M.F.

Lipid Metabolism in Plants

Edited by

Thomas S. Moore, Jr., Ph.D.

Professor
Department of Botany

and

Adjunct Professor
Department of Biochemistry

Louisiana State University
Baton Rouge, Louisiana

CRC Press
Boca Raton Ann Arbor London Tokyo

Library of Congress Cataloging-in-Publication Data

Lipid metabolism in plants / [edited by] Thomas S. Moore, Jr.
  p. cm.
Includes bibliographical references and index.
ISBN 0-8493-4907-9
1. Plant lipids—Metabolism I. Moore, Thomas S.
QK898.L56L55 1993
581.1′3346—dc20   92-46760
                                                                       CIP

Direct all inquiries to CRC Press, Inc., 2000 Corporate Blvd., N.W., Boca Raton, Florida 33431.

© 1993 by CRC Press, Inc.

International Standard Book Number 0-8493-4907-9

Library of Congress Card Number 92-46760
Printed in the United States 1 2 3 4 5 6 7 8 9 0
Printed on acid-free paper

PREFACE

When I took on the task of preparing this book I had the option of either writing it personally or inviting and editing the contributions of other authors. I chose the latter because I felt the field of Plant Lipid Metabolism was moving so fast that a single author could not do it justice. After reading these chapters, I am more convinced than ever. I have been working in this field for about 20 years and have seen an explosion of knowledge during that time.

Thus, I owe each of the authors a great debt of gratitude. Not one person who agreed to contribute defaulted, and the quality of chapters, in my opinion, is outstanding. I am especially appreciative of authors who have written chapters in areas in which they have worked, but may not currently be directly active, or who have contributed a general chapter covering a variety of areas. In these cases, I felt that individuals still could make major and special contributions despite this situation, and I feel they have done so. My special thanks to them.

It might interest some readers to know that this entire book was generated using a variety of word processing programs on different makes of computers, but was relatively easily converted to a single program and submitted on diskette, with the exception of the figures (most or all of which could have been treated similarly). Things have changed dramatically in this area!

Finally, all the authors are indebted to their respective granting agencies, industrial sources, or governmental laboratory support. Such sources have been cited by the respective authors and are due our appreciation. I especially would like to thank the National Science Foudation (NSF) for support over the years, especially for my current grant, IBN-9003817, which provided support for some of my efforts on this book, and also the U.S. Department of Agriculture for additional support for my research (current grant number 91-37304-6437).

Thomas S. Moore, Jr., Ph.D.

THE EDITOR

Thomas S. Moore, Jr., Ph.D., is a Professor in the Department of Botany and an Adjunct Professor in the Department of Biochemistry at Louisiana State University, Baton Rouge, Louisiana.

Dr. Moore began his training at the University of Arkansas, where he received his B.S. in 1964 and his M.S. in 1966. He earned a Ph.D. from Indiana University in 1970. After doing postdoctoral work at the University of California at Santa Cruz, he was appointed an Assistant Professor of Botany at the University of Wyoming, Laramie, Wyoming, in 1973. In 1978 he became an Associate Professor and Chairman of the Department of Botany. He moved to Louisiana State University in 1982 as Professor and Chairman of the Department of Botany, in which capacity he served until 1988. From 1986 to 1988 he was also Interim Chairman of the Department of Microbiology.

Dr. Moore is a member of the American Society of Plant Physiologists, the American Association for the Advancement of Science, and the honorary societies of Sigma Xi and Phi Beta Kappa. He received an Alexander Humboldt Fellowship in 1981.

Dr. Moore is the author of approximately 60 papers and has been the co-author on one previous book. His current major research interests relate to the biosynthesis and utilization of phospholipids in plant cells.

CONTRIBUTORS

Maryse A. Block, Ph.D.
Chargé de Recherche CNRS
Département de Biologie
  Moléculaire et Structurale
Laboratoire de Physiologie
  Cellulaire Végétale
Centre d'Etudes Nucléaires de
  Grenoble
Grenoble, France

Rodney B. Croteau, Ph.D.
Fellow and Professor of
  Biochemistry
Institute of Biological Chemistry
  and Department of Biochemistry
  and Biophysics
Washington State University
Pullman, Washington

Farrist G. Crumley
Biological Laboratory Technician
Microbial Products Research Unit
Richard B. Russell Research
  Center
Athens, Georgia

Roland Douce, Ph.D.
Département de Biologie
  Moléculaire et Structurale
Laboratoire de Physiologie
  Cellulaire Végétale
Centre d'Etudes Nucléaires de
  Grenoble
and
Professor
Université Joseph Fourier
Grenoble, France

Brian G. Fox, Ph.D.
Postdoctoral Research Associate
Department of Chemistry
Carnegie Mellon University
Pittsburgh, Pennsylvania

Margrit Frentzen
Institut für Allgemeine Botanik
Universität Hamburg
Hamburg, Germany

Bernt Gerhardt
Professor
Institut für Botanik
Westfälische Wilhelms-Universität
Münster, Germany

Jonathan Gershenzon, Ph.D.
Assistant Scientist
Institute of Biological Chemistry
Washington State University
Pullman, Washington

Ernst Heinz
Professor
Institut für Allgemeine Botanik
Universität Hamburg
Hamburg, Germany

Anthony H. C. Huang, Ph.D.
Professor
Department of Botany and Plant
  Sciences
University of California
Riverside, California

Jan G. Jaworski, Ph.D.
Professor
Department of Chemistry
Miami University
Oxford, Ohio

Jacques Joyard, Ph.D.
Directeur de Recherche CNRS
Département de Biologie
  Moléculaire et Structurale
Laboratoire de Physiologie
  Cellulaire Végétale
Centre d'Etudes Nucléaires de
  Grenoble
Grenoble, France

Jean-Claude Kader, Ph.D.
Research Director
Laboratory of Plant Cell and
  Molecular Physiology
Université Pierre et Marie Curie
Paris, France

Anthony J. Kinney, D.Phil.
Principal Investigator
Agricultural Products
DuPont Experimental Station
Wilmington, Delaware

**Kathryn F. Kleppinger-Sparace,
  Ph.D.**
Plant Science Department
Macdonald Campus, McGill
  University
Ste. Anne de Bellevue
Quebec, Canada

Hartmut K. Lichtenthaler
Professor
Director of Botanical Institute
Chair of Plant Physiology and
  Plant Biochemistry
University of Karlsruhe
Karlsruhe, Germany

Daniel V. Lynch, Ph.D.
Assistant Professor
Department of Biology
Thompson Biological Laboratory
Williams College
Williamstown, Massachusetts

Agnès Malherbe
Département de Biologie
  Moléculaire et Structurale
Laboratoire de Physiologie
  Cellulaire Végétale
Centre d'Etudes Nucléaires de
  Grenoble
Grenoble, France

Eric Maréchal
Département de Biologie
  Moléculaire et Structurale
Laboratoire de Physiologie
  Cellulaire Végétale
Centre d'Etudes Nucléaires de
  Grenoble
Grenoble, France

W. David Nes, Ph.D.
Research Leader
Plant and Fungal Lipid Group
Microbial Products Research Unit
Richard B. Russell Research
  Center
Athens, Georgia

John B. Ohlrogge, Ph.D.
Professor
Department of Botany and Plant
  Pathology
Michigan State University
East Lansing, Michigan

Stephen R. Parker, Ph.D.
Research Chemist
Plant and Fungal Lipid Group
Microbial Products Research Unit
Richard B. Russell Research
  Center
Athens, Georgia

Dusty Post-Beittenmiller, Ph.D.
Genetics Section Head
The Samuel Roberts Noble
  Foundation
Ardmore, Oklahoma

Samir A. Ross, Ph.D.
Research Institute of
  Pharmaceutical Sciences
School of Pharmacy
University of Mississippi
University, Mississippi

Chris Somerville, Ph.D.
Professor
Michigan State University–
  Department of Energy
  Plant Research Laboratory
Michigan State University
East Lansing, Michigan

Salvatore A. Sparace, Ph.D.
Associate Professor
Plant Science Department
Macdonald Campus, McGill
  University
Ste. Anne de Bellevue
Quebec, Canada

Guy A. Thompson, Jr., Ph.D.
Professor
Department of Botany
University of Texas at Austin
Austin, Texas

Frank J. van de Loo
Michigan State University–
  Department of Energy
  Plant Research Laboratory
Michigan State University
East Lansing, Michigan

Brady A. Vick, Ph.D.
Research Leader
Oilseeds Research
Northern Crop Science
  Laboratory
U. S. Department of
  Agriculture
Agricultural Research Service
Fargo, North Dakota

**Penny M. von Wettstein-
  Knowles, Ph.D.**
Lektor, Associated Professor
Department of Physiology
Carlsberg Laboratory
and
Institute of Genetics
University of Copenhagen
Copenhagen, Denmark

Xuemin Wang, Ph.D.
Assistant Professor
Department of Biochemistry
Kansas State University
Manhattan, Kansas

TABLE OF CONTENTS

SECTION I
ACYL UNIT BIOSYNTHESIS

SECTION II
COMPLEX ACYL LIPIDS

SECTION VI
INDEX

Lipid Metabolism
in Plants

SECTION I

ACYL UNIT BIOSYNTHESIS

Lipids are a basic necessity of life, being involved in such far-ranging biological activities as maintaining cell and organelle integrity and composition, carbon and energy storage, and regulation of development.

Plant lipid metabolism is a complex, highly regulated, and multicompartmented series of reactions for the synthesis and degradation of fatty acids, complex acyl lipids (e.g., phospholipids, triacylglycerol, sphingolipids), and terpenoid lipids (e.g., sterols, carotenoids, phytols). All research points to a very sophisticated network of pathways and transport, coupled with a high degree of regulation. In recent years, major advances have been made in our understanding of these pathways, but many challenges remain.

The field is a difficult one, and in many ways one of the true frontiers of biology. Yet the rewards are many, and the field is not as difficult as some nonlipid researchers seem to feel. Modern analytical techniques provide for sensitive and precise measurement of the lipid products, and enzyme assays with radioactive precursors are quite reliable. Cell fractionation methods allow working with a variety of organelles. On the other hand, enzyme purification of the many membrane-bound enzymes offers to keep the work from being dull! Mutants have been used for lipid investigations for many years, but recently, coupled with molecular genetics techniques, have had a major impact on the field, especially in fatty acid biosynthesis and modification.

This first section concentrates on the biosynthesis and modification of acyl units. This synthesis begins in the plastids with acetyl-CoA and moves out into other organelles for further modifications (desaturations, hydroxylations, elongations, etc.). The production of these basic units of the acyl lipids and their modification are described in Chapters 1 through 5.

Chapter 1

DE NOVO FATTY ACID BIOSYNTHESIS

John B. Ohlrogge, Jan G. Jaworski, and Dusty Post-Beittenmiller

TABLE OF CONTENTS

ABBREVIATIONS

Not defined in text: ACCase — acetyl-CoA carboxylase; ACP — acyl carrier protein; CoA — coenzyme A; FAS — fatty acid synthase; KAS — 3-ketoacyl-ACP synthase; OAA — oxaloacetate; PDC — pyruvate dehydrogenase complex; PEP—phosphoenolpyruvate; PGA — phosphoglyceric acid.

I. INTRODUCTION

The pathway of fatty acid biosynthesis involves the repeated incorporation of two-carbon units derived from acetyl-CoA to initiate and then extend an acyl group to 16 or 18 carbons in length. The main enzymes involved in this synthesis are ACCase (Figure 1, Reaction 1) and FAS (see Figure 1). FAS is a collection of individual enzyme activities (Figure 1, Reactions 2 to 9) catalyzing the conversion of acetyl-CoA and malonyl-CoA to 16:0 and 18:0. In addition, the essential protein co-factor ACP is a component of the FAS. This chapter will provide an overview of current knowledge on fatty acid biosynthesis, including the source of carbon, the enzymes of synthesis and termination, and their regulation. A number of reviews have been published recently on this topic to which the reader is referred for more comprehensive coverage.[1-5]

II. THE SOURCE OF ACETYL-CoA

Fatty acids are assembled by sequential additions of two carbon units which are derived from acetate. To serve as "primer" for the first condensation reaction of fatty acid synthesis acetate must first be activated to form acetyl-CoA. ACCase then converts acetyl-CoA to malonyl-CoA, and it is the malonyl moiety which donates each two carbons for addition to the primer. Because acetyl-CoA does not cross membranes, it must be synthesized inside the plastid in order to provide the carbon precursors for fatty acid synthesis. What is the biosynthetic origin of the acetate and acetyl-CoA used for plastid FAS? This question has been addressed by many studies, but remains one of the major unsolved problems of plant fatty acid biosynthesis.[6,7] Thus, we do not yet have a clear understanding of the path of carbon flow into fatty acid biosynthesis. Several alternatives for the production of acetyl-CoA for plastid fatty acid synthesis have been suggested (Figure 2).

A. PLASTID PYRUVATE DEHYDROGENASE

Acetyl-CoA may be produced from pyruvate and CoA by the pyruvate dehydrogenase reaction. The PDC has been extensively studied in plant mitochondria, (for review, see Randall et al.[8]), but much less is known about PDC in plastids, particularly in chloroplasts. PDC activity has been clearly demonstrated in pea chloroplasts,[9] and there is some evidence for PDC activity

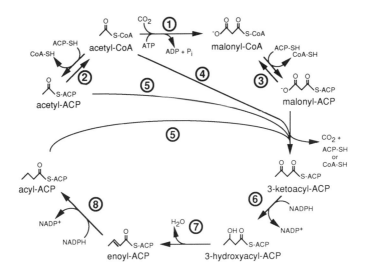

FIGURE 1. Reactions of fatty acid synthesis. 1 — ACCase; 2 — acetyl-CoA:ACP transacylase; 3 — malonyl-CoA:ACP transacylase; 4 — KAS III; 5 — KAS I; 6 — 3-ketoacyl-ACP reductase; 7 — 3-hydroxyacyl-ACP dehydrase; and 8 — enoyl-ACP reductase.

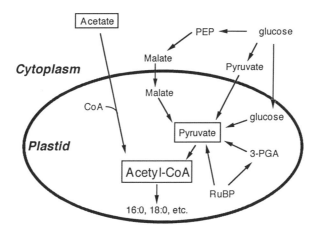

FIGURE 2. A simplified scheme showing several alternative sources of acetyl-CoA production for plastid fatty acid synthesis.

in maize and spinach chloroplasts.[10] From the extensive work of Randall and co-workers, we know that mitochondrial and plastidial PDC are distinctively different enzymes. Although both enzymes are strongly inhibited by their reaction products, and to a lesser extent by metabolites such as adenosine diphosphate (ADP), only the mitochondrial PDC is regulated by reversible phosphorylation, whereas chloroplastic PDC is not. In addition, chloroplastic

PDC is stimulated by changes in pH and divalent cation concentrations which occur in the chloroplast stroma upon illumination. In contrast, mitochondrial PDC is inhibited by changes accompanying illumination.

Since the PDC reaction also produces NADH, it can provide approximately one half of the reducing power needed for fatty acid synthesis. Pyruvate in the chloroplast is derived from the final steps of the glycolytic pathway. Thus, 3-PGA, produced in the initial reaction of the Calvin cycle, can be sequentially converted to 2-PGA, PEP, and, finally, to pyruvate. However, phosphoglycero-mutase, required to convert 3-PGA to 2-PGA, may have insufficient activity in chloroplasts, and therefore this step may require participation of a cytosolic phosphoglyceromutase.[6]

An alternative source of pyruvate in the chloroplast may be as a direct product of ribulose-bisphosphate carboxylase. Andrews and Kane have recently reported that pyruvate represents 0.7% of the products of ribulose-bisphosphate carboxylase from a variety of organisms.[11] However, spinach leaves synthesize fatty acids at rates of 2 to 3 µmol carbon per hour per milligram chlorophyll and total CO_2 fixation is 100 to 200 µmol/mg chlorophyll. This comparison suggests that ribulose-bisphosphate carboxylase production of pyruvate may not be sufficient to provide all the pyruvate precursor needed for lipid synthesis.

B. CYTOPLASMIC ACETATE

Acetyl-CoA may be synthesized from acetate which is imported into the plastid. Chloroplastic pyruvate as a source of acetyl-CoA for *de novo* fatty acid synthesis has been questioned on several grounds. First, the rates of fatty acid biosynthesis from either pyruvate or bicarbonate by isolated chloroplasts are inadequate to account for *in vivo* rates of lipid synthesis.[12,13] In contrast, acetate is incorporated approximately threefold faster than pyruvate, and these rates are sufficient to account for the *in vivo* rates. (The free acetate imported into chloroplasts is ligated to CoA by a very active acetyl-CoA synthetase.[13,14]) Second, although pea chloroplasts have an active PDC, chloroplasts isolated from spinach leaves may not have sufficient pyruvate dehydrogenase activity to account for *in vivo* rates of fatty acid synthesis.[13] Third, pyruvate supplied to chloroplasts appears to be preferentially incorporated into the isoprenoid pathway, whereas label from acetate is almost exclusively found in fatty acids.[15] These results suggest chloroplasts may contain two distinct pools of acetyl-CoA involved either in fatty acid or isoprenoid synthesis. Together, the above considerations have led to suggestions that acetate for fatty acid synthesis is first formed outside the plastid. Since mitochondria are known to have an active pyruvate dehydrogenase, it has further been proposed[16,17] that acetyl-CoA may be produced in the mitochondria, hydrolyzed to free acetate,[17] translocated to the plastid, and then reattached to CoA inside the plastid. This rather circuitous route for chloroplast acetyl-CoA production accounts for a number of observations, but as of yet has not been supported by *in vivo* observations.

Acetyl-carnitine has also been suggested to be a precursor of acetyl-CoA for pea chloroplast fatty acid synthesis.[18,19] Isolated chloroplasts of pea utilized acetyl-carnitine at rates several fold higher than acetate or pyruvate and also at rates twice the highest reported *in vivo* rates of fatty acid synthesis. However, Roughan and co-workers were unable to confirm these results in spinach, pea, amaranthus, or corn, or to detect sufficient activity for the carnitine acetyl-transferase in chloroplasts (which would be required for acetyl-carnitine utilization).[20] Consequently, it is uncertain what role, if any, that acetyl-carnitine plays in chloroplast fatty acid synthesis.

Although acetate is clearly superior to pyruvate as a precursor to fatty acids in isolated chloroplasts, this is not the case for leukoplasts of developing castor endosperm.[21] Plastids of oilseeds appear to have a complete glycolytic pathway which together with pyruvate dehydrogenase can result in the conversion of glucose to acetyl-CoA for fatty acid biosynthesis.[22,23] In addition, sufficient pentose phosphate pathway activities are available to generate NADPH needed for fatty acid synthesis.[24] Thus, compared to chloroplasts, the plastids of developing oilseeds may use different pathways to provide the substrates and co-factors for fatty acid biosynthesis.

Although a complete glycolytic pathway together with pyruvate dehydrogenase in leukoplasts appears sufficient to account for carbohydrate conversion to acetyl-CoA, an alternative pathway has recently been proposed. Smith and co-workers found that rates of fatty acid synthesis from malate by leukoplasts from developing castor endosperm were 4.5- and 120-fold higher than from pyruvate or acetate.[25] They propose that sucrose imported into developing endosperm is converted to malate in the cytoplasm. Thus, cytoplasmic glycolysis could convert glucose to PEP, which is then carboxylated to OAA, which in turn is converted to malate. Malate then enters the plastid where malic enzyme converts it to pyruvate followed by production of acetyl-CoA by pyruvate dehydrogenase. The reactions catalyzed by malic enzyme and pyruvate dehydrogenase each produce NAD(P)H which then can provide sufficient reducing equivalents for fatty acid synthesis.

Finally, it should be noted that in animals and oleaginous yeasts, acetyl-CoA for fatty acid synthesis is provided by ATP:citrate lyase which may be the rate-limiting step for oil production in some yeasts.[26] This enzyme has been reported to be present in developing soybean seeds, germinating castor endosperm, mango fruit, and pea leaves.[27-29] The possible involvement of this enzyme in providing acetyl-CoA for fatty acid synthesis has received very little attention and warrants reexamination.

It is clear from the above that a consensus is not yet available concerning the origin of acetyl-CoA for fatty acid synthesis. Although several pathways can be supported based on *in vitro* data, *in vivo* evidence is lacking in almost all cases. Because acetyl-CoA is perhaps the most central and widely used of all metabolic intermediates, plants may maintain a variety of pathways for its production. Such multiplicity may mean that under different conditions, or in different tissues, the source(s) of acetyl-CoA for fatty acid biosynthesis may differ.

III. COMPONENTS OF FATTY ACID SYNTHESIS

A. ACETYL-CoA CARBOXYLASE (ACCase)

ACCase catalyzes the first step of fatty acid biosynthesis as shown in Figure 1 (Reaction 1). This step is often considered to be the committed step for plastidial *de novo* fatty acid biosynthesis because the malonyl-CoA produced in the plastid is used predominately, if not exclusively, for the synthesis of fatty acids. Malonyl-CoA is synthesized from acetyl-CoA and bicarbonate via a two-step ping-pong reaction as illustrated in the equations below.

$$ATP + HCO_3^- + biotin - E \leftrightarrow ADP + Pi + CO_2 - biotin - E$$

$$CO_2 - biotin - E + acetyl - CoA \leftrightarrow biotin - E + malonyl - CoA$$

The first step involves the adenosine triphosphate (ATP)-dependent carboxylation of the biotinated protein. The second step is the transfer of the carboxyl group from the enzyme to acetyl-CoA, resulting in the formation of malonyl-CoA. Each step is fully reversible and inhibited to varying degrees by its reaction products.

There are two types of ACCase found in nature. The bacterial ACCase is similar to the Type II FAS discussed below in that it is composed of individual and dissociable polypeptides, each with a single enzyme activity or function. The bacterial ACCase is composed of four subunits: biotin carboxylase, biotin carboxyl carrier protein, and two biotin transcarboxylase subunits.[30]

The other type of ACCase is a large (200 to 233 kDa), single polypeptide containing the same three functional domains as the bacterial ACCase. In animals, ACCase is most active as a polymer which is formed from dephosphorylated (active) ACCase subunits. There are various agents which can cause the activation and polymerization of ACCase *in vitro*, including citrate, limited proteolysis, CoA, and the dephosphorylation of inactive ACCase.[31] In addition to its regulation by covalent modification (phosphorylation/dephosphorylation), the animal ACCase is allosterically regulated by its reaction product (malonyl-CoA), by citrate, and by palmitoyl-CoA, an end product of fatty acid biosynthesis. There are also complex transcriptional and posttranscriptional controls of ACCase expression in animals which involve transcriptional induction and differential splicing of the ACCase mRNA in response to diet (liver ACCase) and developmental state (lactating mammary glands).[32,33]

ACCase has also been studied in the photosynthetic diatom, *Cyclotella cryptica*. In this organism, native ACCase (740 kDa) is believed to be a tetramer composed of identical subunits, each containing all three functions,[34] and therefore it resembles animal ACCase in its organization. The diatom enzyme is inhibited by palmitoyl-CoA and, to a lesser extent, by malonyl-CoA, ADP, and orthophosphate.[34] *Cyclotella* accumulates lipids (primarily triacylglycerols) in response to silicon deficiency.[35] An increase in the level of ACCase protein and activity is believed to be, at least in part, responsible for the increased lipid accumulation.[36] Recently, a full length genomic clone for *Cyclotella* ACCase has been isolated and found to have significant (>65%) sequence similarity with the rat and chicken ACCase sequences.[37]

Although higher plants have a dissociated Type II FAS similar to bacteria, plant ACCase has a Type I or associated organization, and therefore it is similar to the animal and diatom ACCases. The maize ACCase appears to function as a dimer.[38] There are reports of plant ACCase isozymes which may not be localized in the plastid.[38] Presumably, additional ACCase isozymes would be responsible for malonyl-CoA production in the cytosol to meet the requirements for the synthesis of flavonoids,[39] phytoalexins,[40] ethylene,[41] and for fatty acyl chain elongation in epidermal wax biosynthesis and seed triacylglycerol biosynthesis. There are reports of smaller molecular weight proteins with ACCase activity in carrot[42] and maize[43] which may represent a Type II ACCase or, alternatively, may be due to limited proteolysis during their isolation.

B. FATTY ACID SYNTHASE (FAS)

FAS refers to reactions 2 through 9 of Figure 1 and, in addition, includes ACP. While the reactions of FAS are essentially the same for all organisms, there are two distinctly different types of FAS found in nature. In animals and yeast, the FAS is referred to as Type I and consists of a multifunctional enzyme complex. Thus, Type I FAS is characterized by large subunits (250 kDa), each capable of catalyzing several different reactions. In contrast, plants and most bacteria have a Type II FAS which characteristically has each enzyme activity and ACP associated with an individual protein which is readily separated from the other activities of FAS. In that sense, the Type II FAS functions much like a metabolic pathway, whereas Type I FAS functions like a large protein complex such as pyruvate dehydrogenase. For the remainder of this chapter, we will limit our discussion to the Type II FAS found in plants.

The metabolic strategy of the fatty acid biosynthetic pathway involves first the energy dependent "activation" of acetyl-CoA by carboxylation to form malonyl-CoA. The decarboxylation of malonyl-CoA is then used to drive a condensation reaction for chain length extension with the formation of a 3-ketoacyl-ACP. Finally, a sequence of three reactions is used to form the fully reduced acyl-ACP. The latter sequence of progressing stepwise from a 3-ketoacyl group to a saturated acyl group is a common reaction series found in biochemical pathways. For example, both β-oxidation and the citric acid cycle use the same series of reactions, but in the reverse order.

1. Acyl Carrier Protein (ACP)

All of the reactions of fatty acid biosynthesis leading from malonyl-CoA to palmitate and oleate require the involvement of ACP (for reviews, see References 2 and 44). ACP is a small (*circa* 9000 Da), acidic protein which has a phosphopantetheine prosthetic group attached to a serine residue near

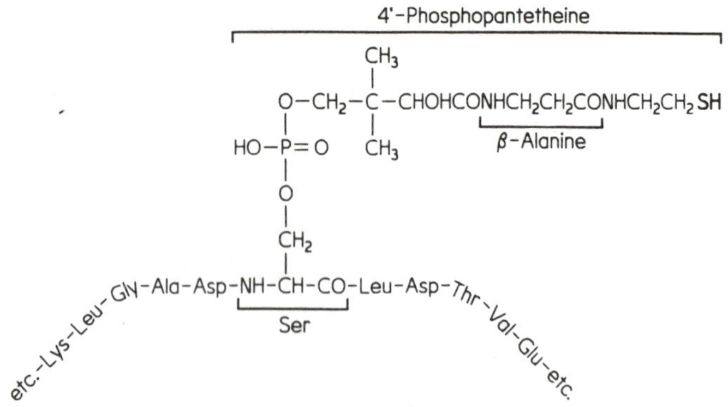

4'-Phosphopantetheine

FIGURE 3. Prosthetic group structure of ACP. A partial amino acid sequence of spinach ACP I which surrounds the serine attachment site for the prosthetic group is shown. The β-alanine component of 4'phosphopantetheine is indicated.

the middle of the polypeptide chain. In many ways, it is the functional equivalent of CoA which has an identical phosphopantetheine group. The acyl groups involved in fatty acid biosynthesis are attached as thioesters to the sulfhydryl at the end of this prosthetic group on ACP (see Figure 3).

Because of their central role in fatty acid biosynthesis, ACPs were the first component of plant fatty acid synthesis to be purified and the first for which cDNA and genomic clones became available. Thus, these proteins have been particularly well-studied. Amino acid sequences are now available for over 15 ACPs from a number of dicot and monocot species, as well as from photosynthetic and nonphotosynthetic bacteria. A 19 amino acid sequence surrounding the prosthetic group attachment site is highly conserved (>90%) in all these proteins. Overall, the sequences from different higher plant species are about 65 to 75 % identical, a level which is somewhat less than that found in comparisons of other sequenced proteins of fatty acid metabolism.

From *in vitro* mRNA translations, it was observed that ACPs were synthesized as larger molecular weight precursors, suggesting that they contained chloroplast transit peptides.[45] This was later confirmed from the cDNA sequence of ACPs from several species which all contained a 50 to 59 amino acid N-terminal extension which had a primary structure characteristic of such transit peptides.[46-51] Together, the above observations confirmed that ACPs are nuclear-encoded proteins which must be imported into the plastid from their site of synthesis on cytoplasmic ribosomes. Lamppa and co-workers have studied the import of pre-ACP by isolated chloroplasts and have shown that this can occur without the presence of the phosphopantetheine prosthetic group[52] and that prosthetic group attachment can occur after uptake of the preapo-ACP into chloroplasts.

Purification of ACPs from spinach and barley leaves led to the separation of two or more isoforms of this protein.[53,54] The major isoforms purified from leaves differed in their N-terminal amino acid sequences and thus appeared to be products of distinct genes. In addition, it is now clear that the different isoforms of ACP exhibit tissue-specific expression patterns. In several dicot species, the most abundant form of ACP observed in leaves is expressed at low or undetectable levels in other tissues. Thus, this form of ACP appears to be leaf specific. Other forms of ACP can be considered to be constitutive, i.e., they are expressed in all the tissues that have been examined (roots, seeds, leaves, stems, and flowers). In *Brassica napus*[55] and *Arabidopsis*,[50] additional forms of ACP have been found to be specifically expressed in seeds.

The existence of multiple forms of ACP is widespread in the plant kingdom and includes all multicellular photosynthetic eukaryots which have been examined, including ferns, and multicellular algae.[56] What is the functional significance of these different isoforms? The two forms of ACP found in spinach leaves appear to act identically with the reactions of fatty acid synthesis both *in vitro*[57] and *in vivo*.[58] However, *in vitro* assays indicate they differ in their activity with the oleoyl-ACP hydrolase and the oleoyl-ACP glycerol-3-phosphate acyltransferase reactions which determine the partitioning of oleic acid between plastidial and cytoplasmic glycerolipid metabolism.[57] Thus, individual ACPs might have specialized functions in directing acyl metabolism toward different metabolic steps. Alternatively, multiple ACP isoforms might afford a higher level of tissue-specific and developmental control over fatty acid synthesis.

Although the great majority of ACP in leaves is found in plastids,[59] recent evidence suggests that plant mitochondria also contain ACP. Chuman and Brody observed proteins in purified pea and potato mitochondria which cross-reacted with antibodies to ACP and whose mobility during polyacrylamide gel electrophoresis (PAGE) was changed upon deacylation.[60] In addition, a clone for a mitochondrial ACP was recently isolated from an *Arabidopsis* cDNA library.[61] It is not yet known whether ACP functions in a complete fatty acid biosynthesis pathway in plant mitochondria or whether it has other functions. In *Neurospora*[62] and bovine heart mitochondria,[63] acyl-ACP has been found to be tightly associated with the complex I NADH:ubiquinone oxidoreductase, but the physiological role of this connection between an electron transport complex and ACP has not been explained.

Genomic sequences are now available for ACPs from *Arabidopsis*,[64,65] *B. napus*,[55] and barley.[66] In all ACP genes so far sequenced, there are three introns which occur in similar locations: one within the transit peptide, one near the junction of the transit peptide to the mature protein, and a third just after the prosthetic group attachment site. The three cloned barley ACP genes have been mapped to chromosomes 1 and 7. Barley *Acl1* is on the short arm of chromosome 7, while *Acl2* and *Acl3* are on the long and short arms of chromosome 1. The promoters of ACP genes are just beginning to undergo scrutiny. In barley, the promoter for *Acl1*, which is probably most highly expressed in leaves, is quite different from the promoter for *Acl3* whose message is found in all tissues examined.[66] The *Acl3* promoter does not appear to have a TATA box at the transcriptional start, is higher in GC content, and contains three GGGCGG elements that are not present in the *Acl1* promoter.[67,68] These promoter characteristics of *Acl3* are similar to those found in mammalian housekeeping genes and are consistent with the expression of *Acl3* mRNA in several tissues.

2. Acetyl-CoA:ACP Transacylase

Acetyl transacylase (Reaction 2 in Figure 1) catalyzes the transfer of the acetyl moiety from CoA to ACP. This reaction proceeds via an acetyl-enzyme intermediate in the following sequence:

$$Acetyl - CoA + enzyme \leftrightarrow acetyl - enzyme + CoA$$
$$Acetyl - enzyme + ACP \leftrightarrow acetyl - ACP + enzyme$$

The covalent acetyl-enzyme intermediate is probably an oxygen ester of serine. The related enzyme from yeast[69] and a similar enzyme, malonyl transacylase, from *Escherichia coli*[70] are known to form covalent intermediates with serine. This enzyme has been purified to homogeneity from *E. coli* and partially purified from spinach. The *E. coli* enzyme is a homodimer whose M_r is 61,000, and the M_r is 29,000 for the subunit.[71] In contrast, the M_r of the spinach transacylase is 49,000.[72]

The role of the transacylase is to produce acetyl-ACP, which was until recently regarded as the primer for fatty acid synthesis. Because acetyl transacylase activity is relatively low compared to other FAS enzymes, it was also considered to be potentially rate limiting and a candidate as a site of regulation for fatty acid synthesis in plants.[72] However, the discovery of KAS III (Reaction 3) in plants and *E. coli*, an enzyme which uses acetyl-CoA directly and thus bypasses the requirement for acetyl-ACP, suggested that acetyl transacylase may not play a central role in fatty acid biosynthesis. Recent analysis of acetyl-ACP utilization by isolated spinach chloroplasts substantiated that acetyl-ACP is only a minor intermediate in fatty acid synthesis, indicating that the role of the acetyl transacylase is also minor. To further complicate our understanding of this enzyme, it has been suggested that the acetyl transacylase may not be a separate enzyme at all, but rather a partial reaction of one of the 3-ketoacyl-ACP synthases, viz., KAS III. This suggestion will be discussed in more detail below.

3. Malonyl-CoA:ACP Transacylase

This reaction (Reaction 3) is analogous to the acetyl transacylase reaction, catalyzing the transfer of a malonyl moiety from CoA to ACP by a similar mechanism:

$$Malonyl - CoA + enzyme \leftrightarrow malonyl - enzyme + CoA$$
$$Malonyl - enzyme + ACP \leftrightarrow malonyl - ACP + enzyme$$

As noted above, this mechanism includes a covalent intermediate with malonate esterified to a serine. Malonyl transacylase has been purified to varying levels of homogeneity from avocado,[73] barley,[74] spinach,[75] soybean,[76] and leek.[77] Kinetic analysis of the spinach enzyme revealed that its mechanism is random sequential.[75] Two forms of this enzyme have been found in both soybean[76] and leek,[77] although functional significance to these isoforms has not as yet been assigned.

4. 3-Ketoacyl-ACP Synthase

These synthases (Reactions 4 and 5) catalyze a condensation reaction, and as a result these enzymes have commonly been referred to as condensing enzymes. Usually, there are three isozymes of the synthases, KAS I, II, and III, each of which are distinguished by their substrate specificity. The general reaction for KAS is the condensation of an acyl-ACP and malonyl-ACP to produce a 3-ketoacyl-ACP and release of CO_2 and ACP-SH. The mechanism for this reaction is

$$Acyl - ACP + enzyme \leftrightarrow acyl - S - enzyme + ACP$$
$$Acyl - S - enzyme + malonyl - ACP \leftrightarrow 3 - ketoacyl - ACP + CO_2 + ACP - SH$$

The condensing enzymes from plants are homodimers with an M_r equal to 43,000 to 45,000 per subunit, while the native enzyme has an M_r equal to 86,700.[78,79]

The substrate specificity of each of the KAS isozymes suggests the role each plays in fatty acid biosynthesis. KAS I is most active with C_4–C_{14} acyl-ACPs[80] and displays small but significant activity with acetyl-ACP.[81] KAS II is active only with longer chain (C_{10}–C_{16}) acyl-ACPs.[80] Finally, the most recently discovered isozyme, KAS III, uses exclusively acetyl-CoA in place of an acyl-ACP.[81] Thus, these *in vitro* activities suggest that KAS III initiates fatty acid biosynthesis, using acetyl-CoA as the primer of synthesis. KAS I is then used to extend the acyl chain to C_{12}–C_{16}, and, finally, KAS II completes the synthesis to C_{18}.

An interesting question that has arisen as a result of the characterization of KAS III concerns the role of acetyl-ACP in fatty acid biosynthesis. Since KAS III can use acetyl-CoA directly at rates several fold greater than acetyl transacylase,[81,82] the favored direction of fatty acid synthesis is apparently to bypass acetyl-ACP, i.e., bypass Reactions 2 and 5 (Figure 1). In fact, data recently obtained from isolated spinach chloroplasts clearly demonstrate that acetyl-ACP is not a major intermediate of fatty acid synthesis.[83] However, significant quantities of acetyl-ACP are found in plants, and it is actually the major form of acyl-ACP that accumulates when spinach plants are placed in the dark. Why then does it accumulate and what is the physiological significance of the acetyl-ACP? One possible explanation is that it accumulates primarily because it is *not* a major intermediate of fatty acid synthesis. Slow utilization of acetyl-ACP by condensing enzymes results in its accumulation as a result of a slightly higher acetyl transacylase activity. This explanation poses a further dilemma. If acetyl-ACP is a minor intermediate of limited importance, why has acetyl transacylase activity been maintained through evolution? One explanation has been suggested as a result of analysis of KAS III in *E. coli* and an understanding of the mechanism of condensing enzymes.[84,85] Each of their partial reactions are reversible, and because acyl-CoA is also a substrate (albeit, poor) for the KASs, the first partial reaction, shown above, allows a condensing enzyme to act as an acyl-CoA-ACP transacylase:

$$Acyl - CoA + ACP \leftrightarrow acyl - ACP + CoA$$

Rather than cells having a distinct enzyme with acetyl transacylase activity, cells may use condensing enzymes to catalyze this reaction. This suggestion has to some extent been verified with purified KAS III from both *E. coli*[86] and spinach.[81] Both of these condensing enzymes have acetyl transacylase activity, although it is 90- to 200-fold less than KAS III activity. It is not known at this time if all the *in vivo* acetyl transacylase activity can be accounted for by condensing enzymes or if the transacylases purified from *E. coli*[71] and spinach[72] were actually condensing enzymes.

5. 3-Ketoacyl-ACP Reductase

The initial reductive step of fatty acid biosynthesis (Reaction 6) is the conversion of a 3-ketoacyl-ACP to a 3-hydroxyacyl-ACP:

$$3 - ketoacyl - ACP + NADPH \leftrightarrow 3 - hydroxyacyl - ACP + NADP^+$$

The native enzyme from avocado has a M_r of 130 kDa and a subunit M_r of 28 kDa, suggesting that the native protein is a tetramer.[87] This subunit size is in good agreement with that reported for spinach[88] and *Arabidopsis*.[89]

At least two isoforms of this reductase are found in avocado, an NADPH-dependent and an NADH-dependent form.[73] The predominant form is NADPH dependent and can account for all the required 3-ketoacyl-ACP reductase activity for fatty acid synthesis. The metabolic role of the NADH-dependent form has not been elucidated. Recently, a cDNA clone for this enzyme was isolated from *Cuphea*.[90]

6. 3-Hydroxyacyl-ACP Dehydrase

The removal of water (Reaction 7 in Figure 1) from the 3-hydroxyacyl-ACP to form the 2,3-*trans*-enoyl-ACP is catalyzed by a dehydrase:

$$3 - hydroxyacyl - ACP \leftrightarrow Enoyl - ACP + H_2O$$

The purified dehydrase from spinach has a M_r of 85 kDa and a subunit size of 19 kDa. Thus, it appears to be a homotetramer. Its substrate specificity is very broad, with high activity demonstrated for acyl groups ranging from C_4–C_{16}.[88]

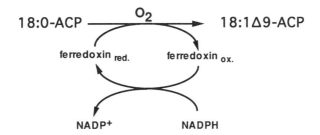

FIGURE 4. Aerobic biosynthesis of oleic acid catalyzed by stearoyl-ACP desaturase.

7. Enoyl-ACP Reductase

The final reduction step (Reaction 8) reduces the 2,3-*trans*-enoyl-ACP to the corresponding saturated acyl-ACP:

$$Enoyl - ACP + NADH \leftrightarrow acyl - ACP + NAD^+$$

There are two isoforms of this enzyme. The major form is specific for NADH and has been purified from avocado mesocarp,[73] spinach leaf,[88] and oilseed rape.[91] This reductase exists as a homotetramer with a native M_r of 115 to 140 kDa and a subunit M_r of 32.5 to 34.8 kDa. A second isoform found in safflower seed uses either NADPH or NADH and is specific for longer chain ($\geq C_{10}$) enoyl-ACPs.[92] A cDNA clone for this enzyme has been isolated from rapeseed.[93]

C. MODIFICATION AND TERMINATION REACTIONS

1. Stearoyl-ACP Desaturase

The enzyme that catalyzes the insertion of a *cis* double bond at the Δ^9 position of stearoyl-ACP is the stearoyl-ACP desaturase. Localized in the stroma of the plastid, it is the only desaturase studied to date that is not associated with a membrane.[94,95] As a consequence, it was the first of the plant desaturases purified to homogeneity.[96] This desaturase has now been purified from several sources and its cDNA cloned.[97-101] The desaturase is a homodimer, with the M_r of the native protein 68 kDa and the subunit M_r 36 to 38 kDa.[96,98]

This desaturase is highly specific for stearoyl-ACP. The desaturase activity for palmitoyl-ACP and stearoyl-CoA are only 1 and 5% compared to stearoyl-ACP. In addition, stearoyl-ACP desaturase requires molecular oxygen and a source of electrons[94,96,102] (Figure 4). The immediate donor of electrons to the desaturase is ferredoxin, which in turn can receive electrons from either NADPH via the NADPH-ferredoxin-NADP+ oxidoreductase system or directly from photosystem I in chloroplasts.[102]

2. *n*-12 Acyl-ACP Desaturase

Although oleic acid is the major 18-carbon moneneoic fatty acid found in most plants, exceptions occur. Several *Umbelliferae* species accumulate large

(>75% of fatty acids) levels of Δ^6-18:1 in their seed lipids. Recently, it has been found that the enzyme responsible for producing Δ^6-18:1 in these plants is an acyl-ACP desaturase which is related to the stearoyl-ACP desaturase which produces oleic acid.[103] A cDNA coding for this desaturase was cloned by screening with antibodies to the avocado stearoyl-ACP desaturase. It had about 65% sequence identity with the Δ^9 producing desaturase. Transformation of tobacco with this clone resulted in production of both Δ^6-18:1 and Δ^4-16:1 in approximately equal proportions. Thus, the cDNA may encode an *n*-12 desaturase or a Δ^4 desaturase.

3. Hydrolases and Acyltransferase

In almost all plant species and tissues, the predominant fatty acids have chain lengths of 18 (60 to 95%) or 16 (5 to 40%) carbons. This is also true for essentially all animal and microbial species and is believed to occur because these chain lengths provide optimal physical properties of bilayer membranes. How do cells determine the chain length of fatty acids they produce? It is likely that the specificity of several enzymes act together to ensure production of 16- and 18-carbon acyl groups. The KAS I is most active in producing fatty acids of up to 16 carbons, whereas the KAS II is specific for elongation of 14:0 and 16:0. KAS II is much less active for the elongation of 18:0-ACP to 20:0-ACP, and therefore the condensing enzymes together produce primarily 16- and 18-carbon acyl chains (attached to ACP). However, release of these fatty acids from the ACP requires the action of either an acyl-ACP hydrolase (thioesterase) or an acyltransferase. Most plants contain an acyl-ACP thioesterase which has highest activity with oleoyl-ACP.[96,104-106] Therefore, in order to produce a substrate for removal of fatty acid from the ACP, the acyl chain must first be desaturated by the stearoyl-ACP desaturase.[96] This enzyme is specific for 18-carbon chains (compared to 16), and therefore the desaturase also contributes to assuring that 18-carbon chains predominate. Thus, the specificities of the KAS II, the stearoyl-ACP desaturase, and the oleoyl-ACP thioesterase all act together to ensure that the 18-carbon oleic acid is the major product of plastid fatty acid synthesis.

Chain termination can also occur when fatty acids are transferred from ACP to glycerol-3-phosphate by acyltransferases. There are two such enzymes in the plastid. The first is a soluble enzyme which prefers oleoyl-ACP as substrate and transfers the oleoyl moiety to the *sn*-1 position of glycerol-3-phosphate.[107] The second enzyme is localized on the inner membrane of the chloroplast envelope,[108] transfers fatty acids onto the *sn*-2 position of 1-acyl-glycerol-3-phosphate, and is specific for palmitoyl-ACP.[107] Thus, this second acyltransferase is a major mechanism which determines that 16-carbon fatty acids are components of plant membranes. Indeed, lipid biochemists sometimes categorize different higher plants as either "16:3" or "18:3" type plants based on the proportions of 16:3 and 18:3 in their leaf lipids.[109] Those which are of the 16:3 type have a higher flux through the chloroplast acyltransferase reaction which selects palmitoyl-ACP from the available pools of potential acyl-ACP substrates.

Although 18- and 16-carbon chain lengths are the rule in nature, exceptions occur. The storage triacylglycerols found in plant seeds frequently contain unusual fatty acid structures, including shorter or longer chain lengths (see Chapter 3). Certain species of *Cuphea,* for example, produce high proportions of lauric (12:0) or other medium-chain length fatty acids in their seeds.[110] Recently, it has been found that a primary mechanism for the production of these shorter chain products is the existence of an additional acyl-ACP thioesterase which prefers substrates such as lauroyl-ACP.[111] This type of acyl-ACP thioesterase has now been cloned from California Bay trees, and its expression in transgenic plants has led to laurate accumulation in the seeds of such plants.[112] Some plant species also produce high proportions of longer chain (usually C_{20}–C_{22}) fatty acids in their seeds. However, the elongation of these fatty acids beyond 18 carbons does not occur as part of the plastid fatty acid biosynthetic machinery, but rather involves export of oleic from the plastid and its elongation by membrane-bound enzymes associated with the endoplasmic reticulum (see Chapter 3).

IV. MOLECULAR ORGANIZATION OF FATTY ACID SYNTHESIS

Increasing evidence indicates that components of many metabolic pathways, previously thought to act separately, may function as multienzyme complexes. If plant FAS functions as a complex *in vivo,* it is relevant to fatty acid biosynthesis in two respects. First, the *in vivo* levels of the individual components of a complex are frequently coordinately regulated at both transcriptional and translational levels. Slabas and co-workers measured the activities of four FAS proteins, ACP, enoyl-ACP reductase, and KAS I and II, on similar populations of rapeseed.[79,91,113] These studies indicate that these four proteins have essentially identical temporal patterns of accumulation, suggesting that their expression and/or steady state levels may be coordinately regulated by the same developmental signals. Second, if plant FAS functions as a multienzyme complex, there may be certain steric constraints on the constituent proteins. This may have specific implications on design strategies for the introduction of new or genetically altered FAS components into oilseed crops. Foreign lipid biosynthetic enzymes may not be conformationally compatible with or have access to substrates within an endogenous FAS complex. Therefore, if plant FAS functions as a complex, it may be important to determine which proteins constitute the complex.

There have been no studies directly demonstrating whether or not plant FAS proteins function as a complex. However, there have been three *in vivo* studies using spinach ACP,[114] California Bay medium-chain acyl-ACP thioesterase,[112] and coriander acyl-ACP desaturase[103] clones to transform plants. In all cases, the transgenic plants expressed the foreign FAS protein, and it was functional in the endogenous FAS machinery. These data suggest that if tobacco FAS functions as a complex *in vivo,* then the nature of this complex is sufficiently flexible to accommodate the foreign FAS proteins.

V. REGULATION OF FATTY ACID SYNTHESIS

The reactions of fatty acid biosynthesis from acetyl-CoA are now well-understood, and most of the enzymes in the pathway are now cloned. In contrast, much is yet to be learned about how this pathway is regulated. It is probably essential to all cells that the flux of carbon through the fatty acid biosynthetic pathway be tightly regulated. Overproduction of fatty acids is not only wasteful of carbon and energy, but free fatty acids are also potentially toxic to cells. In addition, organisms appear to coordinate the production of membrane lipids with membrane proteins such that a constant lipid to protein ratio in membranes is maintained. To avoid accumulation of free fatty acids, utilization of acyl chains for glycerolipid synthesis must be coordinated with their production by the FAS. How is the pathway of plant fatty acid biosynthesis regulated?

The flux through biochemical pathways is frequently controlled by regulating the activity of a "rate-determining" step in the pathway. Often this rate-determining reaction is the first committed step of the pathway or at least one of the early reactions. (Although control of a biosynthetic pathway can sometimes be attributed to a single enzymatic step, in many cases, more than one enzyme is involved and it is more accurate to consider a rate-limiting *system*.) In plant fatty acid biosynthesis, several reactions have been considered as potentially important points of regulation.

In animal fatty acid biosynthesis, ACCase is the most important regulatory and rate-determining step.[31] In animals, malonyl-CoA is used only for fatty acid synthesis, and therefore ACCase is also the first committed step in the pathway. Based on analogy to animals, ACCase has frequently been suggested as rate determining for plant fatty acid synthesis. In contrast to animals, however, malonyl-CoA in plants is utilized in several additional reactions which include chalcone synthase (flavonoid pathway), amino acid malonation, and formation of malonyl-ACC. Nevertheless, the preceding three reactions probably occur outside the plastid, and therefore, within the plastid, the utilization of malonyl-CoA may be committed exclusively for fatty acid production.

One of the enzymes of FAS could also be rate determining. Acetyl transacylase was suggested as a rate-limiting step of fatty acid biosynthesis based on its very low activity in assays of the component enzymes of FAS.[72] However, as discussed earlier, it is now clear that this enzyme plays, at best, a minor role in plant fatty acid biosynthesis. The condensing enzymes (KAS) also have low activity (compared to other FAS components) when assayed *in vitro*[72] and thus are potentially rate limiting. Furthermore, examination of the pools of acyl-ACPs in spinach leaf or seed indicates that the acyl groups are present predominantly in the saturated form rather than as 3-keto-, 3-hydroxy-, or 2-enoyl-ACP intermediates.[58] These observations indicate that the reductases and dehydratase activities *in vivo* are sufficiently higher than that of the condensing enzyme such that the unsaturated intermediates of FAS are not observed.

A strategy useful for identifying potential regulatory steps in a biosynthetic pathway is to examine the pools of substrates and products for individual reactions in the pathway.[115] In order to perform a regulatory role, an enzyme must have a ratio of substrates to products *in vivo* which is substantially higher than the ratio at thermodynamic equilibrium. In the fatty acid biosynthetic pathway, the relevant substrates and intermediates are acetyl-CoA, malonyl-CoA, and the C_3–$C_{18:1}$ acyl-ACP intermediates and products of the pathway. Analysis of these components is complicated by their low level (<1 μM for most acyl-ACPs), rapid turnover ($t_{1/2}$ < 5 s), and potential lability during analysis. However, methods have recently been developed which have allowed measurement of the *in vivo* concentrations of essentially all the substrates and intermediates of plastid fatty acid biosynthesis.[58,116] These methods have taken advantage of the ability to separate acyl-ACPs with different acyl groups using native polyacrylamide electrophoresis gels containing urea and to detect nanogram levels of these acyl-ACPs on immunoblots probed with antibodies to ACP. An example of such analyses is shown in Figure 5. In addition, HPLC methods have been used to measure *in situ* levels of acetyl-CoA and malonyl-CoA in isolated chloroplasts. These studies have led to the following observations:

1. Acyl-ACPs can be detected with all saturated chain lengths from 2:0 to 18:0. The intermediates between 4:0-ACP and 14:0-ACP occur at similar levels.
2. In both young leaves in the light and in developing seeds, approximately 60% of the ACP is unesterified ACP-SH.
3. The ACP I and ACP II isoforms of spinach leaf have indistinguishable profiles of acyl chains, suggesting that these isoforms are utilized similarly in *de novo* FAS reactions.
4. Analysis of concentrations of the substrates and products of ACCase and the condensing enzymes indicate that both of these reactions are substantially displaced from equilibrium and therefore potentially regulatory.

Additional evidence regarding the nature of regulation of FAS in leaf was obtained by examining changes in the acyl-ACP and acyl-CoA pools during the transition between light and dark conditions. Fatty acid synthesis is five- to sixfold lower in leaves in the dark compared to the light,[117] and differences in the acyl-ACP and acyl-CoA pools are observed under these conditions.[58,116] When spinach leaves or chloroplasts were transferred from light to dark, acetyl-ACP levels increased four- to fivefold, and malonyl-ACP levels decreased. In dark-incubated chloroplasts, malonyl-CoA and malonyl-ACP levels were undetectable, but increased when chloroplasts were incubated in the light. Together, these observations provide strong *in situ* evidence that regulation of ACCase activity is responsible, at least in part, for light/dark regulation of leaf fatty acid synthesis. As ACCase activity and malonyl-CoA

A.

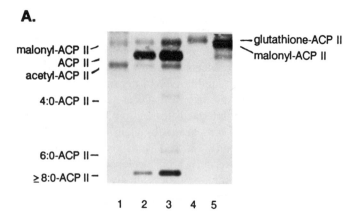

B.

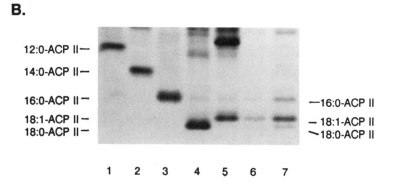

FIGURE 5. Identification of acyl-ACPs in extracts from developing spinach seed using immunoblot analyses of 1 and 5 *M* urea gels. Aliquots of seed extracts were separated on (A) 1 *M* or (B) 5 *M* urea gels, and the ACPs were visualized by immunoblotting techniques using antibodies to spinach ACP. (A) Lanes 1, 4, and 5 contain acetyl-ACP II, a glutathione adduct of ACP II, and malonyl-ACP II, respectively. Lanes 2 and 4 contain 3 and 9 μl of seed extract, respectively. Free ACP II and the chain lengths of the acyl-ACPs identified in the seed extract are indicated at the *left*. Glutathione-ACP II and malonyl-ACP II are also indicated at the right. (B) Long-chain acyl-ACP II standards (12:0-ACP II, 14:0-ACP II, 16:0-ACP II, 18:0-ACP II, and 18:1-ACP II) are shown in Lanes 1 through 5, respectively. Lanes 6 and 7 contain 8 and 24 μl of seed extract, respectively.

synthesis decrease, the rate of the condensing enzyme reactions decrease (due to lack of malonyl-ACP). Consequently, acetyl-ACP levels increase as its rate of utilization by KAS I decreases.

ACCase activity *in vitro* is influenced by ATP to ADP ratios,[118] by pH, and by magnesium.[43] Each of these factors change in light vs. dark chloroplasts such that at least part of the *in vivo* regulation of ACCase may be through modification of the stromal environment. However, other regulation of ACCase is also likely to be important. When light-incubated chloroplasts are rapidly lysed and the ACCase activity determined, the activity is two- to fourfold higher than from dark-incubated chloroplasts.[119] During this *in vitro* assay, the ACCase activity from light-incubated chloroplasts did not change over 4 min, while the activity from dark-incubated chloroplasts increased with time and reached light levels within 2 to 3 min.[149] Thus, ACCase activity appears to be transiently inactivated or inhibited in the dark. Experiments with a number of photosynthetic electron transport inhibitors has indicated that this modulation of ACCase activity is dependent on photosystem I.[149] A similar conclusion was made some years ago regarding the light regulation of fatty acid synthesis in isolated chloroplasts,[120] and photosystem I involvement is also supported by data on mutants of *Chlamydomonas*.[121] The nature of the interaction(s) between photosystem I and ACCase is not yet known. Although thioredoxin is responsible for light/dark control of several stromal enzymes, this does not appear to be true for ACCase.[149]

Although strong evidence now implicates ACCase as a point of regulation of plant FAS in at least some tissues, a major question which remains is how the activity of ACCase is coordinated with the utilization of fatty acids for glycerolipid synthesis. In animals, long-chain acyl-CoA acts as a feedback inhibitor of ACCase. This system serves to assure that production of fatty acids does not exceed their utilization (as acyl-CoA) by acyltransferases. However, long-chain acyl-CoA has only minor inhibition of plant ACC,[34] and furthermore, long-chain acyl-CoAs are not believed to be intermediates within the plastid. Since long-chain acyl-ACPs are the immediate product of plastid FAS, these metabolites might be expected to perform a similar feedback function. However, attempts to observe inhibition of plant ACC by long-chain acyl-ACP have not succeeded.[150] Thus, at present, there is no information about the nature of any feedback system which plant cells might use to coordinate the utilization of fatty acids with their production by plastid FAS.

VI. INHIBITORS OF FATTY ACID SYNTHESIS

Several inhibitors of plant fatty acid synthesis are known, and these have been useful in revealing additional information about fatty acid metabolism. Cerulenin is an antibiotic produced by the fungus *Cephalosporium caerulens*.[122] Cerulenin binds irreversibly to the active site of KAS I, thereby inhibiting fatty acid synthesis at the early condensation steps.[123] KAS II is much less sensitive to cerulenin inhibition and does not appear to form a covalent linkage to the antibiotic. KAS III is apparently not inhibited at all by cerulenin. Thus, the three ketoacyl-ACP synthases exhibit a wide range of sensitivities to cerulenin, which suggests that the active sites of the condensing enzymes may be structurally quite different.

ACCase is inhibited by three classes of inhibitors. The aryloxyphenoxypropionates include herbicides such as fluazifop or haloxyfop and are active at concentrations of 10^{-7} to $10^{-6} M$.[124] The cyclohexanediones include sethoxydim or tralkoxydim and are slightly less potent with I_{50} values of 10^{-6} to 10^{-5} M.[125-127] Recently, another grass-selective herbicide, PP600 (3-isopropyl-6-(N-[2,2-dimethylpropyl]-acetamido-1,3,5-triazine-2,4-(1H,3H)dione), has also been shown to act on ACCase.[128] However, the concentration required to inhibit *in vitro* assays was 20- to 30-fold higher than *in vivo* inhibition of acetate incorporation into lipids, suggesting its mode of action may require metabolism to a more active form. Interestingly, these three types of ACCase inhibitors are structurally quite different, but all are much more potent with ACCase from the Graminacea than from other plant families.

Allicin, a major flavor ingredient of garlic and a long-known antimicrobial agent, has been reported to specifically inhibit plant acetyl-CoA synthetase.[129] However, more recent data now indicate that the inhibition reported by allicin was due to contaminating acetate in the antibiotic preparations.[130]

VII. MUTATIONS AFFECTING FATTY ACID SYNTHESIS

Fatty acids are essential components of all membrane phospholipids, and therefore mutations affecting fatty acid production or chain length in vegetative tissues are likely to be lethal. Unlike bacteria and yeast, it is not possible to maintain higher plants which are fatty acid auxotrophs. However, some variations in fatty acid desaturation and composition within and between lipid classes is tolerated. Consequently, a large number of leaf lipid biosynthetic mutants have been isolated in *Arabidopsis* (for review, see Ohlrogge et al.[2]), but no mutations affecting leaf *de novo* fatty acid synthesis have been isolated.

Although it is unlikely that fatty acid biosynthetic mutants would be viable in vegetative tissues, they can be obtained in developing seeds. Because seeds produce fatty acids for storage oils, they can tolerate variations in fatty acid composition of the triacylglycerols and therefore present a distinct advantage for identification of fatty acid biosynthetic mutants without lethal consequences. In most cases, altered seed oils have been the result of breeding programs. For example, soybean has been bred for changes in either saturated (stearic and palmitic) or unsaturated (oleic, linoleic, and linolenic) content.[131] Often traits are identified by analyzing the seed oil, and then the character is followed by traditional breeding programs without any knowledge of which enzyme(s) may be responsible for the character. For example, several mutant lines of *Cuphea* have been isolated which have changes in the proportion of medium-chain fatty acids in the seed triacylglycerols.[132] Many of these lines segregate as if a single gene is responsible for the character. However, there is no biochemical evidence indicating what enzyme activity may be involved in generating the mutant oil composition. There has been speculation that either a medium-chain-specific thioesterase or a condensing enzyme with limited substrate specificity may be involved. Thioesterases which are specific for medium-chain fatty acids have been characterized and/or cloned from a number of sources (see accompanying sections in this chapter, as well as other relevant chapters in this book). Therefore, the existence of such activities is documented in these species. Condensing enzymes are known to have preferred chain length specificities also. For example, KAS I will elongate from C_2 to C_{16}, while KAS II is quite specific for C_{14} or C_{16} to C_{18} elongation. Furthermore, condensing enzymes are differentially effected by the antibiotic cerulenin, and one study has shown that in *Cuphea*, fatty acyl chain length can be effected by cerulenin.[133] However, no condensing enzymes have been characterized which have a substrate specificity which could account for the predominance of medium-chain fatty acids in these *Cuphea* lines.

VIII. SUMMARY AND FUTURE PERSPECTIVES

As summarized in Table 1, much recent research has focused on the enzymes of plant fatty acid biosynthesis, and in the past several years most of the component proteins have been purified and/or cloned. Thus, we now understand many of the characteristics and details of the structure of the members of this pathway. Nevertheless, major questions remain unanswered, and many of these deal with how the individual components of the fatty acid biosynthetic pathway *interact* with each other and with other components of lipid metabolism. We consider the following as important topics for future research:

1.   What is the source of plastidial acetyl-CoA? Does the pathway for acetyl-CoA production vary with different tissues and during their development?
2.   Is there a supramolecular organization of the lipid biosynthetic components such as a complex of enzymes in a "metabolon". If a complex of FAS enzymes exists, is the stoichiometry of the components always the same?
3.   What determines how much fatty acid is produced by a cell? Is control primarily through the level of FAS gene expression or through metabolic control over enzyme activity?
4.   How are the genes for FAS components regulated? Are all the members of the pathway regulated by the same developmental and environmental signals?
5.   What is the role of isoforms in fatty acid biosynthesis? Do all components of FAS have isoforms which are expressed in a tissue-specific manner?
6.   Is ACCase the major regulatory enzyme for FAS in all plant tissues?
7.   How is the synthesis of fatty acids in the plastid coordinated with the utilization of fatty acids outside the plastid? Are there signals that communicate between the endoplasmic reticulum and the plastid FAS, and if so, what are these signals?

TABLE 1
An Overview of Publications on the Purification and Cloning
of Plant FAS Proteins

FAS protein	Purification	Ref.	Cloning	Ref.
ACCase	Avocado	134	*Cyclotella*	37
	Cyclotella	34		
	Maize	38, 43		
	Parsley	135		
	Rapeseed	136		
	Soybean	137		
	Spinach	134		
	Wheat	135		
ACP	Avocado	138	*Arabidopsis*	
	Barley	54, 139	cDNA	50
	Rapeseed	113	genomic	64, 65
	Spinach	53, 138	Barley	
			cDNA	47, 140
			genomic	67, 140
			Rapeseed	
			cDNA	48, 51
			genomic	55
			Spinach (cDNA)	46, 49
Malonyl-CoA:ACP	Barley	74		
transacylase	Leek	77		
	Spinach	75		
	Soybean	76		
KAS I	Barley	78	Barley	78
	Rapeseed	79		
	Safflower	92		
	Spinach	141		
KAS II	Safflower	92		
	Spinach	80		
KAS III	Spinach	81		
3-ketoacyl-ACP reductase	Avocado	73, 87	*Arabidopsis*	89
	Spinach	88	Cuphea	90
			Rapeseed	89
3-hydroxyacyl-ACP	Spinach	88		
dehydratase				
Enoyl-ACP reductase	Avocado	73	Rapeseed	93
	Rapeseed	91, 142, 143		
	Safflower	92		
	Spinach	88		
Stearoyl-ACP desaturase	Avocado	98	Castor	98, 99
	Safflower	144	Cucumber	97
			Safflower	100
Acyl-ACP thioesterase	Avocado	145	California Bay	112
	Cuphea	146	Safflower	148
	California Bay	111, 147		
	Rapeseed	105		
	Squash	106		

Note: The selection of references is meant to be representative rather than comprehensive.

8. What is the nature of the metabolic regulation of the FAS pathway. Do some molecules such as acyl-ACP act as feedback regulators for this pathway?

The substantial recent progress summarized in this chapter and in Table 1 may now make it possible to address the questions listed above. Plants are a potentially rich source of a wide variety of useful chemicals, as exemplified by the storage triacylglycerols in oilseeds. To exploit the wide range of potential application of genetic engineering techniques toward the production of useful new plant lipids, it will be necessary to develop a detailed mechanistic understanding of most aspects of plant lipid metabolism. Developing answers to the above questions will be an important step toward obtaining such understanding.

REFERENCES

1. **Harwood, J. L.,** Fatty acid metabolism, *Annu. Rev. Plant Physiol. Plant Mol. Biol.,* 39, 101, 1988.
2. **Ohlrogge, J. B., Browse, J., and Somerville, C. R.,** The genetics of plant lipids, *Biochim. Biophys. Acta,* 1082, 1, 1991.
3. **Slabas, A. R. and Fawcett, T.,** The biochemistry and molecular biology of plant lipid biosynthesis, *Plant Mol. Biol.,* 19, 169, 1992.
4. **Post-Beittenmiller, D., Ohlrogge, J. B., and Somerville, C. R.,** Regulation of plant lipid biosynthesis: an example of developmental regulation superimposed on a ubiquitous pathway, in *Control of Plant Gene Expression,* Verma, D. P. S., Ed., CRC Press, Boca Raton, FL, 1992.
5. **Browse, J. and Somerville, C.,** Glycerolipid synthesis: biochemistry and regulation, *Annu. Rev. Plant Physiol. Plant Mol. Biol.,* 42, 467, 1991.
6. **Givan, C.V.,** The source of coenzyme A in chloroplasts of higher plants, *Physiol. Plant.,* 57, 311, 1983.
7. **Liedvogel, B.,** Acetyl coenzyme A and isopentenylpyrophosphate as lipid precursors in plant cells — biosynthesis and compartmentation, *Plant Physiol.,* 124, 211, 1986.
8. **Randall, D. D., Miernyk, J. A., Fang, T. K., Budde, R. J. A., and Schuller, K. A.,** Regulation of the pyruvate dehydrogenase complexes in plants, *Annu. NY Acad. Sci.,* 573, 192, 1989.
9. **Camp, P. J. and Randall, D. D.,** Purification and characterization of the pea chloroplast pyruvate dehydrogenase complex, *Plant Physiol.,* 77, 571, 1985.
10. **Treede, H.-J. and Heise, K.-P.,** Purification of the chloroplast pyruvate dehydrogenase complex from spinach and maize mesophyll, *Z. Naturforsch.,* 41, 1011, 1986.
11. **Andrews, T. J. and Kane, H. J.,** Pyruvate is a by-product of catalysis by ribulose-bisphosphate carboxylase/oxygenase., *J. Biol. Chem.,* 266, 9447, 1991.
12. **Roughan, P., Holland, R., Slack, C., and Mudd, J.,** Acetate is the preferred substrate for long-chain fatty acid synthesis in isolated spinach chloroplasts, *Biochem. J.,* 184, 565, 1978.
13. **Roughan, P. G., Holland, R., and Slack, C. R.,** On the control of long-chain-fatty acid synthesis in isolated intact spinach (*Spinacia oleracea*) chloroplasts, *Biochem. J.,* 184, 193, 1979.
14. **Kuhn, D., Knauf, M., and Stumpf, P.,** Subcellular localization of acetyl CoA synthetase in leaf protoplasts of *Spinacea oleracea, Arch. Biochem. Biophys.,* 209, 441, 1981.

15. **Springer, J. and Heise, K.-P.,** Comparison of acetate- and pyruvate-dependent fatty acid synthesis by spinach chloroplasts, *Planta,* 177, 417, 1989.
16. **Murphy, D. J. and Stumpf, P. K.,** The origin of chloroplastic acetyl coenzyme A, *Arch. Biochem. Biophys.,* 212, 730, 1981.
17. **Liedvogel, B. and Stumpf, P.,** Origin of acetate in spinach leaf cell, *Plant Physiol.,* 69, 897, 1982.
18. **Masterson, C., Wood, C., and Thomas, D. R.,** L-acetylcarnitine, a substrate for chloroplast fatty acid synthesis, *Plant Cell Environ.,* 13, 749, 1990.
19. **Masterson, C., Wood, C., and Thomas, D. R.,** Inhibition studies on acetyl group incorporation into chloroplast fatty acids, *Plant Cell Environ.,* 13, 767, 1990.
20. **Roughan, G., Post-Beittenmiller, D., Ohlrogge, J., and Browse, J.,** Acetylcarnitine and fatty acid synthesis by isolated chloroplasts, 1993, *Plant Physiol.,* in press.
21. **Miernyk, J. A. and Dennis, D. T.,** The incorporation of glycolytic intermediates into lipids by plastids isolated from the developing endosperm of castor oil seeds, *J. Exp. Bot.,* 34, 712, 1983.
22. **Miernyk, J. and Dennis, D.,** Isozymes of the glycolytic enzymes in endosperm from developing castor oil seeds, *Plant Physiol.,* 69, 825, 1982.
23. **Dennis, D. and Miernyk, J.,** Compartmentation of nonphotosynthetic carbohydrate metabolism, *Annu. Rev. Plant Physiol.,* 33, 27, 1982.
24. **Agrawal, P. and Canvin, D.,** The pentose phosphate pathway in relation to fat synthesis in the developing castor oil seed, *Plant Physiol.,* 47, 672, 1971.
25. **Smith, R. G., Gauthier, D. A., Dennis, D. T., and Turpin, D. H.,** Malate- and pyruvate-dependent fatty acid synthesis in leucoplasts from developing castor endosperm, *Plant Physiol.,* 98, 1233, 1992.
26. **Botham, P. A. and Ratledge, C.,** A biochemical explanation for lipid accumulation in candida 107 and other oleaginous micro-organisms, *J. Gen. Microbiol.,* 114, 361, 1979.
27. **Nelson, D. R. and Rinne, R. W.,** Citrate cleavage enzyme from developing soybean cotyledons, *Plant Physiol.,* 55, 69, 1975.
28. **Fritsch, H. and Beevers, H.,** ATP citrate lyase from germinating castor bean endosperm, *Plant Physiol.,* 63, 687, 1979.
29. **Kaethner, T. M. and Rees, A. T.,** Intracellular location of ATP citrate lyase in leaves of Pisum sativum L., *Planta,* 163, 290, 1985.
30. **Guchhait, R. B., Polakis, S. E., Dimroth, P., Stoll, E., Moss, J., and Lane, M. D.,** Acetyl coenzyme A carboxylase system of *Escherichia coli, J. Biol. Chem.,* 249, 6633, 1974.
31. **Kim, K.-H., Lopez-Casillas, F., Bai, B. H., Luo, X., and Pape, M. E.,** Role of reversible phosphorylation of acetyl-CoA carboxylase in long-chain fatty acid synthesis, *FASEB J.,* 3, 2250, 1989.
32. **Lopez-Casillas, F. and Kim, K.-H.,** Heterogeneity at the 5′ end of rat acetyl-coenzyme A carboxylase mRNA, *J. Biol. Chem.,* 264, 7176, 1989.
33. **Luo, X., Park, K., Lopez-Casillas, F., and Kim, K.-H.,** Structural features of the acetyl-CoA carboxylase gene: mechanisms for the generation of mRNAs with 5′ end heterogeneity, *Proc. Natl. Acad. Sci. U.S.A.,* 86, 4042, 1989.
34. **Roessler, P. G.,** Purification and characterization of acetyl-CoA carboxylase from the diatom cyclotella cryptica, *Plant Physiol.,* 92, 73, 1990.
35. **Roessler, P. G.,** Effects of silicon deficiency on lipid composition and metabolism in the diatom cyclotella cryptica, *J. Phycol.,* 24, 394, 1988.
36. **Roessler, P. G.,** Changes in the activities of various lipid and carbohydrate biosynthetic enzymes in the diatom *Cyclotella cryptica* in response to silicon deficiency, *Arch. Biochem. Biophys.,* 267, 521, 1988.
37. **Roessler, P. G. and Ohlrogge, J. B.,** Characterization of the gene for acetyl-CoA carboxylase from the alga *Cyclotella cryptica, Plant Physiol.,* 99, 113, 1992.
38. **Egli, M. A., Gengenbach, B. G., Gronwald, J. W., Somers, D. A., and Wyse, D. L.,** Purification of maize leaf acetyl CoA carboxylase, *Maize Gen. Coop. Newsl.,* 65, 95, 1991.

39. **Ebel, J. and Hahlbrock, K.**, Enzymes of flavone and flavonol-glycoside biosynthesis: coordinated and selective induction in cell-suspension cultures of *Petroselinum hortense*, *Eur. J. Biochem.*, 75, 201, 1977.

40. **Gehlert, R., Schoppner, A., and Kindl, H.**, Stilbene synthase from seedlings of *Pinus sylvestris*: purification and induction in response to fungal infection, *Mol. Plant Micr. Interact.*, 3, 444, 1990.

41. **Gamburg, K. Z., Glusdo, O. V., and Rekoslavskaya, N. I.**, The content of N-malonyl-D-tryptophan in seeds and seedlings of plants, *Plant Sci.*, 77, 149, 1991.

42. **Nikolau, B. J., Croxdale, J., Ulrich, T. H., and Wurtele, E. S.**, Acetyl-CoA carboxylase and biotin-containing proteins in carrot somatic embryogenesis, in *The Metabolism, Structure, and Function of Plant Lipids*, Stumpf, P. K. and Mudd, J. B., Eds., Plenum Press, New York., 1987, 517.

43. **Nikolau, B. J. and Hawke, J.C.**, Purification and characterization of maize leaf acetyl-coenzyme A carboxylase, *Arch. Biochem. Biophys.*, 228, 86, 1984.

44. **Ohlrogge, J. B.**, Biochemistry of plant acyl carrier proteins, in *The Biochemistry of Plants*, Stumpf, P.K., Ed., Academic Press, Orlando, FL, 1987, 137.

45. **Ohlrogge, J. B. and Kuo, T. M.**, Spinach acyl carrier protein: primary structure, mRNA translation and immunoelectrophoretic analysis, in *Structure, Function, and Metabolism of Plant Lipids*, Siegenthaler, P. A. and Eichenberger, W., Eds., Elsevier, Amsterdam, 1984, 63.

46. **Scherer, D. E. and Knauf, V. C.**, Isolation of a cDNA clone for the acyl carrier protein-I of spinach, *Plant Mol. Biol.*, 9, 127, 1987.

47. **Hansen, L.**, Three cDNA clones for barley leaf acyl carrier proteins I and II, *Carlsberg Res. Commun.*, 52, 381, 1987.

48. **Safford, R., Windust, J. H. C., Lucas, C., Silva, J.D., James, C. M., Hellyer, A., Smith, C. G., Slabas, A. R., and Hughes, S. G.**, Plastid-localised seed acyl-carrier protein of *Brassica napus* is encoded by a distinct, nuclear multigene family, *Eur. J. Biochem.*, 174, 287, 1988.

49. **Schmid, K. M. and Ohlrogge, J. B.**, A root acyl carrier protein-II from spinach is also expressed in leaves and seeds, *Plant Mol. Biol.*, 15, 765, 1990.

50. **Hlousek-radojcic, A., Post-Beittenmiller, D., and Ohlrogge, J. B.**, Expression of constitutive and tissue-specific acyl carrier protein isoforms in *Arabidopsis*, *Plant Physiol.*, 98, 206, 1992.

51. **Rose, R. E., DeJesus, C. E., Moylan, S. L., Ridge, N. P., Scherer, D. E., and Knauf, V. C.**, The nucleotide sequence of a cDNA clone encoding acyl carrier protein (ACP) from *Brassica campestris* seeds, *Nucl. Acids Res.*, 15, 7197, 1987.

52. **Fernandez, M. D. and Lamppa, G. K.**, Acyl carrier protein import into chloroplasts — both the precursor and mature forms are substrates for phosphopantetheine attachment by a soluble chloroplast holo-acyl carrier protein synthase, *J. Biol. Chem.*, 266, 7220, 1991.

53. **Ohlrogge, J. B. and Kuo, T. M.**, Plants have isoforms of acyl carrier proteins that are expressed differently in different tissues, *J. Biol. Chem.*, 260, 8032, 1985.

54. **Hoj, P. and Svendson, I.**, Barley chloroplasts contain two acyl carrier proteins coded for by different genes, *Carlsberg Res. Commun.*, 49, 483, 1984.

55. **DeSilva, J., Loader, N., Jarman, C., Windust, J., Hughes, S., and Safford, R.**, The isolation and sequence analysis of two seed-expressed acyl carrier protein genes from *Brassica napus*, *Plant Mol. Biol.*, 14, 537, 1990.

56. **Battey, J. F. and Ohlrogge, J. B.**, Evolutionary and tissue-specific control of expression of multiple acyl carrier protein isoforms in plants and bacteria, *Planta*, 180, 352, 1990.

57. **Guerra, D. J., Ohlrogge, J. B., and Frentzen, M.**, Activity of acyl carrier protein isoforms in reactions of plant fatty acid metabolism, *Plant Physiol.*, 82, 448, 1986.

58. **Post-Beittenmiller, D., Jaworski, J. G., and Ohlrogge, J. B.**, *In vivo* pools of free and acylated acyl carrier proteins in spinach — evidence for sites of regulation of fatty acid biosynthesis, *J. Biol. Chem.*, 266, 1858, 1991.

59. **Ohlrogge, J. B., Kuhn, D. N., and Stumpf, P. K.,** Subcellular localization of acyl carrier protein in leaf protoplasts of *Spinacia oleracea, Proc. Natl. Acad. Sci. U.S.A.,* 76, 1194, 1979.

60. **Chuman, L. and Brody, S.,** Acyl carrier protein is present in the mitochondria of plants and eucaryotic micro-organisms, *Eur. J. Biochem.,* 184, 643, 1989.

61. **Shintani, D. and Ohlrogge, J. B.,** The cloning and characterization of a putative mitochondrial acyl carrier protein from *Arabidopsis thaliana, Plant Physiol.,* 99, 88, 1992.

62. **Sackmann, U., Zensen, R., Rohlen, D., Jahnke, U., and Weiss, H.,** The acyl-carrier protein in *Neurospora crassa* mitochondria is a subunit of NADH:ubiquinone reductase (complex I), *Eur. J. Biochem.,* 200, 463, 1991.

63. **Runswick, M. J., Fearnley, I. M., Mark, S. J., and Walker, J. E.,** Presence of an acyl carrier protein in NADH:ubiquinone oxidoreductase from bovine heart mitochondria, *FEBS,* 286, 121, 1991.

64. **Post-Beittenmiller, M. A., Hlousek-Radojcic, A., and Ohlrogge, J. B.,** DNA sequence of a genomic clone encoding an *Arabidopsis* acyl carrier protein, *Nucl. Acids Res.,* 17, 1777, 1989.

65. **Lamppa, G. and Jacks, C.,** Analysis of two linked genes coding for the acyl carrier protein (ACP) from *Arabidopsis thaliana* (Columbia), *Plant Mol. Biol.,* 16, 469, 1991.

66. **Hansen, L. and von Wettstein-Knowles, P.,** The barley genes Acl1 and Acl3 encoding acyl carrier proteins I and III are located on different chromosomes, *Mol. Gen. Genet.,* 229, 467, 1991.

67. **Devos, K. M., Chinoy, C. N., Atkinson, M. D., Hansen, L., von Wettstein-Knowles, P., and Gale, M. D.,** Chromosomal location in wheat of the genes coding for the acyl carrier protein-I and protein-III, *Theor. Appl. Genet.,* 82, 3, 1991.

68. **von Wettstein-Knowles, P.,** Molecular genetics of lipid synthesis in barley, in *Barley Genetics VI,* Munek, L., Ed., Munksgaard International Publishers Ltd., Copenhagen, 1992, 753.

69. **Stoops, J. K., Singh, N., and Wakil, S. J.,** The yeast fatty acid synthase: pathways for transfer of the acetyl group from coenzyme A to the cys-SH of the condensation site, *J. Biol. Chem.,* 265, 16971, 1990.

70. **Joshi, V. C. and Wakil, S. J.,** Studies on the mechanism of fatty acid synthesis: purification and properties of malonyl-coenzyme A-acyl carrier protein transacylase of *Escherichia coli, Arch. Biochem. Biophys.,* 143, 493, 1971.

71. **Lowe, P. N. and Rhodes, S.,** Purification and characterization of [acyl-carrier-protein] acetyltransferase from *Escherichia coli, Biochem. J.,* 250, 789, 1988.

72. **Shimakata, T. and Stumpf, P. K.,** The purification and function of acetyl coenzyme A: acyl carrier protein transacylase, *J. Biol. Chem.,* 258, 3592, 1983.

73. **Caughey, I. and Kekwich, R. G. O.,** The characteristics of some components of the fatty acid synthetase system in the plastids from the mesocarp of avocado fruit, *Eur. J. Biochem.,* 123, 553, 1982.

74. **Hoj, P. and Mikkelsen, J.,** Partial separation of individual enzyme activities of an ACP-dependent fatty acid synthetase from barley chloroplasts, *Carlsberg Res. Commun.,* 47, 119, 1982.

75. **Stapleton, S. R. and Jaworski, J. G.,** Characterization and purification of malonyl-coenzyme A: [acyl-carrier-protein] transacylases from spinach and *Anabaena variabilis, Biochim. Biophys. Acta,* 51658, 1, 1983.

76. **Guerra, D. J. and Ohlrogge, J. B.,** Partial purification and characterization of two forms of malonyl-CoA:ACP transacylase from soybean leaf tissue, *Arch. Biochem. Biophys.,* 246, 274, 1986.

77. **Lessire, R. and Stumpf, P. K.,** Nature of fatty acid synthetase systems in parenchymal and epidermal cells of *Allium porrum* L. leaves, *Plant Physiol.,* 73, 614, 1982.

78. **Siggaard-Andersen, M., Kauppinen, S., and von Wettstein-Knowles, P.,** Primary structure of a cerulenin-binding beta-ketoacyl-(acyl carrier protein) synthase from barley chloroplasts, *Proc. Natl. Acad. Sci. U.S.A.,* 88, 4114, 1991.

79. **MacKintosh, R. W., Hardie, D. G., and Slabas, A. R.**, A new assay procedure to study the induction of β-ketoacyl-ACP synthase I and II, and the complete purification of β-ketoacyl-ACP synthase I from developing seeds of oilseed rape (*Brassica napus*), *Biochim. Biophys. Acta,* 1002, 114, 1989.

80. **Shimakata, T. and Stumpf, P. K.**, Isolation and function of spinach leaf β-ketoacyl-[acyl-carrier-protein] synthases, *Proc. Natl. Acad. Sci. U.S.A.,* 79, 5808, 1982.

81. **Clough, R. C., Matthis, A. L., Barnum, S. R., and Jaworski, J. G.**, Purification and characterization of 3-ketoacyl-acyl carrier protein synthase from spinach: a condensing enzyme utilizing acetyl-CoA to initiate fatty acid synthesis, *J. Biol. Chem.,* 267, 20992, 1992.

82. **Jaworski, J. G., Clough, R. C., and Barnum, S. R.**, A cerulenin insensitive short chain 3-ketoacyl-acyl carrier protein synthase in *Spinacia oleracea* leaves, *Plant Physiol.,* 90, 41, 1989.

83. **Jaworski, J. G., Post-Beittenmiller, D., and Ohlrogge, J. B.**, Acetyl-acyl carrier protein is not a major intermediate of fatty acid biosynthesis in plants, *Eur. J. Biochem.,* in press.

84. **Jackowski, S. and Rock, C. O.**, Acetoacetyl-acyl carrier protein synthase, a potential regulator of fatty acid biosynthesis in bacteria, *J. Biol. Chem.,* 262, 7927, 1987.

85. **Jackowski, S., Murphy, C. M., Cronan, J. E., and Rock, C. O.**, Acetoacetyl-acyl carrier protein synthase: a target for the antibiotic thiolactomycin, *J. Biol. Chem.,* 264, 7624, 1989.

86. **Tsay, J., Oh, W., Larson, T. J., Jackowski, S., and Rock, C. O.**, Isolation and characterization of the 3-ketoacyl-acyl carrier protein synthase III gene (*fabH*) from *Escherichia coli* K-12, *J. Biol. Chem.,* 267, 6807, 1992.

87. **Sheldon, P. S., Keckwick, R. G. O., Sidebotton, C. G., and Slabas, A. R.**, 3-Oxoacyl-(acyl carrier protein) reductase from avocado (*Persea americana*) fruit mesocarp, *Biochem. J.,* 271, 713, 1990.

88. **Shimakata, T. and Stumpf, P. K.**, Purification and characteristics of β-ketoacyl-[acyl-carrier-protein] reductase, β-hydroxyacyl-[acyl-carrier-protein] dehydrase, and enoyl-[acyl-carrier-protein] reductase from *Spinacia oleracea* leaves, *Arch. Biochem. Biophys.,* 218, 77, 1982.

89. **Slabas, A. R., Chasen, D., Nishida, I., Murata, N., Sidebottom, C., Safford, R., Sheldon, P. S., Kekwick, R. G. O., Hardie, D. G., and Mackintosh, R. W.**, Molecular cloning of higher plant 3-oxoacyl-(acyl carrier protein) reductase, *Biochem. J.,* 283, 321, 1992.

90. **Klein, B., Pawlowski, K., Horicke-Grandpierre, C., Schell, J., and Topfer, R.**, Isolation and characterization of a cDNA from *Cuphea lanceolata* endocing a β-ketoacyl-ACP reductase, *Mol. Gen. Genet.,* 233, 122, 1992.

91. **Slabas, A., Sidebottom, C., Hellyer, A., Kessell, R., and Tombs, M.**, Induction, purification and characterization of NADH-specific enoyl acyl carrier protein reductase from developing seeds of oil seed rape (*Brassica napus*), *Biochim. Biophys. Acta,* 877, 271, 1986.

92. **Shimakata, T. and Stumpf, P. K.**, The procaryotic nature of the fatty acid synthetase of developing *Carthamus tinctorius* L. seeds, *Arch. Biochem. Biophys.,* 217, 144, 1982.

93. **Kater, M. M., Koningstein, G. M., Nijkamp, H. J. J., and Stuitje, A. R.**, cDNA cloning and expression of *Brassica napus* enoyl-acyl carrier protein reductase in *Escherichia coli,* *Plant Mol. Biol.,* 17, 895, 1991.

94. **Nagai, J. and Bloch, K.**, Enzymatic desaturation of stearoyl acyl carrier protein, *J. Biol. Chem.,* 243, 4626, 1968.

95. **Jacobson, B., Jaworski, J., and Stumpf, P.**, Fat metabolism in higher plants LXII. Stearoyl-acyl carrier protein desaturase from spinach chloroplasts, *Plant Physiol.,* 54, 484, 1974.

96. **McKeon, T. and Stumpf, P.**, Purification and characterization of the stearoyl-acyl carrier protein desaturase and the acyl-acyl carrier protein thioesterase from maturing seeds of safflower, *J. Biol. Chem.,* 257, 12141, 1982.

97. **Shanklin, J., Mullins, C., and Somerville, C.**, Sequence of a complementary DNA from *Cucumis sativus* L. encoding the stearoyl-acyl-carrier protein desaturase, *Plant Physiol.*, 97, 467, 1991.

98. **Shanklin, J. and Somerville, C.**, Stearoyl-acyl-carrier-protein desaturase from higher plants is structurally unrelated to the animal and fungal homologs, *Proc. Natl. Acad. Sci. U.S.A.*, 88, 2510, 1991.

99. **Knutzon, D. S., Scherer, D. E., and Schreckengost, W. E.**, Nucleotide sequence of a complementary DNA clone encoding stearoyl-acyl carrier protein desaturase from castor bean, *Ricinus communis*, *Plant Physiol.*, 96, 344, 1991.

100. **Thompson, G. A., Scherer, D. E., Foxall-van Aken, S., Kenny, J. W., Young, H. L., Shintani, D. K., Kridl, J. C., and Knauf, V. C.**, Primary structures of the precursor and mature forms of stearoyl-acyl carrier protein desaturase from safflower embryos and require-ment of ferredoxin for enzyme activity, *Proc. Natl. Acad. Sci. U.S.A.*, 88, 2578, 1991.

101. **Nishida, I., Beppu, T., Matsuo, T., and Murata, N.**, Nucleotide sequence of a cDNA clone encoding a precursor to stearoyl-(acyl-carrier-protein) desaturase from spinach, *Spinacia oleracea*, *Plant Mol. Biol.*, 19, 711, 1992.

102. **Jaworski, J. and Stumpf, P.**, Fat metabolism in higher plants LIX. Properties of a soluble stearoyl-ACP desaturase from maturing *Carthamus tinctorius*, *Arch. Biochem. Biophys.*, 162, 158, 1974.

103. **Cahoon, E. B., Shanklin, J., and Ohlrogge, J. B.**, Expression of a novel coriander desaturase results in petroselinic acid production in transgenic tobacco, *Proc. Natl. Acad. Sci. U.S.A.*, 89, 11184, 1992.

104. **Ohlrogge, J., Shine, W., and Stumpf, P.**, Fat metabolism in higher plants, *Arch. Biochem. Biophys.*, 189, 382, 1978.

105. **Hellyer, A. and Slabas, A. R.**, Acyl-[acyl-carrier protein] thioesterase from oil seed rape — purification and characterization, in *Plant Lipid Biochemistry, Structure and Utilization*, Quinn, B. J. and Harwood, J. L., Eds., Portland Press, London, 1990, 157.

106. **Imai, H., Mishida, I., and Murata, N.**, Acyl-[acyl-carrier-protein] hydrolase of squash cotyledons: purification and characterization, in *Plant Lipid Biochemistry, Structure and Utilization*, Quinn, B. J. and Harwood, J. L., Eds., Portland Press, London, 1990, 160.

107. **Frentzen, M., Heinz, E., McKeon, T. A., and Stumpf, P. K.**, Specificities and selec-tivities of glycerol-3-phosphate acyltransferase and monoacylglcerol-3-phosphate acyltransferase from pea and spinach chloroplasts, *Eur. J. Biochem.*, 129, 629, 1983.

108. **Andrews, J., Ohlrogge, J. B., and Keegstra, K.**, Final step of phosphatidic acid synthesis in pea chloroplasts occurs in the inner envelope membrane, *Plant Physiol.*, 78, 459, 1985.

109. **Roughan, P. and Slack, C.**, Cellular organization of glycerolipid metabolism, *Annu. Rev. Plant Physiol.*, 33, 97, 1982.

110. **Thompson, A. E. and Kleiman, R.**, Effect of seed maturity on seed oil, fatty acid and crude protein content of eight cuphea species, *J. Am. Oil Chem. Soc.*, 65, 139, 1988.

111. **Pollard, M. R., Anderson, L., Fan, C., Hawkins, D. J., and Davies, H. M.**, A specific acyl-ACP thioesterase implicated in medium-chain fatty acid production in immature cotyledons of *Umbellularia californica*, *Arch. Biochem. Biophys.*, 284, 306, 1991.

112. **Voelker, T. A., Worrell, A. C., Anderson, L., Bleibaum, J., Fan, C., Hawkins, D. J., Radke, S. E., and Davies, H. M.**, Fatty acid biosynthesis redirected to medium chains in transgenic oilseed plants, *Science*, 257, 72, 1992.

113. **Slabas, A. R., Harding, J., Hellyer, A., Roberts, P., and Bambridge, H. E.**, Induction, purification and characterization of acyl carrier protein from developing seeds of oil seed rape (*Brassica napus*), *Biochim. Biophys. Acta*, 921, 50, 1987.

114. **Post-Beittenmiller, M. A., Schmid, K. M., and Ohlrogge, J. B.**, Expression of holo and apo forms of spinach acyl carrier protein-I in leaves of transgenic tobacco plants, *Plant Cell*, 1, 889, 1989.

115. **Rolleston, F. S.**, A theoretical background to the use of measured concentrations of intermediates in study of the control of intermediary metabolism, *Curr. Top. Cell. Regul.*, 5, 47, 1972.

116. **Post-Beittenmiller, D., Roughan, G., and Ohlrogge, J. B.,** Regulation of plant fatty acid biosynthesis: analysis of acyl-CoA and acyl-ACP substrate pools in chloroplasts from spinach and pea, *Plant Physiol.,* 100, 923, 1992.

117. **Browse, J., Roughan, P., and Slack, C.,** Light control of fatty acid synthesis and diurnal flucuations of fatty acid composition in leaves, *Biochem. J.,* 196, 347, 1981.

118. **Eastwell, K. C. and Stumpf, P. K.,** Regulation of plant acetyl-CoA carboxylase by adenylate nucleotides, *Plant Physiol.,* 72, 50, 1983.

119. **Sauer, A. and Heise, K.-P.,** Regulation of acetyl-coenzyme A carboxylase and acetyl-coenzyme A synthetase in spinach chloroplasts, *Z. Naturforsch.,* 39c, 268, 1984.

120. **Nakamura, Y. and Yamada, M.,** Fatty acid synthesis by spinach chloroplasts III. Relationship between fatty acid synthesis and photophosphorylation, *Plant Cell Physiol.,* 16, 163, 1975.

121. **Picaud, A., Creach, A., and Tremolieres, A.,** Light-stimulated fatty acid synthesis in Chlamydomonas whole cells, in *Plant Lipid Biochemistry, Structure and Utilization,* Quinn, B. and Harwood, J., Eds., Portland Press, London, 1990, 393.

122. **Omura, S.,** The antibiotic cerulenin, a novel tool for biochemistry as an inhibitor of fatty acid synthesis, *Bacteriol. Rev.,* 40, 681, 1976.

123. **Kauppinen, S., Siggaard-Anderson, M., and von Wettstein-Knowles, P.,** β-Ketoacyl-ACP synthase I of *Escherichia coli*: nucleotide sequence of the *fabB* gene and identification of the cerulenin binding residue, *Carlsberg Res. Commun.,* 53, 357, 1988.

124. **Kobek, K., Focke, M., and Lichtenthaler, H. K.,** Fatty-acid biosynthesis and acetyl-CoA carboxylase as a target of diclofop, fenoxaprop and other aryloxy-phenoxy-propionic acid herbicides, *Z. Naturforsch.,* 43c, 47, 1988.

125. **Walker, K. A., Ridley, S. M., Lewis, T., and Harwood, J. L.,** Fluazifop, a grass-selective herbicide which inhibits acetyl-CoA carboxylase in sensitive plant species, *Biochem. J.,* 254, 307, 1988.

126. **Focke, M. and Lichtenthaler, H. K.,** Inhibition of the acetyl-CoA carboxylase of barley chloroplasts by cycloxydim and sethoxydim, *Z. Naturforsch.,* 42, 1361, 1987.

127. **Burton, J. D., Gronwald, J. W., Somers, D. A., Connelly, J. A., Gengenbach, B. G., and Wyse, D. L.,** Inhibition of plant acetyl-coenzyme A carboxylase by the herbicides sethoxydim and haloxyfop, *Biochem. Biophys. Res. Commun.,* 148, 1039, 1987.

128. **Walker, K. A., Ridley, S. M., Lewis, T., and Harwood, J. L.,** A new class of herbicide which inhibits acetyl-CoA carboxylase in sensitive plant species, *Phytochemistry,* 29, 3743, 1990.

129. **Focke, M., Feld, A., and Lichtenthaler, H.,** Allicin, a naturally occurring antibiotic from garlic, specifically inhibits acetyl-CoA synthetase, *FEBS Lett.,* 261, 106, 1990.

130. **Lichtenthaler, H. K., Focke, M., Golz, A., Hoffmann, S., Kobek, K., and Motel, A.,** Investigation on the starting enzymes of fatty acid biosynthesis and their inhibition, in *The Proceedings of the Tenth International Symposium on Plant Lipids,* Cherif, A., Ben Miled-Daoud, D., Marzouk, B., Smaoui, A., and Zarrouk, M., Eds., Centre National Pédagogique, Tunisia, 1992, 103.

131. **Anon.,** Economic Implications of Modified Soybean Traits, Iowa Soybean Promotion Board, Ames, IA, 1990.

132. **Knapp, S. J.,** Modifying the seed oils of Cuphea — a new commercial source of caprylic, capric, lauric, and myristic acid, in "Seed oils for the future", Mackenzie, S. L. and Taylor, D. C., Eds., *J. Am. Oil. Chem. Soc.,* in press.

133. **Heise, K.-P. and Fuhrmann, J.,** Medium-chain fatty acid synthesis by plastids from *Cuphea* embryos, in *The Proceedings of the Tenth International Symposium on Plant Lipids,* Cherif, A., Ben Miled-Daoud, D., Marzouk, B., Smaoui, A., and Zarrouk, M., Eds., Centre National Pédagogique, Tunisia, 1992, 125.

134. **Mohan, S. B. and Kekwick, R. G. O.,** Acetyl-CoA carboxylase from avocado plastids and spinach chloroplasts, *Biochem. J.,* 187, 667, 1980.

135. **Egin-Buhler, B., Loyal, R., and Ebel, J.,** Comparison of acetyl-CoA carboxylases from parsley cell cultures and wheat germ, *Arch. Biochem. Biophys.,* 203, 90, 1980.

136. **Slabas, A. R. and Hellyer, A.**, Rapid purification of a high molecular weight subunit polypeptide form of rape seed acetyl CoA carboxylase, *Plant Sci.*, 39, 177, 1985.

137. **Charles, D. J. and Cherry, J. H.**, Purification and characterization of acetyl-CoA carboxylase from developing soybean seeds, *Phytochemistry*, 25, 1067, 1986.

138. **Simoni, R., Criddle, R., and Stumpf, P.**, Fat metabolism in higher plants XXXI. Purification and properties of plant and bacterial acyl carrier proteins, *J. Biol. Chem.*, 212, 573, 1967.

139. **Hoj, P. and Svendsen, I.**, Barley acyl carrier protein: its amino acid sequence and assay using purified malonyl-CoA:ACP transacylase, *Carlsberg Res. Commun.*, 48, 285, 1983.

140. **Hansen, L. and Kauppinen, S.**, Barley acyl carrier protein II: nucleotide sequence of cDNA clones and chromosomal location of the Ac12 gene, *Plant Physiol.*, 97, 472, 1991.

141. **Shimakata, T. and Stumpf, P. K.**, Purification and characterization of β-ketoacyl-ACP synthetase I from *Spinacia oleracea* leaves, *Arch. Biochem. Biophys.*, 220, 39, 1983.

142. **Slabas, A. R., Cottingham, I., Austin, A., Fawcett, T., and Sidebottom, C. M.**, Amino acid sequence analysis of rape seed (*Brassica napus*) NADH-enoyl ACP reductase, *Plant Mol. Biol.*, 17, 911, 1991.

143. **Cottingham, I. R., Austin, A., Sidebottom, C., and Slabas, A. R.**, Purified enoyl-[acyl-carrier-protein] reductase from rape seed (*Brassica napus*) contains two closely related polypeptides which differ by a six-amino-acid N-terminal extension, *Biochim. Biophys. Acta*, 954, 201, 1988.

144. **McKeon, T. and Stumpf, P.**, Purification and characterization of the stearoyl-acyl carrier protein desaturase and the acyl-acyl carrier protein thioesterase from maturing seeds of safflower, *J. Biol. Chem.*, 257, 12141, 1982.

145. **Ohlrogge, J. B., Shine, W. E., and Stumpf, P. K.**, Fat metabolism in higher plants. Characterization of plant acyl-ACP and acyl-CoA hydrolases, *Arch. Biochem. Biophys.*, 189, 382, 1978.

146. **Dormann, P., Spener, F., and Ohlrogge, J. B.**, Characterization of two acyl-ACP thioesterases specific for medium chain acyl-ACPs and oleoyl-ACP from developing *Cuphea* seeds, *Planta*, in press.

147. **Davies, H. M., Anderson, L., Fan, C., and Hawkins, D. J.**, Developmental induction, purification, and further characterization of 12:0-ACP thioesterase from immature cotyledons of *Umbellularia californica*, *Arch. Biochem. Biophys.*, 290, 37, 1991.

148. **Knutzon, D. S., Bleibaum, J. L., Nelsen, J., Kridl, J. C., and Thompson, G. A.**, Isolation and characterization of two safflower oleoyl-acyl carrier protein thioesterase cDNA clones, *Plant Physiol.*, 100, 1751, 1992.

149. **Nakahira, K. and Ohlrogge, J. B.**, unpublished data, 1992.

150. **Roessler, P. G. and Ohlrogge, J. B.**, unpublished data, 1991.

Chapter 2

BIOSYNTHESIS OF POLYUNSATURATED FATTY ACIDS

Ernst Heinz

TABLE OF CONTENTS

0-8493-4907-9/93/$0.00+$.50
© 1993 by CRC Press Inc.

ABBREVIATIONS

Not defined in the text: ACP — acyl carrier protein; CoA — coenzyme A; DAG and TAG — di- and triacylglycerol; ER — endoplasmic reticulum; MGD, DGD, and SQD — monogalactosyl-, digalactosyl-, and sulfoquinovosyl diacylglycerol; PA — phosphatidic acid; PC, PE, PG, and PI — phosphatidyl choline, -ethanolamine, -glycerol, and -inositol.

I. INTRODUCTION

In the plant kingdom, a multiplicity of fatty acids with different chain lengths and one, two, or more double bonds in *cis(Z)*- or *trans(E)*-configuration, and even triple bonds are found.[1] Fatty acids are called polyunsaturated when they have two or more double bonds, and this chapter will place emphasis on the biochemistry and enzymology of the introduction of the second and third *cis* double bond into acyl groups. But only a few polyunsaturated fatty acids can be dealt with because only a limited number have been included in detailed studies up to now. These are the unsaturated C_{16} and C_{18} acyl substituents commonly found in membrane lipids and reserve TAGs of higher plants, which have abandoned the diversity elaborated by more primitive organisms, particularly by the various groups of eukaryotic algae.[2]

In view of the various systems used to name and abbreviate fatty acid structures, a short note on nomenclature is necessary. As exemplified with α-linolenic acid, a systematic name such as (9Z, 12Z, 15Z)-octadecatrienoic acid is too long for repeated use, and common or trivial names do not exist for many fatty acids. Therefore, abbreviations will be used which vary depending on the precision required in a given context. They may give only the carbon and double bond numbers separated by a colon (i.e., 18:3) or include location of double bonds with counting their position, usually from the carboxyl group (9,12,15-18:3), or in addition indicate the configuration of double bonds with *cis* for Z and *trans* for E, along with "all", if all double bonds have the same configuration (all *cis*-9,12,15-18:3). Occasionally, counting of the first carbon atom of a double bond starts from the methyl end (*n*-3,6,9-18:3) to clarify correlations not obvious when counting from the carboxyl group. The two different ways of counting have also been used to specify desaturases[3] which recognize the position of the future double bond by measuring the distance either from the carboxyl group (indicated by the prefix Δ^x desaturase) or from the methyl end (*n*-x desaturase).

cis Double bonds are required for fluidity and mobility in membranes and other hydrophobic aggregates. The introduction of *cis* double bonds into saturated fatty acids lowers their melting points and the phase transition temperatures of the corresponding membrane lipids.[204] The drop in melting point is largest for the first *cis* double bond when placed into the middle of the chain. The second and third double bonds result in increasingly smaller differences, whereas the fourth effects a large drop again when it disrupts the saturated

TABLE 1
Some Common Saturated and Unsaturated Fatty Acids
with 16 and 18 Carbon Atoms

Common name as acid	Abbreviation of structure	Mp °C
Stearic	18:0	70
Oleic	9-*cis*-18:1	13
Elaidic	9-*trans*-18:1	45
Stearolic	9-*yn*-18:1	48
Petroselinic	6-*cis*-18:1	32
Vaccenic	11-*cis*-18:1	15
Linoleic	9,12-*cis*-18:2	−5
—	6,9-*cis*-18:2	−11
α-linolenic	9,12,15-*cis*-18:3	−11
γ-linolenic	6,9,12-*cis*-18:3	not determined
Stearidonic	6,9,12,15-*cis*-18:4	−57
Palmitic	16:0	63
Palmitoleic	9-*cis*-16:1	0
—	3-*trans*-16:1	53

Note: The melting points (taken from Ref. 204) may vary slightly depending on the reference.

segment extending toward the carboxyl end (Table 1). The phase transition temperatures of the corresponding diacyl lipids are even lower as evident from a comparison of the melting point of oleic acid (+13°C) and the phase transition temperature of dioleoyl PC (−22°C). Cells make use of these differences to maintain a proper fluidity in membranes at a given temperature by incorporating appropriate mixtures of acyl groups into polar membrane lipids. But reserve lipids also contain polyunsaturated fatty acids, apparently because they are handled more easily in a fluid state during deposition, storage, and consumption, despite the fact that this requires additional enzymes for β-oxidation (see Chapter 16). On the other hand, there are no obvious reasons why marine algae produce large quantities of highly unsaturated acids such as all *cis*-5,8,11,14,17-eicosapentaenoic acid[2] with a melting point of −54°C in a heat-buffered environment, while, on the other hand, higher plants can cope with a temperature span of nearly 100°C with a simple set of just two or three polyunsaturated fatty acids. The prevailing components of this latter set are linoleic and α-linolenic acid, both of which are derived from oleic acid by additional methylene-interrupted *cis* double bonds placed into the C-terminal segment of the parent chain in positions which are not in reach of animal desaturases. Only some insects,[4] such as the cockroach, can produce linoleic from oleic acid. Therefore, in mammals, linoleic and linolenic acid are essential fatty acids and required for an appropriate balance in the human diet.[5] Nevertheless, the proportion of linolenic acid is kept at a minimum in edible oils to increase their chemical stability. In contrast, oleic acid has a very good shelf life and is stable for several thousand years if protected from microbial attack; it is the only unsaturated fatty acid which has survived in remnants of olive oil in Roman pottery, in ancient maize kernels harvested by American Indians 1700 years ago, in hazel nuts preserved for 5000 years after fire roasting, or in meat from mammoth kept frozen for 40,000 years.[6,7]

II. STEREOCHEMISTRY OF HYDROGEN ABSTRACTION

The first experiments on desaturation were concerned with the demonstration that various fatty acids added exogenously to different organisms (bacteria, yeasts, algae, leaves, or leaf homogenates) were desaturated. These experiments showed that, in fact, exogenous [1-^{14}C]stearic acid was desaturated in aerobic reactions sequentially via 18:1 and 18:2 to 18:3. Corresponding results were obtained with [1-^{14}C]oleic acid, which in contrast to stearate is better accepted by leaves.[3] Although, based on current knowledge (see below), one would hesitate to ascribe the above-mentioned sequences to a single compartment, these investigations enabled subsequent experiments on the stereochemistry of hydrogen removal in the course of double bond formation. The method was elucidated while investigating the conversion of stearic to oleic acid. A series of deuterium- and tritium-labeled stearic acids was prepared and incubated,[8,9] including racemic erythro- and threo-9,10-dideuterostearates, as well as 9-L, 9-D, 10-L, and 10-D-tritiostearate. The analysis of tracer retained in the resulting oleic acid indicated that *Corynebacterium* and *Chlorella* (the only organisms investigated so far) remove enantioselectively the D-hydrogen atoms (pro-R) from the prochiral centers at C-9 and C-10. Therefore, the introduction of the first double bond represents a *cis* elimination of the D-erythro pair of hydrogens.

To investigate the same problem during the formation of the second double bond, further substrates were synthesized,[9] including racemic erythro- and threo-12,13-dideutero oleic acid, as well as 12-D and 12-L-tritiostearate. Analysis of the linoleic acid recovered after desaturation showed that the second double bond is also formed by an enantioselective *cis* removal of a pair of hydrogens from the D-erythro-12,13-positions. Similar experiments with further substrates[9] labeled in the 15,16-position demonstrated that also the third desaturation converting linoleic to linolenic acid proceeds via *cis* abstraction of an erythro pair of hydrogens from the 15,16-positions. In this case, an absolute assignment to the D- or L-pairs has not yet been carried out. But we may conclude that most likely all three double bonds are formed following the same stereochemistry, i.e., *cis* removal of an erythro pair of D-hydrogens.

During the conversion of labeled stearic to oleic acid by *Corynebacterium*,[8] a kinetic isotope effect was observed for the 9-D, but not for the 10-D position. From this, it was concluded that the 9-D-hydrogen is removed first in the rate-limiting step. In contrast, in the experiments with *Chlorella*,[9] kinetic isotope effects were observed at both positions of the future double bond. This was interpreted as an indication of a concerted abstraction of both hydrogen atoms. In the staggered zig-zag conformation of a saturated segment of an alkyl chain, hydrogens in the D-position at neighboring C atoms are on the opposite side of the chain. A simultaneous *cis* elimination would be favored when both hydrogens are close together in an eclipsed position.[10] This requires a rotation of the 9,10-carbon bond by 180° to bring the 10-D-hydrogen into the same plane with the 9-D-proton. This rotation produces a bend in the carbon chain which is typical and permanent for *cis* double bonds in contrast to this energetically unfavored conformer in saturated acids. These stereochemical and kinetic considerations may be extended to the other double bonds, but further experiments have not been carried out. It will be interesting to see whether the recently crystallized 18:0-ACP desaturase[11] has a complementary structure in its surface which is able to bind the entire carboxyl part of the acyl chain up to C-10 and to accommodate the bend in the C-9–C-10 segment, or whether it is the function of the protein part of the substrate, i.e., the ACP, to stabilize and offer this conformer. A model proposed for acyl-ACP suggests that the thioester-linked acyl group is bound in the extended form by the hydrophobic internal surface of α-helices only up to C-8, so that the distal segment can adopt other conformations.[12]

As pointed out above, it is not clear which compartment was involved in desaturation by *Chlorella* cells and whether chloroplast and extraplastidic desaturases follow the same stereochemistry and mechanism. In this context, it should be mentioned that the aerobic introduction of double bonds into preformed acyl chains as observed in *Corynebacterium* is considered atypical for bacteria, the majority of which produce unsaturated fatty acids, mainly monoenoic acids, during fatty acid biosynthesis at the level of a C-10 or C-12 intermediate in an anaerobic isomerization reaction. On the other hand, an increasing number of different genera is found (*Alcaligenes, Bacillus, Brevibacterium, Corynebacterium, Flexibacter, Leptospira, Mycobacterium, Pseudomonas, Vibrio*), members of which can carry out oxygen-dependent desaturation of preformed fatty acids at different positions along the hydrocarbon chain.[13]

III. DOUBLE BOND POSITION AND CHAIN LENGTH SPECIFICITY

Another question which was investigated by classical biochemistry with *Chlorella* cells and slices of developing *Ricinus* seeds concerns the specificity of the desaturation reactions with respect to the location of double bonds in fatty acids of different chain length.[10,14,15] Due to the lack of investigations with additional organisms, the results obtained with these two systems are considered as representative for all plants. So far, reactions resulting in the first and second double bond have been investigated, and the results indicated the existence of different desaturases. With regard to the introduction of the first double bond into [1-^{14}C]-labeled saturated acyl chains ranging from C_{14} to C_{19}, two different activities were deduced. The pattern of double bond location shown in Figure 1a indicates that a Δ^7 desaturase introduces a *cis* double bond into exogenously added, radiolabeled fatty acids with chain lengths from C_{14} to C_{16}, whereas longer homologs are not accepted. In support of this, fatty acid analyses of *Euglena* may be mentioned which showed that Δ^7 double bonds in monoenoates were present only in C_{14}, C_{15}, and C_{16}, but not in C_{18} and C_{19} fatty acids.[16] It should be pointed out that the experiments of Figure 1a are the only evidence to suggest that the first double bond in 7,10,13-16:3 is introduced by a Δ^7 and not by a *n*-9 desaturase (see below). The operation of the Δ^7 desaturase in *Ricinus* (Figure 1a) is somewhat surprising, since leaves of castor bean do not express this activity.

The same series of experiments showed that apart from this Δ^7 desaturase, an additional Δ^9 desaturase is operating. This activity has an overlapping but wider acceptance of chain lengths, since C_{14} to C_{19} fatty acids were converted in the same experiments to additional Δ^9 monoenoates (Figure 1b). Both desaturases introduce double bonds into a position determined by its distance from the carboxyl group, whereas the length of the distal part toward the methyl end is less important, particularly in the case of the Δ^9 desaturase, which also operates in the presence of a *cis*-12 double bond. In view of these specificities, the desaturases were appropriately named as Δ^7 and Δ^9 desaturases,[3] whereas the chemical status of the carboxyl group was not known or looked at in these experiments. In view of the present knowledge,[17,18] one would assume that the Δ^7 desaturase is the one which normally prefers the *sn*-2-bound palmitoyl group of prokaryotic MGD. But recent experiments have shown that in spinach leaves, exogenous 16:0 and 18:0 have only access to the eukaryotic *sn*-1 position of this glycolipid where they are desaturated.[19] The identity of the enzyme desaturating 18:0 in this position is particularly interesting (see below).

Some of the experiments concerned with the introduction of the second double bond[15] are shown in Figures 1c and d, whereas many negative results with unaccepted substrates are not included in the figure, but are given in the legend because they are of equal importance for deducing the structural requirements of the desaturase reactions. The data are evidence for the existence of two different desaturases introducing the second double bond into a monoenoic acid. The criteria used for this differentiation are accepted chain length, position of compatible first double bond, distance between first and second double bond, location of the second double bond with respect to the carboxyl group, and location of the second double bond with respect to the methyl end of the resulting dienoic acid.

From the results of Figure 1c, it is obvious that the insertion of the second double bond into the substrates is governed by measuring the distance of the existing as well as of the newly formed double bond from the methyl end. Therefore, this desaturation is ascribed to a *cis*-*n*-6 desaturase which requires a *cis-n*-9 substrate. Both conditions are strictly observed as demonstrated by the nonacceptance of several 16:1, 17:1, and 18:1 isomers with a *cis*, *trans*, or triple bond in different positions as specified in the legend of Figure 1c. On the other hand, this desaturase is relatively unspecific regarding the length of the carboxyl part of the fatty acid, although the minimum length required is that of a C_{16} fatty acid. From recent knowledge, we would ascribe this activity to the chloroplast enzyme which converts *n*-9 acyl groups esterified in both positions of membrane lipids to *n*-6,9 acyl groups, particularly to 7,10-hexadecadienoyl and linoleoyl residues. Recent enzymatic studies with the *n*-6 desaturase solubilized from chloroplast envelopes[20] support this assumption as it desaturated *n*-9 substrates with 16, 18, and 22 carbon atoms (see below).

The second pattern of double bond formation (Figure 1d) was evidence for the operation of another desaturase. It is obvious that this activity requires a Δ^{12} desaturase measuring from the carboxyl group. Also, in this case, the existing *cis* double bond has to be located in a fixed position from the reference point, i.e., in the Δ^9 position, since *cis* double bonds in other positions or a *trans* and triple bond in the Δ^9 position abolish the substrate character of the corresponding monoenoic acid as specified in the legend of Figure 1d. The length and structure of the methyl end is rather uncritical as shown by the chain length variation and the acceptance of 9,18-19:2. In view of current knowledge, one would identify this activity as the extraplastidic enzyme converting lipid-bound oleoyl to linoleoyl groups. It is only the lack of other Δ^9 monoenoates in the extraplastidic compartment which explains the absence of other dienoic acids such as 9,12-16:2.

The exact recognition of isomeric monoenoic fatty acids is not confined to desaturases introducing the second double bond. Similarly high specificities have been observed with other proteins such as lipases and repressor systems. One isoform of the extracellular lipase from *Geotrichum* showed a nearly absolute specificity for the *cis-Δ^9*-isomer of 13 different *cis* octadecenoates.[21] Similarly, the transcription of the stearoyl-CoA desaturase gene in *Saccharomyces* is specifically repressed by *cis*-9-18:1, but not by the *cis*-5- or *cis*-11-isomers.[22]

As mentioned above, experiments on the structural requirements for the introduction of the third double bond have not been carried out. But in view of the structure of the resulting trienoic acids, as well as on the basis of recent genetic experiments,[18] there is evidence that again two different desaturases operate in microsomes and plastids from which one could be a Δ^{15} and the other an *n*-3 desaturase. It is tempting to expect that again both have common and different properties. They measure precisely their site of action with respect to the existing double bond(s) and the reference point, but are relatively sloppy with regard to that part of their substrate which is not involved in this regard. Therefore, the *n*-3 desaturase will not care too much about the length and the structure of the carboxyl part extending beyond *n*-6 (and/or *n*-9), whereas the Δ^{15} desaturase may require a carboxyl part of defined structure

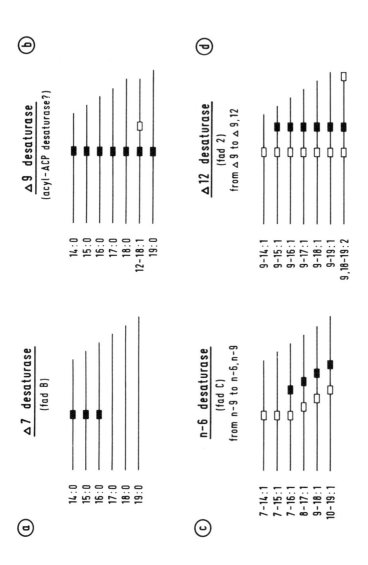

FIGURE 1. Specificities of plant desaturase activities regarding accepted chain length, position of newly introduced double bond (■), and location of accepted preexisting double bond (☐). The results have been obtained by incubating labeled substrates (indicated at the left of each line) with cells of *Chlorella* and slices of *Ricinus* seeds. In view of the present differentiation between plastid and microsomal desaturases, a tentative attribution is made with the resulting difficulties outlined (further details in the text, *fad* loci are defined in Figure 2). (a) Δ^7 Desaturase. This pattern is the only evidence for Δ^7 and not *n*-9 desaturation of 16:0 at *sn*-1 and *sn*-2 of MGD. The Δ^7 desaturase (*fad*B locus) is characteristic for 16:3-plants. $C_{17:0}$–$C_{19:0}$ were not desaturated at this position. (b) Δ^9 Desaturase. The attribution to the plastidial 18:0-ACP desaturase may be questionable, since the applied exogenous fatty acids should have access to the ACP track, whereas in leaves exogenous 16:0 and 18:0 are desaturated as part of eukaryotic MGD in the *sn*-1 position at Δ^7 and Δ^9, respectively. On the other hand, the pattern shown is perfectly followed by isolated 18:0-CoA desaturase from rat liver microsomes. (c) *n*-6 Desaturase. This series can be attributed to the plastidial *n*-6 desaturase (*fad*C locus and possibly *des*A from cyanobacteria). In leaves, exogenous substrates will be desaturated in both *sn* positions of eukaryotic chloroplast lipids. An *n*-6 double bond was not introduced into the following *cis*-monoenoic fatty acids: 9-16:1, 9-17:1, 7- or 8- or 10- or 11- or 12-18:1, 9-19:1; 9-*trans*-18:1, 9-*yn*-18:1, assuming that all substrates were incorporated into lipids. (d) Δ^{12} Desaturase. This profile is attributed to the microsomal enzyme (*fad*2 locus), which has access to these substrates after incorporation into PC. A Δ^{12} double bond was not introduced into the following *cis*-monoenoic fatty acids: 7-14:1, 7-15:1, 8-17:1, 7- or 8- or 10- or 11-18:1; 9-*trans*-18:1, 9-*yn*-18:1 (details in References 9, 10, 14, 15, 19, and 175).

and can be liberal only toward the methyl end beyond the third double bond at Δ^{15}. This difference between the two desaturases may explain why the fourth double bond converting γ-linolenic 6,9,12-18:3 into stearidonic acid 6,9,12,15-18:4 can only be inserted by the n-3, but not by the Δ^{15} desaturase (see below). The very low proportions of 7,13-16:2 and 9,15-18:2 in chloroplast lipids of an (n-6) desaturase mutant from *Arabidopsis*[23] demonstrate that the n-6 double bond is required for optimal activity of the n-3 desaturase.

Similar to the formation of linoleic from oleic acid by the action of the n-6 and the Δ^{12} desaturases, the n-3 and Δ^{15} desaturases will both produce linolenic from linoleic acid as substrate. But only the n-3 desaturase, which according to present knowledge operates in plastids, is able to convert 7,10-16:2 into 7,10,13-16:3, which is the characteristic trienoic acid of 16:3-plants.[24,25]

In contrast to the common double bond positions described so far, members of the *Boraginaceae*[26] are characterized by polyunsaturated fatty acids with a Δ^6 double bond such as γ-linolenic (6,9,12-18:3) and stearidonic acid (6,9,12,15-18:4). Seeds from *Borago* produce γ-linolenic from linoleic acid, and therefore just one further microsomal Δ^6 desaturase is required in addition to the above-described enzymes.[27] Also blue-green algae have a Δ^6 desaturase (see below), and accordingly this activity, in principle, could be expected to be expressed in chloroplasts of higher plants, but so far, a γ-18:3 producing Δ^6 desaturase has not been detected in these organelles. The microsomal Δ^6 desaturase can use linoleic and α-linolenic acid[28] and are different from that which produces petroselinic acid (*cis*-6-18:1) in the fruits of the *Apiaceae*. This particular enzyme is most likely localized in plastids and uses a saturated ACP thioester of unknown chain length.[29] Although the origin of stearidonic acid, which is characteristic for leaves of *Boraginaceae*, has not been investigated in detail, the knowledge of its biosynthesis may contribute to a differentiation between the n-3 and the Δ^{15} desaturases (see above). 6,9,12,15-18:4 may be formed from 6,9,12-18:3 by the n-3 desaturase in plastids which has no strict requirement for a defined carboxyl part, whereas a microsomal Δ^{15} desaturase cannot convert 6,9,12-18:3 into 6,9,12,15-18:4, since the structure of the saturated C-1–C-9 segment is severely altered by the Δ^6 double bond and therefore does not meet the requirements for substrate binding. This is supported by experiments with linseed cotyledons,[30] in which the microsomal Δ^{15} desaturase can convert linoleic to α-linolenic, but not γ-linolenic to stearidonic acid.

Even a fifth double bond can be introduced toward the carboxyl end, resulting in all *cis*-3,6,9,12,15-18:5 which is found in marine algae.[31] A common and accordingly conserved property of desaturases producing the prevailing polyenoic acids, whether measuring from the carboxyl or methyl end, results in the characteristic divinyl methane blocks because the additional double bond is always introduced in the same fixed distance from the existing one. This phylogenetically conserved pattern is only occasionally violated by an increased distance between adjacent double bonds as, for example, in all *cis*-5,9,12,15-18:4 found in *Chlamydomonas*.[32]

IV. LIPID-LINKED DESATURATION IN EUKARYOTIC CELLS

Parallel to investigations with emphasis on the chemical questions, research was carried out aiming at an elucidation of the biochemical and subcellular aspects of desaturation. The first experiments led to the suggestion that in plants the acyl groups of membrane lipids represent the actual substrates in many desaturation reactions, whereas animal desaturases use acyl-CoA thioesters. It was proposed that two different desaturase systems may be operating in eukaryotic, photosynthetically active cells: PC linked in microsomes and MGD linked in chloroplasts.[33-35] The experimental evidence for the new concept was the changing pattern of radioactively labeled but lipid-bound fatty acids following a typical precursor/product relationship at a (ideally) constant level of radioactivity in the particular lipid. This could be explained by direct desaturation of acyl groups covalently bound as ester groups in lipids and was further supported by analysis of molecular species.[36,37] In both lipids (PC and MGD), the pattern of molecular species differed significantly when analyzed at the beginning or at the end of the experiment, and this observation was repeatedly confirmed in subsequent studies; at the beginning the species containing monoenoic and saturated acids predominate, whereas with increasing time, the proportion of more unsaturated species increases. On the other hand, if desaturation had been limited to thioesters, then these lipid precursors should display variation in unsaturation, whereas the actual lipids would become increasingly more labeled, but with an invariant set of unsaturated acyl groups. A severe handicap of such experiments with whole cells and leaves is the fact that rigorous pulse/chase conditions as required for an unambiguous proof of this mechanism are difficult to realize. A complete removal of precursor is nearly impossible, which is particularly critical when fatty acids such as oleic acid are used as precursors. For example, in the experiments with oleic acid-labeling of *Chlorella*, cells were incubated with this acid in the dark under anaerobic conditions, during which time only part of the oleic acid was incorporated into PC. The remainder, as explicitly stated,[37] was present as free oleic acid in the cells and could not be removed. Subsequent incubation in the light in the presence of oxygen resulted in the appearance of linoleate in PC, but no data on absolute quantities of free or PC-bound acyl groups were given, and a search for thioester-bound linoleate was not carried out at that time. Therefore, it is not surprising that the provocative concept of lipid-linked desaturation was received with scepticism and that proof by *in vitro* experiments was called for.

There was some controversy regarding the actual lipid substrate used for desaturation in leaves, but this discrepancy was resolved when it was found that plants show differences in labeling and desaturation kinetics depending on their systematic position.[17] In 18:3-plants (having 18:3, but no 16:3 in MGD), extraplastidic PC is the first most heavily labeled lipid, and the oleate of this lipid is most rapidly desaturated. In 16:3-plants (which have both 18:3

and varying proportions of 16:3 in MGD), MGD in chloroplasts is similarly well-labeled and desaturated. In particular, it is only in this lipid that 18:3 and 16:3 become labeled, whereas labeled 18:3 can hardly be detected in PC. In similar experiments with different plants and seed cotyledons from oil plants including double-labeling experiments with glycerol, acetate and phosphate, a large body of details was accumulated. The evidence for lipid-linked desaturation and, in particular, for PC-bound fatty acids in oil-producing cotyledons became more substantial and more accepted though the number of arguments could hardly be increased by such *in vivo* experiments. In combination with subcellular fractionation studies, the data suggested the operation of two different sets of desaturation reactions in chloroplasts and microsomes. The most relevant data from acetate-labeling kinetics of leaves in correlation with desaturation were the following:

1. 18:0 does not show up in significant proportions in lipids, whereas 16:0 does.
2. 18:1 is the first C_{18} fatty acid incorporated into lipids.
3. The sequence from 18:1 via 18:2 to 18:3 is observed in many, if not all, polar lipids, but with different rates and to different extents.
4. After short-term labeling, unsaturated C_{18} fatty acids are found in both the *sn*-1 and *sn*-2 position of PC and MGD of 18:3-plants, whereas in MGD of 16:3-plants, unsaturated C_{18} acids are confined to the *sn*-1 position.
5. 16:0 shows up in nearly all lipids, and the lowest proportion is found in MGD of 18:3-plants.
6. 16:0 in extraplastidic lipids is confined to the *sn*-1 position and not desaturated.
7. Lipids synthesized in chloroplasts incorporate 16:0 mainly into the *sn*-2 position.
8. The sequence of 16:0 via 16:1 and 16:2 to 16:3 is confined to MGD of 16:3-plants, where these desaturations are usually confined to the *sn*-2 position.
9. The same sequence from 16:0 to 16:3 and even from 18:0 to 18:3 is found in the *sn*-1 position of MGD after labeling with exogenous 16:0 or 18:0.
10. 16:0 in other glycolipids is not desaturated.
11. 16:0 in the *sn*-2 position of chloroplast PG is desaturated to 3-*trans*-16:1.
12. The predominating molecular species of chloroplast lipids after short-term labeling of 16:3-plants are 18:1/16:0-combinations.
13. The molecular species of envelope lipids from spinach chloroplasts, as well as their change with time, do not differ from thylakoid counterparts.
14. In 18:3-plants, chloroplast lipids are characterized by C_{18}/C_{18}-combinations of different degree of unsaturation, depending on sampling time and other parameters.

This mass of information was a challenge for enzymological studies which eventually should explain these kinetic data in terms of specificities and selectivities of acyltransferases and desaturases. And, in fact, many of these details can now be attributed to individual enzymatic steps (see Chapters 1 and 6).

A surprising result of such work was the discovery[38] of a correlation between loss of labeled C_{18} fatty acids as complete DAG from extraplastidic PC and their concomitant incorporation into MGD from 18:3-plants. This was explained by a permanent exchange of fatty acids and DAG between plastids and ER and the cooperation of these compartments in lipid synthesis and desaturation and was supported by double labeling and additional experiments with isolated chloroplasts (see below). This cooperative pathway hypothesis[17] describes a dynamic subcellular lipid trafficking, upon which is superimposed the insertion of double bonds into appropriate substrates during passage to and from the different compartments and membranes. The essentials of this scheme as required in the context of desaturation are outlined briefly below (for details see References 17 and 18).

The *de novo* biosynthesis of fatty acids in chloroplasts provides 16:0 and 18:1, which are esterified with glycerol-3-phosphate to yield finally PG, MGD, DGD, and SQD with 18:1, 16:0/16:0 pairings of fatty acids. These DAG moieties are called prokaryotic, due to their similarity with fatty acid arrangements in lipids from cyanobacteria. In addition, 18:1 and 16:0 are exported into the nucleocytoplasmic compartment. In the membranes of the ER (experimentally defined as microsomes), these fatty acids are incorporated into phospholipids to give 18:1, 16:0/18:1-combinations, which are typical for this compartment and therefore named eukaryotic. The term cooperative refers to the fact that eukaryotic DAG originating from microsomal PC are continuously reimported by plastids and used for additional glycolipid synthesis. Therefore, glycolipids in plastids have both pro- and eukaryotic DAG backbones which are recognized by the presence of C_{16} or C_{18} fatty acids in the *sn*-2 position, respectively. The trienoic acids in the prokaryotic lipids from chloroplasts (16:3 in the *sn*-2 position of MGD and 18:3 in the *sn*-1 position of all) have never left this organelle and have received their double bonds as permanent residents of plastids. On the other hand, the three double bonds of α-linolenic acid in eukaryotic plastid lipids may be considered as different souvenirs brought home by acyl chains from their subcellular excursion: Δ^9 was introduced into 18:0-ACP in the chloroplast stroma, Δ^{12} into the *sn*-1- or *sn*-2-bound oleoyl group of PC in the ER, and the final *n*-3 was inserted again in the chloroplast after return of the linoleoyl group as part of a DAG to be used for glycolipid formation. The interaction of lipid synthesis and desaturation between plastids and the nucleocytoplasmic compartment is emphasized in Figure 2, which shows the situation for a 16:3-plant.

Recent analysis of desaturation mutants from *Arabidopsis* (see below) have suggested a large degree of flexibility in reciprocal lipid exchange between ER and plastids. There may not be just an export of oleic acid from chloroplasts into the ER, but also of polyunsaturated fatty acids or even of DAG. This release could actually use the reverse route of eukaryotic DAG import. Whether this reversibility is a normal phenomenon or just a response to complement

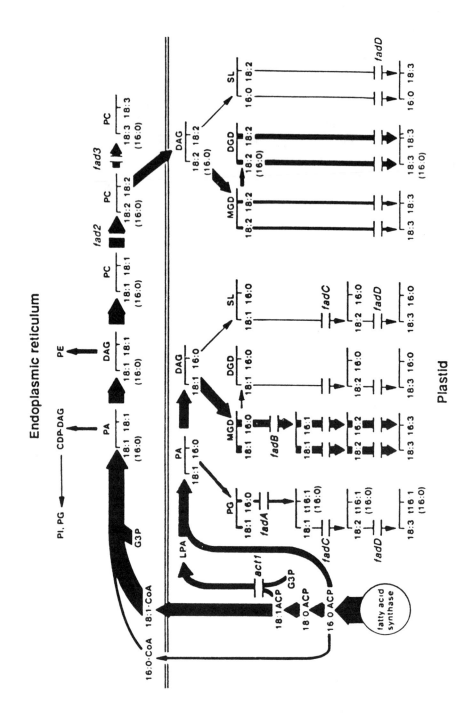

FIGURE 2. Cooperation of subcellular compartments in the biosynthesis of lipids and the desaturation of fatty acids in leaf cells of *Arabidopsis*. Fatty acids are exclusively synthesized in chloroplasts and used for lipid assembly in pro- (18/16) or eukaryotic (18/18) arrangement in chloroplasts or ER membranes, respectively. The capacity to introduce first double bonds (Δ^3-*trans*, Δ^7-*cis*, Δ^9-*cis*) is restricted to chloroplasts, whereas subsequent desaturation steps resulting in *n*-6,9-dienoic and *n*-3,6,9-trienoic acids are catalyzed by two different sets of membrane-bound desaturases, which in both compartments operate with intact lipids as actual substrates. In the above figure, chloroplasts export free fatty acids (16:0, 18:1) and reimport DAG (not necessarily with two 18:2 residues as shown). The widths of the arrows represent the flux rates at the various steps which are controlled by products (desaturase enzymes) of *fad* loci (indicated by breaks). The study of the various *fad* mutants has indicated additional possibilities for mutual exchange of fatty acids or DAG between compartments, and, in particular, the 18:3 content of PC in the ER may depend on polyenoic acid biosynthesis in chloroplasts. The proportion of pro- and eukaryotic DAG used for MGD synthesis is highly flexible and responds to temperature, developmental signals, and desaturase deficiencies in both compartments (for further details see text.) (From Somerville, C. and Browse, J., *Science*, 252, 80, 1991. With permission.)

the deficiency of polyunsaturated fatty acids in the nucleocytoplasmic compartment of mutants is unknown. The subcellular complementation uses pro- and eukaryotic DAG species, although the incorporation of 16:3 into extraplastidic PC is prevented. A similar use of galactolipid-derived DAG for PC synthesis has recently been demonstrated during the process of foliar senescence. It was suggested that 18:3/18:3-species transiently incorporated into PC were finally used for gluconeogenesis in microbodies,[39] which are induced in senescent leaves.

The knowledge that plant cell cultures can incorporate exogenous glyceryl ethers into phospholipids[40] and that lipid-linked desaturation in the ER will start with Δ^9 monoenoates, followed by a subsequent export into plastids, prompted experiments[41] in which 1-*O*-octadec-9'-enyl-glycerol (selachyl alcohol) and its positional isomer 2-*O*-octadec-9'-enyl-glycerol were incubated with photoautotrophic cell cultures from tomato. As expected, these monoacyl glycerol analogs were incorporated into PC and MGD. Analysis by HPLC and mass spectroscopy after 2 weeks showed that the original octadecenyl residues had been desaturated more or less completely to octadecadienyl groups in PC and to octadecatrienyl residues in MGD. Since such cell cultures do not contain endogenous ether lipids and likewise do not resynthesize the ether bond between the alkenyl and glyceryl residue, these results are considered as firm evidence for PC-linked Δ^{12} desaturation of 9-octadecenyl residues at both positions of glycerol. They further indicate that the microsomal Δ^{12} desaturase does not require the carbonyl oxygen at C-1 of the substrate chain for its binding. As repeatedly observed in experiments on fatty acid labeling of leaf lipids, the ether experiments revealed a low proportion of 18:3-ether in PC and a high percentage in MGD. At present, it cannot be decided whether the third double bond was introduced at the level of PC and then transferred as 1-octadecatrienyl-2-acyl-glycerol to chloroplasts for MGD formation so that the PC pool was depleted of this species or whether the third double bond in the ether residue was inserted after import into chloroplasts at the level of MGD. This would show that also the chloroplast *n*-3 desaturase can accept ether-bound substrate chains and thus would be proof for this lipid-linked desaturation in this organelle. Previous experiments with exogenously added intact MGD were not successful[42] because this polar lipid was not taken up and thus could not reach the desaturases in the chloroplasts (see below).

V. LIPID-LINKED DESATURATION IN CYANOBACTERIA

Blue-green algae are considered to represent descendents of phylogenetic precursors of chloroplasts, and numerous investigations have established the close similarity in membrane lipid composition between these prokaryotic organisms and thylakoids.[43] They contain the typical thylakoid lipids (MGD, DGD, SQD, PG), whereas PC, PE, and PI are missing, although the first exceptions to this pattern have been found.[44] The general absence of PC in cyanobacteria is in line with the recent finding that PC in chloroplasts is confined to the outer leaflet of the outer envelope membrane.[45] The fatty acid composition of cyanobacteria differs from bacterial patterns which are normally very simple with regard to unsaturation because anaerobically produced vaccenic (11-18:1) and palmitoleic acid (9-16:1) predominate. In contrast, blue-green algae produce a variety of polyunsaturated fatty acids, including linoleic, α-, and γ-linolenic, as well as 6,9,12,15-octadecatetraenoic acid, all of which are confined to the *sn*-1 position of lipids as typical for prokaryotic combinations, whereas fatty acids with 20 or more carbon atoms have not been found. Also these organisms were used for studies on fatty acid desaturation in experiments with acetate labeling or after application of exogenous 16:0 and 18:0, and it became evident that also in these organisms, desaturation occurs at the level of lipid-linked acyl groups. Some differences or missing links between cyanobacterial and chloroplast lipid and fatty acid metabolism do exist, however.

The gap in lipid metabolism between blue-green algae and chloroplasts would be narrowed if cyanobacterial members could be found which displayed the following features[43] (unless all of them have been developed during the domestication of the guest cell to become the future chloroplast): MGD synthesis by galactosylation of DAG instead of epimerization of glucosyl diacylglycerol; desaturation of acyl-ACP; production of 3-*trans*-16:1 at *sn*-2 of PG; and desaturation of 16:0 at *sn*-2 of MGD to Δ^7-*cis*-16:1. None of the corresponding chloroplast features seem to be necessarily correlated with the new situation as a symbiotic guest, and only the desaturation of 18:0 as an ACP thioester instead of its desaturation as acyl component of lipids can be interpreted as a contribution to the new partnership, since free 18:1 may be exported more easily into the ER immediately after its synthesis than after its incorporation into membrane lipids. In the context of desaturation, it is surprising that apart from one organism, all cyanobacteria investigated so far do not produce Δ^7-16:1 despite the fact that it is the positioning of this fatty acid in chloroplast lipids (as well as of 7,10-16:2 and 7,10,13-16:3) which led to the discovery of the prokaryotic fatty acid pairing and its phylogenetic relevance. The blue-green algae normally produce palmitoleic acid (Δ^9-16:1), including the chlorophyll b-containing *Prochloron*[46] and *Prochlorothrix*.[47] This organism has, in addition, the unusual 4-16:1 and 4,9-16:2, whereas the common 9,12-16:2 is found in several genera. The other free living prochlorophyte *Prochlorococcus*[48] has not yet been investigated in this regard. *Cyanocyta*, the photosynthetically active organelle of *Cyanophora*,[49,50] contains large quantities of polyunsaturated C$_{20}$ acids (20:4 and 20:5) in MGD and thus is more reminiscent of red than of blue-green algae. The only cyanobacterium in which Δ^7-16:1 has been detected together with about equal quantities of palmitoleic acid is *Oscillatoria limnetica*,[51] which may be particularly useful to trace the phylogenetic origin of the Δ^7 desaturase from 16:3-plants. On the other hand, as mentioned above, all these features could have evolved in the process of eukaryotization, and therefore it may be in vain to search for them in present-day cyanobacteria.

Another question which should be addressed in this context is the phylogenetic origin of blue-green algae themselves and accordingly the origin of their characteristic set of lipids and fatty acids. Linear photosynthetic electron transport with two photosystems may have evolved by the coupling of two originally separate systems by joining or gene transfer between two photosynthetic prokaryotes each having only one photosystem.[52] In this model, photosystem II originated from a cyclic system as found in *Rhodobacter* and other purple bacteria, whereas photosystem I would originate from green sulfur bacteria such as *Chlorobium*, which can carry out linear electron transport with one photosystem. If one poses questions concerning the phylogenetic origin of the two cyanobacterial photosystems, one can also ask which organism provided which lipids for the common lipid matrix to accommodate the two photosynthetic units of different origin. Purple bacteria do not contain galactolipids, but SQD has been found in these organisms.[53] On the other hand, green sulfur bacteria[54] contain MGD, as well as SQD, and even Δ^7-16:1 may be present.[55] In this context, it should be stressed that all these organisms are anaerobes, and also *O. limnetica* can produce Δ^7-16:1 and Δ^9-18:1 in the absence of oxygen by a presently unknown mechanism. It should be recalled that anaerobic synthesis of monoenoic acids in bacteria normally results in vaccenic (11-18:1) and palmitoleic acid (9-16:1, both *n*-7), due to the activity of a specific β-D-hydroxydecanoyl-ACP dehydrase, which has additional isomerase activity. In *Brevibacterium*, a similar isomerization is catalyzed at the C_{12} level by a specific β-hydroxydodecanoyl thioester dehydrase.[56] Subsequent chain elongation results in oleic acid, but if stopped at C_{16}, it would produce Δ^7-16:1 (both *n*-9).

As mentioned above, a satisfactory demonstration of lipid-linked desaturation in experiments with intact eukaryotic organisms is difficult to obtain, and efforts in this direction were extended to cyanobacteria as simpler organisms which obviously desaturate lipids by a similar mechanism. But since these organisms do not have acyl-ACP desaturase activity,[57,58] the two saturated fatty acids resulting from *de novo* synthesis (16:0 and 18:0) are first incorporated into lipids (for example into MGD) before both are desaturated: 18:0 in the *sn*-1 position of all lipids and 16:0 in the *sn*-2 position of MGD only. The first double bond introduced in each case is in the Δ^9 position (oleic and palmitoleic acid). To demonstrate lipid-linked desaturation in these organisms, a different approach with ^{13}C labeling and analysis of heavy isotope incorporation by mass spectrometry was used.[59] In contrast to dual labeling with a radioactive tracer, this method answers the question whether different parts of the same molecule (in this case glycerol and fatty acid) are present in the same combination, i.e., without exchange before and after desaturation. For this purpose, *Anabaena* was labeled at 38°C with $[^{13}C]CO_2$ for fatty acid and lipid synthesis which was stopped by addition of cerulenin without blocking desaturation. Subsequent incubation was continued at 22°C at which temperature desaturation is favored. Three samples of MGD were isolated (before $^{13}CO_2$ addition, after cerulenin addition, and after subsequent desaturation at 22°C) and converted to derivatives suitable for mass spectroscopical analysis.

The interpretation of all the resulting data lead to the conclusion that 16:0 is desaturated to 16:1 when it is esterified in the *sn*-2 position of MGD. A release of 16:0 for desaturation and replacement by desaturated but unlabeled 16:1 can be excluded. On the other hand, a tightly coupled deacylation-desaturation-reacylation sequence, i.e., release of labeled 16:0, followed by its desaturation and reincorporation, would not be differentiated by this method and cannot be excluded. In addition, the difficulties of pulse/chase experiments with whole cells become evident, since the rise in the desaturation product is not matched by a parallel decrease in its substrate which indicates that at the beginning of the desaturation not all labeled precursors had been assembled to MGD.

VI. DESATURATION OF ACETATE-LABELED LIPIDS BY ISOLATED CHLOROPLASTS

Experiments with chloroplasts require purified and highly active organelles. In a series of experiments (summarized in Reference 17), the characteristic features of chloroplast lipid assembly were elucidated with organelles which were obtained from the 16:3-plant spinach. In contrast to organelles from 18:3-plants, spinach chloroplasts have a high capacity to synthesize prokaryotic MGD which is required for the biosynthesis of polyunsaturated fatty acids. The experiments on light-dependent acetate labeling were carried out in different modes,[60] depending on the substrate added in addition to acetate. Without any further additions, acetate was mainly incorporated into free fatty acids. Additional glycerol-3-phosphate resulted in a large proportion of label in DAG, whereas complementation with glycerol-3-phosphate and UDP-galactose resulted in a shift to MGD as the major product. In contrast, many previous experiments had been carried out with chloroplasts from 18:3-plants, and supplementing substrates required for glycolipid synthesis had not been included. It was the awareness of lipid-linked desaturation and the expectation that particularly the MGD mode would be useful for desaturation which led to the demonstration of linolenic and hexadecatrienoic acid biosynthesis in isolated spinach chloroplasts. The advantage of this highly active system in comparison to leaves is the fact that it allows the analysis of fatty acid desaturation after very short times in parallel in the thioester fraction (i.e., before incorporation into lipids), as well as in the whole series of lipid intermediates. A complete analysis at the end of such an experiment requires separation of acyl-ACP from lipids and the separate analysis of fatty acids from acyl-ACP, free fatty acids, LPA, PA, DAG, and MGD. The resultant data can be complemented by the analysis of the positional distribution of labeled fatty acids in PA and MGD, as well as by the resolution of labeled molecular species of PA, DAG, and MGD. From many experiments of this kind carried out by several groups, the following sequence in fatty acid and lipid synthesis and the subsequent desaturation steps have been demonstrated: in acyl-ACP, the only long-chain acyl thioesters in chloroplasts, 16:0, 18:0, and 18:1, but no polyenoic acids or unsaturated C_{16} acids are found (the occasionally observed small proportion of linoleic acid[61] does not fit into the scheme and is neglected). 18:1 and 16:0 also appear in the free fatty acid fraction or, when CoA and ATP were included, as acyl-CoA thioesters in the medium. After labeling in the MGD mode, they are found in LPA, PA, DAG, and MGD in prokaryotic arrangement with 18:1 and 16:0 at *sn*-1 and 16:0 at *sn*-2. All precursors of MGD, such as LPA, PA, and DAG, are not in reach of desaturases as evident from the absence of polyunsaturated C_{18} and unsaturated C_{16} acids in these lipids. Only after attachment of a galactose residue the DAG moiety of the resultant MGD is used as desaturase substrate. At times when acyl-ACP does not contain polyunsaturated fatty acids, 18:2 and 18:3 are formed at *sn*-1 of MGD, and palmitic acid at *sn*-2 is desaturated via 16:1 and 16:2 to 16:3. Desaturation of 18:1 at *sn*-1 has also been observed in newly synthesized PG, SQD, and DGD, whereas no *cis* double bonds are introduced into 16:0 at *sn*-2 of these lipids.[62,63] Therefore, polyunsaturated fatty acids of chloroplasts are not synthesized as thioesters, but as oxygen ester components of membrane lipids. As mentioned above, these desaturation reactions can be demonstrated with chloroplasts from 16:3-plants such as spinach and nightshade, whereas isolated 18:3 chloroplasts show very low desaturase activity after acetate labeling, even in cases when high labeling of prokaryotic PG was obtained which should represent an acceptable desaturase substrate.[64]

The desaturation of acetate-labeled MGD in chloroplasts requires oxygen, is independent of light, and is osmotically sensitive because breakage or even swelling of organelles abolishes desaturation.[62,65] From a variety of inhibitors tested, only cyanide showed some selectivity, since the Δ^7-16:0 desaturase (as well as the stearoyl-ACP desaturase) proved to be more resistant than the *n*-6 and *n*-3 desaturases. This difference results in the accumulation of 18:1/16:1-MGD and allows its isolation for studies with solubilized enzymes (see below). The introduction of four *trans* double bonds into phytoene also occurs in chloroplasts, and, similar to fatty acid desaturation, double bonds are inserted into saturated segments of long-chain hydrocarbon substrates. However, these reactions follow a different mechanism, since in anaerobic dehydrogenase steps NAD(P) (or FAD in bacteria and fungi) is used as the electron acceptor.[66]

Desaturation of 16:0 is usually restricted to the *sn*-2 position of MGD. In rare cases, when MGD with 16:1 at *sn*-2 is used for DGD synthesis, the resultant 16:1 in DGD is further desaturated, since the *n*-6 and *n*-3 desaturases operate with all lipids. This has been shown in labeling experiments with the alga *Dunaliella*[67] and may also explain the unusually high proportion of 16:3 in DGD from the moss *Ceratodon*[68] (30% as compared to 48.5% in MGD). But as mentioned above, in some cases, 16:0 at the *sn*-1 position can be desaturated as well. In spinach leaves, high proportions of labeled 16:2 and 16:3 are produced from exogenous 16:0, which enters specifically the eukaryotic *sn*-1 position of PC and MGD.[19] But only the MGD-bound 16:0 is desaturated, whereas 16:0 in PC remains untouched. This C_{16} polyenoic acid may have the characteristic 7,10,13-structure. Its occurrence in the *sn*-1 position shows that the Δ^7 desaturase is not absolutely specific for the glycerol position in MGD and will accept 16:0 in both positions, though 16:0 in *sn*-2 is preferred. It should be mentioned that in contrast to leaves, green algae such as *Dunaliella*[67] and *Chlamydomonas*[32] have the ability to incorporate exogenous 16:0 into the prokaryotic *sn*-2 position of chloroplast lipids, whereas in cyanobacteria 16:0 and 18:0 may be specifically inserted into the *sn*-1 position of MGD by an acyl-ACP:lyso-MGD acyltransferase.[69] On the other hand, exogenous unsaturated C_{16} and C_{18} acids are incorporated into both positions of cyanobacterial lipids.[70]

At this point, it is appropriate to discuss the observation that also exogenous 18:0, similarly incorporated exclusively into the eukaryotic *sn*-1 position of PC and MGD in spinach leaves, was desaturated in MGD.[19] This has to be interpreted in view of the facts that plants apparently have no extraplastidic 18:0-CoA Δ^9 desaturase, that in chloroplasts 18:0 is exclusively desaturated in the form of 18:0-ACP, and that exogenous fatty acids cannot reenter the ACP track in chloroplasts. Although this dogma was questioned by the recent demonstration of low rates of 16:0-ACP formation from exogenous 16:0 in *Spirodela* plants,[71] the exogenous 16:0 entered the *sn*-1 position of spinach MGD, which argues against its incorporation via plastidic 16:0-ACP. Therefore, it is an open question as to which enzyme in chloroplasts can introduce a Δ^9 double bond into *sn*-1-bound 18:0 of MGD and probably into the other fatty acids from Figure 1b as well. On the other hand, cyanobacteria desaturate 18:0 not as ACP thioester, but after incorporation into the *sn*-1 position of MGD. One could speculate that this cyanobacterial Δ^9 desaturase has become the chloroplast 18:0-ACP desaturase to prevent futile lipid synthesis before export of 18:1 into the cytoplasm. This enzyme may recall its ancient properties and also desaturate 18:0 in MGD, although in terms of protein structure some additions or rearrangements would have been required to convert a membrane-bound enzyme into a soluble desaturase accepting hydrophilic ACP thioesters instead of lipophilic membrane lipids. In the near future, DNA-derived amino acid sequences will help to answer such questions. A completely different way to explain the introduction of the Δ^7 double bond would make use of the recently detected shifting of double bonds;[72] the originally inserted Δ^9 could be isomerized to a Δ^7 double bond. In this case, questions concerning the phylogenetic origin of the isomerase remain unanswered.

In summary, the acetate-labeling studies with isolated chloroplasts have confirmed the operation of four *cis* desaturation steps in these organelles (Δ^7, Δ^9, *n*-6, *n*-3), which result in the characteristic C_{16} and C_{18} trienoic acids of these organelles. The only step not yet demonstrated with isolated organelles is the formation of 3-*trans*-16:1 at *sn*-2 of PG. This desaturation is also believed to involve the lipid-linked acyl group, and, in fact, this specific type of lipid-linked desaturation was the first one to be proposed.[73] The formation of a *trans* double bond in close proximity to the ester group may be compared to the desaturation of *N*-acyl sphinganine to *N*-acyl *trans*-4-sphingenine.[74] Recent *in vivo*-labeling studies with the green alga *Dunaliella*[75] have suggested that the PG-specific Δ^3 desaturase is localized in thylakoid membranes and not within the envelope, where PG assembly occurs.

VII. ENZYMATIC STUDIES WITH CHLOROPLAST DESATURASES

The high rates of fatty acid desaturation in isolated, acetate-labeled chloroplasts encouraged experiments on an enzymatic characterization of these desaturases, and it is obvious that such investigations could only be successful with suborganellar preparations. First steps toward the development of pure *in vitro* assays for the separate measurement of individual chloroplast desaturases were experiments[76,77] in which isolated chloroplasts from 18:3-plants such as lettuce, pea, and oat were incubated with exogenously added radiolabeled MGD or PC in the presence of pure (or nonpurified) lipid exchange proteins, resulting in desaturation of preformed and lipid-linked 18:1 and 18:2. In addition, it was demonstrated that desaturase activity could be measured even in ruptured chloroplasts.[78] These studies circumvented the osmotic sensitivity observed before and, in fact, may explain the experiments described above[76] in which all chloroplasts turned out to be broken at the end of the incubations. Spinach chloroplasts were ruptured by a detergent, then diluted and further incubated with various substrates, including different [1-^{14}C]-labeled and -unlabeled fatty acids, glycerol-3-phosphate, and sugar nucleotides. This resulted in the *in situ* assembly of prokaryotic species of MGD containing labeled 16:0, 18:1, or 18:2 in the corresponding prokaryotic positions. Similarly, [1-^{14}C]18:1/16:0-SQD and [1-^{14}C]18:1-PC (labeled in the *sn*-2 position by acyl exchange from 18:1-CoA) were produced *in situ*. All these substrates were desaturated, and highest rates of 18:2- and 18:3-formation were observed with MGD, whereas in SQD and PC only 18:2 was formed. 16:0 in MGD was desaturated to a significantly lower proportion via 16:1 and 16:2 to 16:3.

When it was shown[79] that pure envelope membranes could be used as an enzyme source to desaturate 18:1 to 18:2, it became possible to search for soluble co-factors, and it was found that the desaturation was absolutely dependent on NADPH and ferredoxin. Preliminary evidence for the involvement of NADPH:ferredoxin oxidoreductase (FNR) was obtained by the use of a specific anti-FNR-IgG. In addition, catalase was required, apparently to destroy the inhibitory hydrogen peroxide which forms by rapid autooxidation of reduced ferredoxin.

Similar experiments were carried out with chloroplasts from *Arabidopsis* (another 16:3-plant) and soybean (an 18:3-plant), which were broken by a freeze/thaw cycle.[80] Pro- and eukaryotic MGD (18:2/18:3 and 18:2/16:2), as well as 18:2/18:3-PC, were accepted by the *n*-3 desaturase which produced 18:3 and 16:3. NADPH and catalase were required, and, as above, 16:0 in MGD was desaturated to 16:3. In addition, the substituted pyridazinone herbicide BASF 13-338 (= SAN 9785) inhibited desaturation to 50% at a concentration of about 10^{-7} *M*. This compound may be specific for the *n*-3 desaturase in chloroplasts.

These data show that lipid-linked *n*-6 and *n*-3 desaturases in chloroplasts (and most likely the Δ^7 desaturase as well) use the same co-factors as the 18:0-ACP desaturase in the stroma, i.e., O_2 as final electron acceptor and reduced ferredoxin as the source for the additional two electrons required to reduce O_2 to H_2O, although this detail has not yet been confirmed. Since ferredoxin delivers one electron at a time, the desaturase has to oxidize two reduced ferredoxins, probably in succession, and store the first electron before the double bond can be formed. The similarity in co-factor requirement could suggest a mechanistic and phylogenetic relationship between the *n*-x and the 18:0-ACP desaturase as already speculated above, and, in fact, the cyanobacterial enzymes desaturating MGD-bound 18:0 and 18:1 require reduced ferredoxin and O_2 as recently shown by *in vitro* assays with a membrane fraction from *Synechocystis*.[81]

The localization of desaturase activities in envelopes adds another important function in lipid synthesis to this membrane system, but at present it cannot be decided whether these activities are absent from thylakoids to which the PG-specific Δ^3 desaturase has been attributed.

The desaturases, including the soluble ACP desaturase, increase the number of enzymes in plastids which use reduced ferredoxin and must compete with nitrite reductase, glutamate synthase, and enzymes of sulfate reduction for reducing equivalents. The K_m values of these enzymes for ferredoxin are in the low micromolar range,[82] and a similar ferredoxin concentration was sufficient for half maximal desaturation with envelope membranes.[78] The interaction of ferredoxin with the soluble enzymes involves electrostatic forces between negative ferredoxin carboxylates (glutamates) encircling its buried FeS cluster and positive charges lining the docking sites on the acceptors (lysine and arginine in the case of FNR).[83] Very similar interactions govern the electron transport in microsomal systems (see below). FNR, glutamate synthase, and nitrite reductase show immunological cross-reactions which are limited to the ferredoxin binding sites as indicated by the loss of this cross-reaction when ferredoxin-cross-linked enzymes are used.[84,85] The ferredoxin binding sites[86] of these enzymes are part of protein domains which additionally must accommodate the permanently enzyme-bound electron acceptors such as flavin nucleotides or FeS clusters which accept electrons from ferredoxin. It will be interesting to see which kind of redox centers are used by chloroplast desaturases and whether their ferredoxin docking sites share structural similarities with the above-mentioned enzymes beyond positively charged patches. The knowledge of amino acid and nucleotide sequences together with the availability of crystal data for FNR,[86] ferredoxin,[87] and stearoyl-ACP desaturase[11] should answer this question for the soluble desaturase in the near future.

Under nonphotosynthetic conditions, ferredoxin can be reduced by NADPH via FNR, and it has been shown that nongreen tissues contain ferredoxin[88] and

FNR,[82] both of which can occur in slightly different forms coded by different genes. The roles of these electron carriers and their roles in nonphotosynthetic plastids have been investigated[89-94] and are described in detail in Chapter 17.

In the experiments with envelopes,[79] it was observed that not only MGD, but also other lipids such as DAG, PA, and PC were desaturated. This is in contrast to the *in vivo* situation and was explained by a possible change in the accessibility of desaturases caused by envelope preparation. Similar effects may have contributed to the desaturation of linoleoyl-CoA by thylakoid membranes,[95] which are difficult to separate from residual envelopes. On the other hand, this unspecificity was the basis for a simple enzymatic assay with free oleic acid as substrate including ferredoxin, FNR, NADPH, and catalase. This assay was used to follow the purification of the *n*-6 desaturase after solubilization from spinach envelope membranes.[96] The solubilization provided the possibility to incubate enzyme and labeled galactolipids both in solubilized form and to demonstrate in an *in vitro* system that plant desaturases do, in fact, act on complex membrane lipids.[20] 18:1/16:1-MGD served as a substrate of prokaryotic structure, 18:1/18:1-MGD represented eukaryotic galactolipids, and the esterless 18:1/18:1-ether analog of MGD should prove unambiguously that the desaturation of lipid-linked acyl groups does not even involve their transient release. The pattern of molecular species resolved after incubation of these substrates (Figure 3) demonstrated that all lipids were accepted and that the substrate chains in both glycerol positions were desaturated. Accordingly, the *n*-6 desaturase accepts acyl groups at *sn*-1 and *sn*-2 with equal efficiency, and even ether-bound *cis*-9-octadecenyl residues are desaturated. In addition, free fatty acids including erucic acid (13-*cis*-22:1 = *n*-9-22:1) were desaturated, which is further proof that this desaturase is a *n*-6 desaturase as discussed above.

An apparent molecular weight of 40 kDa was found for the *n*-6 desaturase, which was isolated in sufficient quantity to also determine its N-terminal amino acid sequence. The size of this desaturase is very similar to the corresponding enzyme cloned from blue-green algae (see below). With these experiments, the *in vitro* evidence for lipid-linked desaturation has been provided, which can be considered as a characteristic difference between plant and animal lipid metabolism.

VIII. DESATURATION IN MICROSOMAL MEMBRANES

Most microsomal preparations are highly heterogenous due to their definition as a membrane fraction sedimenting between about 20,000 and 100,000 × *g*. In order to resolve such mixtures, separation methods such as free flow electrophoresis, aqueous phase partitioning, and gradient centrifugation are required,[97] but such highly purified membranes have not yet been used for studies of desaturase activities.

A suitable source for microsomal membranes highly active in desaturation are developing oilseeds. The desaturase activities in these membranes are

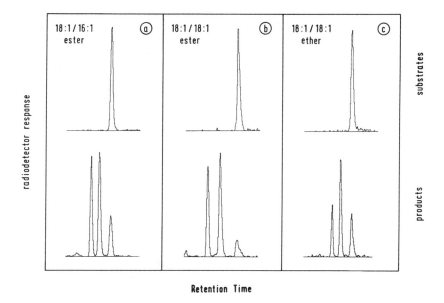

Retention Time

FIGURE 3. Desaturation of intact MGD by solubilized *n*-6 desaturase. The radioactive substrates were incubated for 2 h in detergent solution with the enzyme obtained from envelope membranes of spinach chloroplasts and separated by reversed-phase radio-HPLC into molecular species.[20] The more unsaturated species elute first. The *n*-9-*cis*-monoene substrate chains with 16 and 18 carbon atoms were linked in defined *sn* positions as indicated with *sn*-1 named first. a and b represent MGD with acyl esters in pro- and eukaryotic combination, respectively. c is an esterless ether analog of MGD with two 9-*cis*-octadecenyl residues. In this substrate, a transient de/reacylation during desaturation can be excluded. Fatty acid analysis after desaturation confirmed the production of dienoic acids in a and b, whereas such an analysis is not possible with c.

rather stable and have been extensively studied in many laboratories (summaries in References 17,18, and 98). It took some time for the identification of the actual substrate of desaturation, PC- or CoA-bound oleate, to be conclusive due to technical difficulties with straightforward chemical measurements with mixed products.[99] Therefore, the identification of the desaturation substrate necessitated the separation of acyl-CoA from lipids by phase partitioning and hydrophobic chromatography[100] before analysis of their acyl groups. This methodology was used to follow the fate of the acyl group from [14C]18:1-CoA in microsomal membranes. The acyl-CoA was depleted within the first 5 to 10 min,[101] and applied thioester-bonded 18:1 was recovered in the lipid fraction in which PC was the most heavily labeled component. The oleate was incorporated more than 90% into the *sn*-2 position of PC and, only after longer incubation times, was up to 20 to 30% found at *sn*-1, although this selectivity for the *sn*-1 or *sn*-2 position may vary with the plant.[102] In the presence of NADH and oxygen, the PC-bound oleate was converted to 18:2. Repeated analysis of acyl-CoA samples taken during this desaturation phase showed an unchanged fatty acid composition and no signs of desaturation in the rapidly declining thioester fraction.[101] This is typical for lipid-linked desaturation as already described for chloroplasts: a constant desaturation level in the thioester pool, but increasing unsaturation in the lipid-bound acyl groups. Investigations of microsomal desaturation have the convenience that *in situ* preparation of substrate can be separated from the actual desaturation by omitting NADH during the initial acyl transfer phase of the experiment. The enzyme catalyzing the incorporation of oleate into PC, the acyl-CoA:lyso-PC acyltransferase, has been studied in detail [98,103-105] and is discussed in Chapter 6. The equilibrium of the reaction favors the acylation of lyso-PC, and, despite its reversibility, the rates measured *in vitro* for the reverse formation of acyl-CoA were only a few percent of the forward reaction (Table 2).

Whereas the lyso-PC acyltransferase is of general importance for enrichment of the acyl-CoA pool with unsaturated fatty acids, developing oilseeds provide an alternative possibility for increasing the degree of unsaturation in TAG. This mechanism is based on the reversibility of the CDP-choline:DAG phosphocholine transferase reaction.[108] If, for example, oleoyl groups have missed the first desaturation round initiated by the lyso-PC acyltransferase and instead have been used for glycerol-3-phosphate acylation, there is an additional chance for desaturation following the formation of DAG via the Kennedy pathway. DAG is incorporated into PC which is again subject to desaturation. As outlined above, the unsaturated PC may drain off acyl groups into the acyl-CoA pool, but it can also release its complete DAG moiety, with its increased degree of unsaturation due to the reversibility of the phosphocholine transferase reaction. This reaction continuously remodels the DAG pool toward higher unsaturation, and only the final acylation of DAG with acyl-CoA to TAG represents a point of no return regarding the degree of unsaturation. The phosphocholine transferase in microsomes of different origin has no selectivity toward DAG species as measured for forward and backward reactions. The exact equilibrium of this reaction has not been determined, but the release of DAG is about one order of magnitude slower than the incorporation into PC (Table 2). Both rates depend on the local concentrations of substrates, which have been determined in only a few cases. In linseed cotyledons,[108] overall quantities in terms of nanomoles per gram fresh weight were 2 to 10 for CDP-choline, 84 to 90 for phosphorylcholine, 4200 for DAG, and 2100 for PC, whereas data for CMP are not available. Despite its small size, the pool of CDP-choline was hardly accessible to exogenous labeling, which was explained by assuming that it was in equilibrium with the thousandfold excess of unlabeled PC. In addition, part of this CDP-choline is bound to microsomal membranes in a form resistant to release by washing. The same is true for CMP which is required for the reverse reaction. Also this water-soluble compound is present in washed microsomal membranes, and addition of excess CMP (1.5 mM) or CDP-choline (0.8 mM representing 20× K_m) could, at most, only double the rates for release or incorporation of DAG from or into microsomal PC.[109] Therefore, one has to assume that the water-soluble substrates required for recycling of DAG through PC are present at the microsomal reaction sites in small, but effectively placed pools. The two reactions just described, i.e., cycling of acyl-CoAs via the lyso-PC acyltransferase and cycling of DAG via the phosphocholine transferase, represent the routes by which acyl chains are

TABLE 2
Enzymatic Activities (in nmol/min/mg Microsomal Protein)
Related to Desaturation in Microsomal Membranes
from Developing Safflower *Achenes*

Enzyme	EC number	Activity
Acyl-CoA:*sn*-glycerol-3-phosphate acyltransferase[205]	2.3.1.15	20
Acyl-CoA:1-acyl-glycerol-3-phosphate acyltransferase[205]	2.3.1.51	200
Phosphatidic acid phosphohydrolase[206]	3.1.3.4	18
Acyl-CoA:1,2-diacylglycerol acyltransferase[205]	2.3.1.20	3
Triacylglycerol synthesis (overall reaction)[207]	—	4
Acyl-CoA:1-acyl-glycerol-3-phosphorylcholine acyltransferase[208]	2.3.1.23	340(11)
CDP-choline:diacylglycerol cholinephosphotransferase[109]	2.7.8.2	10(1)
NADH:ferricytochrome b_5 oxidoreductase (a)[124]	1.6.2.2	4000
NADH:ferricytochrome c oxidoreductase[125]	—	117
NADPH:ferricytochrome P-450 oxidoreductase (a)[124]	1.6.2.4	400
Cytochrome b_5 (pmol/mg microsomal protein)[124]	—	300
Oleoyl-phosphatidylcholine Δ^{12} desaturase (b)[106]	1.3.1.35	7
O_2 (μM in fluid bilayer) (c)[121]	—	700
Phosphatidylcholine (nmol/mg microsomal protein)[208]	—	230

Note: a — rate with ferricyanide as acceptor; b — rate for high 18:2-variety; c — 3 to 4 times
the concentration in aqueous media. Numbers in parentheses are rates for reverse reactions.

incorporated into and withdrawn from PC as the only substrate accepted by microsomal desaturase systems. Therefore, the extent of acyl chain desaturation will not only depend on the activity of the desaturases, but also on the relative rates of the ancillary reactions and, in particular, on the release of unsaturated acyl-CoA or DAG in the reverse direction. In cells which do not produce TAG, this will be of minor relevance, since there is no drain of acyl groups into a metabolically inaccessible pool. But in oilseeds, a rapid flow through the Kennedy pathway into TAG would leave acyl and DAG groups little time to cycle in the form of PC repeatedly through the desaturation reactions. The capacity of the Kennedy pathway, and, in particular, of the final DAG acyltransferase in relation to the desaturation reactions, in conjunction with the two ancillary transferases determines the extent to which acyl groups can escape desaturation (or hydroxylation) and end up unmodified in TAG. A comparison of the individual reaction rates is given in Table 1.

Biochemical and genetic evidence indicate the existence of several microsomal desaturases, but detailed enzymological studies have been prevented by an inability to solubilize these membrane-bound enzymes. However, some questions regarding substrate specificity, co-factor requirements, and interference by inhibitors can be analyzed with PC-labeled microsomal membranes. A first question concerns the acceptance of acyl groups in different positions of the glycerol backbone of PC. Positional analysis of desaturation products showed that both *sn*-1- and *sn*-2-bound oleoyl groups function as substrates, from which the *sn*-2-bound residue may have a slightly higher affinity. The formation of PC-bound linoleate has been most widely investigated because oleate desaturation is most active in the microsomes from different sources. It has also been demonstrated to work with both positions of exogenously added PC[110,111] and with a 1-*O*-octadecenyl ether[41] introduced via an analog of lyso-PC. Radio-HPLC of molecular species of [14C]18:1-labeled PC[112] has shown the desaturation of 18:1 as a constituent of different PC species. Convincing evidence that oleate groups at both *sn*-1 and *sn*-2 positions of extraplastidic PC are desaturated by the same enzyme was provided by studies with the *fad*2 mutant of *Arabidopsis*, which carries a defect at a single nuclear locus.[113]

In contrast, the Δ^6 desaturase from *Boraginaceae* accepts only *sn*-2-bound linoleic and α-linolenic acid[114] (whereas the corresponding enzyme from cyanobacteria is confined in its action to the *sn*-1 position of glycolipids). Despite the fact that γ-18:3 is formed exclusively at *sn*-2 of PC, it is found in all three positions of TAG from plants which produce this fatty acid (borage, evening primrose, black currant).[115] This points to at least some reversibility of the lyso-PC acyltransferase reaction as discussed above. On the other hand, the *sn*-1 position of all these TAGs contains significantly less γ-18:3 than the *sn*-2 and *sn*-3 positions (by factors of 3 to 10), which was explained by a high selectivity of the DAG acyltransferase for γ-18:3-CoA. In contrast, γ-18:3 in the TAG produced by a fungus was present in similar proportions at all three positions.

The Δ^{15} desaturase proved to be the most critical one to measure, but recently this activity has been characterized in linseed cotyledons.[116] *In vitro* measurement was only possible with homogenates and not with microsomal membranes. Linoleate desaturation to 18:3 occurs at both *sn*-1 and *sn*-2 positions of PC, but *sn*-2-bound 18:2 is desaturated better. A closer analysis of the desaturation reactions in two mutants suggested that flax cotyledons may express two different 18:2-desaturases, one responsible for conversion of *sn*-1-bound 18:2 and another for desaturation at both positions. It may be repeated that these Δ^{15} desaturases in linseed microsomes could not desaturate γ-linolenic acid despite the fact that it was incorporated into PC.[28]

The microsomal desaturases require O_2 as a final electron acceptor, which so far has not been replaced by an artificial oxidant. The desaturation of oleate by safflower microsomes was saturated above 86 μM O_2 in the incubation medium,[117] and similar data (100 μM) have been reported for soybean microsomes.[111] For comparison, a K_m of 56 μM governs the O_2 affinity of the stromal stearoyl-ACP desaturase, whereas the 20:0-CoA desaturase from microsomes of *Limnanthes*[118] (see below) was found to have a lower K_m of 23 μM. Therefore, the decrease of O_2 solubility with increasing temperature (at 37°C still 210 μM in buffer)[117] will not limit desaturation in these systems. But O_2 solubility not only decreases with increasing temperature, but a similar decrease is also observed with increasing (negative) water potential and, in addition, accentuated by the nature of the solute.[119] In the cell, these parameters will act in combination and result in O_2 concentrations below values usually given for buffers of low osmolarity.

Therefore, the experiments on the desaturation in intact safflower cotyledons and cultured sycamore cells as a function of O_2 availability are of particular relevance, since in these systems the microsomal membranes are surrounded by their natural environment of complex composition and unknown water potential. Under these conditions, the desaturation was invariant above 63 μM O_2 in sycamore cells[120] and above 205 μM O_2 (the lowest value tested) in safflower cotyledons,[117] again suggesting that a rise in temperature will not limit the O_2 supply for desaturation.

On the other hand, an evaluation of the O_2 affinity of the microsomal (and plastidial) desaturases should take into account the partition coefficient of O_2 between water and lipid bilayers.[121] Below the phase transition temperature of membrane lipids, O_2 is about equally soluble in the gel matrix of the membrane and the surrounding buffer. Above the phase transition, the lipophilic O_2 partitions preferentially into the liquid membrane, where its concentration exceeds that of the surrounding incubation medium by factors of 3 to 4. Therefore, the O_2 concentration at the reaction sites of the desaturases in the hydrophobic part of the membrane (if operating there, see below) will be significantly higher, and the actual affinity of the enzymes for O_2 may be lower than the values quoted above. Furthermore, there is a steep gradient of oxygen solubility and diffusion across the bilayer.[122] At the membrane surface, O_2 solubility is considered to be lowest due to the unavailability of hydration water, which in this area is immobilized by the polar head groups of lipids. O_2 solubility increases toward the interior with increasing fluidity and is highest in the central core sandwiched between the regions of Δ^9 double bonds in the two adjacent leaflets. Therefore, the insertion of double bonds into lipid segments near the polar head groups (Δ^3 of 16:0 in the bent *sn*-2 ester of PG may be particularly close to this region, Δ^4 in sphinganine, Δ^5 in 20:0-CoA) may have to operate at lower O_2 concentrations than the desaturation at positions $\Delta^{9,12,\text{ or }15}$ in the O_2-rich membrane interior. But as discussed below, the exact location of the active sites of membrane-bound desaturases may not necessarily reflect the most frequent position of the acyl group segments to be desaturated.

IX. ELECTRON TRANSPORT FOR MICROSOMAL DESATURATION

A. OVERVIEW

It is reasonable to assume that the oxygen which is required for desaturation is completely reduced to water, although this detail has not been confirmed. Therefore, two additional electrons are required per double bond formed. A corresponding electron transport sequence involved in microsomal desaturation was first elucidated with animal and yeast microsomes[123] during conversion of stearoyl- and palmitoyl-CoA to oleoyl- and palmitoleoyl-CoA. Recently, spectroscopic and immunological experiments have confirmed the operation of a similar sequence in plant microsomal desaturation.[124,125] The immediate

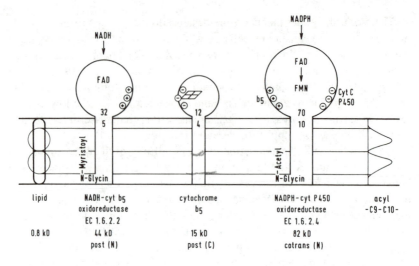

FIGURE 4. Catalytic and membrane-bound domains of microsomal electron transport components required for desaturation. A simplified model is drawn to scale regarding relative size of proteins (0.74 ml/g) and bilayer dimensions (50 Å thick). Cytoplasmically exposed catalytic domains are represented by spheres, membrane-anchoring parts by cylinders (50 Å long, varying diameter). Acylation of membrane-bound N-termini and interactions between protein heads by complementary charge pairing are indicated. The masses of the different parts (in kilodaltons) result from a proteolytic cut between both domains at the membrane surface. A recent model of the yeast 18:0-CoA desaturase ascribes also a large portion (about 80%) of this enzyme (57 kDa) to a cytoplasmic head. The molecular weights (in kilodaltons, second line from bottom) are given for the intact plant proteins. The lowest line indicates the mode of membrane integration which occurs without proteolytic processing: post- (post) or cotranslational (cotrans) insertion with N- or C-terminal parts representing the membrane anchor. The diameter of the protein cylinders in the membrane (11 to 18 Å) may be only twice as large as the diameter (8.6 Å) of the cylinder formally occupied by a single lipid molecule (at the left, liquid-crystalline dioleoyl-PC covering a bilayer surface of 60 Å^2). The polar, water-containing surface layer of each leaflet ends at the C-1 atoms of the acyl groups and extends into a depth of about 7 Å into each leaflet (thin line). The actual volumes accessible to both acyl groups of a lipid molecule are represented by spheres (at the left) rather than by cylinders. The most frequent localization of C-9–C-10 segments of acyl groups and their transversal displacement are represented at the right. The orientation of the heme group (parallel or perpendicular to the membrane) when cytochrome b_5 is donating its electron to the various desaturases is unknown, but important with respect to the distance between the metal center of the desaturase and the eclipsed ethylene segments to be desaturated. The molar ratio of PC/cyt b_5 is about 10^3 (see Table 1). According to a recent model, the b_5 anchor may reach only to the midplane of the membrane. Details and references are in the text.

donor of additional electrons for the desaturase is cytochrome b_5 (b_5), which in turn can receive electrons from two different flavoproteins (Figure 4). These flavoproteins are specific for NADH (EC 1.6.2.2) or NADPH (EC 1.6.2.4), and accordingly desaturation may be supported by both nucleotides. Individual components or combined steps of this electron transport sequence can be measured with the artificial electron acceptors NADH:ferricytochrome b_5 oxidoreductase (NBR, with ferricyanide), NADPH:ferricytochrome P-450 oxidoreductase (NPR, with cytochrome c), and NADH:cytochrome c oxidoreductase, which includes both the NADH-dependent flavoprotein and b_5. In general, the NADH-specific flavoprotein is about ten times more active than the NADPH-dependent one, and both exceed the desaturase activity by several orders of magnitude (Table 1). Therefore, equal rates of desaturation are obtained with 1 to 3 mM NADH or NADPH. Due to the large excess of electron transport capacity compared to desaturase activity, it is not surprising that inhibition of desaturation by antibodies against b_5 is effective only at high antibody concentrations.[125] A small proportion of residual, noncomplexed b_5 will still provide adequate electron flow to support full desaturation.

Recently, dual wavelength spectroscopy has been used to demonstrate directly the involvement of b_5 in desaturation.[124] The steady state level of reduced b_5 observed in NADH-treated microsomes was transiently changed to a lower level by the addition of a limited quantity of oleoyl-CoA. Linoleoyl-CoA was without effect as expected, since these microsomes do not produce linolenic acid. In addition, the rapid response in the level of reduced b_5 confirms that the activity of lyso-PC acyltransferase is not rate limiting in this desaturase system. Cyanide and azide did not interfere with reduction and autoxidation of b_5, so that the inhibition of desaturation by these compounds is specific for the desaturase itself; CO, antimycin, and rotenone were without effect in desaturation and the associated electron transport.

As outlined above, microsomal membranes are a heterogeneous mixture, and the electron transport components of this fraction have been attributed to several different membrane systems purified from this mixture. Therefore, one can assume that NBR, NPR, and b_5 are present in all membrane surfaces covered by the cytoplasm with both donor and acceptor sites facing this compartment. Only chloroplast envelopes[126] and glyoxysomal[127] and outer mitochondrial[128,129] membranes are exceptions which appear not to play a role in desaturation. On the other hand, it is not known exactly which membranes carry out desaturation and whether the ER is truly the subcellular site of this reaction. ER is difficult to purify, and investigations dedicated to this problem were not concerned with desaturation.[97] The electron transport capacity in any of the membranes contributing to the microsomal fraction is sufficient to support desaturation, and it has been shown that lipid-synthesizing enzymes are also present in membranes other than the ER as, for example, in Golgi and plasma membranes.[130]

The abundance of the individual electron transport components in the different membranes varies with plant, developmental stage, and cell type, and therefore a kinetic characterization of reactions involving two components such as NADH-cyt c-oxidoreductase may lead to different results. In general, NBR is about ten times more active than NPR.[124] It should be mentioned that both activities occur also in the soluble fraction,[131] but whether these are truly soluble enzymes comparable to the soluble NBR isoenzyme in erythrocytes (see below) or just proteolytic artifacts is not known. For a reliable measurement of both activities in crude extracts or membrane preparations, the removal of low molecular weight compounds has been recommended.[132] The content of

b_5 and of other b-type cytochromes, including P-450, has been repeatedly determined in microsomal and purified membrane systems, and largely varying contents and ratios have been found.[133-135] In plasma membranes, other b-type cytochromes with higher redox potential (about 160 mV) predominate. In oilseeds, the content of b_5 is strongly dependent on seed development. At optimal stages, a content of 300 pmol/mg microsomal protein corresponding to about 0.4% of the membrane proteins decreases to nearly zero levels at the end toward seed maturation.[124,125]

B. INDIVIDUAL COMPONENTS

The most important entrance of reducing equivalents for desaturation is the NBR. This enzyme contains one bound FAD and can use b_5 and ferricyanide as acceptors, but not cytochrome c. The size determined for the plant enzyme varies from 32 to 44 kDa. Antibodies raised against the purified 44-kDa microsomal enzyme cross-react with a 43-kDa protein from plasma membrane.[136] NADH K_m values for the glyoxysomal, ER, and plasma membrane activities were found to be 7 to 12 μM with cyt c (including b_5) as acceptor and somewhat higher (25 to 97 μM) with ferricyanide as acceptor.[137,138]

The enzyme from pig liver has been crystallized in two forms[139] and serves as a useful model. One form represents the intact protein obtained by detergent solubilization of microsomal membranes and therefore including the hydrophobic membrane anchor (Figure 4), and the other represents the water-soluble catalytic domain released from membranes by a single proteolytic cut. The complex between the reductase and b_5 is stabilized by electrostatic forces, due to complementary charge pairing between positive lysyl residues on the reductase[140] and negative charges on b_5 involving glutamate carboxylates and one exposed heme propionate (Figure 4). The positive residues lining the electron exit on the reductase prevent an interaction with cyt c whose electron transfer site has a similar ring of positively charged lysyl residues surrounding the edge of the buried heme group.[141] The sequenced rat gene [17 kilobasepairs (kbp)] codes for 301 amino acids on nine exons,[142] and only one gene codes for the identical enzymes located in ER and outer mitochondrial membranes,[143] and even a soluble NBR restricted to erythrocytes is expressed from the same gene. In both membrane-bound, the N-terminal glycine carries a myristoyl group (Figure 4) which increases the hydrophobicity.[144] The enzyme is synthesized on free ribosomes and inserted posttranslationally without processing into its target membranes.[145]

The electrons from NADPH supporting desaturation enter via NPR, which has about twice the size as the NBR, ranging from 72 to 82 kDa in plants.[146] The largest form which is considered as the undegraded enzyme was used for antibody production.[147] In contrast to NBR, the NPR can reduce not only ferricyanide and b_5, but also cyt c and P-450, the last one being the best fitting acceptor.[148] The interaction with a wider selection of differently charged acceptors has been ascribed to the fact that the exit area on this reductase is divided into two separate domains, one being charged negatively by carboxylates and enabling the interaction with positively charged cyt c and P-450, whereas the other part with its lysyl residues is positively charged and allows additional electron transfer to the negatively charged b_5 (Figure 4). The K_m values for NADPH reported for the plant enzymes cover an order of magnitude from 6 to 77 μM with both cyt c or ferricyanide as acceptors.[149] The sequenced rat gene for NPR (20 kbp) codes for 678 amino acids on 15 exons,[150] from which the first one codes for the N-terminal membrane domain of 60 amino acids with the N-terminal glycine being acetylated. The enzyme is synthesized on membrane-bound ribosomes and inserted cotranslationally without processing.[145] A single proteolytic cut divides the protein into a membrane anchor of 56 to 60 amino acids (6 to 10 kDa) and a ten times larger fragment representing the water-soluble catalytic domain carrying FAD and FMN.

The central electron carrier connecting the two reductases with the desaturase is b_5. Its iron in the noncovalently bound protoheme is liganded by two additional histidines[141] which prevent access and inhibition by cyanide, azide, and CO. The low redox potential (about -45 mV) of this b-type cytochrome is ascribed to these histidines, whereas other b-type cytochromes with a different sixth ligand such as methionine (b-types in the plasma membranes) or cysteine (P-450) have different redox potentials and redox difference spectra. Cytochrome b_5 proteins have been purified from different plants[151] with molecular weights always close to 15 to 16 kDa, and recently a cDNA from cauliflower b_5 has been sequenced which codes for 134 amino acids.[152] Therefore, this carrier is the smallest of the whole sequence. Previous results considered to indicate a b_5 involvement in β-oxoacyl-CoA reduction during microsomal elongation of fatty acids are now interpreted controversially.[153]

In animal cells, b_5 from ER and outer mitochondrial membranes showed only a 58% identity in amino acid sequence,[141] differing from NBR which was identical in both membranes. The animal b_5 is synthesized on free ribosomes[145] and inserted posttranslationally, without processing, into its target membranes with the C-terminal segment of 30 to 40 amino acids, from which the final 4 to 6 extend to the cytoplasm.[154] In a model differing from the picture shown in Figure 4, the hydrophobic part of b_5 does not penetrate beyond the bilayer midplane into the membrane.[155] Membrane binding[156,157] and the crystalline structure of the catalytic domain[158] have been studied. Recently, an animal desaturase has been expressed in enzymatically active form in the ER of yeast[159] and transgenic tobacco plants[160] which may suggest a general inter- action of b_5 and desaturases of different organisms. The electron acceptor on the desaturase is a nonheme iron in a catalytic site of unknown structure, and it will be interesting to see which metal centers are used by plant desaturases in ER and envelope membranes. This site should be in reach of the heme edge of docked b_5. It may be that the head of b_5 reorients in such a way that the heme is brought from a membrane-parallel position when accepting an electron from the reductase[155] to a more perpendicular orientation when docked to the desaturase; a useful model may be the perpendicularly fixed cytochrome heme interacting with the membrane-bound electron acceptor in the photosynthetic reaction center of bacteria.[161] The active site of the desaturase with the metal ion as acceptor should also have access to the prospective double bond positions of membrane lipids. The location of this site on the desaturase may be a compromise between the accessibility of both the hydrophilic b_5 heme edge and the hydrophobic ethylene segments to be desaturated at various positions along the acyl chains. Therefore, the flexibility of these segments and, in particular, their displacement within the bilayer are of interest in this respect.

Recent analyses of liquid-crystalline bilayers of dioleoyl-PC and 1-oleoyl-2-(9′,10′-dibromostearoyl)-PC by X-ray and neutron scattering have yielded useful information on this question.[162,163] Data on the residential probability of both the double bond and the dibromoethylene segment at C-9–C-10 also reflect the mobility of close by segments relevant for desaturation. Therefore, these two acyl chains can be considered as representing the vertical flexibility of monoenoic and saturated acyl groups of lipids in a fluid membrane into which second (at C-12–C-13 in all lipids) and first double bonds (at C-7–C-8 in MGD) must be inserted. Both segments are found with highest frequency in the middle of the hydrophobic slab of each monolayer (Figure 4), and their presence decreases to low values at both the methyl end in the middle of the bilayer and the polar region on the membrane surface beginning from the inside with the C-1 atoms of the acyl groups at a distance of about 7 Å from the bilayer surface. The vertical displacement of internal acyl group segments due to thermal motion is large enough and their collision with the polar surface high compared to enzymatic rates and, accordingly, sufficient to support desaturation even in this region. This picture is supported by experiments on fluorescence quenching of reporter groups at different positions of the acyl chains[164] and by modeling of the volumes occupied by fluid acyl groups in bilayers which turned out to have more or less round contours along the bilayer normal.[165] Therefore, the vertical insertion of the metal-binding site of the desaturase may be limited primarily by its distance from the heme edge of b_5 and extend only superficially into the hydrophobic part of the membrane. On the other hand, the probability that an acyl chain from the opposite leaflet of the bilayer comes into contact with the active site of the desaturase may be rather low, and an exchange of lipids between leaflets may be required to achieve a similar degree of unsaturation in both membrane halves.[166]

In view of the different interactions of the various components of the desaturation sequence, it is not surprising that it is easily disturbed by alterations of the membrane matrix. Addition of detergent interrupts the NADH-cyt c-oxidoreductase activity, which depends on the mobility of two membrane-bound proteins, whereas the NADH:ferricyanide oxidoreductase is hardly affected. Similar effects result from lipid depletion of the membrane, and the reconstitution of NADH-cyt c-oxidoreductase activity by addition of different lipids was dependent on the nature of the lipid headgroups, which may influence the lateral and rotational mobility of differently charged electron transfer proteins in the membrane.[167]

X. ACYL-CoA DESATURASES

Extensive work[168] has been carried out on the different acyl-CoA desaturases from animal and yeast microsomes (Figure 5), and their comparison with plant desaturases reveals very interesting phylogenetic correlations (Figure 5). The Δ^9-18:0- and the Δ^6-18:2-CoA desaturases were the first to be isolated in functional form with molecular weights varying from 41 kDa (mammalian 18:0-CoA desaturase) and 33 kDa (chicken 18:0-CoA desaturase) to 66 kDa (mammalian 18:2-CoA desaturase). It should be recalled that plants (normally) do not have these and other CoA-dependent enzymes, whereas several insects have a Δ^{12}-18:1-CoA desaturase which produces linoleic acid essential for most other animals.[4] In yeast and fungi, second and third double bonds may be introduced into lipid-bound acyl groups,[169] representing a switch to the plant-type of desaturation. Accordingly, plants and yeasts may follow similar strategies in producing polyunsaturated fatty acids; the first double bond is inserted into a thioester-bound acyl group (ACP, CoA) and the following ones into lipid-linked substrate chains.

cDNA and genomic sequences are known for the mammalian[170,171] (15 kbp, six exons, possibly more than one gene) and the yeast 18:0-CoA desaturase[159] (intron-free gene coding for a 57 kDa protein). The mammalian desaturase is synthesized on free ribosomes which may also apply to the yeast enzyme[159] and inserted posttranslationally without processing into ER membranes. In addition, enzymatic activity has been successfully reconstituted by insertion of the isolated proteins into phospholipid liposomes. Accordingly, the nucleotide sequences for a complete microsomal desaturation set including NBR, NPR, and b_5, as well as their modes of membrane insertion *in vivo*, are known.

Initial studies on structural similarities of the different desaturases used immunological techniques,[168] and differences were found among different desaturases of the same organism[172] and cDNA-derived amino acid sequences have also been demonstrated to differ among organisms.[173,174] Recent modeling studies[159] have suggested a rather large cytoplasmic domain of the yeast enzyme from which only about 90 amino acids from the middle of the total sequence of 510 were placed as a hydrophobic anchor into the membrane. The three most conserved sequences came to lie in the cytoplasmic domain which also should expose the arginines and other positive charges to pair with b_5 and the phosphate groups of the substrate CoA thioesters. The active site involves an iron ion which is said to be bound by tyrosine and easily removed after NADH reduction by metal chelators.[168]

It should be emphasized again that the Δ^9 desaturase in native and reconstituted form requires the CoA thioesters of fatty acids as substrates. Free fatty acids are not accepted,[175] which is a significant difference from the solubilized chloroplast enzyme. The only plant acyl-CoA desaturase activity recognized so far and investigated in some detail is the Δ^5 desaturase present in developing *Limnanthes* seeds.[118] The major fatty acids in the reserve TAG of this plant are 5-*cis*-20:1 (70%), erucic acid (13-*cis*-22:1, 16%), and 5,13-*cis*-22:2 (10%).

The Δ^5 desaturase activity is unstable and rapidly lost during cell fractionation, so that studies had to be carried out with crude homogenates, and a subcellular attribution to microsomes remains uncertain. It is most likely that this enzyme is one which desaturates on the acyl-CoA level with O_2 and NADH as co-factors. The substrates accepted *in vitro* are CoA thioesters of $C_{16:0}$, $C_{18:0}$, and $C_{20:0}$, whereas *n-9-cis*-18:1, 20:1, and 22:1 were not accepted. On the other hand, the occurrence of 5,13-22:2 in the seed suggests the use of erucoyl-CoA as substrate *in vivo*. According to the high proportion of 5-20:1 in the TAG, the Δ^5 desaturase contributes significantly to the desaturation of reserve lipids in this plant and functionally replaces the 18:0-ACP desaturase in other oilseeds.

XI. INVESTIGATIONS WITH MUTANTS

A. DESATURASE MUTANTS AND FIRST GENES

A detailed knowledge of plant desaturases and their ancillary systems in terms of protein structure and molecular biology was limited until recently to the soluble chloroplast components (ferredoxin, FNR, stearoyl-ACP desaturase, see below). This situation has changed with the isolation of a cDNA[152] for microsomal b_5 and the rapid progress in isolating genes for membrane-bound desaturases from blue-green algae and higher plants. Similar results can be predicted for the near future for NBR and NPR, for which antibodies are already available.[136,147]

The gene for the *cis*-Δ^{12}-18:1 desaturase from blue-green algae was the first one to be isolated from photosynthetic organisms, and was accomplished by complementation of desaturase-deficient mutants of *Synechocystis*. The mutants were generated by chemical mutagenesis and selected by their reduced growth at lower temperature.[176] At 22°C, wild-type cells grew to death in the presence of ampicillin, whereas low-temperature-sensitive mutants survived because they do not multiply at this low temperature and resume growth after a subsequent shift to 34°C. From 26 mutants selected, only two turned out to be desaturase mutants. *Fad*12 had lost the ability to introduce a *cis*-12-double bond into lipid-linked *cis*-9-18:1, whereas *Fad*6 was incapable of *cis*-6-desaturation of the *sn*-1-bound Δ^9- and $\Delta^{9,12}$-unsaturated fatty acids of glycolipids.

The transformation of the *Fad*12 mutant with a genomic library of wild-type *Synechocystis* DNA constructed in a suitable vector resulted in the isolation of a transformant which grew at lower temperature and had a wild-type fatty acid profile.[177] The corresponding DNA insert was characterized as the *des*A gene which codes for a 40.5 kDa protein of 351 amino acids. Sequence analysis of the mutant allele in *Fad*12 showed that one T to A substitution had generated a stop codon in the coding region. Recent expression of *des*A in functional form in *Escherichia coli* has demonstrated unambiguously that it codes for the desaturase,[209] and introduction of *des*A into *Anacystis*, which is a blue-green alga incapable of producing polyunsaturated fatty acids, resulted in the accumulation of 9,12-dienoic acids in the transformed organism. In the derived amino acid sequence of the desaturase, two putative membrane-spanning regions are located in positions widely separated along the sequence which may suggest that a hydrophilic domain is anchored twice in the membrane. On the protein level, the sequence similarity between *des*A and acyl-CoA or acyl-ACP desaturases is below 10%. It may be mentioned that the yeast 18:0-CoA desaturase gene was similarly isolated by complementation of a desaturase-defective mutant.[159]

A cDNA coding for an *n*-6 desaturase from spinach chloroplasts was isolated via purification of the 40 kDa protein from envelope membranes,[96] N-terminal amino acid sequencing, and construction of corresponding nucleotide primers. Their use in the polymerase chain reaction resulted in the isolation of a 1.5 kbp cDNA coding for the entire protein from spinach leaves. The partial amino acid sequence deduced so far has similarity with the *des*A-coded protein and suggests a phylogenetic and functional relationship between these two *n*-6 (or Δ^{12}) desaturases.

The approach to gain access to desaturase genes by genetic methods after mutagenesis was initiated in higher plants with the isolation of the desaturase mutant *fad*A of *Arabidopsis*,[178] which cannot desaturate 16:0 at *sn*-2 of PG to 3-*trans*-16:1. In the meantime, further mutants affecting additional chloroplast (*fad*B, C, D = Δ^7, *n*-6, *n*-3) and ER desaturation steps (*fad*2, 3 = Δ^{12}, Δ^{15}) have been isolated and characterized (summarized in References 18 and 179; see Figure 2). These mutants cover all desaturation reactions required for the biosynthesis of the normal *n*-6,9 and *n*-3,6,9 dienoic and trienoic fatty acids in leaves, roots, and seeds, whereas 18:0-ACP desaturase mutants, as expected from their viability problems, were not found in this plant.

Desaturase mutants have no visible phenotype under normal growth conditions, and so screening by selection as applied with cyanobacteria is not possible. Single leaves from individual plants are directly transesterified to obtain fatty acid methyl esters for analysis by GLC. About ten thousand analyses have been carried out by different laboratories, and several independent mutants for some of the *fad* loci mentioned above have been collected.[113,180,181] Crosses between different mutants of the same locus did not lead to complementation, which demonstrates that the block in the particular desaturation step is controlled by a single nuclear-coded locus. Since all these mutant loci express recessive (*fad*B,C,2) or at most intermediate (*fad*A,D,3) phenotypes, only the rare homozygous mutants crossing out in M2 populations will be recognized with certainty (see below).

For gene isolation, classical and molecular genetics have to be combined to map the mutant locus on a specific chromosome and correlate it with genetic and restriction fragment length polymorphism (RFLP) markers. This is required for the subsequent chromosome walking after selection of a suitable yeast artificial chromosome (YAC) from a corresponding library.[182] The first success of this tedious, but in the long run successful, approach is the isolation of the *fad*3 and *fad*D genes, although in both cases actual chromosome walking was circumvented.[183,184] The inserts of YACs identified by the closest RFLP marker were used as probes to screen a cDNA library from *Arabidopsis* seeds. The sequenced positive clones turned out to code for proteins which had homology on the amino acid level with the *des*A-coded Δ^{12} desaturase from *Synechocystis*.[177] The *fad*3 and *fad*D clones code for proteins of 38 and 51 kDa, respectively. Since both complement the original desaturation defects, they can be considered to represent the microsomal Δ^{15} and the plastidial *n*-3 desaturase, respectively. In a similar approach, T-DNA insertional mutants from *Arabidopsis* were screened for fatty acid alterations. So far, one line homozygous for a mutation at the *fad*B locus (Δ^7-16:0-MGD desaturase) has been isolated.[185] A search for the mutated gene with T-DNA as the probe would be expected to be easier than chromosome walking and results may become available soon.

The similarity of the microsomal Δ^{15} desaturase with the chloroplast *n*-6 and *n*-3 desaturases, as well as with the cyanobacterial Δ^{12} desaturase, is surprising. It places the membrane-bound plant desaturases from cyanobacteria, chloroplasts, and the nucleocytoplasmic compartment into a common large gene family (see Figure 5). This group is responsible for polyunsaturation and differs from acyl-ACP-Δ^9 desaturases,[186] which in turn show no Δ^9 acyl-CoA desaturases. Future results will have to confirm whether the desaturases known so far are unexpectedly different. At present, they can be separated according to the form of their substrate chain into lipid-, CoA-, and ACP-linked enzymes, and classification according to Δ^x or *n*-x measurement seems to be less important in this context. This raises interesting questions regarding phylogenetic relationships and, in particular, the position of the cyanobacterial Δ^9 desaturase and the origin of plant ER desaturases. One could imagine that the desaturases inserting the first double bond into a saturated chain are the oldest ones, and, according to presently known sequence data for CoA and ACP desaturases, they are not related. The enzymes producing additional double bonds in acyl-CoAs or lipids may all share common structural motifs and thus could represent a group of common origin. Therefore, the sequence data from animal Δ^5, Δ^6, and Δ^{12} desaturases will be particularly interesting, and the bacterial genera (see above) capable of aerobic production of different monoenoic fatty acids may be very useful for finding phylogenetic roots. But so far, only one enzyme from this group has been studied enzymatically. It is the 24:0-CoA desaturase from *Mycobacterium*,[187] which requires O_2 and reduced ferredoxin to produce *cis*-15-24:1.

B. METABOLIC CHANGES IN *ARABIDOPSIS* DESATURASE MUTANTS

There is no definite proof yet, but a straightforward and most probable interpretation assumes that the *Arabidopsis* desaturase mutants affect the structural and not just regulatory genes. Apart from *Arabidopsis* and *Synechocystis*, other plants, and in particular all the common oilseeds,[188] have been used to produce or select desaturase mutants for agricultural reasons, i.e., for changing the quality of the oil with regard to its degree of unsaturation. Even if these mutants will not be used for gene isolation, as is most likely, they are very useful for complementing and extending the studies carried out with the *Arabidopsis* mutants. These investigations included a detailed analysis of lipid patterns and of their labeling in different organs such as leaves, roots, and seeds, and in addition the response of mutant plants to various stress situations has been studied. The results have demonstrated the enormous potential of these *Arabidopsis* mutants for understanding lipid metabolism and the function of particular lipids and individual fatty acids by confirming the existence of two different sets of individual desaturase activities as well as their subcellular sites of operation, tissue expression, gene number, and substrate specificity. Furthermore, the mutants have significantly refined our understanding of a mutual exchange of lipids between plastids and nucleocytoplasmic membrane systems and opened first insights into the flexibility of this subcellular cooperation and its regulation in response to defects and external or developmental signals. Most of these results would not have been obtained without the use of these mutants. The most important consequences of those studies not yet discussed in detail are given below, for both chloroplast (*fad*A,B,C,D) and ER (*fad*2,3) desaturase mutants.[18,179,188,189]

The chloroplast desaturase mutants have effects on lipids from leaves, but not from roots or seeds, which reflects the importance of this group of desaturases in different tissues. On the other hand, a close inspection of MGD fatty acids and not just of total fatty acids may reveal minor effects also in nongreen tissues. The more or less complete block of each particular desaturation step in homozygous mutants indicates that in *Arabidopsis* only one gene codes for each of these plastidial desaturases. For the recessive loci *fad*B,C (and *fad*2), transcripts from one copy of this gene in heterozygous F1 leaves are sufficient to establish wild-type desaturation. This applies despite the fact that desaturation of *n*-9 and *n*-6 intermediates is not complete. This would indicate that the particular desaturase gene even in its homozygous wild-type form does not provide excess activity. The regulatory mechanisms involved in such phenomena are not understood, as is evidenced by the fact that desaturase mutants affecting the production of end products (*fad*A,D and *fad*3) produce intermediate phenotypes in heterozygous forms, but not recessive ones such as *fad*B,C and *fad*2. The effects of *fad*B and *fad*C mutants are expressed at all temperatures, whereas the *n*-3 desaturase defect caused by a mutation in *fad*D is expressed only at 32°C and not at 18°C. This could indicate that the *fad*D mutant is temperature sensitive and not expressed at 18°C, but it could also point to the existence of two different *n*-3 desaturases, from which *fad*D codes for a housekeeping gene expressed at all temperatures, whereas a different, cold-induced desaturase is not affected by the *fad*D mutation. The leakiness observed in several other mutants could have similar reasons.

A further important result was the observation of a shift in the biosynthesis of MGD from pro- to eukaryotic species in response to decreasing desaturase activity in chloroplasts from *fad*B, C, and D mutants, which may even contribute to a differentiation between 18:3- and 16:3-plants. Only in *fad*D mutants (lacking the *n*-3 desaturase) do the less desaturated prokaryotic tetraene MGD

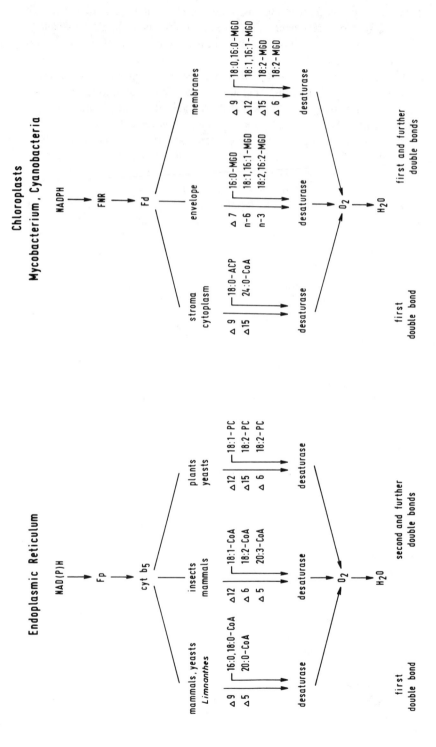

FIGURE 5. Putative electron flow during aerobic desaturation of fatty acids by different soluble and membrane-bound desaturases in various organisms and compartments. All desaturases withdraw electrons from acyl chains and receive additional electrons from soluble or membrane-bound one-electron carriers (ferredoxin = Fd or cytochrome b_5), which in turn are reduced by soluble (NADPH:ferredoxin oxidoreductase = FNR) or membrane-bound flavoproteins (Fp = NADH: cytochrome b_5- and NADPH:cytochrome P-450 oxidoreductase). Oxygen as the final electron acceptor is (most likely) reduced to water. The acyl chains can be linked in thioester form with CoA and ACP or in oxygen ester form with MGD, PC, and other lipids. Animals introduce all double bonds into acyl-CoA thioesters of various degrees of unsaturation using ER-bound acyl-CoA desaturases (left two columns), but Δ^{12} and Δ^{15} desaturases are missing. As a rare exception the Δ^{12} desaturase is found in insects. Yeasts and fungi use similarly ER-bound enzymes, but only the first Δ^9 double bond is inserted by an acyl-CoA desaturase, whereas all others are formed (most likely) from lipid (PC)-bound unsaturated acyl groups. Cyanobacteria also use exclusively membrane-bound enzymes, but the substrates are always lipids (for example MGD, far right). Higher plants have the most complex collection of desaturases: a soluble 18:0-ACP thioester desaturase for the first Δ^9 double bond in C_{18} chains and two membrane-bound sets in chloroplast envelopes (n-6, n-3) and in the ER (Δ^{12}, Δ^{15}, Δ^6) for further double bonds in lipid-bound (MGD, PC) acyl groups of various degrees of unsaturation. The 16:0-MGD Δ^7 desaturase in chloroplasts is characteristic for 16:3-plants. The Δ^3-*trans* desaturase is not included. As exception in plants, *Limnanthes* seeds contain a microsomal acyl-CoA desaturase, and a unique soluble acyl-CoA desaturase is found in the cytoplasm of *Mycobacterium*. The protein sequences known so far suggest three independent groups comprising the 18:0-CoA desaturases (animals, yeasts), the 18:0-ACP desaturases (plants), and the 18:1 and 18:2 desaturases (from chloroplasts, ER, and cyanobacteria), respectively.

species accumulate, and the mutants have the same ratio of pro-/eukaryotic MGD species as wild-type plants. Their degree of unsaturation is apparently sufficient to maintain chloroplast functions at normal and even at low growth temperature. On the other hand, in *fad*C and *fad*B mutants, an increasing replacement of the more saturated prokaryotic by hexaene eukaryotic MGD species is observed, which was explained by the insufficient degree of unsaturation in these mutants. Therefore, the prokaryotic proportion of MGD in a plant is not fixed and may respond flexibly to environmental or developmental signals. In fact, physiologically induced variations of this kind have been demonstrated in leaves and fruits of several 16:3-plants, in which a shift to eukaryotic MGD species was induced by increasing temperature[190,191] and ripening.[94]

In *fad*B mutants, even a slight increase of 16:0 at *sn*-2 of PC was observed and interpreted as resulting from export of nondesaturated prokaryotic DAG into the nucleocytoplasmic compartment. In addition, *fad*D and *fad*C mutants not only had decreased levels of polyunsaturated fatty acids in chloroplast lipids, but in both mutants 18:3 also was significantly reduced in the ER-synthesized PC, PE, and PI. In addition, in *fad*D, the decrease of 18:3 in extraplastidic phospholipids is nearly exactly matched by an increase of 18:2. A straightforward interpretation would see this as resulting from a block in the normal export of polyunsaturated fatty acids from plastids into the ER which is supported by results from *fad*3 mutants (see below). This extends the original picture of the cooperative pathway considerably and may have consequences in planning strategies for genetic manipulation of oilseeds (see below).

The flexibility in switching between pro- and eukaryotic MGD indicates that the desaturation of this lipid may be critical and therefore kept within narrow limits. This contrasts with PG which retains its prokaryotic structure irrespective of the degree of unsaturation. This could simply be due to the fact that the eukaryotic intermediate (DAG) imported into chloroplasts has passed the form (PA) usable for PG synthesis.

A comparison of protein patterns separated by 2D gel electrophoresis claimed that *fad*D mutants lacked a 90 kDa protein ascribed to the stroma/envelope fraction of wild-type chloroplasts. Similarly, the abundance of this protein was reduced in wild-type callus after treatment with the herbicide SAN 9785 known to interfere with 18:2 desaturation.[192] The correlation between this protein and desaturation remains to be demonstrated, particularly in view of the size (51 kDa) of the *n*-3 desaturase.

Studies on the resistance of the various mutants to stress at high or low temperature have shown that *fad*A and *fad*D do not differ from wild type, whereas the less desaturated *fad*B and *fad*C are more resistant at higher and less resistant at lower temperature where growing leaves may even become chlorotic.[193] Accordingly, these mutants have a visible phenotype only when growing at low temperature. The general conclusion is that a high degree of polyunsaturation is required for the biogenesis of chloroplasts of normal size and ultrastructure, but not for photosynthetic electron transport in thylakoid membranes which is hardly affected by the variation in unsaturation as found in the different mutants. Similarly, investigations of temperature effects on different photosynthetic activities and their thermal stability using *Synechocystis* wild type, the *Fad*12 mutant and a *Fad*12/*des*A retransformed strain did not reveal any significant differences.[194] But when both Δ^6 and Δ^{12} desaturases were knocked out, the resultant cells contain only saturated and monounsaturated fatty acids and demonstrated enhanced photoinhibition.[195]

The *fad*2,3[113,181] mutants of microsomal desaturation steps are expressed in leaves, roots, and seeds with largest effects in the nongreen tissues. In roots and seeds, increased levels of the substrate blocked and decreased proportions of the derived products (18:2 and 18:3 in *fad*2, 18:3 in *fad*3) are observed, whereas significantly different patterns were revealed in leaves. Detailed analyses have been carried out so far only for *fad*2, from which four noncomplementing lines have been isolated.[113] Chloroplast-made lipids of this mutant show only minor effects (see below), whereas in all ER-synthesized phospholipids the expected variation (blocked substrate-up; subsequently formed products-down) is only partially seen. 18:1 (*fad*2) and to a lesser extent 18:2 (*fad*3) increase, but 18:3 is hardly reduced in both mutants. Direct *in vitro* measurement of 18:1-PC desaturase in microsomes from *fad*2 leaves did not detect activity, and acetate labeling of *fad*2 leaves confirmed that during the first hours the usually rapid conversion of PC-bound 18:1 to 18:2 was not detectable. These results are best explained by an export of 18:3 from chloroplasts into the ER in *fad*2,3 mutants. Furthermore, in both mutants, the proportion of 16:3 in leaves was increased by about the same percentage, and the detailed analysis of *fad*2 MGD revealed a 36% reduction of eukaryotic species and a corresponding increase of prokaryotic MGD. This is a reversion of the response seen in *fad*B,C mutants and represents the complementary aspect of species switching in MGD synthesis to provide optimal degrees of unsaturation; if desaturation is insufficient in chloroplasts (*fad*B,C), prokaryotic species are replaced by eukaryotic ones; if microsomal desaturation is impaired (*fad*2,3), eukaryotic species are reduced and balanced by prokaryotic ones. The availability of both systems for MGD synthesis would provide greater flexibility, but 18:3-plants have apparently abandoned this possibility.

C. WELL-STUDIED DESATURASE MUTANTS FROM OILSEEDS

Only a few oilseed desaturation mutants have been investigated in detail with respect to fatty acid alterations in lipids and tissues other than TAG from fat-storing organs, but it may be assumed that most affect microsomal desaturases and therefore are homologous to the *fad*2,3 series of *Arabidopsis*. Several of these mutants result in a partial block of one individual desaturation step, which accordingly can be assumed to be controlled by more than one gene. This differs from most *Arabidopsis* mutants as outlined above.

From soybean, mutants with defects in 18:1 (A5) and 18:2 desaturation (C1640) have been investigated in more detail.[111,196] It was found that partial blocks were expressed in cotyledons and roots, but not in leaves, which differs from *Arabidopsis fad*2,3. *In vitro* measurements showed that in A5 cotyledons, 18:1-PC desaturase was reduced, whereas lyso-PC acyltransferase had wild-type activity. This indicates tissue-specific expression of at least two genes for each of these desaturases. From the effects on fatty acid proportions, one has to conclude that even the expression of two genes in wild-type plants is not sufficient to provide enough desaturase activity for complete conversion of minor intermediates such as 18:1 or 18:2 to the subsequent products. Relaxed adjustment of desaturase activities in relation to TAG deposition may be a general characteristic of fat-storing tissues resulting from the absence of a selection pressure. In leaves and, in particular, in the case of chloroplast MGD, the situation is different.

A further specification of desaturases was revealed by detailed genetic and biochemical analyses of two chemically induced flax mutants defective in 18:2 desaturation.[116] Two independent genes *Ln*1 and *Ln*2 on different chromosomes reduce the 18:3 content of seeds each by about half, and a combination of both results in a "zero" line. At times when the 18:3 content of seed PC is already severely reduced, the seed MGD of this zero line is hardly affected. Similar to *fad*3 and *fad*D, both mutants cause intermediate levels of 18:3 in heterozygous F1 plants, and accordingly, in this case, a single intact gene is not sufficient to provide wild-type desaturation. *Ln*2 is rather specific for desaturation at *sn*-2 of PC, whereas *Ln*1 accepts 18:2 in both *sn*-1 and *sn*-2 with about equal efficiency. Similar *sn*-positional specificities of desaturases have been pointed out above (Δ^3-*trans*, Δ^6-*cis*, Δ^7-*cis*), but in view of the flax data such a complication has to be taken into consideration with the "normal" desaturases as well. The *Ln*1 and *Ln*2 defects are not expressed in leaves,[197] which again is an example for tissue- and organelle-specific expression of desaturase genes or of their regulatory systems. Furthermore, these mutants do not interfere with the rate of TAG accumulation, which is particularly encouraging for future genetic manipulation in this field (see below).

As a final example of an oilseed desaturase mutant which has been studied in more detail, the results of high oleic sunflower[188] will be discussed. Wild-type plants accumulate high proportions of 18:2 which vary significantly with growth temperature (increased 18:1 at higher temperature). The chemically induced Pervenets mutant produces a temperature-independent high proportion of 18:1 (80 to 90%). This trait is inherited by a single dominant locus, although subsequent studies suggested a more complicated pattern with two additional modifying loci.[198] A detailed analysis of PC and galactolipids from fruit cotyledons, leaves, and roots showed that the 18:1-desaturation block is expressed only in fruit lipids.[199] On the other hand, the plastids of this organ, though flooded with eukaryotic 18:1-DAG, retain some 18:1- and 18:2-desaturase activity as evident from the small but significant proportions of 18:2 and 18:3 in fruit galactolipids. *In vitro* measurement of enzymatic activities showed that neither electron transport (NBR, NPR) nor lyso-PC acyltransferase were affected, with their activities being comparable to safflower systems. The only defect was a severely reduced or absent 18:1-PC desaturase activity[199] which became obvious at about 11 d after flowering at a time when in wild type this activity starts to rise sharply.[200] These data were explained by the assumption that the fruit-specific expression system of a desaturase gene is blocked, but the dominant character of this phenomenon suggests a complicated interference in a presumed derepression cascade. That this block in 18:1 desaturation does indeed affect regulatory loci and not the structural desaturase gene was concluded from the fact that the same cotyledonary tissue resumed desaturase activity when its development was changed to become green. This was observed during two independent occasions: when fruit development in the head on the plant is interrupted by an unknown mechanism resulting in an outgrowth of cotyledons through the fruit hull[200] or when seeds germinate and cotyledons start to become green.[199] In both cases, the cotyledonary tissue rapidly expressed 18:1 desaturation, and in particular 18:2 and 18:3 were found in PC and galactolipids at germination stages when cotyledons where still white and had not yet left the hulls.

Additional genetic factors as mentioned above and influencing desaturation have also been observed in other oilseeds. The biochemical basis for this interference is not known, but it has been discussed[188] that defects in ancillary systems such as lyso-PC acyltransferase and cholinephosphotransferase, particularly in view of their low rates in reverse reactions (Table 1), could cause such effects. But as pointed out, the lyso-PC acyltransferase was not affected in the examples investigated.[111,199]

In summary, the mutants have shown different types of inheritance and organ expression of desaturase activities. A desaturase can be controlled by one or more loci. A block can be expressed in dominant, intermediate, or recessive form. Its expression can show up in seeds; in seeds and roots; or in seeds, roots, and leaves prohibiting any generalizations.

XII. PERSPECTIVE

Several points have been indicated above as examples where research efforts will be concentrated in the future. In biochemistry, a characterization of the various desaturases (Figure 5) with particular emphasis on metal centers, membrane integration, and interaction with the different substrates will be of prime interest. The elucidation of these problems will be supported by the availability of cloned genes and their expression in native or mutated form in different host organisms. These studies will also contribute to the clarification of the interesting phylogenetic relations between the different desaturase systems. Furthermore, a complete subcellular localization of plastid and microsomal desaturases should be possible with the use of antibodies prepared against recombinant enzymes. With the availability of nucleotide sequences and corresponding probes, well-known physiological phenomena such as tissue expression, differences between 16:3- and 18:3-plants, and variation in response to the multiplicity of external factors can be studied at a new level. Questions on the role of unsaturation in function and biogenesis of membranes will be addressed in a more conclusive way with cyanobacteria[70,195] and higher plants once the double mutants of *Arabidopsis*, which block both ER and chloroplast desaturase genes, have been constructed.[113]

Furthermore, cloned desaturase genes will undoubtedly be used for transformation of oilseeds to suit different demands. The economic consequences of such experiments will alter the innocence of previous investigations and speed up the research in this field, but not necessarily the dissemination of data, which may be discussed with lawyers and submitted to patent granting authorities, before editors of scientific journals and colleagues become aware of the progress. Therefore, an efficient literature coverage in this field should include submitted patent applications.

The future transformation of oilseeds can be divided into groups of increasing complexity. For human consumption, two obvious goals are elimination of linolenic acid and the production of a high oleic oil for deep-fat frying, which has additional interest for chemical use. From the biochemistry of desaturation, both goals can by realized by "simple" antisense strategies to block just one desaturation step catalyzed by 18:2- or 18:1-desaturase activities, respectively. In view of the complications with the export of polyunsaturated fatty acids from chloroplasts into the ER,[189] both chloroplast and ER enzymes should be blocked for safety purposes. This aspect may be of particular relevance in oilseeds such as rape, in which fat-storing tissues are green for a long time.[201] This and the possible existence of multiple genes for each compartment may contribute to the difficulties in reducing 18:3 from the oil of amphidiploid rapeseed by mutation and conventional breeding.[202]

The use of seed oils in the chemical industry requires annual production of large quantities of TAG being homogeneous in the constituent fatty acids with regard to chain length and functional groups (double bonds, hydroxyl and epoxide groups, etc.). Apart from the genes required for hydroxylation or epoxidation which are not available yet, the production of accordingly modified oils of high homogeneity will require the additional use of desaturase genes as outlined by two examples. New oilseed varieties resulting from such transformations will only be successful commercially when the resulting oil is homogeneous and the proportion of the desired fatty acid is approaching 90%. If ricinoleic acid, for example, which is formed by hydroxylation of 18:1 in the ER,[203] should be produced in transgenic rapeseed, it is important that the 18:1 exported from the plastids be hydroxylated before further desaturation to 18:2 occurs, since 18:2 is not hydroxylated and decreases the homogeneity of the product. Therefore, the additional expression of the hydroxylase in the above-described high oleic rapeseed would be a useful strategy. On the other hand, the production of a high vernolic oil, to be mentioned as second example of higher complexity, requires that 18:1, on the contrary, be completely converted to 18:2 for subsequent 12,13-epoxidation, and further desaturation of 18:2 to 18:3 has to be prevented.

These two examples may be sufficient to demonstrate the involvement of desaturases in this field, in which success will depend on the overexpression of modifying enzymes in combination with similarly effective sense and antisense manipulation of desaturation to block unwanted alternatives, which will deteriorate the homogeneity of the product. As mentioned above, the farmers will not grow scientifically interesting prototypes. Rather, what is required are new varieties which produce homogeneous seed oils in unaltered hectare yields and not available from other sources. Therefore, in the short term, it will be easier to produce transgenic low linolenic and high oleic lines because an antisense block of one reaction may work and the resulting change in desaturation should not have any influence on the yield-determining rates of TAG assembly as shown by the zero-18:3 line of flax[116] and the data of Table 1.

ACKNOWLEDGMENTS

The work in the author's laboratory was financially supported by grants from the Bundesministerium für Forschung und Technologie and the Fonds der Chemischen Industrie. The author thanks Dr. H. P. Schmidt and P. Sperling for their engagement and contribution in this research.

REFERENCES

1. **Wolff, I. A.**, Seed lipids, *Science,* 154, 1140, 1966.
2. **Wagner, H. and Pohl, P.**, Eine These: Fettsäurebiosynthese und Evolution bei pflanzlichen und tierischen Organismen, *Phytochemistry,* 5, 903, 1966.
3. **Gurr, M. I.**, The biosynthesis of polyunsaturated fatty acids in plants, *Lipids,* 6, 266, 1971.
4. **De Renobales, M., Cripps, C., Stanley-Samuelson, D. W., Jurenka, R. A., and Blomquist, G. J.**, Biosynthesis of linoleic acid in insects, *Trends Biochem. Sci.,* 12, 364, 1982.
5. **Sardesai, V. M.**, Nutritional role of polyunsaturated fatty acids, *J. Nutr. Biochem.,* 3, 154, 1992.
6. **Rottländer, R. C. A. and Schlichtherle, H.**, Analyse frühgeschichtlicher Gefäßinhalte, *Naturwissenschaften,* 70, 33, 1983.
7. **Priestley, D. A., Galinat, W. C., and Leopold, A. C.**, Preservation of polyunsaturated fatty acid in ancient Anasazi maize seed, *Nature,* 292, 146, 1981.
8. **Schroepfer, G. J. and Bloch, K.**, The stereospecific conversion of stearic acid to oleic acid, *J. Biol. Chem.,* 240, 54, 1965.
9. **Morris, L. J.**, Mechanisms and stereochemistry in fatty acid metabolism; the fifth Colworth medal lecture, *Biochem. J.,* 118, 681, 1970.
10. **Brett, D., Howling, D., Morris, L. J., and James, A. T.**, Specificity of the fatty acid desaturases. The conversion of saturated to monoenoic acids, *Arch. Biochem. Biophys.,* 143, 535, 1971.
11. **Schneider, G., Lindqvist, Y., Shanklin, J., and Somerville, C.**, Preliminary crystallographic data for stearoyl-acyl carrier protein desaturase from castor seed, *J. Mol. Biol.,* 225, 561, 1992.

12. **Mayo, K. H. and Prestegard, J. H.,** Acyl carrier protein from *Escherichia coli*. Structural characterization of short-chain acylated acyl carrier protein by NMR, *Biochemistry*, 24, 7834, 1985.

13. **Wada, M., Fukunaga, N., and Sasaki, S.,** Mechanism of biosynthesis of unsaturated fatty acids in *Pseudomonas* sp. strain E-3, a psychrotrophic bacterium, *J. Bacteriol.*, 171, 4267, 1989.

14. **Howling, D., Morris, L. J., and James, A. T.,** The influence of chain length on the dehydrogenation of saturated fatty acids, *Biochim. Biophys. Acta*, 152, 224, 1968.

15. **Howling, D., Morris, L. J., Gurr, M. I., and James, A. T.,** The specificity of fatty acid desaturases and hydroxylases. The dehydrogenation and hydroxylation of monoenoic acids, *Biochim. Biophys. Acta*, 260, 10, 1972.

16. **Korn, E. D.,** The fatty acids of *Euglena gracilis*, *J. Lipid Res.*, 5, 352, 1964.

17. **Roughan, P. G. and Slack, C. R.,** Cellular organization of glycerolipid metabolism, *Annu. Rev. Plant Physiol.*, 33, 97, 1982.

18. **Browse, J., and Somerville, C.,** Glycerolipid synthesis: biochemistry and regulation, *Annu. Rev. Plant Physiol. Plant Mol. Biol.*, 42, 467, 1991.

19. **Roughan, P. G., Thompson, G. A., Jr., and Cho, S. H.,** Metabolism of exogenous long-chain fatty acids by spinach leaves, *Arch. Biochem. Biophys.*, 259, 481, 1987.

20. **Schmidt, H. and Heinz, E.,** Direct desaturation of intact galactolipids by a desaturase solubilized from chloroplast envelopes, *Biochem. J.*, 289, 777, 1993.

21. **Charton, E. and Macrae, A. R.,** Substrate specificities of lipases A and B from *Geotrichum candidum* CMICC 335426, *Biochim. Biophys. Acta*, 1123, 59, 1992.

22. **McDonough, V. M., Stukey, J. E., and Martin, C. E.,** Specificity of unsaturated fatty acid-regulated expression of the *Saccharomyces cerevisae* OLE1 gene, *J. Biol. Chem.*, 267, 5931, 1992.

23. **Browse, J., Kunst, L., Anderson, S., Hugly, S., and Somerville, C.,** A mutant of *Arabidopsis* deficient in chloroplast 16:1/18:1 desaturase, *Plant Physiol.*, 90, 522, 1989.

24. **Radunz, A.,** Über die Lipide der Pteridophyten-1. Die Isolierung und Identifizierung der Polyensäuren, *Phytochemistry*, 6, 399, 1967.

25. **Jamieson, G. R. and Reid, E. H.,** The occurrence of hexadeca-7,10,13-trienoic acid in the leaf lipids of angiosperms, *Phytochemistry*, 10, 1837, 1971.

26. **Jamieson, G. R and Reid, E. H.,** The leaf lipids of some members of the Boraginaceae family, *Phytochemistry*, 8, 1489, 1969.

27. **Griffiths, G., Stobart, A. K., and Stymne, S.,** Δ^6- and Δ^{12}-desaturase activities and phosphatidic acid formation in microsomal preparations from the developing cotyledons of common borage (*Borago officinalis*), *Biochem. J.*, 252, 641, 1988.

28. **Griffiths, G., Brechany, E. Y., Christie, W. W., Stymne, S., and Stobart, K.,** Synthesis of octadecatetraenoic acid (OTA) in borage (*Borago officinalis*), in *Biological Role of Plant Lipids,* Biacs, P. A., Gruiz, K., and Kremmer, T., Eds., Plenum Press, New York, 1989, 151.

29. **Cahoon, E. B., Shanklin, J., and Ohlrogge, J.,** Expression of a coriander desaturase results in petroselinic acid production in transgenic tobacco, *Proc. Natl. Acad. Sci. U.S.A.*, 89, 11184, 192.

30. **Stymne, S., Green, A., and Tonnet, M. L.,** Lipid synthesis in developing cotyledons of linolenic acid deficient mutants of linseed, in *Biological Role of Plant Lipids,* Biacs, P. A., Gruiz, K., and Kremmer, T., Eds., Plenum Press, New York, 1989, 147.

31. **Kuklev, D. V., Aizdaicher, N. A., Imbs, A. B., Bezuglov, V. V., and Latyshev, N. A.,** All-*cis*-3,6,9,12,15-octadecapentaenoic acid from the unicellular alga *Gymnodinium knowalevskii*, *Phytochemistry*, 31, 2401, 1992.

32. **Giroud, C. and Eichenberger, W.,** Lipids of *Chlamydomonas reinhardtii*. Incorporation of [^{14}C]acetate, [^{14}C]palmitate and [^{14}C]oleate into different lipids and evidence for lipid-linked desaturation of fatty acids, *Plant Cell Physiol.*, 30, 121, 1989.

33. **Nichols, B. W., James, A. T., and Breuer, J.,** Interrelationships between fatty acid biosynthesis and acyl-lipid synthesis in *Chlorella vulgaris*, *Biochem. J.*, 104, 486, 1967.

34. **Gurr, M. I., Robinson, M. P., and James, A. T.,** The mechanism of formation of polyunsaturated fatty acids by photosynthetic tissue. The tight coupling of oleate desaturation with phospholipid synthesis in *Chlorella vulgaris, Eur. J. Biochem.,* 9, 70, 1969.
35. **Roughan, P. G.,** Turnover of the glycerolipids of pumpkin leaves. The importance of phosphatidylcholine, *Biochem. J.,* 117, 1, 1970.
36. **Nichols, B. W. and Moorhouse, R.,** The separation, structure and metabolism of monogalactosyl diglyceride species in *Chlorella vulgaris, Lipids,* 4, 311, 1969.
37. **Gurr, M. I. and Brawn, P.,** The biosynthesis of polyunsaturated fatty acids by photosynthetic tissue. The composition of phosphatidyl choline species in *Chlorella vulgaris* during the formation of linoleic acid, *Eur. J. Biochem.,* 17, 19, 1970.
38. **Roughan, P. G.,** Phosphatidylcholine: donor of 18-carbon unsaturated fatty acids for glycerolipid biosynthesis, *Lipids,* 10, 609, 1975.
39. **Wanner, L., Keller, F., and Matile, Ph.,** Metabolism of radiolabeled galactolipids in senescent barley leaves, *Plant Sci.,* 78, 199, 1991.
40. **Mangold, H. K., Apte, S. S., and Weber, N.,** Biotransformation of alkylglycerols in plant cell cultures: production of platelet activating factor and other biologically active ether lipids, *Lipids,* 26, 1086, 1991.
41. **Sperling, P., Stöcker, S., Mühlbach, H.-P., and Heinz, E.,** Alkenyl ether analogues as substrates for microsomal acylation and desaturation systems, in *Metabolism, structure and utilization of plant lipids,* Cherif, A., Miled-Daoud, D. B., Marzouk, B., Smaoui, A., and Zarrouk, M., Eds., Cent. Natl. Péd., Tunis, 1992, 133.
42. **Heinz, E., Siebertz, H. P., Linscheid, M., Joyard, J., and Douce, R.,** Investigations on the origin of diglyceride diversity in leaf lipids, in *Advances in the Biochemistry and Physiology of Plant Lipids,* Appelqvist, L.-A. and Liljenberg, C., Eds., Elsevier, Amsterdam, 1979, 99.
43. **Murata, N. and Nishida, I.,** Lipids of blue-green algae (Cyanobacteria), in *The Biochemistry of Plants, a Comprehensive Treatise, Vol. 9, Lipids: Structure and Function,* Stumpf, P. K., Ed., Academic Press, Orlando, FL, 1987, 315.
44. **Sallal, A.-K., Ghannoum, M. A., Al-Hasan, R. H., Nimer, N. A., and Radwan, S. S.,** Lanosterol and diacylglycerophosphocholines in lipids from whole cells and thylakoids of the cyanobacterium *Chlorogloeopsis fritschii, Arch. Microbiol.,* 148, 1, 1987.
45. **Dorne, A.-J., Joyard, J., and Douce, R.,** Do thylakoids really contain phosphatidylcholine?, *Proc. Natl. Acad. Sci. U.S.A.,* 87, 71, 1990.
46. **Murata, N. and Sato, N.,** Analysis of lipids in *Prochloron* sp.: occurrence of monoglucosyl diacylglycerol, *Plant Cell Physiol.,* 24, 133, 1983.
47. **Gombos, Z. and Murata, N.,** Lipids and fatty acids of *Prochlorothrix hollandica, Plant Cell Physiol.,* 32, 73, 1991.
48. **Urbach, E., Robertson, D. L., and Chisholm, S. W.,** Multiple evolutionary origins of prochlorophytes within the cyanobacterial radiation, *Nature,* 355, 267, 1992.
49. **Kleinig, H., Beyer, P., Schubert, C., Liedvogel, B., and Lütke-Brinkhaus, F.,** *Cyanophora paradoxa*: fatty acids and fatty acid synthesis *in vitro, Z. Naturforsch.,* 41c, 169, 1986.
50. **Zook, D., Schenk, H. E. A., Frank, H. G., Thiel, D., Poralla, K., and Härtner, T.,** Lipids in *Cyanophora paradoxa.* II. Arachidonic acid in the lipid fractions of the endocyanelle, *Endocyt. C. Res.,* 3, 99, 1986.
51. **Jahnke, L. L., Lee, B., Sweeney, M. J., and Klein, H. P.,** Anaerobic biosynthesis of unsaturated fatty acids in the cyanobacterium, *Oscillatoria limnetica, Arch. Microbiol.,* 152, 215, 1989.
52. **Mathis, P.,** Compared structure of plant and bacterial photosynthetic reaction centers. Evolutionary implications, *Biochim. Biophys. Acta,* 1018, 163, 1990.
53. **Russell, N. J. and Harwood, J. L.,** Changes in the acyl lipid composition of photosynthetic bacteria grown under photosynthetic and non-photosynthetic conditions, *Biochem. J.,* 181, 339, 1979.

54. Knudsen, E., Jantzen, E., Bryn, K., Ormerod, J. G., and Sirevag, R., Quantitative and structural characteristics of lipids in *Chlorobium* and *Chloroflexus*, *Arch. Microbiol.*, 132, 149, 1982.

55. Kenyon, C. N. and Gray, A. M., Preliminary analysis of lipids and fatty acids of green bacteria and *Chloroflexus aurantiacus*, *J. Bacteriol.*, 120, 131, 1974.

56. Saito, K., Kawaguchi, A., Seyama, Y., Yamakawa, T., and Okuda, S., Stereochemistry of β-hydroxydodecanoyl thioester dehydration catalyzed by fatty acid synthetase from *Brevibacterium ammoniagenes*, *Tetrahedron Lett.*, 23, 1689, 1982.

57. Lem, N. W. and Stumpf, P. K., *In vitro* fatty acid synthesis and complex lipid metabolism in the cyanobacterium *Anabaena variabilis*. I. Some characteristics of fatty acid synthesis, *Plant Physiol.*, 74, 134, 1984.

58. Stapleton, S. R. and Jaworski, J. G., Characterization of fatty acid biosynthesis in the cyanobacterium *Anabaena variabilis*, *Biochim. Biophys. Acta*, 794, 249, 1984.

59. Sato, N., Seyama, Y., and Murata, N., Lipid-linked desaturation of palmitic acid in monogalactosyl diacylglycerol in the blue-green alga (cyanobacterium) *Anabaena variabilis* studied *in vivo*, *Plant Cell Physiol.*, 27, 819, 1986.

60. Roughan, P. G., Holland, R., and Slack, C. R., The role of chloroplasts and microsomal fractions in polar-lipid synthesis from [1-^{14}C]acetate by cell-free preparations from spinach (*Spinacia oleracea*) leaves, *Biochem. J.*, 188, 17, 1980.

61. Sanchez, J. and Mancha, M., Separation and analysis of acylthioesters from higher plants, *Phytochemistry*, 19, 817, 1980.

62. Andrews, J. and Heinz, E., Desaturation of newly synthesized monogalactosyldiacyl-glycerol in spinach chloroplasts, *J. Plant Physiol.*, 131, 75, 1987.

63. Heemskerk, J. W. M., Schmidt, H., Hammer, U., and Heinz, E., Biosynthesis and desaturation of prokaryotic galactolipids in leaves and isolated chloroplasts from spinach, *Plant Physiol.*, 96, 144, 1991.

64. Roughan, P. G., Cytidine triphosphate-dependent, acyl-CoA-independent synthesis of phosphatidylglycerol by chloroplasts isolated from spinach and pea, *Biochim. Biophys. Acta*, 835, 527, 1985.

65. Andrews, J., Schmidt, H., and Heinz, E., Interference of electron transport inhibitors with desaturation of monogalactosyl diacylglycerol in intact chloroplasts, *Arch. Biochem. Biophys.*, 270, 611, 1989.

66. Fraser, P. D., Misawa, N., Linden, H., Yamano, S., Kobayashi, K., and Sandmann, G., Expression in *Escherichia coli*, purification, and reactivation of the recombinant *Erwinia uredovora* phytoene desaturase, *J. Biol. Chem.*, 267, 19891, 1992.

67. Cho, S. H. and Thompson, G. A., Jr., On the metabolic relationship between monogalactosyldiacylglycerol and digalactosyldiacylglycerol molecular species in *Dunaliella salina*, *J. Biol. Chem.*, 262, 7586, 1987.

68. Aro, E.-M., Somersalo, S., and Karunen, P., Membrane lipids in *Ceratodon purpureus* protonemata grown at high and low temperatures, *Physiol. Plant.*, 69, 65, 1987.

69. Chen, H.-H., Wickrema, A., and Jaworski, J. G., Acyl-acyl-carrier protein: lysomonogalactosyldiacylglycerol acyltransferase from the cyanobacterium *Anabaena variabilis*, *Biochim. Biophys. Acta*, 963, 493, 1988.

70. Williams, J. P., Maissan, E., Mitchell, K., and Khan, M. U., The manipulation of the fatty acid composition of glycerolipids in cyanobacteria using exogenous fatty acids, *Plant Cell Physiol.*, 31, 495, 1990.

71. Mattoo, A. K., Callahan, F. E., Mehta, R. A., and Ohlrogge, J. B., Rapid *in vivo* acylation of acyl carrier protein with exogenous fatty acids in *Spirodela oligorrhiza*, *Plant Physiol.*, 89, 707, 1989.

72. Shibahara, A., Yamamoto, K., Takeoka, M., Kinoshita, A., Kajimoto, G., Nakayama, T., and Noda, M., Novel pathways of oleic and *cis*-vaccenic acid biosythesis by an enzymatic double-bond shifting reaction in higher plants, *FEBS Lett.*, 264, 228, 1990.

73. Nichols, B. W., Harris, P., and James, A. T., The biosynthesis of *trans*-Δ^3-hexadecenoic acid by *Chlorella vulgaris*, *Biochem. Biophys. Res. Commun.*, 21, 473, 1965.

74. **Merill, A. H., Jr. and Wang, E.,** Biosynthesis of long-chain (sphingoid) bases from serine by LM cells. Evidence for introduction of the 4-*trans*-double bond after *de novo* biosynthesis of *N*-acylsphinganine(s), *J. Biol. Chem.,* 261, 3764, 1986.
75. **Ohnishi, M. and Thompson, G.A., Jr.,** Biosynthesis of the unique *trans*-Δ^3-hexadecenoic acid component of chloroplast phosphatidylglycerol: evidence concerning its site and mechanism of formation, *Arch. Biochem. Biophys.,* 288, 591, 1991.
76. **Jones, V. M. and Harwood, J. L.,** Desaturation of linoleic acid from exogenous lipids by isolated chloroplasts, *Biochem. J.,* 190, 851, 1980.
77. **Ohnishi, J.-I. and Yamada, M.,** Glycerolipid synthesis in *Avena* leaves during greening of etiolated seedlings III. Synthesis of α-linolenoyl-monogalactosyl diacylglycerol from liposomal linoleoyl-phosphatidylcholine by *Avena* plastids in the presence of phosphatidylcholine-exchange protein, *Plant Cell Physiol.,* 23, 767, 1982.
78. **Schmidt, H. and Heinz, E.,** Involvement of ferredoxin in desaturation of lipid-bound oleate in chloroplasts, *Plant Physiol.,* 94, 214, 1990.
79. **Schmidt, H. and Heinz, E.,** Desaturation of oleoyl groups in envelope membranes from spinach chloroplasts, *Proc. Natl. Acad. Sci. U.S.A.,* 87, 9477, 1990.
80. **Norman, H. A., Pillai, P., and St. John, J. B.,** *In vitro* desaturation of monogalactosyldiacylglycerol and phosphatidylcholine molecular species by chloroplast homogenates, *Phytochemistry,* 30, 2217, 1991.
81. **Wada, H., Schmidt, H., Heinz, E., and Murata, N.,** Ferredoxin-dependent desaturation *in vitro* of fatty acids in cyanobacterial thylakoid membranes, *J. Bacteriol.,* 175, in press.
82. **Hirasawa, M., Chang, K.-T., and Knaff, D. B.,** Characterization of a ferredoxin:NADP$^+$ oxidoreductase from a nonphotosynthetic plant tissue, *Arch. Biochem. Biophys.,* 276, 251, 1990.
83. **Zanetti, G., Morelli, D., Ronchi, S., Negri, A., Aliverti, A., and Curti, B.,** Structural studies on the interaction between ferredoxin and ferredoxin-NADP$^+$ reductase, *Biochemistry,* 27, 3753, 1988.
84. **Gotor, C., Pajuelo, E., Romero, L. C., Marquez, A. J., and Vega, J. M.,** Immunological studies of ferredoxin-nitrite reductases and ferredoxin-glutamate synthases from photosynthetic organisms, *Arch. Microbiol.,* 153, 230, 1990.
85. **Hirasawa, M., Chang, K.-T., and Knaff, D. B.,** The interaction of ferredoxin and glutamate synthase: cross-linking and immunological studies, *Arch. Biochem. Biophys.,* 286, 171, 1991.
86. **Karplus, P. A., Daniels, M. J., and Herriott, J. R.,** Atomic structure of ferredoxin-NADP$^+$ reductase: prototype for a structurally novel flavoenzyme family, *Science,* 251, 60, 1991.
87. **Rypniewski, W. R., Breiter, D. R., Benning, M. M., Wesenberg, G., Oh, B.-H., Markley, J. L., Rayment, I., and Holden, H. M.,** Crystallization and structure determination to 2.5-Å resolution of the oxidized [2Fe-2S]ferredoxin isolated from *Anabaena* 7120, *Biochemistry,* 30, 4126, 1991.
88. **Hase, T., Kimata, Y., Yonekura, K., Matsumara, T., and Sakakibara, H.,** Molecular cloning and differential expression of the maize ferredoxin gene familiy, *Plant Physiol.,* 96, 77, 1991.
89. **Yuan, X.-H. and Andersen, L. E.,** Changing activity of glucose-6-phosphate dehydrogenase from pea chloroplasts during photosynthetic induction, *Plant Physiol.,* 85, 598, 1987.
90. **Smith, R. G., Gauthier, D. A., Dennis, D. T., and Turpin, D. H.,** Malate- and pyruvate-dependent fatty acid synthesis in leucoplasts from developing castor endosperm, *Plant Physiol.,* 98, 1233, 1992.
91. **Browse, J., Roughan, P. G., and Slack, C. R.,** Light control of fatty acid synthesis and diurnal fluctuations of fatty acid composition in leaves, *Biochem. J.,* 196, 347, 1981.
92. **Alban, C., Joyard, J., and Douce, R.,** Comparison of glycerolipid biosynthesis in non-green plastids from sycamore (*Acer pseudoplatanus*) cells and cauliflower (*Brassica oleracea*) buds, *Biochem. J.,* 259, 775, 1989.

93. **Alban, C., Dorne, A. J., Joyard, J., and Douce, R.,** [^{14}C]Acetate incorporation into glycerolipids from cauliflower proplastids and sycamore amyloplasts, *FEBS Lett.,* 249, 95, 1989.

94. **Whitaker, B. D.,** Fatty-acid composition of polar lipids in fruit and leaf chloroplasts of "16:3"- and "18:3"-plant species, *Planta,* 169, 313, 1986.

95. **Grechkin, A. N., Gatarova, T. E., and Tarchevsky, I. A.,** Linoleate Δ^{15}-desaturase activity of pea leaf chloroplasts is localized in thylakoids, in *Structure, Function and Metabolism of Plant Lipids,* Siegenthaler, P. A. and Eichenberger, W., Eds., Elsevier, Amsterdam, 1984, 51.

96. **Schmidt, H. and Heinz, E.,** n-6-Desaturase from chloroplast envelopes: purification and enzymatic characteristics, in *Metabolism, structure and utilization of plant lipids,* Cherif, A., Miled-Daoud, D. B., Marzouk, B., Smaoui, A., and Zarrouk, M., Eds., Cent. Natl. Péd., Tunis, 1992, 140.

97. **Yoshida, S. and Kawata, T.,** Isolation of smooth endoplasmic reticulum and tonoplast from mung bean hypocotyls (*Vigna radiata* [L.] Wilczek) using a Ficoll gradient and two-polymer phase partition, *Plant Cell Physiol.,* 29, 1391, 1988.

98. **Stymne, S. and Stobart, A. K.,** Triacylglycerol biosynthesis, in *The Biochemistry of Plants, a Comprehensive Treatise, Vol. 9, Lipids: Structure and Function,* Stumpf, P. K., Ed., Academic Press, Orlando, FL, 1987, 175.

99. **Nichols, B. W. and Safford, R.,** Conversion of lipids to fatty alcohols and lysolipids by $NaBH_4$, *Chem. Phys. Lipids,* 11, 222, 1973.

100. **Stymne, S. and Glad, G.,** Acyl exchange between oleoyl-CoA and phosphatidylcholine in microsomes of developing soya bean cotyledons and its role in fatty acid desaturation, *Lipids,* 16, 298, 1981.

101. **Stymne, S. and Appelqvist, L.-Å.,** The biosynthesis of linoleate from oleoyl-CoA via oleoyl-phosphatidylcholine in microsomes of developing safflower seeds, *Eur. J. Biochem.,* 90, 223,1978.

102. **Serghini-Caid, H., Demandre, C., Justin, A.-M., and Mazliak, P.,** Oleoyl-phosphatidylcholine molecular species desaturated in pea leaf microsomes—possible substrates of oleate desaturase in other green leaves, *Plant Sci.,* 54, 93, 1988.

103. **Hares, W. and Frentzen, M.,** Properties of the microsomal acyl-CoA:sn-1-acyl-glycerol-3-phosphate acyltransferase from spinach (*Spinacia oleracea* L.) leaves, *J. Plant Physiol.,* 131, 49, 1987.

104. **Frentzen, M.,** Comparison of certain properties of membrane bound and solubilized acyltransferase activities of plant microsomes, *Plant Sci.,* 69, 39, 1990.

105. **Stymne, S., Stobart, A. K., and Glad, G.,** The role of the acyl-CoA pool in the synthesis of polyunsaturated 18-carbon fatty acids and triacylglycerol production in the microsomes of developing safflower seeds, *Biochim. Biophys. Acta,* 752, 198, 1983.

106. **Stobart, A. K. and Stymne, S.,** The regulation of the fatty-acid composition of the triacylglycerol in microsomal preparations from avocado mesocarp and the developing cotyledons of safflower, *Planta,* 163, 119, 1985.

107. **Sanchez, J., del Cuvillo, M. T., and Harwood, J. L.,** Glycerolipid biosynthesis by microsomal fractions from olive fruits, *Phytochemistry,* 31, 129, 1992.

108. **Slack, C. R., Campbell, L. C., Browse, J. A., and Roughan, P. G.,** Some evidence for the reversibility of the choline-phosphotransferase-catalyzed reaction in developing linseed cotyledons *in vivo, Biochim. Biophys. Acta,* 754, 10, 1983.

109. **Slack, C. R., Roughan, P. G., Browse, J. A., and Gardiner, S. E.,** Some properties of cholinephosphotransferase from developing safflower cotyledons, *Biochim. Biophys. Acta,* 833, 438, 1985.

110. **Slack, C. R., Roughan, P. G., and Browse, J.,** Evidence for an oleoyl phosphatidylcholine desaturase in microsomal preparations from cotyledons of safflower (*Carthamus tinctorius*) seed, *Biochem. J.,* 179, 649, 1979.

111. **Martin, B. A. and Rinne, R. W.,** A comparison of oleic acid metabolism in the soybean (*Glycine max* [L.] Merr.) genotypes Williams and A5, a mutant with decreased linoleic acid in the seed, *Plant Physiol.,* 81, 41, 1986.

112. **Demandre, C., Trémolières, A., Justin, A. M., and Mazliak, P.,** Oleate desaturation in six phosphatidylcholine molecular species from potato microsomes, *Biochim. Biophys. Acta,* 877, 380, 1986.

113. **Miquel, M. and Browse, J.,** *Arabidopsis* mutants deficient in polyunsaturated fatty acid synthesis. Biochemical and genetic characterization of a plant oleoyl-phosphatidylcholine desaturase, *J. Biol. Chem.,* 267, 1502, 1992.

114. **Stymne, S. and Stobart, A. K.,** Biosynthesis of γ-linolenic acid in cotyledons and microsomal preparations of the developing seeds of common borage (*Borago officinalis*), *Biochem. J.,* 240, 385, 1986.

115. **Lawson, L. D. and Hughes, B. G.,** Triacylglycerol structure of plant and fungal oils containing γ-linolenic acid, *Lipids,* 23, 313, 1988.

116. **Stymne, S., Tonnet, M. L., and Green, A. G.,** Biosynthesis of linolenate in developing embryos and cell-free preparations of high-linolenate linseed (*Linum usitatissimum*) and low-linolenate mutants, *Arch. Biochem. Biophys.,* 294, 557, 1992.

117. **Browse, J. and Slack, C. R.,** The effects of temperature and oxygen on the rates of fatty acid synthesis and oleate desaturation in safflower (*Carthamus tinctorius*) seed, *Biochim. Biophys. Acta,* 753, 145, 1983.

118. **Moreau, R. A., Pollard, M. R., and Stumpf, P. K.,** Properties of a Δ^5-fatty acyl-CoA desaturase in the cotyledons of developing *Limnanthes alba, Arch. Biochem. Biophys.,* 209, 376, 1981.

119. **Zander, R.,** The distribution space of physically dissolved oxygen in aqueous solutions of organic substances, *Z. Naturforsch.,* 31c, 339, 1976.

120. **Rebeille, F., Bligny, R., and Douce, R.,** Rôle de l'oxygène et de la temperature sur la composition en acides gras des cellules isolées d'érable (*Acer pseudoplatanus* L.), *Biochim. Biophys. Acta,* 620, 1, 1980.

121. **Smotkin, E. S., Moy, F. T., and Plachy, W. Z.,** Dioxygen solubility in aqueous phosphatidylcholine dispersions, *Biochim. Biophys. Acta,* 1061, 33, 1991.

122. **Windrem, D. A. and Plachy, W. Z.,** The diffusion-solubility of oxygen in lipid bilayers, *Biochim. Biophys. Acta,* 600, 655, 1980.

123. **Jansson, I. and Schenkman, J. B.,** Studies on three microsomal electron transfer enzyme systems. Specificity of electron flow pathways, *Arch. Biochem. Biophys.,* 178, 89, 1977.

124. **Smith, M. A., Cross, A. R., Jones, O. T. G., Griffiths, W. T., Stymne, S., and Stobart, K.,** Electron-transport components of the 1-acyl-2-oleoyl-*sn*-glycero-3-phosphocholine Δ^{12}-desaturase (Δ^{12}-desaturase) in microsomal preparations from developing safflower (*Carthamus tinctorius* L.) cotyledons, *Biochem. J.,* 272, 23, 1990.

125. **Kearns, E. V., Hugly, S., and Somerville, C.,** The role of cytochrome b_5 in Δ^{12} desaturation of oleic acid by microsomes of safflower (*Carthamus tinctorius* L.), *Arch. Biochem. Biophys.,* 284, 431, 1991.

126. **Douce, R., Holtz, R. B., and Benson, A. A.,** Isolation and properties of the envelope of spinach chloroplasts, *J. Biol. Chem.,* 248, 7215, 1973.

127. **Luster, D. G. and Donaldson, R. P.,** Orientation of electron transport activities in the membrane of intact glyoxysomes isolated from castor bean endosperm, *Plant Physiol.,* 85, 796, 1987.

128. **Moreau, F.,** Electron transport between outer and inner membranes in plant mitochondria, *Plant Sci. Lett.,* 6, 215, 1976.

129. **Dizengremel, P., Kader, J.-C., Mazliak, P., and Lance, C.,** Electron transport and fatty acid synthesis in microsomes and outer mitochondrial membranes of plant tissues, *Plant Sci. Lett.,* 11, 151, 1978.

130. **Sauer, A. and Robinson, D. G.,** Subcellular localization of enzymes involved in lecithin biosynthesis in maize roots, *J. Exp. Bot.,* 36, 1257, 1985.

131. **Thompson, J. E.,** The behaviour of cytoplasmic membranes in *Phaseolus vulgaris* cotyledons during germination, *Can. J. Bot.,* 52, 534, 1974.

132. **Pohl, U. and Wiermann, R.,** NADH/NADPH-cytochrome reductase assay: interference by nonenzymatic cytochrome c reducing activity in plant extracts, *Anal. Biochem.,* 116, 425, 1981.

133. **Askerlund, P., Larsson, C., and Widell, S.,** Cytochromes of plant plasma membranes. Characterization by absorbance difference spectrophotometry and redox titration, *Physiol. Plant.,* 76, 123, 1989.

134. **Hendry, G. A. F., Houghton, J. D., and Jones, O. T. G.,** The cytochromes in microsomal fractions of germinating mung beans, *Biochem. J.,* 194, 743, 1981.

135. **Asard, H., Venken, M., Caubergs, R., Reijnders, W., Oltmann, F. L., and de Greef, J. A.,** b-Type cytochromes in higher plant plasma membranes, *Plant Physiol.,* 90, 1077, 1989.

136. **Askerlund, P., Laurent, P., Nakagawa, H., and Kader, J. C.,** NADH-Ferricyanide reductase of leaf plasma membranes. Partial purification and immunological relation to potato tuber microsomal NADH-ferricyanide reductase and spinach leaf NADH-nitrate reductase, *Plant Physiol.,* 95, 6, 1991.

137. **Luster, D. G., Bowditch, M. I., Eldridge, K. M., and Donaldson, R. P.,** Characterization of membrane-bound electron transport enzymes from castor bean glyoxysomes and endoplasmic reciculum, *Arch. Biochem. Biophys.,* 265, 50, 1988.

138. **Askerlund, P., Larsson, C., and Widell, S.,** Localization of donor and acceptor sites of NADH dehydrogenase activities using inside-out and right-side-out plasma membrane vesicles from plants, *FEBS Lett.,* 239, 23, 1988.

139. **Miki, K., Kaida, S., Kasai, N., Iyanagi, T., Kobayashi, K., and Hayashi, K.,** Crystallization and preliminary x-ray crystallographic study of NADH-cytochrome b5 reductase from pig liver microsomes, *J. Biol. Chem.,* 262, 11801, 1987.

140. **Strittmatter, P., Kittler, J. M., Coghill, J. E., and Ozols, J.,** Characterization of lysyl residues of NADH-cytochrome b_5 reductase implicated in charge-pairing with active-site carboxyl residues of cytochrome b_5 by site-directed mutagenesis of an expression vector for the flavoprotein, *J. Biol. Chem.,* 267, 2519, 1992.

141. **Mathews, F. S.,** The structure, function and evolution of cytochromes, *Prog. Biophys. Mol. Biol.,* 45, 1, 1985.

142. **Zenno, S., Hattori, M., Misumi, Y., Yubisui, T., and Sakaki, Y.,** Molecular cloning of a cDNA encoding rat NADH-cytochrome b_5 reductase and the corresponding gene, *J. Biochem.,* 107, 810, 1990.

143. **Kuwahara, S.-I., Okada, Y., and Omura, T.,** Evidence for molecular identity of microsomal and mitochondrial NADH-cytochrome b_5 reductase of rat liver, *J. Biochem.,* 83, 1049, 1978.

144. **Ozols, J., Carr, S. A., and Strittmatter, P.,** Identification of the NH_2-terminal blocking group of NADH-cytochrome b_5 reductase as myristic acid and the complete amino acid sequence of the membrane-binding domain, *J. Biol. Chem.,* 259, 13349, 1984.

145. **Okada, Y., Frey, A. B., Guenthner, T. M., Oesch, F., Sabatini, D. D., and Kreibich, G.,** Studies on the biosynthesis of microsomal membrane proteins. Site of synthesis and mode of insertion of cytochrome b_5, cytochrome b_5 reductase, cytochrome P-450 reductase and epoxide hydrolase, *Eur. J. Biochem.,* 122, 393, 1982.

146. **Donaldson, R. P. and Luster, D. G.,** Multiple forms of plant cytochromes P-450, *Plant Physiol.,* 96, 669, 1991.

147. **Benveniste, I., Lesot, A., Hasenfratz, M.-P., and Durst, F.,** Immunological characterization of NADPH-cytochrome P-450 reductase from Jerusalem artichoke and other higher plants, *Biochem. J.,* 259, 847, 1989.

148. **Nadler, S. G. and Strobel, H. W.,** Identification and characterization of an NADPH-cytochrome P450 reductase derived peptide involved in binding to cytochrome P450, *Arch. Biochem. Biophys.,* 290, 277, 1991.

149. **Fujita, M. and Asahi, T.,** Purification and properties of sweet potato NADPH-cytochrome c (P-450) reductase, *Plant Cell Physiol.,* 26, 397, 1985.

150. **Porter, T. D., Beck, T. W., and Kasper, C. B.,** NADPH-cytochrome P-450 oxidoreductase gene organization correlates with structural domains of the protein, *Biochemistry,* 29, 9814, 1990.

151. **Bonnerot, C., Galle, A. M., Jolliot, A., and Kader, J.-C.,** Purification and properties of plant cytochrome b_5, *Biochem. J.,* 226, 331, 1985.

152. **Kearns, E. V., Keck, P., and Somerville, C.,** Primary structure of cytochrome b_5 from cauliflower (*Brassica oleracea* L.) deduced from peptide and cDNA sequences, *Plant Physiol.,* 99, 1254, 1992.

153. **Cinti, D. L., Cook, L., Nagi, M. N., and Suneja, S. K.,** The fatty acid chain elongation system of mammalian endoplasmic reticulum, *Prog. Lipid Res.,* 31, 1, 1992.

154. **Ozols, J.,** Structure of cytochrome b_5 and its topology in the microsomal membrane, *Biochim. Biophys. Acta,* 997, 121, 1989.

155. **Chester, D. W., Skita, V., Young, H. S., Mavromoustakos, T., and Strittmatter, P.,** Bilayer structure and physical dynamics of the cytochrome b_5 dimyristoyl-phosphatidylcholine interaction, *Biophys. J.,* 61, 1224, 1992.

156. **George, S. K., Najera, L., Sandoval, R. P., Countryman, C., Davis, R. W., and Ihler, G. M.,** The hydrophobic domain of cytochrome b_5 is capable of anchoring β-galactosidase in *Escherichia coli* membranes, *J. Bacteriol.,* 171, 4569, 1989.

157. **Mathews, F. S., Levine, M., and Argos, P.,** Three-dimensional Fourier synthesis of calf liver cytochrome b_5 at 2.8 Å resolution, *J. Mol. Biol.,* 64, 449, 1972.

158. **Salemme, F. R.,** An hypothetical structure for an intermolecular electron transfer complex of cytochromes c and b_5, *J. Mol. Biol.,* 102, 563, 1976.

159. **Stukey, J. E., McDonough, V. M., and Martin, C. E.,** The *OLE1* gene of *Saccharomyces cerevisiae* encodes the Δ^9 fatty acid desaturase and can be functionally replaced by the rat stearoyl-CoA desaturase, *J. Biol. Chem.,* 265, 20144, 1990.

160. **Grayburn, W. S., Collins, G. B., and Hildebrand, D. F.,** Fatty acid alteration by a Δ^9-desaturase in transgenic tobacco tissue, *Bio/Technology,* 10, 675, 1992.

161. **Deisenhofer, J. and Michel, H.,** High-resolution structure of photosynthetic reaction centers, *Annu. Rev. Biophys. Biophys. Chem.,* 20, 247, 1991.

162. **Wiener, C. M. and White, S. H.,** Structure of a fluid dioleoyl phosphatidylcholine bilayer determined by joint refinement of x-ray and neutron diffraction data III. Complete structure, *Biophys. J.,* 61, 434, 1992.

163. **Wiener, M. C. and White, S. H.,** Transbilayer distribution of bromine in fluid bilayers containing a specifically bromine analogue of dioleoylphosphatidylcholine, *Biochemistry,* 30, 6997, 1991.

164. **Wardlaw, J. R., Sawyer, W. H., and Ghiggino, K. P.,** Vertical fluctuations of phospholipid acyl chains in bilayers, *FEBS Lett.,* 223, 20, 1987.

165. **Pastor, R. W., Venable, R. M., and Karplus, M.,** Model for the structure of the lipid bilayer, *Proc. Natl. Acad. Sci. U.S.A.,* 88, 892, 1991.

166. **Devaux, P. F.,** Protein involvement in transmembrane lipid asymmetry, *Annu. Rev. Biophys. Biomol. Struct.,* 21, 417, 1992.

167. **Mazliak, P., Jolliot, A., Justin, A.-M., and Kader, J.-C.,** Stimulation of potato microsomal NADH-cytochrome c reductase by acidic phospholipids, *Phytochemistry,* 24, 1163, 1985.

168. **Holloway, P. W.,** Fatty acid desaturation, in *The Enzymes,* Vol. 16, 3rd ed., Boyer, P. D., Ed., Academic Press, New York, 1983, 63.

169. **Ferrante, G. and Kates, M.,** Pathways for desaturation of oleoyl chains in *Candida lipolytica, Can. J. Biochem. Cell. Biol.,* 61, 1191, 1983.

170. **Thiede, M. A., Ozols, J., and Stittmatter, P.,** Construction and sequence of cDNA for rat liver stearyl coenzyme A desaturase, *J. Biol. Chem.,* 261, 13230, 1986.

171. **Mikara, K.,** Structure and regulation of rat liver microsomal stearoyl-CoA desaturase gene, *J. Biochem.,* 108, 1022, 1990.

172. **Fujiwara, Y., Okayasu, T., Ishibashi, T., and Imai, Y.,** Immunochemical evidence for the enzymatic difference of Δ^6-desaturase from Δ^9- and Δ^5-desaturase in rat liver microsomes, *Biochem. Biophys. Res. Commun.,* 110, 36, 1983.

173. **Ntambi, J. M., Buhrow, S. A., Kaestner, K. H., Christy, R. J., Sibley, E., Kelly, T.J., Jr., and Lane, M. D.,** Differentiation-induced gene expression in 3T3-L1 preadipocytes. Characterization of a differentially expressed gene encoding stearoyl-CoA desaturase, *J. Biol. Chem.,* 263, 17291, 1988.

174. **Strittmatter, P., Thiede, M. A., Hackett, C. S., and Ozols, J.,** Bacterial synthesis of active rat stearyl-CoA desaturase lacking the 26-residue amino-terminal amino acid sequence, *J. Biol. Chem.,* 263, 2532, 1988.

175. **Enoch, H. G., Catala, A., and Strittmatter, P.,** Mechanism of rat liver microsomal stearyl-CoA desaturase. Studies of the substrate specificity, enzyme-substrate interactions, and the function of lipid, *J. Biol. Chem.,* 251, 5095, 1976.

176. **Wada, H. and Murata, N.,** *Synechocystis* PCC6803 mutants defective in desaturation of fatty acids, *Plant Cell Physiol.,* 30, 971, 1989.

177. **Wada, H., Gombos, Z., and Murata, N.,** Enhancement of chilling tolerance of a cyanobacterium by genetic manipulation of fatty acid desaturation, *Nature,* 347, 200, 1990.

178. **Browse, J., McCourt, P., and Somerville, C.,** A mutant of *Arabidopsis* lacking a chloroplast-specific lipid, *Science,* 227, 763, 1985.

179. **Somerville, C. and Browse, J.,** Plant lipids: metabolism, mutants, and membranes, *Science,* 252, 80, 1991.

180. **James, D. W., Jr. and Dooner, H. K.,** Isolation of EMS-induced mutants in *Arabidopsis* altered in seed fatty acid composition, *Theor. Appl. Genet.,* 80, 241, 1990.

181. **Lemieux, B., Miquel, M., Somerville, C., and Browse, J.,** Mutants of *Arabidopsis* with alterations in seed lipid fatty acid composition, *Theor. Appl. Genet.,* 80, 234, 1990.

182. **Grill, E. and Somerville, C.,** Construction and characterization of a yeast artificial chromosome library of *Arabidopsis* which is suitable for chromosome walking, *Mol. Gen. Genet.,* 226, 484, 1991.

183. **Arondel, V., Lemieux, B., Hwang, J., Gibson, S., Goodman, H. M., and Somerville, C.,** MAP-based cloning of a gene controlling omega-3 fatty acid desaturation in *Arabidopsis, Science,* 258, 1353, 1992.

184. **Iba, K., Gibson, S., Hugly, S., Nishimura, M., and Somerville, C.,** Chromosome walking in the region of *Arabidopsis fad*D locus using yeast artificial chromosomes, in *Research in Photosynthesis,* Vol. 3, Murata, N., Ed., Kluwer Academic, Dordrecht, 1992, 55.

185. **Browse, J., Miquel, M., and Somerville, C.,** Genetic approaches to understanding plant lipid metabolism, in *Plant Lipid Biochemistry, Structure and Utilization,* Quinn, P. J. and Harwood, J. L., Eds., Portland Press, London, 1990, 431.

186. **Sato, A., Becker, C. K., and Knauf, V. C.,** Nucleotide sequence of a complementary DNA clone encoding stearoyl-acyl carrier protein desaturase from *Simmondsia chinensis, Plant Physiol.,* 99, 363, 1992.

187. **Kikuchi, S. and Kusaka, T.,** Isolation and partial characterization of a very long-chain fatty acid desaturation system from the cytosol of *Mycobacterium smegmatis, J. Biochem.,* 99, 723, 1986.

188. **Ohlrogge, J. B., Browse, J., and Somerville, C.,** The genetics of plant lipids, *Biochim. Biophys. Acta,* 1082, 1, 1991.

189. **Browse, J., Kunst, L., Hughly, S., and Somerville, C.,** Modifications to the two pathway scheme of lipid metabolism based on studies of *Arabidopsis* mutants, in *Biological Role of Plant Lipids,* Biacs, P. A., Gruiz, K., and Kremmer, T., Eds., Plenum Press, New York, 1989, 335.

190. **Pearcy, R. W.,** Effect of growth temperature on the fatty acid composition of the leaf lipids in *Atriplex lentiformis* (Torr.) Wats., *Plant Physiol.,* 61, 484, 1978.

191. **Johnson G. and Williams, J. P.,** Effect of growth temperature on the biosynthesis of chloroplastic galactosyldiacylglycerol molecular species in *Brassica napus* leaves, *Plant Physiol.,* 91, 924, 1989.

192. **Brockman, J. A., Norman, H. A., and Hildebrand, D. F.,** Effects of temperature, light and chemical modulator on linolenate biosynthesis in mutant and wild type *Arabidopsis* calli, *Phytochemistry,* 29, 1447, 1990.

193. **Hugly, S. and Somerville, C.,** A role for membrane lipid polyunsaturation in chloroplast biogenesis at low temperature, *Plant Physiol.,* 99, 197, 1992.

194. **Gombos, Z., Wada, H., and Murata, N.,** Direct evaluation of effects of fatty acid unsaturation on the thermal properties of photosynthetic activities, as studied by mutation and transformation of *Synechocystis* PCC6803, *Plant Cell Physiol.,* 32, 205, 1991.

195. **Wada, H., Gombos, Z., Sakamoto, T., and Murata, N.,** Genetic manipulation of the extent of desaturation of fatty acids in membrane lipids in the cyanobacterium *Synechocystis* PCC6803, *Plant Cell Physiol.,* 33, 535, 1992.

196. **Wang, X.-M., Norman, H. A., St. John, J. B., Yin, T., and Hildebrand, D. F.,** Comparison of fatty acid composition in tissues of low linolenate mutants of soybean, *Phytochemistry,* 28, 411, 1989.

197. **Tonnet, M. L. and Green, A. G.,** Characterization of seed and leaf lipids of high and low linolenic acid flax genotypes, *Arch. Biochem. Biophys.,* 252, 646, 1987.

198. **Fernandez-Martinez, J., Jimenez, A., Dominguez, J., Garcia, J. M., Garcés, S. R., and Mancha, M.,** Genetic analysis of the high oleic acid content in cultivated sunflower (*Helianthus annuus* L.), *Euphytica,* 41, 39, 1989.

199. **Sperling, P., Hammer, U., Friedt, W., and Heinz, E.,** High oleic sunflower: studies on composition and desaturation of acyl groups in different lipids and organs, *Z. Naturforsch.,* 45c, 166, 1990.

200. **Garcés, R. and Mancha, M.,** Oleate desaturation in seeds of two genotypes of sunflower, *Phytochemistry,* 28, 2593, 1989.

201. **Thies, W.,** Der Einfluß der Chloroplasten auf die Bildung von ungesättigten Fettsäuren in reifenden Rapssamen, *Fette Seifen Anstrichm.,* 73, 710, 1971.

202. **Röbbelen, G. and Nitsch, A.,** Genetical and physiological investigations on mutants for polyenoic fatty acids in rapeseed, *Brassica napus* L. I. Selection and description of new mutants, *Z. Pflanzenzücht.,* 75, 93, 1975.

203. **Bafor, M., Smith, M. A., Jonsson, L., Stobart, K., and Stymne, S.,** Ricinoleic acid biosynthesis and triacylglycerol assembly in microsomal preparations from developing castor-bean (*Ricinus communis*) endosperm. *Biochem. J.,* 280, 507, 1991.

204. **Gunstone, F. D., Harwood, J. L., and Padley, F. B., Eds.,** *The Lipid Handbook,* Chapman and Hall, London, 1986.

205. **Ichihara, K., Asahi, T., and Fujii, S.,** 1-Acyl-*sn*-glycerol-3-phosphate acyltransferase in maturing safflower seeds and its contribution to the non-random fatty acid distribution of triacylglycerol, *Eur. J. Biochem.,* 167, 339, 1987.

206. **Ichihara, K.,** The action of phosphatidate phosphatase on the fatty acid composition of safflower triacylglycerol and spinach glycerolipids, *Planta,* 183, 353, 1991.

207. **Stobart, A. K., Stymne, S., and Höglund, S.,** Safflower microsomes catalyse oil accumulation *in vitro*: a model system, *Planta,* 169, 33, 1986.

208. **Stymne, S. and Stobart, K.,** Evidence for the reversibility of the acyl-CoA: lysophosphatidylcholine acyltransferase in microsomal preparations from developing safflower (*Carthamus tinctorius* L.) cotyledons and rat liver, *Biochem. J.,* 223, 305, 1984.

209. **Wada, H.,** personal communication, 1992.

Chapter 3

UNUSUAL FATTY ACIDS

Frank J. van de Loo, Brian G. Fox, and Chris Somerville

TABLE OF CONTENTS

0-8493-4907-9/93/$0.00+$.50
© 1993 by CRC Press Inc.

I. INTRODUCTION

Extensive surveys of the fatty acid composition of seed oils from different species of higher plants have resulted in the identification of more than 210 naturally occurring fatty acids which can be broadly classified into 1 of 18 structural classes (Table 1). The classes are defined by the number and arrangement of double or triple bonds and various functional groups, such as hydroxyls, ketones, epoxys, cyclopentenyl or cyclopropyl groups, furans, or halogens. This level of structural diversity is similar to that of some of the least diverse families of plant secondary metabolites, a class of compounds which have been estimated to contain as many as 100,000 different structures.[3] A summary of the range of structures to be found in plant fatty acids and a small number of examples of representative fatty acids are presented in Table 2. Extensive lists of the amounts and sources of plant fatty acids are available in earlier reviews.[1,2,7-11] Because of the magnitude of this extensive and scattered literature which can no longer be adequately represented by traditional compilations, the development of a computer database of the fatty acid composition of all characterized plant species would be very desirable.

The most commonly occurring fatty acids which may occur in both membrane and storage lipids are a small family of 16- and 18-carbon fatty acids which may have from zero to three methylene-interrupted *cis* unsaturations. All members of the family are descended from the fully saturated species as the result of a series of sequential desaturations which begin at the ω-9 carbon and progress in the direction of the ω-3 carbon.[12] Fatty acids which cannot be described by this simple algorithm are generally considered "unusual" even though several, such as lauric (12:0), erucic (22:1), and ricinoleic (12-OH, 18:1), are of significant commercial importance.

Much of the research to date concerning unusual plant fatty acids has been focused on the identification of new structures or cataloging the composition of fatty acids found in various plant species. Relatively little is known about the mechanisms responsible for the synthesis and accumulation of unusual fatty acids or of their significance to the fitness of the plants which accumulate them. However, the infusion of molecular genetics into most aspects of plant biology and new methods of protein purification have provided new motivations and methods for undertaking detailed mechanistic investigations of the synthesis of unusual fatty acids. In this review, we attempt to provide an overview of the major questions concerning unusual fatty acid biosynthesis. Also, in order to stimulate thinking about the properties of the enzymes which are thought to participate in many of the characteristic modifications, we present an overview of certain aspects of contemporary enzymology which are relevant to the important class of enzymes, such as desaturases and hydroxylases, which participate in O_2-dependent modifications of fatty acids.

TABLE 1
Minimum Summary of the Kinds of Fatty Acids Found in Plants

Type	Number of structures
Saturated	14
Monounsaturated	11
Diunsaturated[a]	9
Triunsaturated[a]	9
Tetraunsaturated[a]	6
Pentaunsaturated[a]	3
Hexaunsaturated[a]	1
Nonconjugated ethylenic[b]	32
Conjugated ethylenic	15
Acetylenic	26
Monohydroxy	33
Polyhydroxy	12
Keto	5
Epoxy	9
Cyclopentenyl	15
Cyclopropanoid	6
Furanoid	2
Halogenated	2

Note: The numbers were compiled by counting the numbers of distinct fatty acids described in compilations.[1,2]

[a] All double bonds are methylene-interrupted *cis* isomers.
[b] Configurations other than methylene-interrupted *cis* double bonds.

II. BIOLOGY

A. TAXONOMIC RELATIONSHIPS

The taxonomic relationships between plants having similar or identical kinds of unusual fatty acids have been examined.[1,7] In some cases, particular fatty acids occur mostly or solely in related taxa. For example, the cyclopentenyl fatty acids have been found only in the family Flacourtiaceae, although the presence of cyclopentenylglycine, the biosynthetic precursor of the cyclopentenyl fatty acids, in the Passifloraceae and Turneraceae suggests that these acids may also be found in these other families of the order Violales.[13,14] Petroselinic acid is most commonly found in the related families Apiaceae, Araliaceae, and Garryaceae, but has also been observed in unrelated families.[15] In other cases, there does not appear to be a direct link between taxonomic relationships and the occurrence of unusual fatty acids. For example, lauric acid is prominent in both the unrelated families Lauraceae and Arecaceae. Similarly, ricinoleic acid has now been identified in 12 genera from 10 families.[16-27] However, in

TABLE 2
Diversity of Unusual Plant Fatty Acids[a]

Unusual property	Name and structure	Source[b] (seed oil unless otherwise noted)
A. Chain length		
1. Medium chain	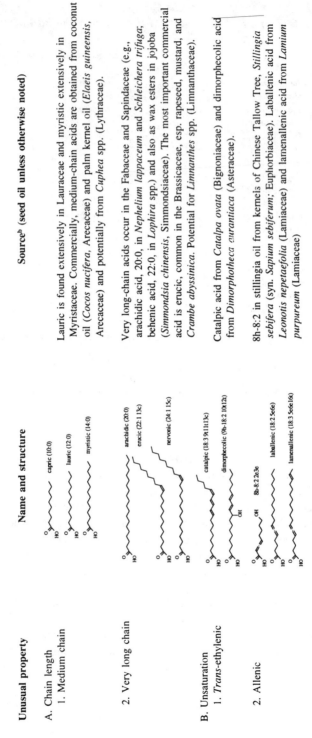 capric (10:0) lauric (12:0) myristic (14:0)	Lauric is found extensively in Lauraceae and myristic extensively in Myristaceae. Commercially, medium-chain acids are obtained from coconut oil (*Cocos nucifera*, Arecaceae) and palm kernel oil (*Elaeis guineensis*, Arecaceae) and potentially from *Cuphea* spp. (Lythraceae).
2. Very long chain	arachidic (20:0) erucic (22:1 13c) nervonic (24:1 15c)	Very long-chain acids occur in the Fabaceae and Sapindaceae (e.g., arachidic acid, 20:0, in *Nephelium lappaceum* and *Schleichera trifuga*; behenic acid, 22:0, in *Lophira* spp.) and also as wax esters in jojoba (*Simmondsia chinensis*, Simmondsiaceae). The most important commercial acid is erucic, common in the Brassicaceae, esp. rapeseed, mustard, and *Crambe abyssinica*. Potential for *Limnanthes* spp. (Limnanthaceae).
B. Unsaturation		
1. *Trans*-ethylenic	catalpic (18:3 9t11t13c) dimorphecolic (9h-18:2 10t12) OH	Catalpic acid from *Catalpa ovata* (Bignoniaceae) and dimorphecolic acid from *Dimorphotheca aurantiaca* (Asteraceae).
2. Allenic	8h-8:2 2e3e OH laballenic (18:2 5e6e) lamenallenic (18:3 5e6e16e)	8h-8:2 in stillingia oil from kernels of Chinese Tallow Tree, *Stillingia sebifera* (syn. *Sapium sebiferum*; Euphorbiaceae). Laballenic acid from *Leonotis nepetaefolia* (Lamiaceae) and lamenallenic acid from *Lamium purpureum* (Lamiaceae)

Commercially in tung oil, *Aleurites fordii* (Euphorbiaceae). Various other examples, generally also containing *trans* double bonds (e.g., catalpic acid, above). Industrially, conjugated diene (9,11) acids obtained by dehydration of ricinoleic acid from castor oil, yielding also nonconjugated diene (9,12) acids.

Tariric acid from *Picramnia* spp. (Simaroubaceae) and crepenynic acid from *Crepis* spp. (Asteraceae). Various more complex acetylenic acids occur, often in the Olacaceae and Santalaceae.

20:1 5c from *Limnanthes douglasii*, *L. alba* (Limnanthaceae). Petroselinic acid widespread in the Apiaceae and Araliaceae, e.g., coriander, carrot. *Cis*-vaccenic acid a relatively common minor acid but at higher levels in some seed oils (e.g., *Entandraphragma* spp., Meliaceae[4]) and tropical fruits (e.g., *Diospyros kaki*[5]). Various other examples, including erucic acid, above.

Ricinoleic acid from castor oil, *Ricinus communis* (Euphorbiaceae); isoricinoleic acid from *Strophanthus* spp., *Wrightia* spp. (Apocynaceae); and auricolic acid from *Lesquerella auriculata* (Brassicaceae). Some of these acids occur in various other species; castor oil is the most important commercial source. ω and ω-1 hydroxy acids are common components of

3. Conjugated

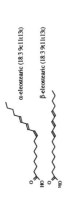

α-eleostearic (18:3 9c11t13t)

β-eleostearic (18:3 9t11t13t)

4. Acetylenic

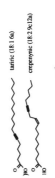

tariric (18:1 6a)

crepenynic (18:2 9c12a)

5. Unusual positions

20:1 5c

petroselinic (18:1 6c)

cis-vaccenic (18:1 11c)

C. Oxygenated
1. Hydroxy

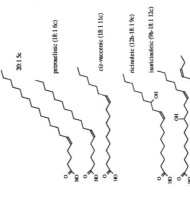

ricinoleic (12h-18:1 9c)

isoricinoleic (9h-18:1 12c)

auricolic (14h-20:2 11c17c)

TABLE 2 (continued)
Diversity of Unusual Plant Fatty Acids[a]

Unusual property	Name and structure	Source[b] (seed oil unless otherwise noted)
2. Epoxy	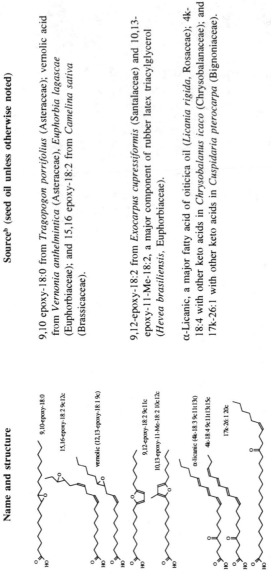	9,10 epoxy-18:0 from *Tragopogon porrifolius* (Asteraceae); vernolic acid from *Vernonia anthelmintica* (Asteraceae), *Euphorbia lagascae* (Euphorbiaceae); and 15,16 epoxy-18:2 from *Camelina sativa* (Brassicaceae).
3. Furanoid	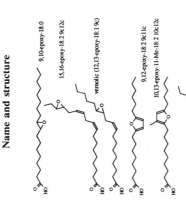	9,12-epoxy-18:2 from *Exocarpus cupressiformis* (Santalaceae) and 10,13-epoxy-11-Me-18:2, a major component of rubber latex triacylglycerol (*Hevea brasiliensis*, Euphorbiaceae).
4. Keto		α-Licanic, a major fatty acid of oiticica oil (*Licania rigida*, Rosaceae); 4k-18:4 with other keto acids in *Chrysobalanus icaco* (Chrysobalanaceae); and 17k-26:1 with other keto acids in *Cuspidaria pterocarpa* (Bignoniaceae).

D. Other

1. Branched chain

angelic (2-Me-4:1 2c)

tiglic (2-Me-4:1 2t)

16-Me-17:0

Angelic acid from roots of *Angelica archangelica* (Apiaceae) and tiglic acid from roots and seeds of *Croton tiglium* (Euphorbiaceae). 16Me-17:0 with other branched acids in vegetative and generative tissues of plastome mutants of *Antirrhinum majus* (Scrophulariaceae).[6] Branched-chain acids are rare in plants and are much more common in bacteria.

2. Cyclic

sterculic

dihydrosterculic

hydnocarpic

Sterculic acid from *Sterculia foetida* (Sterculiaceae); dihydrosterculic acid from *Euphoria longana* (Sapindaceae); and hydnocarpic acid from chaulmoogra oil, *Hydnocarpus wightiana* (Flacourtiaceae). Cyclopentenyl acids in chaulmoogra oil formerly used in the treatment of leprosy.

3. Fluoro

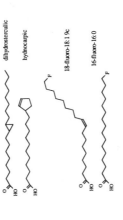

18-fluoro-18:1 9c

16-fluoro-16:0

Very rare acids found in seeds of *Dichapetalum toxicarium* (Dichapetalaceae).

a   This table is intended to give an impression of the diversity of plant fatty acid structures, with a small number of examples of both the fatty acids and their sources only. Reviews cited in the text give a comprehensive coverage of the occurrence of unusual fatty acids.

b   Compiled chiefly from References 1 and 2. Original references are given where not available from these sources.

c   Abbreviations: c— *cis*-ethylenic; t — *trans*-ethylenic; e — ethylenic; a — acetylenic; h — hydroxy; k — keto; and Me — methyl.

the large genus *Linum*, ricinoleic acid was found in the seed oil of all species tested from the section *Syllinum*, but in no species from other sections of the genus.[17] If any conclusion can be drawn from the taxonomic distribution of unusual fatty acids, it would seem to be that the ability to synthesize some unusual fatty acids appears to have evolved several times independently, while for others it may have evolved only once.

B. TISSUE LOCALIZATION OF UNUSUAL FATTY ACIDS

A feature of unusual fatty acids is that they are generally confined to seed triacylglycerols.[28-33] Analyses of vegetative tissues have generated few reports of unusual fatty acids, other than those occurring in the cuticle. A small number of exceptions exist in which unusual fatty acids are found in tissues other than the seed. The Δ^6-desaturated acids γ-linolenic ($\Delta^{6,9,12}$-18:3) and octadecatetraenoic ($\Delta^{6,9,12,15}$-18:4) are found in leaves and seeds of some plants of the Boraginaceae.[34,35] The cyclopropenoid fatty acids of the order Malvales are not restricted to the seeds.[36-38] The cyclopentenyl fatty acids of Flacourtiaceae seed oils are also found in smaller proportions in the phospholipids and glycolipids of various tissues.[39] Acetylenic fatty acids are found in root, stem, and leaf in the Santalaceae.[11,36] The small triacylglycerol fraction of rape leaves and siliques contains erucic acid.[36] Similarly, petroselinic acid is found in the pericarp (a maternal tissue), as well as within the seeds of ivy (*Hedera helix*).[36] Latex of the rubber tree, *Hevea brasiliensis*, contains triacylglycerol in which a furanoid acid (Table 2) is the major component,[40] while seeds of the same tree contain no unusual fatty acids.[41] An interesting observation is that branched-chain fatty acids accumulate in the yellow-white chloroplast-deficient parts of leaves of plastome mutants of snapdragon and tobacco, which have reduced levels of unsaturated fatty acids. The branched-chain fatty acids are completely absent from normal leaves and normal green parts of mosaic leaves.[6]

C. POTENTIAL PHYSIOLOGICAL ROLES

Since the ability to synthesize various unusual fatty acids must have evolved independently, the common feature of confinement to seed triacylglycerol indicates some selective constraint or functional significance. One possible function of unusual fatty acids is that by being toxic or indigestible they protect the seed against herbivory. Some unusual fatty acids may be inherently toxic, such as the acetylenic fatty acids or some of their metabolites described below which have antibiotic properties.[11] Other unusual fatty acids are toxic upon catabolism by the herbivore, such as the ω-fluoro fatty acids of *Dichapetalum toxicarium*.[42] One of these acids (threo-18-fluoro-9,10-dihydroxystearic acid) was lethal to rats when injected intraperitoneally at 25 mg/kg, probably due to catabolism to toxic fluoroacetate. The cyclopentenyl fatty acids were long used in the treatment of leprosy, and activity of hydnocarpic acid (Table 2) against many *Mycobacterium* species has been demonstrated.[39,43] These acids were also deleterious to the patients, causing a range of side effects. The cyclopropenoid fatty acids also appear to have biological activities, possibly due to the accumulation in animal tissues of partial catabolites containing the cyclopropene ring, which inhibits β-oxidation.[39] Three effects of cyclopropenoid fatty acids have been described, but are poorly understood: there is some alteration of the properties of membranes, there is an inhibition of fatty acid desaturase activity, and there is a carcinogenic effect or a co-carcinogenic effect with aflatoxins.[39,44,45] The tumor-promoting effect of cyclopropenoid fatty acids may be dependent upon their incorporation into membranes.[46] Interestingly, malvalic and sterculic acids inhibit the growth of seed-eating lepidopteran larvae and may be part of the defense of cotton plants against these insects.[47] These fatty acids may also be effective antifungal agents, inhibiting the growth of some plant pathogenic fungi at concentrations that appear biologically relevant.[38,48] The most intensely studied of the unusual fatty acids from a dietary viewpoint is erucic acid, due to fears that the consumption of rapeseed oil may be detrimental to human health. Chronic feeding of erucic acid to experimental animals has a range of deleterious effects,[49] but whether these are sufficiently severe to propose a herbivore-defense role for erucic acid in seeds is questionable.

It is possible that the use of unusual fatty acids as a carbon source may require adaptations of lipases or β-oxidation enzymes not present in the herbivore, so that it cannot catabolize them and remains unrewarded for eating seeds in which unusual fatty acids make up a large component of stored carbon. The purgative properties of castor oil, in which triricinolein is the predominant lipid, are well-known. However, the significance of ricinoleic acid in this context is unclear, since the castor seed seems already well-protected by the presence of toxic and allergenic proteins.

The question of why so many unusual fatty acids (Table 1) have evolved in plants may be considered as a subset of the same question concerning the extreme diversity of plant secondary metabolites.[3,50] Indeed, in some cases, unusual fatty acids may be starting points for plant secondary metabolism. For example, crepenynic acid (Table 2) is believed to be the precursor of most of the large range of polyacetylenes synthesized by a small group of plant families. These polyacetylenes may be involved in plant-plant or plant-animal interactions, and some have toxic or antibiotic properties. However, the presence of acetylenic fatty acids in other plant families such as the Santalaceae seems not to be accompanied by the accumulation of polyacetylenes.[15]

It is apparent that the accumulation of unusual fatty acids in lipid bodies of the seed does not incur any selective disadvantage. By contrast, some unusual fatty acids might be expected to disrupt membrane structure and function if incorporated into membrane-forming lipids. There is relatively little direct biological evidence for a disruptive effect of unusual fatty acids upon membrane structure. The few available studies have exploited mutants of *Escherichia coli* and of yeast which are incapable of fatty acid desaturation and require exogenous unsaturated fatty acids for growth. The fatty acids supplemented to the growth medium become incorporated in the membranes and thus provide a technique for correlating fatty acid structure with ability to augment saturated fatty acids in the formation of functional membranes.

Data from such experiments are somewhat contradictory, perhaps due to toxic impurities in some of the fatty acids used. In yeast,[51-54] it appeared that *cis* polyunsaturates supported growth regardless of double bond positions. *Cis* monounsaturates appeared to be effective only for certain desaturation positions, though this was due to an incompatibility with long stretches of contiguous saturation rather than an incompatibility with certain bond positions. *Trans* unsaturation was less effective than *cis*. Certain hydroxy fatty acids (e.g., ricinoleic) were compatible with growth, though they were less effective than the nonoxygenated analogs (e.g., oleic acid). The use of some unusual fatty acids in membranes was associated with their modification, such as the acetylation of hydroxystearic acids.[55] Results of studies with *E. coli* are broadly similar; branched-chain, brominated, and *trans* unsaturated fatty acids could support growth, as could cyclopropanoid acids which are in fact "usual" fatty acids for this organism[56-58] (see below). These studies show that at least some unusual fatty acids can exist in some functional membranes, but it remains quite possible that their inclusion in normal plant membranes would be deleterious.

A new technique for modifying plant membrane fatty acid composition *in vivo* has been described[59] which might be used to address the question of the compatibility of unusual fatty acids with membrane function. When fatty acids (15:0, 17:0, 17:1, 18:1) were applied as their Tween esters to leaves or other organs, they were extensively incorporated into membrane lipids. Techniques also exist for the study of physical properties of multilamellar liposomes containing unusual fatty acids generated *in vitro*.[60,61] One such study[60] showed that cyclopropanoid fatty acids are, in fact, eminently suited to the formation of functional membranes stable over a broad temperature range. As mentioned above, such fatty acids are normal components of some bacterial membranes.[62,63]

III. BIOCHEMISTRY

As discussed above, the majority of unusual fatty acids characterized from plants are found in seed oils, sometimes accumulating in substantial quantities. Various substituted acids also occur in cutin and suberin; these are considered separately (Chapter 4) as is fatty acid oxygenation by lipoxygenases (Chapter 5). In addition, a number of other fatty acid modifications have been described in vegetative tissues.[64-69] These typically appear to involve cytochrome P-450 enzymes which produce minor quantities of secondary metabolites with uncertain roles in the plant and are not considered further here.

A. BIOSYNTHESIS OF UNUSUAL FATTY ACIDS
1. Ricinoleic Acid
The biosynthesis of ricinoleic (12-hydroxy-*cis*-9-octadecenoic) acid from oleic acid in the developing endosperm of castor (*Ricinus communis*) is rela-

tively well-studied. Morris[70] established in elegant double-labeling studies that hydroxylation occurs directly by hydroxyl substitution rather than via an unsaturated, keto, or epoxy intermediate. Developing endosperm slices were incubated with a mixture of [1-^{14}C]oleic acid and *erythro*-12,13-ditritio-oleic acid. The ^{3}H:^{14}C ratio of ricinoleate synthesized was found to be 75% of that of the substrate mixture, and upon chemical oxidation to 12-keto-oleate the ratio was 55%, close to a predicted 50%. These results can only be obtained by a hydroxyl substitution mechanism (^{3}H:^{14}C ratios in the ricinoleate and 12-keto-oleate would be, respectively, for a keto intermediate: 50 and 50%; for an unsaturated intermediate: 50 and 25%; and for an epoxy intermediate: also 50 and 25%). Hydroxylation using oleoyl coenzyme A (CoA) as precursor can be demonstrated in crude preparations or microsomes and requires NAD(P)H and molecular oxygen.[71] NADPH supports lower rates of hydroxylation than NADH.[71,72] The incorporation of $^{18}O_2$ into the hydroxyl of ricinoleic acid has been shown by mass spectrometry.[73] Data from a study of the substrate specificity of the hydroxylase show that all substrate parameters (i.e., chain length and double bond position with respect to both ends) are important; deviations in these parameters caused reduced activity relative to oleic acid.[74] The position at which the hydroxyl was introduced, however, was determined by the position of the double bond, always being three carbons distal. The substrate for hydroxylation *in vivo* is most likely oleate esterifed in the *sn-2* position of phosphatidylcholine, from which ricinoleate is released as the free acid before activation and incorporation into triacylglycerol.[75] Antibodies raised against purified plant cytochrome b$_5$ inhibit the hydroxylation reaction, indicating that cytochrome b$_5$ is the electron donor to the hydroxylase.[76] Carbon monoxide (CO) does not inhibit hydroxylation, suggesting that a cytochrome P-450 is not involved.[71,72] These characteristics of the hydroxylase are generally similar to those of the microsomal desaturases (Chapter 2).

2. Isoricinoleic Acid

Investigations of the biosynthesis of isoricinoleic acid (9-hydroxy-*cis*-12 octadecenoic acid) in *Wrightia* spp.[77,78] have provided no conclusive evidence relating to the pathway of biosynthesis. Interestingly, 9-OH-18:0 is detected in the seed oil and can be labeled after incubation of developing seed halves with [^{14}C]acetate, indicating that the substrate for hydroxylation may be stearate (by direct hydroxylation) or oleate (by double bond hydration). These workers[78] favored the latter possibility because under their conditions [^{14}C]18:1 and [^{14}C]18:2 could act as precursors for [^{14}C]9-OH-18:1, and this conversion was greater under anaerobic conditions. However, it seems that these anaerobic conditions may not have been rigorous, and incubation times were very long; indeed the conversion of [^{14}C]18:2 to [^{14}C]9-OH-18:0 and [^{14}C]9-OH-18:1 at equivalent rates must be viewed with some concern.

3. Δ^5 Desaturation

A family of Δ^5-unsaturated fatty acids have been reported in the triacylglycerols of meadowfoam (*Limnanthes alba*). These were Δ^5-18:1, Δ^5-

20:1, Δ^5-22:1, and $\Delta^{5,13}$-22:2; Δ^{11}-20:1 and Δ^{13}-22:1 were also present.[79] Δ^5-Desaturated fatty acids, namely $\Delta^{5,9}$-18:2, $\Delta^{5,9,12}$-18:3, $\Delta^{5,9,12,15}$-18:4, $\Delta^{5,11}$-20:2, $\Delta^{5,11,14}$-20:3, and $\Delta^{5,11,14,17}$-20:4, were also identified in 20 species of gymnosperms.[80] The range of fatty acids identified is consistent with the Δ^5 desaturation of the common fatty acids or their C_2 elongation products. Activity of a Δ^5 desaturase was reported in experiments with developing cotyledons and cell-free extracts of *L. alba*.[81,82] Activity required molecular oxygen, preferred NADH over NADPH, and was not inhibited by CO. Palmitoyl-CoA, stearoyl-CoA, and eicosanoyl-CoA could all be desaturated, consistent with the range of fatty acids found in the seed. Unfortunately, the enzyme was unstable after fractional centrifugation, so that only crude extracts could be studied. This precluded a direct assessment of the cellular location of desaturation, but the desaturation of chain-elongated fatty acids suggested that desaturation occurs in the same compartment as elongation, believed to be the cytoplasm. With eicosanoyl-CoA as substrate, newly synthesized Δ^5-20:1 accumulated in the acyl-CoA fraction earlier than in the complex lipid fractions, suggesting that desaturation occurs on an acyl-CoA substrate. However, such results are to be interpreted with caution in studies using crude extracts.

4. Petroselinic Acid

Desaturation of plant fatty acids at the Δ^6 position seems to occur by two unrelated pathways. One pathway is widespread in seeds of the Apiaceae, Araliaceae, and Garryaceae which accumulate petroselinic (*cis*-6-octadecenoic) acid.[83] Crude homogenates of endosperm from developing coriander seeds could synthesize petroselinic acid from [^{14}C]malonyl-CoA. However, no desaturation of [^{14}C]18:0 or [^{14}C]16:0 or the corresponding acyl-ACPs could be demonstrated in the same extracts.[84] An important development was the discovery that antibodies raised against the 38-kDa stearoyl-ACP Δ^9 desaturase of avocado[85] recognized an additional 36-kDa protein which was only present in those tissues which synthesize petroselinic acid.[86] This allowed the isolation of a coriander cDNA clone which, when expressed in transgenic tobacco callus, caused the synthesis of petroselinic acid not otherwise found in tobacco. This clone has a high degree of sequence similarity at the deduced amino acid level to the Δ^9 desaturase clones. These results demonstrate that the biosynthesis of petroselinic acid involves a soluble plastidic enzyme which is expected to have similar biochemical properties to the Δ^9 desaturase.[85,87] Interestingly, the transgenic tobacco callus also synthesized Δ^4-16:1, suggesting two possible biosynthetic pathways for petroselinic acid. Either petroselinic acid could be synthesized by C_2 elongation of Δ^4-16:1 (analogous to the biosynthesis of *cis*-vaccenic acid discussed below) or both 16:0 and 18:0 could be desaturated at the ω-12 position. The latter possibility would be similar to the specificity of the membrane-bound desaturases of the plastid rather than that of the homologous soluble Δ^9 desaturase.[12]

5. γ-Linolenic and Octadecatetraenoic Acids

The second plant Δ^6 desaturation pathway is involved in the synthesis of γ-linolenic acid ($\Delta^{6,9,12}$-18:3) which occurs in a few species, notably borage (*Borago officinalis*), evening primrose (*Oenothera biennis*), and currant (*Ribes* spp.), and octadecatetraenoic acid ($\Delta^{6,9,12,15}$-18:4) found, for example, in borage leaves. Microsomes prepared from developing borage cotyledons actively desaturated [^{14}C]linoleate to γ-linolenate and [^{14}C]γ-linolenate to octadecatetraenoate. This activity was dependent upon NADH, and the substrate appeared to be esterified to the *sn*-2 position of phosphatidylcholine.[88-90] This microsomal Δ^6 desaturase, therefore, has substantially different properties to that involved in petroselinic acid biosynthesis. Furthermore, the borage Δ^6 desaturase, like the *Limnanthes* Δ^5 desaturase, is peculiar in that it desaturates at a position between the carboxyl group and preexisting double bonds, whereas other plant enzymes desaturate sequentially toward the methyl group. The Δ^6-desaturated acids are also found in the chloroplast lipids of borage leaves,[90] presumably as a result of lipid transport from the endoplasmic reticulum.

6. cis-Vaccenic Acid

cis-Vaccenic (*cis*-11-octadecenoic) acid is thought to be relatively common as a minor component of plant fatty acids,[4,91] but is considered here because it occurs at significant levels in some seed oils[1,4] and in the pulp lipids of some tropical fruits.[5,92] It has long been suggested[93-95] that the major pathway of *cis*-vaccenic acid biosynthesis is by elongation of palmitoleic acid, as is the case in animals and bacteria. In developing *Sinapis alba* seed, where *cis*-vaccenic acid accounted for a few percent of total fatty acids, labeling of *cis*-vaccenic acid with [^{14}C]malonyl-CoA was mainly due to elongation of palmitoleic acid rather than *de novo* synthesis.[96] The elongation pathway has been confirmed in fruit pulp of *Diospyros kaki* where *cis*-vaccenic acid accounted for 29% of total fatty acids.[5] These workers found an NADPH-dependent elongation of [2,2-^{2}H$_2$]palmitoleoyl-CoA with malonyl-CoA to dideutero-*cis*-vaccenic acid. An unexpected result was that some [2,2-^{2}H$_2$]palmitoleate was converted into oleate. It was subsequently shown that label in *cis*-vaccenic acid could be recovered in oleic acid and vice versa. This indicates that palmitoleate can be elongated to *cis*-vaccenate, but that *cis*-vaccenate can also be made by isomerization of oleate and vice versa.[97] No evidence has been found for a specific Δ^{11}-desaturase.

7. Erucic Acid

The biosynthesis of erucic (*cis*-13-docosenoic) acid and other very long-chain fatty acids has been studied in seeds of a number of species[81,98-101] and has been reviewed previously.[102-104] It has been clearly established that erucate is formed by elongation of oleate, involving two successive condensations with malonyl-CoA in the presence of NAD(P)H.[96,105] It is not clear whether the same or different enzymes are involved in the different elongation steps, since in *Brassica juncea* the reactions from 20:1 to 22:1 showed a preference for NADPH and a sensitivity to inhibitors not found for the reactions from 18:1 to 20:1.[106] The subcellular location of seed elongase activity is not clear; a particulate fraction (15,000 × *g* pellet) was more active than soluble, microsome, or oil body fractions.[107] Partial solubilization of elongase activity[108] and preliminary identification of the expected elongation intermediates, that is β-ketoacyl-CoA, β-hydroxyacyl-CoA, and *trans*-2-enoyl-CoA, have been reported.[105]

A microsomal acyl-CoA elongase of leaf epidermal cells has been characterized and partially purified.[106,109-111] This enzyme complex elongates saturated acyl-CoAs for biosynthesis of waxes, reviewed by von Wettstein-Knowles (Chapter 4).

8. Cyclic Fatty Acids

We are not aware of any biosynthetic studies concerning the cyclopropenoid and cyclopentenyl fatty acids in plants, since this was reviewed by Mangold and Spener.[39] Only a brief summary is given here. The cyclopropanoid acid, dihydrosterculic acid (Table 2), is formed by reaction of methionine, most probably as *S*-adenosylmethionine, with the nine and ten carbons of oleic acid. Dihydrosterculic acid is desaturated in an oxygen-requiring reaction to sterculic acid. Dihydromalvalic and malvalic acids appear to be derived from dihydrosterculic and sterculic acids respectively by α-oxidation.

The cyclopentenyl fatty acids are formed by elongation of aleprolic acid (cyclopentenyl carboxylic acid) by sequential two-carbon additions as for the biosynthesis of straight-chain acids. The nonprotein amino acid cyclopentenylglycine is a precursor of aleprolic acid and the first detectable intermediate containing the cyclopentenyl ring. Thus, the cyclopentenyl fatty acids are synthesized by priming a chain elongating system with the cyclopentenyl group rather than by ring closure of straight-chain acids as might otherwise be envisaged. The dependency of cyclopentenyl fatty acid biosynthesis upon CoA or acyl carrier protein (ACP) has not yet been shown.

9. Medium-Chain Fatty Acids

Possible mechanisms for the synthesis of medium-chain fatty acids have been reviewed recently.[112] At that time (1988), little experimental support was available for any of the possible mechanisms described. Recent studies with *Umbellularia californica*[113] have demonstrated the existence of a medium-chain-specific isozyme of acyl-ACP thioesterase in the seed plastids. This activity could be separated by partial purification from long-chain-specific acyl-ACP thioesterase activity and medium-chain acyl-CoA thioesterase activity. The implication is that the medium-chain thioesterase cleaves ACP from the growing acyl chain when it reaches a certain length. The free fatty acids would then be exported from the plastid and converted to acyl-CoA by an acyl-CoA synthetase associated with the plastid envelope.[12] The activity of the medium-chain acyl-ACP thioesterase increased with maturation of the seeds, consistent with a possible role in determining the acyl composition of the seed oil.

10. Other Fatty Acids

Enzymes involved in the synthesis of epoxy fatty acids in seed oils have not been studied. Mechanistic considerations (below) suggest, however, that epoxidases may be similar enzymes to hydroxylases. Indeed, some hydroxylases may also have epoxidase activity when presented with the unsaturated substrate analog.[114-116] This suggests that the epoxidases may be homologous enzymes to the hydroxylases, such as that involved in the biosynthesis of ricinoleic acid described above. Alternatively, the enzymes active in some seeds may have been recruited from a normal role in cutin biosynthesis. Fatty acid epoxidases (and hydroxylases) involved in cutin biosynthesis or secondary metabolism appear to be cytochromes P-450, since they are inhibited by carbon monoxide, often photoreversibly.[65,69,117]

As with epoxidases, the enzymes involved in synthesis of other unusual fatty acid types, such as the keto and acetylenic fatty acids, have not been studied. It might be speculated that keto fatty acids are derived from hydroxy fatty acids by a secondary alcohol dehydrogenase. It might also be speculated that the acetylenic fatty acids could be derived from epoxy fatty acids by the action of an epoxide hydrolase, yielding the dihydroxy fatty acid, followed by two dehydration steps, yielding first the enol and then the acetylenic fatty acid.

B. GENERAL CONSIDERATIONS

In addition to the presence of enzymes involved directly in the synthesis of unusual fatty acids, plants which accumulate unusual fatty acids may require other specialized proteins. Germination of the seeds in which they occur requires that the catabolic enzymes, such as lipases and the enzymes of β-oxidation, must be able to accept the unusual fatty acids and that unusual structures formed during β-oxidation can be processed. We are aware of only one pertinent study. When [^{14}C]ricinoleate was catabolized by homogenates of germinating pea and castor seeds, β-oxidation was blocked at the C_{10} level in pea, but went to completion in castor.[118] The C_{10} product identified in pea, 4-keto-decanoic acid, was presumably not further metabolizable, causing the arrest of β-oxidation at this point. A pathway was proposed for the degradation in castor of 4-hydroxy-decanoic acid via 2-hydroxy-octanoic acid or 4-keto-decanoic acid, 2-keto-octanoic acid, and heptanoic acid, but the operation of this pathway has not been verified.

As mentioned above, unusual fatty acids accumulate almost exclusively in the triacylglycerol fraction and are in some way excluded from the polar lipids. This is particularly intriguing, since diacylglycerol is a precursor of both triacylglycerol and polar lipid. With castor microsomes, there was some indication that the pool of ricinoleoyl-containing polar lipid is minimized by a preference of diacylglycerol acyltransferase for ricinoleate-containing diacylglycerols.[75] A similar result was obtained with *Cuphea lanceolata* microsomes[119] where the diacylglycerol acyltransferase is highly active and selective for diacylglycerol containing medium-chain fatty acids. In addition, the lysophosphatidic acid acyltransferase was selective for both donor and acceptor acyl groups such that didecanoyl and dioleoyl species of phosphatidic acid accounted for the majority synthesized. The lysophosphatidic acid acyltransferase of palm (*Syagrus cocoides*) microsomes also preferentially acylates lysophosphatidic acid containing medium-chain (12:0) acyl groups with medium-chain acyl-CoAs, again favoring dilauroylglycerol over mixed (12:0, 18:1) diacylglycerol.[120] In borage (*Borago officinalis*), γ-linolenic acid may be efficiently acylated to the *sn*-1 and *sn*-2 positions of glycerol-3-phosphate, but, in fact, accumulates at the *sn*-3 position of triacylglycerol. The diacylglycerol acyltransferase preferably uses γ-18:3-CoA, and this may minimize the pool size of γ-18:3-CoA, such that γ-18:3 is concentrated in triacylglycerol.[89] Data accruing from a number of other studies[121-123] tend to strengthen the view obtained from those described above — that targeting of unusual fatty acids to triacylglycerol can be at least partly explained by the relative activities and selectivities of the acyltransferases.

In some cases, the discrimination of acyltransferases actually limits the unusual fatty acid content of triacylglycerol. A well-known example is the "66% barrier" to erucate content in triacylglycerol of *Brassica* seeds. In this case, the lysophosphatidic acid acyltransferase does not accept erucoyl-CoA as an acyl donor, limiting erucate to the *sn*-1 and *sn*-3 positions of triacylglycerol. This specificity is observed in several species, but the enzyme from meadowfoam (*L. alba*) has been recognized as an exception.[124] An understanding of enzyme specificities and the ability to exploit those enzymes with desirable properties, such as the meadowfoam lysophosphatidic acid acyltransferase, will be important facets in the future manipulation of oil crops through molecular techniques. Likewise, it is important to investigate the metabolic fate of unusual fatty acids introduced (*in vitro* and eventually through molecular techniques) into plants lacking unusual fatty acids. One such study[125] showed that the targeting of unusual fatty acids to triacylglycerol is maintained, but the rate of esterification of the alien fatty acids was slow.

C. MECHANISTIC CONSIDERATIONS RELATED TO O_2-DEPENDENT TRANSFORMATIONS OF PLANT FATTY ACIDS

A number of the modifications of plant fatty acids described in this review involve O_2-dependent desaturation or hydroxylation reactions. A common feature of these reactions is the energetically demanding requirement for the cleavage of an unactivated, aliphatic C-H bond. At present, two classes of biological cofactors have been found sufficiently reactive to catalyze this type of O_2-dependent chemistry. The first class, consisting of the heme-containing oxygenases including cytochrome P-450 and peroxidase, has been extensively characterized.[126-137] In contrast, the second class has only recently been identified in the soluble bacterial enzyme methane monooxygenase and is thus relatively less well-characterized.[138-141] Through consideration of the properties of each of these two classes, a rationalization of the mechanistic constraints associated with O_2-dependent hydrocarbon oxidation can be made. As described below, these constraints reasonably apply to the O_2-dependent transformations of plant fatty acids as well.

1. Heme Oxygenase Mechanisms

For the P-450 oxygenases, the reductive activation of O_2 occurs via two discrete single electron transfers.[142] In P-450$_{cam}$, binding of substrates to the resting state iron(III) enzyme causes E° for the iron(III)/iron(II) redox couple to increase by ~130 mV,[143] providing thermodynamic control of electron transfer. An oxygenated intermediate formally equivalent to coordinated superoxide, Fe(III)-O_2^-, is formed when O_2 binds to the iron(II) state.[144-146] The iron(II) state of P-450 also forms a strongly inhibitory complex with CO that is commonly used as a diagnostic for this type of oxygenase.[147,148] A subsequent single electron transfer to Fe(III)-O_2^- initiates catalysis by reduction of this adduct to the peroxide level Fe(III)-OO^{2-}.[149] An apparently equivalent species can be transiently attained by complexation of H_2O_2[150] or organic peroxides[150-152] to the resting state Fe(III) P-450 enzyme, leading to "peroxide shunt" catalysis. The observation of "peroxide shunt" catalysis for P-450 supports both the requirement for two-electron reduction of O_2 prior to catalysis and the formation of a catalytically relevant peroxide level adduct. The Fe(III)-OO^{2-} adduct can be cleaved by either homolytic[153,154] (RO:OH ↔ RO·+·OH) or heterolytic[153,155] (RO:OH ↔ RO$^+$ + $^-$·OH) mechanisms. A generally accepted feature of the mechanism of heme oxygenase chemistry is that Fe(III)-OO^{2-} undergoes heterolytic cleavage to generate water and a reactive intermediate formally described as Fe(V)=O.[155,156] Spectroscopic characterizations of horseradish peroxidase,[157-159] cytochrome c peroxidase,[160-162] chloroperoxidase,[163] and synthetic models for peroxidase[164] and P-450[155] indicate this reactive intermediate is best formulated as (L$^{·+}$)-Fe(IV)=O, where (L$^{·+}$) represents either a porphyrin π cation radical[158,164] or an active site tryptophanyl cation radical.[160,161] The presence of an oxidizable ligand species is proposed to provide stabilization of the formal Fe(V)=O moiety via intramolecular electron transfer.

No spectroscopic evidence for the presence of (L$^{·+}$)-Fe(IV)=O or for the identity of the putative radical species has been obtained for P-450, presumably due to the inherent high reactivity of this species. However, the P-450 reactive intermediate has been proposed to abstract a hydrogen atom from the substrate to generate an enzyme bound [(L$^{·+}$)-Fe(III)=O·RH] diradical caged pair.[114] The involvement of radical intermediates is required to account for the epimerization[165] and the scrambling of regiochemistry caused by allylic migration.[166] Through the study of the distribution of rearrangement products obtained from the oxidation of highly strained cyclic hydrocarbons,[167-169] the lifetime of the diradical caged pair has been recently estimated to be ~10^{10}/s.[170] The hydroxylation reaction is completed by recombination of the diradical pair, which has been called the "oxygen rebound" mechanism.[171]

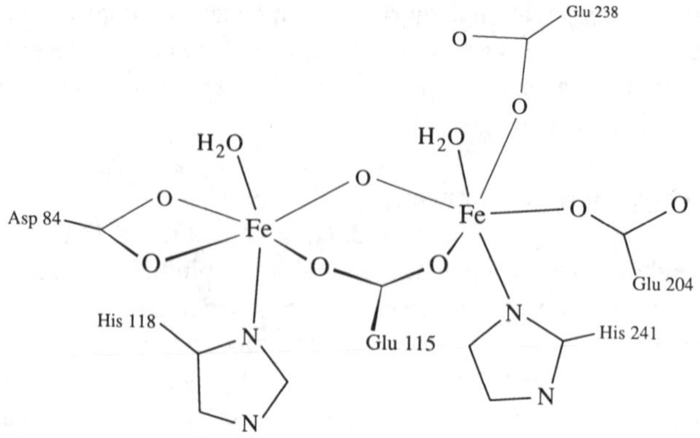

FIGURE 1. Structure of the FeOFe cluster present in ribonucleotide reductase.[183]

The extensive research literature available on the heme oxygenases has provided a basis for understanding the general strategies by which biological systems generate and utilize high valent oxoiron species to effect hydrocarbon oxidation catalysis. An important aspect of this research has been the realization of the energetic constraints imposed by the thermodynamic stability of unactivated, aliphatic C-H bonds. These constraints demand that intermediates with similar formal oxidation state and reactivity must be created for oxidative attack on unactivated hydrocarbons regardless of the cofactor utilized to effect the oxidation.

2. Introduction of a New Oxygenase Cofactor

Recently, a second biological cofactor competent for the O_2-dependent activation of aliphatic C-H bonds has been identified in methane monooxygenase (MMO). MMO is a three-protein component bacterial enzyme that catalyzes the O_2- and NADH-dependent hydroxylation of methane to yield methanol.[139,172] In addition, a wide variety of other hydrocarbons act as adventitious substrates for the purified enzyme.[172,173] The cofactor contained in the terminal hydroxylase component of MMO is a diiron cluster (FeOFe).[138,140,141] As evidenced by the presence of similar, but not identical, clusters in a wide variety of other proteins including hemerythrin,[174-178] purple acid phosphatases,[179-181] rubrerythrin,[182] and the B$_2$ subunit of ribonucleotide reductase,[183,184] this structural unit appears to be widely distributed in biological systems. The FeOFe cluster consists of a pair of spin-coupled iron atoms containing protein-derived nitrogen or oxygen ligands. An example of this structure, based on the 2.2-Å X-ray crystal structure[183] of ribonucleotide reductase, is shown in Figure 1.

For the MMO hydroxylase, quantitation of the iron content indicates the most active holoprotein contains 4 mol of iron;[185] lower iron content is associated with reduced enzymatic activity.[186,187] Electron paramagnetic resonance (EPR) and Mössbauer spectroscopic characterizations indicate all iron is contained in FeOFe clusters.[185,188] No mononuclear iron species, other metals, no organic co-factors, or catalytically active stable free radical species are present.[185] Thus, the catalytically active MMO hydroxylase (245 kDa $(\alpha\beta\gamma)_2$ holoprotein structure) contains two binuclear iron clusters.[185,188,189] Based on a combination of spectroscopic,[190] biochemical,[189,191] and genetic[192] evidence, the FeOFe cluster has been localized within the α-subunit of the MMO hydroxylase. For ribonucleotide reductase and methemerythrin, resonance Raman studies have demonstrated the bridging oxygen atom is derived from water.[193,194] In contrast, the origin of the oxygen atom in the MMO FeOFe cluster, whether solvent or protein derived, has not been defined. Both Mössbauer[188] and Extended X-ray Absorption Fine Structure (EXAFS)[195] studies of the FeOFe cluster of the MMO hydroxylase suggest the bridging oxygen moiety is protonated or otherwise triply coordinated at all accessible pH values.

Of the above-mentioned group of proteins containing FeOFe clusters, presently, only MMO[185,196] and ribonucleotide reductase[197-200] are known to catalyze oxygenase chemistry. It is therefore noteworthy that recent optical and Mössbauer spectroscopic characterization[201] of the recombinant stearoyl-ACP Δ^9 desaturase from *Ricinus communis* has conclusively shown FeOFe clusters are present in this enzyme as well. Unlike most P-450 enzymes, both MMO[202] and the plant fatty acid desaturases[203] and hydroxylases[71,72] are not significantly inhibited by CO, further implicating a similar nonheme reactive center. Based on presently available biochemical and spectroscopic evidence, a catalytic cycle for the FeOFe cluster of MMO has been formulated as shown in Figure 2.[141,196,204] The following discussion will present evidence for this cycle in order to promote a more general awareness of the catalytic potential of this newly recognized oxygenase cofactor.

3. Reduction of the FeOFe Cluster and Interaction with O_2

Three stable redox states can potentially be obtained for the FeOFe cluster:[141,205-207] a resting diferric state; a one-electron reduced mixed valence state consisting of a spin-coupled ferrous and ferric pair; and a two-electron reduced diferrous state. Thus, a key structural difference between the FeOFe cluster and heme is the ability of the FeOFe cluster to store two reducing equivalents prior to interactions with O_2. A variety of experimental evidence supports the cycle of two-electron reduction of the FeOFe cluster of MMO by either protein-mediated or chemical processes prior to interaction with O_2 (see Figures 2A and B). First, EPR studies of a novel integer-spin resonance associated with a spin S = 4 state of the diferrous FeOFe cluster have shown this redox state is quantitatively produced[208] and is readily reactive[185] with O_2. In contrast, the S = $1/2$ EPR signal characteristic of the mixed valence FeOFe cluster is produced in relatively low yield (~10 to 30% of the cluster concentration) and is unreactive in the presence of O_2.[185,209] Second, anaerobic chemical conversion of the MMO hydroxylase FeOFe cluster to the diferrous state in the absence of the other protein components supports single turnover catalytic hydroxylation and epoxidation reactions upon readmission of O_2 (see Figures 2B through E).[185] Similar experiments where the MMO hydroxylase

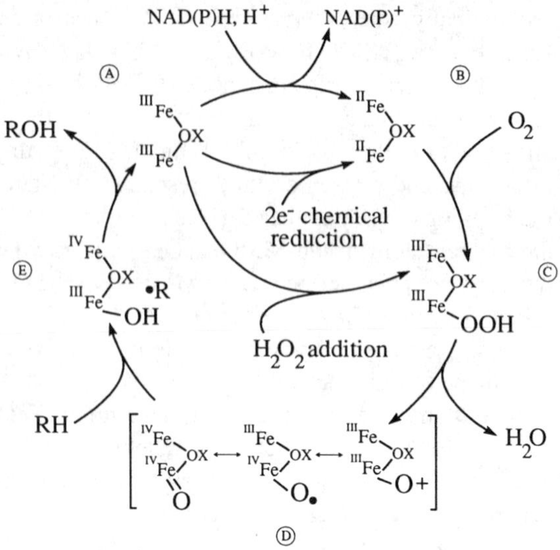

Lipid Metabolism in Plants

FIGURE 2. Proposed catalytic cycle for the FeOFe cluster of methane monooxygenase.

FeOFe cluster was converted to the mixed valence state did not support single turnover catalysis.[185] Taken together, these observations demonstrate that the diferrous FeOFe cluster is a catalytically competent structure and that all catalytic entities required for O_2-dependent hydroxylation reactions are inherent in the MMO hydroxylase protein.

4. Heterolytic Cleavage of Reduced O_2

Coordination of O_2 to the diferrous FeOFe cluster of MMO presumably yields a transitory FeOFe-OOH adduct (see Figures 2B and C). The protonation state of this adduct is presently unknown; that shown here is for convenience only. The formation of this type of adduct is strongly supported by the ability of the diferric MMO hydroxylase to catalyze hydroxylation and epoxidation reactions in the presence of exogenous H_2O_2 (see Figures 2A and C through E).[204] However, reduced oxygen species such as enzyme-bound superoxide or H_2O_2 alone are insufficiently reactive to catalyze the cleavage of unactivated C-H bonds[210,211] observed for MMO chemistry and potentially required for the stereospecific hydroxylation or the position-specific desaturation of plant fatty acids. Moreover, freely diffusible reactive species such as hydroxyl radical[212,213] (Fenton-like chemistry) generated by homolytic cleavage of a peroxide-level species yield reaction products that are inconsistent with the selectivity observed for the enzymatic transformations catalyzed by MMO and the plant FeOFe desaturases. Thus, in accord with the results obtained from study of the heme oxygenases, it is highly probable that the FeOFe-OOH adduct undergoes heterolytic cleavage to generate a tightly enzyme-bound high valent oxoiron reactive species.

The creation of a high valent oxoiron reactive species in MMO catalysis has been substantiated by the observation of intramolecular atomic migrations during the oxidation of haloalkenes,[196] strained cyclic compounds,[214,215] and aromatic compounds.[216] These products are consistent with the generation of an electrophilic oxidizing agent in the enzyme-active site during MMO catalysis. These intramolecular atomic migrations are directly analogous to the 1,2 shift of deuterium originally observed during P-450 oxidations of aromatic hydrocarbons,[217] subsequently called the "NIH shift". For MMO, the 32% inversion[218] of stereochemistry observed during the oxidation of chirally labeled ethane effectively eliminates a concerted mechanism for the aliphatic hydrocarbon hydroxylation reaction. Rather, this result provides strong experimental evidence for the stepwise creation of a substrate-based radical intermediate consistent with the "oxygen rebound" mechanism proposed for P-450.[171] The observation of products derived from rearrangements of substrate radicals formed during the oxidation of 1,1-dimethylcyclopropane,[214] allylic rearrangements during the oxidation of cyclohexene,[219,220] and epimerization during the hydroxylation of norbornane[219] provide additional support for a stepwise rather than concerted oxidation. An *intramolecular* isotope effect $k_{H,exo}/k_{D,exo} \geq 5.5$ was determined for the MMO-catalyzed oxidation of *exo,exo,exo,exo* -2,3,5,6-d_4 norbornane, supporting the abstraction of a hydrogen atom during the catalytic reaction.[219] However, the *intermolecular* isotope effect on V_m was small (approximately 1.8) for the oxidation of CH_4 and CD_4,[219] suggesting enzymatic processes other than C-H bond cleavage were rate limiting in the reconstituted enzyme reaction.

No structural evidence for a high valent oxidizing species has presently been obtained for MMO. However, study of a structural[221,222] and catalytic model[223] for the FeOFe cluster of MMO has recently allowed the identification of an intermediate with potential relevance to the enzymatic process.[224] Combined optical, resonance Raman, EPR, and Mössbauer studies have shown that treatment of the synthetic cluster with H_2O_2 at –40°C caused cleavage of the FeOFe dimer and formation of a *monomeric* high valent species. The catalytic potential of this species was shown by correlation of the hydroxylation of cumene with the disappearance of the optical spectrum of the intermediate at –40°C. The magnetic spectral properties of the intermediate were rationalized by ferromagnetic coupling[159,163] between a spin S = 1 ferryl species (Fe(IV)=O)[225] and a radical. The radical, denoted (L⁻⁺), has been proposed to be an oxidized state of the tris(2-pyridylmethyl)amine ligand. This study provides the first direct spectroscopic evidence for the existence of a high valent oxoiron species outside of a heme environment that is also capable of hydrocarbon oxidation. Moreover, the electronic structure of this intermediate is formally analogous to the (L⁻⁺)-Fe(IV)=O species proposed for P-450 and peroxidase chemistry. This analogy is highly supportive of the proposal that a common chemical strategy will be utilized for *stabilization* of reactive high valent oxoiron in-termediates. In the case of MMO, delocalizable electron density derived from resonance rearrangements within the spin-coupled FeOFe co-factor (see Figure

2D) has been proposed to promote stability of the oxidizing species.[141,196] With respect to enzymatic catalysis, the potential relevance of the cleavage of the FeOFe model cluster during the *generation* of the reactive intermediate has not yet been addressed.

5. Modulation of the Reactivity of $(L^{\cdot +})$ – Fe(IV) = O

The ability of the heme oxygenases to direct catalysis from a common intermediate state must reside in the specific electronic and structural properties of the enzyme active site.[128,130,136,137,155,226-231] Similar considerations may also be true for modulation of the reactivity of the FeOFe cluster. However, the mechanisms of modulating the reactive intermediate in the FeOFe oxygenases are presently unknown. The reactivity of the FeOFe cluster may be controlled by composition of the protein ligands to the cluster iron atoms, the nature of the bridging oxygen atom, the protein environment surrounding the active site, and the spatial orientation of hydrocarbon substrates relative to the FeOFe cluster. By analogy with the reactivity of MMO, it is evident that the FeOFe cluster is sufficiently reactive to effect both the hydroxylation and epoxidation reactions observed in unusual plant fatty acids. It is also likely that C-H bond cleavage required for fatty acid desaturation[232-234] will require the participation of a reactive intermediate such as the $(L^{\cdot +})$-Fe(IV)=O species proposed here. Due to the apparent requirement for two-electron oxidation without oxygen atom transfer,[55] fatty acid desaturation may also exhibit mechanistic aspects in common with a peroxidase-type reaction. While no peroxidase enzyme containing an FeOFe cluster has been presently described, the FeOFe cluster of ribonucleotide reductase does catalyze the O_2-dependent single electron oxidation of tyrosine to form a tyrosyl cation radical without oxygen atom transfer.[198,199] Surprisingly, site-specific mutagenesis of Phe 208 to Tyr resulted in conversion of the enzyme to an oxygen transfer catalyst.[235] Upon purification of the mutant protein, Tyr 208 was shown to be hydroxylated in the *meta* position and was acting as a bidentate ligand to one iron of the FeOFe cluster.

6. Potential Implications for Plant Fatty Acid Transformations

All meaningful proposals for enzyme reaction mechanisms are necessarily based on the correct structural identification of the cofactors participating in the reaction. The conclusive identification of an FeOFe cluster in the stearoyl-ACP Δ^9 desaturase[201] finally offers the key to opening the mechanistic "black box" surrounding the mechanism of aerobic fatty acid desaturation first declared by Bloch in 1969[236] and subsequently outlined by others.[237,238] By analogy to the reactivity of MMO and ribonucleotide reductase, hydroxylation, epoxidation, and desaturation reactions can be reasonably anticipated for plant enzymes containing FeOFe clusters. However, significant experimentation is still required to define the intimate details of these reactions. All known proteins which contain FeOFe clusters and which interact with O_2 are reduced to the diferrous state[239] prior to interaction with O_2. At present, the roles of protein-protein and protein-substrate interactions in the control of the timing of electron transfers, in the coordination of hydrocarbon and O_2 binding, and in the outcome of catalysis in the plant FeOFe oxygenases are unknown. In MMO, these interactions have been shown to modify the observed percentage of the mixed valence state of the FeOFe cluster,[185,189,240] to change the rate of catalysis by greater than 100-fold,[185,186,189] and to modulate the reaction specificity with complex substrates.[241] Heterolytic cleavage of an FeOFe-OOH adduct would be required to generate a tightly enzyme-bound oxidative intermediate sufficiently reactive to catalyze the specific hydroxylation, epoxidation, or desaturation reactions observed in higher plant tissues. It is an intriguing possibility that the position-specific transformations of plant fatty acids described in this review may arise from isozyme-like variations of the substrate binding site relative to a structurally and functionally conserved FeOFe cluster unit which is ultimately responsible for oxidative catalysis.

IV. CONCLUDING REMARKS

As in most other aspects of biology, the study of plant lipid metabolism is currently being transformed by the application of the methods and approaches of molecular genetics. The broad utility of the molecular approach is evident in the recent progress in understanding the mechanistic and structural basis of catalysis by desaturases. Until recently, it had not been possible to purify significant quantities of these soluble proteins. However, following the cloning and expression in microorganisms of genes for stearoyl-ACP desaturase,[85,87] it was possible to obtain the gram quantities of pure protein required for structural studies by X-ray crystallography[242] and spectroscopic methods.[201] The availability of the gene also permitted the isolation of a related gene for the "Δ^6 desaturase" from coriander.[86] In conjunction with the determination of the tertiary structure of the Δ^9 desaturase, comparison of the closely related structures of the Δ^9 and Δ^6 desaturases should provide an unambiguous insight into the mechanisms which regulate the insertion of the double bond.

By extrapolation from the recent progress made in understanding desaturases, it seems likely that comparable progress may soon be made, by similar means, in understanding the mechanisms responsible for many of the other modifications of fatty acids which give rise to the diversity of structures summarized in Table 2. Indeed, the formulation of probable reaction mechanisms for the oxygen-dependent modifications, described here, may assist in the development of strategies for the identification of genes encoding the relevant enzymes. For instance, in view of the apparent similarities between desaturases and the hydroxylase responsible for ricinoleate biosynthesis in castor, we anticipate substantial structural homology will be found in the primary structures of the desaturases and the hydroxylases. Similarly, because of the coincidence of occurrence of epoxy and hydroxy fatty acids in certain genera, the observation that various substitutions occur at characteristic carbons and on the basis of proposed mechanistic considerations, we may also expect structural homology between desaturases or hydroxylases and epoxidases.

The eventual isolation of a family of genes for fatty acid modifying enzymes may also be expected to create new opportunities for the agricultural production of novel fats and oils. It seems likely that many unusual fatty acids could be directly useful in nonfood applications if they could be obtained in adequate quantities and substantial purity at costs comparable to the cost of edible oils. As the genes for modifying enzymes become available, it should be possible to transfer these genes into oilseed species with good agronomic characteristics and produce modified fatty acids as specialty oil crops. This may require not only knowledge of the biosynthesis and catabolism of unusual fatty acids, but also detailed knowledge of the mechanisms by which the source species generally exclude unusual fatty acids from membrane lipids. Thus, although there appear to be many attractive opportunities for applying basic knowledge of plant lipid biochemistry to the creation of useful new agricultural commodities, it will be necessary to fill in many large gaps in our current knowledge. It is to be hoped that the opportunities will invigorate interest in the study of unusual fatty acids.

REFERENCES

1. **Hitchcock, C. and Nichols, B. W.,** *Plant Lipid Biochemistry,* Academic Press, London, 1971, chap. 1.
2. **Gunstone, F. D., Harwood, J. L., and Padley, F. B.,** *The Lipid Handbook,* Chapman and Hall, London, 1986, chaps. 1 and 3.
3. **Wink, M.,** Plant breeding: importance of plant secondary metabolites for protection against pathogens and herbivores, *Theor. Appl. Genet.,* 75, 225, 1988.
4. **Kleiman, R. and Payne-Wahl, K. L.,** Fatty acid composition of seed oils of the Meliaceae, including one genus rich in cis-vaccenic acid, *J. Am. Oil Chem. Soc.,* 61, 1836, 1984.
5. **Hibahara, A., Yamamoto, K., Takeoka, M., Kinoshita, A., Kajimoto, G., Nakayama, T., and Noda, M.,** Application of a GC-MS method using deuterated fatty acids for tracing cis-vaccenic acid biosynthesis in kaki pulp, *Lipids,* 24, 488, 1989.
6. **Radunz, A.,** On the function of methyl-branched chain fatty acids in phospholipids of cell membranes in higher plants, in *The Metabolism, Structure, and Function of Plant Lipids,* Stumpf, P. K., Mudd, J. B., and Nes, W. D., Eds., Plenum Press, New York, 1987, 197.
7. **Hitchcock, C.,** Structure and distribution of plant acyl lipids, in *Recent Advances in the Chemistry and Biochemistry of Plant Lipids,* Galliard, T. and Mercer, E. I., Eds., Academic Press, London, 1975, chap. 1.
8. **Smith, C. R., Jr.,** Occurrence of unusual fatty acids in plants, in *Progress in the Chemistry of Fats and Other Lipids,* Vol. 11 (Part 1), Holman, R.T., Ed., Pergamon Press, London, 1970.
9. **Markley, K. S.,** *Fatty Acids; Their Chemistry, Properties, Production, and Uses,* Vol. 1, 2nd ed., Robert E. Krieger Publishing, Malabar, FL, 1983.
10. **Smith C. R., Jr.,** Unusual seed oils and their fatty acids, in *Fatty Acids,* Pryde, E. H., Ed., American Oil Chemists Society, Champaign, IL, 1979, chap. 2.
11. **Hopkins, C. Y.,** Fatty acids with conjugated unsaturation, in *Topics in Lipid Chemistry,* Vol. 3, Gunstone, F.D., Ed., John Wiley & Sons, New York, 1972, chap. 2.
12. **Browse, J. and Somerville, C.,** Glycerolipid synthesis: biochemistry and regulation, *Annu. Rev. Plant Physiol. Plant Mol. Biol.,* 42, 467, 1991.

13. **Spener, F. and Tober, I.,** Cyclopentenylfettsauren im Brennpunkt, *Fette Seifen Anstrichm.,* 83, 401, 1981.

14. **Tober, I. and Conn, E. E.,** Cyclopentenylglycine, a precursor of deidaclin in Turnera ulmifolia, *Phytochemistry,* 24, 1215, 1985.

15. **Harborne, J. B. and Turner, B. L.,** *Plant Chemosystematics,* Academic Press, London, 1984, chap. 8.

16. **Brieskorn, C. H. and Kabelitz, L.,** Hydroxyfettsauren aus dem cutin des blattes von *Rosmarinus officinalis, Phytochemistry,* 10, 3195, 1971.

17. **Green, A. G.,** The occurrence of ricinoleic acid in *Linum* seed oils, *J. Am. Oil Chem. Soc.,* 61, 939, 1984.

18. **Garg, S. P., Sherwani, M. R. K., Arora, A., Agarwal, R., and Ahmad, M.,** Ricinoleic acid in *Vinca rosea* seed oil, *J. Oil Technol. Assoc. India,* 19, 63, 1987.

19. **Ahmad, M. U., Husain, S. K., and Osman, S. M.,** Ricinoleic acid in *Phyllanthus niruri* seed oil, *J. Am. Oil Chem. Soc.,* 58, 673, 1981.

20. **Daulatabad, C. D. and Mirajkar, A. M.,** Ricinoleic acid in *Artocarpus integrifolia* seed oil, *J. Am. Oil Chem. Soc.,* 66, 1631, 1989.

21. **Daulatabad, C. D., Desai, V. A., Hosamani, K. M., and Jamkhandi, A. M.,** Novel fatty acids in *Azima tetracantha* seed oil, *J. Am. Oil Chem. Soc.,* 68, 978, 1991.

22. **Daulatabad, C. D. and Hosamani, K. M.,** Unusual fatty acids in *Brunfelsia americana* seed oil: a rich source of oil, *J. Am. Oil Chem. Soc.,* 68, 608, 1991.

23. **Yunusova, S. G., Gusakova, S. D., and Glushenkova, A. I.,** Lipids from the seeds of *Securinega suffruticosa, Khim. Prir. Soedin. (Tashk.),* 3, 277, 1986.

24. **Kleiman, R., Spencer, G. F., Earle, F. R., Nieschlag, H. J., and Barclay, A. S.,** Tetra-acid triglycerides containing a new hydroxy eicosadienoyl moiety in *Lesquerella auriculata* seed oil, *Lipids,* 7, 660, 1972.

25. **Sreenivasan, B., Kamath, N. R., and Kane, J. G.,** Studies on castor oil. 1. Fatty acid composition of castor oil, *J. Am. Oil Chem. Soc.,* 33, 61, 1956.

26. **Badami, R. C. and Kudari, S. M.,** Analysis of *Hiptage madablota* seed oil, *J. Sci. Food Agric.,* 21, 248, 1970.

27. **Nikolin, A., Nikolin, B., and Jankovic, M.,** Ipopurpuroside, a new glycoside from *Ipomoea purpurea, Phytochemistry,* 17, 451, 1978.

28. **Rao, S. V., Paulose, M. M., and Vijayalakshmi, B.,** A note on the fatty acids present in oilseed phospholipids, *Lipids,* 2, 88, 1967.

29. **Persmark, U.,** Main constituents of rapeseed lecithin, *J. Am. Oil Chem. Soc.,* 45, 742, 1968.

30. **Vijayalakshmi, B. and Rao, S. V.,** Fatty acid composition of phospholipids in seed oils containing unusual acids, *Chem. Phys. Lipids,* 9, 82, 1972.

31. **Kondoh, H. and Kawabe, S.,** Phospholipids in castor seeds, *Agric. Biol. Chem.,* 39, 745, 1975.

32. **Nasirullah, Werner, G., and Seher, A.,** Fatty acid composition of lipids from edible parts and seeds of vegetables, *Fette Seifen Anstrichm.,* 86, 264, 1984.

33. **Prasad, R. B. N., Rao, Y. N., and Rao, S. V.,** Phospholipids of palash (*Butea monosperma*), papaya (*Carica papaya*), jangli badam (*Sterculia foetida*), coriander (*Coriandrum sativum*) and carrot (*Daucus carota*) seeds, *J. Am. Oil Chem. Soc.,* 64, 1424, 1987.

34. **Jamieson, G. R. and Reid, E. H.,** Analysis of oils and fats by gas chromatography. V. Fatty acid composition of the leaf lipids of *Myosotis scorpioides, J. Sci. Food Agric.,* 19, 628, 1968.

35. **Jamieson, G. R. and Reid, E. H.,** The leaf lipids of some members of the Boraginaceae family, *Phytochemistry,* 8, 1489, 1969.

36. **Appelqvist, L.-A.,** Biochemical and structural aspects of storage and membrane lipids in developing oilseeds, in *Recent Advances in the Chemistry and Biochemistry of Plant Lipids,* Galliard, T. and Mercer, E. I., Eds., Academic Press, London, 1975, chap. 8.

37. **Shenstone, F. S. and Vickery, J. R.,** Occurrence of cyclo-propene fatty acids in some plants of the order Malvales, *Nature,* 190, 168, 1961.

38. **Schmid, K. M. and Patterson, G. W.,** Distribution of cylclopropenoid fatty acids in malvaceous plant parts, *Phytochemistry,* 27, 2831, 1988.
39. **Mangold, H. K. and Spener, F.,** Biosynthesis of cyclic fatty acids, in *The Biochemistry of Plants,* Vol. 4, Stumpf, P. K., Ed., Academic Press, London, 1980, chap. 19.
40. **Lie Ken Jie, M. S. F. and Sinha, S.,** Fatty acid composition and the characterization of a novel dioxo C_{18}-fatty acid in the latex of *Hevea brasiliensis, Phytochemistry,* 20, 1863, 1981.
41. **Gandhi, V. M., Cherian, K. M., and Mulky, M. J.,** Nutritional and toxicological evaluation of rubber seed oil, *J. Am. Oil Chem. Soc.,* 67, 883, 1990.
42. **Harper, D. B. and Hamilton, J. T. G.,** Identification of threo-18-fluoro-9,10-dihydroxystearic acid: a novel ω-fluorinated fatty acid from *Dichapetalum toxicarium* seeds, *Tetrahedron Lett.,* 31, 7661, 1990.
43. **Abdel-Moety, E. M.,** Cyclopentenylfettsauren als ausgangsmaterial zur gewinnung neuer wirkstoffe, *Fette Seifen Anstrichm.,* 83, 65, 1981.
44. **Greenberg, A. and Harris, J.,** Cyclopropenoid fatty acids, *J. Chem. Educ.,* 59, 539, 1982.
45. **Schuch, R. and Ahmad, F.,** Structure and biological significance of triacylglycerols containing cyclopropene acyl moieties, *Fat Sci. Technol.,* 89, 338, 1987.
46. **Pawlowski, N. E., Hendricks, J. D., Bailey, M. L., Nixon, J. E., and Bailey G. S.,** Structural-bioactivity relationship for tumor promotion by cyclopropenes, *J. Agric. Food Chem.,* 33, 767, 1985.
47. **Chan, B. G., Waiss, A. C., Jr., Binder, R. G., and Elliger, C. A.,** Inhibition of lepidopterous larval growth by cotton constituents, *Entomol. Exp. Appl.,* 24, 294, 1978.
48. **Schmid, K. M. and Patterson, G. W.,** Effects of cyclopropenoid fatty acids on fungal growth and lipid composition, *Lipids,* 23, 248, 1988.
49. **Kramer, J. K. G., Sauer, F. D., and Pigden, W. J.,** *High and Low Erucic Acid Rapeseed Oils,* Academic Press, Toronto, 1983, chaps. 11–21.
50. **Williams, D. H., Stone, M. J., Hauck, P. R., and Rahman, S. K.,** Why are secondary metabolites (natural products) biosynthesized?, *J. Nat. Prod.,* 52, 1189, 1989.
51. **Proudlock, J. W., Haslam, J. M., and Linnane, A. W.,** Biogenesis of mitochondria 19. The effects of unsaturated fatty acid depletion on the lipid composition and energy metabolism of a fatty acid desaturase mutant of *Saccharomyces cerevisiae, Bioenergetics,* 2, 327, 1971.
52. **Wisnieski, B. J. and Kiyomoto, R. K.,** Fatty acid desaturase mutants of yeast: growth requirements and electron spin resonance spin-label distribution, *J. Bacteriol.,* 109, 186, 1972.
53. **Lands, W. E. M., Sacks, R. W., Sauter, J., and Gunstone, F.,** Selective effects of fatty acids upon cell growth and metabolic regulation, *Lipids,* 13, 878, 1978.
54. **Nes, W. D., Adler, J. H., and Nes, W. R.,** A structure-function correlation for fatty acids in *Saccharomyces cerevisiae, Exp. Mycol.,* 8, 55, 1984.
55. **Light, R. J., Lennarz, W. J., and Bloch, K.,** The metabolism of hydroxystearic acids in yeast, *J. Biol. Chem.,* 237, 1793, 1962.
56. **Silbert, D. F., Ruch, R., and Vagelos, P. R.,** Fatty acid replacements in a fatty acid auxotroph of *Escherichia coli, J. Bacteriol.,* 95, 1658, 1968.
57. **Silbert, D. F., Ladenson, R. C., and Honegger, J. L.,** The unsaturated fatty acid requirement in *Escherichia coli.* Temperature dependence and total replacement by branched-chain fatty acids, *Biochim. Biophys. Acta,* 311, 349, 1973.
58. **Machtiger, N. A. and Fox, C. F.,** Biochemistry of bacterial membranes, *Annu. Rev. Biochem.,* 42, 575, 1973.
59. **Terzaghi, W. B.,** Manipulating membrane fatty acid compositions of whole plants with tween-fatty acid esters, *Plant Physiol.,* 91, 203, 1989.
60. **Dufourc, E. J., Smith, I. C. P., and Jarrell, H. C.,** Role of cyclopropane moieties in the lipid properties of biological membranes: a ^2H NMR structural and dynamical approach, *Biochemistry,* 23, 2300, 1984.
61. **Isaacson, Y., Riehl, T. E., and Stenson, W. F.,** Nonelectrolyte permeability of liposomes of hydroxyfatty acid-containing phosphatidylcholines, *Biochim. Biophys. Acta,* 986, 295, 1989.

62. **Grogan, D. W. and Cronan, J. E.,** Cloning and manipulation of the *Escherichia coli* cyclopropane fatty acid synthase gene: physiological aspects of enzyme overproduction, *J. Bacteriol.,* 158, 286, 1984.

63. **Grogan, D. W. and Cronan, J. E.,** Characterization of *Escherichia coli* mutants completely defective in synthesis of cyclopropane fatty acids, *J. Bacteriol.,* 166, 872, 1986.

64. **Ferrante, G. and Kates, M.,** Characteristics of the oleoyl- and linoleoyl-CoA desaturase and hydroxylase systems in cell fractions from soybean cell suspension cultures, *Biochim. Biophys. Acta,* 876, 429, 1986.

65. **Salaun, J.-P., Weissbart, D., Durst, F., Pflieger, P., and Mioskowski, C.,** Epoxidation of cis and trans Δ^9-unsaturated lauric acids by a cytochrome P-450-dependent system from higher plant microsomes, *FEBS Lett.,* 246, 120, 1989.

66. **Blee, E. and Schuber, F.,** Stereochemistry of the epoxidation of fatty acids catalyzed by soybean peroxygenase, *Biochem. Biophys. Res. Commun.,* 173, 1354, 1990.

67. **Grechkin, A. N., Kukhtina, N. V., Gafarova, T. E., and Kuramshin, R. A.,** Oxidation of [1-^{14}C]linoleic acid in isolated microsomes from pea leaves, *Plant Sci.,* 70, 175, 1990.

68. **Janistyn, B.,** Evidence for conversion of arachidonic acid to hydroxyeicosatetraenoic acids by a cell-free homogenate of maize seedlings, *Phytochemistry,* 29, 2453, 1990.

69. **Fahlstadius, P.,** Absolute configurations of 9,10-epoxydodecanoic acids biosynthesized by microsomes from jerusalem artichoke tubers, *Phytochemistry,* 30, 1905, 1991.

70. **Morris, L. J.,** The mechanism of ricinoleic acid biosynthesis in *Ricinus communis* seeds, *Biochem. Biophys. Res. Commun.,* 29, 311, 1967.

71. **Galliard, T. and Stumpf, P. K.,** Fat metabolism in higher plants XXX. Enzymatic synthesis of ricinoleic acid by a microsomal preparation from developing *Ricinus communis* seeds, *J. Biol. Chem.,* 241, 5806, 1966.

72. **Moreau, R. A. and Stumpf, P. K.,** Recent studies of the enzymic synthesis of ricinoleic acid by developing castor beans, *Plant Physiol.,* 67, 672, 1981.

73. **Underhill, E. W., van de Loo, F. J., and Somerville, C. R.,** unpublished data, 1989.

74. **Howling, D., Morris, L. J., Gurr, M. I., and James, A. T.,** The specificity of fatty acid desaturases and hydroxylases. The dehydrogenation and hydroxylation of monoenoic acids, *Biochim. Biophys. Acta,* 260, 10, 1972.

75. **Bafor, M., Smith, M. A., Jonsson, L., Stobart, K., and Stymne, S.,** Ricinoleic acid biosynthesis and triacylglycerol assembly in microsomal preparations from developing castor-bean (*Ricinus communis*) endosperm, *Biochem. J.,* 280, 507, 1991.

76. **van de Loo, F. J., Kearns, E. V., and Somerville, C. R.,** unpublished data, 1991.

77. **Ahmad, F., Schiller, H., and Mukherjee, K. D.,** Lipids containing isoricinoleoyl (9-hydroxy-*cis*-12-octadecenoyl) moieties in seeds of Wrightia species, *Lipids,* 21, 486, 1986.

78. **Ahmad, F. and Mukherjee, K. D.,** Biosynthesis of lipids containing isoricinoleic (9-hydroxy-cis-12-octadecenoic) acid in seeds of Wrightia species, *Z. Naturforsch.,* 43c, 505, 1988.

79. **Nikolova-Damyanova, B., Chrisie, W. W., and Herslof, B.,** The structure of the triacylglycerols of meadowfoam oil, *J. Am. Oil Chem. Soc.,* 67, 503, 1990.

80. **Takagi, T. and Itabashi, Y.,** *Cis*-5-olefinic unsaturated fatty acids in seed lipids of Gymnospermae and their distribution in triacylglycerols, *Lipids,* 17, 716, 1982.

81. **Pollard, M. R. and Stumpf, P. K.,** Biosynthesis of C_{20} and C_{22} fatty acids by developing seeds of *Limnanthes alba*: chain elongation and Δ^5 desaturation, *Plant Physiol.,* 66, 649, 1980.

82. **Moreau, R. A., Pollard, M. R., and Stumpf, P. K.,** Properties of a $\Delta5$-fatty acyl-CoA desaturase in the cotyledons of developing *Limnanthes alba, Arch. Biochem. Biophys.,* 209, 376, 1981.

83. **Kleiman, R. and Spencer, G. F.,** Search for new industrial oils: XVI. Umbelliflorae — seed oils rich in petroselinic acid, *J. Am. Oil Chem. Soc.,* 59, 29, 1982.

84. **Cahoon, E. B. and Ohlrogge, J. B.,** Deposition and synthesis of petroselinic acid in seeds of Umbelliferae, *Inform,* 2 (Abstr.), 342, 1991.

85. **Shanklin, J. and Somerville, C.,** Stearoyl-acyl-carrier-protein desaturase from higher plants is structurally unrelated to the animal and fungal homologs, *Proc. Natl. Acad. Sci. U.S.A.,* 88, 2510, 1991.

86. **Cahoon, E. B., Shanklin, J., and Ohlrogge, J. B.,** Expression of a coriander desaturase results in petroselinic acid production in transgenic tobacco, *Proc. Natl. Acad. Sci. U.S.A.*, 89, 11184, 1992.

87. **Thompson, G. A., Scherer, D. E., Foxall-van Aken, S., Kenny, J. W., Young, H. L., Shintani, D. K., Kridl, J. C., and Knauf, V. C.,** Primary structures of the precursor and mature forms of stearoyl-acyl carrier protein desaturase from safflower embryos and requirement of ferredoxin for enzyme activity, *Proc. Natl. Acad. Sci. U.S.A.*, 88, 2578, 1991.

88. **Stymne, S. and Stobart, A. K.,** Biosynthesis of γ-linolenic acid in cotyledons and microsomal preparations of the developing seeds of common borage (*Borago officinalis*), *Biochem. J.*, 240, 385, 1986.

89. **Griffiths, G., Stobart, A. K., and Stymne, S.,** Δ^6- and Δ^{12}-desaturase activities and phosphatidic acid formation in microsomal preparations from the developing cotyledons of common borage (*Borago officinalis*), *Biochem. J.*, 252, 641, 1988.

90. **Griffiths, G., Brechany, E. Y., Christie, W. W., Stymne, S., and Stobart, K.,** Synthesis of octadecatetraenoic acid (OTA) in borage (*Borago officinalis*), in *Biological Role of Plant Lipids*, Biacs, P. A., Gruiz, K., and Kremmer, T., Eds., Plenum Press, New York, 1989, 151.

91. **Seher, A. and Gundlach, U.,** Isomere monoensauren in pflanzenolen, *Fette Seifen Anstrichm.*, 84, 342, 1982.

92. **Yamamoto, K., Shibahara, A., Sakuma, A., Nakayama, T., and Kajimoto, G.,** Occurrence of n-5 monounsaturated fatty acids in jujube pulp lipids, *Lipids*, 25, 602, 1990.

93. **Hawke, J. C. and Stumpf, P. K.,** Fat metabolism in higher plants. XXVII. Synthesis of long-chain fatty acids by preparations of *Hordeum vulgare* L. and other Graminae, *Plant Physiol.*, 40, 1023, 1965.

94. **Kuemmel, D. F. and Chapman, L. R.,** The 9-hexadecenoic and 11-octadecenoic acid content of natural fats and oils, *Lipids*, 3, 313, 1968.

95. **Grosbois, M.,** Biosynthese des acides gras au cours du developpement du fruit et de la graine du lierre, *Phytochemistry*, 10, 1261, 1971.

96. **Mukherjee, K. D.,** Elongation of (n-9) and (n-7) cis-monounsaturated and saturated fatty acids in seeds of *Sinapis alba*, *Lipids*, 21, 347, 1986.

97. **Shibahara, A., Yamamoto, K., Takeoka, M., Kinoshita, A., Kajimoto, G., Nakayama, T., and Noda, M.,** Novel pathways of oleic and cis-vaccenic acid biosynthesis by an enzymatic double-bond shifting reaction in higher plants, *FEBS Lett.*, 264, 228, 1990; see errata 268, 306, 1990.

98. **Appleby, R. S., Gurr, M. I., and Nichols, B. W.,** Studies on seed-oil triglycerides; factors controlling the biosynthesis of fatty acids and acyl lipids in subcellular organelles of maturing *Crambe abyssinica* seeds, *Eur. J. Biochem.*, 48, 209, 1974.

99. **Ohlrogge, J. B., Pollard, M. R., and Stumpf, P. K.,** Studies on biosynthesis of waxes by developing jojoba seed tissue, *Lipids*, 13, 203, 1978.

100. **Pollard, M. R., McKeon, T., Gupta, L. M., and Stumpf, P. K.,** Studies on biosynthesis of waxes by developing jojoba seed. II. The demonstration of wax biosynthesis by cell-free homogenates, *Lipids*, 14, 651, 1979.

101. **Pollard, M. R. and Stumpf, P. K.,** Long chain (C_{20} and C_{22}) fatty acid biosynthesis in developing seeds of *Tropaeolum majus*, *Plant Physiol.*, 66, 641, 1980.

102. **Stumpf, P. K. and Pollard, M. R.,** Pathways of fatty acid biosynthesis in higher plants with particular reference to developing rapeseed, in *High and Low Erucic Acid Rapeseed Oils*, Kramer, J. K. G., Sauer, F. D., and Pigden, W. J., Eds., Academic Press, Toronto, 1983, chap. 5.

103. **Stymne, S. and Stobart, A. K.,** Triacylglycerol biosynthesis, in *The Biochemistry of Plants*, Vol. 9, Stumpf, P.K., Ed., Academic Press, Orlando, FL, 1987, chap. 8.

104. **Griffiths, G., Hakman, I., Tillberg, E., Hellman, M., Stymne, S., and Stobart, A. K.,** The biosynthesis of triacylglycerols in oil-seeds with a perspective view on the role of plant growth regulators, in *Plant Lipids: Targets for Manipulation, Monograph 17*, Pinfield, N. J. and Stobart, A. K., Eds., British Plant Growth Regulator Group, Bristol, 1988, 11.

105. **Fehling, E. and Mukherjee, K. D.,** Acyl-CoA elongase from a higher plant (*Lunaria annua*): metabolic intermediates of very-long-chain acyl-CoA products and substrate specificity, *Biochim. Biophys. Acta,* 1082, 239, 1991.

106. **Agrawal, V. P. and Stumpf, P. K.,** Elongation systems involved in the biosynthesis of erucic acid from oleic acid in developing *Brassica juncea* seeds, *Lipids,* 20, 361, 1985.

107. **Murphy, D. J. and Mukherjee, K. D.,** Biosynthesis of very long chain monounsaturated fatty acids by subcellular fractions of developing seeds, *FEBS Lett.,* 230, 101, 1988.

108. **Murphy, D. J. and Mukherjee, K. D.,** Elongases synthesizing very long chain monounsaturated fatty acids in developing oilseeds and their solubilization, *Z. Naturforsch.,* 44c, 629, 1989.

109. **Bessoule, J.-J., Lessire, R., and Cassagne, C.,** Partial purification of the acyl-CoA elongase of *Allium porrum* leaves, *Arch. Biochem. Biophys.,* 268, 475, 1989.

110. **Lessire, R., Bessoule, J.-J., and Cassagne, C.,** Involvement of a β-ketoacyl-CoA intermediate in acyl-CoA elongation by an acyl-CoA elongase purified from leek epidermal cells, *Biochim. Biophys. Acta,* 1006, 35, 1989.

111. **Bessoule, J.-J., Lessire, R., and Cassagne, C.,** Theoretical analysis of the activity of membrane-bound enzymes using amphiphilic or hydrophobic substrates. Application to the acyl-CoA elongases from *Allium porrum* cells and to their purification, *Biochim. Biophys. Acta,* 983, 35, 1989.

112. **Harwood, J. L.,** Fatty acid metabolism, *Annu. Rev. Plant Physiol. Plant Mol. Biol.,* 39, 101, 1988.

113. **Pollard, M. R., Anderson, L., Fan, C., Hawkins, D. J., and Davies, H. M.,** A specific acyl-ACP thioesterase implicated in medium-chain fatty acid production in immature cotyledons of *Umbellularia californica, Arch. Biochem. Biophys.,* 284, 306, 1991.

114. **Ortiz de Montellano, P. R.,** Oxygen activation and transfer, in *Cytochrome P-450: Structure, Mechanism, and Biochemistry,* Ortiz de Montellano, P. R., Ed., Plenum Press, New York, 1986, 217.

115. **Ruettinger, R. T. and Fulco, A. J.,** Epoxidation of unsaturated fatty acids by a soluble cytochrome P-450-dependent system from *Bacillus megaterium, J. Biol. Chem.,* 256, 5728, 1981.

116. **Oliw, E. H.,** Biosynthesis of $18(R_D)$-hydroxyeicosatetraenoic acid from arachidonic acid by microsomes of monkey seminal vesicles; some properties of a novel fatty acid ω3-hydroxylase and ω3-epoxygenase, *J. Biol. Chem.,* 264, 17845, 1989.

117. **Kolattukudy, P. E.,** Cutin, suberin and waxes, in *The Biochemistry of Plants,* Vol. 4, Stumpf, P. K., Ed., Academic Press, London, 1980, chap. 18.

118. **Hutton, D. and Stumpf, P. K.,** Fat metabolism in higher plants LXII. The pathway of ricinoleic acid catabolism in the germinating castor bean (*Ricinus communis* L.) and pea (*Pisum sativum* L.), *Arch. Biochem. Biophys.,* 142, 48, 1971.

119. **Bafor, M., Jonsson, L., Stobart, A. K., and Stymne, S.,** Regulation of triacylglycerol biosynthesis in embryos and microsomal preparations from the developing seeds of *Cuphea lanceolata, Biochem. J.,* 272, 31, 1990.

120. **Oo, K.-C. and Huang, A. H. C.,** Lysophosphatidate acyltransferase activities in the microsomes from palm endosperm, maize scutellum, and rapeseed cotyledon of maturing seeds, *Plant Physiol.,* 91, 1288, 1989.

121. **Mukherjee, K. D. and Kiewitt, I.,** Changes in fatty acid composition of lipid classes in developing mustard seed, *Phytochemistry,* 23, 349, 1984.

122. **Cao, Y.-Z. and Huang, A. H. C.,** Acyl coenzyme A preference of diacylglycerol acyltransferase from the maturing seeds of Cuphea, maize, rapeseed, and canola, *Plant Physiol.,* 84, 762, 1987.

123. **Fehling, E. and Mukherjee, K. D.,** Biosynthesis of triacylglycerols containing very long chain mono-unsaturated fatty acids in seeds of *Lunaria annua, Phytochemistry,* 29, 1525, 1990.

124. **Cao, Y.-Z., Oo, K.-C., and Huang, A. H. C.,** Lysophosphatidate acyltransferase in the microsomes from maturing seeds of meadowfoam (*Limnanthes alba*), *Plant Physiol.,* 94, 1199, 1990.

125. **Battey, J. F. and Ohlrogge, J. B.**, A comparison of the metabolic fate of fatty acids of different chain lengths in developing oilseeds, *Plant Physiol.*, 90, 835, 1989.
126. **Guengerich, F. P.**, Cytochrome P-450: advances and prospects, *FASEB J.*, 6, 667, 1992.
127. **Coon, M. J., Ding, X., Pernecky, S. J., and Vaz, A. D. N.**, Cytochrome P-450: progress and predictions, *FASEB J.*, 6, 669, 1992.
128. **Poulos, T. J. and Ragg, R.**, Cytochrome P-450$_{cam}$: crystallography, oxygen activation, and electron transfer, *FASEB J.*, 6, 674, 1992.
129. **Hollenberg, P. F.**, Mechanisms of cytochrome P-450 and peroxidase-catalyzed xenobiotic metabolism, *FASEB J.*, 6, 686, 1992.
130. **Johnson, E. F., Kronbach, T., and Hsu, M.-H.**, Analysis of the catalytic specificity of cytochrome P-450 enzymes through site-directed mutagenesis, *FASEB J.*, 6, 700, 1992.
131. **Guengerich, F. P.**, Reactions and significance of cytochrome P-450 enzymes, *J. Biol. Chem.*, 266, 10019, 1991.
132. **Porter, T. D. and Coon, M. J.**, Cytochrome P-450: multiplicity of isoforms, substrates, and catalytic and regulatory mechanisms, *J. Biol. Chem.*, 266, 13469, 1991.
133. **Dawson, J. H.**, Probing structure-function relations in heme-containing oxygenases and peroxidases, *Science*, 240, 433, 1988.
134. **Ortiz de Montellano, P. R., Ed.**, *Cytochrome P-450: Structure, Mechanism, and Biochemistry*, Plenum Press, New York, 1986, 556.
135. **Marnett, L. J., Weller, P., and Battista, J. R.**, Comparison of the peroxidase activity of hemoproteins and cytochrome P-450, in *Cytochrome P-450: Structure, Mechanism, and Biochemistry*, Ortiz de Montellano, P. R., Ed., Plenum Press, New York, 1986, 29.
136. **Dunford, H. B., Araiso, T., Job, D., Ricard, J., Rutter, R., Hager, L. P., Wever, R., Kast, W. M., Boelens, R., Ellfolk, N., and Rönnberg, M.**, Peroxidases, in *The Biological Chemistry of Iron*, Vol. 89, Dunford, H. B., Dolphin, D., Raymond, K., and Sieker, L., Eds., D. Reidel, Boston, 1982, 337.
137. **Hewson, W. D. and Hager, L. P.**, Peroxidases, catalases, and chloroperoxidase, in *The Porphyrins*, Vol. 7, Dolphin, D., Ed., Academic Press, New York, 1979, 295.
138. **Froland, W. F., Andersson, K. K., Lee, S.-K., Liu, Y., and Lipscomb, J. D.**, Oxygenation by methane monooxygenase: oxygen activation and component interactions, in *Applications of Enzyme Biotechnology*, Kelly, J. W. and Baldwin, T. O., Eds., Plenum Press, New York, 1991, 39.
139. **Dalton, H.**, Structure and mechanism of the enzyme(s) involved in methane oxidation, in *Applications of Enzyme Biotechnology*, Kelly, J. W. and Baldwin, T. O., Eds., Plenum Press, New York, 1991, 55.
140. **Rosenzweig, A. C., Feng, X. D., and Lippard, S. J.**, Studies of methane monooxygenase and alkane oxidation model complexes, in *Applications of Enzyme Biotechnology*, Kelly, J. W. and Baldwin, T. O., Eds., Plenum Press, New York, 1991, 69.
141. **Fox, B. G. and Lipscomb, J. D.**, Methane monooxygenase: a novel biological catalyst for hydrocarbon oxidations, in *Biological Oxidation Systems*, Vol. 1, Reddy, C. C., Hamilton, G. A., and Madyastha, K. M., Eds., Academic Press, New York, 1990, 367.
142. **Peterson, J. A. and Prough, R. A.**, Cytochrome P-450 reductase and cytochrome b_5 in cytochrome P-450 catalysis, in *Cytochrome P-450: Structure, Mechanism, and Biochemistry*, Ortiz de Montellano, P. R., Ed., Plenum Press, New York, 1986, 89.
143. **Sligar, S. G. and Gunsalus, I. C.**, A thermodynamic model of regulation: modulation of the redox equilibria in camphor monooxygenase, *Proc. Natl. Acad. Sci. U.S.A.*, 73, 1078, 1976.
144. **Bangcharoenpaurpong, O., Rizos, A. K., Champion, P. M., Jollie, D., and Sligar, S. G.**, Resonance Raman detection of bound dioxygen in cytochrome P-450$_{cam}$, *J. Biol. Chem.*, 261, 8089, 1986.
145. **Sharrock, M., Münck, E., Debrunner, P. G., Marshall, V., Lipscomb, J. D., and Gunsalus, I. C.**, Mössbauer studies of cytochrome P-450$_{cam}$, *Biochemistry*, 12, 258, 1973.

146. **Ishimura, Y., Ullrich, V., and Peterson, J. A.,** Oxygenated cytochrome P-450 and its possible role in enzymatic hydroxylation, *Biochem. Biophys. Res. Commun.,* 42, 140, 1971.

147. **Raag, R. and Poulos, T. L.,** Crystal structure of the carbon monoxide-substrate-cytochrome P-450$_{cam}$ ternary complex, *Biochemistry,* 28, 7586, 1989.

148. **Omura, T. and Sato, R.,** The carbon monoxide-binding pigment of liver microsomes. I. Evidence for its hemoprotein nature, *J. Biol. Chem.,* 239, 2370, 1964.

149. **Bonfils, C., Balny, C., and Maurel, P.,** Direct evidence for electron transfer from ferrous cytochrome b_5 to the oxyferrous intermediate of liver cytochrome P-450 LM$_2$, *J. Biol. Chem.,* 256, 9457, 1981.

150. **Hyrcay, E. G., Gustafsson, J.-Å., Ingelman-Sundberg, M., and Ernster, L.,** Sodium periodate, sodium chlorite, organic hydroperoxides, and H$_2$O$_2$ as hydroxylating agents in steroid hydroxylation reactions by the partially purified cytochrome P-450, *Biochem. Biophys. Res. Commun.,* 66, 209, 1975.

151. **Hyrcay, E. G., Gustafsson, J.-Å., Ingelman-Sundberg, M., and Ernster, L.,** The involvement of cytochrome P-450 in hepatic microsomal steroid hydroxylation reactions supported by sodium periodate, sodium chlorite, and organic hydroperoxides, *Eur. J. Biochem.,* 61, 43, 1976.

152. **Nordblom, G. D., White, R. E., and Coon, M. J.,** Studies on hydroperoxide dependent substrate hydroxylation by purified liver microsomal cytochrome P-450, *Arch. Biochem. Biophys.,* 175, 524, 1976.

153. **White, R. E. and Coon, M. J.,** Oxygen activation by cytochrome P-450, *Annu. Rev. Biochem.,* 49, 315, 1980.

154. **White, R. E., Sligar, S. G., and Coon, M. J.,** Evidence for a homolytic mechanism of peroxide oxygen-oxygen bond cleavage during substrate hydroxylation by cytochrome P-450, *J. Biol. Chem.,* 255, 11108, 1980.

155. **Groves, J. T. and Watanabe, Y.,** Oxygen activation by metalloporphyrins related to peroxidase and cytochrome P-450. Direct observation of the O-O bond cleavage step, *J. Am. Chem. Soc.,* 108, 7834, 1986.

156. **McMurray, T. J. and Groves, J. T.,** Metalloporphyrin models for cytochrome P-450, in *Cytochrome P-450: Structure, Mechanism, and Biochemistry,* Ortiz de Montellano, P. R., Ed., Plenum Press, New York, 1986, 1.

157. **Rutter, R., Valentine, M., Hendrich, M. P., Hager, L. P., and Debrunner, P. G.,** Chemical nature of the porphyrin cation radical in horseradish peroxidase compound I, *Biochemistry,* 22, 4769, 1983.

158. **Roberts, J. E., Hoffman, B. M., Rutter, R., and Hager, L. M.,** Electron-nuclear double resonance of horseradish peroxidase compound I: detection of the porphyrin π-cation radical, *J. Biol. Chem.,* 256, 2118, 1981.

159. **Schultz, C. E., Devany, P. W., Winkler, H., Debrunner, P. G., Doan, N., Chiang, R., Rutter, R., and Hager, L. P.,** Horseradish peroxidase compound I: evidence for spin coupling between the heme iron and the "free" radical, *FEBS Lett.,* 103, 102, 1979.

160. **Fishel, L. A., Farnum, M. F., Mauro, J. M., Miller, M. A., Kraut, J., Liu, Y., Tan, C.-L., and Scholes, C. P.,** Compound I radical in site-directed mutants of cytochrome *c* peroxidase as probed by electron paramagnetic resonance and electron-nuclear double resonance, *Biochemistry,* 30, 1986, 1991.

161. **Mauro, J. M., Fishel, L. A., Hazzard, J. T., Meyer, T., Tollin, G., Cusanovich, M. A., and Kraut, J.,** Tryptophan 191 to phenylalanine, a proximal-side mutation in yeast cytochrome *c* peroxidase that strongly affects the kinetics of ferrocytochrome *c* oxidation, *Biochemistry,* 27, 6243, 1988.

162. **Hoffman, B. M., Roberts, J. E., Kang, C. H., and Margolish, E.,** Electron paramagnetic and electron nuclear double resonance of the hydrogen peroxide compound of cytochrome *c* peroxidase, *J. Biol. Chem.,* 256, 6556, 1981.

163. **Rutter, R., Hager, L. P., Dhonau, H., Hendrich, M. P., Valentine, M., and Debrunner, P. G.,** Chloroperoxidase compound I: electron paramagnetic resonance and Mössbauer studies, *Biochemistry,* 23, 6809, 1984.

164. **Dolphin, D., Forman, A., Borg, D. C., Fajer, J., and Felton, R. H.,** Compounds I of catalase and horse radish peroxidase: π-cation radicals, *Proc. Natl. Acad. Sci. U.S.A.,* 68, 614, 1971.

165. **Groves, J. T., McClusky, G. A., White, R. E., and Coon, M. J.,** Aliphatic hydroxylation by highly purified liver microsomal cytochrome P-450: evidence for a carbon radical intermediate, *Biochem. Biophys. Res. Commun.,* 81, 154, 1978.

166. **Groves, J. T. and Subramanian, D. V.,** Hydroxylation by cytochrome P-450 and metalloporphyrin models. Evidence for allylic rearrangement, *J. Am. Chem. Soc.,* 106, 2177, 1984.

167. **Bowry, V. W., Lustzk, J., and Ingold, K. U.,** Calibration of a new horology of fast radical "clocks". Ring-opening rates for ring- and a-alkyl-substituted cyclopropylcarbinyl radicals and for the bicyclo[2.1.0]pent-2-yl radical, *J. Am. Chem. Soc.,* 113, 5687, 1991.

168. **Ortiz de Montellano, P. R. and Stearns, R. A.,** Timing of the radical recombination step in cytochrome P-450 catalysis with ring-strained probes, *J. Am. Chem. Soc.,* 109, 3415, 1987.

169. **Griller, D. and Ingold, K. U.,** Free-radical clocks, *Acc. Chem. Res.,* 13, 317, 1980.

170. **Bowry, V. W. and Ingold, K. U.,** A radical clock investigation of microsomal cytochrome P-450 hydroxylation of hydrocarbons. Rate of oxygen rebound, *J. Am. Chem. Soc.,* 113, 5699, 1991.

171. **Groves, J. T. and McCluskey, G. A.,** Aliphatic hydroxylation *via* oxygen rebound: oxygen transfer catalyzed by iron, *J. Am. Chem. Soc.,* 98, 859, 1976.

172. **Dalton, H.,** Oxidation of hydrocarbons by methane monooxygenase from a variety of microbes, *Adv. Appl. Microbiol.,* 26, 71, 1980.

173. **Higgins, I. J., Best, D. J., and Hammond, R. C.,** New findings in methane-oxidizing bacteria highlight their importance in the biosphere and their commercial importance, *Nature,* 286, 561, 1980.

174. **McCormick, J. M., Reem, R. C., and Solomon, E. I.,** Chemical and spectroscopic studies of the mixed-valent derivatives of the non-heme iron protein hemerythrin, *J. Am. Chem. Soc.,* 113, 9066, 1991.

175. **Reem, R. C. and Solomon, E. I.,** Spectroscopic studies of the binuclear ferrous active site of deoxyhemerythrin: coordination number and probable bridging ligands for the native and ligand bound forms, *J. Am. Chem. Soc.,* 109, 1216, 1987.

176. **Pearce, L. L., Kurtz, D. M., Jr., Xia, Y.-M., and Debrunner, P. G.,** Reduction of the binuclear iron site in octameric methemerythrins. Characterizations of intermediates and a unifying reaction scheme, *J. Am. Chem. Soc.,* 109, 7286, 1987.

177. **Stenkamp, R. E., Sieker, L. C., and Jensen, L. H.,** Binuclear iron complexes of methemerythrin and azidomethemerythrin at 2.0-Å resolution, *J. Am. Chem. Soc.,* 106, 618, 1984.

178. **Okamura, M. Y., Klotz, I. M., Johnson, C. E., Winter, M. R. C., and Williams, R. J. P.,** The state of iron in hemerythrin. A Mössbauer study, *Biochemistry,* 8, 1951, 1969.

179. **Sage, J. T., Xia, Y.-M., Debrunner, P. G., Keough, D. T., de Jersey, J., and Zerner, B.,** Mössbauer analysis of the binuclear iron site in purple acid phosphatase from pig allantoic fluid, *J. Am. Chem. Soc.,* 111, 7239, 1989.

180. **Averill, B. A., Davis, J. C., Burman, S., Zirino, T., Sanders-Loehr, J., Loehr, T. M., Sage, J. T., and Debrunner, P. G.,** Spectroscopic and magnetic studies of the purple acid phosphatase from bovine spleen, *J. Am. Chem. Soc.,* 109, 3760, 1987.

181. **Davis, J. C. and Averill, B. A.,** Evidence for a spin-coupled binuclear iron unit at the active site of the purple acid phosphatase from beef spleen, *Proc. Natl. Acad. Sci. U.S.A.,* 79, 4623, 1982.

182. **LeGall, J., Prickril, B. C., Moura, I., Xavier, A. V., Moura, J. J. G., and Huynh, B.-H.,** Isolation and characterization of rubrerythrin, a non-heme iron protein from *Desulfovibrio vulgaris* that contains rubredoxin centers and a hemerythrin-like binuclear iron cluster, *Biochemistry*, 27, 1636, 1988.

183. **Nordlund, P., Sjöberg, B.-M., and Eklund, H.,** Three-dimensional structure of the free radical protein of ribonucleotide reductase, *Nature*, 345, 593, 1990.

184. **Lynch, J. B., Juarez-Garcia, C., Münck, E., and Que, L., Jr.,** Mössbauer and EPR studies of the binuclear iron center in ribonucleotide reductase from *Escherichia coli*, *J. Biol. Chem.*, 264, 8091, 1989.

185. **Fox, B. G., Froland, W. A., Dege, J. E., and Lipscomb, J. D.,** Methane monooxygenase from *Methylosinus trichosporium* OB3b: purification and properties of a three-component system with high specific activity from a Type II methanotroph, *J. Biol. Chem.*, 264, 10023, 1989.

186. **Fox, B. G. and Lipscomb, J. D.,** Purification of a high specific activity methane monooxygenase hydroxylase component from a Type II methanotroph, *Biochem. Biophys. Res. Commun.*, 154, 165, 1988.

187. **Woodland, M. P. and Dalton, H.,** Purification and characterization of component A of the methane monooxygenase from *Methylococcus capsulatus* (Bath), *J. Biol. Chem.*, 259, 53, 1984.

188. **Fox, B. G., Surerus, K. K., Münck, E., and Lipscomb, J. D.,** Evidence for a μ-oxo-bridged binuclear iron cluster in the hydroxylase component of methane monooxygenase: Mössbauer and EPR studies, *J. Biol. Chem.*, 263, 10553, 1988.

189. **Fox, B. G., Liu, Y., Dege, J. E., and Lipscomb, J. D.,** Complex formation between the protein components of methane monooxygenase from *Methylosinus trichosporium* OB3b: identification of the sites of component interactions, *J. Biol. Chem.*, 266, 540, 1991.

190. **Hendrich, M. P., Fox, B. G., Andersson, K. K., Debrunner, P. G., and Lipscomb, J. D.,** Ligation of the diiron site of the hydroxylase component of methane monooxygenase: an electron nuclear double resonance study, *J. Biol. Chem.*, 267, 261, 1992.

191. **Prior, S. D. and Dalton, H.,** Acetylene as a suicide substrate and active site probe for methane monooxygenase from *Methylococcus capsulatus* (Bath), *FEMS Micro. Lett.*, 29, 105, 1985.

192. **Cardy, D. L. N., Laidler, V., Salmond, G. P. C., and Murrel, J. C.,** Molecular analysis of the methane monooxygenase (MMO) gene cluster of *Methylosinus trichosporium* OB3b, *Mol. Microbiol.*, 5, 335, 1991.

193. **Sjöberg, B.-M., Loehr, T. M., and Sanders-Loehr, J.,** Raman spectral evidence for a μ-oxo bridge in the binuclear iron center of ribonucleotide reductase, *Biochemistry*, 21, 96, 1982.

194. **Shiemke, A. K., Loehr, T. M., and Sanders-Loehr, J.,** Resonance Raman study of oxyhemerythrin and hydroxymethemerythrin. Evidence for hydrogen bonding of ligands to the Fe-O-Fe center, *J. Am. Chem. Soc.*, 108, 2437, 1986.

195. **DeWitt, J. G., Bentsen, J. G., Rosenzweig, A. C., Hedman, B., Green, J., Pilkington, S., Papaefthymiou, G. C., Dalton, H., Hodgson, K. O., and Lippard, S. J.,** X-ray absorption, Mössbauer, and EPR studies of the dinuclear iron center in the hydroxylase component of methane monooxygenase, *J. Am. Chem. Soc.*, 113, 9219, 1992.

196. **Fox, B. G., Borneman, J. G., Wackett, L. P., and Lipscomb, J. D.,** Haloalkene oxidation by the soluble methane monooxygenase from *Methylosinus trichosporium* OB3b: mechanistic and environmental implications, *Biochemistry*, 29, 6419, 1990.

197. **Bollinger, J. M., Jr., Edmondson, D. E., Huynh, B. H., Filley, J., Norton, J. R., and Stubbe, J.,** Mechanism of assembly of the tyrosyl radical dinuclear iron cluster cofactor of ribonucleotide reductase, *Science*, 253, 292, 1991.

198. **Bollinger, J. M., Jr., Stubbe, J., Huynh, B. H., and Edmondson, D. E.,** Novel differic radical intermediate responsible for tyrosyl radical formation in assembly of the cofactor of ribonucleotide reductase, *J. Am. Chem. Soc.*, 113, 6289, 1991.

199. **Elgren, T. E., Lynch, J. B., Juarez-Garcia, C., Münck, E., Sjöberg, B.-M., and Que, L., Jr.,** Electron transfer associated with oxygen activation in the B2 protein of ribonucleotide reductase from *Escherichia coli, J. Biol. Chem.,* 266, 19265, 1991.

200. **Sahlin, M., Sjöberg, B.-M., Backes, G., Loher, T., and Sanders-Loehr, J.,** Activation of the iron-containing B2 protein of ribonucleotide reductase by hydrogen peroxide, *Biochem. Biophys. Res. Commun.,* 167, 813, 1990.

201. **Fox, B. G., Shanklin, J., Somerville, C., and Münck, E.,** Stearoyl-acyl carrier protein Δ^9-desaturase from *Ricinus communis* is a diiron-oxo protein, *Proc. Natl. Acad. Sci. U.S.A.,* in press.

202. **Stirling, D. I. and Dalton, H.,** The fortuitous oxidation and cometabolism of various carbon compounds by whole-cell suspensions of *Methylococcus capsulatus* (Bath), *FEMS Micro. Lett.,* 5, 315, 1979.

203. **Jaworski, J. G. and Stumpf, P. K.,** Fat metabolism in higher plants. Properties of a soluble stearyl-acyl carrier protein desaturase from maturing *Carthamus tinctorius, Arch. Biochem. Biophys.,* 162, 158, 1974.

204. **Andersson, K. K., Froland, W. A., Lee, S.-K., and Lipscomb, J. D.,** Dioxygen independent oxygenation of hydrocarbons by methane monooxygenase hydroxylase component, *New. J. Chem.,* 15, 411, 1991.

205. **Kurtz, D. M., Jr.,** Oxo- and hydroxobridged diiron complexes: a chemical perspective on a biological unit, *Chem. Rev.,* 90, 585, 1990.

206. **Que, L., Jr. and True, A. E.,** Dinuclear iron- and manganese-oxo sites in biology, in *Progress in Inorganic Chemistry: Bioinorganic Chemistry,* Vol. 38, Lippard, S. J., Ed., John Wiley & Sons, New York, 1990, 97.

207. **Que, L., Jr. and Scarrow, R. C.,** Active sites of binuclear iron-oxo proteins, in *ACS Symposium Series No. 372 Metal Clusters in Proteins,* Que, L., Jr., Ed., American Chemical Society, Washington, D.C., 1988, 152.

208. **Hendrich, M. P., Münck, E., Fox, B. G., and Lipscomb, J. D.,** Integer-spin EPR studies of the fully reduced methane monooxygenase hydroxylase component, *J. Am. Chem. Soc.,* 112, 5861, 1990.

209. **Woodland, M. P., Patil, D. S., Cammack, R., and Dalton, H.,** ESR studies of protein A of the soluble methane monooxygenase from *Methylococcus capsulatus* (Bath), *Biochem. Biophys. Acta,* 873, 237, 1986.

210. **Hamilton, G. A.,** Mechanisms of biological oxidation reactions involving oxygen and reduced oxygen derivatives, in *Biological Oxidation Systems,* Vol. 1, Reddy, C. C., Hamilton, G. A., and Madyastha, K. M., Eds., Academic Press, New York, 1990, 3.

211. **Hamilton, G. A.,** Chemical models and mechanisms for oxygenases, in *Molecular Mechanisms of Oxygen Activation,* Hayaishi, O., Ed., Academic Press, New York, 1974, 405.

212. **Sheu, C., Sobkowaik, A., Zhang, L., Ozbalik, N., Barton, D. H. R., and Sawyer, D. T.,** Iron-hydroperoxide-induced phenylselenization of hydrocarbons (Fenton chemistry), *J. Am. Chem. Soc.,* 111, 8030, 1989.

213. **Walling, C.,** Fenton's reagent revisited, *Acc. Chem. Res.,* 8, 125, 1975.

214. **Ruzicka, F., Huang, D.-S., Donnelly, M. I., and Frey, P. A.,** Methane monooxygenase catalyzed oxidation of 1,1-dimethylcyclopropane. Evidence for radical and carbocationic intermediates, *Biochemistry,* 29, 1696, 1990.

215. **Green, J. and Dalton, H.,** Substrate specificity of soluble methane monooxygenase: mechanistic implications, *J. Biol. Chem.,* 246, 17698, 1989.

216. **Jezequel, S. G. and Higgins, I. J.,** Mechanistic aspects of biotransformations by the monooxygenase system of *Methylosinus trichosporium* OB3b, *J. Chem. Tech. Biotech.,* 33B, 139, 1983.

217. **Jerina, D. M. and Daly, J. W.,** Arene oxides: a new aspect of drug metabolism, *Science,* 185, 573, 1974.

218. **Priestly, N. D., Floss, H. D., Froland, W. A., Lipscomb, J. D., Williams, P. G., and Morimoto, H.,** Cryptic stereochemistry of the methane monooxygenase reaction, *J. Am. Chem. Soc.,* 114, 7561, 1992.

219. **Rataj, M. J., Knauth, J. E., and Donnelly, M. I.,** Oxidation of deuterated compounds by high specific activity methane monooxygenase from *Methylosinus trichosporium* OB3b, *J. Biol. Chem.,* 266, 18684, 1991.

220. **Leak, D. J. and Dalton, H.,** Studies on the regioselectivity and stereoselectivity of the soluble methane monooxygenase from *Methylococcus capsulatus* (Bath), *Biocatalysis,* 1, 23, 1987.

221. **Norman, R. E., Holz, R. C., Menage, S., Que, L. J., Zhang, Z. H., and O'Connor, C. J.,** Structures and properties of dibridged (μ-oxo)diiron(III) complexes. Effects of Fe-O-Fe angle, *Inorg. Chem.,* 29, 4629, 1990.

222. **Norman, R. E., Yan, S., Que, L., Jr., Backes, G., Ling, J., Sanders-Loehr, J., Zhang, J. H., and O'Connor, C. J.,** (μ-Oxo)(μ-carboxylato)diiron(III) complexes with distinct iron sites. Consequences of the inequivalence and its relevance to dinuclear iron-oxo proteins, *J. Am. Chem. Soc.,* 112, 1554, 1990.

223. **Leising, R. A., Norman, R. E., and Que, L., Jr.,** Alkane functionalization by nonporphyrin iron complexes: mechanistic insights, *Inorg. Chem.,* 29, 2553, 1990.

224. **Leising, R. L., Brennan, B. A., Que, L., Jr., Fox, B. G., and Münck, E.,** Models for nonheme iron oxygenases: a high valent iron-oxo intermediate, *J. Am. Chem. Soc.,* 113, 3988, 1991.

225. **Oosterhuis, W. T. and Lang, G.,** Magnetic properties of the t^4_{2g} configuration in low symmetry crystal fields, *J. Chem. Phys.,* 58, 4757, 1973.

226. **Bangcharoenpaurpong, O., Champion, P. M., Hall, K. S., and Hager, L. P.,** Resonance Raman studies of isotopically labelled chloroperoxidase, *Biochemistry,* 25, 2374, 1986.

227. **Sligar, S. G., Filipovic, D., and Stayton, P. S.,** Mutagenesis of cytochromes P-450$_{cam}$ and b_5, *Methods Enzymol.,* 206, 31, 1991.

228. **Martinis, S. A., Atkins, W. A., Stayton, P. S., and Sligar, S. G.,** A conserved residue of cytochrome P-450 is involved in heme-oxygen stability and activation, *J. Am. Chem. Soc.,* 111, 9252, 1989.

229. **Sligar, S. G., Egeberg, K. D., Sage, J. T., Morikis, D., and Champion, P. M.,** Alteration of heme axial ligands by site-directed mutagenesis: a cytochrome becomes a catalytic demethylase, *J. Am. Chem. Soc.,* 109, 7896, 1987.

230. **Atkins, W. A. and Sligar, S. G.,** The roles of active site hydrogen bonding in cytochrome P-450$_{cam}$ as revealed by site-directed mutagenesis, *J. Biol. Chem.,* 263, 18842, 1988.

231. **Stayton, P. S. and Sligar, S. G.,** The cytochrome P-450$_{cam}$ binding surface as defined by site-directed mutagenesis and electrostatic modeling, *Biochemistry,* 29, 7381, 1990.

232. **Morris, L. J.,** Mechanisms and stereochemistry in fatty acid metabolism, *Biochem. J.,* 118, 681, 1970.

233. **Morris, L. J., Harris, R. V., Kelly, W., and James, A. T.,** The stereochemistry of desaturations of long-chain fatty acids in *Chlorella vulgaris, Biochem. J.,* 109, 673, 1968.

234. **Schroepfer, G. J., Jr. and Bloch, K.,** The stereospecific conversion of stearic acid to oleic acid, *J. Biol. Chem.,* 240, 54, 1965.

235. **Ormö, M., deMaré, S., Regnström, K., Aberg, A., Sahlin, M., Ling, J., Loehr, T. M., and Sanders-Loehr, J.,** Engineering of the iron site in ribonucleotide reductase to a self-hydroxylating monooxygenase, *J. Biol. Chem.,* 267, 8711, 1992.

236. **Bloch, K.,** Enzymatic synthesis of monounsaturated fatty acids, *Acc. Chem. Res.,* 2, 193, 1969.

237. **Holloway, P. W.,** Fatty acid desaturation, in *The Enzymes,* Vol. 16, 3rd ed., Boyer, P. D., Ed., Academic Press, New York, 1983, 63.

238. **Gurr, M. I.,** The biosynthesis of unsaturated fatty acids, in *MTP International Review of Science,* Vol. 4, Goodwin, T. W., Ed., Butterworths, London, 1974, 181.

239. **Sanders-Loehr, J.,** Involvement of oxo-bridged binuclear iron centers in oxygen transport, oxygen reduction and oxygenation, *Prog. Clin. Biol. Res.,* 274, 193, 1988.

240. **Liu, K. E. and Lippard, S. J.,** Redox properties of the hydroxylase component of methane monooxygenase from *Methylococcus capsulatus* (Bath), *J. Biol. Chem.,* 266, 12836, 1991.

241. **Froland, W. A., Andersson, K. K., Lee, S.-K., Liu, Y., and Lipscomb, J. D.,** Methane
      monooxygenase component B and reductase alter the regioselectivity of the hydroxylase
      component catalyzed reactions: a novel role for protein-protein interactions in oxygenase
      mechanisms, *J. Biol. Chem.,* 267, 17588, 1992.
242. **Schneider, G., Lindqvist, Y., Shanklin, J., and Somerville, C.,** Preliminary crystallo-
      graphic data for stearoyl-acyl carrier protein desaturase from castor seed, *J. Mol. Biol.,*
      225, 561, 1992.

Chapter 4

WAXES, CUTIN, AND SUBERIN

Penny M. von Wettstein-Knowles

TABLE OF CONTENTS

0-8493-4907-9/93/$0.00+$.50
© 1993 by CRC Press Inc.

I. SCOPE

Following a very brief introduction to the location, structure, composition, and function of the three groups of plant lipids, this chapter will delve into the mechanisms whereby they are synthesized. Although they are generally treated as three quite distinct groups of lipids, they will be seen to be very closely related. That is, in each case, they are synthesized by a set of membrane-localized enzymes each belonging to a small family, with different members of each family participating in various combinations. The result is three groups of overlapping lipid classes present in different proportions. Moreover, other members of the pertinent gene families encode isoenzymes which function in synthesis of the membrane and seed storage lipids. The synthetic reactions are divided into two groups, those producing the carbon skeletons and those modifying them.

II. NATURE AND FUNCTION OF PLANT WAXES, CUTIN, AND SUBERIN

Plant waxes consisting of very long-chain, relatively nonpolar lipid molecules are associated primarily with the cuticle which extends in a continuous sheet exterior to the walls of the epidermal cells of aerial tissues. Two subgroups of these lipids soluble in organic solvents are recognized. (1) The epicuticular waxes forming the outermost layer of the cuticle exist as an amorphous lipid film on which may be found a wide variety of structured bodies. (2) The intracuticular waxes embed cutin, a rather insoluble lipid polymer which is the main structural component of the cuticular membrane. The relatively thin outer layer of the cuticular membrane, or primary cuticle, generally appears amorphous, but sometimes has distinct lamellae. The much thicker inner portion of the cuticular membrane, or secondary cuticle, also containing wax and cutin often displays a reticulate appearance. It is deposited between the primary cuticle and wall after cell expansion is finished. In underground tissues (roots, storage organs), stems undergoing secondary growth, and wound healing sites, waxes are associated with the suberin matrix, a polymer related to cutin which has in addition to an aliphatic domain also an aromatic one. Suberin is located between the plasma lemma and walls of the outermost one or two cell layers. Microscopic examination of the suberized regions reveals a lamellar structure interpreted to represent alternating layers of wax and polymer. While numerous observations support this contention,[1] it has not been definitively demonstrated.[2] Waxes are also found at internal sites having specialized functions where they can be recognized by their characteristic lamellate structure.[3] An exception to the generalization that internal waxes are associated with suberin occurs in the seeds of some plants which store their major energy reserves as waxes rather than as triacylglycerols.[4-6]

An extraordinary diversity of aliphatic lipid classes has been identified in epicuticular waxes, including numerous types of hydrocarbons, ketones, β-diketones, esters, estolides (polyesters), alcohols, aldehydes, and free fatty acids.[7] Generally, a series of homologs encompassing ten carbons occurs dominated by either the even or odd members. While chains 20–35 carbons long are most frequently encountered, fatty acids and hydrocarbons with fewer than 20 and esters with more than 60 are known. The waxes associated with cutin and suberin polymers, by comparison, are much less variable, being limited to *n*-alkanes, fatty acids, primary alcohols, and esters. Usually, the chain length distributions are broader, have a greater preponderance of shorter members, and the relative prominence of either even or odd chain lengths is less marked than in the epicuticular waxes. Only esters and hydrocarbons have been reported in wax-storing seeds. Other lipids such as terpenoids and sterols which occur in epicuticular waxes are the subjects of Chapters 11 and 12.

The building blocks of cutin and the aliphatic domain of suberin consist of fatty acids with one or more additional substitutions either at the opposite that is ω-end or internally circa midchain.[7] Cutin monomers, which are derivatives of C_{16} and $C_{18:1}$, normally have midchain substitutions (predominantly hydroxy and epoxy). By comparison, those of suberin, which generally range from C_{16}–C_{24} including $C_{18:1}$, but may be as long as C_{30}, are more likely to have ω-substitutions (hydroxy and oic). Since terminal groups are used in construction of linear polymers, while midchain groups participate in cross-linking, the resulting cutin polymer is the more rigid one. Monomers of suberin's aromatic domain include the phenolic aldehydes *p*-hydroxybenzaldehyde and vanillin with small amounts of syringaldehyde. The latter is an important constituent of the lignin polymer. A structure for the aromatic domain of suberin similar to that of lignin has been envisaged[8] which is covalently attached on one side to the cell wall and on the other to its aliphatic domain. The tentative nature of the model has been emphasized.[3,7,9]

The primary function of the polymer-associated waxes is to prohibit loss of water and/or other molecules by diffusion through the cutin and suberin matrices.[10,11] The plant or a tissue thereof is not only protected against desiccation, but can enclose specified regions to shield against the internal movement of unwanted solutes or to prevent decay.[3,12] In addition to providing a structural matrix in and on which the waxes are located, the polymers have been implicated as a line of defense against pathogens which invade by direct penetration of the cuticle. For fungi with an active extracellular cutinase, that is an esterase belonging to the class of serine hydrolases, penetration of the cutin polymer presents no difficulties.[13] That this enzyme is required in some cases has been well-documented, for example, by inhibiting penetration using specific antibodies and restoring penetration to cutinase defective mutants by adding cutinase. Thus, a few molecules of the cutinase are released in a fluid of unknown composition by the fungal spore when it lands on the plant. The enzyme attacks the polymer, releasing by hydrolysis a few monomers which have been shown to induce transcription of the enzyme in the fungus, thereby ensuring an adequate supply of cutinase for complete penetration. The mechanism whereby the cutin monomers on the plant surface trigger transcription in fungal nuclei is at present an enigma, but with the progress being made in dissecting the promoter region of the cutinase structural gene,[14] an answer should be forthcoming. Another interesting question is why the epicuticular wax structures apparently dissolve when the fluid is released by the conidium upon landing, since most of the wax lipids, as exemplified by barley leaves, lack an ester bond, the substrate for the cutinase.[15,16] To what extent a battery of cutinase enzymes with different properties can be used to clarify the detailed structure of the polymers remains to be determined.

Because of their location on the outer surface of the cuticle in direct contact with the environment, the epicuticular waxes provide a diversity of other protective functions for the plant, many of which are analogous to those assigned to surface waxes of insects.[17] Wax bodies exist in a fascinating array of forms, and in a limited number of cases a direct correlation between structure and chemical composition has been established. Each aerial organ, tissue, or even cell type has its own peculiar morphology plus density of wax bodies which helps determine how effective the protection is. For example, the wax coat is a major factor in determining whether or not moisture is trapped[18,19] and/or retained on a cuticle surface as illustrated in Figure 1. Water on the surface for extended periods of time provides a suitable atmosphere for germination of fungal spores[20-22] and leads to leaching of nutrients.[23] Some wax coats result in an airspace above the epidermal cells which has been implicated as a factor contributing to frost hardiness, since nucleating ice crystals are not in direct contact with the cuticular membrane.[24-26] The nature of the wax layer also determines what type of insects as adults or larvae can adhere to or move on the surface[27-29] and can even assist a plant in trapping insects.[30] Additional potential interactions between plant epicuticular waxes and insects have been described.[31,32] Finally, light reflection and refraction from the cuticle surface is affected by the structure of the wax bodies. While plant temperature may be influenced,[33] and perhaps also photosynthetic efficiency,[34] the phenotypic modification resulting from the changed wax morphology permitted the isolation of numerous mutants with altered wax composition[35] (Figure 1). The mutations all affect the specific lipids making major contributions to the structure of the wax coats. Studies of a number of them have contributed to unraveling various facets of the wax biosynthetic pathways. Moreover, a few have been designated regulatory genes.[35-37] In most cases, no indication as to the nature of the mutated gene could be discerned.

III. SYNTHESIS OF THE CARBON SKELETONS

A. THE CONDENSATION ELONGATION MECHANISM, ELONGASES, AND POLYKETIDE SYNTHASES

Elongases are enzyme complexes which repetitively condense short activated carbon chains to an activated primer and prepare the growing chain for the next addition. Included in this general definition is the soluble fatty acid

FIGURE 1. Epicuticular waxes function as raincoats. Circa half an hour after raining, water drops are gone from the leaf blades of the wild type Bonus barley (*Hordeum vulgare*), although they still hang from the plastic divider (top left) and adhere to the leaf surfaces on *cer-j* mutants (top right). Transmission electron micrographs of shadowed carbon replicas of the cuticle surfaces reveal that this difference is due to the presence of small lobed plates, diagnostic for predominating amounts of primary alcohols in the wax, on the wild type (bottom left) and their absence on *cer-j* mutants (bottom right) which have few, small smooth plates scattered among thin plates appressed to the surface. Bar = 1 μm.

synthetase (FAS) complex, which joins eight activated C_2 units together to give 16-carbon chains. An additional elongation step carried out by another soluble complex known as C_{16} elongase gives 18-carbon acyl chains, which are in turn desaturated by the soluble stearoyl desaturase giving $C_{18:1}$. This plastid-localized pathway supplying the 16- and 18-carbon acyl chains characterizing plant membranes is the subject of Chapter 1. The same or an analogous condensation elongation pathway presumably provides the carbon chains for the C_{16} and $C_{18:1}$ families of cutin monomers and the substrates for the membrane-localized elongases whose coordinated action results in the longer carbon skeletons characterizing the wax lipids and many of the suberin monomers. It should be noted that C_{18} rather than $C_{18:1}$ chains generally serve as the primer for these

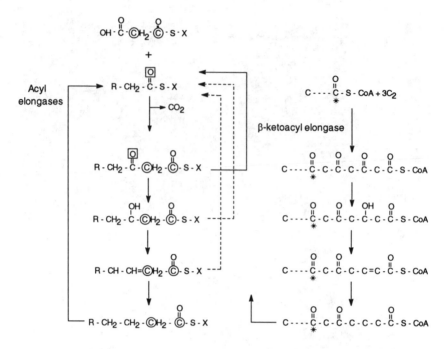

FIGURE 2. The basic condensation elongation mechanism synthesizes the carbon skeletons of wax acyl lipids and cutin plus suberin aliphatic monomers. In syntheses carried out by acyl elongases including FAS, the 3-oxo group introduced by the condensation reaction is removed in three steps by the reactions shown before the next addition takes place (far left). Polyketide synthesis is characterized by the omission of one or more of these three reactions in specified cycles of elongation (center), with the solid arrow illustrating the β-ketoacyl elongase functioning in the formation of wax acyl lipids. Three successive condensations carried out by the β-ketoacyl elongase result in a tetraoxo acyl chain from which the 3-oxo group is then removed in the same manner as in elongation carried out by acyl elongases (right). The primer and donor units used by the microsomal acyl and β-ketoacyl elongases are activated (X) by coenzyme A (CoA), whereas those used by plastid-localized complexes and most polyketide synthases are activated by an acyl carrier protein (ACP), with the exception that in the first condensation carried out by FAS, a CoA primer serves.

elongases. Whether one considers the wax lipids or the suberin monomers, the primary elongated products of the elongases in the form of free fatty acids are usually minor components, since most of them serve as substrates for the associated enzymes discussed in Section IV. The total length attained during elongation, however, can be deduced from the chain lengths of the members of the various lipid classes.

As illustrated in Figure 2 (left), each condensation carried out by an elongase introduces a β-keto group into the growing chain. This keto group is normally removed by a series of three reactions: a β-keto reduction, a β-hydroxy dehydration, and an enoyl reduction. Thus, an elongase requires a minimum of four activities to carry out each elongation step, if one excludes from consideration the enzyme providing the donor unit. Variations of this basic condensation elongation biosynthetic mechanism are well-known, however, which give rise to compounds classified as polyketides.[38] Their modified carbon chains synthesized by polyketide synthase enzyme complexes can be recognized by the presence of keto groups, hydroxy groups, or double bonds that were not removed before the next condensation took place (Figure 2, center). For example, plastid-localized 6-methylsalicylic acid syntase (6-MSA) joins four activated C_2 units, leaving keto groups after the first and third rounds of elongation and a double bond after the second. That polyketide synthases, as well as elongases, participate in construction of the wax and polymer acyl chains increases the number of possible carbon skeleton structures. Use of different primer and donor units expands the repetoire still further as illustrated below.

1. Straight, Branched, and Unsaturated Chains

Kolattukudy[39-42] obtained the original data supporting the elongation thesis by analyzing the alkanes which are major components of the *Brassica*, pea (*Pisum sativum*), spinach (*Spinacia oleracea*), *Senecio*, and tobacco (*Nicotiana tobacum*) waxes. Single- and double-labeled C_{12}–C_{18} fatty acids presented to whole leaf and/or tissue slices were recovered intact in the expected C_{29}, C_{31}, and/or C_{33} alkanes. Tissue slices of leek (*Alliun porrum*) leaves were later shown to incorporate lignoceric acid into C_{26}–C_{32} fatty acids and the corresponding C_{25}–C_{31} alkanes.[43] That elongation was by C_2 units was indicated by feeding 1- and 2-labeled acetate to *Brassica* leaves, followed by isolation and degradation of nonacosan-15-one. The data revealed that the methyl group of acetate preferentially became the oxygen-bearing carbon.[44]

Similar radio tracer experiments demonstrated that use of different primers and/or donor units by the elongases gave rise to other carbon skeletons. Primer units derived from valine and isoleucine led to the synthesis of the most frequently encountered branched plant wax lipids. That is, the former amino acid was incorporated by tobacco leaves into 2-methyl C_{15}–C_{25} fatty acids and 2-methyl C_{28}–C_{32} alkanes, whereas the latter gave rise to 3-methyl C_{16}–C_{26} fatty acids and 3-methyl C_{29}–C_{31} alkanes.[40] More recently, detailed investigations of the synthesis of branched hydrocarbons in insect waxes pinpointed activated methylmalonic acid as a primary source of methyl branches. The latter compound functioned both as an initial primer unit giving 3-methyl branches and/or as a donor unit yielding other internal methyl branches. While the same role for methylmalonic acid is well-documented for given vertebrate tissues[45] and in prokaryotes such as *Mycobacterium tuberculosis*,[46] its potential contribution to the synthesis of 3-methyl and internal methyl branches in plant wax lipids has not been investigated. These experiments established the origin of the major homologs in the lipid classes and inferred that the minor ones arose from a C_3 primer unit, presumably activated propionic acid. The prominence of the minor homologs would depend on the ratio of C_2 to C_3 primer units used.

Some of the double bonds in the wax and monomer lipids arise from use of an unsaturated primer by a membrane elongase. For example, adding C_2

units to $C_{18:1}$ chains results in $C_{20:1}$, $C_{22:1}$, and $C_{24:1}$ monounsaturated chains stored as the acyl moieties of seed ester and triacylglycerols and would explain the origin of the up to C_{37} alkenes in waxes.[47]

2. β-Diketo Chains

To ascertain the origin of the β-diketone carbon skeletons, acetate was fed to barley spikes and hentriacontan-14,16-dione isolated. The C_{14} and C_{16} fatty acids derived from the two ends of the molecule by hydrolysis were subjected to α-oxidation and the specific activities of the individual carbons determined by radio-GC. The results implied elongation proceeding from C_{31} to C_1.[48] When exposure to arsenite preceeded the acetate, label was found solely in the C_{1-14} end, indicating that only endogenous precursors of 18 or more carbons could be elongated[49] in accord with the known inhibitory effect of arsenite on C_{16} elongase. Stearic acid did not function as a β-diketone precursor.[49] Feeding pentadecanoic acid resulted in a novel C_{30} β-diketone with the label in the C_{16-30} end.[50] These results confirmed the proposal[51] that the two oxygens were built into the carbon chain during elongation (Figure 2, right). In other words, β-diketones can be classified as polyketides. The steps at which the introduction of the oxygen occurs is quite variable considering the known range of β-diketones. For example, synthesis of hentriacontan-8,10-dione in *Buxus sempervirens*[52] would require retention of the 3-oxo group in either a C_{10} or C_{24} acyl chain depending on the direction of elongation vs. a C_{18} acyl chain in barley. The condensation elongation system responsible for synthesis of the β-diketone carbon skeletons was named a β-ketoacyl elongase after the 3-oxoacyl primers it uses to distinguish it from the elongases using acyl primers in constructing the other wax and polymer acyl chains (Figure 2).

B. DISSECTING THE ELONGATION CONDENSATION PATHWAYS WITH MUTANTS AND INHIBITORS

The term elongation system as used herein encompasses not only FAS, elongase, and polyketide synthase complexes, but also complementary factors such as substrate availability and compartmentalization, while excluding the associated enzyme systems receiving the acyl chains. Mutations and inhibitors (photoperiodic and chemical) are unlikely to alter the composition of the epicuticular waxes and acyl chains of seed esters and triacylglycerols plus cutin and suberin monomers in a manner allowing easy recognition of the primary-induced change. Nevertheless, analyses of differential effects on waxes and seeds often in conjunction with radiotracer experiments have uncovered a surprising number of parallel and sequential elongation systems.

1. Sequential Systems

One of the earliest indications for the existence of sequential systems came from pea experiments in which the photoperiod was varied.[53] The results together with those from a later study in barley[16] revealed that in epicuticular waxes on light-grown seedling leaves one chain length or group thereof predominated, whereas on dark-grown leaves an additional shorter prominent chain length or group thereof was also present. An analogous differential in the ability of the yellow vs. green segments of developing maize leaves to incorporate acetate into waxes also results in a uni- vs. bimodal distribution of homologs.[54,55]

Mutations in genes whose products function in elongation can potentially be recognized by an increase in the amount of shorter homologs. Table 1 lists those meeting this criterium. The four maize genes and five *Brassica* mutations each distinguish a minimum of three systems; in the *Brassica* genus, for example, elongation up to C_{26}, from C_{26}–C_{28}, and from C_{28}–C_{30}. Interestingly, of the nine *gl* genes which have been mapped in maize (*Zea mays*) and their effect on the wax composition determined, the four involved in elongation are located in chromosome 2 (*gl*$_2$ and *gl*$_{11}$) and chromosome 4 (*gl*$_3$ and *gl*$_4$), while the others are scattered among five different chromosomes.[56] Isolation and characterization of the four pertinent genes by chromosome tagging is feasible and would reveal whether any of them represent duplicated genes.[57] The *wa* gene in peas also acts in the terminal step of elongation.[58] Mutations in the *FAE1* gene in *Arabidopsis thaliana* inhibit three elongation systems, namely $C_{18:0}$–$C_{20:0}$, $C_{18:1}$–$C_{20:1}$, and $C_{20:1}$–$C_{22:1}$.[59] The gene has been mapped on chromosome 4 with respect to restriction fragment length polymorphism (RFLP) markers, and walking is in progress[60,61] with the goal of isolating it. Determining if *FAE1* or any of the four maize *gl* genes codes for an elongase component should be possible by comparison to the relevant gene sequences becoming available for FAS components.[62-65]

Of the numerous reports describing inhibitory effects of selected chemicals on wax synthesis, relatively few present detailed analyses of the wax composition. Trichloroacetic acid and various thiocarbamates were not only among the earliest studied (see, for example, References 44 and 66), but are also those which have received most attention. While the results are most readily interpreted as an inhibition of elongation starting with the terminal step and progressing toward shorter chain lengths, a closer examination reveals inconsistencies (References 54 and 67 and references therein). The difficulty in interpreting the data reflects the facts that the lipid classes occur in markedly different proportions and that the degree of elongation differs from one class to the other.[68] Thus, neither trichloroacetic acid nor the various thiocarbamates have been useful in discriminating among elongation systems. By comparison, arsenite, cyanide, and 2-mercaptoethanol identified a minimum of four sequential systems contributing to elongation of C_{18}–C_{32} chains present in barley waxes[69] (Table 1). Extrapolation from the known effect of arsenite[70] infers that the condensing moiety (a β-ketoacyl-CoA synthase) of the C_{20}–C_{22} system is more closely related to that of C_{16} elongase than to those of FAS or C_{18} elongase. The effects of arsenite and cerulenin are reciprocal with respect to the plastid acyl carrier protein (ACP) condensing enzymes β-ketoacyl-ACP synthase I and II (KAS I, II) (Table 1). Given the differences in sensitivity to arsenite of the CoA elongases, it would be of interest to carry out parallel experiments with cerulenin. Should one of the acyl-CoA elongases be very sensitive,

TABLE 1
Sites of Blocks in Elongation Induced by Gene Mutations and Chemical Inhibitors

Blocked in addition to carbon[a]	Mutations					Chemicals[e]			
	Epicuticular waxes				Seeds				
	Zea mays	Pisum sativum	Brassicas[c]	Hordeum vulgare	Arabidopsis[d]	Arsenite	Mercaptoethanol	Cyanide	Cerulenin
4–14						↔ KAS I			⇕ KAS I
12				cer-q					
14				cer-q					
16				cer-q		⇕ KAS II			↔ KAS II
18, 18:1					fae1				
20, 20:1					fae1	⇕			
22, 22:1					★		⇕		
26	gl$_{11}$		gl$_3$ K						
28	gl$_3$		gl$_1$ BS					⇕	
			gl$_2$ BS						
			gl$_3$ BS						
			gl$_5$ C ★						
30	gl$_2$, gl$_4$ ★[b]	wa ★							
32				★					

[a] Unless specified, applies to saturated acyl chain.
★ — Longest significant chain *in vivo*.
[b] ★
[c] *B. oleracea*; K — kale; BS — brussel sprouts; C — cauliflower.
[d] *A. thaliana*, impaired in elongation of 18:0, 18:1, and 20:1.
[e] ↔ — somewhat sensitive; ⇕ — very sensitive; KAS I and II — β-ketoacyl-ACP synthase I and II, respectively.

isolation of its condensing activity by radio tagging should be feasible, since cerulenin binds covalently to the active site cysteine.[71,72]

Neither genes nor inhibitors have been identified that modify the chain length of the β-diketones.

2. Parallel Systems

The best characterized example was unveiled with the aid of the *eceriferum* (*cer*) mutants in barley, especially those belonging to the *cer-cqu* gene.[35] Alkanes (primarily C_{31}) and primary alcohols (primarily C_{26}) are synthesized by acyl system(s), while β-diketones (primarily C_{31}) and esterified alkan-2-ols (primarily C_{13} and C_{15}) arise from the β-ketoacyl system (Figure 3). Initial indications of parallel elongation included: (1) Cyanide stimulated alkane formation at concentrations inhibiting β-diketone synthesis, whereas 2-mercaptoethanol essentially eliminated alkane synthesis without affecting formation of β-diketones.[49] (2) Stearic acid served as a primer for the lipids derived from the acyl, but not the β-ketoacyl system.[49,50] (3) Only the β-diketone and esterified alkan-2-ol lipids were altered by *cer-cqu* mutations.[78,79] Subsequent studies revealed that the *cer-cqu* gene codes for a multifunctional polypeptide having at least three functional domains, corresponding to complementation groups *cer-c*, *-q*, and *-u*.[35] That the function determined by the *cer-q* complementation group is the condensing activity of the β-ketoacyl elongase was ascertained by feeding appropriately labeled and activated substrates for three successive steps in a cycle of elongation, namely C_{16} acyl, 3-oxoacyl, and 3-hydroxyacyl-CoAs.[80] Formation of both β-diketone and esterified alkan-2-ol lipids is inhibited by arsenite, intimating that the condensing activity of the β-ketoacyl elongase is more similar to that of C_{16} elongase than those of FAS.[49,69,81] Cyanide inhibits β-diketone synthesis from 3-oxoacyl-CoAs, but not that of the esterified alkan-2-ols.[49,81] Whether this implies that an additional elongase participates in some of the later steps of β-diketone chain synthesis is unknown. Cyanide impedes addition of C_2 units to C_{26} chains by an acyl elongase[69] (Table 1).

The existence of parallel acyl systems have also been deduced. For example, six *cer* genes in barley have been identified in which elongation of alkanes is impeded, but that of the alkenes is not.[47] In the *Brassica* genus, branched homologs are present only in some of the lipid classes, suggesting that they arise from systems using branched primers which the other systems do not use. The gl_n mutation in rape (*B. napus*) and the gl_6 mutation in marrow stem kale (*B. oleracea*) modify the system(s) giving solely *n*-chains in the wild types, so that lipid class(es) arising therefrom also contain branched homologs.[82,83] In maize, the occurrence of two acyl systems has been proposed to account for the differences in wax composition on the dull leaves of young plants vs. the glossy appearing ones on mature plants. The authors suggest that one elongation system functions specifically in young leaves (ED-1), while the other (ED-2) functions throughout the life of the plant,[84] although an alternative interpretation of the data is possible.[35]

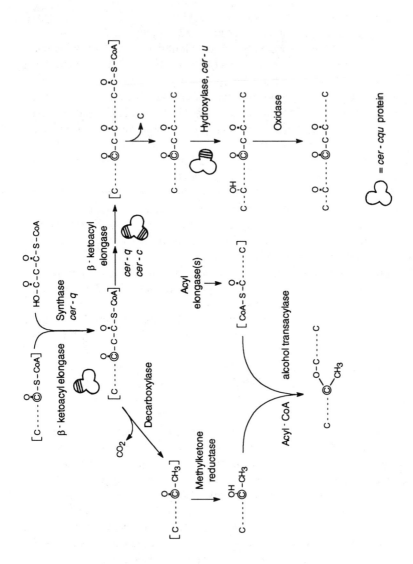

FIGURE 3. The β-ketoacyl elongase pathways. The β-ketoacyl elongase produces a pool of 3-oxoacyl-CoA intermediates. Some of these (left branch) are converted into methylketones by a decarboxylase, which in turn are acted upon by a methylketone reductase to give alkan-2-ols, which can be esterified with acyl-CoAs yielding alkan-2-ol esters. After two oxo groups are introduced into other intermediates as illustrated in Figure 2, additional C_2 units are added (right branch). Removal of a carbon, possibly by decarbonylation (Figure 5), gives β-diketones. A hydroxyl group can then be inserted and oxidized to give hydroxy- and oxo-β-diketones, respectively. While β-diketone lipids have been reported only in epicuticular waxes, the alkan-2-ol lipids also occur in suberin waxes[73] and covalently attached to the cutin and suberin monomers.[74,75] The lipids synthesized by these basic pathways may serve as precursors for other reactions. For example, in the grass *Agropyron elongatum*, insertion of another hydroxyl group yields hydroxyoxo-β-diketones,[76] or in barley insertion of an oxygen forms oxo-alkan-2-ol esters.[77] In barley the multifunctional protein determined by the *cer-cqu* gene plays three roles in this pathway at the sites indicated. The q domain carries out the condensing (synthase) activity of the β-ketoacyl elongase, the u domain has hydroxylase activity, but the function of the c domain remains to be ascertained. Brackets enclose lipids that have not been found in waxes or the cutin and suberin monomers.

C. MEMBRANE ELONGASES
1. Progress Toward Isolation

Early experiments with peeled tissue from *Senecio*[40] and leek[85] demonstrated that epicuticular waxes were specifically synthesized in epidermal cells. Thereafter, work with pea and leek epidermal cells revealed that elongase activities producing 20- and longer carbon chains were associated with membrane fractions.[86,87] A similar correlation between synthesis of long acyl chains and membrane preparations was derived at the same time for systems from pea cotyledons[88] plus germinating peas[89,90] and, more recently, for those from developing oilseeds,[91-95] although the first results from differential centrifugation experiments with jojoba (*Simmondsia chinensis*)[94] and brown mustard (*B. juncea*)[91] suggested that the floating wax pad also had significant elongating activity. This has not been substantiated, but was caused by adhering membranes.[94,95] While all the membrane systems were similar in that malonyl-CoA donated C_2 units to primers in the absence of exogenous ACP, they differed in other respects such as the nature of the reductant, primer, or product. Despite the obvious difficulties in isolating elongases from membranes, considerable progress has been made using the leek epidermal peel and honesty (*Lunaria annua*) developing seed systems, thereby revealing the striking similarity between them.

In 1984, the existence of a microsomal ATP-dependent elongation in leek epidermal cells was deduced.[96] Endogenous primers added to microsomes, that is membranes from a 13,000 to 100,000 × g pellet, in the presence of ATP and NADPH or NADP, were elongated up to 28 carbons. In the absence of ATP, saturated, exogenous C_{10}–C_{22}-CoAs resulted in a series of higher homologs related to the chain length of the presented primer, while $C_{18:1}$-CoA did not. These results provided an explanation as to how leek epidermal cells which had previously been shown to lack a C_{16} elongase[97] were able to synthesize 18- and longer carbon chains. Treatment of the microsomes with low concentrations of Trition X-100 to solubilize proteins followed by sucrose density gradient centrifugation or Sephacryl S-300 chromatography led to the separation of two ATP-independent elongating activities differing in their ability to elongate the two tested CoAs, C_{18} and C_{20}.[98] In the honesty system, elongation from $C_{18:1}$ up to $C_{24:1}$ was associated with the particulate fraction pelleted at 15,000 × g,[92] whereas the microsomal fraction was less active. While NADPH or NADH were equally effective for $C_{20:1}$ formation in mustard (*Sinapsis alba*), another oilseed, NADPH functioned best as co-factor for the subsequent step giving $C_{22:1}$, thereby distinguishing two elongation steps[93] analogous to those identified in the leek system and to two revealed by the inhibitor experiments (Table 1). As in the case of C_{18}-CoA elongase from leek,[99] the initial products of elongation in the honesty and mustard particulate systems are CoA esters.[100] In contrast with the leek microsomal system, the honesty particulate fraction[101] used both saturated and monounsaturated C_{14}–C_{22}-CoAs as primers. Identification of the metabolic intermediates demonstrated for the first time in a plant membrane system the presence of the four predicted elongase activities required for elongation of $C_{18:1}$: β-ketoacyl-CoA synthase, β-ketoacyl-CoA reductase, β-hydroxyacyl-CoA dehydrase, and enoyl-CoA reductase (Figure 2, left).

Partial purification of the C_{18}-CoA elongating activity from the leek system was obtained by subjecting Triton X-100 treated microsomes to anion exchange chromatography on diethylaminoacetyl (DEAE) cellulose plus gel filtration chromatography on Ultrogel 34 AcA.[99,102,103] The resulting fraction, containing three or four prominent proteins upon sodium dodecyl sulfate polyacrylamide gel electrophoresis (SDS-PAGE) analysis, not only elongated C_{18} chains, but also displayed β-ketoacyl-ACP synthase and malonyl-CoA:ACP transacylase activities. The latter, a characteristic of plastid-localized, soluble FAS, and C_{16} elongase, is not one of the expected membrane elongase activities. Preliminary results support the presence of the expected dehydrase, since in Western blots antibodies[104] raised against β-hydroxyacyl-CoA dehydrase from rat liver microsomal acyl-CoA elongase react with one band. Elongation was stimulated upon addition of phosphatidylcholine, the major phospholipid of the microsomes which had been almost totally depleted during purification. No evidence for covalent binding of C_{18}-CoA or malonyl-CoA to the enzyme complex was adduced. For the honesty $C_{18:1}$ elongating activity, the procedure was modified by inclusion of a polyethylene glycol precipitation before the DEAE chromatography.[105] A major protein fraction with an apparent M_r greater than 300 kDa from the Ultrogel column had β-ketoacyl synthase and β-ketoacyl reductase activities. SDS-PAGE revealed the presence of at least 10 to 12 proteins in the 12 to 60 kDa range. A second smaller peak with significant $C_{18:1}$ elongating activity was not further investigated. Despite the noted differences, the C_{18} and $C_{18:1}$-CoA elongases from the two plants appear very similar to the rat liver microsomal acyl-CoA elongase.[106]

2. Molecular Organization and Programming

Observations from the experiments to purify membrane elongases are in accord with those from the genetic and inhibitor studies in implying that many elongation systems can participate in the synthesis of the long acyl carbon skeletons. This is not unique for plants, as demonstrated by the recent observations in *Saccharopolyspora erythraea*.[107] In this species six tandem modules of DNA, each coding for a set of elongase activities, function in the six successive elongation steps required for synthesis of the macrolide antibiotic erythromycin. One of the interesting questions for the coming years will be to determine the physical relationship of these enzyme complexes and whether this is the only programming factor involved. Initial experiments in this direction have been carried out with the leek system. Sucrose gradient separation and phase partitioning of the microsomal membranes from etiolated seedlings combined with appropriate assays led to the deductions that (1) the C_{18} and C_{20}-CoA elongases were present in the endoplasmic reticulum and Golgi apparatus-enriched fractions, respectively; (2) the ATP-independent elongase was in a third uncharacterized fraction; and (3) all three elongases were lacking in the plasmalemma-enriched fraction.[108,109] These results were intepreted to imply that C_{20} acyl chains synthesized in the endoplasmic reticulum are transferred to the Golgi apparatus, where they are elongated by one or two steps before being transferred to the plasmalemma and moved onto the cuticle surface. If this scenario is true, then additional elongases using 22- and longer carbon substrates must exist in the plasmalemma and/or wall in order to account for the leek epidermal wax skeletons which are predominantly C_{26}–C_{32}.

The known molecular organization of the functional units participating in elongase and polyketide syntase pathways range from multifunctional proteins coded for by single genes as in mammalian FAS and *Penicillium* 6-MSA to complexes of proteins coded for by different genes such as in plastid FAS and *Streptomyces* polyketide synthases.[38] On the basis of the observations summarized above, the leek $C_{18:0}$-CoA elongase is considered[102] to consist of discrete polypeptides. This membrane elongase differs from the soluble elongases and polyketide synthases, excluding chalcone synthase, in using CoA- rather than ACP-activated units for elongation. The limited and presently available primary sequences in databases support the notion that the genes encoding all proteins and domains with a similar function have evolved from the same ancestral gene. The only apparent exception is chalcone synthase vs. all the other thus far characterized condensing moieties.[110] Will this generalization still hold when the genes for the membrane elongase components have been isolated and characterized?

IV. REACTIONS OF THE ASSOCIATED PATHWAYS

The carbon chains produced by the coordinated action of elongases normally serve as substrates for associated pathways, thereby generating the diverse range of lipid classes occurring in waxes, cutin, and suberin. Four basic pathways have been distinguished (Table 2). The dominant homologs of the lipid classes arising from the reductive and hydroxylative + epoxidative pathways have even numbered carbon chains vs. those from the decarbonylative/ decarboxylative pathways with odd numbered ones. The primary lipid classes from the hydroxylative + epoxidative pathway, in contrast to those from the other three pathways, occur most frequently as monomers of wax estolides and of the cutin and suberin aliphatic polymers. Acyl chains not entering one of the pathways are found in waxes as free fatty acids. The overlap among the synthetic mechanisms participating in these four pathways is extensive, for example, hydroxylations and transacylations occur in all. One of the few possible unique reactions is the decarboxylation of β-oxoacyl-CoAs to methylketones, precursors of the alkan-2-ols. To emphasize the commonality, the individual reactions rather than the pathways are focused on below.

TABLE 2
The Associated Pathways Use Carbon Skeletons Built by Elongases to Synthesize the Diverse Lipid Classes Via Four Basic Pathways

Associated pathways	Substrates from	Primary lipid classes
Decarbonylative Figure 5	Acyl-CoA elongase	Aldehydes, hydrocarbons, secondary alcohols, ketones, esters
Decarbonylative/ decarboxylative? Figure 3	β-ketoacyl-CoA elongase	β-diketones, hydroxy-β-diketones, oxo-β-diketones, alkan-2-ols, esters
Reductive Figure 4	Acyl-CoA elongase	Aldehydes, primary alcohols, esters
Hydroxylative + epoxidative Figure 6	Acyl-CoA elongase	ω- and midchain hydroxy acids, di- and trihydroxy acids, dicarboxylic acids, epoxy-ω-hydroxy acids

A. REDUCTION

One fate of the acyl-CoA products of elongases is reduction. A cell-free system with this capacity was described in broccoli (*B. oleracea*) in 1971.[111,112] Starting with an acetone powder extract of young leaves, ammonium sulfate fractionation plus gel filtration chromatography on Sepadex G-100 separated two reducing activities. The first one ($\approx$70,000 kDa) required NADP to reduce an acyl-CoA to the corresponding aldehyde, while the second ($\approx$35,000 kDa) preferred NADPH to reduce the aldehyde to primary alcohol (Figure 4). Although in these experiments 16- or 18-carbon substrates were used, the results were assumed to simulate the *in vivo* situation where 32-carbon substrates serve and infer a lack of chain length specificity by the enzymes for their substrates. This work was not followed up, and the only related observations in a higher plant carried out using the floating wax pad from jojoba seeds implicated microsomal enzymes (see Section III.C.1). In the presence of NADPH, $C_{20:1}$ and $C_{22:1}$ primary alcohols were generated from the corresponding labeled acyl-CoAs. Whether or not an aldehyde intermediate is formed during the reduction is unknown, but if so it does not accumulate. The latter observations resemble those in a range of organisms including *Euglena gracilis* and birds (see Reference 112), honey bees (*Apis mellifera*),[114] and various mammals[115,116] in which microsomal systems have been shown to generate long-chain alcohols from acyl-CoAs.

Very strong support in favor of an analogous system in wax synthesis comes from analyses of the composition of the epicuticular waxes on wild type and "*gl$_5$*" maize leaves.[117] The percent of the primary alcohols and aldehydes on leaves of these two genotypes was 62.7 and 20.4 vs. 8.7 and 83.5, respectively. Subsequently, the "*gl$_5$*" plants analyzed were shown to carry two mutations at different loci that have been designated *gl$_5$* and *gl$_{20}$*.[56] Analyses of wax from

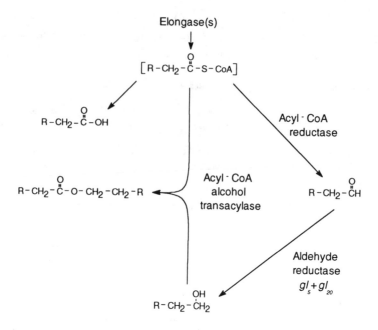

FIGURE 4. The reductive pathway. Acyl-CoAs synthesized by acyl elongases may be converted by acyl-CoA reductase to aldehydes which can serve as substrates for an aldehyde reductase leading to formation of primary alcohols (right). The latter can be esterified with other acyl-CoAs by an acyl-CoA alcohol transacylase (center). Finally, the acyl-CoAs may be hydrolyzed to free acids (left). The lipids synthesized by this basic pathway can serve as precursors for other reactions. For example, ω-hydroxylation (Figure 6) of hexacosanol is presumed to give the 1,16-diol that is a major component of oat (*Avena sativum*) leaf wax.[113] Esterification of the diols yields diesters. In maize when both gl_5 and gl_{20} are mutated, reduction of aldehydes to primary alcohols is impaired. Brackets enclose lipids that have not been found in waxes or the cutin and suberin polymers.

gl_5 and gl_{20} mutants have not been made, but the glossy mutant phenotype is obtained only in plants homozygous for both mutations, suggesting that gl_5 and gl_{20} are duplicated genes. This infers that the gene products function as interchangeable monomers, homomultimers, or as heteromers. Whether gl_5 and gl_{20} are structural genes for the maize aldehyde reductase can be determined using molecular biological techniques. Namely, the isolated genes could be expressed individually and/or simultaneously in *Escherichia coli*, and the resulting proteins could be assayed for activity.

B. DECARBONYLATION AND DECARBOXYLATION

Approximately 60 years ago, several theories to explain the origin of the hydrocarbon constituents in plant and insect waxes were debated.[118-120] Thereafter, the subject remained essentially dormant until the mid-1960s when Kolattukudy using tissue slices and C_2–C_{18}-labeled precursors rapidly adduced the first evidence that long-chain fatty acids with *n*-carbons were precursors of alkanes with *n*-1 carbons (see Reference 121 for review). Additional support

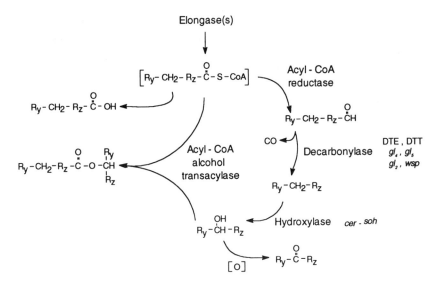

FIGURE 5. A decarbonylative pathway. The pool of acyl-CoAs synthesized by the acyl elongases serves as substrates for synthesis in sequence (right) of aldehydes by acyl-CoA reductase, hydrocarbons by a decarbonylase, secondary alcohols by a midchain hydroxylase, and ketones by oxidation. The acyl-CoAs can also (center) be esterified with the secondary alcohols to give esters and sometimes (far left) hydrolyzed into free fatty acids. The lipids synthesized by this basic pathway may serve as precursors for other reactions. For example, midchain hydroxylation of nonacosan-10-ol in waxes of pines is presumed to give diols such as 5,10-C_{29} and 10,13-C_{29}.[127] Esterification of the diols with various fatty acids yields estolides with three or more units, the dominating component of wax on mature pine needles.[128] The pea gene *wsp*, the three *Brassica gl* genes, and the thiols DTE and DTT block the decarbonylase reaction. In barley *cer-*soh inhibits the hydroxylase reaction. Brackets enclose lipids that have not been found in waxes or the cutin and suberin polymers.

for this contention was obtained when even longer (C_{24}–C_{32}) exogenous fatty acids were shown to be incorporated by developing flower petals[122] and tissue slices of leaves from cabbage[123] and leek[43] into the appropriate alkanes. A decarboxylation mechanism was presumed to carry out the conversion as originally proposed.[94] Extending the studies to a cell-free system from epidermal cells of pea leaves intimated that α-oxidation was involved.[124] However, in the pea microsomal system, α-hydroxy C_{18} fatty acid was not a good precursor of C_{17} alkanes,[125] although α-hydroxy C_{24} fatty acid was for C_{23} alkanes in a 10,000 × *g* supernatant fraction from termites (*Zootermopsis angusticollis*).[126]

That alkanes might be closely related biosynthetically to aldehydes was suggested a number of times during the 1970s, based on the effects of mutants and inhibitors on wax composition (Figure 5). In peas and the *Brassica* genus, alkanes and their derivatives are the major wax classes. These lipids are markedly reduced in the mutant *wsp* in peas,[53,58] gl_4 in marrow stem kale,[129,130] gl_1 in rape,[82] and gl_2 in brussel sprouts.[131] Although an accompanying increase of the appropriate chain length aldehydes occurs, only in the case of gl_2 does the absolute amount even begin to compensate for the decrease of the specified lipids. That is, in gl_2, the C_{30} aldehydes account for *circa* 14.5 $\mu g/cm^2$ and the C_{29} alkanes, secondary alcohols, and ketones for 3.9 $\mu g/cm^2$, whereas in the wild type the corresponding values are trace and 45. Another indication that aldehydes might be alkane precursors in pea waxes came from studying the effect of dithioerythritol (DTE) on the incorporation of labeled acetate by tissue slices into wax lipids.[132] From the data presented, the ratio of the increase in label in all 32 carbon chains to the decrease in those with 31 is 0.23. This estimate is a minimum, since data for all lipid classes is not available.[69] The same potential aldehyde/precursor-alkane/product relationship was also detected using the related thiol dithiothreitol (DTT) and studying wax biosynthesis on barley spikes, where neither the alkanes, primary alcohols, nor derivatives of these two lipids are predominant wax components.[49,69] These observations imply action of the four genes and the two thiols in the conversion of aldehydes to alkanes. Decarbonylation was proposed (Figure 5) as a mechanism for this reaction in 1984.[125] This was a novel idea as no enzymatic analog had then been described. The pea microsomal system was immediately exploited in pursuit of supporting evidence.[133] Two sucrose density gradient centrifugation steps were employed to obtain a relatively heavy particulate fraction that lacked acyl-CoA reductase and α-oxidation activity, but converted [1-^{3}H,1-^{14}C]-C_{17} aldehyde to ^{3}H-C_{17} alkane. Close to but somewhat less than the stoichiometric amount of ^{14}CO produced during the reaction was trapped by the rhodium complex $RhCl[(C_6H_6)_3P]_3$, but no $^{14}CO_2$ was trapped by hyamine hydroxide. The reaction was not inhibited by imidazol or stimulated by ascorbate, but was sensitive to metal ion chelating agents and trypsin. Surprisingly, DTE had little effect.

Thereafter, observations were extended to microsomal preparations from the uropygial gland of the eared grebe (*Podiceps nigricollis*)[134] and the green colonial alga (*Botryococcus braunii* strain A).[135] Work has continued with the latter, as it is a much more favorable source of the enzyme.[136] Solubilization with octyl β-glucoside, followed by gel filtration chromatography on superose 6, ion exchange on Mono Q, and SDS-PAGE identified a heteromer of 66 and 55 kDa subunits. Electron probe analysis identified cobalt. When the green alga was grown in the presence of $^{57}CoCl_2$, label followed the protein during purification. Combined with absorption spectra analysis, the decarbonylase appears to be a cobalt-porphyrin (or corrin) containing enzyme. A function for cobalt in plants has been identified if the same holds true for the plant enzyme.

In addition to the alkanes, two other odd chain lipid classes, β-diketones and methylketones, are thought to be synthesized by loss of a carbon from an even chain acyl precursor (Figure 3). (1) While decarboxylation was initially proposed as a mechanism for the origin of the β-diketones,[49] decarbonylation of a β-keto aldehyde intermediate is an alternative possibility. Although no direct experimental data is available, the following can be noted. Synthesis in barley of β-diketones, in contrast to that of the alkanes, is insensitive to DTT, as is that of the primary alcohols.[49] Inhibition of the decarbonylation of aldehydes to alkanes by the thiols DTT and DTE was characteristic for all crude microsomal systems studied excepting that from termites as noted above. On the other hand, after separation of acyl-CoA reduction from decarbonylation activity in particulate preparations from peas, a marked decrease in sensitivity of the latter activity to DTE was observed.[133] If a decarbonylase is involved in β-diketone synthesis, then either it is a different enzyme system than that giving alkanes or two different acyl-CoA reductase systems must exist: a thiol insensitive one in primary alcohol and β-diketone syntheses and a sensitive one in alkane formation. Only a single system of either is required if decarbonylation is not a step in β-diketone production. The acyl-CoA products of the elongases serving as precursors for barley alkanes and β-diketones have not been detected, suggesting a tight coupling of the pertinent reactions. Furthermore, in barley no β-keto aldehydes have been detected in the waxes from the 37 surveyed out of 54 *cer* complementation groups which are characterized by a decrease in β-diketone synthesis.[35] (2) Methylketones have not been described *in vivo*, but have been invoked as intermediates in synthesis of alkan-2-ols. Results of tissue slice experiments with barley suggested that esterified alkan-2-ols were derived from an elongase condensation product. That is, feeding 3-oxo C_{16}-CoA led to accumulation of labeled pentadecan-15-one, and a distribution of alkan-2-ol esters synthesized which matched their *in vivo* distribution. Exogenous methylketones were rapidly reduced and esterified.[80] In addition, an NADPH-specific methylketone reductase was partially purified.[137] The consistent detection of α-oxidation activity in the experiments leading to the identification of the decarbonylation mechanism[124,125,133] also supports the idea that methylketones arise by decarboxylation.

C. HYDROXYLATION
1. Midchain
Epicuticular waxes containing alkanes may have secondary alcohols and sometimes even ketones of the same chain lengths. Generally, the secondary alcohols occur free, rarely esterified as in some roses (Rosaceae)[138] and pines (*Pinus radiata*).[121] Early experiments, summarized in Reference 121, to elucidate the biosynthetic relationships of these three lipids led to feeding labeled C_{29} substrates to tissue slices of broccoli.[139,140] ^{3}H-alkane was incorporated into both the other lipids, while ^{3}H-secondary alcohol was converted solely into the ketone (Figure 5). The introduction of the hydroxyl group required oxygen. Partial reversion by Fe^{2+} of phenanthroline inhibition implicated a monooxygenase.[140]

Into what lipid insertion of the hydroxyl group takes place has not been studied. Whether it is significant that neither 16 nor 15 hydroxylated C_{30} fatty acids or aldehydes have been identified in the intensively studied *Brassica* waxes is unknown. While this occurs adjacent to or in the middle of the predominant chain in *Brassica*[82,140-142] and unrelated *Clarkia elegans*[143] giving

13-, 14-, and 15-ol isomers of C_{29} and in peas[58] giving 14-, 15-, and 16-ol isomers of C_{31}, the apparent specificity of the enzymes in other plants varies widely. For example, in some roses the C_{25}–C_{33} secondary alcohols are hydroxylated in positions 4, 5, and 6, whereas in others 7- and 10-isomers of C_{29} plus the 9-isomer of C_{31} have been identified.[138,144] In pines only hentriacontan-10-ol has been reported,[128] but in barley a mixture of heptacosan-9-ol, nonacosan-10- and -11-ols, and hentriacontan-10-, -11-, -12-, and -13-ols occurs.[145] These last two examples illustrate the most frequently encountered situation,[141] namely a mixture of isomers among which the 10-ol predominates. Isoforms of lauric acid cytochrome P-450, a monooxygenase with analogous differences in site and degree of specificity of hydroxylation, have been characterized.[146]

Results of radiotracer studies with leaf slices and a cell-free system from broad bean (*Vicia faba*) inferred that an analogous enzyme participates in conversion of 16-hydroxyhexadecanoic acid into 10,16-dihydroxy C_{16} acid, a major cutin monomer[147] (Figure 6). While the latter is most frequent in nature, the 9,16- as well as 8,16- and 7,16-isomers also occur.[7] The insertion hypothesis was strengthened by the demonstration that only one hydrogen was lost.[147] Thereafter, the activity associated with a crude microsomal preparation was subjected to sucrose density gradient centrifugation, and a fraction recovered that catalyzed synthesis of both 9- and 10-isomers.[151] NADPH and oxygen were required for the activity which was inhibited by thiols, N_3^-, and metal ion chelators. Sensitivity to CO and the degree of photoreversibility were studied. These observations led to the deduction that a cytochrome P-450 monooxygenase was carrying out the midchain hydroxylation.

Two genes have been identified in barley which play a role in insertion of hydroxyl groups. The wild type allele of *cer-*[soh] results in synthesis of the just described mixture of secondary alcohols specifically on the awns.[152] Mutations in the *cer-u* complementation group of the *cer-cqu* gene result in a decrease or absence of the hydroxy (primarily C_{25})-β-diketones accompanied by a proportional increase in β-diketones.[153] Monooxygenases carrying out hydroxylations are complexes consisting of at least a haem protein (cytochrome P-450) plus a flavoprotein (NADPH-cytochrome P-450 reductase). How the u domain of the *cer-cqu* determined protein, representing one component of the monooxygenase, interacts with the other component is intriguing. The *cer-cqu* encoded protein is by no means small, since it carries out a condensation reaction (q domain) and at least one other as yet unidentified reaction (c domain). With the continually improving techniques for isolating these monooxygenase membrane-localized proteins,[146,154-156] antibody and cDNA probes will soon be available to start exploring the nature of the *cer-*[soh] gene product. For example, will the ability of a preparation from wild type awns to synthesize secondary alcohols be impeded in the presence of antibodies raised against the Jerusalem artichoke (*Helianthus tuberosus*) cytochrome P-450 reductase.

2. Omega

Most cutin and a substantial proportion of suberin monomers are hydroxylated in the ω-position. Using the same experimental approach as for midchain insertions, ω-hydroxylation of C_{18} was observed in a number of tissue slice systems from apple, grape, and *Senecio odoris*.[157] Thereafter, a microsomal cytochrome P-450 monooxygenase was implicated in synthesis of ω-hydroxy C_{16} from C_{16}[151,158] (Figure 6). The degree of sensitivity to CO and lack of photoreversibility by light, however, implied that a different member of the enzyme family was involved than that participating in midchain hydroxylation. Recent investigations with a microsomal system from etiolated seedlings of *V. sativa* treated with clofibrate to induce the hydroxylase confirmed that ω-hydroxylation was carried out by a cytochrome P-450 monooxygenase.[159] The enzyme used not only $C_{18:1}$, but also its epoxide and diol derivatives as substrates (Figure 6). Hydroxylation was restricted to the ω-position, and antibodies raised against a NADPH-cytochrome P-450 reductase from Jerusalem artichokes inhibited the reaction.

3. Epoxidation + Hydration

The most prominent members of the C_{18} family of cutin monomers are the 9,10-epoxy-18-hydroxy and 9,10,18-trihydroxy compounds. Precursor studies showed that oleic acid was readily converted into both of the specified lipids[157] (Figure 6). After differential centrifugation, a $3,000 \times g$ pellet from spinach which contained cutin was found to have the highest epoxidase activity.[160] Its characterization again implicated a cytochrome P-450-dependent enzyme, which in contrast to those mentioned above was inhibited by CN^- and had an absolute requirement for ATP and CoA. Oleic acid and its ω-acetylated derivative were not readily epoxidized compared to the ω-hydroxy derivative, suggesting a significant substrate specificity. The $3,000 \times g$ pellet was unable to carry out ω-hydroxylation of oleic acid. Since cutin from apple skins is rich in the trihydroxy C_{18} monomers compared to that from spinach, epoxide hydrolase activity was studied in a $3,000 \times g$ pellet prepared therefrom.[161] The enzyme which was active with 9,10-epoxy-18-hydroxy C_{18} as a substrate was sensitive to thiols and trichloropropylene oxide and did not require any co-factors or metal ions.

Very recent studies with a *V. sativum* microsomal system[159] rather than a $3,000 \times g$ pellet gave somewhat different results. An epoxidase was identified that required neither ATP nor CoA. Addition of β-mercaptoethanol to the microsomes inhibited epoxidation, but not the ω-hydroxylation dependent on cytochrome P-450 in the presence of NADPH. The enzyme was suggested (data not given) to be a peroxygenase carrying out a peroxide rather than an oxygen-dependent reaction (Figure 6). Of interest in this connection is the recent characterization of a microsome-localized peroxygenase from soybeans (*Glycine max*) which in the presence of alkylhydroperoxides as co-substrates catalyzes epoxidation of *cis* oleic and linoleic acids.[162]

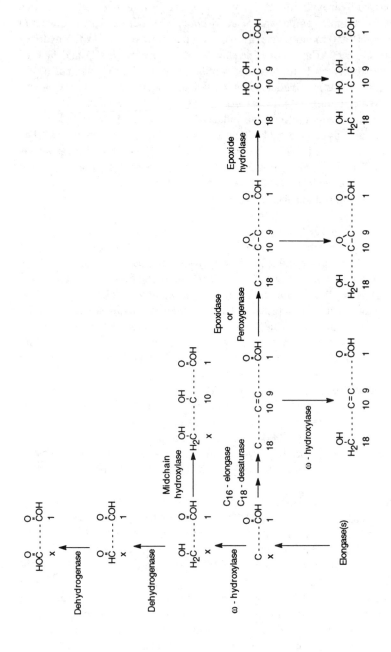

FIGURE 6. The hydroxylative + epoxidative pathways. Subcellular localization and characterization of the elongases and stearoyl desaturase giving the starting C_{16} and $C_{18:1}$ acyl chains have not been carried out. The acyl chains are shown as free acids, since the literature is not clear as to whether they are activated. ω-Hydroxylation of palmitic acid (x = 16) followed (1) by midchain hydroxylation results in a characteristic cutin monomer, namely 10,16-dihydroxy C_{16} and (2) by two oxidation steps yields 1,16-dioic C_{16} representative of suberin (top and left). Action of additional acyl elongases as illustrated in Figures 4 and 5 is required to give the substrates (x = 18 to 24) for the same series of reactions giving rise to the ω-hydroxy and dioic fatty acids that are frequent in suberin. The 9,10-epoxy-10-hydroxy and 9,10,16-trihydroxy C_{18} fatty acids characterizing the C_{18} family of cutin monomers are synthesized by the coordinated action of a ω-hydroxylase, epoxidase, or peroxygenase (see text) plus epoxide hydrolase (center and right). The lipids synthesized by these basic pathways frequently serve as precursors for the cutin and suberin acyl transacylases building the respective polymers. Analogous reactions, however, presumably take place in synthesis of (1) the ω-hydroxy, midchain hydroxy, and dioic fatty acid monomers of the estolides characterizing conifer epicuticular waxes,[148,149] and (2) the midchain hydroxy and epoxy acyl moieties of some seed storage lipids.[150]

D. OXIDATION

Dicarboxylic fatty acids are common but minor cutin monomers and prominent suberin monomers. Experiments to determine their synthetic origin were carried out with cell-free extracts of *V. faba* epidermis. These showed that ω-hydroxy C_{16} was converted into both 1,16-dioic and 16-oxo C_{16} fatty acids, while feeding the latter resulted only in synthesis of 1,16-dioic C_{16}[163] (Figure 6). Differential centrifugation revealed interestingly that most of the activity, which preferred NADP as co-factor and did not require CoA or ATP, was present in the $105,000 \times g$ supernatant. Thiol reagents inhibited the reaction. These observations were extended by using potato tubers in which suberin synthesis had been induced by wounding. Starting with an ammonium sulfate fraction from an acetone powder, two dehydrogenase activites were resolved by Sepharose 6B gel filtration chromatography.[164] One converted the C_{16} ω-hydroxy substrate into the ω-oxo derivative, and the second converted the latter into the dioic fatty acid. Only the ω-hydroxy acid dehydrogenase was induced (*circa* 16-fold) by wounding. Its purification identified a dimer of 31,000 kDa subunits. A detailed characterization revealed that this enzyme was quite similar to other dehydrogenases and led to a proposed model of the active site.[165] Repetition of the purification today for amino acid sequence data should readily lead to isolation of the gene.

In comparison to the ω-hydroxy groups, midchain hydroxy groups of cutin and suberin monomers are rarely if ever oxidized (Figure 6), whereas ketone derivatives of secondary alcohols are well-known constituents of some epicuticular waxes (Figure 5). The enzymes in broccoli and *C. elegans* exhibit a marked substrate specificity in that they selectively oxidize the 15- vs. the 14-ol isomer of nonacosane.[140,143] The two isomers occur *in vivo* and were synthesized in a ratio of *circa* 60:40, respectively, whereas 92% of the ketones were substituted on carbon 15. Whether this specificity applies to the dehydrogenases converting secondary alcohols to ketones in other plant waxes is unknown as the latter have not been studied in enough detail. An intriguing question for the future is to explain the synthetic mechanism giving rise to the 13- and 14-hydroxynonacosan-15-ones and 15-hydroxynonacosan-14- and -16-ones in the *Brassica* genus.[166]

E. TRANSACYLATION

Esterification reactions in higher plants have received very little attention. Initially, a protein fraction from gel filtration on Sephadex G-100 was shown to esterify endogenous alcohols (primarily C_{18}) with labeled C_{16}-CoA, while the starting acetone powder from broccoli leaves also used free fatty acids and the acyl chains of phospholipids in synthesis of esters (reviewed in Reference 121). Exogenous C_{16}–C_{24} primary alcohols were transferred to C_{12}–C_{18} fatty acyl chains by microsomes from barley leaves.[67] If the two studied systems are those producing the wax esters *in vivo*, then the observations infer a marked lack of specificity for their alcohol substrates, since *in vivo* the ester alcohol moieties are much longer. More extensive analysis with the floating wax pad from jojoba seeds demonstrated esterification of $C_{20:1}$ primary alcohols and CoA-activated, but not free $C_{18:1}$–$C_{22:1}$ acyl chains.[94] Most likely, the activity can be contributed to the contaminating membranes as shown for the elongating activity.[95] Analyses of the composition of esters in plant epicuticular waxes have uncovered examples of esterification of alcohol and acyl moieties that are approximately random, as well as those that are nonrandom. To illustrate, in silver ferns (*Cyathea dealbata*) C_{22}–C_{34} primary alcohols are randomly combined with C_{16}–C_{24} fatty acids to give C_{40}–C_{52} esters.[167] On pine needles the C_{36}–C_{42} esters are composed of a C_6–C_{12} moiety combined with one that is C_{26}–C_{32}.[168] These sparse observations from higher plants are in accord with those from a wide variety of other organisms[35] supporting the existence of the acyl-CoA:transacylases diagrammed in Figures 3 to 5.

Construction of the linear and/or cross-linked polymers characterizing wax estolides and cutin plus suberin matrices requires di- and multisubstituted acyl chains, respectively. A single study[169] carried out almost 20 years ago showed that a 3,000 × g pellet from *V. faba* could add various C_{10}–C_{18} fatty acids to the hydroxyl residues of the cutin present in the pellet. Palmitic acid and its ω-hydroxy and 10,16-dihydroxy derivatives were the preferred substrates, and ATP and CoA were required. Solubilization by sonication of the pellet was possible. Activity then also depended upon addition of cutin as a primer whose efficiency could be increased by chemical and enzymatic treatments that increased the number of available hydroxyl groups. Cellulose, hexadecanol, and other potential hydroxyl acceptor compounds did not serve as primers.

V. SUBCELLULAR LOCALIZATION OF THE SYNTHESIZING MACHINERY

Elongases and the associated enzyme complexes synthesizing the acyl components of waxes, cutin, and suberin are indubitably associated with particulate fractions (see Section III.C.1 and Table 3). The unsolved enigma is which membrane(s) and to what extent the cell wall may be involved. One can visualize acyl chains being synthesized in plastids and transferred to the endoplasmic reticulum and plasmalemma wherein further elongations and the associated reactions take place. The resulting lipids are then transported to their final destinations. For the cutin monomers and epicuticular waxes, this entails passing through plasmalemma and inner layers to an outer layer of the epidermal cell wall or onto the external surface. Polymerization and cross-linking of the cutin monomers would take place thereafter. At the other end of the spectrum, the complete carbon skeletons could be synthesized, and the associated reactions take place in the plasmalemma and juxtaposed cell wall or even entirely within the cell wall. In the latter case, the only lipids requiring transport would be the waxes destined for the cuticular surface. As of today, only fragmentary information is available bearing on the question of the site of synthesis vs. final destination. No experimental results are available for suberin and its associated waxes, but its localization between the plasmalemma and cell wall predicts pull**154**   *Lipid Metabolism in Plants*

TABLE 3
Fractionation by Differential Centrifugation of Enzyme Activites in the Associated Pathways Participating in Wax, Cutin, and Suberin Biosynthesis

Enzyme (figure)	Plant	Organ	Present in (cutin)[a]	Ref.
Acyl-CoA reductase (4)	*P. sativum*	Very young leaves	Microsomal (+)	133
Decarbonylase (5)	*P. sativum*	Very young leaves	Microsomal (+)	133
Acyl transacylase				
Alcohol (4)	*H. vulgare*	Seedling leaves	Microsomal	67
Cutin	*V. faba*	Young leaves, flower buds	3,000 × g pellet (+)	169
	S. odoris	Leaves	3,000 × g pellet (+)	169
Hydroxylase				
C_{16} ω (6)	*V. faba*	Dark grown shoots	Microsomal	158
$C_{18:1}$ ω (6)	*V. sativa*	Dark grown shoots	Microsomal	159
Midchain (6)	*V. faba*	Young leaves	Microsomal	151
Dehydrogenase				
ω-hydroxy (6)	*V. faba*	Young leaves	Microsomal	163
ω-aldehyde (6)	*V. faba*	Young leaves	Microsomal	163
Epoxidase (6)	*S. oleracea*	Mature leaves	3,000 × g pellet (+)	160
	S. odoris	Leaves	3,000 × g pellet (+)	160
	M. pumila	Young fruits	3,000 × g pellet (+)	160
Peroxygenase (6)	*V. sativa*	Dark grown shoots	Microsomal	159
Epoxide hydrolase (6)	*V. sativa*	Dark grown shoots	Microsomal	159
	M. pumila	Young fruits	3,000 × g pellet (+)	161
	S. odoris	Leaves	3,000 × g pellet (+)	161
	S. oleracea	Mature leaves	3,000 × g pellet (+)	161

[a] Microsomal = pellet from centrifugation in range 9,000 to 110,000 × g. + = Diagnostic cutin monomers present.

a similar scenario as for cutin and its associated waxes. With the exception of the analysis of the *cer-cqu* gene in barley, the relevant observations are derived from cell fractionation experiments employing centrifugation. The same dilemma is not encountered with respect to the seed waxes which remain within the cell, and hence they are excluded from the following presentation.

A. CELL FRACTIONATION
Several studies have been made in which microsomal preparations were fractionated and the presence of more than one activity tracked. In some cases, this was an indirect result of experiments designed for other purposes. For example, with the goal of solubilizing the hydrocarbon synthesizing component, pea microsomes were treated with trypsin. Loss of their elongating activity accompanied by retention of that for hydrocarbon synthesis was interpreted to imply that the former activity was on the cytoplasmic surface and the latter embedded within a membrane.[125] Shortly thereafter, the same laboratory characterized, after two rounds of sucrose density gradient centrifugation, a

heavy particulate fraction from the pea microsomes capable of carrying out the decarbonylation reaction, but with a markedly decreased ability to reduce acyl-CoAs. All other fractions from the gradient, however, had both activities.[133] Rather than assay for the presence of diagnostic activites in the various fractions, they were examined in the electron microscope. The heavy fraction of interest lacked membrane vesicles, but contained structures resembling cell wall and/ or cuticle fragments whose identity was confirmed by demonstrating the presence of the major pea cutin monomer 10,16-dihydroxydecanoic acid. The results were interpreted to simulate the *in vivo* situation where elongases and acyl-CoA reductase were associated with membranes whose products were secreted to the wall for decarbonylation. Identification of a fraction containing only vesicles and elongase/reductase activities would have strongly supported the hypothesis.

After sucrose density gradient centrifugations, both midchain and ω-hydroxylase activities from *V. faba*[151] were present in the same fraction containing an endoplasmic reticulum marker activity well-separated from the mitochondrial membrane marker. Since activities diagnostic for membranes such as the plasmalemma and Golgi apparatus having similar densities to the endoplasmic reticulum were not assayed, it appears today to have been somewhat premature to have concluded that these hydroxylases were localized in the endoplasmic reticulum. Whether or not cutin is a component of the microsomal preparations from the other plants listed in Table 3, in addition to peas, is unknown as the appropriate chemical analyses were not carried out. Cutin was identified in the $3,000 \times g$ pellet which served as a source for three of the associated enzyme activities (Table 3). Not surprisingly, it was required as a primer for the cutin transacylase, that is polymerase activity.[169] This aspect was not investigated with respect to the epoxidase and epoxide hydrolase activities. Why, if the latter two enzymes were in the cell wall,[160] the midchain and ω-hydroxylases should occur in the endoplasmic reticulum[151] is curious (see Figure 6). The very recent observations that double bonds can be converted to epoxide rings by a microsomally localized peroxygenase again emphasizes the need for a detailed investigation.

The more recent work with leek microsomes (see Section III.C.2) led to the suggestion that C_{18} and C_{20} elongases are present and active in the endoplasmic reticulum, but lacking in the plasmalemma. At first, this proposal appears incompatible with the earlier report that protoplasts of epidermal cells from barley leaf sheaths can synthesize C_{18} chains, but are essentially unable to elongate them further.[170] However, simultaneously prepared mesophyll protoplasts were characterized by a reduced ability to desaturate $C_{18:1}$, an activity generally assumed to occur in the endoplasmic reticulum. The same effects were observed in a tissue slice system treated with the enzyme preparations used in protoplast preparation, inferring that one or more of the components of the latter may enter the cell and impede endoplasmic reticulum functions. If so, then the leek results are in accord with those from barley. Another way to reconcile the observations is to invoke feedback inhibition.

For example, with protoplasts and plasmalemma fractions lacking a cell wall, feedback inhibition normally arising from the receipt of the elongated chains would also be absent. This aspect may be important to consider in future experiments designed to assess the physical relationship among the participating enzymes or complexes.

Another aspect to be taken into account when considering subcellular localization of the activities is that fatty acyl chains not entering one of the four basic pathways occur in the waxes as free acids. Although generally comprising a small proportion of the wax, they may be as much as 90% of the epicuticular wax, as on leaf sheaths of sorghum.[171] Their presence in waxes betokens that the final elongase products have access to the cuticle surface just as the primary and final lipids arising via the associated pathways. How this trafficking of lipids onto the surface is accomplished is unknown.

B. THE *cer-cqu* GENE

The blue-grey color of wild type barley spikes and uppermost leaf sheaths plus internodes results from the presence of a wax coating of long thin tubes on the cuticle surface correlated with the presence of β-diketones.[153,172] The most frequently induced mutations modifying the color to green and yellow-green have been assigned to the *cer-cqu* gene (522/900). Of these 204, 157 and 148 belong to either complementation group *cer-c*, *-q*, or *-u*. The other 13 belong to more than one of these, that is *-cu*, *-cq*, *-qu*, or *-cqu*. Mutations in these four complementation groups plus those in *cer-c* and *-q* result in wax coats devoid of or with a reduced number of tubes, while those of *cer-u* are characterized by the presence of lobed plates among the tubes. That all three complementation groups belong to the same gene was demonstrated by the recovery of partial and wild type revertants from all 13 "multiple" mutants after NaN_3 mutagenesis. The genetic data reveals that *cer-cqu* determines a polypeptide having at least three functional domains corresponding to *cer-c*, *-q*, and *-u*.[35]

Two of the three functions are known. Analyses of spike waxes led to the deduction that barley esterified alkan-2-ols (primarily C_{13} and C_{15}, but also C_{11} and C_{17}) and β-diketones are synthesized from a common precursor by a β-ketoacyl elongase independently of acyl elongase-derived wax lipids.[78] The inhibition of synthesis of all β-ketoacyl elongase-derived lipids by *cer-q* mutations intimated that the function of this complementation group was previous to or at the branch point in the pathway (Figure 3). By contrast, *cer-c*-induced blocks of β-diketone formation were accompanied by increases of esterified alkan-2-ols, inferring a site of action in the branch leading to the β-diketones. Subsequently, appropriately activated and labeled substrates for three successive steps in a cycle of elongation, that is, C_{16} acyl-, 3-oxoacyl-, and 3-hydroxyacyl-CoAs, were fed to tissue slices prepared from *cer-c*[36], *-q*[42], and *-u*[69] plants.[80] The 3-hydroxyacyl-CoA was ineffective as a precursor in all cases. The acyl- and 3-oxoacyl-CoAs served as substrates for esterified alkan-2-ols in *cer-c*[36] tissue and for both lipids in *cer-u*[69] tissue. While neither of these compounds was incorporated into β-diketones by *cer-q*[42] tissue, the 3-oxoacyl-CoA but not the acyl-CoA proved a good substrate for esterified alkan-2-ols. That the 3-oxoacyl-CoA failed as a β-diketone precursor in *cer-q*[42] tissue is presumably due to the fact that, in contrast to alkan-2-ol synthesis, two successive rounds of 3-oxo conservation are required in β-diketone synthesis[173] (Figure 2). These results defined the function of complementation group *q* as the condensing activity of the β-ketoacyl elongase. This activity which is sensitive to arsenite participates relatively early in synthesis of the carbon skeletons; that is, C_{12}-, C_{14}-, and C_{16}-acyl-CoAs are elongated by the β-ketoacyl elongase to the C_{14}-, C_{16}-, and C_{18}-3-oxoacyl-CoAs. The first two are channeled preferentially into the alkan-2-ol branch and the latter into the β-diketone branch of the pathway (Table 1, Figure 3). By comparison, C_{12} and C_{14} chains are intermediates in the synthesis of palmitic acid by plastid FAS which is relatively insensitive to arsenite. Combined, these observations suggest that if *de novo* synthesis of β-ketoacyl elongase-derived lipids takes place in plastids, then C_{12} and C_{14} intermediates, as well as the C_{16} product of FAS, are selectively transported out of the organelle. That this is feasible is illustrated by the oilseeds storing C_{12} chains in triacylglycerols. The function of the *u* complementation group, a midchain hydroxylation of the completed β-diketone chain (section IV.C.1), is attributable to an associated reaction. Thus, the *cer-cqu* gene product, which on the basis of complementation analyses acts as a multimer with a minimum of two units, carries out one of the earliest elongation reactions and the final associated reaction before the lipid is conveyed onto the cuticle surface.

Where is the *cer-cqu* determined protein located? My working hypothesis is in the plasmalemma and/or cell wall, since (1) the decarbonylase which carries out an associated reaction has been isolated in a cutin containing membrane vesicle deficient fraction,[133] and (2) cutin serves as a primer for the associated transacylation reactions producing the polymer.[169] If this hypothesis is true, then a similar location can be predicted for the other elongases and associated enzymes participating in synthesis of the aliphatic components of cutin, suberin, and their waxes. Definitive answers to these questions will become possible with the cloning of the structural genes. The first steps toward the *cer-cqu* gene have been successfully completed with the identification of a tightly linked RFLP[35] marker and the cloning of FAS condensing enzymes in barley.[71,174] The genes *cer-cqu* and *Kas12*, which codes for β-ketoacyl-ACP synthase I isoenzyme 2, as well as two fragments hybridizing to *Chs1*, encoding chalcone synthase, have all been localized to chromosome 2,[35,175,176] revealing that this chromosome harbors a number of members of the condensing enzyme gene family which has a minimum of eight members participating in acyl lipid synthesis in barley.

REFERENCES

1. **Soliday, C. L., Kolattukudy, P. E., and Davis, R.W.,** Chemical and ultrastructural evidence that waxes associated with the suberin polymer constitute the major diffusion barrier to water vapor in potato tuber (*Solanum tuberosum* L.), *Planta,* 146, 607, 1979.

2. **Holloway, P. J.,** Structure and histochemistry of plant cuticular membranes: an overview, in *The Plant Cuticle,* Cutler, D. F., Alvin, K. L., and Price, C. E., Eds., Academic Press, London, 1982, 1.

3. **Kolattukudy, P. E. and Espelie, K. E.,** Biosynthesis of cutin, suberin, and associated waxes, in *Biosynthesis and Biodegradation of Wood Components,* Higuchi, T., Ed., Academic Press, New York, 1985, 161.

4. **Kartha, A. R. and Singh, S. P.,** The *in vivo* 'quantum' synthesis of reserve waxes in seeds of *Murraya koenigii* (Linn.) Spreng, *Chem. Ind.,* 1342, 1969.

5. **Miwa, T.,** Jojoba oil wax esters and derived fatty acids and alcohols: gas chromatographic analyses, *J. Am. Oil Chem. Soc.,* 48, 259, 1971.

6. **Spencer, G. F., Plattner, R. D., and Miwa, T.,** Jojoba oil analysis by high pressure liquid chromatography and gas chromatography/mass spectrometry, *J. Am. Oil Chem. Soc.,* 54, 187, 1977.

7. **Walton, T. J.,** Waxes, cutin and suberin, *Methods Plant Biochem.,* 4, 105, 1990.

8. **Kolattukudy, P. E.,** Biochemistry and function of cutin and suberin, *Can. J. Bot.,* 62, 2918, 1984.

9. **Matzke, K. and Riederer, M.,** A comparative study into the chemical constitution of cutins and suberins from *Picea abies* (L.) Karst., *Quercus robur* L., and *Fagus sylvatica* L., *Planta,* 185, 223, 1991.

10. **Schönherr, J.,** Resistance of plant surfaces to water loss: transport properties of cutin, suberin and associated lipids, in *Encyclopedia of Plant Physiology New Series Vol. 12B,* Lange, O. L., Nobel, P. S., Osmond, C. B., and Ziegler, H., Eds., Springer-Verlag, Berlin, 1982, 153.

11. **Riederer, M. and Schneider, G.,** The effect of the environment on the permeability and composition of *Citrus* leaf cuticles. II. Composition of soluble cuticular lipids and correlation with transport properties, *Planta,* 180, 154, 1990.

12. **Kolattukudy, P. E.,** Biopolyester membranes of plants: cutin and suberin, *Science,* 208, 990, 1980.

13. **Köller, W.,** The plant cuticle. A barrier to be overcome by fungal plant pathogens, in *The Fungal Spore and Disease Initiation in Plants and Animals,* Cole, G. T. and Hoch, H. C., Eds., Plenum Press, New York, 1991, 219.

14. **Bajar, A., Podila, G. K., and Kolattukudy, P. E.,** Identification of a fungal cutinase promoter that is inducible by a plant signal via a phosphorylated trans-acting factor, *Proc. Natl. Acad. Sci. U.S.A.,* 88, 8208, 1991.

15. **Kunoh, H., Nicholson, R. L., Yosioka, H., Yamaoka, N., and Kobayashi, I.,** Preparation of the infection court by *Erysiphe graminis:* degradation of the host cuticle, *Physiol. Mol. Plant Pathol.,* 36, 397, 1990.

16. **Giese, B. N.,** Effects of light and temperature on the composition of epicuticular wax of barley leaves, *Phytochemistry,* 14, 921, 1975.

17. **Hadley, N. F.,** The arthropod cuticle. *Sci. Am.,* 254, 104, 1986.

18. **Ahmad, I. and Wainwright, S. J.,** Ecotype differences in leaf surface properties of *Agrostis stolonifera* from salt marsh, spray zone and inland habitats, *New Phytol.,* 76, 361, 1976.

19. **Leyton, L. and Juniper, B. E.,** Cuticle structure and water relations of pine needles, *Nature,* 198, 770, 1963.

20. **Tewari, J. P. and Skoropad, W. P.,** Relationship between epicuticular wax and blackspot caused by Alternaria brassicae in three lines of rapeseed, *Can. J. Plant Sci.,* 56, 781, 1976.

21. **Sutton, J. C., Rowell, P. M., and James, T. D.,** Effects of leaf wax, wetness duration and temperature on infection of onion leaves by *Botrytis squamosa, Phytoprotection,* 65, 65, 1984.

22. **Marois, J. J., Nelson, J. K., Morrison, J. C., Lile, L. S., and Bledsoe, A. M.,** The influence of berry contact within grape clusters on the development of *Botrytis cinerea* and epicuticular wax, *Am. J. Enol. Vitic.,* 37, 293, 1986.

23. **von Wettstein-Knowles, P.,** unpublished data, 1971.

24. **Thomas, D. A. and Barber, H. N.,** Studies on leaf characteristics of a cline of *Eucalyptus urnigera* from Mount Wellington, Tasmania. I. Water repellency and the freezing of leaves, *Aust. J. Bot.,* 22, 501, 1974.

25. **Single, W. V. and Marcellos, H.,** Studies on frost injury to wheat. IV. Freezing of ears after emergence from the leaf sheath, *Aust. J. Agric. Res.,* 25, 679, 1974.

26. **Harwood, C. E.,** Frost resistance of subalpine *Eucalyptus* species. I. Experiments using a radiation frost room, *Aust. J. Bot.,* 28, 587, 1980.

27. **Stork, N. E.,** Role of waxblooms in preventing attachment to brassicas by the mustard beetle, *Phaedon cochleariae, Entomol. Exp. Appl.,* 28, 100, 1980.

28. **Edwards, P. B.,** Do waxes on juvenile *Eucalyptus* leaves provide protection from grazing insects?, *Aust. J. Ecol.,* 7, 347, 1982.

29. **Eigenbrode, S. D. and Shelton, A. M.,** Behavior of neonate diamondback moth larvae (Lepidoptera: Plutellidae) on glossy-leafed resistant *Brassica oleracea* L., *Environ. Entomol.,* 19, 1566, 1990.

30. **Juniper, B. E. and Burras, J. K.,** How pitcher plants trap insects, *New Sci.,* 13, 75, 1962.

31. **Woodhead, S. and Chapman, R. F.,** Insect behaviour and the chemistry of plant surface waxes, in *Insects and the Plant Surface,* Juniper, B. E. and Southwood, R., Eds., Edward Arnold Publisher, London, 1986, 123.

32. **Städler, E.,** Oviposition and feeding stimuli in leaf surface waxes, in *Insects and the Plant Surface,* Juniper, B. and Southwood, R., Eds., Edward Arnold Publisher, London, 1986, 105.

33. **Jefferson, P. G., Johnson, D. A., and Asay, K. H.,** Epicuticular wax production, water status and leaf temperature in triticeae range grasses on contrasting visible glaucousness, *Can. J. Plant Sci.,* 69, 513, 1989.

34. **Cameron, R. J.,** Light intensity and the growth of *Eucalyptus* seedlings. II. The effect of cuticular waxes on light absorption in leaves of *Eucalyptus* species, *Aust. J. Bot.,* 18, 275, 1970.

35. **von Wettstein-Knowles, P.,** Biosynthesis and genetics of waxes, in *Waxes,* Hamilton, R. J., Ed., Oily Press, Ayr, Scotland, 1993, in press.

36. **Bianchi, A., Bianchi, G., Avato, P., and Salamini, F.,** Biosynthetic pathways of epicuticular wax of maize as assessed by mutation, light, plant age and inhibitor studies, *Maydica,* 30, 179, 1985.

37. **Lundqvist, U. and von Wettstein-Knowles, P.,** Dominant mutations at *Cer-yy* change barley spike wax into leaf blade wax, *Carlsberg Res. Commun.,* 47, 29, 1982.

38. **Hopwood, D. A. and Sherman, D. H.,** Molecular genetics of polyketides and its comparison to fatty acid biosynthesis, *Annu. Rev. Genet.,* 4, 37, 1990.

39. **Kolattukudy, P. E.,** Biosynthesis of wax in *Brassica oleracea.* Relation of fatty acids to wax, *Biochemistry,* 5, 2265, 1966.

40. **Kolattukudy, P. E.,** Further evidence for an elongation-decarboxylation mechanism in the biosynthesis of paraffins in leaves, *Plant Physiol.,* 43, 375, 1968.

41. **Kolattukudy, P. E.,** Tests whether a head to head condensation mechanism occurs in the biosynthesis of n-hentriacontane, the paraffin of spinach and pea leaves, *Plant Physiol.,* 43, 1466, 1968.

42. **Kolattukudy, P. E.,** Biosynthesis of surface lipids. Biosynthesis of long-chain hydrocarbons and waxy esters is discussed, *Science,* 159, 498, 1968.

43. **Cassagne, C. and Lessire, R.,** Studies on alkane biosynthesis in epidermis of *Allium porrum* L. leaves, *Arch. Biochem. Biophys.,* 165, 274, 1974.

44. **Kolattukudy, P. E.,** Biosynthesis of wax in *Brassica oleracea, Biochemistry,* 4, 1844, 1965.

45. **Buckner, J. S., Kolattukudy, P. E., and Rogers, L.,** Synthesis of multimethyl-branched fatty acids by avian and mammalian fatty acid synthetase and its regulation by malonyl-CoA decarboxylase in the uropygial gland, *Arch. Biochem. Biophys.,* 186, 152, 1978.

46. **Rainwater, D. L. and Kolattukudy, P. E.,** Fatty acid biosynthesis in *Mycobacterium tuberculosis* var. *bovis Bacillus Calmette-Guérin.* Purification and characterization of a novel fatty acid synthase, mycocerosic acid synthase, which elongates *n*-fatty acyl-CoA with methylmalonyl-CoA, *J. Biol. Chem.,* 260, 616, 1985.

47. **von Wettstein-Knowles, P.,** The origin of the double bond in the C_{23}-C_{41} alkenes of barley epicuticular wax, in *Structure, Function and Metabolism of Plant Lipids,* Sigenthaler, P.-A. and Eichenberger, W., Eds., Elsevier Science Publishers, Amsterdam, 1984, 521.

48. **Netting, A. G. and von Wettstein-Knowles, P.,** Biosynthesis of the β-diketones of barley spike epicuticular wax, *Arch. Biochem. Biophys.,* 174, 613, 1976.

49. **Mikkelsen, J. D. and von Wettstein-Knowles, P.,** Biosynthesis of β-diketones and hydrocarbons in barley spike epicuticular wax, *Arch. Biochem. Biophys.,* 188, 172, 1978.

50. **Mikkelsen, J. D.,** Structure and biosynthesis of β-diketones in barley spike epicuticular wax, *Carlsberg Res. Commun.,* 44, 133, 1979.

51. **von Wettstein-Knowles, P. and Netting, A. G.,** Esterified alkan-1-ols and alkan-2-ols in barley epicuticular wax, *Lipids,* 11, 478, 1976.

52. **Dierickx, P. J.,** New β-diketones from *Buxus sempervirens, Pytochemistry,* 12, 1498, 1973.

53. **Macey, M. J. K. and Barber, H. N.,** Chemical genetics of wax formation on leaves of *Pisum sativum, Phytochemistry,* 9, 5, 1970.

54. **Avato, P., Mikkelsen, J. D., and von Wettstein-Knowles, P.,** Effect of inhibitors on synthesis of fatty acyl chains present in waxes on developing maize leaves, *Carlsberg Res. Commun.,* 45, 329, 1980.

55. **von Wettstein-Knowles, P.,** Elongases and epicuticular wax biosynthesis, *Physiol. Vég.,* 20, 797, 1982.

56. **Coe, E. H., Hoisington, D. A., and Neuffer, M. G.,** Linkage map of corn (maize) (*Zea mays* L.) (2N = 20), in *Genetic Maps,* O'Brien, S. J., Ed., Cold Spring Harbor Laboratory Press, Cold Spring Harbor, New York, 1990, 6.39.

57. **von Wettstein-Knowles, P.,** Genes, elongases and associated enzyme systems in epicuticular wax synthesis, in *The Metabolism, Structure and Function of Plant Lipids,* Stumpf, P. K., Mudd, J. B., and Nes, W. D., Eds., Plenum Press, New York, 1987, 489.

58. **Holloway, P. J., Hunt, G. M., Baker, E. A., and Macey, M. J. K.,** Chemical composition and ultrastructure of the epicuticular wax in four mutants of *Pisum sativum* (L), *Chem. Phys. Lipids,* 20, 141, 1977.

59. **Kunst, L., Taylor, D. C., and Underhill, E. W.,** Fatty acid elongation in developing seeds of *Arabidopsis thaliana, Plant Physiol. Biochem.,* 30, 425, 1992.

60. **Kunst, L. and Underhill, E.,** Molecular genetic analysis of fatty acid elongation in seeds of *Arabidopsis,* in Int. Soc. Plant Mol. Biol. Congress, Tucson, Arizona, 1991, Abstr. 718.

61. **Lemieux, B., Hauge, B., and Somerville, C.,** RFLP mapping and chromosome walking to the *Arabidopsis thaliana FAE1* locus, in Int. Soc. Plant Mol. Biol. Congress, Tuscon, Arizona, 1991, Abstr. 727.

62. **Kater, M. M., Koningstein, G. M., Nijkamp, J. J., and Stuitje, A. R.,** cDNA cloning and expression of *Brassica napus* enoyl-acyl carrier protein reductase in *Escherichia coli, Plant Mol. Biol.,* 17, 895, 1991.

63. **Siggaard-Andersen, M., Kauppinen, S., and von Wettstein-Knowles, P.,** Primary structure of a cerulenin-binding β-ketoacyl-[acyl carrier protein] synthase from barley chloroplasts, *Proc. Natl. Acad. Sci. U.S.A.,* 88, 4114, 1991.

64. Klein, B., Pawlowski, K., Höricke-Grandpierre, C., Schell, J., and Töpfer, R., Isolation and characterization of a cDNA from *Cuphea lanceolata* encoding a β-ketoacyl-ACP reductase, *Mol. Gen. Genet.*, 233, 122, 1992.

65. Slabas, A.R., Chase, D., Nishida, I., Murata, N., Sidebottom, C., Safford, R., Sheldon, P.S., Kekwick, R., Hardie, D.G., and Mackintosh, R.W., Molecular cloning of higher-plant 3-oxoacyl-(acyl carrier protein) reductase. Sequence identities with the *nodG*-gene product of the nitrogen-fixing soil bacterium, *Biochem. J.*, 283, 321, 1992.

66. Still, G. G., Davis, D. G., and Zander, G. L., Plant epicuticular lipids: alteration by herbicidal carbamates, *Plant Physiol.*, 46, 307, 1970.

67. Avato, P., Synthesis of wax esters by a cell-free system from barley (*Hordeum vulgare* L.), *Planta*, 162, 487, 1984.

68. von Wettstein-Knowles, P., Genetics and biosynthesis of plant epicuticular waxes, in *Advances in the Biochemistry and Physiology of Plant Lipids*, Appelqvist, L.-Å. and Liljenberg, C., Eds., Elsevier/North-Holland Biomedical Press, Amsterdam, 1979, 1.

69. Mikkelsen, J. D., The effects of inhibitors on the biosynthesis of the long chain lipids with even carbon numbers in barley spike epicuticular wax, *Carlsberg Res. Commun.*, 43, 15, 1978.

70. Shimakata, T. and Stumpf, P. K., Isolation and function of spinach leaf β-ketoacyl-[acyl-carrier-protein] synthases, *Proc. Natl. Acad. Sci. U.S.A.*, 79, 5808, 1982.

71. Kauppinen, S., Siggaard-Andersen, M., and von Wettstein-Knowles, P., β-ketoacyl synthase I of Escherichia coli: nucleotide sequence of the *fabB* gene and identification of the cerulenin binding residue, *Carlsberg Res. Commun.*, 53, 357, 1988.

72. Funabashi, H., Kawaguchi, A., Tomoda, H., Ōmura, S., Okuda, S., and Iwasaki, S., Binding site of cerulenin in fatty acid synthetase, *J. Biochem.*, 105, 751, 1989.

73. Espelie, K. E., Sadek, N. Z., and Kolattukudy, P. E., Composition of suberin-associated waxes from the subterranean storage organs of seven plants, *Planta*, 148, 468, 1980.

74. Espelie, K. E. and Kolattukudy, P. E., Composition of the aliphatic components of 'suberin' from the bundle sheaths of *Zea mays* leaves, *Plant Sci. Lett.*, 15, 225, 1979.

75. Espelie, K. E., Dean, B. B., and Kolattukudy, P. E., Composition of lipid-derived polymers from different anatomical regions of several plant species, *Plant Physiol.*, 64, 1089, 1979.

76. Tulloch, A. P., Epicuticular waxes from *Agropyron dasystachyum*, *Agropyron riparium* and *Agropyron elongatum*, *Phytochemistry*, 22, 1605, 1983.

77. von Wettstein-Knowles, P. and Madsen, J. Ø., 7-Oxopentadecan-2-ol esters — a new epicuticular wax lipid class, *Carlsberg Res. Commun.*, 49, 57, 1984.

78. von Wettstein-Knowles, P., Biosynthetic relationships between β-diketones and esterified alkan-2-ols deduced from epicuticular wax of barley mutants, *Mol. Gen. Genet.*, 144, 43, 1976.

79. von Wettstein-Knowles, P. and Søgaard, B., The *cer-cqu* region in barley: gene cluster or multifunctional gene, *Carlsberg Res. Commun.*, 45, 125, 1980.

80. Mikkelsen, J. D., Biosynthesis of esterified alkan-2-ols and β-diketones in barley spike epicuticular wax: synthesis of radioactive intermediates, *Carlsberg Res. Commun.*, 49, 391, 1984.

81. von Wettstein-Knowles, P., Effects of inhibitors on synthesis of esterified alkan-2-ols in barley spike epicuticular wax, *Carlsberg Res. Commun.*, 50, 239, 1985.

82. Holloway, P. J., Brown, G. A., Baker, E. A., and Macey, M. J. K., Chemical composition and ultrastructure of the epicuticular wax in three lines of *Brassica napus* (L), *Chem. Phys. Lipids*, 19, 114, 1977.

83. Netting, A. G., Macey, M. J. K., and Barber, H. N., Chemical genetics of a sub-glaucous mutant of *Brassica oleracea*, *Phytochemistry*, 11, 579, 1972.

84. Bianchi, G., Avato, P., and Salamini, F., Surface waxes from grain, leaves, and husks of maize (*Zea mays* L.), *Cereal Chem.*, 61, 45, 1984.

85. Cassagne, C., Les Hydrocarbares Végétaux: Biosynthèse et Localisation Cellulaire, Ph.D. thesis, University of Bordeaux, France, 1970.

86. **Cassagne, C. and Lessire, R.,** Biosynthesis of saturated very long chain fatty acids by purified membrane fractions from leek epidermal cells, *Arch. Biochem. Biophys.,* 191, 146, 1978.
87. **Kolattukudy, P. E. and Buckner, J. S.,** Chain elongation of fatty acids by cell-free extracts of epidermis from pea leaves (*Pisum sativum*), *Biochem. Biophys. Res. Commun.,* 46, 801, 1972.
88. **Macey, M. J. K. and Stumpf, P. K.,** Fat metabolism in higher plants XXXVI: long chain fatty acid synthesis in germinating peas, *Plant Physiol.,* 43, 1637, 1968.
89. **Bolton, P. and Harwood, J. L.,** Fatty acid biosynthesis by a particulate preparation from germinating pea, *Biochem. J.,* 168, 261, 1977.
90. **Harwood, J. L. and Stumpf, P.K.,** Fat metabolism in higher plants XLIII. Control of fatty acid synthesis in germinating seeds, *Arch. Biochem. Biophys.,* 142, 281, 1971.
91. **Agrawal, V. P. and Stumpf, P. K.,** Elongation systems involved in the biosynthesis of erucic acid from oleic acid in developing *Brassica juncea* seeds, *Lipids,* 20, 361, 1985.
92. **Murphy, D. J. and Mukherjee, K. D.,** Biosynthesis of very long chain monounsaturated fatty acids by subcellular fractions of developing seeds, *FEBS Lett.,* 230, 101, 1988.
93. **Murphy, D. J. and Mukherjee, K. D.,** Elongases synthesizing very long chain monounsaturated fatty acids in developing oilseeds and their solubilization, *Z. Naturforsch.,* 44C, 629, 1989.
94. **Pollard, M. R., McKeon, T., Gupta, L. M., and Stumpf, P.K.,** Studies on biosynthesis of waxes by developing jojoba seed. II. The demonstration of wax biosynthesis by cell-free homogenates, *Lipids,* 14, 651, 1979.
95. **Lardans, A. and Trémolières, A.,** Fatty acid elongation activities in subcellular fractions of developing seeds of *Limnanthes alba, Phytochemistry,* 31, 121, 1992.
96. **Agrawal, V. P., Lessire, R., and Stumpf, P. K.,** Biosynthesis of very long chain fatty acids in microsomes from epidermal cells of *Allium porrum* L., *Arch. Biochem. Biophys.,* 230, 580, 1984.
97. **Lessire, R. and Stumpf, P. K.,** Nature of the fatty acid synthetase systems in parenchymal and epidermal cells of *Allium porrum* L. leaves, *Plant Physiol.,* 73, 614, 1982.
98. **Lessire, R., Bessoule, J.-J., and Cassagne, C.,** Solubilization of C18-CoA and C20-CoA elongases from *Allium porrum* L. epidermal cell microsomes, *FEBS Lett.,* 187, 314, 1985.
99. **Lessire, R., Bessoule, J.-J., and Cassagne, C.,** Involvement of a β-ketoacyl-CoA intermediate in acyl-CoA elongation by an acyl-CoA elongase purified from leek epidermal cells, *Biochim. Biophys. Acta,* 1006, 35, 1989.
100. **Fehling, E., Murphy, D. J., and Mukherjee, K. D.,** Biosynthesis of triacylglycerols containing very long chain monounsaturated acyl moieties in developing seeds, *Plant Physiol.,* 94, 492, 1990.
101. **Fehling, E. and Mukherjee, K. D.,** Acyl-CoA elongase from a higher plant (*Lunaria annua*): metabolic intermediates of very-long-chain acyl-CoA products and substrate specificity, *Biochim. Biophys. Acta,* 1082, 239, 1991.
102. **Bessoule, J.-J., Lessire, R., and Cassagne, C.,** Partial purification of the acyl-CoA elongase of *Allium porrum* leaves, *Arch. Biochem. Biophys.,* 268, 475, 1989.
103. **Lessire, R., Bessoule, J.-J., and Cassagne, C.,** Structural organization of plant icosanoyl-CoA synthase, in *Plant Lipid Biochemistry, Structure and Utilization,* Quinn, P. J. and Harwood, J. L., Eds., Portland Press, London, 1990, 142.
104. **Osei, P., Suneja, S. K., Laguna, J. C., Nagi, M. N., Cook, L., Prasad, M. R., and Cinti, D. L.,** Topography of rat hepatic microsomal enzymatic components of the fatty acid chain elongation system, *J. Biol. Chem.,* 264, 6844, 1989.
105. **Fehling, E., Lessire, R., Cassagne, C., and Mukherjee, K. D.,** Solubilization and partial purification of constituents of acyl-CoA elongase from *Lunaria annua, Biochim. Biophys. Acta,* 1126, 88, 1992.
106. **Bernert, J. T. and Sprecher, H.,** The isolation of acyl-CoA derivatives as products of partial reactions in the microsomal chain elongation of fatty acids, *Biochim. Biophys. Acta,* 573, 436, 1979.

107. **Donadio, S., Staver, M. J., McAlpine, J. B., Swanson, S. J., and Katz, L.,** Modular organization of genes required for complex polyketide biosynthesis, *Science,* 252, 675, 1991.
108. **Moreau, P., Bertho, P., Juguelin, H., and Lessire, R.,** Intracellular transport of very long chain fatty acids in etiolated leek seedlings, *Plant Physiol. Biochem.,* 2, 173, 1988.
109. **Moreau, P., Juguelin, H., Lessire, R., and Cassagne, C.,** Plasma membrane biogenesis in higher plants: *in vivo* transfer of lipids to the plasma membrane, *Phytochemistry,* 27, 1631, 1988.
110. **Lanz, T., Tropf, S., Marner, F.-J., Schröder, J., and Schröder, G.,** The role of cysteines in polyketide synthases. Site-directed mutagenesis of resveratrol and chalcone synthases, two key enzymes in different plant-specific pathways, *J. Biol. Chem.,* 266, 9971, 1991.
111. **Kolattukudy, P. E.,** Enzymatic synthesis of fatty alcohols in *Brassica oleracea, Arch. Biochem. Biophys.,* 142, 701, 1971.
112. **Kolattukudy, P. E., Rogers, L., and Larson, J. D.,** Enzymatic reduction of fatty acids and α-hydroxy fatty acids, *Methods Enzymol.,* 71, 263, 1981.
113. **Tulloch, A. P. and Hoffman, L. L.,** Leaf wax of oats, *Lipids,* 8, 617, 1973.
114. **Blomquist, G. J. and Ries, M. K.,** The enzymatic synthesis of wax esters by a microsomal preparation from the honeybee *Apis mellifera* L., *Insect Biochem.,* 9, 183, 1979.
115. **Kolattukudy, P. E. and Rogers, L.,** Acyl-CoA reductase and acyl-CoA: fatty alcohol acyl transferase in the microsomal preparation from the bovine meibomian gland, *J. Lipid Res.,* 27, 404, 1986.
116. **Moore, C. and Snyder, F.,** Properties of microsomal acyl coenzyme A reductase in mouse preputial glands, *Arch. Biochem. Biophys.,* 214, 489, 1982.
117. **Bianchi, G., Avato, P., and Salamini, F.,** Glossy mutants of maize. VIII: Accumulation of fatty aldehydes in surface waxes of gl_5 maize seedlings, *Biochem. Genet.,* 16, 1015, 1978.
118. **Channon, H. J. and Chibnall, A. C.,** XXII. The ether-soluble substances of cabbage leaf cytoplasm. V. The isolation of *n*-nonacosane and di-*n*-tetradecyl ketone, *Biochem. J.,* 23, 168, 1929.
119. **Chibnall, A. C. and Piper, S. H.,** CCLXXXVIII. The metabolism of plant and insect waxes, *Biochem. J.,* 28, 2209, 1934.
120. **Clenshaw, E. and Smedley-Maclean, I.,** XV. The nature of the unsaponifiable fraction of the lipoid matter extracted from green leaves, *Biochem. J.,* 23, 107, 1929.
121. **Kolattukudy, P. E.,** Biosynthesis of cuticular lipids, *Annu. Rev. Plant Physiol.,* 21, 163, 1970.
122. **Kolattukudy, P. E., Croteau, R., and Brown, L.,** Structure and biosynthesis of cuticular lipids. Hydroxylation of palmitic acid and decarboxylation of C_{28}, C_{30}, and C_{32} acids in *Vicia faba* flowers, *Plant Physiol.,* 54, 670, 1974.
123. **Kolattukudy, P. E., Buckner, J. S., and Brown, L.,** Direct evidence for a decarboxylation mechanism in the biosynthesis of alkanes in *B. oleracea, Biochem. Biophys. Res. Commun.,* 47, 1306, 1972.
124. **Khan, A. A. and Kolattukudy, P. E.,** Decarboxylation of long chain fatty acids to alkanes by cell free preparations of pea leaves (*Pisum sativum*), *Biochem. Biophys. Res. Commun.,* 61, 1379, 1974.
125. **Bognar, A. L., Paliyath, G., Rogers, L., and Kolattukudy, P. E.,** Biosynthesis of alkanes by particulate and solubilized enzyme preparations from pea leaves (*Pisum sativum*), *Arch. Biochem. Biophys.,* 235, 8, 1984.
126. **Chu, A. J. and Blomquist, G. J.,** Decarboxylation of tetracosanoic acid to *n*-tricosane in the termite *Zootermopsis angusticollis, Comp. Biochem. Physiol.,* 66B, 313, 1980.
127. **Franich, R. A., Gowar, A. P. and Volkman, J. K.,** Secondary diols of *Pinus radiata* needle epicuticular wax, *Phytochemistry,* 18, 1563, 1979.
128. **Franich, R. A., Wells, L. G., and Holland, P. T.,** Epicuticular wax of *Pinus radiata* needles, *Phytochemistry,* 17, 1617, 1978.

129. **Macey, M. J. K.,** Wax synthesis in *Brassica oleracea* as modified by trichloroacetic acid and glossy mutations, *Phytochemistry,* 13, 1353, 1974.

130. **Macey, M. J. K. and Barber, H. N.,** Chemical genetics of wax formation on leaves of *Brassica oleracea, Phytochemistry,* 9, 13, 1970.

131. **Baker, E. A.,** The influence of environment on leaf wax development in *Brassica oleracea* var. *gemmifera, New Phytol.,* 73, 955, 1974.

132. **Buckner, J. S. and Kolattukudy, P. E.,** Specific inhibition of alkane synthesis with accumulation of long chain compounds by dithioerythritol, dithiothreitol, and mercaptoethanol in *Pisum sativum, Arch. Biochem. Biophys.,* 156, 34, 1973.

133. **Cheesbrough, T. M. and Kolattukudy, P. E.,** Alkane biosynthesis by decarbonylation of aldehydes catalyzed by a particulate preparation from *Pisum sativum, Proc. Natl. Acad. Sci. U.S.A.,* 81, 6613, 1984.

134. **Cheesbrough, T. M. and Kolattukudy, P. E.,** Microsomal preparation from an animal tissue catalyzes release of carbon monoxide from a fatty aldehyde to generate an alkane, *J. Biol. Chem.,* 263, 2738, 1988.

135. **Dennis, M. W. and Kolattukudy, P. E.,** Alkane biosynthesis by decarbonylation of aldehyde catalyzed by a microsomal preparation from *Botryococcus braunii, Arch. Biochem. Biophys.,* 287, 268, 1991.

136. **Dennis, M. and Kolattukudy, P. E.,** A cobalt-porphyrin enzyme converts a fatty aldehyde to a hydrocarbon and CO, *Proc. Natl. Acad. Sci. U.S.A.,* 89, 5306, 1992.

137. **Mikkelsen, J. D. and von Wettstein-Knowles, P.,** Biosynthesis of esterified alkan-2-ols in barley spike epicuticular wax, in *Structure, Function and Metabolism of Plant Lipids,* Siegenthaler, P.-A. and Eichenberger, W., Eds., Elsevier Science Publishers, Amsterdam, 1984, 517.

138. **Mladenova, K. and Stoianova-Ivanova, B.,** Composition of neutral components in flower wax of some decorative roses, *Phytochemistry,* 16, 269, 1977.

139. **Kolattukudy, P. E. and Liu, T. J.,** Direct evidence for biosynthetic relationships among hydrocarbons, secondary alcohols and ketones in *Brassica oleracea, Biochem. Biophys. Res. Commun.,* 41, 1369, 1970.

140. **Kolattukudy, P. E., Buckner, J. S., and Liu, T. J.,** Biosynthesis of secondary alcohols and ketones from alkanes, *Arch. Biochem. Biophys.,* 156, 613, 1973.

141. **Holloway, P. J., Jeffree, C. E., and Baker, E. A.,** Structural determination of secondary alcohols from plant epicuticular waxes, *Phytochemistry,* 15, 1768, 1976.

142. **Netting, A. G. and Macey, M. J. K.,** The composition of ketones and secondary alcohols from *Brassica oleracea* waxes, *Phytochemistry,* 10, 1917, 1971.

143. **Hunt, G. M, Holloway, P. J., and Baker, E. A.,** Ultrastructure and chemistry of *Clarkia elegans* leaf wax: a comparative study with Brassica leaf waxes, *Plant Sci. Lett.,* 6, 353, 1976.

144. **Wollrab, V.,** Secondary alcohols and paraffins in the plant waxes of the family Rosaceae, *Phytochemistry,* 8, 623, 1969.

145. **von Wettstein-Knowles, P. and Netting, A. G.,** Composition of epicuticular waxes on barley spikes, *Carlsberg Res. Commun.,* 41, 225, 1976.

146. **Durst, F., Benveniste, I., Salaün, J. P., and Werck-Reichhart, D.,** Function, mechanism and regulation of cytochrome *P-450* enzymes in plants, *Biochem. Soc. Trans.,* 20, 353, 1992.

147. **Walton, T. J. and Kolattukudy, P. E.,** Enzymatic conversion of 16-hydroxypalmitic acid into 10,16-dihydroxypalmitic acid in *Vicia faba* epidermal extracts, *Biochem. Biophys. Res. Commun.,* 46, 16, 1972.

148. **Schulten, H.-R., Simmleit, N., and Rump, H. H.,** Soft ionization mass spectrometry of epicuticular waxes isolated from coniferous needles, *Chem. Phys. Lipids,* 41, 209, 1986.

149. **Tulloch, A. P. and Bergter, L.,** Epicuticular wax of *Juniperus scopulorum, Phytochemistry,* 20, 2711, 1981.

150. **Smith, M. A., Stobart, K., Bafor, M., Jonsson, L., and Stymne, S.,** The biosynthesis of ricinoleic and vernolic acid in microsomal preparations from developing endosperm of castor bean and *Euphorbia lagascae,* in *Metabolism, Structure and Utilization of Plant Lipids,* Cherif, A., Miled-Daoud, D., Marzouk, B., Smaoui, A., and Zarrouk, M., Eds., Centre National Pédagogique, Tunis, 1992, 148.

151. **Soliday, C. L. and Kolattukudy, P. E.,** Midchain hydroxylation of 16-hydroxypalmitic acid by the endoplasmic reticulum fraction from germinating *Vicia faba, Arch. Biochem. Biophys.,* 188, 338, 1978.

152. **Placing, S., Kannangara, C. G., Mikkelsen, J. D., Simpson, D., and von Wettstein-Knowles, P.,** Presence of *cer-* soh conditions the synthesis of secondary alcohols in barley epicuticular wax, *Barley Genet. Newslett.,* 9, 75, 1979.

153. **von Wettstein-Knowles, P.,** Genetic control of β-diketone and hydroxy-β-diketone synthesis in epicuticular waxes of barley, *Planta,* 106, 113, 1972.

154. **Benveniste, I., Lesot, A., Hasenfratz, M.-P., and Durst, F.,** Immunochemical characterization of NADPH-cytochrome *P*-450 reductase from Jerusalem artichoke and other higher plants, *Biochem. J.,* 259, 847, 1989.

155. **Benveniste, I., Lesot, A., Hasenfratz, M.-P., Kochs, G., and Durst, F.,** Multiple forms of NADPH-cytochrome P450 reductase in higher plants, *Biochem. Biophys. Res. Commun.,* 177, 105, 1991.

156. **Werck-Reichhart, D., Benveniste, I., Teutsch, H., Durst, F., and Gabriac, B.,** Glycerol allows low-temperature phase separation of membrane proteins solubilized in Triton X-114: application to the purification of plant cytochromes P-450 and b_5, *Anal. Biochem.,* 197, 125, 1991.

157. **Kolattukudy, P. E., Walton, T. J., and Kushwaha, R. P.,** Biosynthesis of the C_{18} family of cutin acids: ω-hydroxyoleic acid, ω-hydroxy-9,10-epoxystearic acid, 9,10,18-trihydroxystearic acid, and their Δ^{12}-unsaturated analogs, *Biochemistry,* 12, 4488, 1973.

158. **Soliday, C. L. and Kolattukudy, P. E.,** Biosynthesis of cutin. ω-hydroxylation of fatty acids by a microsomal preparation from germinating *Vicia faba, Plant Physiol.,* 59, 1116, 1977.

159. **Pinot, F., Salaün, J.-P., Bosch, H., Lesot, A., Mioskowski, C., and Durst, F.,** ω-hydroxylation of Z9-octadecenoic, Z9,10-epoxystearic and 9,10-dihydroxystearic acids by microsomal cytochrome P450 systems from *Vicia sativa, Biochem. Biophys. Res. Commun.,* 184, 183, 1992.

160. **Croteau, R. and Kolattukudy, P. E.,** Biosynthesis of hydroxyfatty acid polymers. Enzymatic epoxidation of 18-hydroxyoleic acid to 18-hydroxy-*cis*-9,10-epoxystearic acid by a particulate preparation from spinach (*Spinacia oleracea*), *Arch. Biochem. Biophys.,* 170, 61, 1975.

161. **Croteau, R. and Kolattukudy, P. E.,** Biosynthesis of hydroxyfatty acid polymers. Enzymatic hydration of 18-hydroxy-*cis*-9,10-epoxystearic acid to *threo*-9,10,18-trihydroxystearic acid by a particulate preparation from apple (*Malus pumila*), *Arch. Biochem. Biophys.,* 170, 73, 1975.

162. **Blée, E. and Schuber, F.,** Efficient epoxidation of unsaturated fatty acids by a hydroperoxide-dependent oxygenase, *J. Biol. Chem.,* 265, 12887, 1990.

163. **Kolattukudy, P. E., Croteau, R., and Walton, T. J.,** Biosynthesis of cutin. Enzymatic conversion of ω-hydroxy fatty acids to dicarboxylic acids by cell-free extracts of *Vicia faba* epidermis, *Plant Physiol.,* 55, 875, 1975.

164. **Agrawal, V. P. and Kolattukudy, P. E.,** Purification and characterization of a wound-induced ω-hydroxyfatty acid:NADP oxidoreductase from potato tuber disks (*Solanum tuberosum* L.), *Arch. Biochem. Biophys.,* 191, 452, 1978.

165. **Agrawal, V. P. and Kolattukudy, P.E.,** Mechanism of action of a wound-induced ω-hydroxyfatty acid: NADP oxidoreductase isolated from potato tubers (*Solanum tuberosum* L.), *Arch. Biochem. Biophys.,* 191, 466, 1978.

166. **Holloway, P. J. and Brown, G. A.**, The ketol constituents of *Brassica* epicuticular waxes, *Chem. Phys. Lipids,* 19, 1, 1977.
167. **Franich, R. A., Goodin, S. J., and Hansen, E.**, Wax esters of the New Zealand silver fern, *Cyathea dealbata, Phytochemistry,* 24, 1093, 1985.
168. **Franich, R. A., Goodin, S. J., and Volkman, J. K.**, Alkyl esters from *Pinus radiata* foliage epicuticular wax, *Phytochemistry,* 24, 2949, 1985.
169. **Croteau, R. and Kolattukudy, P. E.**, Biosynthesis of hydroxyfatty acid polymers. Enzymatic synthesis of cutin from monomer acids by cell-free preparations from the epidermis of *Vicia faba* leaves, *Biochemistry,* 13, 3193, 1974.
170. **Mikkelsen, J.D.**, Synthesis of lipids by epidermal and mesophyll protoplasts isolated from barley leaf sheaths, in *Biogenesis and Function of Plant Lipids,* Mazliak, P., Benveniste, P., Costes, C., and Douce, R., Eds., Elsevier/North-Holland Biomedical Press, Amsterdam, 1980, 285.
171. **Bianchi, G., Avato, P., Bertorelli, P., and Mariani, G.**, Epicuticular waxes of two sorghum varieties, *Phytochemistry,* 17, 999, 1978.
172. **Simpson, D. and von Wettstein-Knowles, P.**, Structure of epicuticular waxes on spikes and leaf sheaths of barley as revealed by a direct platinum replica technique, *Carlsberg Res. Commun.,* 45, 465, 1980.
173. **von Wettstein-Knowles, P.**, Role of *cer-cqu* in epicuticular wax biosynthesis, *Biochem. Soc. Trans.,* 14, 576, 1986.
174. **Wissenbach, M., Siggaard-Andersen, M., Kauppinen, S., and von Wettstein-Knowles, P.**, Condensing enzymes of barley in *Metabolism, Structure and Utilization of Plant Lipids,* Cherif, A., Miled-Daoud, D., Marzouk, B., Smaoui, A., and Zarrouk, M., Eds., Centre National Pédagogique, Tunis, 1992, 393.
175. **Kauppinen, S. K.**, Structure and expression of the *Kas12* gene encoding a β-ketoacyl-ACP synthase I isozyme from barley, *J. Biol. Chem.,* 267, 23999, 1992.
176. **von Wettstein-Knowles, P.**, Barley (Hordeum vulgare) 2N = 14, in *Genetic Maps,* O'Brien, S. J., Ed., Cold Spring Harbor Laboratory Press, Cold Spring Harbor, New York, 1992, in press.

Chapter 5

OXYGENATED FATTY ACIDS OF THE LIPOXYGENASE PATHWAY

Brady A. Vick

TABLE OF CONTENTS

0-8493-4907-9/93/$0.00+$.50
© 1993 by CRC Press Inc.

167

I. INTRODUCTION

Among the promising fields of plant lipid research which have received special attention during the past decade, the study of oxygenated fatty acids stands out as an exciting example of a discipline in which important discoveries are yet to be revealed. Like adventurers deciphering a faded and torn treasure map, researchers of oxygenated fatty acid metabolism in plants have now identified important landmarks to serve as a basis for further exploration. The mysteries surrounding the physiological function of these diverse metabolites, the clues suggesting their roles as defense signals or metabolic regulators, and the similarities of the plant compounds to potent intracellular mediators of mammalian metabolism all combine to make the plant oxygenated fatty acids a fascinating topic for plant biochemists and physiologists. The "treasure" is yet to be discovered.

Fatty acid oxidation in plants occurs by at least four separate pathways. Two of these, α- and β-oxidation, result in the removal of one or two carbon units, respectively, from the carboxyl terminus of the fatty acid carbon chain and are described in detail in Chapter 16. Another pathway, ω-oxidation, results in oxidation at the methyl end of the fatty acid molecule. All three pathways are generally associated with the catabolism of fatty acids. The fourth pathway of oxidative metabolism, and the focus of this chapter, is the direct incorporation of dioxygen into a polyunsaturated fatty acid in a reaction catalyzed by the enzyme lipoxygenase. As we shall discover, the fatty acid hydroperoxide products of this reaction are predominantly metabolized by one of two divergent pathways in plants. The distinguishing enzymes of these two pathways are hydroperoxide lyase and hydroperoxide dehydrase. Together with the lipoxygenase reaction, these metabolic routes constitute the lipoxygenase pathway of plants. This oxidative pathway is not usually regarded as a catabolic pathway in the healthy, nonsenescing plant, but rather as a biosynthetic pathway leading to important mediators of plant metabolism.

Over the years, lipoxygenase has undergone name changes. The enzyme was first observed in legumes in the 1920s because of its ability to bleach carotenes in wheat flour and was initially termed carotene oxidase. After further studies showed that the oxidation of carotene was dependent upon the simultaneous oxidation of unsaturated lipids, the enzyme was renamed lipoxidase.[1] In the early 1970s, the term lipoxygenase (linoleate: oxygen oxidoreductase; EC 1.13.11.12) came into use to more accurately describe the reaction catalyzed. Several reviews which emphasize various aspects of the lipoxygenase pathway have been published in recent years.[2-8]

II. LIPOXYGENASE

A. LIPOXYGENASE REACTION

Lipoxygenase catalyzes the incorporation of dioxygen into polyunsaturated fatty acids containing a 1Z,4Z-pentadiene structure. In plants, the two most abundant fatty acids with this feature are linoleic and α-linolenic acids. The reaction may be diagrammed as:

$$\textit{cis} \qquad\qquad \textit{cis}$$

$$R - CH = CHCH_2CH = CH - R'$$

$$O_2 \;\downarrow\; \text{Lipoxygenase}$$

$$OOH$$

$$| \qquad \textit{trans} \qquad\qquad \textit{cis}$$

$$R - CH - CH = CH - CH = CH - R'$$

$$+$$

$$OOH$$

$$\textit{cis} \qquad\qquad \textit{trans} \qquad |$$

$$R - CH = CH - CH = CH - CH - R'$$

The oxygen molecule usually combines with the fatty acid at either the *n*-6 or the *n*-10 carbon, and the resulting hydroperoxide group assumes the *S*-stereoconfiguration. The *cis* double bond at the point of attack isomerizes to the *trans* configuration and moves into conjugation with the neighboring *cis* double bond. The formation of a conjugated double bond system in the product allows for a convenient spectrophotometric assay of the enzyme because the conjugated product absorbs strongly at 234 nm. Another common method to measure lipoxygenase activity is to monitor oxygen uptake through the use of an oxygen-specific electrode.

The regiospecificity (the specific point of oxygen attack) of lipoxygenases is dependent upon the source of the enzyme and upon reaction conditions. Some plant sources, such as soybean,[9] flaxseed,[10] apples,[11] and tea leaves,[12] have lipoxygenases that catalyze the incorporation of oxygen predominantly at C-13 of linoleic or linolenic acid. Lipoxygenases from other sources, such as tomato,[13] potato,[14] and corn,[15] oxygenate primarily at C-9. Some plants have several isozymes of lipoxygenase, and each can vary in its regiospecificity. The pH at which the reaction is conducted also influences the 9:13 ratio of the products.

B. STRUCTURAL PROPERTIES

Soybean seed lipoxygenase was one of the first enzymes to be purified and crystallized[16] and is the best-characterized of all plant lipoxygenases. It is composed of three isozymes, L-1, L-2, and L-3, which can be separated by anion exchange chromatography.[17,18] The three isozymes show differences in regiospecificity, pH optimum, and enzymatic properties. L-1 catalyzes the incorporation of oxygen predominantly at C-13 of linoleic and linolenic acids and has its optimum activity at pH 9, higher than most plant lipoxygenases. L-2 and L-3, which have their maximum activities at neutral pH, do not show positional selectivity for oxygen incorporation and catalyze the formation of approximately equal proportions of 9- and 13-hydroperoxide isomers. All three isozymes contain one atom of nonheme iron at the active site of the enzyme molecule.

The complete amino acid sequence of each soybean seed lipoxygenase isozyme has been deduced from the nucleotide sequence of its respective cDNA.[19-21] All three isozymes are composed of single polypeptide chains of similar lengths. L-1 has a M_r of 94,038 and is composed of 838 amino acids. L-2, the largest of the three, has a M_r of 97,036 and contains 865 amino acid residues. L-3 has 859 amino acids with a calculated M_r of 96,541. L-1 and L-3 share 81 and 74% homology, respectively, with the L-2 isozyme, with the highest homology occurring in the carboxyl-terminal half of their polypeptide chains. Within this section, there is a region known as the "histidine region", which has particularly high identity among the three soybean isozymes. This 44-amino acid segment begins at amino acids 489, 518, and 509 of L-1, L-2, and L-3, respectively. The same region is also highly conserved in rat and human lipoxygenases.[22-24] The highly conserved nature of this locus suggests that it participates at the active site surrounding the iron atom.[20]

The environment around the metal ion has been the subject of intense study. It is presumed that certain amino acids serve as ligands to the iron atom. Imidazole nitrogen atoms of histidine are common ligands of iron in many nonheme proteins, and so are oxygen atoms such as the hydroxyl of tyrosine or the carboxyl of acidic amino acids. Examination of the amino acids in the highly conserved region reveals six histidines; two tyrosines; and two acidic amino acids, aspartic and glutamic, which are possible ligands. Although tyrosine is a common ligand of iron, spectroscopic studies of soybean lipoxygenase rule out iron-tyrosine complexes. Thus, the environment around the iron atom is most likely an octahedral coordination sphere composed of a planar arrangement of four imidazole nitrogens of histidine with two carboxylate oxygens coordinated in the axial positions (Figure 1).[20,25,26] A water molecule has also been suggested as a possible ligand in place of one of the carboxylate anions.[27] The precise geometry of the coordination sphere around

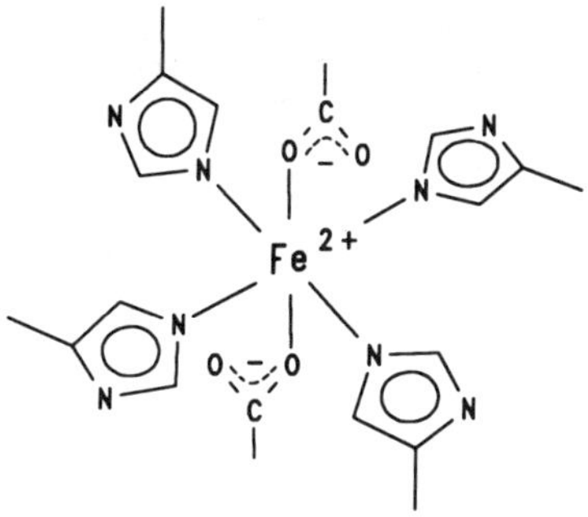

FIGURE 1. Model for the ligand coordination of iron in the active site of soybean lipoxygenase. Imidazole ligands are from histidine residues. (Redrawn from Navaratnam, S., et al., *Biochim. Biophys. Acta*, 956, 70, 1988. With permission.)

the iron atom in the active site will have to await the results of X-ray crystallographic studies, which are currently in the initial stage.[28,29]

Pea lipoxygenase is the only other plant lipoxygenase whose cDNA sequence has been reported. A strong similarity exists between the two major isozymes of pea and the L-2 and L-3 isozymes of soybean. The L-2- and L-3-like pea isozymes have 81 and 83% homology, respectively, with the DNA of the soybean isozymes, and their predicted M_r are 97,134 and 97,628.[30] Pea lipoxygenases also possess the conserved "histidine region" characteristic of soybean lipoxygenase, with five of the six histidines retained, as well as aspartic and glutamic acid.

C. REACTION MECHANISM
1. The Aerobic Reaction

The iron atom in lipoxygenase alternates between the Fe(II) and the Fe(III) states during catalysis. Native soybean seed L-1 is in the high spin Fe(II) state, and in this state the enzyme is largely inactive (Figure 2). Activation of the native E-Fe(II) enzyme requires oxidation of the iron atom from Fe(II) to Fe(III) by the reaction product, 13(*S*)-hydroperoxylinoleic acid. The addition of hydroperoxide product also results in visible color changes to the enzyme in solution. When a molar equivalent of 13(*S*)-hydroperoxylinoleic acid is added to the native enzyme to oxidize it to E-Fe(III), the activated enzyme takes on a yellow appearance and is referred to as the "yellow enzyme". Because of the product activation requirement, the oxygenation reaction exhibits a characteristic initial lag period. The source of the 13(*S*)-hydroperoxylinoleic acid required for activation is unclear, but probably results from oxygenation catalyzed by small amounts of active E-Fe(III) in equilibrium with native E-Fe(II).

Another colored lipoxygenase species, a "purple enzyme", results when a molar excess of 13(*S*)-hydroperoxylinoleic acid is added to soybean L-1. The purple enzyme constitutes only a small proportion of the active species, but it masks the color of the predominant yellow enzyme. It has an altered iron

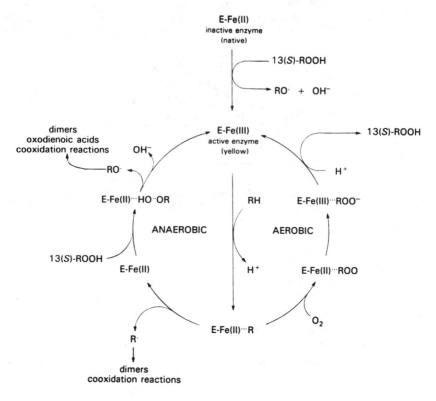

FIGURE 2. Proposed reaction mechanism for soybean lipoxygenase under aerobic and anaerobic conditions. (Adapted from Veldink, G. A., et al., *Prog. Chem. Fats Other Lipids,* 15, 131, 1977. With permission.)

coordination sphere, is unstable, and ultimately reverts back to the yellow form.

The initial interaction between the active enzyme and linoleic acid substrate is the loss of a hydrogen atom from C-11 of linoleic acid to form a linoleoyl radical. This occurs by the stereospecific abstraction of a proton from C-11, possibly by a basic amino acid, and the transfer of an electron back to the iron atom to reduce it to E-Fe(II). Molecular oxygen reacts with the linoleoyl radical to form a peroxyl radical, which subsequently accepts an electron from Fe(II) and acquires a proton to complete the hydroperoxidation. Finally, linoleoyl hydroperoxide dissociates from the enzyme which has returned to its active E-Fe(III) state.

The regiospecificity of oxygen addition to the linoleoyl radical appears to depend upon the ability of the enzyme to recognize the carboxyl or methyl terminal end of the substrate. At pH9, soybean L-1 recognizes and orients the methyl end of the linoleic acid. With Fischer projections as the reference, the enzyme abstracts the D-hydrogen from C-11 and directs oxygen substitution at C-13 from the opposite side of the fatty acid molecule, so that the hydroperoxide group is oriented in the L-stereoconfiguration.[31] Using the *R,S* convention for stereoconfiguration, this corresponds to abstraction of the pro(*S*) hydrogen and formation of the 13*S*-hydroperoxide. Some lipoxygenase enzymes, such as wheat,[32] recognize the carboxyl group, abstract the L-hydrogen from C-11, and catalyze the formation of $9D_S$-hydroperoxylinoleic acid.

2. The Anaerobic Reaction

Soybean L-1 also undergoes an anaerobic reaction which results in the formation of a variety of products. When the oxygen supply is depleted, the activated enzyme abstracts a hydrogen atom from C-11 as usual to form a linoleoyl radical, and the enzyme returns to the native E-Fe(II) form. Because no oxygen atoms are available to react with the radical, it dissociates from the enzyme to form a free radical. Some of the radicals recombine with other radicals to form fatty acid dimers. The native E-Fe(II) enzyme is reactivated by oxidation with 13-hydroperoxylinoleic acid product remaining from the aerobic reaction. Reduction products of the hydroperoxide are hydroxyl ions and alkoxy radicals, which rearrange or combine to form oxodienoic acids, dimers, and pentane. The alkyl and alkoxy radicals produced during the anaerobic reaction are thought to be responsible for the pigment bleaching activity of lipoxygenase, such as the carotene oxidizing activity originally attributed to the enzyme.

D. DOUBLE DIOXYGENATION

Some lipoxygenases can catalyze the incorporation of a second oxygen molecule into a fatty acid hydroperoxide if the reaction conditions are properly manipulated. The formation of the conjugated triene, 9,16-dihydroperoxy-10*E*,12*Z*,14*E*-octadecatrienoic acid, from α-linolenic acid has been reported in the case of soybean[33] and potato tuber lipoxygenases.[34] Double dioxygenation of α-linolenic acid seems to occur only if the initial oxygenation is at C-9.

Although arachidonic acid (5*Z*,8*Z*,11*Z*,14*Z*-icosatetraenoic acid) is not normally present in higher plants, it can serve as a substrate for double dioxygenation by plant lipoxygenases *in vitro*. Under suitable conditions, soybean lipoxygenase catalyzes the formation of 8*S*,15*S*-dihydroperoxy-5*Z*,9*E*,11*Z*,13*E*-icosatetraenoic acid[35] or 5*S*,15*S*-dihydroperoxy-6*E*,8*Z*,11*Z*,13*E*-icosatetraenoic acid.[36]

E. DISTRIBUTION OF LIPOXYGENASES
1. Distribution Among Species

Lipoxygenase was originally discovered in leguminous plants, and early work on the enzyme centered on its possible role in metabolic pathways peculiar to the leguminosae. It soon became clear, however, that lipoxygenases are widespread throughout the plant kingdom, including algae[37,38] and bryophytes.[39] Among the vascular plants, most angiosperms express lipoxygenase during some stage of development.[40,41] Lipoxygenase has been demonstrated in both monocotyledonous and dicotyledonous plants.

Remarkably, there have been few demonstrations of lipoxygenase in gymnosperms. Within this class, only the leaves of ginkgo, a primitive gymnosperm, have been reported to exhibit lipoxygenase activity, albeit low.[42] There are no accounts in the literature of the presence of lipoxygenase in conifers. It would be premature, however, to conclude that this class of plants is devoid of the enzyme. Attempts to demonstrate lipoxygenase in gymnosperms probably have not yet been exhaustive.

2. Distribution Among Plant Organs

The variety of plant organs in which lipoxygenase has been identified has been so diverse among species that it offers no useful clues concerning the functional role of the enzyme in plant metabolism. Seeds of numerous species, such as peas, soybeans, and flaxseed, are rich sources of lipoxygenase. In contrast, many plants do not express lipoxygenase until after the seed has germinated. For example, the cotyledons of many dicotyledonous oil-bearing plants, such as sunflower, cotton, and watermelon, have virtually no lipoxygenase activity in the dry seed, but exhibit high levels of activity after 1 week of growth. Hypocotyls of these plants are also good sources of lipoxygenase.

Leaves of some dicots, such as soybean, alfalfa, and tea, have considerable lipoxygenase activity. Frequently, however, leaves are poor sources of lipoxygenase. Buffered extracts of sunflower and spinach leaves, for example, show negligible lipoxygenase activity, even though they possess high activity of subsequent enzymes in the lipoxygenase pathway. It is not clear whether these leaves are inherently low in lipoxygenase activity or whether inhibitory substances, such as chlorophyll, interfere with the assay. In contrast, the coleoptiles and young leaves of many monocots, such as corn, barley, and wheat, are high in lipoxygenase activity. Numerous fruits, such as apple, pear, and tomato, have significant amounts of lipoxygenase. Potato tuber is a particularly rich source of the enzyme.

3. Subcellular Distribution

Many attempts have been made to identify a precise subcellular location of lipoxygenase activity. Unfortunately, the results of these studies have not been consistent across species or plant organs. Many reports have concluded that lipoxygenase is primarily a soluble, cytoplasmic enzyme. In several photosynthetic tissues, lipoxygenase has been demonstrated to be predominantly associated with the chloroplasts.[43,44] Vacuoles and lipid bodies have also been reported as sites of lipoxygenase sequestration.

F. LIPOXYGENASE INHIBITORS

A number of useful lipoxygenase inhibitors have been identified, and these have been important tools in the characterization of the lipoxygenase reaction mechanism. A few of the classical inhibitors, as well as some interesting new inhibitors, will be discussed briefly here. A thorough treatment of the current status of lipoxygenase inhibitors in mechanistic studies can be found in a recent review by Veldink and Vliegenthart.[8]

Classical inhibitors of lipoxygenase include acetylenic fatty acids, catechols, hydroxamic acids, and phenylhydrazine/phenylhydrazones. The 20-carbon fatty acid, 5,8,11,14-icosatetraynoic acid (ETYA), is a good example of the acetylenic fatty acid inhibitors. When ETYA is added to a solution containing lipoxygenase, the enzyme abstracts the n-8 methylene hydrogen and the resulting allene reacts irreversibly with the enzyme to inactivate it. Among the catechol inhibitors, nordihydroguaiaretic acid (NDGA), catechol, and n-propylgallate are frequently used as lipoxygenase inhibitors. These compounds, well-known as antioxidants, were originally thought to inhibit lipoxygenase by their ability to scavenge free radicals that are proposed to be intermediates in the catalytic mechanism. However, recent studies have shown that NDGA and catechol inactivate soybean L-1 by reduction of its iron atom from Fe(III) to Fe(II).[45]

Hydroxamic acids, such as salicylhydroxamic acid (SHAM), cause a strong inhibition of lipoxygenase by binding to the enzyme. Phenylhydrazine and various phenylhydrazones are thought to act as noncompetitive inhibitors by binding of the compound (or its autoxidation product) in the vicinity of the active site.[46] Other recognized inhibitors, especially of mammalian lipoxygenases, are nonsteroidal antiinflammatory drugs such as indomethacin and BW 755C.

G. PHYSIOLOGICAL ROLE OF HYDROPEROXIDE PRODUCTS

One of the great mysteries about plant lipoxygenase has been the question of the physiological role of its fatty acid hydroperoxide products, which are potentially toxic to the plant. Fatty acid hydroperoxides readily form hydroperoxy radicals that are known to disrupt membrane integrity, as well as inactivate proteins and amino acids by reacting with sulfhydryl groups.[47] DNA is also susceptible to degradation as a result of hydroperoxide attack on guanine nucleotides.[48] Why, then, have plants evolved the capacity to produce these potentially self-destructive products?

The reactive properties attributed to hydroperoxides have led to speculation that lipoxygenase participates in the senescence process in plants. However, conflicting experimental evidence leaves this question unanswered for the moment. Studies with peas have demonstrated that lipoxygenase activity increases as the pea leaf senesces, and the addition of lipoxygenase inhibitors retards senescence.[49] Likewise, lipoxygenase activity increases in senescing cotyledons of *Phaseolus vulgaris* concomitant with an increase in lipid hydroperoxides in the microsomal membranes, which ultimately leads to membrane disruption.[50] Both examples suggest a role for lipoxygenase in plant senescence. On the other hand, lipoxygenase activity in soybean cotyledons decreases as the cotyledons turn yellow and senesce. When senescence is reversed by removal of the seed pods, the cotyledons rejuvenate, and lipoxygenase activity increases.[51] Thus, the correlation between regreening cotyledons and elevated lipoxygenase activity argues against a role for lipoxygenase in senescence and instead supports a role for lipoxygenase in active plant growth.

Lipoxygenase has also been proposed to be involved in the regulation of the Calvin cycle.[43] According to this hypothesis, fatty acid hydroperoxide products oxidize thioredoxin, which is necessary to keep thiol groups of carbon fixation enzymes in their active, reduced state. Thus, an elevated lipoxygenase activity would result in increased oxidation of thioredoxin and a corresponding decrease in carbon fixation.

It is possible that many of the hypotheses put forward about the role of fatty acid hydroperoxides in plant metabolism have merit, even though results with various plant species appear to be contradictory. Conceivably, different plant species could have evolved unique uses for specific lipoxygenase isozymes. A more universal role for plant lipoxygenase, however, is its role in providing fatty acid hydroperoxide substrates for two divergent enzyme systems designated as the hydroperoxide lyase and hydroperoxide dehydrase pathways. These two metabolic routes, which are widespread in the plant kingdom, are described in the following section.

III. HYDROPEROXIDE METABOLISM

A. HYDROPEROXIDE LYASE PATHWAY
1. Hydroperoxide Lyase

Hydroperoxide lyase catalyzes the cleavage of fatty acid hydroperoxides into aldehyde and oxoacid fragments (Figure 3). The existence of the enzyme was first predicted from studies on the biogenesis of volatiles from bananas and other fruits.[52] Both 13- and 9-hydroperoxides of linoleic or linolenic acid serve as substrates for the enzyme, but each species may have a different specificity toward the positional isomers and toward linoleic or linolenic acid. Lyases isolated from leaf tissue typically have higher specificity toward linolenic acid-derived hydroperoxides than those from linoleic acid. In the majority of plants, hydroperoxide lyase is specific for 13-hydroperoxides. However, the enzyme from a few plants, such as cucumber and kidney bean leaves, utilizes both 9- and 13-hydroperoxide isomers. Pear fruit is an example where only the 9-hydroperoxide isomer is a suitable substrate for hydroperoxide lyase.[53]

It is now well-established that the volatile aldehydes produced by hydroperoxide lyase are important components of the characteristic aromas and flavors of many fruits and vegetables. When 13-hydroperoxy-9Z,11E,15Z-octadecatrienoic acid (originating from linolenic acid) is the substrate, the hydroperoxide lyase products are 3Z-hexenal and 12-oxo-9Z-dodecenoic acid. In many plants, these two metabolites are quickly isomerized to 2E-hexenal and 12-oxo-10E-dodecenoic acid. Whether the isomerization is catalyzed by enzyme action or occurs nonenzymically is not entirely clear for most of the

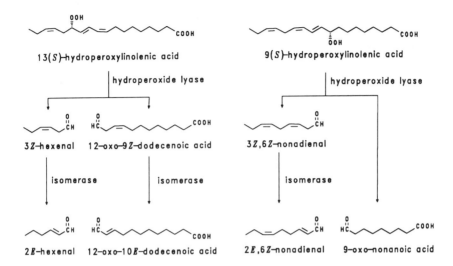

FIGURE 3. Reactions of the hydroperoxide lyase pathway with 13- and 9-hydroperoxides of linolenic acid. Similar reactions occur with 13- and 9-hydroperoxides of linoleic acid, resulting in hexanal and 2E-nonenal as the aldehyde products.

plants studied. Only in the case of cucumber has a *cis*-3:*trans*-2-enal isomerase been described.[54] In watermelon seedlings, the *cis:trans* isomerization activity could not be separated from the cleavage activity, and both activities were concluded to occur as a concerted mechanism.[55] It is possible, however, that isomerization could have occurred nonenzymically because *cis:trans* isomerization readily occurs under favorable reaction conditions, such as pH extremes.

Likewise, 3Z,6Z-nonadienal and 9-oxo-nonanoic acid are the products of hydroperoxide lyase-catalyzed cleavage of 9-hydroperoxy-10E,12Z,15Z-octadecatrienoic acid (Figure 3). As with the six-carbon unsaturated aldehyde product, double bond isomerization of the nine-carbon aldehyde proceeds readily to form 2E,6Z-nonadienal. Analogous reactions occur with 13- and 9-hydroperoxides of linoleic acid, resulting in hexanal, 3Z-nonenal, 2E-nonenal, and 9-oxo-nonanoic acid as the products. Isozymes of hydroperoxide lyase apparently exist in plants as evidenced by differences in substrate specificities toward hydroperoxides of linoleic or linolenic acids. The enzyme in nonphotosynthetic plant tissues is typically more active with hydroperoxides of linoleic acid, whereas hydroperoxide lyases in photosynthetic organs usually show higher activity with the hydroperoxides of linolenic acid.[56] Some plants also possess alcohol dehydrogenases that reduce the aldehyde products to alcohols. One of these products, 3Z-hexenol, also known as leaf alcohol, imparts a characteristic odor frequently associated with green leaves.

Hydroperoxide lyase is also present in algae, but the products are different. In algae, the enzyme cleaves the 13-hydroperoxide of linoleic or linolenic acid to produce 5- and 13-carbon fragments, rather than 6- and 12-carbon fragments as observed in higher plants. The freshwater alga *Chlorella pyrenoidosa* possesses a hydroperoxide lyase that converts the 13-hydroperoxide of linoleic or linolenic acid to pentane (from linoleic) or pentene (from linolenic) and 13-oxo-9Z,11E-tridecadienoic acid.[57] A similar cleavage takes place with hydroperoxide lyase from the blue-green alga *Oscillatoria* sp., except that the five-carbon product is pentanol rather than pentane.[58]

Hydroperoxide lyases have been partially purified and characterized from several plant sources, mostly dicots.[55,59-62] In each species, hydroperoxide lyase has a high molecular weight, usually in the range of 200,000 to 250,000. The molecular weight of native soybean hydroperoxide lyase, estimated by gel filtration, is approximately 250,000,[62] but SDS gel electrophoresis of the purified enzyme reveals a protein band with a M_r of only 62,000, suggesting that soybean hydroperoxide lyase exists as a tetramer in its native state. It is likely that other high molecular weight hydroperoxide lyases are also composed of multiple subunits. The enzyme from tea leaves (M_r 55,000)[63] is probably monomeric in its native state, as is the enzyme from lower plant forms such as *C. pyrenoidosa* (M_r 48,000)[57] and *Oscillatoria* sp. (M_r 56,000).[58]

2. Subcellular Distribution

Most hydroperoxide lyases appear to be membrane bound. Detergents such as Triton X-100 are generally required in the tissue extraction medium in order to solubilize the enzyme. Because they are membrane-bound proteins, hydroperoxide lyases are highly hydrophobic and tend to aggregate. Thus, detergents must also be present in the eluants during chromatographic procedures in order to suppress hydrophobic aggregation of the enzyme. In the flesh of cucumber fruit, a nonphotosynthetic tissue, hydroperoxide lyase is associated with three membrane fractions: plasma membrane, Golgi, and endoplasmic reticulum. The enzyme in photosynthetic tissues is typically found in chloroplasts where it is bound to the thylakoid membrane.[59,64]

3. Physiological Role of Hydroperoxide Lyase Pathway Products

The function of hydroperoxide lyase in plant metabolism is unresolved, although most hypotheses focus on a role for the enzyme in plant defense and wound repair. One of the hydroperoxide lyase products originating from linolenic acid, 2E-hexenal, is an effective fungicide,[65,66] bactericide,[67] and insecticide.[68] The other cleavage product, 12-oxo-10E-dodecenoic acid, is the active component of traumatin, the so-called "wound hormone".[69] The putative function of plant wound hormone, first proposed in the early part of this century,[70] is to stimulate cell division near the wound site, resulting in the formation of a protective callus around the wound. Thus, in its proposed role as a protective enzyme, hydroperoxide lyase simultaneously provides two active defense agents: (1) a short-chain aldehyde to destroy or inhibit the attacking pest and (2) a wound-healing agent to protect and mend the damaged tissue. The 9-carbon aldehydes resulting from cleavage of 9-hydroperoxy fatty acids probably have similar functions; however, no role has yet been proposed for 9-oxo-nonanoic acid.

A recent hypothesis has been proposed in which lipoxygenase and hydroperoxide lyase participate in oxidative catabolism of triacylglycerols during fat mobilization in germinating seedlings.[71] According to this proposal, lipoxygenase present in the lipid bodies of germinating cucumber seedlings may utilize triacylglycerol directly as a substrate and catalyze the oxygenation of polyunsaturated fatty acids at carbon 9 while they are still covalently bound to glycerol. Hydroperoxide lyase could cleave the hydroperoxides to produce 9-carbon aldehydes, which could be oxidized to the acid and shuttled to the glyoxysomes for β-oxidation, bypassing the need for fatty acid hydrolysis by lipase.

It should be emphasized that the aforementioned roles are still only proposals. The lack of solid evidence for the function of hydroperoxide lyase in plants reaffirms the conviction that exciting opportunities still exist for imaginative new hypotheses about its role, and for innovative ideas and methods to test and assess their validity.

B. HYDROPEROXIDE DEHYDRASE PATHWAY: (+)7-ISO-JASMONIC ACID BIOSYNTHESIS

1. Enzymes of the Hydroperoxide Dehydrase Pathway

a. Hydroperoxide Dehydrase

Hydroperoxide dehydrase (EC 4.2.1.92), also called allene oxide synthase, is the first enzyme of a divergent pathway of hydroperoxide metabolism that leads to the biosynthesis of (+)7-iso-jasmonic acid, a plant growth regulator. The enzyme catalyzes the dehydration of a fatty acid hydroperoxide to form an allene oxide (Figure 4).[72,73] Discovered by Zimmerman,[74] it was initially named "hydroperoxide isomerase" because at that time the participation of a short-lived allene oxide intermediate in the pathway was not known. The specificity of hydroperoxide dehydrase toward the 13- or 9-hydroperoxide of linoleic or linolenic acid is species dependent. In flaxseed and cotton, there is a strong preference for the 13-hydroperoxide isomer,[75,76] whereas the enzyme from corn germ shows no preference.[77] In this chapter, we will focus on hydroperoxide dehydrase conversion of 13-hydroperoxy-9Z,11E,15Z-octadecatrienoic acid because only this substrate leads to the formation of (+)7-iso-jasmonic acid. The product of the hydroperoxide dehydrase reaction is the allene oxide, 12,13(S)-epoxy-9Z,11,15Z-octadecatrienoic acid (Figure 4). It is extremely unstable and has a half-life of only 26 sec at 0°C.[78] Two types of products result from nonenzymatic transformation of the transitory allene oxide: ketols and a cyclopentenyl metabolite. The ketols are produced by spontaneous hydrolysis of the allene oxide to form an α-ketol, 12-oxo-13-hydroxy-9Z,15Z-octadecadienoic acid, and a γ-ketol, 12-oxo-9-hydroxy-10E,15Z-octadecadienoic acid (Figure 4). Of the two ketols formed, the α-

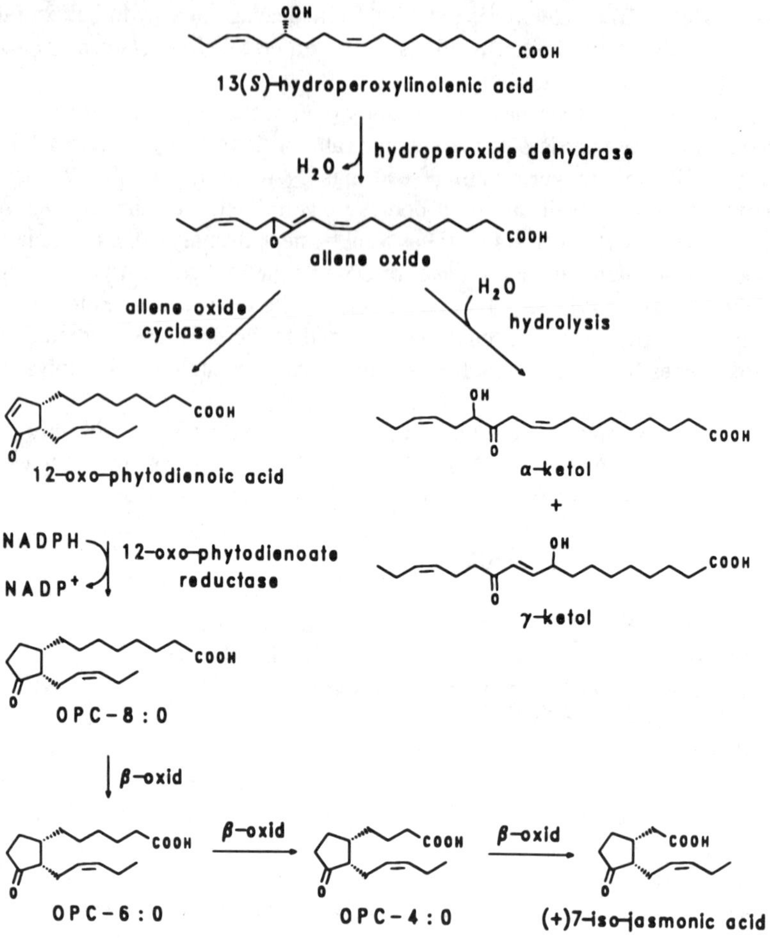

FIGURE 4. Reactions of the hydroperoxide dehydrase pathway with 13-hydroperoxylinolenic acid as substrate. Other substrates, such as 13-hydroperoxylinoleic acid, 9-hydroperoxylinoleic acid, or 9-hydroperoxylinolenic acid, lead only to the formation of ketols and do not participate in the biosynthesis of jasmonic acid.

ketol is the predominant product. It was the formation of ketols that first led to the description of the enzyme as a "hydroperoxide isomerase".

The second type of product, a cyclopentenyl compound, results from the spontaneous cyclization of the allene oxide to form 8-[2-(cis-2'-pentenyl)-3-oxo-cyclopent-4-enyl]octanoic acid (Figure 4), which was given the common name 12-oxo-phytodienoic acid (12-oxo-PDA). The double bond at carbon 15 of the allene oxide intermediate appears to facilitate the cyclization reaction, because in its absence the cyclization rate is low. The alkyl side chains are in the *cis* configuration, but in the case of spontaneous cyclization a racemic mixture (9S,13S and 9R,13R) is formed. As we shall see later, the cyclization reaction can also occur with stereospecificity through enzyme catalysis. The original identification of the cyclic product occurred prior to the discovery of the allene oxide intermediate, and consequently the term "hydroperoxide cyclase" was used for this reaction in the early literature.[79] The names "hydroperoxide isomerase" and "hydroperoxide cyclase" have now been discontinued.

Hydroperoxide dehydrase from flaxseed has been purified and characterized as a cytochrome P-450 enzyme with a M_r of 55,000.[80] It is a hydrophobic, membrane-bound protein which usually requires the presence of a detergent for complete solubilization. Hydroperoxide dehydrase from other plant sources frequently has a much higher M_r, typically 220,000 to 250,000,[59,76] suggesting that it probably exists as a tetramer. While it is commonly accepted that hydroperoxide dehydrase is associated with a particulate fraction, the specific organelle or membrane location is not known in most cases. However, in spinach leaves, the enzyme has been conclusively demonstrated in chloroplast membranes, but this has not been confirmed as a general property of dehydrases in all photosynthetic tissues.

b. Allene Oxide Cyclase

Allene oxide cyclase (EC 5.3.99.6) catalyzes the stereospecific cyclization of 12,13(*S*)-epoxy-9Z,11,15Z-octadecatrienoic acid (allene oxide) to 9*S*,13*S*-12-oxo-phytodienoic acid.[78] The enzyme has been detected in several plant species, but is best-characterized from corn kernels and potato tubers. Corn allene oxide cyclase has a M_r of 45,000, whereas the potato enzyme is slightly larger at 50,000. Differential centrifugation of enzyme preparations from both sources shows that the cyclase enzyme is found predominantly in the 105,000 *g* supernatant fraction, indicating that allene oxide cyclase is a soluble enzyme.[78]

c. 12-Oxo-Phytodienoate Reductase

12-Oxo-phytodienoate reductase (EC 1.3.1.42) catalyzes the reduction of the Δ^{10} double bond of 12-oxo-phytodienoic acid to produce (1*S*,2*S*)3-oxo-2-(2′pentenyl)cyclopentaneoctanoic acid (abbreviated OPC-8:0, where 8:0 refers to the eight-carbon side chain).[81] OPC-8:0 retains the same *cis* stereochemistry of the side chains as its precursor. The enzyme has been observed in many species,[82] including algae,[57] but its partial purification has only been described in corn kernel. In corn, the enzyme has a M_r of 54,000 and utilizes NADPH as the preferred reductant. The K_m of 12-oxo-phytodienoic acid is 190 μM, and for NADPH it is 13 μM. The subcellular location of 12-oxo-phytodienoate reductase has not been systematically investigated.

d. β-Oxidation Enzymes

The existence of β-oxidation enzymes in the (+)7-iso-jasmonic acid pathway has only been inferred on the basis of the observed products, which are OPC-6:0, OPC-4:0, and OPC-2:0 [(+)7-iso-jasmonic acid]. Each product retains the *cis* configuration of the side chains. Because no metabolites with odd-numbered side chains were identified, the enzymes responsible for the suc-

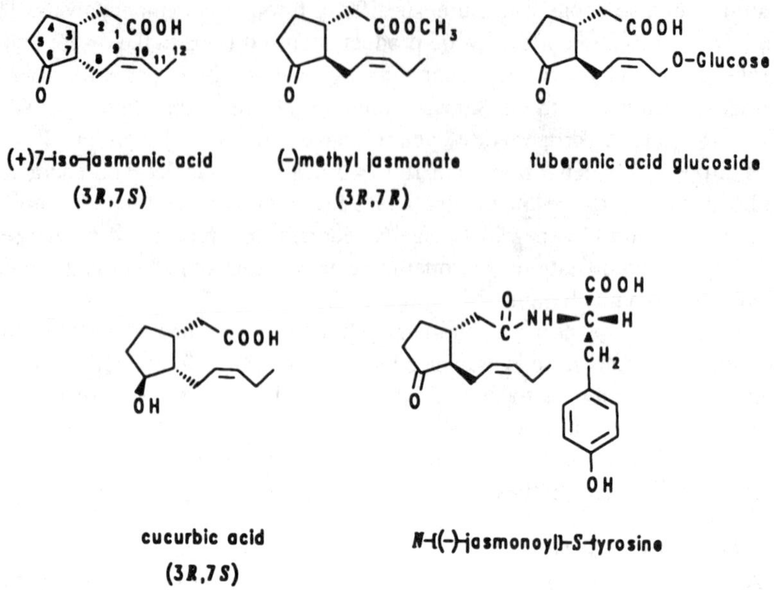

(+)7-iso-jasmonic acid
(3*R*,7*S*)

(–)methyl jasmonate
(3*R*,7*R*)

tuberonic acid glucoside

cucurbic acid
(3*R*,7*S*)

N-(–)jasmonoyl)-*S*-tyrosine

FIGURE 5. Examples of some naturally occurring derivatives of (+)7-iso-jasmonic acid.

cessive two-carbon losses from OPC-8:0 were concluded to be due to β-oxidation. It is not known whether enzymes of the normal β-oxidation pathway are involved or whether a specialized β-oxidation pathway exists for (+)7-iso-jasmonic acid biosynthesis. The intracellular site of the chain-shortening process is unknown.

2. Jasmonate Stereochemistry and Metabolism

A brief review of the stereochemistry of (+)7-iso-jasmonic acid and its related metabolites (jasmonates) is relevant to the discussion of their biological activity in Section III.B.4. In this chapter, we will use an informal, but often used, numbering system for the carbon skeleton of the 12-carbon jasmonates, with the carboxyl group designated as C-1 and the bridgehead carbons as C-3 and C-7 (Figure 5). (–)Methyl jasmonate was the first of the jasmonates to be characterized and was originally isolated from the essential oil of *Jasminum grandiflorum* L.[83] It has a pleasant fragrance and serves as an important component of perfumes. The side chains of (–)methyl jasmonate have a *trans* configuration (3*R*,7*R*) in contrast to the *cis* (3*R*,7*S*) form of (+)7-iso-jasmonic acid. Its acid form, (–)jasmonic acid, was first purified and characterized as a plant growth inhibitor from the fungus *Lasiodiplodia theobromae*.[84]

Based on its biosynthetic pathway, (+)7-iso-jasmonic acid (also called epijasmonic or *cis*-jasmonic acid) is generally accepted as the primary stereoisomer product of the hydroperoxide dehydrase pathway. However, this compound slowly epimerizes at carbon 7, and even more rapidly under heat, acid, or alkaline conditions, to form (–)jasmonic acid. Because most extraction and purification techniques employ at least one of these conditions, some epimerization at carbon 7 is inevitable. It is not known with certainty whether (+)7-iso-jasmonic acid is the only physiologically active compound *in vivo* or whether (–)jasmonic acid and other stereoisomers are also active. This issue has remained ambiguous because most studies on the biological activity of jasmonates have utilized chemically synthesized (±)methyl jasmonate or (±)jasmonic acid, which are stereoisomeric mixtures in the approximate ratio of 92:8, (±)jasmonate:(±)7-iso-jasmonate. The recent synthesis of stereochemically pure jasmonates should help to answer this question. Preliminary evidence suggests that the various stereoisomers differ in biological activity in different bioassays,[85] although (+)7-iso-jasmonic acid is frequently the most effective isomer.[86]

Many derivatives of (+)7-iso-jasmonic acid have been isolated as natural products from plants and fungi. While none of the interconversions of the jasmonates have yet been characterized, the probable routes are predictable and involve ketone reductions, ω-oxidations, glycosylations, and amide formation. Some examples of derivatives of (+)7-iso-jasmonic acid and (–)jasmonic acid are illustrated in Figure 5.

3. Physiological Role of Ketols

The ketol products of hydroperoxide dehydrase have no known physiological role. In fact, it is quite possible that ketols are not formed in significant amounts during normal *in vivo* metabolism in unstressed plants. Ketols may only exist as products of *in vitro* allene oxide metabolism or in wounded plants where extensive cell disruption has occurred. Cell disruption would likely force allene oxide metabolites into an environment more favorable for hydrolysis than for stereospecific cyclization by allene oxide cyclase.

Recently, it has been shown that ketols resulting from linolenic acid metabolism can be further dioxygenated *in vitro* by lipoxygenase to form hydroperoxides of α-ketols.[87] Whether this is a significant reaction *in vivo* remains to be established. This recent discovery opens up new opportunities for investigation of the physiological effects of multioxygenated fatty acids in plant metabolism.

4. Physiological Role of Jasmonates
a. Growth Inhibition and Promotion

A remarkable surge in research concerning the physiological relevance of (+)7-iso-jasmonic acid and other jasmonates has taken place within the past 2 years. Excellent reviews on the status of jasmonates in plant metabolism have been written,[88,89] but the dynamic nature of the field requires vigilant attention to the literature to keep abreast. In recent years, the jasmonates have come to be regarded by many as a new class of plant growth regulator. Both inhibitory and promotive effects have been attributed to the jasmonates in plant bioassays. Examples of inhibitory activities include retardation of seedling growth and tissue culture growth, suppression of seed and pollen germination, and inhibition of pigment formation. Jasmonates frequently promote processes associated with plant stress, such as stomatal closure, senescence, chlorophyll degradation, and respiration.[88] In this respect, jasmonates share many similarities to abscisic acid. An example of growth promotion is the induction of tuber formation in potatoes by several jasmonates, including jasmonic acid, methyl jasmonate, cucurbic acid, and tuberonic acid (Figure 5).[90] Jasmonic acid also stimulates the development of axillary buds and adventitious roots on potato stem cuttings.[91]

b. Chemical Signaling

An intriguing role proposed for jasmonic acid that has received wide attention is its function as a signal, i.e., a chemical messenger, in response to certain stressors. Anderson[92] has presented a thoughtful review of the pre-1988 literature on the role of fatty acid-derived products as second messengers. Since then, further evidence has strengthened the hypothesis that jasmonates function as chemical signals. For example, water stress (desiccation) induces the accumulation of several characteristic polypeptides in barley leaves, as well as the leaves of many other plant species. Interestingly, the application of methyl jasmonate to healthy, unstressed leaves results in induction of the same polypeptides, which are referred to as jasmonate-induced polypeptides (JIPs).[93] A similar phenomenon occurs in soybean leaves[94,95] when either the seed pods are removed, phloem export is blocked, or the plant is subjected to water deficit. All three stress treatments induce the synthesis of two soybean leaf polypeptides (28 and 31 kDa) known as vegetative storage proteins (VSP). VSPs are thought to be repositories for nitrogen, which can be mobilized under conditions favorable for growth and transported to developing plant organs. Application of methyl jasmonate to soybean plants mimics the effects of stressors by inducing VSP synthesis in healthy soybean plants. Thus, the ability of jasmonates to induce JIP and VSP formation suggests that they are chemical signals dispatched from a stress sensor apparatus and transported to the leaves where they induce the synthesis of specific polypeptides.

Similar proposals for chemical signaling have been put forward for jasmonates in tendril coiling. Gaseous methyl jasmonate promotes tendril coiling in *Bryonia dioica* Jacq., and the kinetics of the coiling parallels that of coiling induced by mechanical stimulation.[96] Hence, the suggestion has been made that methyl jasmonate is an endogenous chemical signal produced in mechanically perturbed parts of a tendril that then diffuses throughout the intracellular spaces to activate tendril coiling.

Further strong evidence of a role for jasmonates as chemical signals for plant defense comes from plant cell suspension culture experiments. Cultured cells of *Rauvolfia canescens* and *Eschscholtzia californica* (California poppy) can be induced by a yeast elicitor to synthesize low molecular weight defensive compounds derived from the phenylpropanoid pathway.[97] Within 30 min after exposure to the elicitor, the concentration of jasmonic acid increases sharply, suggesting that the elicitor stimulates the jasmonate biosynthetic pathway to produce jasmonic acid as a chemical signal. Furthermore, when jasmonates are applied to cell cultures in the absence of the yeast elicitor, they independently induce the synthesis of defensive compounds. Thus, jasmonic acid appears to be a chemical messenger, synthesized in response to a pathogen, which turns on a set of defensive genes.

The best-characterized model for chemical signaling by jasmonates is the synthesis of two wound-inducible proteinase inhibitors in tomato leaves.[98] Each proteinase inhibitor is a low molecular weight protein that interferes with insect digestion, prompting a decline in feeding by the insect. Synthesis of the inhibitors is activated either by insects chewing on the leaf or by mechanical damage to the leaf. Accumulation of the inhibitors is both local and systemic. Application of jasmonic acid, or its biosynthetic precursors, to the leaf leads to the same specific induction of proteinase inhibitor synthesis. The jasmonates also stimulate the synthesis of new inhibitor mRNA, indicating that regulation occurs at the level of transcription. In addition to induction by jasmonic acid, exposure of tomato plants to gaseous methyl jasmonate, or to plant species like sagebrush that synthesize methyl jasmonate, also stimulates proteinase inhibitor synthesis. Such a response by one plant to the presence of another plant suggests a mechanism for interplant communication.[99] The similarity between inhibitor induction by wounding and induction by jasmonates has led to the hypothesis that jasmonates are key chemical messengers produced in response to wounding and that they play an important role in plant defense.

Thus, a similar pattern is evident from the various reports on jasmonates as signal molecules. First, a stressor (e.g., insect, mechanical, water deficit) stimulates the biosynthesis of jasmonic acid, possibly by activating a lipase to release the precursor, linolenic acid. Second, jasmonic acid or a related jasmonate acts as a signal to directly or indirectly activate a class of species-specific genes that code for low or high molecular weight proteins with specific functions in plant defense.

C. DIVINYL ETHER FORMATION

Extracts of potato tubers catalyze the conversion of 9-hydroperoxides of linoleic or linolenic acid to divinyl ether fatty acids, named colneleic and colnelenic acids, respectively.[100]

$$CH_3CH_2CH = CHCH_2 - CH = CH - CH = CH - O - CH = CH(CH_2)_6 COOH$$

colnelenic acid

The reaction appears to be unique to potatoes because neither product has been reported as a metabolite of fatty acid hydroperoxides in any other species. Divinyl ethers of fatty acids are relatively unstable, especially in the presence of metal ions, and decompose to aldehyde products similar to those from hydroperoxide lyase.

IV. COMPARISON TO MAMMALIAN PATHWAYS

Plants and animals have obvious differences in the manner in which they metabolize polyunsaturated fatty acids by dioxygenation. In plants, it is the 18-carbon fatty acids which are the predominant substrates for conversion to the octadecanoids, the oxygenated products of linoleic or linolenic acid. In contrast, the principal fatty acid substrate in mammals for dioxygenation is the 20-carbon fatty acid, arachidonic acid, which is metabolized to a wide variety of oxygenated derivatives called the icosanoids.[101] At present, the lipoxygenase pathway is the only pathway known to be operative in plants for fatty acid dioxygenation, whereas animals have two primary dioxygenation pathways: one initiated by lipoxygenase and the other by cycloxygenase. Plant and animal lipoxygenases also have differences in regiospecificity. The lipoxygenases from plants usually catalyze dioxygenation at either carbon 13 or carbon 9 of linoleic or linolenic acid, whereas lipoxygenases from mammals are classified into several types depending upon regiospecificity. Three of these, 15-lipoxygenase, 12-lipoxygenase, and 5-lipoxygenase, are the most important.

The physiological function of the icosanoids in mammalian metabolism is much better understood than is the function of the octadecanoids in plant metabolism. Metabolites of the cyclooxygenase pathway include prostaglandins, thromboxane, and prostacyclin. Thromboxane, found in platelets, is a vasoconstrictor and stimulator of platelet aggregation, whereas prostacyclin has an opposing function as a vasodilator and inhibitor of platelet aggregation. The roles of the 15- and 12-lipoxygenase products are not yet well-understood, but 5-lipoxygenase is recognized as the first step in the biosynthesis of leukotrienes. The leukotrienes are synthesized by a variety of white blood cells and by the cells of several body organs. When conjugated with specific peptides, the peptidoleukotrienes act at very low concentrations to contract respiratory, vascular, and intestinal smooth muscle. They have been implicated as active mediators of asthma, inflammatory reactions, and hypersensitive reactions (allergies).

While the differences between plant and animal systems for the metabolism of polyunsaturated fatty acids are readily apparent, there are also noteworthy parallels. It is remarkable that both plants and animals have retained lipoxygenase over the course of evolution, suggesting that the enzyme was essential to the survival of a primitive ancestor common to both kingdoms. Although a plant cyclooxygenase has not yet been demonstrated, plants have nevertheless acquired the ability, through the lipoxygenase pathway, to produce fatty acids with a cyclopentane ring structure similar to the mammalian prostaglandins. A common theme emerging from studies of the physiological function of oxygenated fatty acids in both mammalian and plant metabolism is their role in injured cells. Many of the mammalian icosanoids have been implicated in the inflammatory response. Similarly, metabolites of the plant octadecanoid pathway, such as jasmonic acid, are thought to have roles as signal molecules in response to mechanical injury and other stress conditions. The next several years will be fascinating times for research in this field of plant lipid metabolism, as contemporary techniques such as molecular cloning are integrated with traditional plant biochemistry and physiology to offer exciting new approaches to the mysteries still associated with the lipoxygenase pathway in plants.

REFERENCES

1. **André, E. and Hou, K.-W.**, The presence of a lipoid oxidase in soy bean, *Glycine soya*, Lieb., *C. R. Hebd. Séances Acad. Sci.*, 194, 645, 1932.
2. **Kühn, H., Schewe, T., and Rapoport, S. M.**, The stereochemistry of the reactions of lipoxygenases and their metabolites. Proposed nomenclature of lipoxygenases and related enzymes, *Adv. Enzymol.*, 58, 273, 1986.
3. **Vick, B. A. and Zimmerman, D. C.**, Oxidative systems for modification of fatty acids: the lipoxygenase pathway, in *The Biochemistry of Plants: A Comprehensive Treatise, Vol. 9, Structure and Function*, Stumpf, P. K. and Conn, E. E., Eds., Academic Press, Orlando, FL, 1987, 53.
4. **Mack, A. J., Peterman, T. K., and Siedow, J. N.**, Lipoxygenase isozymes in higher plants: biochemical properties and physiological role, *Isozymes: Curr. Top. Biol. Med. Res.*, 13, 127, 1987.
5. **Hildebrand, D. F., Hamilton-Kemp, T. R., Legg, C. S., and Bookjans, G.**, Plant lipoxygenases: occurrence, properties and possible functions, *Curr. Top. Plant Biochem. Physiol.*, 7, 201, 1988.
6. **Gardner, H. W.**, Recent investigations into the lipoxygenase pathway of plants, *Biochim. Biophys. Acta*, 1084, 221, 1991.
7. **Siedow, J. N.**, Plant lipoxygenase: structure and function, *Annu. Rev. Plant Physiol. Plant Mol. Biol.*, 42, 145, 1991.
8. **Veldink, G. A. and Vliegenthart, J. F. G.**, Substrates and products of lipoxygenase catalysis, in *Studies in Natural Products Chemistry, Vol. 9, Structure and Chemistry*, Rahman, Atta-ur, Ed., Elsevier Science Publishers, Amsterdam, 1991, 559.
9. **Hamberg, M. and Samuelsson, B.**, Specificity of the oxygenation of unsaturated fatty acids catalyzed by soybean lipoxidase, *J. Biol. Chem.*, 424, 5329, 1967.
10. **Zimmerman, D. C.**, Specificity of flaxseed lipoxidase, *Lipids*, 5, 392, 1970.
11. **Grosch, W., Laskawy, G., and Fischer, K.-H.**, Positions-Spezifität der Peroxidierung von Linol- und Linolensäure durch Homogenate aus Äpfeln und Birnen, *Z. Lebensm. Unters. Forsch.*, 163, 203, 1977.
12. **Kajiwara, T., Nagata, N., Hatanaka, A., and Naoshima, Y.**, Stereoselective oxygenation of linoleic acid to 13-hydroperoxide in chloroplasts from *Thea sinensis*, *Agric. Biol. Chem.*, 44, 437, 1980.
13. **Matthew, J. A., Chan, H. W.-S., and Galliard, T.**, A simple method for the preparation of pure 9-D-hydroperoxide of linoleic acid and methyl linoleate based on the positional specificity of lipoxygenase in tomato fruit, *Lipids*, 12, 324, 1977.
14. **Galliard, T. and Phillips, D. R.**, Lipoxygenase from potato tubers; partial purification and properties of an enzyme that specifically oxygenates the 9-position of linoleic acid, *Biochem. J.*, 124, 431, 1971.
15. **Gardner, H. W.**, Sequential enzymes of linoleic acid oxidation in corn germ: lipoxygenase and linoleate hydroperoxide isomerase, *J. Lipid Res.*, 11, 311, 1970.

16. **Theorell, H., Holman, R. T., and Åkeson, Å.,** Crystalline lipoxidase, *Acta Chem. Scand.,* 1, 571, 1947.

17. **Christopher, J. P., Pistorius, E. K., and Axelrod, B.,** Isolation of an isozyme of soybean lipoxygenase, *Biochim. Biophys. Acta,* 198, 12, 1970.

18. **Christopher, J. P., Pistorius, E. K., and Axelrod, B.,** Isolation of a third isozyme of soybean lipoxygenase, *Biochim. Biophys. Acta,* 284, 54, 1972.

19. **Shibata, D., Steczko, J., Dixon, J. E., Hermodson, M., Yazdanparast, R., and Axelrod, B.,** Primary structure of soybean lipoxygenase-1, *J. Biol. Chem.,* 262, 10080, 1987.

20. **Shibata, D., Steczko, J., Dixon, J. E., Andrews, P. C., Hermodson, M., and Axelrod, B.,** Primary structure of soybean lipoxygenase-2, *J. Biol. Chem.,* 263, 6816, 1988.

21. **Yenofsky, R. L., Fine, M., and Liu, C.,** Isolation and characterization of a soybean (*Glycine max*) lipoxygenase-3 gene, *Mol. Gen. Genet.,* 211, 215, 1988.

22. **Dixon, R. A. F., Jones, R. E., Diehl, R. E., Bennett, C. D., Kargman, S., and Rouzer, C. A.,** Cloning of the cDNA for human 5-lipoxygenase, *Proc. Natl. Acad. Sci. U.S.A.,* 85, 416, 1988.

23. **Matsumoto, T., Funk, C. D., Rådmark, O., Höög, J.-O., Jörnvall, H., and Samuelsson, B.,** Molecular cloning and amino acid sequence of human 5-lipoxygenase, *Proc. Natl. Acad. Sci. U.S.A.,* 85, 26, 1988 and correction 85, 3406, 1988.

24. **Balcarek, J. M., Theisen, T. W., Cook, M. N., Varrichio, A., Hwang, S.-M., Strohsacker, M. W., and Crooke, S. T.,** Isolation and characterization of a cDNA clone encoding rat 5-lipoxygenase, *J. Biol. Chem.,* 263, 13937, 1988.

25. **Dunham, W. R., Carroll, R. T., Thompson, J. F., Sands, R. H., and Funk, M. O., Jr.,** The initial characterization of the iron environment in lipoxygenase by Mössbauer spectroscopy, *Eur. J. Biochem.,* 190, 611, 1990.

26. **Navaratnam, S., Feiters, M. C., Al-Hakim, M., Allen, J. C., Veldink, G. A., and Vliegenthart, J. F. G.,** Iron environment in soybean lipoxygenase-1, *Biochim. Biophys. Acta,* 956, 70, 1988.

27. **Nelson, M. J.,** Evidence for water coordinated to the active site iron in soybean lipoxygenase-1, *J. Am. Chem. Soc.,* 110, 2985, 1988.

28. **Steczko, J., Muchmore, C. R., Smith, J. L., and Axelrod, B.,** Crystallization and preliminary x-ray investigation of lipoxygenase 1 from soybeans, *J. Biol. Chem.,* 265, 11352, 1990.

29. **Boyington, J. C., Gaffney, B. J., and Amzel, L. M.,** Crystallization and preliminary x-ray analysis of soybean lipoxygenase-1, a non-heme iron-containing dioxygenase, *J. Biol. Chem.,* 265, 12,771, 1990.

30. **Ealing, P. M. and Casey, R.,** The cDNA cloning of a pea (*Pisum sativum*) seed lipoxygenase; sequence comparisons of the two major pea seed lipoxygenase isoforms, *Biochem. J.,* 264, 929, 1989.

31. **Veldink, G. A., Vliegenthart, J. F. G., and Boldingh, J.,** Plant lipoxygenases, *Prog. Chem. Fats Other Lipids,* 15, 131, 1977.

32. **Kühn, H., Heydeck, D., Wiesner, R., and Schewe, T.,** The positional specificity of wheat lipoxygenase; the carboxylic group as signal for the recognition of the site of the hydrogen removal, *Biochim. Biophys. Acta,* 830, 25, 1985.

33. **Sok, D.-E. and Kim, M. R.,** Enzymatic formation of 9,16-dihydro(pero)xyoctadecatrienoic acid isomers from α-linolenic acid, *Arch. Biochem. Biophys.,* 277, 86, 1990.

34. **Grechkin, A. N., Kuramshin, R. A., Safonova, E. Y., Yefremov, Y. J., Latypov, S. K., Ilyasov, A. V., and Tarchevsky, I. A.,** Double hydroperoxidation of α-linolenic acid by potato tuber lipoxygenase, *Biochim. Biophys. Acta,* 1081, 79, 1991.

35. **Bild, G. S., Ramadoss, C. S., and Axelrod, B.,** Multiple dioxygenation by lipoxygenase of lipids containing all-*cis*-1,4,7-octatriene moieties, *Arch. Biochem. Biophys.,* 184, 36, 1977.

36. **Van Os, C. P. A., Rijke-Schielder, G. P. M., van Halbeek, H., Verhagen, J., and Vliegenthart, J. F. G.,** Double dioxygenation of arachidonic acid by soybean lipoxygenase-1; kinetics and regio-stereo specificities of the reaction steps, *Biochim. Biophys. Acta,* 663, 177, 1981.

37. **Zimmerman, D. C. and Vick, B. A.,** Lipoxygenase in *Chlorella pyrenoidosa, Lipids,* 8, 264, 1973.

38. **Beneytout, J. L., Andrianarison, R. H., Rakotoarisoa, Z., and Tixier, M.,** Properties of a lipoxygenase in blue-green algae (*Oscillatoria* sp.), *Plant Physiol.,* 91, 367, 1989.

39. **Matsui, K., Narahara, H., Kajiwara, T., and Hatanaka, A.,** Purification and properties of lipoxygenase in *Marchantia polymorpha* cultured cells, *Phytochemistry,* 30, 1499, 1991.

40. **Pinsky, A. S., Grossman, S., and Trop, M.,** Lipoxidase content and antioxidant activity of some fruits and vegetables, *J. Food Sci.,* 36, 571, 1971.

41. **Axelrod, B.,** Lipoxygenases, *ACS Adv. Chem. Ser.,* 136, 324, 1974.

42. **Hatanaka, A., Sekiya, J., and Kajiwara, T.,** Distribution of an enzyme system producing *cis*-3-hexenal and *n*-hexanal from linolenic and linoleic acids in some plants, *Phytochemistry,* 17, 869, 1978.

43. **Douillard, R.,** Hypothèses sur la localisation et le rôle des lipoxygénases des cellules végétales, *Physiol. Veg.,* 19, 533, 1981.

44. **Hatanaka, A., Kajiwara, T., Sekiya, J., Masaya, I., and Inouye, S.,** Participation and properties of lipoxygenase and hydroperoxide lyase in volatile C_6-aldehyde formation from C_{18}-unsaturated fatty acids in isolated tea chloroplasts, *Plant Cell Physiol.,* 23, 91, 1982.

45. **Kemal, C., Louis-Flamberg, P., Krupinsky-Olsen, R., and Shorter, A. L.,** Reductive inactivation of soybean lipoxygenase-1 by catechols: a possible mechanism for regulation of lipoxygenase activity, *Biochemistry,* 26, 7064, 1987.

46. **Wallach, D. P. and Brown, V. R.,** A novel preparation of human platelet lipoxygenase; characteristics and inhibition by a variety of phenyl hydrazones and comparisons with other lipoxygenases, *Biochim. Biophys. Acta,* 663, 361, 1981.

47. **Gardner, H. W.,** Lipid hydroperoxide reactivity with proteins and amino acids: a review, *J. Agric. Food Chem.,* 27, 220, 1979.

48. **Inouye, S.,** Site-specific cleavage of double-strand DNA by hydroperoxide of linoleic acid, *FEBS Lett.,* 172, 231, 1984.

49. **Leshem, Y. Y., Grossman, S., Frimer, A., and Ziv, J.,** Endogenous lipoxygenase control and lipid-associated free radical scavenging as modes of cytokinin action in plant senescence retardation, in *Advances in the Biochemistry and Physiology of Plant Lipids,* Appelqvist, L.-Å. and Liljenberg, C., Eds., Elsevier/North Holland Biomedical Press, Amsterdam, 1979, 193.

50. **Pauls, K. P. and Thompson, J. E.,** Evidence for the accumulation of peroxidized lipids in membranes of senescing cotyledons, *Plant Physiol.,* 75, 1152, 1984.

51. **Peterman, T. K. and Siedow, J. N.,** Behavior of lipoxygenase during establishment, senescence, and rejuvenation of soybean cotyledons, *Plant Physiol.,* 78, 690, 1985.

52. **Tressl, R. and Drawert, F.,** Biogenesis of banana volatiles, *J. Agric. Food Chem.,* 21, 560, 1973.

53. **Kim, I.-S. and Grosch, W.,** Partial purification and properties of a hydroperoxide lyase from fruits of pear, *J. Agric. Food Chem.,* 29, 1220, 1981.

54. **Phillips, D. R., Matthew, J. A., Reynolds, J., and Fenwick, G. R.,** Partial purification and properties of a *cis*-3:*trans*-2-enal isomerase from cucumber fruit, *Phytochemistry,* 18, 401, 1979.

55. **Vick, B. A. and Zimmerman, D. C.,** Lipoxygenase and hydroperoxide lyase in germinating watermelon seedlings, *Plant Physiol.,* 57, 780, 1976.

56. **Sekiya, J., Tanigawa, S., Kajiwara, T., and Hatanaka, A.,** Fatty acid hydroperoxide lyase in tobacco cells cultured *in vitro, Phytochemistry,* 23, 2439, 1984.

57. **Vick, B. A. and Zimmerman, D. C.,** Metabolism of fatty acid hydroperoxides by *Chlorella pyrenoidosa, Plant Physiol.,* 90, 125, 1989.

58. **Andrianarison, R.-H., Beneytout, J.-L., and Tixier, M.,** An enzymatic conversion of lipoxygenase products by a hydroperoxide lyase in blue-green algae (*Oscillatoria* sp.), *Plant Physiol.,* 91, 1280, 1989.

59. **Vick, B. A. and Zimmerman, D. C.,** Pathways of fatty acid hydroperoxide metabolism in spinach leaf chloroplasts, *Plant Physiol.,* 85, 1073, 1987.

60. **Schreier, P. and Lorenz, G.,** Separation, partial purification and characterization of a fatty acid hydroperoxide cleaving enzyme from apple and tomato fruits, *Z. Naturforsch.,* 37c, 165, 1982.

61. **Hatanaka, A., Kajiwara, T., and Sekiya, J.,** Fatty acid hydroperoxide lyase in plant tissues; volatile aldehyde formation from linoleic and linolenic acid, in *Biogeneration of Aromas,* ACS Symp. Ser. No. 317, Parliment, T. H. and Croteau, R., Eds., American Chemical Society, Washington, D.C., 1986, 167.

62. **Olías, J. M., Rios, J. J., Valle, M., Zamora, R., Sanz, L. C., and Axelrod, B. A.,** Fatty acid hydroperoxide lyase in germinating soybean seedlings, *J. Agric. Food Chem.,* 38, 624, 1990.

63. **Matsui, K., Toyota, H., Kajiwara, T., Kakuno, T., and Hatanaka, A.,** Fatty acid hydroperoxide cleaving enzyme, hydroperoxide lyase, from tea leaves, *Phytochemistry,* 30, 2109, 1991.

64. **Götz-Schmidt, E.-M., Wenzel, M., and Schreier, P.,** C_6-Volatiles in homogenates from green leaves: localization of hydroperoxide lyase activity, *Lebensm. Wiss. Technol.,* 19, 152, 1986.

65. **Major, R. T., Marchini, P., and Sproston, T.,** Isolation from *Ginkgo biloba* L. of an inhibitor of fungus growth, *J. Biol. Chem.,* 235, 3298, 1960.

66. **Zeringue, H. J., Jr. and McCormick, S. P.,** Relationships between cotton leaf-derived volatiles and growth of *Aspergillus flavus, J. Am. Oil Chem. Soc.,* 66, 581, 1989.

67. **Schildknecht, H. and Rauch, G.,** Defensive substances of plants. II. Chemical nature of the volatile phytocides of leafy plants, particularly of *Robinia pseudoacacia, Z. Naturforsch.,* B 16b, 422, 1961.

68. **Lyr, H. and Banasiak, L.,** Alkenals, volatile defense substances in plants, their properties and activities, *Acta Phytopathol. Acad. Sci. Hung.,* 18, 3, 1983.

69. **Zimmerman, D. C. and Coudron, C. A.,** Identification of traumatin, a wound hormone, as 12-oxo-*trans*-10-dodecenoic acid, *Plant Physiol.,* 63, 536, 1979.

70. **Haberlandt, G.,** Wundhormone als Erreger von Zellteilungen, *Beitr. Allg. Bot.,* 2, 1, 1921.

71. **Feußner, I. and Kindl, H.,** A lipoxygenase is the main lipid body protein in cucumber and soybean cotyledons during the stage of triglyceride mobilization, *FEBS Lett.,* 298, 223, 1992.

72. **Hamberg, M.,** Mechanism of corn hydroperoxide isomerase: detection of 12,13(S)-oxido-9(Z),11-octadecadienoic acid, *Biochim. Biophys. Acta,* 920, 76, 1987.

73. **Baertschi, S. W., Ingram, C. D., Harris, T. M., and Brash, A. R.,** Absolute configuration of *cis*-12-oxophytodienoic acid of flaxseed: implications for the mechanism of biosynthesis from 13(S)-hydroperoxide of linolenic acid, *Biochemistry,* 27, 18, 1988.

74. **Zimmerman, D. C.,** A new product of linoleic acid oxidation by a flaxseed enzyme, *Biochem. Biophys. Res. Commun.,* 23, 398, 1966.

75. **Feng, P. and Zimmerman, D. C.,** Substrate specificity of flax hydroperoxide isomerase, *Lipids,* 14, 710, 1979.

76. **Vick, B. A. and Zimmerman, D. C.,** Lipoxygenase, hydroperoxide isomerase, and hydroperoxide cyclase in young cotton seedlings, *Plant Physiol.,* 67, 92, 1981.

77. **Gardner, H. W.,** Sequential enzymes of linoleic acid oxidation in corn germ: lipoxygenase and linoleate hydroperoxide isomerase, *J. Lipid Res.,* 11, 311, 1970.

78. **Hamberg, M. and Fahlstadius, P.,** Allene oxide cyclase: a new enzyme in plant lipid metabolism, *Arch. Biochem. Biophys.,* 276, 518, 1990.

79. **Vick, B. A., Feng, P., and Zimmerman, D. C.,** Formation of 12[^{18}O]oxo-*cis*-10,*cis*-15-phytodienoic acid from 13-[^{18}O]hydroperoxylinolenic acid by hydroperoxide cyclase, *Lipids,* 15, 468, 1980.

80. **Song, W.-C. and Brash, A. R.,** Purification of an allene oxide synthase and identification of the enzyme as a cytochrome P-450, *Science,* 253, 781, 1991.

81. **Vick, B. A. and Zimmerman, D. C.,** Characterization of 12-oxo-phytodienoic acid reductase in corn; the jasmonic acid pathway, *Plant Physiol.,* 80, 202, 1986.

82. **Vick, B. A. and Zimmerman, D. C.**, Biosynthesis of jasmonic acid by several plant species, *Plant Physiol.*, 75, 458, 1984.

83. **Demole, E., Lederer, E., and Mercier, D.**, Isolement et détermination de la structure du jasmonate de méthyle, constituant odorant caractéristique de l'essence de jasmin, *Helv. Chim. Acta*, 45, 675, 1962.

84. **Aldridge, D. C., Galt, S., Giles, D., and Turner, W. B.**, Metabolites of *Lasiodiplodia theobromae*, *J. Chem. Soc. C*, 1971, 1623, 1971.

85. **Koda, Y., Kikuta, Y., Kitahara, T., Nishi, T., and Mori, K.**, Comparisons of various biological activities of stereoisomers of methyl jasmonate, *Phytochemistry*, 31, 1111, 1992.

86. **Miersch, O., Meyer, A., Vorkefeld, S., and Sembdner, G.**, Occurrence of (+)-7-iso-jasmonic acid in *Vicia faba* L. and its biological activity, *J. Plant Growth Regul.*, 5, 91, 1986.

87. **Grechkin, A. N., Kuramshin, R. A., Safonova, E. Y., Latypov, S. K., and Ilyasov, A. V.**, Formation of ketols from linolenic acid 13-hydroperoxide via allene oxide. Evidence for two distinct mechanisms of allene oxide hydrolysis, *Biochim. Biophys. Acta*, 1086, 317, 1991.

88. **Parthier, B.**, Jasmonates, new regulators of plant growth and development: many facts and few hypotheses on their actions, *Bot. Acta*, 104, 446, 1991.

89. **Van den Berg, J. H. and Ewing, E. E.**, Jasmonates and their role in plant growth and development, with special reference to the control of potato tuberization: a review, *Am. Potato J.*, 68, 781, 1991.

90. **Koda, Y., Kikuta, Y., Tazaki, H., Tsujino, Y., Sakamura, S., and Yoshihara, T.**, Potato tuber-inducing activities of jasmonic acid and related compounds, *Phytochemistry*, 30, 1435, 1991.

91. **Ravnikar, M., Vilhar, B., and Gogala, N.**, Stimulatory effects of jasmonic acid on potato stem node and protoplast culture, *J. Plant Growth Regul.*, 11, 29, 1992.

92. **Anderson, J. M.**, Membrane-derived fatty acids as precursors to second messengers, in *Second Messengers in Plant Growth and Development*, Boss, W. F. and Morré, D. J., Eds., Alan R. Liss, Inc., New York, 1989, 181.

93. **Weidhase, R. A., Kramell, H., Lehmann, J., Liebisch, H., Lerbs, W., and Parthier, B.**, Methyl jasmonate-induced changes in the polypeptide pattern of senescing barley leaf segments, *Plant Sci.*, 51, 177, 1987.

94. **Mason, H. S. and Mullet, J. E.**, Expression of two soybean vegetative storage protein genes during development and in response to water deficit, wounding and jasmonic acid, *Plant Cell*, 2, 569, 1990.

95. **Staswick, P. E., Huang, J.-F., and Rhee, Y.**, Nitrogen and methyl jasmonate induction of soybean vegetative storage protein genes, *Plant Physiol.*, 96, 130, 1991.

96. **Falkenstein, E., Groth, B., Mithöfer, A., and Weiler, E. W.**, Methyljasmonate and α-linolenic acid are potent inducers of tendril coiling, *Planta*, 185, 316, 1991.

97. **Gundlach, H., Müller, M. J., Kutchan, T. M., and Zenk, M. H.**, Jasmonic acid is a signal transducer in elicitor-induced plant cell cultures, *Proc. Natl. Acad. Sci. U.S.A.*, 89, 2389, 1992.

98. **Farmer, E. E., Johnson, R. R., and Ryan, C. A.**, Regulation of expression of proteinase inhibitor genes by methyl jasmonate and jasmonic acid, *Plant Physiol.*, 98, 995, 1992.

99. **Farmer, E. E. and Ryan, C. A.**, Interplant communication: airborne methyl jasmonate induces synthesis of proteinase inhibitors in plant leaves, *Proc. Natl. Acad. Sci. U.S.A.*, 87, 7713, 1990.

100. **Galliard, T. and Chan, H. W.-S.**, Lipoxygenases, in *The Biochemistry of Plants: A Comprehensive Treatise*, Vol. 4, Lipids: Structure and Function, Stumpf, P. K. and Conn, E. E., Eds., Academic Press, New York, 1980, 131.

101. **Needleman, P., Turk, J., Jakschik, B. A., Morrison, A. R., and Lefkowith, J. B.**, Arachidonic acid metabolism, *Annu. Rev. Biochem.*, 55, 69, 1986.

SECTION II

COMPLEX ACYL LIPIDS

I have included in this category the storage lipids, the major category of membrane lipids (phospholipids), the major membrane lipids of the chloroplasts (galactolipids, as well as sulfolipids), and the sphingolipids.

Most of these lipids are comprised of a glycerol backbone and two acyl units attached to positions 1 and 2, while the third position may have a third acyl unit or a unique polar head group. The first chapter in this section (Chapter 6) deals with the addition of the acyl units to the backbone and modifications of these units by exchange. Two later chapters, 7 and 8, are concerned with addition of polar head groups, which makes them suitable for membrane structures.

Chapter 9 is concerned with the sphingolipids, which are polar modifications of acyl units not involving glycerol, but which also make them suitable for membrane structures. This field has been relatively neglected, but hopefully this situation will change.

The final chapter in this set describes lipid transfer mechanisms. The intricate shuttling of lipids among membranes for synthesis and functioning requires careful regulation, but the means of such movements remain incompletely understood, and surprises continue to come in this field. Chapter 10 describes the current status of this very important field.

Chapter 6

ACYLTRANSFERASES AND TRIACYLGLYCEROLS

Margrit Frentzen

TABLE OF CONTENTS

0-8493-4907-9/93/$0.00+$.50
© 1993 by CRC Press Inc.

ABBREVIATIONS

Not defined in text: ACP — acyl carrier protein; CDP — cytidine diphosphate; CoA — coenzyme A; DAG — sn-1,2-diacylglycerol; DAGAT — acyl-CoA(ACP):sn-1,2-diacylglycerol acyltransferase; ER — endoplasmic reticulum; GPAT — acyl-CoA(ACP):sn-glycerol-3-phosphate acyltranferase; LPA — lysophosphatidic acid; LPAAT — acyl-CoA(ACP):sn-1-acylglycerol-3-phosphate acyltransferase; LPC — lysophosphatidylcholine; LPE — lysophosphatidylethanolamine; LPCAT — acyl-CoA:lysophosphatidylcholine acyltranferase; PA — phosphatidic acid; PC — phosphatidylcholine; PE — phosphatidylethanolamine; PG — phosphatidylglycerol; TAG — triacylglycerol; fatty acids are denoted by number of carbon atoms and double bonds; positions of double bonds are given in parentheses.

I. INTRODUCTION

Glycerolipids, by far the largest group of plant lipids, are structurally based on glycerol. This group includes glycodiacylglycerols, phospholipids, and acylglycerols such as triacylglycerol. Glycodiacylglycerols and phospholipids are the major polar membrane lipids of plants, whereas triacylglycerols serve as storage lipids in most plant species. The properties of the triacylglycerols depend on the fatty acids esterified with each of the three hydroxyl groups of glycerol. Since the glycerol molecule does not have rotational symmetry, the three carbon atoms and, thus, the acyl groups at each carbon atom are readily distinguished from each other. Unlike acylglycerols, glycodiacylglycerols and phospholipids are amphiphilic compounds, and their properties arise from both the two nonpolar acyl groups at the sn-1 (stereochemical numbering) and sn-2 position of the glycerol backbone and the polar head group at the sn-3 position.

The distribution of glycodiacylglycerol and phospholipids among the various membrane systems of plant cells reveals appreciable differences, especially between plastidial and extraplastidial membrane systems. While plastidial membranes are characterized by the glycodiacylglycerols typical of cyanobacteria, phospholipids dominate in the extraplastidial membranes (see Chapters 7 and 8).

Differences exist not only with respect to the composition of the polar head groups, but also with respect to the acyl groups esterified at the sn-1 and sn-2 position of the glycerol backbone. In contrast to the storage lipids (see Chapter 3), the membrane lipids of the various plant species predominantly carry C_{16} and C_{18} acyl groups, which are distributed in a characteristic pattern between the sn-1 and sn-2 position of the glycerol backbone.[1] According to the positionally specific distribution of the acyl groups, glycerolipids can be divided into two major groups.[2,3] One group exclusively carries C_{16} fatty acids at the sn-2 position, whereas the sn-1 position contains predominantly, albeit not exclusively, C_{18} fatty acids. Since this fatty acid distribution reflects the pattern of glycerolipids of cyanobacteria,[1,2] it is termed prokaryotic. The other group of glycerolipids specifically contains C_{18} acyl groups at the *sn*-2 position,while the *sn*-1 position is esterified with both C_{16} and C_{18} fatty acids. Since this pattern is not found in the glycerolipids of cyanobacteria, it is termed eukaryotic.[1,2] This pattern is the typical fatty acid distribution of phospholipids apart from PG, which is a notable exception. The fatty acid distribution of PG of the nucleocytoplasmic part of the cell is eukaryotic like the other phospholipids, while the PG of mitochondria can possess both eukaryotic and prokaryotic patterns.[4] On the other hand, the plastidial PG of plant cells exhibits a prokaryotic fatty acid distribution. Thus, plastids of all plants contain eukaryotic PC but prokaryotic PG. The glycodiacylglycerols, the major glycerolipids of plastids, can show both types of patterns, but the proportion of pro- and eukaryotic glycodiacylglycerols varies in different plants.[1] In many families of Angiospermae, designated as 18:3-plants, the glycodiacylglycerols are eukaryotic. In 16:3-plants, however, which contain not only linolenic acid (18:3), but also hexadecatrienoic acid (16:3) as trienoic fatty acids, the glycodiacylglycerols are prokaryotic in a relatively higher proportion. These 16:3-plants include species of some Angiospermae families such as Chenopodiaceae, Apiaceae, and Brassicaceae. Hence, the prokaryotic fatty acid distribution of the glycodiacylglycerols of cyanobacteria is still preserved in the corresponding membrane lipids of 16:3-plants, but not in the more advanced 18:3-plants.

The *de novo* biosynthesis of the pro- and eukaryotic diacylglycerol moieties of the membrane lipids in the different subcellular compartments are described in the first sections of this chapter. The discovery that two distinct pathways are utilized for the biosynthesis of the plastidial glycerolipids is one of the important discoveries in plant lipid metabolism during the past decade.[2,5] While the so-called prokaryotic pathway within the plastids is used for the biosynthesis of glycerolipids with a prokaryotic pattern, eukaryotic diacylglycerol moieties are formed in the extraplastidial part of the cell, particularly in the ER. Hence, the classification of glycerolipids based on analogies to cyanobacteria clearly emphasizes the prokaryotic nature of plastids with regard to glycerolipid biosynthesis, but due to the probably different phylogenetic origins of plastids and mitochondria, it cannot cover the prokaryotic nature of mitochondria. Consequently, unlike the plastidial glycerolipids, the synthetic origin of those of the mitochondria cannot be directly deduced from their fatty acid patterns.

The last part of this chapter is concerned with the biosynthesis of triacylglycerols, especially in developing seeds. Vegetable oils represent an enormous renewable natural resource. The major part of the world production of plant oils is consumed by humans and animals, but a significant proportion is utilized by the chemical industry. The fatty acid composition of the triacylglycerols determines their properties and thus their usefulness in both food and industrial applications which, however, make quite different demands on their fatty acid composition. Due to the progress in the genetic engineering of several oilseed crops, there is now a widespread interest in altering the fatty acid composition of oils of traditional seed crops by the use of cloned genes and thus in producing new, high quality oils designed for specific market applications.[6,7] These developments have stimulated efforts to obtain detailed knowledge of the pathway of the biosynthesis of triacylglycerol, as well as the enzymes and the regulation mechanisms involved. In recent years, there has been a huge increase in investigations concerning the biosynthesis of triacylglycerol esterified with unusual fatty acids (see Section VI.B). Current models of mechanisms of triacylglycerol biosynthesis are described.

II. GLYCEROL-3-PHOSPHATE PATHWAY AND ITS SUBCELLULAR LOCALIZATION WITHIN PLANT CELLS

The first evidence that plant tissue can synthesize glycerolipids as bacteria and animal cells via the glycerol-3-phosphate pathway came from experiments with avocado mesocarp[8] and spinach leaves.[9,10] The individual steps of this pathway are given in Figure 1. *sn*-Glycerol-3-phosphate is converted into *sn*-1,2-diacylglycerol-3-phosphate, PA, by the action of two discrete acyltransferases which utilize activated fatty acid groups, namely CoA or ACP thioesters, as acyl donors. The first enzyme, GPAT (EC 2.3.1.15) catalyzes the acylation reaction at the *sn*-1 position of *sn*-glycerol-3-phosphate to form *sn*-1-acylglycerol-3-phosphate (LPA). Subsequently, the second enzyme, LPAAT (EC 2.3.1.51, lysophosphatidate acyltransferase) utilizes 1-acylglycerol-3-phosphate as substrate and catalyzes the acylation at its *sn*-2 position so that PA is formed.

As depicted in Figure 1, PA is the key intermediate in the biosynthesis of the various glycerolipids. On the one hand, it serves via CDP-diacylglycerol as precursor for the formation of the acid phospholipids (see Chapter 8). On the other hand, it is converted by a PA phosphatase (EC 3.1.3.4) into diacylglycerol (DAG), the intermediate in the biosynthesis of glycodiacyl-glycerol (see Chapter 7), and the predominant phospholipids PC and PE (see Chapter 8). Furthermore, DAG is used as substrate by DAGAT, which catalyzes the acylation reaction at its *sn*-3 position and thus the ultimate step in the biosynthesis of TAG.

At first, the glycerol-3-phosphate pathway was found to operate in microsomal fractions of plant tissue,[8-10] while the incorporation rates of glycerol-3-phosphate into PA determined in mitochondrial and plastidial fractions were attributed to a contamination by microsomal membranes. Due to the development of reliable methods for preparing pure organelle fractions, it is now well-established that *de novo* biosynthesis of glycerolipids via the glycerol-3-phosphate pathway occurs in different subcellular compartments of plant cells, namely plastids, mitochondria, and microsomes, especially the ER, although certain activity is also present in the Golgi apparatus.[3,11,12] As described below and in the

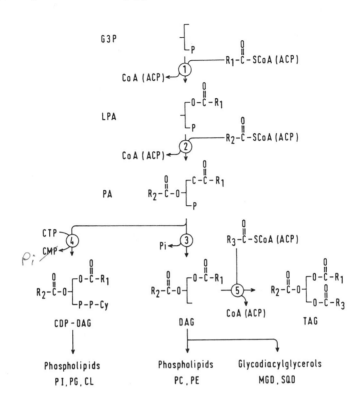

FIGURE 1. Glycerol-3-phosphate pathway in plants. 1 — GPAT; 2 — LPAAT; 3 — PA phosphatase; 4 — CTP:PA cytidyltransferase; 5 — DAGAT; G3P — *sn*-glycerol-3-phosphate; PI — phosphatidylinositol; CL — cardiolipin; MGD — monogalactosyldiacylglycerol; SQD — sulfoquinovosyldiacylglycerol.

subsequent chapters, the activities of these compartments have different importance for the biosynthesis of glycerolipids.

In addition to the acyltransferases given in Figure 1, there are indications for further acyltransferases in various compartments which utilize lysoglycerolipids as substrates.[11-13] But apart from the LPCAT (EC 2.3.1.23), which can play an important role in the biosynthesis of TAG,[13] as discussed below, little is known about the functions and properties of these enzymes.

III. FORMATION OF THE WATER-SOLUBLE PRECURSORS GLYCEROL-3-PHOSPHATE AND ACTIVATED FATTY ACIDS

The glycerol backbone of the various lipids is derived from glycerol-3-phosphate, and there are no indications for glycerolipids of plants being synthesized via the direct acylation of dihydroxyacetone phosphate. Hence,

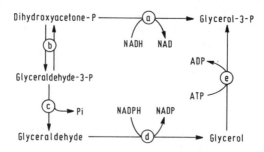

FIGURE 2. Formation of *sn*-glycerol-3-phosphate within plants. a — NAD:glycerol-3-phosphate oxidoreductase; b — triose phosphate isomerase; c — glyceraldehyd-3-phosphate phosphatase; d — NADPH: glyceraldehyde reductase; e — ATP:glycerol kinase.

glycerol-3-phosphate and activated fatty acids are needed for the first steps of *de novo* biosynthesis of glycerolipids (Figure 1).

As described in Chapter 1, *de novo* biosynthesis of fatty acids occurs within plastids and results in the formation of palmitoyl-, stearoyl-, and oleoyl-ACP. These thioesters, which can reach total concentrations of about 3 to 4 µM,[14,15] are either used directly for the plastidial lipid biosynthesis or exported as CoA thioesters into the cytoplasm where they can be utilized directly or after modifications (Chapters 2 and 3) for lipid biosynthesis at extraplastidial sites. Whether acyl-CoA thioesters diffuse in the cytoplasm in free form or bound to specific proteins such as fatty acid binding proteins[16] awaits further clarification. Furthermore, no data are available concerning the concentrations of acyl-CoA thioesters *in vivo*, albeit it is conceivable that the thioesters exist in a small, highly active metabolic pool.

As depicted in Figure 2, glycerol-3-phosphate can be formed by different reaction sequences which are of different importance to the various plant organs and species. It can be produced from dihydroxyacetone phosphate by the action of NAD+:glycerol-3-phosphate oxidoreductase.[17-19] In leaves, this enzyme occurs in both cytoplasm and chloroplasts,[19-21] whereas in developing seeds of castor bean only a cytosolic form was detectable.[18] Alternatively, glycerol-3-phosphate can be formed from dihydroxyacetone phosphate in the cytoplasm via the glycerol kinase reaction, which seems to be the rate-limiting step of this reaction sequence[22,23] (Figure 2).

Unlike oxidoreductase, glycerol kinase activity appears to be present in all plant organs.[1,8,22-25] It plays an important role, for instance, in germinating seeds where it converts glycerol released from TAG by the action of lipases into glycerol-3-phosphate[24,25] (see Chapter 14). High glycerol kinase but no oxidoreductase activity has also been found in developing groundnut seeds.[23] In these seeds, glycerol-3-phosphate is formed from dihydroxyacetone phosphate via glycerol kinase (Figure 2), the activity of which is more than adequate for the observed rates of TAG deposition *in vivo*.[23] However, in seeds of other plant species and in leaves possessing both kinase and oxidoreductase, the activity of the kinase is distinctly lower than that of the oxidoreductase,[13,19,22] suggesting that in these organs glycerol-3-phosphate is predominantly formed by the oxidoreductase. In seeds, this reaction occurs in the cytoplasm,[18] whereas in leaves the oxidoreductase located in the chloroplasts seems to play the major role, and activities of about 10 μmol/h/mg of chlorophyll have been determined.[19-21] This plastidial enzyme comprises up to 80% of the total oxidoreductase activity of leaves and is activated by light, under which conditions the production of the thioester needed for glycerolipid biosynthesis also takes place (see Chapter 1), while the cytosolic form is activated in the dark.[19-21] Hence, it is likely that chloroplasts provide the other compartments not only with their fatty acid biosynthesis products, but also with glycerol-3-phosphate. The level of glycerol-3-phosphate in leaves was found to be more than twice as high in the light as in the dark,[26] and in illuminated spinach leaves concentrations of about 0.15 mM within chloroplasts and about 0.7 mM within the cytoplasm have been determined. Even higher glycerol-3-phosphate concentrations of 0.3 to 0.6 mM have been reported with chloroplasts of other plant species.[26,27]

IV. *DE NOVO* BIOSYNTHESIS OF MEMBRANE GLYCEROLIPIDS

In all compartments of plant cells, a GPAT and an LPAAT catalyze the first two steps in the course of *de novo* biosynthesis of glycerolipids (Figure 1). So far, only one of these acyltransferases, namely the soluble GPAT of chloroplasts,[28-30] has been purified to apparent homogeneity. Therefore, the determination of the properties of the various acyltransferases which have been reported has predominantly been carried out by using subcellular fractions or partially purified acyltransferase fractions as enzyme source. In all compartments, the GPAT possesses a specificity for *sn*-glycerol-3-phosphate and exclusively catalyzes the acylation of the *sn*-1 position of the acyl acceptor[3,31-33] (Figure 1). These properties can thus be considered as general properties of the GPATs of plants. The LPAATs of the various compartments utilize *sn*-1-acylglycerol-3-phosphate as the acyl acceptor for the second acylation reaction (Figure 1), but they are inactive with the respective *sn*-2-acyl isomer.[3,33,34] As outlined below, the isofunctional enzymes of the various compartments, which are probably all coded in the nucleus, display appreciable differences with respect to their abilities to utilize CoA or ACP thioesters and also with respect to their specificities and selectivities for defined acyl groups of the substrates.

A. PLASTIDIAL ACYLTRANSFERASES

The GPAT and LPAAT of plastids have been intensively investigated with respect to their intraorganelle localization and their properties (Table 1). In chloroplasts, as well as in different types of nongreen plastids, the LPAAT is an integral membrane protein of the envelope.[27,35-40] In pea chloroplasts, the enzyme occurs in the inner envelope only,[40] whereas in spinach chloroplasts both the outer and inner envelope membranes contain LPAAT.[41] The local-

TABLE 1
Comparison of Certain Properties of the GPAT and LPAAT
of Different Subcellular Compartments of Vegetative Organs of Plants

Enzyme	Plastids	Mitochondria	Microsomes
GPAT			
Localization	Stroma, envelope	Intermembrane Space, OM, IM	ER, Golgi
pH optimum	7– 8	7–8	7–8
Acyl acceptor	G3P	G3P	G3P
Acyl donor	Acyl-ACP	Acyl-CoA	Acyl-CoA
Positional specificity	*sn*-1	*sn*-1	*sn*-1
FA selectivity	18:1 ≥ 16:0	16:0 > 18:1	16:0 > 18:1
LPAAT			
Localization	Outer, inner envelope	OM, IM	ER, Golgi
pH optimum	7.5–8	8.5–10	9–11
Acyl acceptor	1-16:0-G3P = 1-18:1-G3P	1-16:0-G3P = 1-18:1-G3P	1-18:1-G3P > 1-16:0-G3P
Acyl donor	Acyl-ACP	Acyl-CoA	Acyl-CoA
FA selectivity	16:0 > 18:1	18:1 ≥ 16:0	18:1 > 16:0

Note: OM — outer membrane; IM — inner membrane; G3P — glycerol-3-phosphate; FA — fatty acid; 1-16:0- and 1-18:1-G3P — *sn*-1-palmitoyl- and *sn*-1-oleoylglycerol-3-phosphate. Data are taken from the references given in the text.

ization of LPAAT in the envelopes, as well as in the internal membrane systems, has so far been described for the plastids of cauliflower buds, in which the internal membranes are clearly connected to the inner envelope,[31] and chloroplasts of growing *Chlamydomonas* cells.[42] It is not yet clear whether the enzymes of the internal membrane systems or of the outer envelope display the same properties as those of the inner one.

Apart from a few exceptions,[38,42,43] the GPAT of chloroplasts and nongreen plastids of various plant species has been shown to be a soluble stroma protein.[3,27,28,35,36,44] Its amount is rather low and makes up only 0.04% of the stroma protein of chloroplasts.[45] This enzyme appears to catalyze the rate-limiting step of glycerolipid biosynthesis, at least in certain plants.[12] The soluble GPAT of plastids has been studied in more detail. Depending on the plant species, it has been found to occur in one to three isomeric forms[28,44,46] (Table 2). Both forms of pea show the same catalytic properties,[44] while one of the three isomeric forms of squash chloroplasts, designated AT1, can be distinguished from AT2 and AT3 not only by its isoelectric point and its molecular mass[28] (Table 2), but also by its kinetic data.[47] One isomeric GPAT form of pea and two of squash chloroplasts[28-30] have been purified to apparent homogeneity. A soluble GPAT has also been isolated from cocoa seeds, which has a M_r of 20 kDa.[48] Although the subcellular localization of this enzyme was not studied, in view of the recent results,[49] it is likely that it is a plastidial GPAT as suggested by Joyard and Douce.[12] cDNAs for the GPAT of pea, squash,

TABLE 2
Isoelectric Points and Molecular Masses of the Plastidial GPAT of Various Plant Species

	A			B		C	D
	AT1	**AT2**	**AT3**	**P1**	**P2**		
Isoelectric point	6.6	5.6	5.5	6.3	6.6	nd	5.2
Molecular mass (kDa)							
Mature protein	30	40	40	41	41	41	44
Preprotein	nd		44		51	50	nd

Note: The molecular masses of the mature protein and of the preprotein deduced from the cDNA sequences are given (A — isomeric forms AT1, AT2, AT3 of squash; B — isomeric forms P1, P2 of pea; C — *Arabidopsis*; D — spinach). The data are from References 28–30, 39, 44, 50, and 51. nd — not determined.

cucumber, and *Arabidopsis*[30,50-52] and a genomic DNA for the enzyme of *Arabidopsis*[51] have been sequenced. These data suggest that the plastidial GPAT is synthesized as a preprotein with a transit peptide of about 90 amino acids which is processed to the mature protein (Table 2). The distinctly shorter transit peptide deduced from the cDNA sequence of squash[50] (Table 2) has been found to be incomplete.[51] A comparison of the deduced amino acid sequences of the mature proteins of the four plant species reveals a highly significant and continuous sequence homology of up to 86%.[30,50-52] On the other hand, there is hardly any homology between the sequences of the plastidial GPATs and that of the corresponding enzyme of *Escherichia coli*.[30,50-52] In order to elucidate whether the isomeric forms of the plastidial GPAT represent isoenzymes or result, for instance, from posttranslational modifications, further sequence data are needed.

1. Substrate Specificities and Selectivities

The plastidial acyltransferases of higher plants[31,39,40] unlike those of *Euglena*[53] can use both acyl-CoA and acyl-ACP thioesters as substrates for the biosynthesis of PA. When, however, mixtures of CoA and ACP thioesters are offered, the GPAT, as well as the LPAAT, of plastids of higher plants shows a pronounced preference for acyl-ACP thioesters[31,39] (Table 1). Since ACP thioesters are formed in the stroma and CoA thioesters at the outer envelope membrane[3] (Figure 3), it is now generally accepted that acyl-ACP thioesters are the physiological acyl donors for the plastidial glycerolipid biosynthesis in higher plants.[3,5,12]

Results of fatty acid selectivities of the plastidial GPAT and LPAAT of various plant species are summarized in Table 3. In order to determine such selectivities, enzyme fractions are incubated, for instance, with mixtures of various acyl-labeled acyl-ACP(CoA) thioesters, and subsequently the fatty acid composition of the reaction products is analyzed. The data given in Table 3 clearly reveal that oleoyl groups are highly enriched in *sn*-1-acylglycerol-

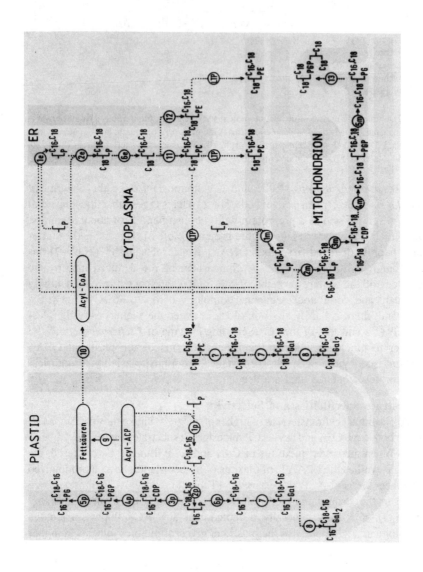

FIGURE 3. *De novo* biosynthesis of membrane glycerolipids in leaves via the prokaryotic plastidial, the eukaryotic microsomal, and the mitochondrial pathway. 1 — GPAT; 2 — LPAAT; 3 — CTP:PA cytidyltransferase; 4 — CDP-DAG:glycerol-3-phosphate phosphatidyltransferase; 5 — phosphatidylglycerophosphate phosphatase; 6 — PA phosphatase; 7 — UDP-galactose:DAG galactosyltransferase; 8 — MGD:MGD galactosyltransferase; 9 — acyl-ACP thioesterase; 10 — acyl-CoA synthetase; 11 — CDP-choline:DAG diacylglycerol cholinephosphotransferase; 12 — CDP-ethanolamine:DAG ethanolaminephosphotransferase; 13 — CDP-DAG:PG phosphatidyltransferase; LTP — lipid transfer protein; e — microsomal; m — mitochondrial; p — plastidial enzymes; polar head groups: C — choline; E — ethanolamine; G — glycerol; Gal — galactose; P — phosphate.

TABLE 3
Fatty Acid Selectivities of Plastidial GPAT and LPAAT of Spinach,
Pea, *Amaranthus*, Squash Leaves, and Cauliflower Buds

	Composition of fatty acid groups of					
	LPA				*sn*-2 of PA	
	16:0/18:1		18:0/18:1		16:0/18:1	
Enzyme source	(%)		(%)		(%)	
Pea	15	85	35	65	97	3
Spinach	10	90	13	87	96	4
Cauliflower	22	78	—	—	95	5
Amaranthus	47	53	38	62	88	12
Squash AT1	29	71	30	70	—	—
AT2	45	56	38	62	—	—
AT3	46	54	38	62	—	—

Note: The fatty acid composition of LPA and of the *sn*-2 position of PA formed by the various enzyme fractions from equimolar mixture of acyl-ACP thioesters (16:0-/18:1-ACP, 18:0-/18:1-ACP) and glycerol-3-phosphate and 1-acylglycerol-3-phosphate, respectively, is given in percent of total according to the data of References 27, 31, 39, and 47.

3-phosphate formed by the GPAT of spinach and pea chloroplasts[39] and of cauliflower leukoplasts[31] in comparison to both palmitoyl and stearoyl groups. Hence, the plastidial GPAT of these plant species is selective for oleoyl-ACP. One of the isomeric forms of the enzyme of squash chloroplasts, namely AT1, also preferentially uses oleoyl-ACP as substrate, although under physiologically relevant conditions its selectivity is less distinct than that of the GPAT of spinach chloroplasts.[47] On the contrary, AT2 and AT3 of squash, as well as the GPAT of *Amaranthus* chloroplasts, utilize the acyl groups to almost the same proportion as they are offered and thus catalyze an unselective *sn*-1 acylation of glycerol-3-phosphate.[27,47] The determination of the kinetic data of the plastidial GPAT shows that the observed fatty acid selectivity for oleoyl-ACP can be attributed to differences in the K_m values.[39,47]

Consequently, in spite of the high sequence homologies between the plastidial GPAT of various plant species, its fatty acid selectivity can vary significantly from plant to plant. The results further demonstrate that disregarding modifications of the esterified acyl groups (see Chapter 2), the fatty acid composition of the glycerolipids of plants can be determined by the substrate specificities and selectivities of the acyltransferases, as well as by the mixtures of acyl donors available to the enzyme. As outlined above, the importance of these two factors can vary appreciably from plant to plant.

In contrast to the plastidial GPAT, the plastidial LPAAT of all plant species studied displays distinctly higher activities with palmitoyl- than with oleoyl-

ACP, while it utilizes 1-palmitoyl- and 1-oleoylglycerol-3-phosphate to the same extent[27,39] (Table 1). The kinetic data of this enzyme have not been determined yet. When the LPAAT is provided with a mixture of palmitoyl- and oleoyl-ACP, the enzyme of the 16:3-plants spinach and cauliflower, as well as of the 18:3-plants pea and *Amaranthus,* almost exclusively directs palmitoyl groups to the *sn*-2 position of 1-acylglycerol-3-phosphate[27,31,39] (Table 3). The results of the enzymic studies are consistent with the results of [14C]acetate-labeling experiments with isolated chloroplasts of various 16:3- and 18:3-plants.[54-56]

In conclusion, due to the properties of the acyltransferases in plastids of both 16:3- and 18:3-plants, PA with a prokaryotic fatty acid pattern is formed; its *sn*-2 position is specifically esterified with palmitic acid, while its *sn*-1 position contains, in addition to the oleoyl groups, a more or less high proportion of saturated acyl groups (Figure 3). In all plants, this PA is used as a precursor for the biosynthesis of PG (see Chapter 8), and in 16:3-plants it is also utilized as a precursor for the biosynthesis of the different glycodiacylglycerols (see Chapter 7). Primary products of the prokaryotic pathway within the plastids are glycerolipid species which are esterified with saturated or monounsaturated fatty acids. These lipid species are subsequently converted by the different plastidial desaturases into the typical highly unsaturated lipid species of the plastidial membranes (see Chapter 2). Glycerolipids with eukaryotic fatty acid pattern, however, cannot be synthesized *de novo* within the plastids, but have to be imported from the ER.[2,3,5]

The prokaryotic plastidial pathway is, thus, well-established for leaf chloroplasts and has been shown to operate in certain nongreen plastids as well. There are, however, indications that plastids of roots[57] (see Chapter 17) and chloroplasts of certain unicellular algae[58] have biosynthetic capacities for glycerolipids different from those of leaf chloroplasts. But it is not yet clear whether the properties of the GPAT and LPAAT of these plastids differ from those of chloroplasts of higher plants.

2. Important Features of the Plastidial Glycerol-3-Phosphate Acyltransferase

As shown by Murata and co-workers,[59,60] chilling sensitivity of several plants is correlated with their proportion of nonfluid, prokaryotic PG species, in which both the *sn*-1 and *sn*-2 position are esterified with a saturated or a *trans* unsaturated fatty acid (see Chapter 18). Since *cis* double bonds cannot be introduced into saturated acyl groups esterified in PG by plant desaturases (see Chapter 2) and since the plastidial LPAAT invariantly directs palmitoyl groups to the *sn*-2 position of the glycerol backbone, the formation of a critical proportion of nonfluid lipid species is determined by the properties of the plastidial GPAT. As outlined above, the enzymes of *Amaranthus* and squash, two chilling-sensitive plants, do not display the pronounced selectivity for oleoyl-ACP characteristic of the GPAT of the chilling-resistant plants such as pea, spinach, and cauliflower.[27,31,39,47] In contrast to the resistant plants, in the sensitive ones a significant proportion of nonfluid prokaryotic lipid species can be formed.

Recently, direct evidence for the significance of the properties of the GPAT in regard to chilling sensitivity has been provided by transformation experiments with tobacco and *Arabidopsis* plants using the cloned genes for the GPAT of *E. coli*, squash, and *Arabidopsis*.[61,62] The functional expression of the genes for the unselective GPAT form of squash[61] and for the palmitate-selective enzyme of *E. coli*[62] within the chloroplasts of the transgenic plants resulted in a significant increase in the proportion of the nonfluid PG species, as well as in an increased sensitivity of the plants to low, nonfreezing temperatures. Corresponding experiments with the gene for the oleate-selective GPAT of *Arabidopsis* yielded opposite although less pronounced effects in transgenic tobacco plants.[61]

Furthermore, the activity of the plastidial GPAT can play a decisive role in controlling the acyl flux via the prokaryotic plastidial pathway.[3,5] As mentioned above, plastids cannot synthesize their eukaryotic glycerolipids, but have to import the respective diacylglycerol moieties formed via the eukaryotic microsomal pathway (Figure 3). According to the different proportions of pro- and eukaryotic glycerolipid species within plastidial membranes of 16:3- and 18:3-plants, the two pathways contribute differently to the biosynthesis of plastidial glycerolipids. Therefore, the biosynthesis rates of both pathways must be correspondingly well-balanced and controlled.

The available data already indicate complex regulatory mechanisms.[3,5] The acyl flux through the two pathways can be directly controlled by the relative activities of the GPAT and the acyl-ACP thioesterase (Figure 3). In the stroma, these two enzymic activities compete for the products of *de novo* fatty acid biosynthesis, where the acyl groups are channeled into the prokaryotic pathway by the activity of the GPAT and into the eukaryotic one by the activity of the thioesterase (Figure 3). This is supported by the results of enzymic studies with stroma fractions of chloroplasts of 16:3- and 18:3-plants using the physiologically relevant substrate concentrations described above.[63] The studies show that the 16:3-plants spinach and mustard have distinctly higher ratios of GPAT vs. thioesterase activities than do the 18:3-plants pea and maize. Within one plant, the ratio of the two enzymic activities can be influenced to a certain degree by the substrate concentrations available in plastids, especially by those of glycerol-3-phosphate[63] and, as shown so far for enzymes of spinach chloroplasts, by those of ACP isoforms, since the GPAT preferentially uses oleoyl-ACP II and the thioesterase oleoyl-ACP I.[64]

The importance of the plastidial GPAT for the biosynthesis of glycerolipids via the prokaryotic pathway has been impressively demonstrated by experiments with mutants of the 16:3-plant *Arabidopsis,* in which the plastidial GPAT activity is reduced to <5% of the activity in the wild type.[65] In these mutants, the amounts of the individual lipids are hardly altered, but the defect in the plastidial GPAT converts the 16:3-plant into an 18:3-plant.

Experiments with the 16:3-plant tobacco indicate that the different regulatory mechanisms are of varying importance or that the glycerolipid biosynthesis is controlled differently in various plant species. For instance, the ratio of the activities of the plastidial GPAT vs. acyl-ACP thioesterase in stroma fractions of tobacco is as low as in those of 18:3-plants.[63] Furthermore, in transgenic tobacco plants, the overexpression of the plastidial GPAT of squash or *Arabidopsis* did not result in an alteration of the acyl flux via the two pathways.[61]

B. MITOCHONDRIAL ACYLTRANSFERASES

Mitochondria, like plastids, are capable of *de novo* biosynthesis of glycerolipids. The semiautonomy of these organelles is also reflected in their lipid biosynthesis, since mitochondria are similar to plastids, in that they can only produce a part of their own membrane lipids, mainly PG and cardiolipin; they have to import glycerolipids formed at the ER[11] (Figure 3). The contribution of mitochondria to the production of glycerolipids within a plant cell is relatively low. In castor bean endosperm,[66] as well as in the phytoflagellate *Euglena*,[43] about 10% of the cellular activity has been found to be associated with mitochondria.

In contrast to the plastidial acyltransferases, there are only a few publications concerning GPAT and LPAAT in purified mitochondrial fractions of plants.[32,66-68] With respect to the submitochondrial localization of the enzymes, different results have been reported (Table 1). The LPAAT is firmly bound to the membranes, but has been demonstrated to occur in the outer membrane of the organelles of potato tubers and pea leaves[32] and in both the outer and inner membrane of mitochondria of castor bean endosperm.[68] The mitochondrial GPAT of potato tubers and castor bean endosperm shows the same submitochondrial localization as the respective LPAAT, whereas the GPAT of the organelles of pea leaves is a soluble protein of the intermembrane space.[32,68] Since the data have been obtained with mitochondrial fractions of different organs of different plant species, it is not yet clear whether they reflect organ- or species-specific differences.

The recent discovery of acyl-ACP in plant mitochondria[69] raises the possibility that ACP thioesters are used for PA biosynthesis as in plastids. However, the mitochondrial acyltransferases display higher activities with acyl-CoA than with the corresponding ACP thioesters[32] (Table 1). According to these results, it is likely that acyl-CoA rather than acyl-ACP thioesters serve as acyl donor for the biosynthesis of PA (Figure 3).

Determinations of the acyl-CoA specificities and selectivities show that the mitochondrial GPAT of pea leaves and potato tubers differ in their specificities, but have similar selectivities, namely a slight preference for saturated acyl-CoA thioesters in comparison to unsaturated ones, particularly oleoyl-CoA[32,70] (Table 1). LPA formed by the GPAT is subsequently used as acyl acceptor by the LPAAT, which does not show a fatty acid specificity with respect to the acyl group of the acyl acceptor[32] (Table 1). But the mitochondrial LPAAT displays an acyl-CoA specificity and selectivity. It discriminates against both palmitoyl- and stearoyl-CoA, while oleoyl- and linoleoyl-CoA are incorporated to almost the same extent. The acyl-CoA selectivity of the LPAAT of pea leaves, however, is less distinct than that of potato tubers.[32] Thus, if the acyltransferases are provided with similar acyl-CoA thioester mixtures, PA formed in the organelles of pea leaves will contain a higher proportion of saturated acyl groups at the *sn*-2 position than that of potato tubers. These results correlate very well with the different fatty acid compositions at the *sn*-2 position of PG synthesized in the mitochondria of the different plant organs. While PG of pea leaf mitochondria carries palmitoyl groups not only at its *sn*-1, but also at its *sn*-2 position, where this acyl group amounts to even more than 20% of the total fatty acids of this lipid,[4] this percentage comprises less than one third in PG of potato tuber mitochondria.[70]

In conclusion, the mitochondrial GPAT and LPAAT differ significantly from the isofunctional enzymes of plastids (Table 1). According to the properties of the acyltransferases, in plastids glycerolipids with prokaryotic fatty acid patterns are formed, whereas mitochondria have the capability of synthesizing glycerolipids with both prokaryotic and eukaryotic patterns. Hence, *de novo* biosynthesis of glycerolipids in mitochondria does not fit into the scheme of either a prokaryotic or a eukaryotic pathway. It rather represents an additional and unique mitochondrial pathway (Figure 3).

C. MICROSOMAL ACYLTRANSFERASES

In the nucleocytoplasmic part of plant cells, the ER is the primary site of *de novo* biosynthesis of phospholipids with eukaryotic fatty acid patterns, but not exclusively since dictyosomes are also capable of *de novo* biosynthesis.[3,11] In the endosperm of castor bean, which contains only few Golgi apparati, microsomal PA biosynthesis has been shown to be located in the ER,[66] whereas in growing maize roots about half of the microsomal GPAT activity is associated with the dictyosomes.[71] *In vivo* labeling experiments, however, cast doubt that the activities determined in isolated dictyosomes play a role *in vivo*.[3] Consequently, *de novo* biosynthesis of phospholipids via the eukaryotic pathway at the ER is relevant to the biogenesis of the entire membrane systems of cells. The ER is also the site of the biosynthesis of triacylglycerols during seed and fruit developement.

The specific activities of the microsomal GPAT and LPAAT vary significantly with respect to the plant tissue and species, but the LPAAT generally shows activities which are several times higher than those of the GPAT.[72-74] The intermediate LPA is therefore not accumulated in the membranes. Both the microsomal GPAT and LPAAT specifically use acyl-CoA thioesters as acyl donors, and both are inactive with the respective ACP thioesters[34,70] (Table 1). The fatty acid specificities and selectivities of the acyltransferases of leaf microsomes can differ appreciably from those of the corresponding enzymes of developing oilseeds, especially if the seeds store triacylglycerols esterified with unusual fatty acids. The properties of the seed enzymes are therefore described separately in the next sections.

The microsomal GPAT of leaves shows a slight preference for palmitoyl groups, whereas the LPAAT displays highest activities with such acyl donors and acceptors which contain unsaturated C_{18} acyl groups[33,34,72] (Table 1). These results have been obtained with both the membrane-bound LPAAT and partially purified enzyme fractions.[34] Thus, the microsomal LPAAT shows a fatty acid specificity with respect to the acyl group of both the acyl donor and the acyl acceptor, although the latter is less pronounced. In this respect, the microsomal enzyme differs from the corresponding enzymic activities of plastids and mitochondria (Table 1). Regardless of the acyl group of the acceptor, the microsomal LPAAT displays a selectivity for linoleoyl-CoA over oleoyl-CoA, and it almost completely excludes palmitoyl- and stearoyl-CoA from the *sn*-2 position of PA.[34,72] Consequently, eukaryotic fatty acid patterns are established by the substrate specificities and selectivities of the microsomal acyltransferases (Figure 3). PA formed in the ER is subsequently utilized as a precursor for the biosynthesis of the various phospholipids via the eukaryotic pathway (see Chapter 8). For instance, PA phosphatase and DAG cholinephosphotransferase convert PA into PC, which serves as a transport metabolite from the ER to the plastids and its eukaryotic diacylglycerol moiety is used for the biosynthesis of glycodiacylglycerols within the envelope membranes (Figure 3). As described in the next sections, PC is also a central intermediate in the biosynthesis of triacylglycerols esterified with polyunsaturated or hydroxylated fatty acids.

V. TRIACYLGLYCEROL AND ITS DEPOSITION IN PLANT CELLS

Almost all plants store lipids as energy reserves in the form of TAG. It is detectable in nearly all plant organs, but it is only deposited in appreciable amounts during seed and fruit development. The seed represents the most important organ for the storage of TAG, which is located in cotyledons or endosperm. Some plant species such as avocado or olive store TAG in the mesocarp of their fruits.

In contrast to the membrane lipids, the structures of fatty acids esterified in TAG have been found to be remarkably diverse (see Chapter 3). However, only a few fatty acids, namely those also found most commonly in the membrane lipids, contribute to more than 90% of world oil production.[6,7] A huge number of plant species, however, have the capacity to synthesize fatty acids which are specifically esterified in TAG, but largely excluded from the membrane lipids. Such fatty acids, which are of great interest for industrial applications, are termed unusual or family specific fatty acids, since they are often prevalent in species of particular plant families[6,7] (see Chapter 3 and below).

During seed and fruit development, TAG accumulation takes place only over a short period during the midstage of development,[13] presumably due to temporal specificity of expression of the genes involved.[7,75] In order to investigate the biosynthesis of TAG, it is therefore important to obtain plant material at the correct stages of development. Unlike plant cell cultures, somatic and gametophytic embryo cultures of oil plants, especially the microspore-derived embryo system, have been shown to be an attractive alternative to zygotic embryos.[76] In microspore-derived embryos, oil synthesis is readily induced, and the fatty acid composition resembles that of zygotic embryos.[76] Recent results, however, indicate that during the development of microspore-derived embryos the timing of synthesis of certain proteins differs from that of zygotic ones.[77-79]

Within the oil-storing cells, TAG accumulates in discrete organelles, the oil bodies. In seeds, they are spherical bodies with diameters of 0.2 to 2.5 μm.[75,80] Their core is made up of reserve lipids surrounded and stabilized by an annulus of phospholipids and proteins of a width of 2 to 4 nm.[75,80] These data are consistent with results which indicate that oil bodies are bounded by a half-unit membrane.[13] The biogenesis of oil bodies has been a contentious field of research.[13] Immense progress has recently been made with respect to the oil body proteins which have been found to show rather simple patterns.[75,80] From experiments with seeds of a wide range of plant species, it is now well-established that the major proteins of oil bodies represent a unique class, termed oleosins, which are not associated with any other subcellular compartments. Oleosins are a structurally related group of interfacially active proteins of low M_r (16 to 26 kDa), which occur throughout the plant kingdom. They possess some of the structural features of the transport apolipoproteins of animals. They have a hydrophilic N-terminal and an amphipathic C-terminal region, whereas the central domain is hydrophobic and shows considerable amino acid sequence similarity between plant species.[75,80-85] It is believed that these proteins stabilize the oil bodies during dehydration of the seeds by preventing coalescence of the oil droplets and thus maintaining an optimal ratio of surface to volume for a rapid mobilization of the storage material during germination[7,75,80] (see Chapters 14 and 16). This hypothesis is supported by the structure of the oil bodies in mesocarp tissue, which does not undergo desiccation and in which oil is deposited to attract animals for seed dispersal and not to nourish the new seedling.[86] The oil bodies of mesocarp tissue are distinctly larger than those of seed tissue and are not stabilized by oleosins.[86]

During seed development of certain plant species, oleosins are synthesized after the main period of TAG accumulation and shortly before the onset of seed dehydration.[75] Thus, TAG is deposited in nascent oil bodies, which contain hardly any protein, as already suggested by Bergfeld et al.[87] In view of these results and those of further biochemical studies, it is conceivable that the *de novo* biosynthesis of TAG determined to occur in isolated oil body fractions was actually due to contamination with membrane fractions.[13,88] It is now generally accepted that the ER is the principal site of TAG biosynthesis,[5] but the final step of this pathway appears to be catalyzed by oil bodies as well.[89] It has been proposed that the hydrophobic TAG molecules will tend to accumulate between the two leaflets of the ER membranes, and nascent oil bodies eventually bud off as oil droplets bounded by a phospholipid mono-layer.[75,80] However, TAGs synthesized *in vitro* by isolated membranes, at rates very similar to the maximal rates of TAG deposition observed in developing seeds *in vivo*, have been found to be accumulated as droplets without a monolayer of phospholipids.[90] Whether the lack of phospholipids is merely due to the membranes being provided with the substrates needed for TAG biosyn-thesis, but not with those needed for phospholipid biosynthesis as suggested by Browse and Somerville,[5] awaits further clarification. As seeds approach maturity, oleosins are synthesized and oil bodies become coated with pro-teins.[75] The transport mechanism of the oleosins has not been elucidated, but recent experiments suggest that the signal for the correct targeting of the proteins to the oil bodies resides in their central, conserved domain.[91]

VI. TRIACYLGLYCEROL BIOSYNTHESIS

In principle, the biosynthesis of membrane lipids has the reaction sequence from glycerol-3-phosphate to DAG in common with that of TAG (Figure 1). The DAGAT, which catalyzes the acylation reaction at the *sn*-3 position of the glycerol backbone, represents the only enzyme unique to the biosynthesis pathway of TAG (Figure 1). In developing seeds and fruits, as well as germinating seeds, most of the total cellular activity of the DAGAT is located in the microsomal membranes,[13,89,92-96] especially in the ER.[97] The enzyme of germinating soybeans has been purified to apparent homogeneity and shown to consist of three different subunits.[96,98]

In contrast to seeds, leaves contain most of the DAGAT activity in the envelope membranes of chloroplasts.[99] This plastidial enzyme displays a specificity for saturated acyl-CoA thioesters[100] and can perhaps utilize ACP thioesters as well.[101] TAG normally accumulates in plastids in low amounts, but can increase significantly under certain conditions, such as stress, under which the DAGAT converts DAG released by degradation of the membrane lipids into TAG.[102-104]

In oil-storing tissue, there are no indications for the plastidial DAGAT being of any importance for the huge accumulation of TAG. In recent years, attention has been focused on the enzymic activities determining the fatty acid pattern of the reserve lipids (see below), but less attention has been drawn toward the mechanism involved in the regulation of the rates of TAG biosynthesis. In seeds as in animal tissues, the step catalyzed by the PA phosphatase is suggested to be rate limiting for the production of reserve lipids.[13] Recently, Ichihara and co-workers[105] provided evidence that in developing safflower seeds as in animal tissue the biosynthetic rate of TAG is controlled by a reversible translocation of the PA phosphatase from the cytoplasm to the ER. In developing seeds of groundnut as in those of safflower, the activity profile of the microsomal PA phosphatase reflects the period of active TAG deposition,[92] but in contrast to safflower seeds, no soluble phosphatase activity was detectable in groundnut seeds.[92] Furthermore, there are indications that in seeds of certain plant species, such as rapeseed, the DAGAT rather than the PA phosphatase catalyzes the rate-limiting step in TAG biosynthesis.[106,107]

It is premature to draw general conclusions, and further work will be required to elucidate the mechanisms controlling the biosynthetic rates of TAG in oil-storing plant tissues.

A. TRIACYLGLYCEROLS WITH UNSATURATED C_{18} FATTY ACIDS

Many plant species, including the major edible oil-producing species, accumulate TAG predominantly esterified with polyunsaturated C_{18} fatty acids. The fatty acid distribution of these reserve lipids is generally nonrandom; the *sn*-2 position carries almost exclusively unsaturated C_{18} acyl groups, while the *sn*-1 and *sn*-3 position also contain different proportions of saturated acyl groups.[13]

The microsomal GPAT and LPAAT of developing seeds and fruits of such plant species resemble the corresponding enzymic activities of leaves. The GPAT displays a selectivity for saturated acyl-CoA thioesters, while the LPAAT has a distinct preference for unsaturated C_{18}-CoA thioesters[74,108-111] and thus can establish the typical fatty acid pattern at the *sn*-1 and *sn*-2 position of TAG. Subsequently, PA is converted into DAG by the PA phosphatase, which displays hardly any specificity for the PA species formed by the acyltransferases.[112,113] The DAGAT of several plant species has been found to display a rather broad acyl-CoA specificity and selectivity,[89,94,97,114] although that is not generally the case. For instance, the microsomal DAGAT of developing sunflower seeds utilizes palmitoyl- and oleoyl-CoA to the same extent, but strongly discriminates against stearoyl-CoA.[115] On the other hand, in palm mesocarp, the DAGAT detected in oil body fractions, unlike the microsomal enzyme, has a preference for oleoyl- over palmitoyl-CoA.[89] Further notable exceptions include the DAGAT of seeds of Boraginaceae which accumulate TAG with γ-linolenic acid (18:3 (6,9,12))[116] or both DAGAT and GPAT of cocoa seeds, which deposit TAG with 1-palmitoyl-2-oleoyl-3-stearoyl-*sn*-glycerol as main molecular species.[49] These acyltransferases show distinct preferences for substrates with γ-linolenoyl and saturated acyl groups, respectively.[49,116]

During the last decade, research on reserve lipids was especially directed toward the elucidation of the mechanisms which make polyunsaturated C_{18} fatty acids synthesized on PC (see Chapter 2) available for TAG biosynthesis.[2,3,5,13] As shown by both labeling experiments and enzymic studies, PC is a central intermediate in TAG biosynthesis, and two enzymic activities are involved in the channeling of polyunsaturated acyl groups from PC to TAG.[2,3,5,13] The results are summarized in Figure 4. One enzymic activity, the cholinephosphotransferase, catalyzes not only the conversion of DAG to PC, but also the backward reaction, without showing any specificity with respect to the molecular lipid species[117,118] (Figure 4). By this enzymic activity, the

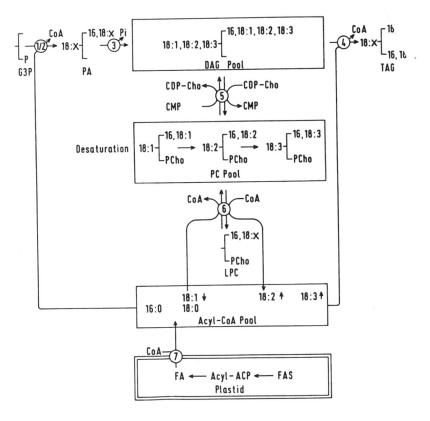

FIGURE 4. Biosynthetic scheme of triacylglycerols with polyunsaturated C_{18} fatty acids in oil-storing tissue. 1/2 — GPAT/LPAAT; 3 — PA phosphatase; 4 — DAGAT; 5 — CDP-choline:DAG cholinephosphotransferase; 6 — LPCAT; 7 — acyl-CoA synthetase; P — phosphate; Cho — choline.

PC and DAG pools can be equilibrated, and DAG species esterified with polyunsaturated C_{18} fatty acids become available to the DAGAT. A further enzymic activity, however, is needed to provide the DAGAT with polyunsaturated C_{18}-CoA thioesters. Such thioesters are produced via a reversible LPCAT reaction.[119] The LPCAT is highly selective for unsaturated C_{18}-CoA thioesters and catalyzes both the acyl-CoA-dependent acylation of LPC and the backward reaction, namely the transesterification of acyl groups from the *sn*-2 position of PC to CoA.[13] The combined forward and backward reactions effect an acyl-exchange between the acyl groups at the *sn*-2 position of PC and those of the acyl-CoA pool (Figure 4). In this way, polyunsaturated C_{18}-CoA thioesters become available to the DAGAT for the *sn*-3 acylation reaction. Furthermore, the acyl-CoA pool also provides the GPAT and LPAAT with substrates, so that linoleoyl-CoA or linolenoyl-CoA can be incorporated directly into PA, preferentially into its *sn*-2 position in accord with the fatty acid selectivities of the GPAT and LPAAT described above (Figure 4).

As argued by Stymne and Stobart,[13] acyl exchange appears to be a common feature of perhaps all LPCAT. It has been demonstrated to occur not only in seed, but also in leaf tissue,[120] so that polyunsaturated C_{18}-CoA thioesters can also become available to the microsomal GPAT and LPAAT of leaves as well. Microsomal LPCATs of developing seeds have been solubilized and partially purified.[121,122] The solubilized fractions, however, catalyze not only the acylation of LPC, but also a wide range of other acyl acceptors.[121] Furthermore, microsomal fractions of sunflower seeds have been found to acylate both the *sn*-2 and the *sn*-1 isomers of LPC,[123] whereas those of soybean seeds acylate LPE at high rates.[124] It is likely that these different acylation reactions are catalyzed by different enzymes, but this awaits clarification.

Both the interconversion of DAG with PC and the acyl-exchange catalyzed by LPCAT (Figure 4) have been demonstrated to operate in developing seeds of various plant species accumulating TAG with polyunsaturated fatty acids, although the relative importance of the two alternative routes appears to vary with the plant species.[13,125,126] On the other hand, oil-storing tissue such as avocado mesocarp or cocoa seeds, which accumulate saturated and monounsaturated rather than polyunsaturated C_{18} acyl groups, do not have detectable acyl exchange or PC-DAG interconversion.[13,49] In such tissues, TAG is synthesized via the glycerol-3-phosphate pathway without the involvement of PC (Figure 4).

B. TRIACYLGLYCEROLS WITH UNUSUAL FATTY ACIDS

Since unusual fatty acids are of particular interest for industrial applications, efforts are now being directed to modifying the genomes of traditional oilseed crops in such a way that they accumulate unusual fatty acids in sufficient quantities and sufficiently pure form so that new, economically valuable oils are produced.[5-7] Application of DNA technology in these approaches requires a complete understanding of the biosynthesis of fatty acids (see Chapters 1 and 2) as well as TAG. Current investigations into TAG biosynthesis are especially concerned with the questions of which enzymic activities unusual fatty acids are effectively incorporated into TAG; which factors determine the fatty acid composition of the reserve lipids; and by which mechanisms unusual fatty acids are channeled into TAG, but excluded from the membrane lipids where they may adversely affect structure and function of the membranes.

In oilseeds which naturally accumulate unusual fatty acids, these acyl groups are essentially restricted to the reserve lipids where they can comprise more than 80% of the fatty acids of TAG, with only a few percent being found in the phospholipids.[6] Thus, on the one hand, these seeds have to possess enzymic activities which can utilize the unusual acyl groups for the biosynthesis of TAG, and on the other hand mechanisms must exist by which the unusual fatty acids are specifically channeled to the reserve lipids. A restriction of these fatty acids to TAG can be achieved if the biosynthesis pathways resulting in the formation of reserve and membrane lipids are spatially separated. However, the available evidence suggests that the biosynthesis of TAG occurs along with that of phospholipids via the glycerol-3-phosphate pathway at the ER and, as described below, the substrate specificities of the enzymes involved rather than spatial separations play the decisive role in the specific targeting of the acyl groups.

1. Hydroxylated Fatty Acids

TAGs with hydroxylated fatty acids are accumulated by a few plant species such as castor bean, the oil of which is known to be rich in ricinoleic acid (18:1(12-OH)). Such fatty acids are particularly useful because they allow many chemical conversions.[6,7] The desaturation step of ricinoleic acid biosynthesis appears to occur on oleoyl groups esterified to PC (see Chapters 2 and 3), suggesting that PC is a central intermediate in the biosynthesis of TAG with both polyunsaturated and hydroxylated fatty acids. According to the recent data of Bafor et al.,[127] the following pathway emerges for the TAG biosynthesis in developing seeds of castor bean. Similar to the biosynthesis of TAG with polyunsaturated fatty acid (Figure 4), the cholinephosphotransferase of castor bean catalyzes DAG-PC interconversion. DAG containing ricinoleic acid does not accumulate in the microsomal membranes, but is rapidly converted into TAG. These results indicate that the DAGAT has a preference for ricinoleoyl-containing DAG species, and this has recently been confirmed by the determination of the substrate specificities and selectivities of the enzyme.[128] Thus, the DAGAT prevents an accumulation of ricinoleoyl-containing DAG moieties in PC. In this respect, the mechanism which makes ricinoleoyl groups as CoA thioesters available to acyltransferases is of even more importance.[127] In contrast to polyunsaturated acyl groups, ricinoleoyl groups are not released from PC as CoA thioesters via acyl exchange (Figure 4), but as free fatty acids. This reaction is presumably catalyzed by a phospholipase A_2 which rapidly digests PC species with hydroxylated, but not unsaturated, acyl groups. The thermodynamically favored degradation of PC to LPC and ricinoleic acids effectively removes PC with hydroxylated fatty acids from the membranes. It is notable that a specific degradation of PC with oxygenated acyl groups is not confined to castor bean seeds, but also occurs in seeds of other plant species such as rape, which do not accumulate ricinoleic acid.[127,129]

In castor bean seeds, LPC is subsequently reacylated to PC by an LPCAT which has a distinct preference for oleoyl-CoA over ricinoleoyl-CoA.[127] Ricinoleic acid, on the other hand, is converted into the CoA thioester by an acyl-CoA synthetase present in the castor bean microsomal fractions, and this thioester can be used as a substrate by the acyltransferases involved in TAG biosynthesis.[127,128,130] The DAGAT is highly selective for ricinoleoyl-CoA. This thioester can also be used for the acylation of glycerol-3-phosphate to PA. The fatty acid specificities and selectivities of the microsomal GPAT and LPAAT of castor bean seeds have not been studied in detail yet. However, the available data already show that a high proportion of ricinoleoyl-CoA within the acyl-CoA pool is needed for the biosynthesis of diricinolein, which can be converted into triricinolein, the main molecular species in castor bean oil.[127] The fatty acid composition of castor bean oil is thus predominantly controlled by the acyl-CoA pool available to the GPAT and LPAAT and by

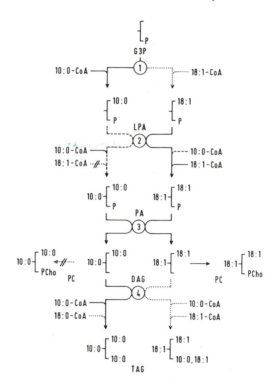

FIGURE 5.   Biosynthetic scheme of triacylglycerols with medium-chain fatty acids in developing seeds of *C. lanceolata* according to Bafor et al.[131] The fatty acid selectivities of the enzymes are symbolized in the figure. (For abbrevations see Figure 4.)

the selectivities of the DAGAT for both acyl-CoA- and DAG-containing ricinoleoyl groups.

In conclusion, TAGs with ricinoleoyl groups are synthesized via a modification of the pathway shown in Figure 4. The main difference lies in the mechanism resulting in the release of the acyl groups from PC. The primary production of free fatty acids and their subsequent conversion into acyl-CoA demands more energy than acyl exchange, but they cause the exclusion of ricinoleoyl groups from the membranes.[127]

2. Medium-Chain Fatty Acids

TAG, rich in medium-chain fatty acids, is especially useful to the detergent industry.[6,7] Such oils are deposited in seeds of coconut, oil palm (palm kernel oil), and certain species of *Cuphea*.[6,7,13] The biosynthesis of TAG in developing *Cuphea* seeds has recently been investigated, especially with respect to the mechanisms controlling the restriction of medium-chain fatty acids to TAG.[131,132] The results of Bafor et al.[131] obtained with seeds of *C. lanceolata* accumulating oil almost exclusively composed of capric acid (10:0) are summarized in Figure 5.

All three microsomal acyltransferases can use medium-chain acyl-CoA thioesters as substrates, which are exported from the plastids (see Chapter 3), and the enzymes have been found to display even higher activities with caproyl-CoA than with oleoyl-CoA. With respect to the acyl acceptor, the LPAAT prefers 1-oleoyl- over 1-caproylglycerol-3-phosphate, but, more importantly, the acyl group of the acyl acceptor affects the acyl-CoA specificity and selectivity of the LPAAT. 1-Oleoylglycerol-3-phosphate can be acylated with either caproyl- or oleoyl-CoA, but oleoyl-CoA is preferentially used, whereas the LPAAT exclusively utilizes medium-chain acyl-CoA thioesters for the acylation of 1-caproylglycerol-3-phosphate (Figure 5). Similar results have been reported for the microsomal LPAAT of seeds of other plant species depositing TAG rich in medium-chain fatty acids.[132-134] Consequently, due to the properties of the microsomal LPAAT, the formation of mixed-chain PA species esterified with both a caproyl and an oleoyl group is largely prevented (Figure 5). The DAGAT displays a specificity and selectivity for dicaprate over dioleate and almost exclusively directs caproyl-CoA to the *sn*-3 position, whereas dioleate is converted into PC by the activity of the cholinephosphotransferase and the LPCAT appears to be inactive with caproyl-CoA.[128,131]

In summary, the results show that *Cuphea* seeds possess acyltransferases with optimal properties for the incorporation of medium-chain fatty acids into TAG and that due to the substrate selectivities of the LPAAT, DAGAT, cholinephosphotransferase, and LPCAT the specific targeting of the acyl groups is achieved.[128,131]

3. Very Long-Chain Fatty Acids

The biosynthesis of TAG esterified with very long-chain fatty acids, which are especially used as lubricants,[6,7] has been predominantly investigated in developing seeds of species of Brassicaceae such as rapeseed. The oil of Brassicaceae, however, is not exclusively esterified with very long-chain fatty acids. Erucic and gondolic acid (20:1(11)) are only found at its *sn*-1 and *sn*-3 position, while the *sn*-2 position contains polyunsaturated C_{18} acyl groups.[13] Therefore, unlike the pathway in *Cuphea* seeds (Figure 5), in rapeseed mixed-chain DAG species have to be channeled into TAG, and PC has to provide the *sn*-2 position of TAG with polyunsaturated C_{18} groups. The available evidence suggests[126] that polyunsaturated C_{18} groups are predominantly directed via acyl exchange rather than PC-DAG interconversion to the *sn*-2 position of rapeseed oil, so that mixed-chain DAG moieties can be largely excluded from PC.

As given in Table 4, the microsomal GPAT of species of Brassicaceae can utilize erucoyl-CoA, but in contrast to the GPAT of *Cuphea* species it does not show a specificity for CoA thioesters with unusual acyl groups.[73,74,135] According to *in vitro* experiments with exogenously added and *in situ*-synthesized acyl-CoA thioesters, the GPAT preferentially uses acyl-CoA thioesters formed in the membranes[88] and thus explains the different incorporation rates previously described.[136-139] In view of these results, it is likely that the products

TABLE 4
Acyl-CoA Specificities of the Microsomal GPAT of Developing Seeds of Various Plant Species

Plant species	Acyl groups of acyl-CoA						
	10:0	12:0	14:0	16:0	18:1	20:1	22:1
Cuphea lanceolata	100	100	47	31	44	—	—
C. procumbens	33	100	72	28	49	—	—
C. wrightii	3	100	33	14	7	—	—
Tropaeolum majus	—	28	—	110	100	67	40
Limnanthes douglasii	—	4	—	79	100	52	24
Brassica napus	—	10	—	103	100	—	51
B. campestris	4	46	—	—	100	39	13
Carthamus tinctorius	3	14	53	69	100	7	1

Note: GPAT activities of the *Cuphea* species are expressed as percentage of the activity determined with lauroyl-CoA according to the data of References 131 and 132. GPAT activities of *T. majus, L. douglasii, B. napus, B. campestris,* and *Carthamus tinctorius* are expressed as percentage of the activity determined with oleoyl-CoA.[74,135,144]

of the elongase systems (see Chapter 3) formed in the membranes are directly channeled to the GPAT and thus can be incorporated into the *sn*-1 position of glycerol-3-phosphate.[138,139] The microsomal DAGAT has been found to be active with erucoyl-CoA.[95,114,135,137] In certain rape varieties, the DAGAT even displays a specificity and selectivity for this thioester in comparison to oleoyl-CoA[95,135] and seems to have a preference for DAG species with very long-chain acyl groups.[137,140] It is well-established that the exclusion of erucic acid from the *sn*-2 position of TAG is governed by the properties of the microsomal LPAAT. This enzyme resembles the corresponding enzyme of leaves or of oilseeds accumulating polyunsaturated C_{18} fatty acids described above. It displays a selectivity for unsaturated C_{18}-CoA thioesters, and completely excludes erucoyl-CoA irregardless of the acyl group of the acyl acceptor[135,141,143] (Table 5). The LPAAT even discriminates against other unusual acyl-CoA thioesters, especially when the acyl acceptor carries the unusual acyl group as well.[88,130,134,142] In order to modify the fatty acid composition of rapeseed oil, the LPAAT is thus an important target enzyme for genetic engineering.[7]

Unlike rapeseed oil, TAG of nasturtium and species of Limnanthaceae contains very long-chain acyl groups at each position of the glycerol backbone. Trierucin is the main molecular species of nasturtium oil, while the oil of *Limnanthes* species is predominantly esterified with *cis*-5-eicosenic acid, but erucic acid is enriched in its *sn*-2 position.[143,144] The microsomal GPAT of developing seeds of nasturtium and *L. douglasii* shows properties very similar to the corresponding enzyme of rapeseed[144] (Table 4). Furthermore, the microsomal DAGAT of nasturtium, like that of certain rape varieties, preferentially uses both acyl-CoA and DAG with very long-chain fatty acids.[144] As

TABLE 5
Substrate Specificities of the Microsomal LPAAT of Developing Seeds of Various Plant Species

Products	A	B	C	D	E	F
18:1/18:1	100%	100%	100%	100%	100%	100%
18:1/20:1	46%	—	20%	—	—	—
18:1/22:1	86%	71%	0%	0%	5%	4%
20:1/18:1	86%	—	22%	47%	—	—
20:1/20:1	98%	—	4%	—	—	—
20:1/22:1	109%	—	0%	0%	—	—
22:1/18:1	50%	23%	24%	25%	150%	111%
22:1/20:1	51%	—	4%	—	—	—
22:1/22:1	65%	60%	0%	0%	5%	1%

Note: 100% corresponds to the LPAAT activity determined with oleoy-CoA and 1-oleoylglycerol-3-phosphate of A — *Limnanthes douglasii*; B — *L. alba*; C — *Tropaeolum majus*; D — *Brassica napus*; E — *Glycine max*; and F — *Zea mays*.[141,143,144]

expected from the fatty acid composition of the oils, developing seeds of both *L. douglasii* and *L. alba* possess a microsomal LPAAT which can utilize erucoyl-CoA[141,143,144] (Table 5), whereas the microsomal LPCAT as that of rapeseed is inactive with this thioester.[135,144] As given in Table 5, the LPAT of *Limnanthes* seeds displays highest activities with erucoyl-CoA if the acyl acceptor contains an eicosenoyl group, the acyl group enriched in the *sn*-1 position of *Limnanthes* oil.[144] Even with eicosenoyl- or erucoylglycerol-3-phosphate, the LPAAT of the *Limnanthes* species, however, shows no preference for the unusual acyl-CoA thioesters.[141,143,144] In this respect, the LPAAT of *Limnanthes* species differs from the corresponding enzyme of *Cuphea* seeds[131,132] described above, but resembles the LPAAT of castor bean seeds[127] or carrot seeds, which accumulate TAG rich in petroselinic acid (18:1 (6)).[142]

Surprising results were obtained with the microsomal LPAAT of developing seeds of nasturtium.[143,144] Microsomal fractions of these seeds synthesize TAG in high rates from glycerol-3-phosphate,[144] but the microsomal LPAAT has been found to be inactive with erucoyl-CoA irrespective of the acyl group of the acyl acceptor[143,144] (Table 5). Although it is not conclusive, the available data suggest that nasturtium has developed mechanisms different from those of the *Limnanthes* species for the incorporation of erucoyl groups into the *sn*-2 position of its TAG.[144] Further work will be required to elucidate the biosynthesis pathway of trierucin in nasturtium seeds.

In summary, the available evidence suggests that apart from the one exception described above, unusual fatty acids are incorporated into TAG via the glycerol-3-phosphate pathway, and a restriction of these acyl groups to the reserve lipids is achieved by the substrate specificities or selectivities of the enzymes involved in TAG and phospholipid biosynthesis. Further experiments are needed to show whether these mechanisms are generally used in oilseeds or whether other plant species have developed different mechanisms as suggested for nasturtium.

According to *in vitro* experiments, even oilseeds of plant species that do not normally synthesize medium- or very long-chain fatty acids can restrict them to TAG.[145] But these oilseeds often incorporate the unusual acyl groups at only low rates and exclusively into the *sn*-1 and *sn*-3 position of TAG,[73,114,134,142,143,145] according to the substrate specificities of acyltransferases described above. On the other hand, oilseeds depositing high proportions of unusual fatty acids possess acyltransferase activities which can effectively incorporate the acyl groups into all three positions of the glycerol backbone. The microsomal LPAAT of these seeds appears to have the unique property that it can accept unusual acyl groups present in both its substrates, so that DAG species with unusual acyl groups in both positions are formed.[130-134,141-144] These DAG species are preferentially used by the DAGAT for the formation of TAG.[127,128,131,140,144] Unlike the DAGAT of these seeds, the GPAT and LPAAT often have no specificities or selectivities for unusual acyl groups,[127,141-144] and so the fatty acid compositions of these oils are often governed by the acyl-CoA pool available to the acyltransferases and the substrate selectivities of the DAGAT.

The properties of the microsomal GPAT and LPAAT of developing seeds can differ significantly from those of the corresponding enzymes of leaves, suggesting a seed-specific expression of these enzymes.[144,146] By now it is not conclusive that only one set of microsomal acyltransferases is involved in both TAG and phospholipid biosynthesis as suggested in Figure 5.[146] Due to the progress recently made with the purification of a microsomal GPAT and LPAAT[147,148] and since a cDNA clone has recently been isolated from an expression library of rapeseed[149] which complements an *Escherichia coli* mutant[150] deficient in its LPAAT, this problem will probably be solved in the near future.

VII. CONCLUDING REMARKS

Considerable progress has been made on plant acyltransferases and triacylglycerol biosynthesis in the past few years, especially with regard to the plastidial GPAT, the biosynthesis of oleosomes, and the biosynthesis of oils with unusual fatty acids. But numerous questions remain unanswered, particularly because most of the enzymes have not been purified and cloned genes are now available only for the plastidial GPAT. The purification of further enzymes and the isolation of further genes for the enzymes will substantially increase our understanding of plant glycerolipid metabolism and the regulatory mechanisms. Moreover, the availability of cloned genes for the enzymes involved in biosynthesis of reserve lipids will allow application of our knowledge of this pathway directly toward production of new, high quality oils designed for specific commercial utilizations. This is certainly a very promising area for future applications of genetic engineering methods.

ACKNOWLEDGMENTS

I am grateful to my colleagues for providing me with preprints of their publications and their recent results not yet published. I wish to thank Dr. F.P. Wolter for critically reading the manuscript. Research in the author's laboratory was supported by the Deutsche Forschungsgemeinschaft and by the Bundesminister für Forschung und Technologie (Förderkennzeichen 031600A and 0319412G).

REFERENCES

1. **Heinz, E.**, Enzymatic reactions in galactolipid biosynthesis, in *Lipids and Lipid Polymers in Higher Plants*, Tevini, M. and Lichtenthaler, H.K., Eds., Springer-Verlag, Berlin, 1977, 102.
2. **Roughan, G. and Slack, C. R.**, Cellular organization of glycerolipid metabolism, *Annu. Rev. Plant Physiol.*, 33, 97, 1982.
3. **Frentzen, M.**, Biosynthesis and desaturation of the different diacylglycerol moieties in higher plants, *J. Plant Physiol.*, 124, 193, 1986.
4. **Dorne, A.-J. and Heinz, E.**, Position and pairing of fatty acids in phosphatidylglycerol from pea leaf chloroplasts and mitochondria, *Plant Sci.*, 60, 39, 1989.
5. **Browse, J. and Somerville, C.**, Glycerolipid biosynthesis: biochemistry and regulation, *Annu. Rev. Plant Physiol. Plant Mol. Biol.*, 42, 467, 1991.
6. **Battey, J. F., Schmid, K. M., and Ohlrogge, J. B.**, Genetic engineering for plant oils: potential and limitations, *Trends Biotechnol.* 7, 122, 1989.
7. **Hills, M. J. and Murphy, D. J.**, Biotechnology of oilseeds, *Biotechnol. Genet. Eng. Rev.*, 9, 1, 1991.
8. **Barron, E. J. and Stumpf, P. K.**, Fat metablism in higher plants XIX. The biosynthesis of triglycerides by avocado-mesocarp enzymes, *Biochim. Biophys. Acta*, 60, 329, 1962.
9. **Cheniae, G. M.**, Phosphatidic acid and glyceride synthesis by particles from spinach leaves, *Plant Physiol.*, 40, 235, 1965.
10. **Sastry, P. S. and Kates, M.**, Biosynthesis of lipids in plants II. Incorporation of glycerophosphate-^{32}P into phosphatides by cell-free preparations from spinach leaves, *Can. J. Biochem.*, 44, 459, 1966.
11. **Moore, T. S., Jr.**, Phospholipid biosynthesis, *Annu. Rev. Plant Physiol.*, 33, 235, 1982.
12. **Joyard, J. and Douce, R.**, Galactolipid synthesis, in *The Biochemistry of Plants*, Vol. 9, Stumpf, P.K. and Conn, E.E., Eds., Academic Press, Orlando, FL, 1987, 215.
13. **Stymne, S. and Stobart, A. K.**, Triacylglycerol biosynthesis, in *The Biochemistry of Plants*, Vol. 9, Stumpf, P.K. and Conn, E.E., Eds., Academic Press, Orlando, FL, 1987, 175.
14. **Soll, J. and Roughan, G.**, Acyl-acyl carrier protein pool sizes during steady-state fatty acid synthesis by isolated spinach chloroplasts, *FEBS Lett.*, 146, 189, 1982.
15. **Roughan, G. and Nishida, I.**, Concentrations of long-chain acyl-acyl carrier proteins during fatty acid synthesis by chloroplasts isolated from pea (*Pisum sativum*), safflower (*Carthamus tinctoris*), and amaranthus (*Amaranthus lividus*) leaves, *Arch. Biochem. Biophys.*, 276, 38, 1990.
16. **Arondel, V., Vergnolle, C., Tchang, F., and Kader, J.-C.**, Bifunctional lipid-transfer: fatty acid-binding proteins in plants, *Mol. Cell. Biochem.*, 98, 49, 1990.
17. **Santora, G. T., Gee, R., and Tolbert, N.E.**, Isolation of a *sn*-glycerol 3-phosphate:NAD oxidoreductase from spinach leaves, *Arch. Biochem. Biophys.*, 196, 403, 1979.
18. **Finlayson, S. A. and Dennis, D. T.**, NAD$^+$-specific glycerol 3-phosphate dehydrogenase from developing castor bean endosperm, *Arch. Biochem. Biophys.*, 199, 179, 1980.

19. **Gee, R. W., Byerrum, R. U., Gerber, D. W., and Tolbert, N. E.,** Dihydroxyacetone phosphate reductase in plants, *Plant Physiol.,* 86, 98, 1988.

20. **Gee, R. W., Byerrum, R. U., Gerber, D. W., and Tolbert, N. E.,** Differential inhibition and activation of two leaf dihydroxyacetone phosphate reductases, role of fructose 2,6-bisphosphate, *Plant Physiol.,* 87, 379, 1988.

21. **Gee, R., Goyal, A., Gerber, D., Byerrum, R. U., and Tolbert, N. E.,** Isolation of dihydroxyacetone phosphate reductase from *Dunaliella* chloroplasts and comparison with isozymes from spinach leaves, *Plant Physiol.,* 88, 896, 1988.

22. **Hippmann, H. and Heinz, E.,** Glycerol kinase in leaves, *Z. Pflanzenphysiol.,* 79, 408, 1976.

23. **Ghosh, S. and Sastry, P. S.,** Triacylglycerol synthesis in developing seeds of groundnut (*Arachis hypogaea*): pathway and properties of enzymes of sn-glycerol 3-phosphate formation, *Arch. Biochem. Biophys.,* 262, 508, 1988.

24. **Huang, A. H. C.,** Enzymes of glycerol metabolism in the storage tissues of fatty seedlings, *Plant Physiol.,* 55, 555, 1975.

25. **Sadava, D. and Moore, K.,** Glycerol metabolism in higher plants: glycerol kinase, *Biochem. Biophys. Res. Commun.,* 143, 977, 1987.

26. **Sauer, A. and Heise, K.-P.,** Control of fatty acid incorporation into chloroplast lipids *in vitro, Z. Naturforsch.,* 39c, 593, 1984.

27. **Cronan, J. E., Jr. and Roughan, P. G.,** Fatty acid specificity and selectivity of the chloroplast sn-glycerol 3-phosphate acyltransferase of the chilling sensitive plant, *Amaranthus lividus, Plant Physiol.,* 83, 676, 1987.

28. **Nishida, I., Frentzen, M., Ishizaki, O., and Murata, N.,** Purification of isomeric forms of acyl-[acyl-carrier-protein]:glycerol-3-phosphate acyltransferase from greening squash cotyledons, *Plant Cell Physiol.,* 28, 1071, 1987.

29. **Douady, D. and Dubacq, J.-P.,** Purification of acyl-CoA: glycerol-3-phosphate acyltranferase from pea leaves, *Biochim. Biophys. Acta,* 921, 615, 1987.

30. **Weber, S., Wolter, F.-P., Buck, F., Frentzen, M., and Heinz, E.,** Purification and cDNA sequencing of an oleate-selective acyl-ACP:sn-glycerol-3-phosphate acyltransferase from pea chloroplasts, *Plant Mol. Biol.,* 17, 1067, 1991.

31. **Alban, C., Joyard, J., and Douce, R.,** Comparison of glycerolipid biosynthesis in non-green plastids from sycamore (*Acer pseudoplatanus*) cells and cauliflower (*Brassica oleracea*) buds, *Biochem. J.,* 259, 775, 1989.

32. **Frentzen, M., Neuburger, M., Joyard, J., and Douce, R.,** Intraorganelle localization and substrate specificities of the mitochondrial acyl-CoA:sn-glycerol-3-phosphate *O*-acyltranferase and acyl-CoA:1-acyl-sn-glycerol-3-phosphate *O*-acyltransferase from potato tubers and pea leaves, *Eur. J. Biochem.,* 187, 395, 1990.

33. **Frentzen, M.,** Comparison of certain properties of membrane bound and solubilized acyltransferase activities of plant microsomes, *Plant Sci.,* 69, 39, 1990.

34. **Hares, W. and Frentzen, M.,** Substrate specificities of the membrane-bound and partially purified microsomal acyl-CoA:1-acylglycerol-3-phosphate acyltransferase from etiolated shoots of *Pisum sativum* (L.), *Planta,* 185, 124, 1991.

35. **Joyard, J. and Douce, R.,** Site of synthesis of phosphatidic acid and diacylglycerol in spinach chloroplasts, *Biochim. Biophys. Acta,* 486, 273, 1977.

36. **Alban, C., Joyard, J., and Douce, R.,** Preparation and characterization of envelope membranes from nongreen plastids, *Plant Physiol.,* 88, 709, 1988.

37. **Liedvogel, B. and Kleinig, H.,** Galactolipid synthesis in chromoplasts *in vitro, Planta,* 144, 467, 1979.

38. **Fishwick, M. J. and Wright, A. J.,** Isolation and characterization of amyloplast envelope membranes from *Solanum tuberosum, Phytochemistry,* 19, 55, 1980.

39. **Frentzen, M., Heinz, E., McKeon, T. A., and Stumpf, P. K.,** Specificities and selectivities of glycerol-3-phosphate acyltransferase and monoacylglycerol-3-phosphate acyltransferase from pea and spinach chloroplasts, *Eur. J. Biochem.,* 129, 629, 1983.

40. **Andrews, J., Ohlrogge, J. B., and Keegstra, K.,** Final step of phosphatidic acid synthesis in pea chloroplasts occurs in the inner envelope membrane, *Plant Physiol.*, 78, 459, 1985.
41. **Block, M. A., Dorne, A.-J., Joyard, J., and Douce, R.,** The phosphatidic acid phosphatase of the chloroplast envelope is located on the inner envelope membrane, *FEBS Lett.*, 164, 111, 1983.
42. **Michaels, A. S., Jelsema, C. L., and Barrnett, R. J.,** Membrane lipid metabolism in *Chlamydomonas reinhardtii* 137⁺ and y-1: II. Cytochemical localization of acyltransferase activities, *J. Ultrastruct. Res.*, 82, 35, 1983.
43. **Boehler, B. A. and Ernst-Fonberg, M. L.,** sn-Glycerol-3-phosphate transacylase activity in *Euglena gracilis* organelles, *Arch. Biochem. Biophys.*, 175, 229, 1976.
44. **Bertrams, M. and Heinz, E.,** Positional specificity and fatty acid selectivity of purified sn-glycerol-3-phosphate acyltransferases from chloroplasts, *Plant Physiol.*, 68, 653, 1981.
45. **Douady, D., Passaquet, C., and Dubacq, J.-P.,** Immunochemical characterisation and in vitro synthesis of glycerol-3-phosphate acyltransferase from pea, *Plant Sci.*, 66, 65, 1990.
46. **Dubacq, J.-P. and Murata, N.,** personal communication, 1988.
47. **Frentzen, M., Nishida, I., and Murata, N.,** Properties of the plastidial acyl-(acyl-carrier-protein):glycerol-3-phosphate acyltransferase from the chilling-sensitive plant squash (*Cucurbita moschata*), *Plant Cell Physiol.*, 28, 1195, 1987.
48. **Fritz, P. J., Kauffman, J. M., Robertson, C. A., and Wilson, M. R.,** Cocoa butter biosynthesis, purification and characterization of a soluble sn-glycerol-3-phosphate acyltransferase from cocoa seeds, *J. Biol. Chem.*, 261, 194, 1986.
49. **Griffiths, G. and Harwood, J. L.,** The regulation of triacylglycerol biosynthesis in cocoa (*Theobroma cacao*) L., *Planta*, 184, 279, 1991.
50. **Ishizaki, O., Nishida, I., Agata, K., Eguchi, G., and Murata, N.,** Cloning and nucleotide sequence of cDNA for the plastid glycerol-3-phosphate acyltransferase from squash, *FEBS Lett.*, 238, 424, 1988.
51. **Nishida, I., Tasaka, Y., Shiraishi, H., and Murata, N.,** The gene and the RNA for the precursor to the plastid-located glycerol-3-phosphate acyltransferase of *Arabidopsis thaliana*, *Plant Mol. Biol.*, 21, 267, 1993.
52. **Johnson, T. C., Schneider, J. C., and Somerville, C.,** Nucleotide sequence of acyl-acyl carrier protein: glycerol-3-phosphate acyltransferase from cucumber, *Plant Physiol.*, 99, 771, 1992.
53. **Herschenson, S., Boehler-Kohler, B. A., and Ernst-Fonberg, M. L.,** Comparison of glycerophosphate acyltransferases from *Euglena* chloroplasts and microsomes, *Arch. Biochem. Biophys.*, 223, 76, 1983.
54. **Heinz, E. and Roughan, P. G.,** Similarities and differences in lipid metabolism of chloroplasts isolated from 18:3 and 16:3 plants, *Plant Physiol.*, 72, 273, 1983.
55. **Gardiner, S. E. and Roughan, P. G.,** Relationship between fatty-acyl composition of diacylgalactosylglycerol and turnover of chloroplast phosphatidate, *Biochem. J.*, 210, 949, 1983.
56. **Gardiner, S. E., Heinz, E., and Roughan, P. G.,** Rates and products of long-chain fatty acid synthesis from [1-¹⁴C]acetate in chloroplasts isolated from leaves of 16:3 and 18:3 plants, *Plant Physiol.*, 74, 890, 1984.
57. **Stahl, R. J. and Sparace, S. A.,** Fatty acid and glycerolipid biosynthesis in isolated pea root plastids, in *Plant Lipid Biochemistry, Structure and Utilization*, Quinn, P. J. and Harwood, J. L., Eds., Portland Press, London, 1990, 154.
58. **Sato, N.,** Lipid in *Crytomonas* CR-1. II. Biosynthesis of betaine lipids and galactolipids, *Plant Cell Physiol.*, 32, 845, 1991.
59. **Murata, N., Sato, N., Takahashi, N., and Hamazaki, Y.,** Compositions and positional distributions of fatty acids in phospholipids from leaves of chilling-sensitive and chilling-resistant plants, *Plant Cell Physiol.*, 23, 1071, 1982.
60. **Murata, N.,** Molecular species composition of phosphatidylglycerols from chilling-sensitive and chilling-resistant plants, *Plant Cell Physiol.*, 24, 81, 1983.

61. **Murata, N., Ishizaki-Nishizawa, O., Higashi, S., Hayashi, H., Tasaka, Y., and Nishida, I.**, Genetically engineered alteration in the chilling sensitivity of plants, *Nature*, 356, 710, 1992.

62. **Wolter, F. P., Schmidt, R., and Heinz, E.**, Chilling sensitivity of *Arabidopsis thaliana* with genetically engineered membrane lipids, *EMBO J.*, 11, 4685, 1992.

63. **Löhden, I. and Frentzen, M.**, Role of plastidial acyl-acyl carrier protein:glycerol-3-phosphate acyltransferase and acyl-acyl carrier protein hydrolase in channelling the acyl flux through the prokaryotic and eukaryotic pathway, *Planta*, 176, 506, 1988.

64. **Guerra, D. J., Ohlrogge, J. B., and Frentzen, M.**, Activity of acyl carrier protein isoforms in reactions of plant fatty acid metabolism, *Plant Physiol.*, 82, 448, 1986.

65. **Kunst, L., Browse, J., and Somerville, C.**, Altered regulation of lipid biosynthesis in a mutant of *Arabidopsis* deficient in chloroplast glycerol-3-phosphate acyltransferase activity, *Proc. Natl. Acad. Sci. U.S.A.*, 85, 4143, 1988.

66. **Vick, B. and Beevers, H.**, Phosphatidic acid synthesis in castor bean endosperm, *Plant Physiol.*, 59, 459, 1977.

67. **Douce, R.**, Incorporation de l'acide phosphatidique dans le cytidine diphosphate diglycéride des mitochondries isolées des inflorescences de choufleur, *C.R. Acad. Sci. Ser. D*, 272, 3146, 1971.

68. **Sparace, S. A. and Moore, T. S., Jr.**, Phospholipid metabolism in plant mitochondria, submitochondrial sites of synthesis, *Plant Physiol.*, 63, 963, 1979.

69. **Chuman, L. and Brody, S.**, Acyl carrier protein is present in the mitochondria of plants and eucaryotic microorganisms, *Eur. J. Biochem.*, 184, 643, 1989.

70. **Frentzen, M.**, unpublished data, 1989.

71. **Sauer, A. and Robinson, D. G.**, Subcellular localization of enzymes involved in lecithin biosynthesis in maize roots, *J. Exp. Bot.*, 36, 1257, 1985.

72. **Hares, W. and Frentzen, M.**, Properties of the microsomal acyl-CoA:*sn*-1-acyl-glycerol-3-phosphate acyltransferase from spinach (*Spinacia oleracea* L.) leaves, *J. Plant Physiol.*, 131, 49, 1987.

73. **Sun, C., Cao, Y.-Z., and Huang, A. H. C.**, Acyl coenzyme A preference of the glycerol phosphate pathway in the microsomes from the maturing seeds of palm, maize and rapeseed, *Plant Physiol.*, 88, 56, 1988.

74. **Bafor, M., Stobart, A. K., and Stymne, S.**, Properties of the glycerol acylating enzymes in microsomal preparations from the developing seeds of safflower (*Carthamus tinctorius*) and turnip rape (*Brassica campestris*) and their ability to assemble cocoa-butter type fats, *J. Am. Oil Chem. Soc.*, 67, 217, 1990.

75. **Murphy, D.**, Storage lipid bodies in plants and other organisms, *Prog. Lipid Res.*, 29, 299, 1990.

76. **Weber, N., Taylor, D. C., and Underhill, E. W.**, Biosynthesis of storage lipids in plant cell and embryo cultures, *Adv. Biochem. Eng. Biotechnol.*, 45, 99, 1992.

77. **Slocombe, S. P., Fairbairn, D., Bowra, S., Taylor, R. D., and Murphy, D. J.**, Investigation of temporal and tissue-specific regulation of the stearoyl-acyl carrier protein (ACP) desaturase in oilseed rape, in *Metabolism, Structure and Utilization of Plant Lipids,* Cherif, A., Miled-Daoud, D. B., Marzouk, B., Smaoui, A., and Zarrouk, M., Eds., Centre National Pédagogique, Tunis, 1992, 436.

78. **Holbrook, L. A., van Rooijen, G. J. H., Wilen, R. W., and Moloney, M. M.**, Oilbody proteins in microspore-derived embryos of *Brassica napus*, hormonal, osmotic, and developmental regulation of synthesis, *Plant Physiol.*, 97, 1051, 1991.

79. **Holbrook, L. A., Magus, J. R., and Taylor, D. C.**, Abscisic acid induction of elongase activity, biosynthesis and accumulation of very long chain monounsaturated fatty acids and oil body proteins in microspore-derived embryos of *Brassica napus* L. cv Reston, *Plant Sci.*, 84, 99, 1992.

80. **Huang, A. H. C.**, Oil bodies and oleosins in seeds, *Annu. Rev. Plant Physiol. Plant Mol. Biol.*, 43, 177, 1992.

81. **Kalinski, A., Loer, D. S., Weisemann, J. M., Matthews, B. F., and Herman, E. M.,** Isoforms of soybean seed oil body membrane protein 24 kDa oleosin are encoded by closely related cDNAs, *Plant Mol. Biol.,* 17, 1095, 1991.

82. **Van Rooijen, G. J. H., Terning, L. I., and Moloney, M. M.,** Nucleotide sequence of an Arabidopsis thaliana oleosin gene, *Plant Mol. Biol.,* 18, 1177, 1992.

83. **Li, M., Smith, L. J., Clark, D. C., Wilson, R., and Murphy, D. J.,** Secondary structures of a new class of lipid body proteins from oilseeds, *J. Biol. Chem.,* 267, 8245, 1992.

84. **Cummins, I. and Murphy, D. J.,** cDNA sequence of a sunflower oleosin and transcript tissue specificity, *Plant Mol. Biol.,* 19, 873, 1992.

85. **Keddie, J. S., Hübner, G., Slocombe, S. P., Jarvis, R. P., Cummins, I., Edwards, E. W., Shaw, C. H., and Murphy, D. J.,** Cloning and characterisation of an oleosin gene from *Brassica napus, Plant Mol. Biol.,* 19, 443, 1992.

86. **Ross, J. H. E. and Murphy, D. J.,** Ultrastructural and immunological studies on storage product accumulation in olive seed (*Olea europaea*), in *Metabolism, Structure and Utilization of Plant Lipids,* Cherif, A., Miled-Daoud, D. B., Marzouk, B., Smaoui, A., and Zarrouk, M., Eds., Centre National Pédagogique, Tunis, 1992, 444.

87. **Bergfeld, R., Hong, Y.-N., Kühne, T., and Schopfer, P.,** Formation of oleosomes (storage lipid bodies) during embryo-genesis and their breakdown during development in cotyledons of *Sinapis alba* L., *Planta,* 143, 297, 1978.

88. **Bernerth, R.,** Enzymatische Untersuchungen zur Triacylglycerinbiosynthese in reifenden Rapssamen, Dissertation, Universität Hamburg, Germany, 1991.

89. **Oo, K. C. and Chew, Y. H.,** Diacylglycerol acyltransferase in microsomes and oil bodies of oil palm mesocarp, *Plant Cell Physiol.,* 33, 189, 1992.

90. **Stobart, A. K., Stymne, S., and Höglund, S.,** Safflower microsomes catalyse oil accumulation *in vitro*: a model system, *Planta,* 168, 33, 1986.

91. **Lee, W. S., Tzen, J. T. C., Kridl, J. C., Radke, S. E., and Huang, A. H. C.,** Maize oleosin is correctly targeted to seed oil bodies in *Brassica napus* transformed with the maize oleosin gene, *Proc. Natl. Acad. Sci. U.S.A.,* 88, 6181, 1991.

92. **Sukumar, V. and Sastry, P. S.,** Triacylglycerol synthesis in developing seeds of groundnut (*Arachis hypogaea*): studies on phosphatidic acid phosphatase and diacylglycerol acyltransferase during seed maturation, *Biochem. Inter.,* 14, 1153, 1987.

93. **Gracia, J. M., Quintero, L. C., and Mancha, M.,** Oil bodies and lipid synthesis in developing soybean seeds, *Phytochemistry,* 27, 3083, 1988.

94. **Ichihara, K., Takahashi, T., and Fujii, S.,** Diacylglycerol acyltransferase in maturing safflower seeds: its influences on the fatty acid composition of triacylglycerol and on the rate of triacylglycerol synthesis, *Biochim. Biophys. Acta,* 958, 125, 1988.

95. **Weselake, R. J., Taylor, D. C., Pomeroy, M. K., Lawson, S. L., and Underhill, E. W.,** Properties of diacylglycerol acyltransferase from microspore-derived embryos of *Brassica napus, Phytochemistry,* 30, 3533, 1991.

96. **Kwanyuen, P. and Wilson, R. F.,** Isolation and purification of diacylglycerol acyltransferase from germinating soybean cotyledons, *Biochim. Biophys. Acta,* 877, 238, 1986.

97. **Cao, Y.-Z. and Huang, A. H. C.,** Diacylglycerol acyltransferase in maturing oil seeds of maize and other species, *Plant Physiol.,* 82, 812, 1986.

98. **Kwanyuen, P. and Wilson, R. F.,** Subunit and amino acid composition of diacylglycerol acyltransferase from germinating soybean cotyledons, *Biochim. Biophys. Acta,* 1039, 67, 1990.

99. **Martin, B. A. and Wilson, R. F.,** Subcellular localization of triacylglycerol synthesis in spinach leaves, *Lipids,* 19, 117, 1984.

100. **Martin, B. A. and Wilson, R. F.,** Properties of diacylglycerol acyltransferase from spinach leaves, *Lipids,* 18, 1, 1983.

101. **Shine, W. E., Mancha, M., and Stumpf, P. K.,** Fat metabolism in higher plants, different incorporation of acyl-coenzymes A and acyl-acyl carrier proteins into plant microsomal lipids, *Arch. Biochem. Biophys.,* 173, 472, 1976.

102. **Browse, J., Somerville, C. R., and Slack, C. R.,** Changes in lipid composition during protoplast isolation, *Plant Sci.,* 56, 15, 1988.

103. **Sakaki, T., Saito, K., Kawaguchi, A., Kondo, N., and Yamada, M.,** Conversion of monogalactosyldiacylglycerols to triacylglycerols in ozone-fumigated spinach leaves, *Plant Physiol.,* 94, 766, 1990.

104. **Sakaki, T., Kondo, N., and Yamada, M.,** Pathway for the synthesis of triacylglycerols from monogalactosyldiacylglycerols in ozone-fumigated spinach leaves, *Plant Physiol.,* 94, 773, 1990.

105. **Ichihara, K., Murota, N., and Fujii, S.,** Intracellular translocation of phosphatidate phosphatase in maturing safflower seeds: a possible mechanism of feedforward control of triacylglycerol synthesis by fatty acids, *Biochim. Biophys. Acta,* 1043, 227, 1990.

106. **Perry, H. J. and Harwood, J. L.,** Studies of lipid metabolism in developing oilseed rape, in *Plant Lipid Biochemistry, Structure and Utilization,* Quinn, P. J. and Harwood, J. L., Eds., Portland Press, London, 1990, 204.

107. **Perry, H. J., Bligny, R., Gout, E., Douce, R., and Harwood, J. L.,** Lipid metabolism in development of oilseed rape seeds, in *Metabolism, Structure and Utilization of Plant Lipids,* Cherif, A., Miled-Daoud, D. B., Marzouk, B., Smaoui, A., and Zarrouk, M., Eds., Centre National Pédagogique, Tunis, 1992.

108. **Ichihara, K.,** *sn*-Glycerol-3-phosphate acyltransferase in a particulate fraction from maturing safflower seeds, *Arch. Biochem. Biophys.,* 232, 685, 1984.

109. **Griffiths, G., Stobart, A. K., and Stymne, S.,** The acylation of *sn*-glycerol 3-phosphate and the metabolism of phosphatidate in microsomal preparations from the developing cotyledons of safflower (*Carthamus tinctorius* L.) seed, *Biochem. J.,* 230, 379, 1985.

110. **Ichihara, K., Asahi, T., and Fujii, S.,** 1-Acyl-*sn*-glycerol-3-phosphate acyltransferase in maturing safflower seeds and its contribution to the non-random fatty acid distribution of triacylglycerol, *Eur. J. Biochem.,* 167, 339, 1987.

111. **Eccleston, V. S. and Harwood, J. L.,** Acylation reactions in developing avocado fruits, in *Plant Lipid Biochemistry, Structure and Utilization,* Quinn, P. J. and Harwood, J. L., Eds., Portland Press, London, 1990, 178.

112. **Ichihara, K., Norikura, S., and Fujii, S.,** Microsomal phosphatidate phosphatase in maturing safflower seeds, *Plant Physiol.,* 90, 413, 1989.

113. **Ichihara, K.,** The action of phosphatidate phosphatase on the fatty-acid composition of safflower triacylglycerol and spinach glycerolipids, *Planta,* 183, 353, 1991.

114. **Cao, Y.-Z. and Huang, A. H. C.,** Acyl coenzyme A preference of diacylglycerol acyltransferase from the maturing seeds of *Cuphea*, maize, rapeseed and Canola, *Plant Physiol.,* 84, 762, 1987.

115. **Stymne, S.,** personal communication, 1992.

116. **Griffiths, G., Stobart, A. K., and Stymne, S.,** Δ^6- and Δ^{12}-desaturase activities and phosphatidic acid formation in microsomal preparations from the developing cotyledons of common borage (*Borago officinalis*), *Biochem. J.,* 252, 641, 1988.

117. **Slack, C. R., Campbell, L. C., Browse J. A., and Roughan, P. G.,** Some evidence for the reversibility of the cholinephosphotransferase-catalyzed reaction in developing linseed cotyledons in vivo, *Biochim. Biophys. Acta,* 754, 10, 1983.

118. **Slack, C. R., Roughan, P. G., Browse, J. A., and Gardiner, S. E.,** Some properties of cholinephosphotransferase from developing safflower cotyledons, *Biochim. Biophys. Acta,* 833, 438, 1985.

119. **Stymne, S. and Stobart, A. K.,** Evidence for the reversibility of the acyl-CoA: lysophosphatidylcholine acyltransferase in microsomal preparations from developing safflower (*Carthamus tinctorius* L.) cotyledons and rat liver, *Biochem. J.,* 223, 305, 1984.

120. **Griffiths, G., Stymne, S., Beckett, A., and Stobart, K.,** Lipid metabolism in immature cotyledons of safflower (*Carthamus tinctorius* L.) exposed to light, in *Regulation of Chloroplast Differentiation,* Akoyunoglou, G. and Senger, H., Eds., Alan R. Liss, Inc., New York, 1986, 147.

121. **Moreau, R. A. and Stumpf, P. K.,** Solubilization and characterization of an acyl-coenzyme A *O*-lysophospholipid acyltransferase from the microsomes of developing safflower seeds, *Plant Physiol.,* 69, 1293, 1982.

122. **Rajasekharan, R., Hitz, W. D., and Kinney, A. J.,** Kinetic characterization of partially purified lysophosphatidyl choline/acyl-CoA acyltransferase from developing soybean seeds, in *Plant Lipid Biochemistry, Structure and Utilization,* Quinn, P. J. and Harwood, J. L., Eds., Portland Press, London, 1990, 187.

123. **Sperling, P., Stöcker, S., Mühlbach, H.-P., and Heinz, E.,** Alkenyl ether analogues as substrates for microsomal acylation and desaturation systems, in *Metabolism, Structure and Utilization of Plant Lipids,* Cherif, A., Miled-Daoud, D. B., Marzouk, B., Smaoui, A., and Zarrouk, M., Eds., Centre National Pédagogique, Tunis, 1992, 133.

124. **Kinney, A. J.,** personal communication, 1992.

125. **Griffiths, G., Stymne, S., and Stobart, A. K.,** The utilization of fatty-acid substrates in triacylglycerol biosynthesis by tissue-slices of developing safflower (*Carthamus tinctorius* L.) and sunflower (*Helianthus annuus* L.) cotyledons, *Planta,* 173, 309, 1988.

126. **Stymne, S., Griffiths, G., and Stobart, K.,** Desaturation of fatty acids on complex-lipid substrates, in *The Metabolism, Structure, and Function of Plant Lipids,* Stumpf, P. K., Mudd, J. B., and Nes, W. D., Eds., Plenum Press, New York, 1987, 405.

127. **Bafor, M., Smith, M. A., Jonsson, L., Stobart, K., and Stymne, S.,** Ricinoleic acid biosynthesis and triacylglycerol assembly in microsomal preparations from developing castor-bean (*Ricinus communis*) endosperm, *Biochem. J.,* 280, 507, 1991.

128. **Wiberg, E., Tillberg, E., and Stymne, S.,** Specificities of diacylglycerol acyltransferases in microsomal fractions from developing oil seeds, in *Metabolism, Structure and Utilization of Plant Lipids,* Cherif, A., Miled-Daoud, D. B., Marzouk, B., Smaoui, A., and Zarrouk, M., Eds., Centre National Pédagogique, Tunis, 1992, 83.

129. **Banas, A., Johansson, I., and Stymne, S.,** Plant microsomal phospholipases exhibit preference for phosphatidylcholine with oxygenated acyl groups, *Plant Sci.,* 84, 137, 1992.

130. **Stymne, S., Bafor, M., Jonsson, L., Wiberg, E., and Stobart, K.,** Triacylglycerol assembly, in *Plant Lipid Biochemistry, Structure and Utilization,* Quinn, P. J. and Harwood, J. L., Eds., Portland Press, London, 1990, 191.

131. **Bafor, M., Jonsson, L., Stobart, A. K., and Stymne, S.,** Regulation of triacylglycerol biosynthesis in embryos and microsomal preparations from the developing seeds of *Cuphea lanceolata, Biochem. J.,* 272, 31, 1990.

132. **Bafor, M. and Stymne, S.,** Substrate specificities of glycerol acylating enzymes from developing embryos of two *Cuphea* species, *Phytochemistry,* 31, 2973, 1992.

133. **Bafor, M., Wiberg, E., and Stymne, S.,** Palm kernel (*Elaeis guineensis*): lipid accumulation, fatty acid changes and acyltransferase activities, in *Plant Lipid Biochemistry, Structure and Utilization,* Quinn, P. J. and Harwood, J. L., Eds., Portland Press, London, 1990, 198.

134. **Oo, K.-C. and Huang, A. H. C.,** Lysophosphatidate acyltransferase activities in the microsomes from palm endosperm, maize scutellum, and rapeseed cotyledon of maturing seeds, *Plant Physiol.,* 91, 1288, 1989.

135. **Bernerth, R. and Frentzen, M.,** Utilization of erucoyl-CoA by acyltransferases from developing seeds of *Brassica napus* (L.) involved in triacylglycerol biosynthesis, *Plant Sci.,* 67, 21, 1990.

136. **Mukherjee, K. D.,** Glycerolipid synthesis by homogenate and oil bodies from developing mustard (*Sinapis alba* L.) seed, *Planta,* 167, 279, 1986.

137. **Taylor, D. C., Weber, N., Barton, D. L., Underhill, E. W., Hogge, L. R., Weselake, R. J., and Pomeroy, M. K.,** Triacylglycerol bioassembly in microspore-derived embryos of *Brassica napus* L. cv Reston, *Plant Physiol.,* 97, 65, 1991.

138. **Fehling, E., Murphy, D. J., and Mukherjee, K. D.,** Biosynthesis of triacylglycerols containing very long chain monounsaturated acyl moieties in developing seeds, *Plant Physiol.,* 94, 492, 1990.

139. **Taylor, D. C., Barton, D. L., Rioux, K. P., MacKenzie, S. L., Reed, D. W., Underhill, E. W., Pomeroy, M. K., and Weber, N.,** Biosynthesis of acyl lipids containing very-long chain fatty acids in microspore-derived and zygotic embryos of *Brassica napus* L. cv Reston, *Plant Physiol.,* 99, 1609, 1992.

140. **Taylor, D. C., Weber, N., Hogge, L. R., Underhill, E. W., and Pomeroy, H. K.,** Formation of trierucoylglycerol (trierucin) from 1,2-dierucoylglycerol by a homogenate of microspore-derived embryos of *Brassica napus* L., *J. Am. Oil Chem. Soc.,* 1993, in press.

141. **Löhden, I., Bernerth, R., and Frentzen, M.,** Acyl-CoA:1-acyl-glycerol-3-phosphate acyltransferase from developing seeds of *Limnanthes douglasii* (R. Br.) and *Brassica napus* (L.), in *Plant Lipid Biochemistry, Structure and Utilization,* Quinn, P. J. and Harwood, J. L., Eds., Portland Press, London, 1990, 175.

142. **Dutta, P. C., Appelqvist, L. A., and Stymne, S.,** Utilization of petroselinate (C18:1$^{\Delta 6}$) by glycerol acylation enzymes in microsomal preparations of developing embryos of carrot (*Daucus carota* L.), safflower (*Carthamus tinctorius* L.) and oil rape (*Brassica napus* L.), *Plant Sci.,* 81, 57, 1992.

143. **Cao, Y.-Z., Oo, K.-C., and Huang, A. H. C.,** Lysophosphatidate acyltransferase in the microsomes from maturing seeds of meadowfoam (*Limnanthes alba*), *Plant Physiol.,* 94, 1199, 1990.

144. **Löhden, I. and Frentzen, M.,** Triacylglycerol biosynthesis in developing seeds of *Tropaeolum majus* L. and *Limnanthes douglasii* R. Br., *Planta,* 188, 215, 1992.

145. **Battey, J. F. and Ohlrogge, J. B.,** A comparison of the metabolic fate of fatty acids of different chain lengths in developing oilseeds, *Plant Physiol.,* 90, 835, 1989.

146. **Laurent, P. and Huang, A. H. C.,** Organ- and development-specific acyl coenzyme A lysophosphatidate acyltransferases in palm and meadowfoam, *Plant Physiol.,* 99, 1711, 1992.

147. **MacKenzie, S. L.,** personal communication, 1992.

148. **Eccleston, V. S. and Harwood, J. L.,** Acyltransferase reactions in avocado microsomes, in *Metabolism, Structure and Utilization of Plant Lipids,* Cherif, A., Miled-Daoud, D. B., Marzouk, B., Smaoui, A., and Zarrouk, M., Eds., Centre National Pédagogique, Tunis, 1992, 79.

149. **Peterek, G., Schmidt, V., Wolter, F.-P., and Frentzen, M.,** Approaches for cloning 1-acylglycerol-3-phosphate acyltransferase genes of oil plants, in *Metabolism, Structure and Utilization of Plant Lipids,* Cherif, A., Miled-Daoud, D. B., Marzouk, B., Smaoui, A., and Zarrouk, M., Eds., Centre National Pédagogique, Tunis, 1992, 401.

150. **Coleman, J.,** Characterization of *Escherichia coli* cells deficient in 1-acyl-*sn*-glycerol-3-phosphate acyltransferase activity, *J. Biol. Chem.,* 265, 17215, 1990.

Chapter 7

ORIGIN AND SYNTHESIS OF GALACTOLIPID AND SULFOLIPID HEAD GROUPS

Jacques Joyard, Maryse A. Block, Agnès Malherbe, Eric Maréchal, and Roland Douce

TABLE OF CONTENTS

0-8493-4907-9/93/$0.00+$.50
© 1993 by CRC Press Inc.

231

I. INTRODUCTION

Galactolipids and sulfolipids are characteristic plant membrane glycerolipids. The first evidence for the occurrence and structure of galactolipids was obtained by Carter et al.[16] They isolated two polar lipids from wheat flour which yielded a mixture of galactosylglycerols after mild alkaline hydrolysis. These compounds were separated and identified by enzymatic cleavage and periodate oxidation as β-D-galactopyranosyl-1-glycerol and α-D-galactopyranosyl-(1→6)-β-D-galactopyranosyl-1-glycerol. After further analyses, the structures of monogalactosyldiacylglycerol (or MGDG) and digalactosyldiacylglycerol (or DGDG) were established respectively as 1,2-diacyl-3-O-(β-D-galactopyranosyl)-*sn*-glycerol and 1,2-diacyl-3-O-(α-D-galactopyranosyl-(1→6)-O-β-D-galacto-pyranosyl)-*sn*-glycerol. Benson et al.[5] also identified higher homologs of DGDG in which an additional D-galactopyranosyl moiety was linked α-(1→6) to the terminal galactose unit of DGDG. Therefore, galactolipids are neutral amphipatic lipids. A plant sulfolipid was characterized and its structure assigned by Benson et al.[6] on the basis of their radiochromatographic studies with [35]S- and ^{14}C-labeled lipid. After deacylation, periodate oxidation, β-galactosidase, and acid hydrolysis, the structure of the polar head group was assigned as 6-sulfo-α-D-quinovopyranosyl-(1→1')-*sn*-glycerol. Therefore, this sulfolipid, or sulfoquinovosyldiacylglycerol (SQDG),* consists of a soluble anionic sugar derivative of sulfonic acid bound to diacylglycerol.

Together, galactolipids and SQDG comprise four fifths of spinach lamellar lipoprotein.[94] Of these lipids, 90% are MGDG and DGDG, whereas SQDG makes up the rest.[8] Their relative quantitative distribution varies among the plant tissues. In photosynthetic tissues, the amount of MGDG consistently exceeds the amount of DGDG. They contain MGDG and DGDG in concentrations ranging from about 0.6 to 15 μmol/g and 0.5 to 7 μmol/g fresh weight, respectively.[81] Ratios of galactolipids (MGDG + DGDG) to SQDG are in the range 5 to 20.[70] Nonphotosynthetic tissues contain fewer glycolipids than photosynthetic tissues and show a preponderance of DGDG. These differences reflect the expansion of plastid membranes where these polar lipids are concentrated. In photosynthetic tissues, chloroplast membranes are the most widely distributed membrane system, and therefore their major constituent, MGDG, is the most prevalent glycerolipid. In nonphotosynthetic tissues, extraplastidial membranes are more developed than plastid membranes which are, in addition, mostly restricted to the envelope membranes, and thus glycolipids only represent between 25 to 30% of the total amount of glycerolipids. Analyses of purified plant membranes clearly demonstrate that galactolipids and SQDG are absent from extraplastidial membranes (mitochondria, peroxi-

---

* Numerous sulfur-containing compounds soluble in organic solvents occur in nature. A nomenclature for sulfolipids based on the form of sulfur contained within the molecule was proposed by Haines.[37] Sulfoquinovosyldiacylglycerol, which contains a sulfonic acid, is a sulfonolipid.

somes, etc.) but are the major membrane constituents from both envelope membranes and thylakoids.[27,29]

Galactolipids in thylakoid and envelope membranes contain a high amount of polyunsaturated fatty acids; up to 95% (in some species) of the total fatty acid is linolenic acid (18:3). In nongreen plastids, 18:3 is still a major component in galactolipids, although appreciable amounts of 18:1 and 18:2 are present. Therefore, the most abundant molecular species of MGDG and DGDG have 18:3 at both *sn*-1 and *sn*-2 positions of the glycerol backbone (Figure 1). Some plants, such as pea, having almost exclusively 18:3 in MGDG are called "18:3 plants". Other plants, such as spinach, contain significant amounts of 16:3 in MGDG and are called "16:3 plants".[45] The positional distribution of 16:3 in MGDG is highly specific: this fatty acid is only present at the *sn*-2 position of glycerol and is almost excluded from the *sn*-1 position. Therefore, two major structures are found in galactolipids: one with C_{18} fatty acids at both *sn* positions and one with C_{18} and C_{16} fatty acids, respectively, at the *sn*-1 and *sn*-2 positions (Figure 1). The first one is typical of "eukaryotic" lipids (such as phosphatidylcholine or PC), and the second one corresponds to a "prokaryotic" structure because it is typical for glycerolipids from cyanobacteria.*

Plant SQDG has a high proportion of saturated fatty acids, much more than galactolipids. According to Mudd and Kleppinger-Sparace,[70] the fatty acid composition of SQDG can be as much as 80% saturated fatty acid (palmitic acid) and usually not less than 40%, and therefore some SQDG molecules contain exclusively 16:0 fatty acids at both *sn*-1 and *sn*-2 positions (see, for instance, Siebertz et al.[86]). In some plant species such as spinach, the predominant molecular species of SQDG, like in MGDG, contain C_{18} and C_{16} fatty acids, respectively, at *sn*-1 and *sn*-2 positions, whereas in other plants, such as cucumber, C_{16} fatty acids were located predominantly at *sn*-1 position, corresponding respectively to prokaryotic and eukaryotic structures.[10]

The purpose of this article is to review the last steps in the biosynthesis of plant glycolipids. As shown above, both galactolipids and SQDG have a very similar structure and their formation should proceed in a very similar way, as suggested almost 30 years ago by Benson.[7] Galactolipids and SQDG result from the addition of a soluble sugar moiety to an appropriate diacylglycerol backbone. Therefore, we will discuss successively the origin of glycolipid precursors, i.e. diacylglycerol, and of the different sugar moieties, and the localization and functioning of the enzymes catalyzing the final steps of glycolipid assembly. In addition, since studies on the regulation of polar lipid synthesis in higher plants are usually limited to determination of the incorporation of radioactivity labeled substrates, we will focus on the mechanisms that could be involved in the regulation of plastid glycerolipid synthesis.

---

* The most characteristic feature of the so-called prokaryotic structure of plant glycerolipids is the presence of C_{16} fatty acids at the *sn*-2 position of the glycerol moiety.

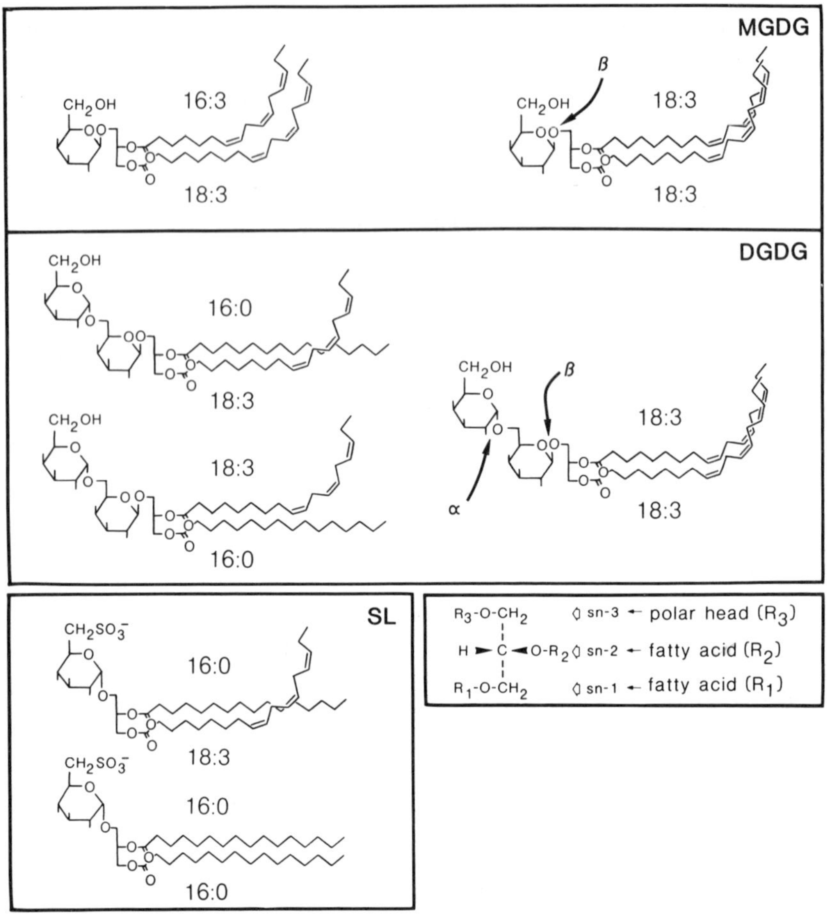

FIGURE 1.   Structure of higher plant galactolipids and sulfolipid.

II. ORIGIN OF GALACTOLIPIDS AND SULFOLIPID PRECURSORS

A. ORIGIN OF THE DIACYLGLYCEROL MOIETY

The differences among galactolipid molecules are mostly confined to the diacylglycerol portion. The same is true for SQDG. As discussed above, at least two major diacylglycerol structures are found in plastid glycerolipids. It is now clearly established that diacylglycerol molecules with C_{16} fatty acids at *sn*-2 position, i.e., with the "prokaryotic" structure, are synthesized within plastids, whereas it is postulated that the biosynthetic pathway for diacylglycerol molecules with a eukaryotic structure could involve several cell compartments.

1. Origin of *sn*-Glycerol-3-Phosphate

For the major plant lipids (i.e., glycerolipids), the precursor for the diacylglycerol backbone of galactolipids and SQDG is *sn*-glycerol-3-phosphate.

In ^{31}P-NMR spectra obtained from neutralized perchloric acid extracts of plant cells and chloroplasts (Figure 2), a sharp resonance peak at 4.2 ppm was identified as a signal from *sn*-glycerol-3-phosphate.[12] However, in spinach leaf pieces, the signal was much lower than those from other phosphomonoesters, whereas in isolated intact chloroplasts the concentration of *sn*-glycerol-3-phosphate could be estimated to about 0.5 mM, an amount sufficient to sustain glycerolipid synthesis in chloroplasts. The accumulation of *sn*-glycerol-3-phosphate within chloroplasts is probably related to the presence of a rather powerful dihydroxyacetone phosphate reductase activity in chloroplasts.[34] This enzyme catalyzes the formation of *sn*-glycerol-3-phosphate at pH 7.0 using NADH, (and not NADPH) as reducing power. Therefore, regulation of dihydroxyacetone phosphate reductase activity, in the cytosol as well as in plastids, probably involves the level of NADH, since the apparent K_m values are 62 and 23 µM, respectively, for the plastid and the cytosolic enzyme. This is very important in chloroplasts, since most plastid reductases use NADPH. Other important features of regulation of plastid dihydroxyacetone phosphate reductase are the rather sharp and low pH optimum (around 7.0) and the effect of reduced thioredoxin. The ratio of reductase activity to that of the reverse reaction (dehydrogenase activity) is 10:1, and therefore the enzyme is probably active *in vivo* for *sn*-glycerol-3-phosphate formation, unless the concentration of this compound reaches considerable values.[34] Plastid dihydroxyacetone phosphate reductase activity represents between 70 and 75% of the total cell activity in pea as well as in spinach, and the remaining activity is present in the cytosol. Therefore, *sn*-glycerol-3-phosphate is formed directly in the major compartments involved in glycerolipid biosynthesis. The substrate for the enzyme involved, dihydroxyacetone phosphate, is a central compound in several essential metabolic pathways, such as photosynthesis and glycolysis, and its physiological concentrations are in the range of 0.1 mM.[89] From these data, it is clear that dihydroxyacetone phosphate reductase activity is probably not a major regulatory step in glycerolipid biosynthesis. In fact, measurements of the pool sizes of glycerol-containing precursors of polar lipids in plants indicate that the pool size of glycerol-3-phosphate is much larger than that of diacylglycerol and phosphatidic acid. This implies that conversion of *sn*-glycerol-3-phosphate to diacylglycerol is rate limiting in the synthesis of polar lipids.

2. Origin of Fatty Acids

Fatty acid synthesis in plant cells is very well-documented. A long series of experiments, starting with Smirnov,[88] has demonstrated that plastids possess all the biosynthetic machinery for fatty acid synthesis (for reviews, see Stumpf[90]). In short, plastids are probably the sole site for fatty acid synthesis within the plant cell. They catalyze high rates of palmitic (16:0) and oleic (18:1) acid synthesis as acyl-ACP (acyl carrier protein) thioesters (16:0-ACP and 18:1-ACP). Then, fatty acids are either used within plastids or exported to the cytosol for the synthesis of complex lipids. A detailed presentation of fatty acid biosynthesis and its regulation is given in Chapter 1.

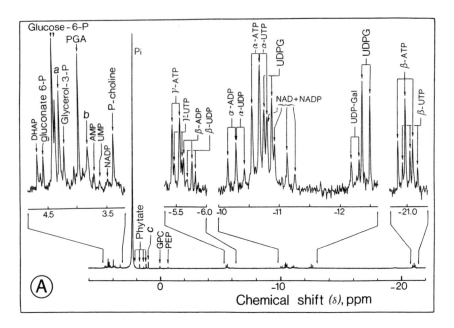

FIGURE 2. Proton-decoupled ^{31}P-NMR spectrum of perchloric acid extracts of spinach leaves (A) and chloroplasts (B). Perchloric extracts were prepared as described by Bligny et al.[12] The samples (2.5 ml) containing 300 μl of ^{2}H$_2$O were analyzed at 25°C with a 10-mm multinuclear probe tuned at 162 MHz. With a 90° pulse angle and a recycling time of 0.4 s under proton decoupling, 2048 Free Induction Decays (FIDs) were acquired. Two levels of broad band ^{1}H-decoupling were employed: 2.5 W during the data acquisition, and 0.5 W during the delay period. The resulting FIDs memorized, over 8000, were Fourier transformed over 32,000 after gaussian multiplication. All the interesting regions are shown on an expanded scale. Peak assignments: a — unidentified peak; b — position of fructose-6-phosphate, ribose-5-phosphate, and phosphoryl-ethanolamine; Pi — total inorganic phosphate; c — unidentified peak (phosphodiester ?); DHAP — dihydroxyacetonephosphate; PGA — glycerate-3-phosphate; GPC — glycerophosphorylcholine, phosphorylcholine; PEP — phosphoenolyruvate; UDPG — UDG-glucose. Note the absence of UDP-galactose and UDP-glucose in chloroplast; in contrast, glycerol-3-phosphate is present in both samples.

3. Biosynthesis of 18:1/16:0 Diacylglycerol

All plastids (chloroplasts, proplastids, amyloplasts, etc.) are able to acylate *sn*-glycerol-3-phosphate to synthesize phosphatidic acid.[58] The first enzyme involved is a soluble glycerol-3-phosphate acyltransferase, closely associated with the inner envelope membrane, which catalyzes the transfer of 18:1 from 18:1-ACP to the *sn*-1 position of glycerol,[2,32,54] producing 1-oleoyl-*sn*-glycerol-3-phosphate (lysophosphatidic acid). Bertrams and Heinz[9] first provided biochemical data for the purified chloroplast glycerol-3-phosphate acyltransferase from pea and spinach. According to the plant species analyzed, the chloroplast stroma contains one to three isomeric forms of glycerol-3-phosphate acyltransferase.[9,24,73] The sequence of the cDNA corresponding to two isoforms of the squash chloroplast protein was determined by Ishizaki et al.[52] The plastid glycerol-3-phosphate acyltransferase is syn-

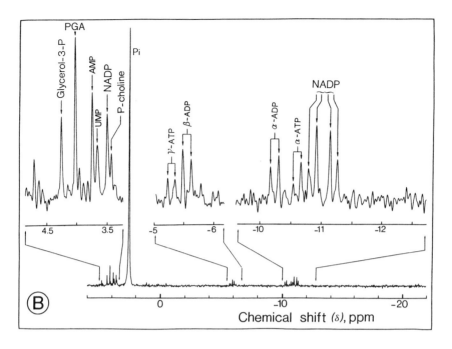

FIGURE 2. (Continued)

thesized as a precursor protein of 396 amino acids, which is processed to become a mature protein of 368 amino acids, losing a transit peptide of 28 amino acids.[66] Interestingly, Kunst et al.[66] have characterized an *Arabidopsis* mutant, named *JB25*, in which the chloroplast glycerol-3-phosphate acyltransferase activity was reduced to less than 4% of the activity in the wild type. Wild-type *Arabidopsis* is a typical 16:3-plant, and therefore its MGDG contains 16:3 and 18:3 fatty acids; in contrast, MGDG from the mutant *JB25* contains only 18:3 and no 16:3. In addition, the MGDG content of the mutant is nearly the same as the wild type. This experiment demonstrates the remarkable flexibility of glycerolipid biosynthesis in plants and provides an insight into the nature of the regulatory mechanisms involved. Since the acyltransferase involved in the biosynthesis of lysophosphatidic acid is strongly reduced, the pathway for the synthesis of glycerolipids with the prokaryotic structure is less active, whereas the activity of the enzymes catalyzing the synthesis of the eukaryotic structure is increased to compensate for the loss of prokaryotic glycerolipids.[66] Finally, the soluble acyltransferase seems to be important for determining the level of phosphatidylglycerol (PG) fatty acid unsaturation.[9,73] For instance, the squash enzyme is responsible for the insertion of a higher proportion of C_{16} fatty acids into lysophosphatidic acid. The gene encoding for squash glycerol-3-phosphate acyltransferase was introduced into tobacco plants, and the transgenic plants obtained contained more saturated PG than the normal tobacco, whereas MGDG, DGDG, and SQDG were almost not affected.[72]

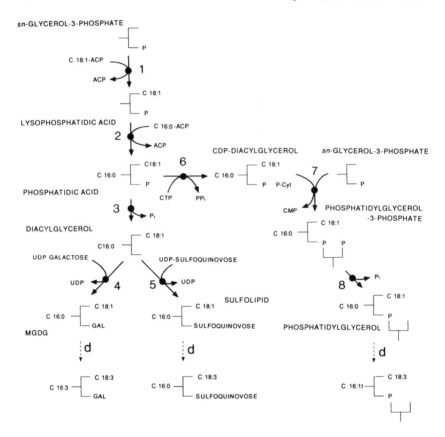

FIGURE 3.   Biosynthesis of plastid glycerolipids by plastid envelope membranes. The enzymes involved are (1) glycerol-3-phosphate acyltransferase, (2) 1-acylglycerol-3-phosphate acyltransferase, (3) phosphatidate phosphatase, (4) monogalactosyldiacylglycerol synthase, (5) sulfolipid synthase, (6) phosphatidate cytidyltransferase, (7) CDP-diacylglycerol-glycerol-3-phosphate 3-phosphatidyltransferase, (8) phosphatidylglycerophosphatase, and (9) desaturases. This biosynthetic pathway leads to the synthesis of monogalactosyldiacylglycerol, sulfolipid, and phosphatidylglycerol with unsaturated C_{18} and C_{16} fatty acids at the *sn*-1 and *sn*-2 positions of glycerol, respectively.

This experiment demonstrates that this acyltransferase is a regulatory point in the biosynthetic pathways for glycolipids and PG.

Lysophosphatidic acid synthesized by the soluble glycerol-3-phosphate acyltransferase is further acylated to form 1,2-diacyl-*sn*-glycerol-3-phosphate (phosphatidic acid) by the action of a 1-acylglycerol-3-phosphate acyltransferase (Figure 3). This enzyme is present in envelope membranes from chloroplasts[56] and nongreen plastids.[2] In spinach chloroplasts, both the outer and the inner envelope membranes contain this acyltransferase,[13] but in pea chloroplasts only the inner membrane does.[3] Since lysophosphatidic acid used for this reaction is esterified at the *sn*-1 position, the enzyme will direct fatty acids, almost exclusively 16:0, to the available *sn*-2 position.[32,54]

Therefore, the two plastid acyltransferases have distinct specificities and selectivities for acylation of *sn*-glycerol-3-phosphate (Figure 3). Together, they lead to the formation of phosphatidic acid having 18:1 and 16:0 fatty acids, respectively, at the *sn*-1 and *sn*-2 positions of the glycerol backbone. As discussed above, this structure is typical of the so-called prokaryotic glycerolipids. In contrast, extraplastidial acyltransferases have distinct localization and properties (nature of the acyl donor, specificities, selectivities) as discussed by Frentzen et al.[31,33] (see also Chapter 6).

Phosphatidic acid synthesized in envelope membranes is further metabolized into either diacylglycerol or PG as shown in Figure 3. At this point probably lies a major regulatory mechanism which controls the biosynthesis of plastid glycerolipids and the final proportions of PG, MGDG, DGDG, and SQDG (see also Murata et al.[72]). Diacylglycerol biosynthesis occurs in the envelope membrane owing to a membrane-bound phosphatidate phosphatase[56,57] exclusively located on the inner envelope membrane.[3,13] In contrast to chloroplasts from 16:3-plants, those from 18:3-plants have a rather low phosphatidate phosphatase activity[47] and cannot deliver diacylglycerol fast enough to sustain the full rate of glycolipid synthesis. The same is true in nongreen plastids from 18:3-plants.[2] These results can explain why 18:3-plants contain only small amounts of galactolipids and SQDG with C_{16} fatty acids at the *sn*-2 position, but contain PG (which is synthesized from phosphatidic acid) with such a structure. It is not yet known whether the reduced level of phosphatidate phosphatase activity is due to lower expression (species specific) of the gene coding for the enzyme or to the presence of regulatory molecules which control the activity of the enzyme. Finally, among the possible mechanisms involved in the regulation of phosphatidate phosphatase activity, feedback inhibition by diacylglycerol probably plays a major role. Using isolated intact chloroplasts and envelope membranes from thermolysin-treated chloroplasts. Malherbe et al.[96] have recently demonstrated that diacylglycerol is a powerful inhibitor of the enzyme; the apparent K_i value was 70 μM (with dioleoylglycerol), whereas the apparent K_m for phosphatidic acid was as high as 600 μM. Obviously, inhibition of phosphatidate phosphatase by diacylglycerol would favor channeling of phosphatidic acid toward PG.

4. Origin of C_{18}/C_{18} Diacylglycerol

The specificities of the envelope acyltransferases do not allow the formation of phosphatidic acid and diacylglycerol with only C_{18} fatty acids. Under natural conditions, prokaryotes are also unable to synthesize such a structure. *In vivo* kinetics of acetate incorporation into chloroplast lipids have suggested that PC could provide the diacylglycerol backbone for plastid glycerolipids. The reader is referred to reviews by Heinz,[45] Douce and Joyard,[28] Roughan and Slack,[78] and Joyard and Douce[58] for detailed presentations of the arguments in favor of this hypothesis. The hypothesis involves (1) the synthesis of PC in extraplastidial membranes, (2) its transfer to the outer envelope membrane, (3) the formation of diacylglycerol, and (4) the integration of diacylglycerol into MGDG or SQDG. Except during the initial steps, the plastid envelope membranes should play a central role in this pathway. However, the complete demonstration that this pathway indeed operates in plants has not yet been provided. A missing link is the conversion of PC into diacylglycerol, since envelope membranes apparently lack a phospholipase C activity which would produce diacylglycerol.[58]

B. ORIGIN OF UDP-GALACTOSE

In a typical proton-decoupled ^{31}P-NMR spectrum of perchloric acid extracts of spinach leaves (Figure 2A), complex multipeak resonance showing absorption bands at approximately −12.4 and −12.6 ppm and corresponding to α- and β-UDP-galactose are clearly visible.[12] In contrast, under the same experimental conditions (Figure 2B), no UDP-galactose could be detected in chloroplast extracts.[12] Figure 2 also demonstrates that UDP-glucose is present in large amounts in the cytosol, but could not be detected in the chloroplast extract. In addition, the enzymes involved in UDP-glucose and UDP-galactose synthesis, i.e., UDP-glucose pyrophosphorylase and UDP-glucose epimerase, are cytosolic enzymes.[65] Therefore, UDP-glucose and UDP-galactose are in permanent equilibrium in the cytosol. All together, these observations provide evidence for an extraplastidial location of UDP-galactose synthesis and accumulation. However, this raises the question of the availability of UDP-galactose for MGDG synthesis in plastids. MGDG synthase is a plastid envelope enzyme[25] and, at least in plants such as spinach, is located on the inner membrane,[13] a membrane shown to be impermeable to UDP-galactose.[40] Therefore, there is no need for UDP-galactose to accumulate in the plastid stroma only if MGDG synthesis occurs on the outer surface of the inner envelope membrane, but this remains to be demonstrated. If this is true, the concentrations of UDP-glucose (about 4 to 5 mM) and UDP-galactose (about 0.2 to 0.5 mM) found in the cytosol by ^{31}P-NMR analyses are high enough to sustain optimal rates of galactolipid biosynthesis (see below), at least under normal physiological conditions.

C. ORIGIN OF THE SULFOQUINOVOSE MOIETY

Almost 30 years after the discovery of sulfolipid by Benson et al.,[6] little is known about the biosynthetic pathway involved in sulfate incorporation into the polar head group of SQDG and especially on the origin of the carbon-sulfur bond occurring in the sulfoquinovose moiety of SQDG. Shibuya et al.[85] have reported identification of a nucleoside diphosphosulfoquinovose (UDP-sulfoquinovose) in extracts of *Chlorella*, but since this time it has been impossible to characterize unambiguously the biosynthetic pathway between sulfate and SQDG. Several reviews have been devoted to this puzzling question.[39,64,70] However, Heinz et al.[49] have recently demonstrated that chemically synthesized UDP-sulfoquinovose can be used by isolated envelope membranes for SQDG synthesis, thus demonstrating that the SQDG biosynthetic pathway proposed by Shibuya et al.[85] could have some physiological significance. But, as discussed by Heinz,[46] one should keep in mind that UDP-sulfoquinovose has not yet been characterized in higher plant chloroplasts and therefore should still be considered as a putative precursor for SQDG. To summarize, two main hypotheses have been proposed for the biosynthesis of UDP-sulfoquinovose; the carbon-sulfur bond can be formed by sulfonation of either a three-carbon compound or a six-carbon compound. Sulfonation of a three-carbon compound requires further elongation to a six-carbon compound and a rather complex enzymatic pathway.[64] Sulfonation of a glucose derivative, first proposed by Zill and Cheniae,[95] could occur on a sugar nucleotide (or on a monoglucosyl-diacylglycerol). Up to now, the only experimental evidence available is in favor of sulfonation of a sugar nucleotide intermediate.

Recently, Benning and Somerville[4] have developed an approach combining biochemistry and molecular biology to characterize enzymes involved in the biosynthesis of SQDG precursors. They have generated a series of mutants from *Rhodobacter sphaeroides*, a photosynthetic bacterium containing SQDG. In these mutants, only very low levels of SQDG were found, and sulfate reduction and activation were not altered, whereas water-soluble sulfate-containing compounds (probably precursors to UDP-sulfoquinovose) accumulated. By complementation with cosmids from a wild-type genomic DNA library, the defect in SQDG was restored. The genes identified are still under investigation, and the corresponding proteins have not yet been identified. One of the proteins has some analogy with UDP-glucose epimerase from bacteria and yeast.[4] When fully characterized and identified, the genes from *R. sphaeroides* will be used for characterization of their higher plant counterparts.

Intracellular localization of the biosynthesis of SQDG precursors is still unknown. Sulfate is a suitable precursor for SQDG synthesis; when intact and purified chloroplasts were incubated in light and the presence of $^{35}SO_4^2$, high rates of SQDG synthesis were demonstrated.[36,59,63] In higher plants, assimilatory sulfate reduction is mostly located in plastids,[82,83] although cysteine synthesis occurs in each subcellular compartment where the synthesis of proteins occurs (i.e., plastids, cytosol, and mitochondria), as shown recently by Lunn et al.[67] and Rolland et al.[77] Adenosine 5′-phosphosulfate (APS) and 3′phosphoadenosine 5′-phosphosulfate (PAPS) are also good precursors for SQDG biosynthesis.[51,64] Similar rates of SQDG synthesis (about 3 nmol/mg of chlorophyll per hour) have been reported when APS, PAPS, and sulfate were used as precursors.[64] The light dependency for SQDG biosynthesis is due to the light requirement for sulfate reduction and activation, i.e., for APS synthesis, since an ATP-generating system allows SQDG synthesis in the dark.[7,64] Finally, reduction beyond the level of sulfite is not necessary for SQDG synthesis in higher plants,[35,64] in contrast with the situation in *Euglena* in which an alternative pathway for SQDG biosynthesis involving cysteic acid is predominant.[21,80] It is also possible that in nongreen plastids, the oxidative pentose phosphate pathway also functions to provide reductant for sulfate assimilation.

As a central metabolite in the chloroplast (see above), dihydroxyacetone phosphate has a major effect on plastid metabolism and was shown to stimulate SQDG synthesis.[64] Several possible roles for dihydroxyacetone phosphate can explain this stimulation. First, being converted into glycerol-3-phosphate by dihydroxyacetone phosphate reductase[34] and therefore as a precursor of the diacylglycerol moiety (see above), dihydroxyacetone phosphate can stimulate the biosynthesis of all plastid glycerolipids. Dihydroxyacetone phosphate is also a precursor for fructose-1,6-bisphosphate and therefore is a possible precursor for the sugar moiety of UDP-sulfoquinovose. Finally, dihydroxy-acetone phosphate is involved in the shuttling of metabolites across the envelope through the phosphate translocator, which allow the indirect transfer across the envelope of reducing equivalents and nucleotides which are used for activation of sulfur compounds.

Finally, one should keep in mind that most nucleotide sugars found within plastids (and involved in starch synthesis) are linked to ADP and usually not to UDP. However, the observations that SQDG synthesis by isolated intact chloroplasts is stimulated by UTP (Kleppinger-Sparace et al.[64]) and that isolated envelope membranes are active in SQDG biosynthesis when supplied with UDP-sulfoquinovose[49,87] provide some support for the possible use (and therefore formation) of UDP-sulfoquinovose by chloroplasts. Up to now, direct evidence is still lacking. In addition, Heinz et al.[49] synthesized different nucleoside 5'-diphospho-sulfoquinovoses and demonstrated that both UDP- and GDP-sulfoquinovose significantly increased SQDG synthesis by spinach chloroplasts and isolated envelope membranes, with UDP-sulfoquinovose being only twice as active as the GDP derivative, thus suggesting that SQDG synthase could use other substrates. The major limitation for the characterization of the *in vivo* sulfoquinovose donor is due to the difficulty in analyzing the sulfur-containing compounds in whole tissues and in intact chloroplasts. Since the pioneering work of Shibuya et al.,[85] numerous sulfur-containing compounds have been shown in plant extracts, but very few have been identified, especially in chloroplasts (for instance, elemental sulfur).[60] Obviously, proper procedures for protecting these compounds from degradation during purification and analyzing them are still lacking.

Together, these observations lead to the conclusion that the precursor of the polar head group for SQDG (probably UDP-sulfoquinovose) is made within plastids, thus suggesting that SQDG synthase is accessible from the stromal side of the inner envelope membrane. In contrast, and as discussed above, UDP-gal is synthesized in the cytosol, and therefore MGDG synthase is accessible from the cytosolic side of the inner envelope membrane. The putative distinct localization, on opposite sides of the same membrane, of these two enzymes manipulating the same substrate is important for understanding their interacting role in plastid glycolipid biosynthesis (see below).

III. GALACTOLIPID BIOSYNTHESIS

The reader is referred to the reviews by Heinz,[45] Harwood,[38] Douce and Joyard,[28] Roughan and Slack,[78] and Joyard and Douce,[58] which describe the early studies on galactolipid formation in whole plants, tissues, organelles or membranes, and on the enzymes involved. In this part, only the final steps in galactolipid biosynthesis will be considered, i.e., the biochemical characterization of MGDG synthase and its purification, and the possible pathway for DGDG synthesis.

A. MGDG SYNTHESIS

The plastid 1,2-diacylglycerol 3-β-galactosyltransferase (or MGDG synthase) transfers a galactose from a water-soluble donor, UDP-galactose, to a hydrophobic acceptor molecule, diacylglycerol, to synthesize MGDG with the release of UDP. Therefore, this enzyme is responsible for the formation of a β-glycosidic bond between galactose and diacylglycerol.

1. Localization

The localization of MGDG synthase in envelope membranes from all chloroplasts and nongreen plastids is now clearly established, and therefore MGDG synthase is considered to be the best marker enzyme for envelope membranes.[29,35] In spinach, MGDG synthase is exclusively located in the inner envelope membrane,[13] whereas it could be also present on the outer membrane in pea chloroplasts[18] (but see Joyard and Douce[58] for a discussion of this point).

2. Purification

Purification of the envelope MGDG synthase is very difficult to achieve. First, envelope membranes are only a minor membrane structure within the plant cell; they do not represent more than 1 to 2% of the plastid proteins; therefore, purification of envelope membranes is a prerequisite for further purification of any envelope protein. Second, envelope membranes have a very high lipid to protein ratio (1.2 to 1.5 mg lipid per milligram protein), and their complete solubilization requires large amounts of detergents that might be harmful to enzymatic activities. In addition, the solubilization conditions for MGDG synthase should also be suitable for the diacylglycerol substrate. Therefore, a compromise between solubilization and optimal activity is difficult to achieve. Third, among the numerous envelope proteins, only a few of them (such as E30, i.e., the phosphate/triose phosphate translocator, and E37) are major components; all the others are only present in limited amounts.[27,29] Therefore, large quantities of purified envelope membranes are necessary for enzyme purification; approximately 100 mg envelope proteins are routinely used for MGDG synthase purification.[68] This enzyme, which is responsible for the synthesis of the most abundant polar lipid on earth, only represents between 0.1 to 0.5% of the total envelope proteins.[58]

Covès et al.[19,20] described suitable conditions for MGDG synthase solubilization and assay; they also achieved a partial purification of this enzyme from spinach chloroplast envelope membranes. Further purification of MGDG synthase activity was recently obtained by two groups,[68,91] but led to characterization of two different polypeptides. Teucher and Heinz[91] characterized a 22-kDa polypeptide. MGDG synthase activity in the purest fraction we have obtained was enriched more than 500-fold and was associated with a 19-kDa polypeptide.[68] Table 1 summarizes the activities and purification factors during this purification, and Figure 4 presents the corresponding polypeptide patterns. No 22-kDa polypeptide could be detected in our purest fractions. This raises the question of the correct identification of MGDG synthase which is hampered by the very low amounts of protein in the purest fractions (about 1 μg). A possible explanation for the discrepancy between the two results is that the enzyme could be associated with a minor component present in both preparations. Therefore, additional work is still needed for unambiguous identification of MGDG synthase. However, despite this major problem, both series of experiments confirm the observation that MGDG synthase is a very minor envelope component. The specific activity of MGDG synthase is extremely high in order to sustain the high rates of MGDG synthesis required for plastid membranes. In addition, the very low level of MGDG synthase in envelope membranes raises several problems. For instance, one can question whether the MGDG synthase is mobile within the inner envelope membrane or associated as a complex with the other envelope enzymes (acyltransferases, phosphatidate phosphatase, SQDG synthase, desaturases, etc.) involved in glycerolipid biosynthesis. In addition, it is not known if MGDG formation is restricted to specific parts of the inner envelope membrane or spread within the whole membrane. Up to now, we cannot provide any answer to these questions.

3. Properties

The galactosyltransferase responsible for MGDG synthesis has a broad pH optimum above 7.5 and up to 9.0.[55,92] A limited stimulation of galactose incorporation into envelope galactolipids by Mg^{2+}, Mn^{2+}, and Ca^{2+} was observed.[18,43,55] In addition, Heemskerk et al.[43] observed a remarkable inhibition of MGDG synthesis in isolated envelope membranes by Zn^{2+}, Fe^{2+}, and Cd^{2+}. Teucher and Heinz[91] and Maréchal et al.[68] took advantage of this observation by using Zn^{2+}-chelating sepharose chromatography for MGDG synthase purification. Furthermore, although addition of salts (NaCl, KCl, KH_2PO_4, etc.) was almost without effect on enzyme activity when measured with native envelope, a high ionic strength in the assay mixture was necessary for optimal activity.[19]

Covès et al.[20] reported a very specific lipid requirement for optimal MGDG synthase activity after partial purification. Neutral lipids (such as galactolipids) and zwitterionic phospholipids (such as PC) were almost without effect on MGDG synthase activity, whereas negatively charged lipids such as SQDG and, more efficiently, PG strongly stimulated MGDG synthase activity.[20] Covès et al.[20] demonstrated that this stimulation was due to the properties of the polar head group rather than being due to a specificity for fatty acids. In contrast, Teucher and Heinz[91] could not demonstrate any stimulation of their purified MGDG synthase activity by added PG. In fact, Teucher and Heinz[91] used a negatively charged detergent, cholate, to solubilize the envelope MGDG synthase, and this probably provides a similar environment to the enzyme. Together, these data suggest that optimal activity of solubilized MGDG synthase

TABLE 1
Purification of MGDG Synthase from Spinach Chloroplast Envelope Membranes

	Protein (μg)	Total activity (nmol galactose (incorporated/h)	Yield (%)	Specific activity (nmol galactose incorporated/ h/mg protein)	Enrichment factor
Solubilized envelope	86,000	7,740	—	90	1
Hydroxyapatite	157	578	7.5	3,682	41
Biogel P-6 DG	138	508	6.5	3,682	41
Blue dextran agarose	47.5	240	3.1	5,046	56
Chelating sepharose-6B	<1[a]	46.4	0.6	>46,400[a]	515

[a] Proteins were estimated by comparison of absorbance at 520 nm of Coomassie blue stained polypeptide bands with (a) that of molecular mass markers (see Figure 4, lane 1) and (b) that of the other envelope fractions (see Figure 4, lanes 2 to 7) whose protein content was known. Both determinations gave very similar results.

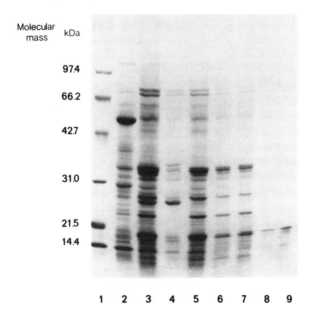

FIGURE 4. SDS-PAGE analysis of envelope polypeptides during MGDG synthase purification. From left to right: (1) molecular mass markers (Bio-Rad), (2) solubilized envelope, (3) fraction from peak 3 of hydroxyapatite chromatography, (4) void volume from Blue Dextran-Agarose, (5) active fraction from Blue Dextran-Agarose, (6,7) void volume from Chelating-Sepharose-6B, and (8,9) active fractions from Chelating-Sepharose-6B. Electrophoresis was done using 7.5 to 15% acrylamide gradient in the presence of sodium dodecyl sulfate. The polypeptides were then stained with Coomassie Blue. Protein load in each well was lanes 2 to 5, 20 μg; lanes 6 and 7, 6 μg; and lanes 8 and 9, estimated to less than 0.5 μg. Total activities loaded in each well were lane 2, 1.8 nmol/h; lane 3, 74 nmol/h; lane 4, 5 nmol/h; lane 5, 101 nmol/h; lanes 6 and 7, 8.5 nmol/h; lane 8, 5.2 nmol/h; and lane 9, 8.2 nmol/h.

requires a negatively charged hydrophobic environment which can be provided either by SQDG, PG, or a charged detergent such as cholate. However, we have demonstrated that addition of PG to our incubation mixture strongly facilitates diacylglycerol solubilization by CHAPS,[97] thus explaining why such a negatively charged environment is necessary for full MGDG synthase activity *in vitro*. In order to gain insight into the modulation of MGDG synthase activity by changing the polar lipid environment, the purified enzyme must be reconstituted into mixed micelles containing its substrate diacylglycerol.[97]

One interesting feature of MGDG synthase is its sulfhydryl nature, established in spinach chloroplast preparations by Chang[17] and Mudd et al.[71] and confirmed with isolated envelope membranes.[43] After solubilization, MGDG synthase is extremely sensitive to sulfhydryl reagents. For instance, incubation of isolated envelope membranes in presence of 5 mM N-ethylmaleimide (NEM) led to complete inhibition of MGDG synthase, whereas the same result was obtained after a brief (15 s) incubation of solubilized envelope in the presence of very low (10 μM) NEM concentrations.[19] Together, the effect of ionic strength, of negatively charged lipids, and of sulfhydryl reagents demonstrate that the optimal activity of envelope MGDG synthase relies on very tight environmental conditions which are met within the inner envelope membrane and should be carefully reconstituted during solubilization and assay.

Kinetic parameters of MGDG synthase were determined with purified enzyme fractions. Covès et al.[20] found an apparent K_m value for UDP-galactose of about 100 μM, a value very close to those previously reported for the enzyme within the envelope (see Joyard and Douce for a review).[58] The amount of UDP-galactose (up to 0.5 mM as demonstrated by ^{31}P-NMR, see Figure 2) present in the cytosol is therefore high enough for an optimal enzyme activity.

The use of purified MGDG synthase fractions devoid of envelope lipids allowed for the first time determination of the affinity of the enzyme for its hydrophobic substrate, diacylglycerol. However, such a study is hampered by several technical difficulties, since diacylglycerol solubilization and control of its availability to the enzyme are not easy to monitor; for instance, apparent K_m values of 500 μg/ml (i.e., about 800 μM) and 180 μM were obtained respectively by Covès et al.[20] and Maréchal et al.[97] Therefore, the affinity of the enzyme for the hydrophobic substrate is not very high,* and clearly diacylglycerol availability at the level of MGDG synthase could be a major regulatory mechanism for MGDG synthesis. This is also true for SQDG and PG, since diacylglycerol will also inhibit the envelope phosphatidate phosphatase (Malherbe et al.[96]). Interestingly, Maréchal et al.[97] found that apparent K_m values are almost identical with all diacylglycerol molecular species analyzed. This observation is in agreement with a series of experiments, using envelope membranes, which have demonstrated that MGDG synthase can use

---

* One should keep in mind, however, that rather high diacylglycerol concentrations can be reached within the membrane and that the surface concentration is probably more appropriate to express this value.[22]

all diacylglycerol molecules available (see Joyard and Douce for a discussion[58]). However, apparent V_{max} values vary in a very wide range according to the different diacylglycerol molecular species used.[97] Interestingly, the lowest activity was obtained with 16:0/16:0 diacylglycerol, a structure which is never found in MGDG.[97]

Finally, the two-substrate kinetic analyses performed by Maréchal et al.[97] (Figures 5 and 6) led to the conclusion that MGDG synthase is a random bireactant system (Figure 7); that is, diacylglycerol and UDP-galactose bind randomly to the enzyme, but the binding of one substrate does not change the dissociation constant for the other substrate. This result demonstrates that the hypothesis of a ping-pong mechanism proposed for MGDG synthase by van Besouw and Wintermans[93] is not valid. This conclusion is also supported by studies of enzyme inhibition by UDP, one of the reaction products. UDP is a competitive inhibitor (K_i = 10 μM[20]) of MGDG synthase when UDP-gal is used as the variable substrate.[20,92] However, at saturating diacylglycerol concentrations, UDP inhibition of MGDG synthesis (with diacylglycerol as the variable substrate) was much more complex[20] and when diacylglycerol concentration was expressed as mole fraction rather than molar value, we found that UDP inhibition was competitive toward diacylglycerol. These observations provide additional evidence for a random bireactant system; MGDG synthase possesses two distinct and independent substrate-binding sites: a hydrophilic one for UDP-galactose and a hydrophobic one for diacylglycerol.

B. FORMATION OF DGDG AND OF HIGHER HOMOLOGS

The galactolipid:galactolipid galactosyltransferase, which is the only DGDG-forming enzyme clearly described in plastids to date, was characterized first on envelope membranes by van Besouw and Wintermans.[92] It catalyzes an enzymatic galactose exchange between galactolipids with the formation of diacylglycerol, but *in vitro* unnatural galactolipids with more than two galactose residues (tri- and tetra-galactosyldiacylglycerol or tri-GDG and tetra-GDG) can be synthesized as follows:

$$2\ MGDG \rightarrow DGDG + 1,2 - diacylglycerol$$
$$DGDG + MGDG \rightarrow tri - GDG + 1,2 - diacylglycerol$$
$$2\ DGDG \rightarrow tetra - GDG + 1,2 - diacylglycerol$$

This enzyme is located on the cytosolic side of the outer envelope membrane because it is destroyed during mild proteolytic digestion (with thermolysin) of intact chloroplasts[23] or nongreen plastids.[1] The localization was further confirmed using purified outer and inner envelope membranes.[42]

The galactolipid:galactolipid galactosyltransferase is more active at pH 6.0.[43,48] Furthermore, its activity is strongly dependent upon the presence of cations such as Mg^{2+} or Ca^{2+} [43,54] and is strongly inhibited by the addition of EDTA and NaF.[54] In contrast, activity in the presence of other cations such

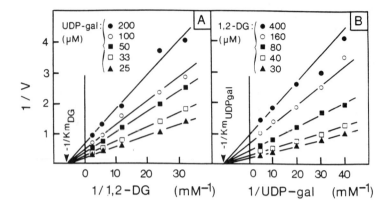

FIGURE 5. Kinetic studies of MGDG synthase as a function of UDP-gal and diacylglycerol concentrations. Representations according to Lineweaver and Burke. A: Diacylglycerol is the variable substrate, and each curve represents a series of experiments at given UDP-gal concentrations (from 25 to 200 μM). B: UDP-gal is the variable substrate, and each curve represents a series of experiments at given diacylglycerol concentrations (from 30 to 400 μM). MGDG synthase activity was determined in 400 μl of incubation mixture containing 6 mM CHAPS; 50 mM MOPS-NaOH, pH 7.8; 1 mM DTT; 250 mM KH_2PO_4; 1 μg PG/ml; 200 μl of MGDG synthase (corresponding to 7 microgram protein); and both substrates, i.e., UDP-gal and diacylglycerol at given concentrations. After 2, 4, and 6 min incubations, 125 μl of the mixture is taken. The reaction is stopped, and the lipids extracted according to Maréchal et al.[97] MGDG synthase activity is expressed as micromole galactose incorporated per hour per milligram protein. The two curves provide several apparent values for V_{max} and one apparent K_m value for each substrate. V_{max} values are then used for the representation in Figure 6.

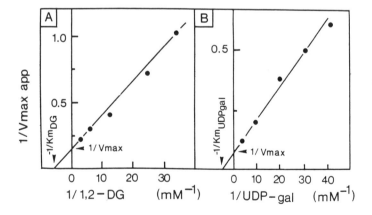

FIGURE 6. Determination of apparent V_{max} for MGDG synthase as a function of UDP-gal and diacylglycerol concentrations. Representations according to Lineweaver and Burke. A: Diacylglycerol is the variable substrate, and V_{max} values were obtained from figure 2B. B: UDP-gal is the variable substrate, V_{max} values were obtained from Figure 2A. The experimental conditions are those from Figure 5. MGDG synthase activity is expressed as micromole galactose incorporated per hour per milligram of protein. V_{max} values are, of course, identical: 11.3 and 12.1 μmol galactose incorporated per hour per milligram of protein, respectively, for A and B.

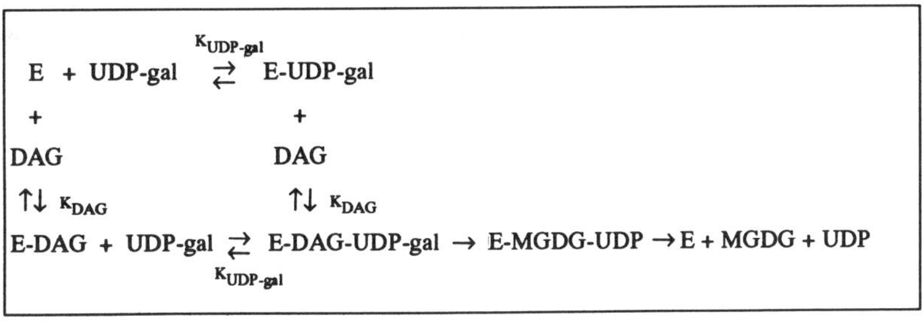

$$
\begin{array}{cc}
 & K_{\text{UDP-gal}} \\
E + \text{UDP-gal} & \rightleftharpoons \quad E\text{-UDP-gal} \\
+ & + \\
\text{DAG} & \text{DAG} \\
\updownarrow K_{\text{DAG}} & \updownarrow K_{\text{DAG}} \\
E\text{-DAG} + \text{UDP-gal} \rightleftharpoons & E\text{-DAG-UDP-gal} \rightarrow E\text{-MGDG-UDP} \rightarrow E + \text{MGDG} + \text{UDP} \\
& K_{\text{UDP-gal}}
\end{array}
$$

FIGURE 7. MGDG synthase is a random bireactant system. Abbreviations used: E — MGDG synthase; DAG — diacylglycerol.

as Co^{2+}, Fe^{2+}, Zn^{2+}, and monovalent cations is almost negligible.[43] Therefore, the biochemical properties of the galactolipid:galactolipid galactosyltransferase are very different from those of MGDG synthase. In addition, one should keep in mind that MGDG synthesis results in the formation of a β-glycosidic bond, whereas an α-glycosidic bond is formed during DGDG synthesis. Heemskerk et al.[42] have demonstrated that the envelope galactolipid:galactolipid galactosyltranferase has a rather low affinity for its substrate, and its activity strongly increases with MGDG concentration and can be saturated only by large quantities of MGDG (above 300 μM). High velocities can thus be obtained (2 to 3 μmol of MGDG converted per hour per milligram of envelope protein), provided that enough MGDG was present in the incubation mixture. In addition, the enzyme is still active at 4°C. The properties of this enzyme have major consequences for the lipid composition of isolated envelope membranes. During the course of envelope membrane isolation and purification (in the presence of 4 mM $MgCl_2$), the outer and inner envelope membranes fuse together, and therefore all the inner membrane MGDG become available for the galactolipid:galactolipid galactosyltransferase of the outer membrane.[23] Consequently, the envelope MGDG is converted into DGDG, tri-GDG, tetra-GDG, and diacylglycerol, and the glycerolipid composition of purified envelope membranes is strongly modified. It is only after destruction of the galactolipid:galactolipid galactosyltransferase (i.e., by thermolysin treatment of isolated intact chloroplasts prior envelope membranes purification) that a normal glycerolipid composition can be determined.[23]

van Besouw and Wintermans,[93] Heemskerk and Wintermans,[41] and Heemskerk et al.[44] have proposed that the galactolipid:galactolipid galactosyltransferase is indeed responsible for DGDG synthesis. For instance, Heemskerk et al.[44] analyzed galactolipid synthesis in chloroplasts or chromoplasts from eight species of 16:3- and 18:3-plants and found that DGDG formation is never stimulated by UDP-gal or any other nucleoside 5'-diphosphodigalactoside; in all cases, DGDG formation was reduced by thermolysin digestion of intact organelles. *In vitro*, the galactolipid:galactolipid galactosyltransferase does not show strong specificity for any MGDG molecular species. However, if this enzyme is indeed the DGDG-synthesizing enzyme, it should discriminate *in vivo* between the various MGDG molecular species that are available, since (1) the proportion of eukaryotic molecular species is higher in DGDG than in MGDG[45] and (2) DGDG contains 16:0 fatty acids (up to 10 to 15%) at both the *sn*-1 and *sn*-2 position and very little 16:3 (in 16:3-plants), whereas MGDG contains little 16:0, but (in 16:3-plants) 16:3 at the *sn*-2 position.[45]

Sakaki et al.[79] have proposed another physiological significance for the envelope galactolipid:galactolipid galactosyltransferase. Using ozone-fumigated spinach leaves, they demonstrated *in vivo* that MGDG was converted into diacylglycerol by the galactolipid:galactolipid galactosyltransferase and then to triacylglycerol (by acylation with 18:3-CoA), owing to a diacylglycerol acyltransferase associated with envelope membranes.[69] Whether the galactolipid:galactolipid galactosyltransferase is involved in DGDG synthesis is the focus of current investigations in several laboratories, and hopefully a definitive answer to this question will be available in the near future.

IV. SULFOLIPID BIOSYNTHESIS

Several comprehensive reviews have described the biosynthesis of SQDG *in vivo* or *in vitro*.[39,64,70] However, very little is known about the final step in SQDG biosynthesis, i.e., on the functioning of SQDG synthase. Most of the information available on this enzyme was obtained with studies on isolated intact chloroplasts[36,59,63] and, recently, with envelope membranes.[49,87] Indeed, the availability of chemically synthesized UDP-sulfoquinovose[50] paved the way for a biochemical characterization of the enzyme system involved in SQDG formation.

A. BIOCHEMICAL CHARACTERIZATION OF SQDG SYNTHASE

All the recent experiments provide evidence that the final step in SQDG synthesis is catalyzed by a 1,2-diacylglycerol 3-β-sulfoquinovosyltransferase (or SQDG synthase), which could transfer a sulfoquinovose from a water-soluble donor, UDP-sulfoquinovose, to a hydrophobic acceptor molecule, diacylglycerol, to synthesize SQDG with the release of UDP. Therefore, the functioning of the SQDG synthase closely mimics that of the MGDG synthase.

As discussed above, SQDG synthesis takes place in plastids, and as expected from our current knowledge of plastid glycerolipid biosynthesis, SQDG synthase is located in envelope membranes;[49] when supplied with UDP-sulfoquinovose, isolated envelope membranes prepared from spinach chloroplasts catalyze the formation of SQDG at rates about 40 nmol/h/mg of protein, a value about 10 times lower than MGDG synthase activity. This is almost in the range of activity of the envelope enzymes involved in phosphatidic acid and diacylglycerol biosynthesis.[58]

The final step in SQDG synthesis is optimal at pH 7.5,[87] a value lower than for MGDG synthase. After the addition of 5 μM of Mg^{2+} to the incubation mixture, the apparent K_m value for UDP-sulfoquinovose decreases from 80 to 10 μM, whereas the apparent V_{max} remains unaffected. An apparent K_m value of 0.7 mM for Mg^{2+} activation of SQDG synthase was determined,[87] a value within the range of Mg^{2+} variations found during dark/light transitions in the chloroplast stroma. Seifert and Heinz[87] therefore proposed that the light-induced release of Mg^{2+} into the stroma could result in an increased efficiency of SQDG synthesis. GDP-sulfoquinovose was used with a much lower affinity (apparent K$_m$ of 400 µM, with 5 mM of Mg^{2+}). Finally, no effect of dithiothreitol on SQDG synthase activity was observed, and therefore there is probably no regulation of SQDG synthase activity by a thiol-dependent oxidation/reduction cycle.

B. SQDG SYNTHASE AND DIACYLGLYCEROL

SQDG has a unique diacylglycerol structure (Figure 1) which could be due to specificities of SQDG synthase. In theory, the different diacylglycerol molecular species available for SQDG synthesis are the same as those which are used for galactolipid biosynthesis. Indeed, competition of SQDG synthase and MGDG synthase for the same pool of diacylglycerol molecules formed *de novo* in the inner envelope membrane was shown using intact chloroplasts by Joyard et al.[59] Under these conditions, the steric placement of these enzymes within the membrane, as well as their kinetic parameters, may be relevant to the outcome of the competition.

Further studies of this competition were done with envelope membranes. Analyses of higher plant membranes, such as the chloroplast envelope,[86] have demonstrated that SQDG contains dipalmitoyl species in addition to the major 18:3/16:0 (or 16:0/18:3). This composition is not found in any galactolipid. When isolated envelope membranes, loaded with 16:0/16:0 and/or 18:1/16:0 diacylglycerol, are incubated in presence of UDP-gal or UDP-sulfoquinovose, both diacylglycerols can be used by MGDG and SQDG synthases, but 16:0/16:0 is incorporated with a much higher efficiency into SQDG than into MGDG (Figure 8).[87] As discussed above, this last observation is supported by kinetic studies of MGDG synthase demonstrating that 16:0/16:0 is used at a very low rate by the purified enzyme.[97] Together, these results demonstrate that SQDG synthase, which can use both 16:0/16:0 and 18:1/16:0 diacylglycerol species with almost the same efficiency, is less selective than MGDG synthase. This observation could explain the occurrence of fully saturated species among SQDG molecules.

V. CONCLUDING REMARKS

In this chapter, we have tried to summarize our current knowledge on the final steps in assembly of galactolipid and SQDG molecules. However, the glycerolipids that are synthesized by MGDG and SQDG synthases contain 16:0 and 18:1, which must be desaturated to polyunsaturated fatty acids. There is evidence that desaturation of C$_{16}$ and C$_{18}$ fatty acids occurs when esterified to glycerolipids[3,58,78] (see Chapter 2) and that the enzymes involved are localized on envelope membranes.[84] Together, all the information concerning plastid glycerolipid formation and desaturation demonstrate the key role of the

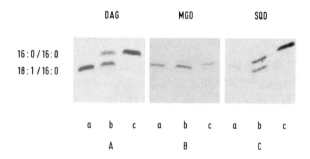

FIGURE 8. Diacylglycerol selectivity of MGDG and SQDG synthases. Thermolysin-treated envelope membranes were preloaded with different molecular species of diacylglycerol (see Seifert and Heinz[87] for experimental details). After 1 h incubation, each sample was analyzed for diacylglycerol species (A) which were mode a, 18:1/16:0; mode b, 16:0/16:0 and 18:1/16:0 in equal proportions; and mode c, 16:0/16:0. The remaining parts were divided into two equal portions for subsequent formation of MGDG and SQDG by addition of UDP-galactose and UDP-sulfoquinovose, respectively. After 1 min (MGDG) or 10 min (SQDG), reactions were stopped and extracted lipids separated by thin layer chromatography (TLC) for isolation of MGDG and SQDG. MGDG molecular species were analyzed by argentation TLC (B), whereas in the case of SQDG transformation into methyl esters was required before argentation TLC (C). Note that SQDG synthase utilizes 16:0/16/0 diacylglycerol molecular species more efficiently than MGDG synthase. (From Seifert, U. and Heinz, E., *Bot. Acta*, 105, 197, 1992. With permission.)

plastid envelope membranes. Indeed, the biochemical characterization of envelope membranes illustrates the dynamics of this unique membrane system as it exists in plant cells.[29,61]

We have also pointed out the major problems and obstacles that have impeded advances in plastid glycerolipid enzymology. Because the purification of all (but one or two) of the biosynthetic enzymes involved has not been achieved, only limited molecular information is available. Esko and Raetz wrote that "the phospholipid-synthesizing system offers considerable challenge to the modern enzymologist, since many of the principles of enzymology that developed out of the study of soluble enzymes do not hold for membrane enzymes that act on membrane-associated substrates."[30] Almost 10 years later, despite major achievements, the same observation holds true for many aspects of lipid synthesis, especially in plants.[11,15]

Consequently, the genes coding for the enzymes of glycerolipid biosynthesis in higher plants are almost unknown. The only plant glycerolipid synthesis gene that has been cloned and sequenced is that of the soluble plastidial acyltransferase involved in lysophosphatidic acid synthesis.[52] In contrast, most of the genes that code for the enzymes of glycerolipid biosynthesis have been characterized in *Escherichia coli*; point mutations that have been characterized in these genes were first isolated, then the genes were mapped, cloned, sequenced, and overexpressed.[75] Similar investigations are now in progress in cyanobacteria, photosynthetic bacteria, and in higher plants (for a review, see Ohlrogge et al.[74]). Browse et al.[14] have initiated a genetic approach to the analysis of lipid composition and synthesis by isolation of a series of mutants of the small crucifer *A. thaliana* with specific alterations in leaf fatty acid or glycerolipid composition. Parenthetically, results suggest that higher plant cells exhibit an unusual tolerance for alteration in their fatty acid-associated polar lipid composition.[76] The analysis of the effects of the mutations on membrane lipid composition complements other methods used for studying leaf lipid metabolism and its regulation.[74]

We have discussed some aspects of the regulation of glycerolipid biosynthesis. MGDG and SQDG synthases are the final enzymes in a pathway which lead to the formation of the main constituents of the lipid matrix of plastid membranes. This is an essential event in plastid biogenesis; glycerolipid requirements are not the same in greening or mature leaves, in chloroplasts, in amyloplasts, or in other specialized plastids. In addition, plants grow under rapidly changing environmental and physiological conditions, and very little is known concerning the mechanisms that regulate the biosynthesis and functioning of pivotal enzymes such as phosphatidate (at the branch point between PG, MGDG, and SQDG synthesis) and the biosynthesis of glycolipid precursors (acyl-ACP, glycerol-3-phosphate, UDP-gal, or UDP-sulfoquinovose). The events involved in plastid glycerolipid biosynthesis rely on gene regulation, but are also dependent upon the biosynthetic pathways of carbohydrate metabolism (photosynthesis, glycolysis, pentose pathway, respiration), which are the major factors which control the plant life.[26] Obviously, new opportunities to study these regulatory mechanisms are offered by genetic techniques applied to the analyses and modification of plant lipid metabolism. Unfortunately, mutants defining regulatory genes involved in polar lipid synthesis have not yet been identified in higher plants. Remaining problems are the characterization of metabolic signals involved in the coordinate regulation of plastid polar lipid synthesis. It is obvious that we are only beginning to understand the complex network of the regulatory mechanism that governs the synthesis and intracellular routing of polar lipids in plastids.

REFERENCES

1. **Alban, C., Joyard, J., and Douce, R.,** Preparation and characterization of envelope membranes from non-green plastids, *Plant Physiol.*, 88, 709–717, 1988.
2. **Alban, C., Joyard, J., and Douce, R.,** Comparison of glycerolipid biosynthesis in non-green plastids from sycamore (*Acer pseudoplatanus*) cells and cauliflower (*Brassica oleracea*) buds, *Biochem. J.*, 259, 775–783, 1989.
3. **Andrews, J., Ohlrogge, J. B., and Keegstra, K.,** Final steps of phosphatidic acid synthesis in pea chloroplasts occurs in the inner envelope membrane, *Plant Physiol.*, 78, 459–465, 1985.
4. **Benning, C. and Somerville, C. R.,** Identification of an operon involved in sulfolipid biosynthesis in *Rhodobacter sphaeroides*, *J. Bacteriol.*, 174, 2352–2360, 1992.
5. **Benson, A. A., Wiser, R., Ferrari, R. A., and Miller, J. A.,** Photosynthesis of galactolipids, *J. Am. Chem. Soc.*, 80, 4740–4750, 1958.

6. **Benson, A. A., Daniel, H., and Wiser, R.,** A sulfolipid in plants, *Proc. Natl. Acad. Sci. U.S.A.,* 45, 1582–1587, 1959.

7. **Benson, A. A.,** The plant sulfolipid, *Adv. Lipid Res.,* 1, 387–394, 1963.

8. **Benson, A. A.,** Lipids of chloroplasts, in *Structure and Function of Chloroplasts,* Gibbs, M., Ed., Springer-Verlag, Berlin, 1971, 129–148.

9. **Bertrams, M. and Heinz, E.,** Experiments on enzymatic acylation of sn-glycerol-3-phosphate acyltransferases from chloroplasts, *Plant Physiol.,* 68, 653–657, 1981.

10. **Bishop, D. G., Sparace, S. A., and Mudd, J. B.,** Biosynthesis of sulfoquinovo-syldiacylglycerol in higher plants: the origin of the diacylglycerol moiety, *Arch. Biochem. Biophys.,* 240, 851–858, 1985.

11. **Bishop, W. R. and Bell, R. M.,** Assembly of phospholipids into cellular membranes: biosynthesis, transmembrane movement and intracellular translocation, *Annu. Rev. Cell Biol.,* 4, 579–610, 1988.

12. **Bligny, R., Gardestrom, P., Roby, C., and Douce, R.,** ^{31}P NMR studies of spinach leaves and their chloroplasts, *J. Biol. Chem.,* 265, 1319–1326, 1990.

13. **Block, M. A., Dorne, A.-J., Joyard, J., and Douce, R.,** Preparation and characterization of membrane fractions enriched in outer and inner envelope membranes from spinach chloroplasts. II. Biochemical characterization. *J. Biol. Chem.,* 258, 13281–13286, 1983.

14. **Browse, J. A., McCourt, P., and Somerville, C. R.,** Fatty acid composition of leaves determined after combined digestion and fatty acid methyl ester formation from fresh tissue, *Anal. Biochem.,* 152, 141–146, 1986.

15. **Carman, G. M. and Henry, S. A.,** Phospholipid biosynthesis in yeast, *Annu. Rev. Biochem.,* 58, 635–669, 1989.

16. **Carter, H. E., McCluer, R. H., and Slifer, E. D.,** Lipids of wheat flour. I. Characterization of galactosylglycerol components, *J. Am. Chem., Soc.* 78, 3735–3738, 1956.

17. **Chang, S. B.,** Sulfhydryl nature of galactosyl transfer enzymes of spinach chloroplasts, *Phytochemistry,* 9, 1947–1948, 1970.

18. **Cline, K. and Keegstra, K.,** Galactosyltransferases involved in galactolipid biosynthesis are located in the outer membrane of pea chloroplast envelopes, *Plant Physiol.,* 71, 366–372, 1983.

19. **Covès, J., Block, M. A., Joyard, J., and Douce, R.,** Solubilization and partial purification of UDP-galactose:diacylglycerol galactosyltransferase activity from spinach chloroplast envelope, *FEBS Lett.,* 208, 401–406, 1986.

20. **Covès, J., Joyard, J., and Douce, R.,** Lipid requirement and kinetic studies of solubilized UDP-galactose:diacylglycerol galactosyltransferase activity from spinach chloroplast envelope membranes, *Proc. Natl. Acad. Sci. U.S.A.,* 85, 4966–4970, 1988.

21. **Davies, W. H., Mercer, E. I., and Goodwin, T. W.,** Some observations on the biosynthesis of the plant sulfolipid by *Euglena gracilis, Biochem. J.,* 98, 369–373, 1966.

22. **Deems, R. A., Eaton, B. R., and Dennis, E. A.,** Kinetic analysis of phospholipase A_2 activity toward mixed micelles and its implications for the study of lipolytic enzymes, *J. Biol. Chem.,* 250, 9013–9020, 1975.

23. **Dorne, A.-J., Block, M. A., Joyard, J., and Douce, R.,** The galactolipid:galactolipid galactosyltransferase is located on the outer membrane of the chloroplast envelope, *FEBS Lett.,* 145, 30–34, 1982.

24. **Douady, D. and Dubacq, J.-P.,** Purification of acyl-CoA:glycerol-3-phosphate acyltransferase from pea leaves, *Biochim. Biophys. Acta,* 921, 615–619, 1987.

25. **Douce, R.,** Site of biosynthesis of galactolipids in spinach chloroplasts, *Science,* 183, 852–853, 1974.

26. **Douce, R.,** *Mitochondria in Higher Plants,* Academic Press, New York, 1985.

27. **Douce, R. and Joyard, J.,** Structure and function of the plastid envelope, *Adv. Bot. Res.,* 7, 1–116, 1979.

28. **Douce, R. and Joyard, J.,** Plant galactolipids, in *The Biochemistry of Plants, Vol. 4, Lipids: Structure and Function,* Stumpf, P. K., Ed., Academic Press, New York, 1980, 321–362.

29. **Douce, R. and Joyard, J.,** Biochemistry and function of the plastid envelope, *Annu. Rev. Cell Biol.,* 6, 173–216, 1990.

30. **Esko, J. D. and Raetz, C. H. R.,** Synthesis of phospholipids in animal cells, in *The Enzymes,* Vol. 16, Boyer, P. D., Ed., Academic Press, New York, 1983, 207–253.

31. **Frentzen, M.,** Biosynthesis and desaturation of the different diacylglycerol moieties in higher plants, *J. Plant Physiol.,* 124, 193–209, 1986.

32. **Frentzen, M., Heinz, E., McKeon, T. A., and Stumpf, P. K.,** Specificities and selectivities of glycerol-3-phosphate acyltransferase and monoacylglycerol-3-phosphate acyltransferase from pea and spinach chloroplasts, *Eur. J. Biochem.,* 129, 629–636, 1983.

33. **Frentzen, M., Neuburger, M., Joyard, J., and Douce, R.,** Intraorganelle localization and substrate specificities of mitochondrial acyl-CoA:*sn*-glycerol-3-phosphate O-acyltransferase and acyl-CoA:1-acyl-*sn*-glycerol-3-phosphate O-acyltransferase from potato tubers and pea leaves, *Eur. J. Biochem.,* 187, 395–402, 1990.

34. **Gee, R. W., Byerrum, R. U., Gerber, D. W., and Tolbert, N. E.,** Dihydroxyacetone phosphate reductase in plants, *Plant Physiol.,* 86, 98–103, 1988.

35. **Gupta, S. D. and Sastry, P. S.,** The biosynthesis of sulfoquinovosyl diacylglycerol: studies with ground nut (*Arachis hypogaea*) leaves, *Arch. Biochem. Biophys.,* 260, 125–133, 1988.

36. **Haas, R., Siebertz, H. P., Wrage, K., and Heinz, E.,** Localization of sulfolipid labeling within cells and chloroplasts, *Planta,* 148, 238–244, 1980.

37. **Haines, T. H.,** Microbial sulfolipids, in *CRC Handbook of Microbiology,* 2nd Ed., Laskin, A. I. and Lechevallier, H. A., Eds., CRC Press, Boca Raton, FL, 1984, 115–123.

38. **Harwood, J. L.,** Plant acyl lipids: structure, distribution and analysis, *Prog. Lipid Res.,* 18, 55–86, 1979.

39. **Harwood, J. L.,** Sulfolipids, in *The Biochemistry of Plants, Vol. 4, Lipids: Structure and Function,* Stumpf, P. K., Ed., Academic Press, New York, 1980, 301–320.

40. **Heber, U.,** Metabolite exchange between chloroplasts and cytoplasm, *Annu. Rev. Plant Physiol.,* 25, 393, 1974.

41. **Heemskerk, J. W. M. and Wintermans, J. F. G. M.,** The role of the chloroplast in the leaf acyl-lipid synthesis, *Physiol. Plant.,* 70, 558–568, 1987.

42. **Heemskerk, J. W. M., Wintermans, J. F. G. M., Joyard, J., Block, M. A., Dorne, A. J., and Douce, R.,** Localization of galactolipid:galactolipid galactosyltransferase and acyltransferase in the outer envelope membrane of spinach chloroplasts, *Biochim. Biophys. Acta.,* 877, 281–289, 1986.

43. **Heemskerk, J. W. M., Jacobs, F. H. H., Scheijen, M. A. M., Heslper, J. P. F. G., and Wintermans, J. F. G. M.,** Characterization of galactosyltransferases in spinach chloroplast envelope, *Biochim. Biophys. Acta,* 918, 189–203, 1987.

44. **Heemskerk, J. H. W., Storz, T., Schmidt, R. R., and Heinz, E.,** Biosynthesis of digalactosyldiacylglycerol in plastids from 16:3 and 18:3 plants, *Plant Physiol.,* 93, 1286–1294, 1990.

45. **Heinz, E.,** Enzymatic reactions in galactolipid biosynthesis, in *Lipids and Lipid Polymers,* Tevini, M. and Lichtenthaler, H. K., Eds., Springer-Verlag, Berlin, 1977, 102–120.

46. **Heinz, E.,** Recent investigations on the biosynthesis of the plant sulfolipid, *Phyton,* in press.

47. **Heinz, E. and Roughan, P. G.,** Similarities and differences in lipid metabolism of chloroplasts isolated from 18:3 and 16:3 plants, *Plant Physiol.,* 72, 273–279, 1983.

48. **Heinz, E., Bertrams, M., Joyard, J., and Douce, R.,** Demonstration of an acyltransferase in chloroplast envelopes, *Z. Pflanzenphysiol.,* 87, 273–279, 1978.

49. **Heinz, E., Schmidt, H., Hoch, M., Jung, K.-H., Binder, H., and Schmidt, R. R.,** Synthesis of different nucleoside 5′-diphospho-sulfoquinovoses and their use for studies on sulfolipid biosynthesis in chloroplasts, *Eur. J. Biochem.,* 184, 445–453, 1989.

50. **Hoch, M., Heinz, E., and Schmidt, R. R.,** Synthesis of 6-deoxy-6-sulfo-α-D-glucopyranosyl phosphate, *Carbohydr. Res.,* 191, 21–28, 1989.

51. **Hoppe, W. and Schwenn, J. D.,** *In vitro* biosynthesis of the plant sulfolipid: on the origin of the sulfonate group, *Z. Naturforsch.,* 36c, 820–826, 1981.

52. **Ishizaki, O., Nishida, I., Agata, K., Eguchi, G., and Murata, N.,** Cloning and nucleotide sequence of cDNA from the plastid glycerol-3-phosphate acyltransferase from squash, *FEBS Lett.*, 238, 424–430, 1988.

53. **Jaworski, J. G.,** Biosynthesis of monoenoic and polyenoic fatty acids, in *The Biochemistry of Plants, Vol. 9, Lipids: Structure and Function*, Stumpf, P. K., Ed., Academic Press, New York, 1987, 159–174.

54. **Joyard, J.,** L'enveloppe des chloroplastes, Thèse de Doctorat d'Etat, Université de Grenoble, France, 1979.

55. **Joyard, J. and Douce, R.,** Mise en évidence et rôle des diacylglycérols de l'enveloppe des chloroplastes d'épinard, *Biochim. Biophys. Acta*, 424, 125–131, 1976.

56. **Joyard, J. and Douce, R.,** Site of synthesis of phosphatidic acid and diacylglycerol in spinach chloroplasts, *Biochim. Biophys. Acta*, 486, 273–285, 1977.

57. **Joyard, J. and Douce, R.,** Characterization of phosphatidate phosphohydrolase activity associated with chloroplast envelope membranes, *FEBS Lett.* 102, 147–150, 1979.

58. **Joyard, J. and Douce, R.,** Galactolipid synthesis, in *The Biochemistry of Plants, Vol. 9, Lipids: Structure and Function*, Stumpf, P. K., Ed., Academic Press, New York, 1987, 215–274.

59. **Joyard, J., Blée, E., and Douce, R.,** Sulfolipid synthesis from $^{35}SO_4^{2-}$ and [1-^{14}C]-acetate in isolated intact spinach chloroplasts, *Biochim. Biophys. Acta*, 879, 78–87, 1986.

60. **Joyard, J., Forest, E., Blée, E., and Douce, R.,** Characterization of elemental sulfur in isolated intact spinach chloroplasts, *Plant Physiol.*, 88, 961–964, 1988.

61. **Joyard, J., Block, M. A., and Douce, R.,** Molecular aspects of plastid envelope biochemistry, *Eur. J. Biochem.*, 199, 489–509, 1991.

62. **Kleppinger-Sparace, K. F.,** Biosynthesis of sulfolipids in higher plants, Ph.D. Dissertation, University of California, Berkeley, 1990.

63. **Kleppinger-Sparace, K. F., Mudd, J. B., and Bishop, D. G.,** Biosynthesis of sulfoquinovosyldiacylglycerol in higher plants: the incorporation of $^{35}SO_4$ by intact chloroplasts, *Arch. Biochem. Biophys.*, 240, 859–865, 1985.

64. **Kleppinger-Sparace, K. F., Mudd, J. B., and Sparace, S. A.,** Biosynthesis of plant sulfolipids, in *Sulfur Nutrition and Sulfur Assimilation in Higher Plants*, Rennenberg, H., Brunold, Ch., Dekok, L. J., and Stulen, I., Eds., SPB Academic Publishing, The Hague, 1990, 77–88.

65. **Königs, B. and Heinz, E.,** Investigation of some enzymatic activities contributing to the biosynthesis of galactolipid precursors in *Vicia faba, Planta*, 118, 159–169, 1974.

66. **Kunst, L., Browse, J., and Somerville, C.,** Altered regulation of lipid biosynthesis in a mutant of *Arabidopsis* deficient in chloroplast glycerol-3-phosphate acyltransferase activity, *Proc. Natl. Acad. Sci. U.S.A.*, 85, 4143–4147, 1988.

67. **Lunn, J. E., Droux, M., Martin, J., and Douce, R.,** Localization of ATP sulfurylase and O-acetylserine(thiol)lyase in spinach leaves, *Plant Physiol.*, 94, 1345–1352, 1990.

68. **Maréchal, E., Block, M. A., Joyard, J., and Douce, R.,** Purification de l'UDP-galactose:1,2-diacylglycérol galactosyltransférase de l'enveloppe des chloroplastes d'épinard, *C.R. Acad. Sci. Paris*, 313, 521–528, 1991.

69. **Martin, B. A. and Wilson, R. F.,** Subcellular localization of triacylglycerol synthesis in spinach leaves, *Lipids*, 19, 117–121, 1984.

70. **Mudd, J. B. and Kleppinger-Sparace, K. F.,** Sulfolipids, in *The Biochemistry of Plants, Vol. 9, Lipids: Structure and Function*, Stumpf, P. K., Ed., Academic Press, New York, 1987, 275–289.

71. **Mudd, J. B., McManus, T. T., Ongun, A., and McCullogh, T. E.,** Inhibition of glycolipid biosynthesis in chloroplasts by ozone and sulfhydryl reagents, *Plant Physiol.*, 48, 335–339, 1971.

72. **Murata, N., Ishizaki-Nishizawa, O., Higashi, S., Hayashi, H., Tasaka, Y., and Nishida, I.,** Genetically engineered alteration in the chilling sensitivity of plants, *Nature*, 356, 710–713, 1992.

73. **Nishida, I., Frentzen, M., Ishizaki, O., and Murata, N.,** Purification of isomeric forms of acyl-[acyl-carrier-protein]:glycerol-3-phosphate acyltransferase from greening squash cotyledons, *Plant Cell Physiol.,* 28, 1071–1079, 1987.

74. **Ohlrogge, J. B., Browse, J. A., and Somerville, C. R.,** The genetics of plant lipids, *Biochim. Biophys. Acta,* 1082, 1–26, 1991.

75. **Raetz, C. R. H. and Dowhan, W.,** Biosynthesis and function of phospholipids in *Escherichia coli., J. Biol. Chem.,* 265, 1235–1238, 1990.

76. **Rebeillé, F., Bligny, R., and Douce, R.,** Rôle de l'oxygène et de la température sur la composition en acides gras des cellules isolées d'érable (*Acer pseudoplatanus* L.), *Biochim. Biophys. Acta,* 620, 1–9, 1980.

77. **Rolland, N., Droux, M., and Douce, R.,** Subcellular distribution of O-acetylserine(thiol)-lyase in cauliflower (*Brassica oleracea* L.) inflorescence, *Plant Physiol.,* 98, 927–935, 1992.

78. **Roughan, P. G. and Slack, C. R.,** Cellular organization of glycerolipid metabolism, *Annu. Rev. Plant Physiol.,* 33, 97–132, 1982.

79. **Sakaki, T., Kondo, N., and Yamada, M.,** Pathway for the synthesis of triacylglycerols from monogalactosyldiacylglycerols in ozone-fumigated spinach leaves, *Plant Physiol.,* 94, 773–780, 1990.

80. **Saidha, T. and Schiff, J. A.,** The role of mitochondria in sulfolipid biosynthesis by *Euglena* chloroplasts, *Biochim. Biophys. Acta,* 1001, 68–273, 1989.

81. **Sastry, P. S.,** Glycosyl glycerides, *Adv. Lipid Res.,* 12, 251–310, 1974.

82. **Schiff, J. A.,** Reduction and other metabolic reactions of sulfate, in *Encyclopedia of Plant Physiology,* New Series, Vol. 15, Pirson, A. and Zimmerman, M. H., Eds., Springer-Verlag, New York, 1983, 401–421.

83. **Schmidt, A.,** Regulation of sulfur metabolism in plants, *Prog. Bot.,* 48, 133–150, 1986.

84. **Schmidt, H. and Heinz, E.,** Desaturation of oleoyl groups in envelope membranes from spinach chloroplasts, *Proc. Natl. Acad. Sci. U.S.A.,* 87, 9477–9480, 1990.

85. **Shibuya, I., Yagi, T., and Benson, A. A.,** Sulfonic acids in algae, in *Microalgae and Photosynthetic Bacteria,* Japan Society of Plant Physiology, Eds., University of Tokyo Press, Tokyo 1963, 627–636.

86. **Siebertz, H. P., Heinz, E., Linscheid, M., Joyard, J., and Douce, R.,** Characterization of lipids from chloroplast envelopes, *Eur. J. Biochem.* 101, 429–438, 1979.

87. **Seifert, U. and Heinz, E.,** Enzymatic characteristics of UDP-sulfoquinovose:diacylglycerol sulfoquinovosyltransferase from chloroplast envelopes, *Bot. Acta,* 105, 197–205, 1992.

88. **Smirnov, B. P.,** The biosynthesis of higher acids from acetate in isolated chloroplasts of *Spinacia oleracea* leaves, *Biochemistry,* 25, 419–426, 1960. Translated from *Biokhimiya* 25, 545–555, 1960.

89. **Stitt, M., Kürzel, B., and Heldt, H. W.,** Control of photosynthetic sucrose synthesis by fructose 2,6-bisphosphate. II. *Plant Physiol.,* 75, 544–560, 1984.

90. **Stumpf, P. K.,** The biosynthesis of saturated fatty acids, in *The Biochemistry of Plants, Vol. 9, Lipids: Structure and Function,* Stumpf, P. K., Ed., Academic Press, New York, 1987, 121–136.

91. **Teucher, T. and Heinz, E.,** Purification of UDP-galactose:diacylglycerol from chloroplast envelopes of spinach (*Spinacia oleracea* L.), *Planta,* 184, 319–326, 1991.

92. **van Besouw, A. and Wintermans, J. F. G. M.,** Galactolipid formation in chloroplast envelopes. I. Evidence for two mechanisms in galactosylation, *Biochim. Biophys. Acta,* 529, 44–53, 1978.

93. **van Besouw, A. and Wintermans, J. F. G. M.,** The synthesis of galactolipids by chloroplast envelopes, *FEBS Lett.,* 102, 33–47, 1979.

94. **Wintermans, J. F. G. M.,** Concentrations of phosphatides and glycolipids in leaves and chloroplasts, *Biochim. Biophys. Acta,* 44, 49–54, 1960.

95. **Zill, L. P. and Cheniae, G. M.,** Lipid metabolism, *Annu. Rev. Plant Physiol.,* 13, 225–264, 1962.

96. **Malherbe, A., Block, M. A., Joyard, J., and Douce, R.,** Feedback inhibition of phosphatidate phosphatase from spinach chloroplast envelope membranes by diacylglycerol, *J. Biol. Chem.,* 267, 23546–23553, 1992.

97. **Maréchal, E., Block, M. A., Dorne, A.-J., Joyard, J., and Douce, R.,** Biochemical characterization of MGDG synthase purified from spinach chloroplast envelope, in *Metabolism, Structure and Utilization of Plant Lipids,* Cherif, A., Miled-Daoud, D., Marzouk, B., Smaoui, A., Zarrouk, M., Eds., Centre National Pédagogique, Tunis, 1992, 10-A.

Chapter 8

PHOSPHOLIPID HEAD GROUPS

Anthony J. Kinney

TABLE OF CONTENTS

0-8493-4907-9/93/$0.00+$.50
© 1993 by CRC Press Inc.

I. INTRODUCTION

The major phospholipids found in plant tissues are similar to those found in all eukaryotes, namely phosphatidylcholine (PC), phosphatidylethanolamine (PE), phosphatidylserine (PS), phosphatidylinositol (PI), phosphatidylglycerol (PG), and diphosphatidylglycerol (DPG). The distribution of these lipids among different organelles of different tissues and different plants has been comprehensively studied.[1,2] The pathways by which these lipids are synthesized have also been studied extensively, but, in contrast to animals and yeasts, the relative contribution of different pathways to the synthesis of phospholipids *in vivo* is far from being understood. Unlike the situation with the bakers' yeast, *Saccharomyces cerevisiae*, for example, there have been no phospholipid biosynthetic mutants described in plants, and very few of the enzymes involved in the synthesis pathways have been purified or their corresponding genes cloned.

In general, phospholipid synthesis may be divided up into three pathway types: phospholipids derived *de novo* from cytidine diphosphate (CDP)-diacylglycerol, phospholipids derived *de novo* from diacylglycerol (Figure 1), and phospholipids derived from head group exchange reactions with other phospholipids. The CDP-diacylglycerol pathway is the only pathway of phospholipid biosynthesis in prokaryotes, namely *Escherichia coli*,[3] which make three major phospholipids, PE, PG, and DPG. The PE of *E. coli* is derived from the decarboxylation of a minor *E. coli* lipid, PS, which itself is synthesized from CDP-diacylglycerol and serine. In animals, the fatty acids of the major phospholipids, PE and PC, are derived from 1,2-diacylglycerol (DAG), PE by the reaction of a CDP aminoalcohol (CDP-ethanolamine) with DAG, and PC by an analogous reaction with CDP-choline or by the sequential methylation of PE.[4] On the other hand, PG, DPG, and PI in animals are derived from CDP-diacylglycerol. Animals are not able to synthesize PS from CDP-diacylglycerol; instead, there is a calcium-dependent base exchange reaction with PE. Yeast, like prokaryotes, can derive all of their phospholipid fatty acyl chains, including those of PS, PE, and PC, from CDP-diacylglycerol. However, under some conditions, they are able to repress the enzymes of this "prokaryotic" pathway and synthesize phospholipids using the fatty acids of DAG.[5] Like yeast, the biosynthetic pathways of plants may also be divided into CDP-diacylglycerol and DAG pathways. In addition, the availability of mutants, pure enzymes, monospecific antibodies, and genes in the yeast *S. cerevisiae* has led to a detailed understanding of the regulation of phospholipid biosynthesis in this organism, and this understanding provides a useful starting point from which to investigate biosynthetic pathways and their regulation in plants. As will be noted, however, phospholipid synthesis in plants has some unique features not found in prokaryotes and any other eukaryotes, and these features must be taken into account by any model of plant phospholipid biosynthesis. A definitive study of the enzymology of plant phospholipid biosynthesis was published in 1982[11] and of yeast phospholipid biosynthesis in 1989.[5] The

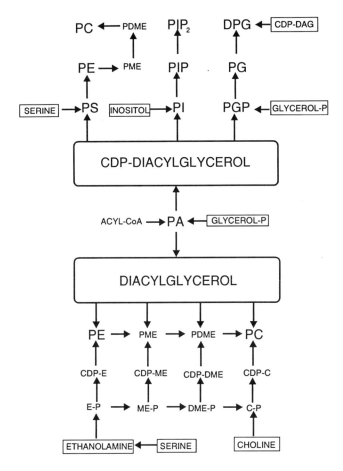

FIGURE 1. Potential pathways of head group incorporation into acylglycerols derived from DAG and CDP-DAG. Every reaction shown has been demonstrated in at least one plant tissue. Co-substrates and co-factors for these reactions are described in the text. *Abbreviations*: E — ethanolamine; M — methyl; DM — dimethyl; C — choline; P — phosphate; CDP — cytidine diphospho; PE — phosphatidylethanolamine; PME — phosphatidylmethylethanolamine; PDME — phosphatidyldimethylethanolamine; PC — phosphatidylcholine; PS — phosphatidylserine; PI — phosphatidylinositol; PIP — phosphatidylinositol phosphate; PIP_2 — phosphatidylinositol bisphosphate; PGP — phosphoglycerolphosphate; PG — phosphatidylglycerol; DPG — diphosphatidylglycerol.

reader is referred to these reviews as a source for original papers published up to these dates.

II. BIOSYNTHESIS OF DIACYLGLYCEROL AND CDP-DIACYLGLYCEROL

The precursor molecule common to the *de novo* synthesis of all phospholipids in prokaryotes and eukaryotes is phosphatidic acid (PA). PA is synthesized by the sequential acylation of glycerol-3-phosphate by two acyltransferases, glycerol-3-phosphate acyltransferase and 1-monoacylglycerol-3-phosphate acyltransferase, both of which utilize acyl-CoA as a source of acyl moieties. These enzymes are described in detail in Chapter 6. PA may be converted to CDP-diacylglycerol by the action of the enzyme CDP-diacylglycerol synthase (CTP:phosphatidate cytidylyltransferase, E.C. 2.7.7.41) or to DAG by the action of the enzyme phosphatidate phosphohydrolase (E.C. 3.1.3.4).

The enzyme CDP-diacylglycerol synthase catalyzes the reaction between PA and CTP to yield CDP-DAG and inorganic phosphate. CDP-diacylglycerol synthase has been purified to homogeneity from the mitochondria of *Saccharomyces*.[6] There is also a yeast mutant which has 25% of the wild-type CDP-diacylglycerol synthase activity.[7] In yeast, this enzyme catalyzes the first step in the primary biosynthetic pathway of all the major phospholipids, and studies with pure enzyme and the mutant[8,9] have indicated that this step is highly regulated. Precursors of the nucleotide pathway such as choline and ethanolamine, in the presence of inositol, repress the activity of the enzyme, and its activity is also repressed by inositol alone. In plants, CDP-diacylglycerol synthase activity has been detected in the inner chloroplast envelope of peas,[10] it being part of the pathway of PG biosynthesis in plastids. CDP-diacylglycerol synthase is also present in the mitochondria of a number of plant tissues[11] and in both the mitochondria and microsomes of castor bean endosperm.[12] The two forms of the castor bean enzyme have different kinetic properties and pH optima and are thus probably separate enzymes. Little else is known about this important enzyme.

The enzyme competing for substrate with CDP-diacylglycerol synthase is phosphatidate phosphohydrolase. This enzyme catalyzes the dephosphorylation of PA to DAG and inorganic phosphate. In yeast, both a microsomal (104 kDa) phosphatidate phosphohydrolase and a form of the enzyme present in both microsomes and mitochondria (45 kDa) have been purified.[13] The enzyme has been detected in chloroplast envelope membranes, microsomes, and soluble fractions of a number of plant tissues.[11,14-16] It seems probable, by analogy with yeast, that there is also a mitochondrial form of phosphatidate phosphohydrolase in plants. It has been postulated that the animal enzyme is regulated by translocation of an inactive form from the cytoplasm to the microsomal membranes where it becomes activated,[17] and there is some evidence that a similar mechanism may operate in plants.[18] This may explain some of the conflicting reports from plant tissues, with some investigators believing it to be a membrane-associated enzyme and some claiming it is soluble.[11,19] In developing oilseeds, the DAG produced by phosphatidate phosphohydrolase is the direct precursor of TAG, and in green plastids DAG is a source of acyl chains for galactolipids and sulfolipid.[19] Indeed the presence or absence of phosphatidate phosphohydrolase in plastids determines whether or not plastid DAG, containing 16-carbon fatty acids at position 2, will be available for galactolipid and sulfolipid synthesis. In the absence of a plastid phosphatidate phosphohydrolase, all of the DAG for synthesis of these lipids is derived from the cytoplasm and contains only 18-carbon fatty acids, hence the presence or absence of a plastid phosphatidate phosphohydrolase marks the differentiation between the so-called "16:3"- and "18:3"- plants.[20] In plant microsomes, phosphatidate phosphohydrolase also provides DAG for the phospholipids, PE and PC, synthesized by the nucleotide pathways discussed below.

In yeast, the 45-kDa phosphatidate phosphohydrolase and the CDP-diacylglycerol synthase are highly regulated in response to inositol in the growth medium.[5,13] The 45-kDa phosphatidate phosphohydrolase is induced in the presence of inositol,[13] whereas CDP-diacylglycerol synthase activity is repressed.[5] A similar effect is observed when yeast enter the stationary phase where CDP-diacylglycerol synthase activity is repressed,[5] and both the 104-kDA and the 45-kDa phosphatidate phosphohydrolases are induced.[13] Thus, the flux of acyl chains through CDP-DAG or DAG in yeast appears to be controlled by the relative activities of CDP-diacylglycerol synthase and phosphatidate phosphohydrolase. It is not clear what the relative contributions of the 45-kDa and 104-kDa forms of phosphatidate phosphohydrolase are to overall phospholipid synthesis rates. In light of the above studies in yeast, further investigation of the plant phosphatidate phosphohydrolases and CDP-diacylglycerol synthase would perhaps shed light on the situation in plant cells.

III. SYNTHESIS OF PHOSPHOLIPIDS FROM CDP-DIACYLGLYCEROL

The attachment of a polar head group to a nonpolar, acylglycerol molecule to yield an amphipathic phospholipid shares a common mechanism in nature. This mechanism is an enzyme-catalyzed, nucleophilic displacement of cytidine monophosphate (CMP) from a CDP ligand by a hydroxyl group of the attacking molecule. The CDP ligand can be an acylglycerol molecule, namely CDP-diacylglycerol, in which case the CMP is displaced by the hydroxyl group of, for example, serine to yield phosphatidylserine. In the case of PE and PC, the hydroxyl group of DAG can displace CMP from a CDP aminoalcohol, either CDP-ethanolamine or CDP-choline, a reaction catalyzed by an aminoalcohol phosphotransferase. In all organisms, PG and DPG are synthesized via CDP-diacylglycerol, and in eukaryotes PI is also synthesized via CDP-diacylglycerol. In addition, this mechanism is used for the synthesis of PS by prokaryotes, yeast, and, most probably, plants.

A. PHOSPHATIDYLGLYCEROL AND DIPHOSPHATIDYLGLYCEROL

PG is the only phospholipid associated with chloroplast thylakoids,[19] although varying amounts of this lipid, from a trace to as much as 15% of total phospholipid, have been found in other plant cell membrane fractions.[2] DPG (sometimes referred to as cardiolipin), on the other hand, is restricted to the inner mitochondrial membrane.[2,19] The enzymes involved in the synthesis of these phospholipids in all organisms, but especially plants and yeast, are the least well-characterized of all phospholipid biosynthetic enzymes. In *E. coli* and yeast, PG is synthesized by a two-step reaction; CDP-diacylglycerol reacts with glycerol phosphate to yield phosphatidylglycerolphosphate and CMP, a reaction catalyzed by the enzyme glycerophosphate:CDP-diacylglycerol phosphatidyltransferase (PGP synthase, E.C. 2.7.8.5). Phosphatidylglycerol-phosphate is then dephosphorylated to PG by the enzyme phosphatidylglycerol-phosphate phosphohydrolase (PGP phosphatase, E.C. 3.1.3.27). In yeast, these reactions are localized exclusively in the inner membranes of mitochondria,[21] and there is evidence that these activities are also present in plant cell mito-chondria.[22] By following the incorporation of ^{14}C-labeled acetate and glycerol-3-phosphate into PG via CDP-diacylglycerol, PG biosynthetic activity has also been measured in the inner envelope membrane of pea chloroplasts,[10] as well as in intact spinach chloroplasts.[23] In contrast to yeast, however, these reactions appear not to be restricted to prokaryotic-type membranes, since there are a number of reports of PG synthesis from labeled CDP-diacylglycerol in plant microsomal membranes.[11] In all the above studies in plants, manganese was found to be a necessary co-factor for PG biosynthetic activity. Although in some of these studies labeled PGP has been detected as well as PG,[10] individual PGP synthase and PGP phosphatase activities have not been characterized in plant tissues. DPG is synthesized in plant mitochondria as it is in yeast by reaction of PG with CDP-diacylglycerol and not by condensation of two PG molecules as in *E.coli*.[3] The reaction is catalyzed by the enzyme PG: CDP diacylglycerol phosphatidltranserase about which very little is known except that the enzyme requires either magnesium or, surprisingly, cobalt, for activity.[11]

B. PHOSPHOINOSITIDES

The enzyme PI synthase (CDP-diacylglycerol:myo-inositol phosphatidyl-transferase, E.C. 2.7.8.11) catalyzes the reaction between free inositol and CDP-diacylglycerol to yield PI and CMP and has an absolute requirement for a divalent cation, either manganese or magnesium. PI synthase was first detected in plant tissue over 20 years ago, and its activity since has been measured in many different plants.[11] There are no reports to date of its pu-rification from a plant source, but PI synthase has been purified from Triton X-100 solubilized yeast microsomes by CDP-diacylglycerol-sepharose affinity chromatography.[5] The purified protein has a subunit M_r of 34 KDa, and like the plant enzyme the yeast PI synthase requires manganese or magnesium for activity. The use of CDP-diacylglycerol affinity has proven very successful in the purification of a number of other yeast phospholipid biosynthetic enzymes[5] and may be of value in the purification of plant PI synthase. In yeast and animals, PI, in the presence of ATP, is converted to PI 4-phosphate (PIP) and AMP, a reaction catalyzed by the enzyme PI 4-kinase (ATP:phosphatidylinositol

4-phosphotransferase, E.C. 2.7.1.67). PIP is in turn converted to PI 4,5-bisphosphate (PIP$_2$) by a second ATP-dependent kinase, PIP kinase (ATP:phosphatidylinositol 4-phosphate 5-phosphotransferase, E.C. 2.7.1.68). A PI 4-kinase, with a M_r of 45 kDa, has been purified from yeast microsomes,[24,25] and monospecific polyclonal antibodies which precipitated PI 4-kinase activity were raised.

It has been well-established that the synthesis and turnover of PIP and PIP$_2$ plays an essential role in the response of animal cells to various agonists.[26] In yeast polyphosphoinositides or their breakdown products, DAG and inositol trisphosphate appear to be involved in the regulation of cell proliferation.[27] In recent years, polyphosphoinositides, and their corresponding lysophospholipids, have been detected in many plant tissues.[28-30] They are present at very low concentrations in plant cells, and PIP and PIP$_2$ represent less than 1% each of the total inositol lipid, whereas PI accounts for about 93%.[30,31] PI itself accounts for less than 10% of total lipid phosphorus in plant cells;[2] consequently, polyphosphoinositides were not discovered in plant tissues until they were specifically looked for in *myo*-[2-H^3]-inositol-labeled cells.[31] Their identification in carrot cultures has been rigorously confirmed by the use of fast-atom bombardment mass spectrometry.[32] Both PI 4-kinase and PIP kinase activities have been measured in plasma membranes isolated from wheat roots,[33] carrot suspension cells,[34] and sunflower hypocotyls,[35] and a soluble form of the carrot PI 4-kinase has been partially purified.[36] Most recently, evidence has been presented for the occurrence of PI 4-kinase in association with the cytoskeleton of carrot suspension cultured cells.[84] The yeast enzyme has a requirement for magnesium (Mg) ions (about 25 mM) which cannot be substituted with manganese (Mn) ions.[24] In contrast, the carrot enzyme required either 20 mM magnesium or 1 mM manganese for activity.[36] This is interesting, since the true substrate for most ATP kinases, including yeast PI kinase, is Mg-ATP.[24,25] Perhaps the solubilization and purification of the membrane PI kinase from a plant source will provide an answer to this paradox. However, unlike the purification of yeast PI synthase, the purification procedure used for the yeast PI 4-kinase involves a considerable number of steps[24,25] and may not be appropriate for the plant enzyme which appears to be more unstable than the yeast kinase. However, it is possible that the antibodies raised against the yeast 45-kDa PI 4-kinase, if they cross-react with the carrot protein, may be of use as an affinity ligand in further purification procedures.

The presence of polyphosphoinositides, and the enzymes responsible for their synthesis, leads to the suggestion that they are involved in signal transduction processes in plants as they are in other eukaryotes. By analogy, it has been postulated that the agonist-induced breakdown of PIP and PIP$_2$ via a membrane receptor acts as a signal transduction mechanism in plant cells.[30] There is some good evidence that this analogy to the animal model is also valid in plant systems.[28,29] However, since polyphosphoinositides and their inositol-containing hydrolysis products are present in such vanishingly small amounts in plant tissues, most attempts to correlate their synthesis or turnover with a cellular response have not been entirely convincing and have led to conflicting claims. For example, a study with *Samanea saman* leaves demonstrated that a 10 to 30 s pulse of white light led to the hydrolysis of ^{3}H-inositol-labeled PIP$_2$ to DAG and inositol trisphosphate.[37] Unfortunately, the increase of ^{3}H label in inositol trisphosphate amounted to only about 10 to 30% of a few hundred counts. In an additional study,[38] the same authors found that, after an initial increase, the amount of DAG, the other hydrolysis product of polyphosphoinositides, decreased, while inositol bis- and trisphosphate levels were still rising. It could be argued that DAG is involved in many other lipid metabolic pathways other than PI metabolism, and this might account for the inconsistency. However, as Memon and Boss[35] have pointed out, a contrasting study with wheat seedlings[39] demonstrated that the enzyme responsible for polyphosphoinositide hydrolysis, phospholipase C, was actually more active in the dark than in the light. In a third study, using sunflower hypocotyls, the *in vitro* enzyme activities of the PI and PIP kinases along with phospholipase C were measured in response to light. The study demonstrated that a 10-s pulse of light resulted in a 50% decrease in PIP kinase activity, resulting in less PIP$_2$, without any change in PIP$_2$ hydrolysis by phospholipase C. Interestingly, a corresponding decrease in the vanadate-sensitive ATPase activity was also observed. Addition of PIP or PIP$_2$ to plasma membranes resulted in a doubling of ATPase activity.[35] The same group has also demonstrated that treatment of carrot suspension cells with *Driselase* species, a cell wall degrading enzyme mix, leads to an increase in plasma membrane PI and PIP kinase activity and a concomitant increase in plasma membrane ATPase activity.[34] The authors of these last two papers speculate that PIP and PIP$_2$, rather than their hydrolysis products, may be involved in the regulation of cellular metabolism in plant cells, possibly via an effect on plasma membrane ATPase activity.[35] Phosphoinositide metabolism is clearly an exciting new field of exploration in plant phospholipid biosynthesis, but, in light of the above reports, clearly one which must be explored with caution.

C. PHOSPHATIDYLSERINE AND ITS METABOLISM

In yeast, PS is synthesized from CDP-diacylglycerol and serine, a reaction catalyzed by the enzyme CDP-diacylglycerol:*L*-serine *O*-phosphatidyltransferase (PS synthase, E.C. 2.7.8.8). The 30-kDa yeast protein has been purified[20] and its structural gene cloned.[5] The regulation of PS synthase activity has been studied extensively in yeast,[5,40] and it has been shown to play a major role in overall phospholipid biosynthesis and cell growth.[5] There is increasing evidence that, unlike animals, PS synthase is also present in plant tissues. Plant PS synthase activity was originally reported in spinach leaf microsomes[11] and has since been detected in castor bean leaves, stems, roots, and cotyledons.[41] No PE-PS exchange reaction was detected in any of these tissues, which lends support to the idea that all the PS in most plant tissue is made via PS synthase. In contrast, in germinating castor bean endosperms, PS appears to be made exclusively via the exchange reaction and not by PS synthase.[11,41] In yeast, PS is the major precursor to PE and PC.[5] PS is decarboxylated to PE by the action of PS decarboxylase (E.C. 4.1.1.65), and PE is sequentially methylated to PC by the action of two N-methyltransferases, which utilize S-adenosylmethionine as the methyl donor. The first of these enzymes PE N-methyltransferase (E.C. 2.1.1.17) catalyzes the methylation on PE to phosphatidylmethylethanolamine (PME), and the other enzyme phospholipid N-methyltransferase catalyzes the two methylations necessary to convert PME to PC.[42] Thus, yeast are able to synthesize all of their major phospholipids using acyl chains derived from CDP-diacylglycerol. Since PS decarboxylase activity has also been measured in plants,[11] the presence of PS synthase would imply that some plant tissues are also capable of synthesizing all of their phospholipids, with the possible exception of PC, without the participation of DAG. There is mixed evidence for the methylation of PE to PC in plant tissue, and this will be discussed later in this chapter, but the pathway appears to operate in some plant tissues,[11,43] and thus PC too may be synthesized without the participation of DAG. In germinating castor bean endosperms, where there may be a large pool of DAG due to mobilization of storage TAG, PS appears to be synthesized from DAG derived from PE via the exchange reaction and not from CDP-diacylglycerol.[41] In tissues where PS synthase activity has been measured, its contribution to the regulation of PE biosynthesis is not known, but if PS decarboxylation were rapid then PS synthase would be rate-limiting for PE biosynthesis. Thus, confirmation of the presence of PS synthase in plants, either by protein purification or by the cloning of its gene, will be awaited with interest. It may be possible to purify the enzyme using the yeast purification protocol, which utilizes CDP-diacylglycerol affinity chromatography, or to clone a PS synthase cDNA by heterologous hybridization with the yeast gene. The plant cDNA might also be cloned by complementation of the yeast PS synthase gene-disruption mutant with a yeast expression vector system containing a plant cDNA library. Successful complementation would allow the mutant to grow in the absence of choline supplementation and could therefore be easily selected for. With the protein and/or cDNA in hand, it would be possible to study the distribution and regulation of expression of PS synthase protein among different tissues and address the role of PS synthase activity in the regulation of phospholipid biosynthesis in plants.

IV. SYNTHESIS OF PHOSPHOLIPIDS FROM DIACYLGLYCEROL

A. NUCLEOTIDE PATHWAYS

In eukaryotic cells, PE and PC can be synthesized by the displacement of CMP from CDP-ethanolamine or CDP-choline by the hydroxyl group of DAG. The CDP aminoalcohol can be synthesized from free ethanolamine or choline by a two-step reaction, the first step being phosphorylation of the free aminoalcohol. The CDP aminoalcohol is synthesized by reaction of the

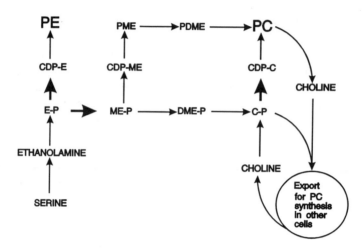

FIGURE 2. Major fluxes through the phosphoaminoalcohol pathway in plants. Flux of methyl groups through phosphorylated ethanolamine compounds vs. phosphatidylethanolamine compounds is plant specific (see text). Potential regulatory steps are shown by the large arrows. See Figure 1 for abbreviations.

phosphoaminoalcohol with CTP with the liberation of inorganic phosphate (Figure 2). The phosphorylation of ethanolamine is catalyzed by the enzyme ethanolamine kinase (ATP:ethanolamine phosphotranferase, E.C. 2.7.1.82) and the phosphorylation of choline by an equivalent enzyme, choline kinase (ATP:cholinephosphotransferase, E.C. 2.7.1.32), both of which are soluble enzymes in animal cells.[44] In yeast, the gene for choline kinase (*CK1*) has been cloned by complementation of a choline kinase mutant and was shown to encode a polypeptide of about 66 kDa,[45] which is similar to the native molecular weight of a partially purified yeast choline kinase.[46] When the *CK1* gene was expressed in *E.coli*, both choline kinase and ethanolamine kinase activities were observed. Disruption of the *CK1* gene in wild-type yeast resulted in the loss of both choline and most of the ethanolamine kinase activities.[45] These results have led some investigators to suggest that both activities are catalyzed by the same enzyme in yeast. In plant cells, ethanolamine and choline kinase activities are found in soluble fractions.[11,47,48] Choline kinase has been purified from soybeans and partially purified from spinach and castor bean endosperm.[11,48] Although the reported native M_r of soybean choline kinase, about 60 kDa, is close to that of the yeast polypeptide encoded by *CK1*,[49] the subunit M_r of the soybean enzyme reported in an earlier study was only 36 kDa.[11] Since the soybean enzyme is very unstable, the 36-kDa polypeptide may have been a proteolytic degradation product of a larger protein, and the actual subunit molecular weight of plant choline kinases may be closer to that of yeast *CK1*, but confirmation will await cloning of the plant cDNA. The kinetic properties of the castor bean endosperm choline kinase have been studied in detail.[48] Unlike the yeast choline kinase, which follows a random Bi-Bi reaction mechanism,[46] the castor bean enzyme catalyzes a sequentially ordered reaction of the enzyme with ATP-Mg^{2+} followed by free choline, with the resulting enzyme-substrate complex requiring free Mg^{2+} for activation and release of choline phosphate.

In contrast to the detailed studies of plant choline kinase reported recently, there have been no detailed studies of plant ethanolamine kinase for over ten years. The native M$_r$ of a partially purified spinach ethanolamine kinase was observed to be 110 kDa, which is about twice the size of the yeast *CK1* subunit.[11] However, the yeast *CK1* gene product appears to be monomeric, and there is no evidence to date that the plant ethanolamine kinase (or the choline kinase for that matter) is a dimer in its native state. Neither the spinach ethanolamine kinase nor a purified ethanolamine kinase from soybean showed any activity toward choline. Likewise, the spinach and soybean choline kinases were not inhibited by ethanolamine.[11] Thus, all the evidence to date indicates that the plant choline and ethanolamine kinases are distinct enzymes and that neither enzyme bears any resemblance to the yeast *CK1* gene product.

The conversion of ethanolamine and choline phosphate to their respective CDP aminoalcohols is catalyzed by a phosphoaminoalcohol cytidylyltransferase. Choline phosphate cytidylyltransferase (CTP:choline phosphate cytidylyltransferase, E.C. 2.7.7.15) has been purified from rat liver,[50,51] and its cDNA has been cloned.[52] The choline phosphate cytidylyltransferase gene (*CCT1*) has also been cloned from yeast by complementation of a choline phosphate cytidylyltransferase mutant,[53] and the gene has been expressed in *E. coli.*[64] *E. coli* lack any endogenous aminoalcohol cytidylyltransferase activity and therefore provide a clean background in which to characterize the yeast protein. The yeast *CCT1* gene product has negligible activity toward ethanolamine phosphate and monomethylethanolamine phosphate when expressed in *E.coli.*[64] The M$_r$ of the rat liver protein is 42 kDA and is thought to be a dimer in its native state,[50,51] whereas the predicted molecular mass of the yeast cDNA product is 49 kDa.[52] The two proteins have divergent N-terminal and C-terminal domains, but have a highly conserved (75% identity) central region of about 160 amino acids which may contain the active site. Choline phosphate cytidylyltransferase is the only plant phospholipid biosynthetic enzyme that has been both purified to homogeneity and its corresponding cDNA cloned.[54-56] The purified pea stem choline phosphate cytidylyltransferase had a reported M$_r$ of 56 kDa[56] and was observed to be monomeric in its native state. In contrast, the corresponding castor bean endosperm choline phosphate cytidylyltransferase had a subunit M$_r$ of 40 kDa[54] and a native mass of 155 kDa, which led the authors to conclude that it exists naturally as a tetramer. Although the cloning of the castor bean choline phosphate cytidylyltransferase cDNA was recently reported, the full cDNA sequence has not yet been published.[55]

In animal cells[4] and in rye roots, pea stems, castor bean endosperm, and celery cell cultures,[47,57-59] there is evidence that choline phosphate cytidylyltransferase plays a key role in regulating the flux through the CDP-choline pathway for PC biosynthesis. The evidence in plants is based on pulse-chase experiments with labeled choline.[58] Additional evidence comes from studies which correlate the change in activity of cytidylyltransferase and changes in the choline phosphate/CDP-choline pool sizes with changes in the rate of choline incorporation into PC.[47,57-59] Like phosphatidate phosphatase, both the animal and the plant choline phosphate cytidylyltransferase appear to be ambiquitous, with their activities having been measured in both soluble and membrane fractions from cell extracts.[4,11,47,60,61] It has been proposed that the activity of the plant enzyme is regulated in a manner similar to that of its animal equivalent, by the lipid-promoted translocation of an inactive cytosolic form to the ER membranes where it becomes activated.[61] There is some evidence for this proposed mechanism in castor bean endosperm cells,[58,61] and stimulation of purified castor bean choline phosphate cytidylyltransferase by both castor bean phospholipids and phosphatidylcholine-oleate vesicles has been demonstrated.[54] The association of the rat liver enzyme with membranes is controlled by the reversible phosphorylation of the cytidylyltransferase protein,[4] and there is additional evidence that castor bean cytidylyltransferse activity is also regulated by phosphorylation.[62] However, it must be noted that the soluble pea enzyme activity is not regulated by phosphorylation.[63] In addition, the pea enzyme is not stimulated by phospholipids, the association of enzyme with membranes is decreased rather than increased by oleate, and membrane-associated pea cytidylyltransferase is *less* active than the soluble form.[63] Analysis of the peptide sequence of the rat liver protein has revealed a 56-residue, amipathic, α-helical region at the C-terminal end of the protein which contains three repeated motifs of 11 amino acids each. It is proposed that the association of the rat cytidylyltransferase with the membrane is mediated by the interaction of the hydrophobic domains of this amipathic helix with the phospholipids of the bilayer in a manner similar to the interaction of apolipoproteins with lipids.[52] Apolipoproteins also contain 11-mer repeat motifs within an amphipathic domain and interact with lipids by intercalation of the hydrophobic side chains. In yeast, the choline phosphate cytidylyltransferase is not known to regulate PC biosynthesis.[5] Furthermore, since more than 90% of the yeast cytidylyltransferase activity is membrane associated, even when the yeast *CCT1* gene is overexpressed in yeast cells,[64] it is probably not ambiquitous and therefore not regulated by a translocation mechanism. In concert with this view, it is of interest to note that the peptide sequence of the yeast cytidylyl-transferase does not contain any amphipathic domains or the repeated motifs observed in the rat peptide. The question of the mechanism of regulation of the plant enzyme therefore may be partially resolved with the publication of the castor cDNA sequence, since, if the protein is regulated by membrane association, the deduced peptide should contain an amphipathic domain, and possibly repeat motifs, similar to those of the rat liver enzyme.

The enzyme ethanolamine phosphate cytidylyltransferase (CTP:ethanolamine phosphate cytidylyltransferase, E.C. 2.7.7.14) has been partially purified from rat liver cells, is a soluble protein which does not associate with membranes, has no lipid requirement for activity,[4] and is thus clearly distinct from the rat liver choline phosphate cytidylyltransferase. A number of attempts to demonstrate the presence of an ethanolamine phosphate cytidylyltransferase in plant cell extracts have been unsuccessful, which has led to the speculation that synthesis of CDP-ethanolamine in plants, if it occurs, may be catalyzed by choline phosphate cytidylyltransferase and not by a separate enzyme.[65,66] The presence of an ethanolamine phosphate cytidylyltransferase in carrot cells was briefly mentioned in a 1956 report (Kennedy and Weiss, cited in Reference 66), but the enzyme was not investigated in any detail. It was also observed that the purified choline phosphate cytidylyltransferase from castor beans has no activity toward ethanolamine phosphate.[54] The uncertainty over the existence of a complete CDP-ethanolamine pathway for PE biosynthesis in plants has been finally resolved by a report which clearly demonstrates the presence of a distinct ethanolamine phosphate cytidylyltransferase in the membrane fractions of mitochondria and ER from castor bean endosperm.[66] Most of the castor bean ethanolamine phosphate cytidylyltransferase activity (80%) appears to be in the mitochondria, whereas the choline phosphate cytidylyltransferase is found mainly in the ER and cytosol. Therefore, although this study resolves one issue, the existence of ethanolamine phosphate cytidylyltransferase in plants, it raises another with regard to cellular compartmentalization. There is no ready explanation for the mitochondrial localization of ethanolamine phosphate cytidylyltransferase in plants, since the enzyme which utilizes its product, ethanolamine phosphotransferase, is associated mainly with the ER and the enzyme which supplies its substrate, ethanolamine kinase, is cytosolic.[11]

The reaction between DAG and a CDP aminoalcohol to yield a phospholipid and CMP is catalyzed by an aminoalcohol phosphotransferase. Both cholinephosphotransferase (CDP-choline:1,2,diacylglycerol cholinephospho-transferase, E.C. 2.7.8.2) and ethanolaminephosphotransferase (CDP-ethano-lamine: 1,2,diacylglycerol ethanolaminephosphotransferase, E.C. 2.7.8.1) activities have been measured in microsomal fractions of plant cells,[11,47,67-69] but like their animal counterparts[4] these enzymes have so far eluded solubi-lization and purification. One possible exception is the preliminary report of the solubilization of cholinephosphotransferase from the microsomes of castor bean endosperm using octyl β-D-glucopyranoside.[70] Cholinephosphotransfer-ase activity has also been detected in castor bean mitochondria.[11] Since amino-alcohol phosphotransferases catalyze a potentially reversible, bisubstrate re-action in which one of the substrates and one of the products is a lipid, any kinetic studies need to be performed in a system where water-insoluble, lipid substrates and products can be efficiently supplied to the enzyme and their concentrations independently varied along with other co-factors and substrates. Perhaps the best overall approach for this type of enzyme is the mixed micelle technique of Lichtenberg et al.,[71] but, since microsomes contain significant levels of DAG, PE, and PC, *any* attempt to study the kinetics of aminoalcohol phosphotransferases requires the use of solubilized and prefer-ably purified enzyme. In the absence of pure or lipid-free enzyme, reports of the kinetic constants of these enzymes are essentially meaningless, except perhaps the observation that all plant aminoalcohol phosphotransferases studied to date have been shown to require a divalent cation, either Mg^{2+} or Mn^{2+}, for activity.[11]

It is possible to study the selectivity of aminoalcohol phosphotransferases for different DAG species by incubating plant microsomes with radiolabeled CDP-ethanolamine and CDP-choline and quantitating the distribution of fatty acids in newly synthesized PE and PC. Use of this technique led initially to the publication of a number of conflicting reports from which it was not possible to draw any generalizations on the selectivity, or lack of selectivity, of plant aminoalcohol phosphotransferases for DAG species.[68,72,73] However, it has now become possible to accurately separate different molecular species of DAG, PE, and PC using reversed-phase HPLC,[74] and recent studies of aminoalcohol phosphotransferases in germinating soybeans, potato tubers, and pea leaves have reported a marked preference by both ethanolamine- and cholinephosphotransferases for 1-palmitoyl-2-linoleoyldiacylglycerol as the lipid substrate.[74] Oleic acid was never observed at position 2 of ^{14}C-labeled PC or PE in these studies. Since PC is the substrate for oleate desaturation in higher plants,[20] the results of these latter studies suggest two possibilities. 1-Palmitoyl-2-oleoyldiacylglycerol could be incorporated into PC and the oleoyl moiety at position *sn*-2 completely desaturated before the first time point of the experiment, 1 min after incubation with ^{14}C-CDP-choline. This is very unlikely, since not only would this desaturation require the addition of exogenous NADH, but desaturation of oleoyl-PC in microsomes has not been observed to go to completion even after 240 min of incubation with radiolabeled substrate.[75] The other explanation is that in these species 18:1-CoA is incorporated into PC for desaturation independently of CDP-choline and the cholinephosphotransferase. The activity of an enzyme capable of catalyzing the exchange of 18:1-CoA with the fatty acid at the *sn*-2 position of PC, acyl-CoA:lysophosphatidylcholine acyltransferase, (E.C. 2.3.1.23), is well-documented in a number of plant tissues.[75] In developing oilseeds, however, there is good evidence that polyunsaturated fatty acids in triacylglycerol can be derived from oleoyl-DAG, which is first incorporated into PC via the cho-linephosphotransferase reaction, desaturated while esterified to PC, and then returned to the DAG pool by the reverse reaction of the cholinephospho-transferase.[75] In these tissues, the relative contributions of the acyl-CoA:lysophosphatidylcholine acyltransferase and cholinephosphotrans-ferase to the desaturated fatty acid composition of TAG is not known. Presumably though, unlike in germinating soybeans, cholinephosphotransferases from developing soybean and other developing oilseeds must preferentially utilize 1-palmitoyl-2-oleoyldiacylglycerol as a substrate in the forward direction and 1-palmitoyl-2-linoleoyl-PC or 1-palmitoyl-2-linolenyl-PC in the reverse direction.

It is clear, however, that in oilseeds which synthesize DAG species with unusual fatty acids, such as capric or erucic acid, the DAG is prevented from being incorporated into membrane lipids by strong selectivity of the aminoalcohol phosphotransferases for DAGs with 16- and 18-carbon fatty acids.[76]

Most studies of ethanolaminephosphotransferase activity in plants have noted that the enzyme is inhibited by CDP-choline and that both aminoalcohol phosphotransferases have similar properties.[11,19] This has inevitably led to the speculation that both aminoalcohol phosphotransferase activities are catalyzed by a single enzyme, and this possibility has been raised whenever these enzymes have been reviewed.[11,19,20] As all of these reviewers point out, however, that purification of the enzyme(s) or cloning of cDNAs which express individual activities is required before a definitive statement can be made. The situation in yeast is instructive in this case. Using the powerful technique of colony autoradiography,[77] mutants for cholinephosphotransferase (*CPT1*) and ethanolaminephosphotransferase (*EPT1*) activity were obtained.[78,79] The corresponding *CPT1* and *EPT1* genes were cloned by complementation[78,79] and a double *CPT1/EPT1* disruption mutant made.[80] This disruption mutant was used to express *CPT1* and *EPT1* separately, so that the individual properties of each aminoalcohol phosphotransferase could be investigated in isolation from the other enzyme. It was discovered that *CPT1* is specific for CDP-choline and dimethylethanolamine, whereas *EPT1* is almost equally effective at utilizing CDP-ethanolamine, CDP-choline, or any methylated aminoalcohol intermediate of the two as substrates. Thus, yeast has two distinct activities: a specific cholinephosphotransferase and a general aminoalcohol phosphotransferase. The two enzymes can be easily distinguished by the observation that *EPT1* is strongly inhibited by CMP, whereas *CPT1* is only very weakly inhibited by this reaction product. While this may provide an interesting model on which to base further investigations of the plant aminoalcohol phosphotransferases and is especially relevant to the discussion of ethanolamine metabolism below, it would probably be more profitable to attempt to clone plant aminoalcohol phosphotransferases by complementation of the yeast double disruption mutant with plant cDNA. If plant aminoalcohol phosphotransferases are expressed in the yeast mutant, they should be detected by colony autoradiography. In addition, it may be possible to clone *CPT1*- and *EPT1*-like cDNAs from plants by heterologous screening with the corresponding yeast genes.

B. ORIGINS OF ETHANOLAMINE AND CHOLINE

The source of ethanolamine and choline for the nucleotide pathways in plants is not entirely clear. In intact castor bean endosperms, radiolabel originating from a 30-min incubation with ^{14}C-serine is rapidly observed in free ethanolamine[81] in support of the notion that ethanolamine is derived by the decarboxylation of free serine. At the end of the pulse period, however, label is also observed in PS and PE; thus, the potential pathway

$$Serine \rightarrow PS \rightarrow PE \rightarrow Ethanolamine$$

cannot be eliminated by these experiments. It must also be noted that serine may also be metabolized into the DAG portions of PS and the methyl groups of PE.[82,83] Consequently, any labeling experiments with radioactive serine as a precursor must take this into account. In a serine-labeling experiment with *Lemna*, PS and PE were isolated and acid hydrolyzed, and the radioactivity in the liberated serine and ethanolamine head groups were determined independently of the diacylglycerol moiety. Flow of label from serine to the methyl groups of PE was prevented by pregrowing the cells in unlabeled methionine, thus providing a large pool of soluble methionine in the cells to trap any labeled methionine derived from serine. Under these conditions, it is clear that *Lemna* cells evolve CO_2 derived from the carboxyl carbon of serine, resulting in ethanolamine formation.[83] While the most likely explanation of the results reported in this study is decarboxylation of free serine, the authors were still not able to completely eliminate the participation of PS and PE as intermediates in this reaction. The free ethanolamine pool of *Lemna* is, however, tightly regulated. When exogenous ethanolamine is supplied to the cells, endogenously synthesized ethanolamine is reduced by an equivalent amount.[98] If excess exogenous ethanolamine (i.e., in amounts greater than the total ethanolamine content of unsupplemented plants) is supplied, the synthesis of endogenous ethanolamine is stopped.[98] The mechanisms of this regulation are unknown, but its existence implies that maintenance of a free ethanolamine pool for PE biosynthesis via the nucleotide pathway is important in plant cells.

The source of choline for the nucleotide pathway of PC biosynthesis is even less clear than the origin of free ethanolamine. Methylation of free ethanolamine to choline has not been conclusively demonstrated in any organism. Nonhepatic animal cells are not able to make choline and are dependent on a supply of choline from the blood. The choline in the blood is obtained from dietary sources and is also synthesized in the liver by methylation of PE to PC and then hydrolysis of the PC to choline.[4] Yeast cells do not normally contain choline and only utilize it for PC biosynthesis if it is supplied in the growth medium.[5] Eukaryotic microorganisms which make PC exclusively by the nucleotide pathway require exogenous choline for growth.[85]

Recently, the methylation of free ethanolamine to methylethanolamine was demonstrated in castor bean endosperm.[105,106] Methylation of the phosphorylated derivatives appeared not to occur in this tissue, although the methylated products were rapidly phosphorylated.[106] These results were confirmed by assaying an enzyme capable of catalyzing the ethanolamine methylation, and the enzyme had a stronger affinity for ethanolamine than the enzyme utilizing phosphoethanolamine for its substrate.[107]

In plants in general, however, the only clearly defined sources of free choline within the cell are the turnover of PC and its metabolites[86,87] or the catabolism of other choline-containing compounds such as sinapoylcholine.[11] On the other hand, many different types of plant tissue and cells, such as *Lemna*, rye roots, castor bean endosperm, wheat protoplasts, and suspension cultures of soybean, carrot, or sycamore, rapidly take up choline from their environment and incorporate it into PC.[11,47,58,88-91] Indeed, in *Lemna*, there is a specific transport system for the uptake of choline, and significant amounts of choline are taken up at external concentrations as low as 0.1 m*M*.[88] The total choline requirement of this plant can be met with an external choline concentration of only 0.65 mM. Also, in sycamore cells, there appears to be a carrier-mediated transport system for choline which facilitates uptake of choline from the media when it is present at concentrations less than 100 mM.[91] At concentrations higher than 100 mM, choline is also taken up by passive diffusion. All of the choline taken up by these cells is immediately converted to choline phosphate, which is stored in the cytosol until it is required for PC biosynthesis.[91,92] The rapid action of choline kinase is thought to act in concert with the choline transporter to trap both choline and inorganic phosphate for lipid biosynthesis.[91,92] Thus, apart from the turnover of preexisting choline-containing compounds, the only clearly demonstrated source of free choline for PC biosynthesis by the nucleotide pathway in plant cells is exogenous choline. Choline phosphate, as well as serving as a cytosolic choline storage pool, may also function to transport choline around the plant, since it is transported in the xylem of tomatoes and barley from the roots to rapidly growing tissues, such as leaves, where it is used for PC biosynthesis.[93] Choline phosphate is broken down on the outside of plant cells by a specific, extra-cellular phosphatase; taken up by the cell as choline and phosphate; and then immediately resynthesized into choline phosphate by the choline kinase.[92] Since choline phosphate is fairly metabolically inert except as a substrate for PC synthesis,[91] it becomes apparent why choline phosphate cytidylyltranferase is observed to regulate the rate of PC biosynthesis by this pathway in many plant tissues. In effect, choline phosphate cytidylyltranferase is the first com-mitted step of PC biosynthesis from choline in plant cells.

C. METHYLATION PATHWAYS

The source of the aminoalcohol in PC when exogenous choline or choline phosphate is not available therefore needs to be considered. In yeast growing in the absence of choline[5] and in animal hepatic cells,[4] PC is synthesized by direct methylation of PE. In animals, the reaction between PE and three *S*-adenosylmethionine (AdoMet) molecules is catalyzed by a single methyl-transferase enzyme,[4] and in yeast it is catalyzed by two separate methyl-transferases, one catalyzing the first rate-limiting reaction and one catalyzing the subsequent two methylations.[42] By analogy with animals and yeast, PC biosynthesis has been conceptually divided into CDP-choline and PE methy-lation pathways by reviewers[11,19] as a basis for experimentally investigating the regulation of PC biosynthesis.[47,94] It has been known for a number of years, however, that in plants these methylation reactions may occur at the level of phosphoaminoalcohols. Methylation of ethanolaminephosphate to cholinephosphate, followed by PC biosynthesis via CDP-choline, was ob-served in barley leaves in addition to direct methylation of PE.[86] In sugar beet leaves, the methylation of ethanolamine phosphate and not the methylation of PE was shown to account for all of the incorporation of ^{14}C-formate into PC.[87] Thus, synthesis of PE and PC from aminoalcohols and DAG cannot be easily divided into two separate pathways, since both may be interacting and operating as a single pathway. In a series of studies of radiolabeled methionine metabolism in *Lemna,* soybean, and carrot, Datko and Mudd[90,95–97] have presented evidence to support this view. In none of these plants have they observed significant methylation of PE, instead the committing step for PC biosynthesis is the methylation of ethanolamine phosphate to monomethylethanolamine phosphate. In *Lemna,* this compound is completely methylated to choline phosphate, which in turn is converted to CDP-choline and PC.[90,95] In soybean, the monomethylethanolamine phosphate is converted to CDP-monomethylethanolamine and then PME. The PME is then methylated to phosphatidyldimethylethanolamine (PDE) and PC.[96,97] In carrot cells, both choline phosphate and PME are formed from monomethylethanolamine, and PC is synthesized via both routes.[96,97] An enzyme activity capable of catalyzing the conversion of ethanolamine phosphate and AdoMet to monomethylethanolamine phosphate was measured in cell-free extracts from all of these tissues.[97] Aminoalcohol cytidylylytransferase activity capable of catalyzing the formation of methylated CDP-ethanolamines was also detected. All of the tissues have *in vitro* activities of microsomal enzymes which catalyzed the conversion of PME to PDE and PDE to PC. None of the plants have a PE methyltranferase activity. Carrot and *Lemna* have soluble activities which can methylate monomethylethanolamine phosphate to dimethylethanolamine phosphate and choline phosphate.[97] The soluble methyltransferases are quite clearly unique enzymes which have not been described for any other organism. The methylaminoalcohol cytidylyltransferase activities, in fact, may be the choline phosphate and ethanolamine phosphate cytidylyltransferases, with wider substrate specificity than was previously assumed. Presumably the CDP-aminoalcohol phosphotransferases have equally broad specificities. In this context, the broad specificity of the yeast *EPT1* gene product described above should be borne in mind.

It seems apparent that in most plant tissues, in the absence of exogenous choline, both PE and PC are derived from ethanolamine. The committing step for PC biosynthesis is the methylation of ethanolamine phosphate to monomethylethanolamine[87,90,96] by the enzyme AdoMet:ethanolamine phosphate *N*-methyltransferase.[97] Subsequent methylations are of either phosphoamino-alcohols, phosphatidylaminoalcohols, or both, depending on the species of plant. The phosphoaminoalcohols are converted to phosphatidylaminoalcohols via CDP intermediates. It is reasonable to assume, therefore, that reactions of the nucleotide and methylation pathways are really two characteristics of a single phosphoaminoalcohol pathway for PE and PC biosynthesis. In all plant tissues where there is evidence for the operation of the phosphoaminoalcohol pathway, the methylation of PE appears to be minimal; in *Lemna,* for example, Mudd and Dakto estimate that the rate of methylation of ethanolamine phosphate predominates over the rate of PE methylation by at least 99:1.[90] The major fluxes through the phosphoaminoalcohol pathway, based on the above studies, are summarized in Figure 2.

D. REGULATION OF THE PHOSPHOAMINOALCOHOL PATHWAY

In yeast cells, biosynthesis of PE and PC from CDP-DAG is repressed when the cells are grown in the presence of aminoalcohol precursors plus inositol. This repression is at the level of gene expression.[5] There is good evidence that the phosphoaminoalcohol pathway in plants is also regulated when exogenous choline or choline phosphate is available for PC biosynthesis. Mudd and Dakto have shown that when intact *Lemna* plants are grown in the presence of 3 mM choline, the flux of methionine groups through methylated derivatives of ethanolamine is reduced by 90 to 95%.[98] Similar reductions in methylation (77 to 98%) are seen when carrot and soybean cells are cultured in the presence of 50 mM choline.[99] In all three plants, exogenous choline reduced the *in vitro* activity of AdoMet:ethanolamine phosphate N-methyltransferase.[98,99] Thus, biosynthesis of PC from ethanolamine phosphate is repressed when a ready supply of choline phosphate is available to the cell, and the key regulatory enzyme in this repression appears to be AdoMet:ethanolamine phosphate N-methyltransferase. Mudd and Datko provide indirect evidence that the enzyme is regulated both at the level of gene expression and by feedback inhibition of AdoMet:ethanolamine phosphate N-methyltransferase activity by choline phosphate.[98] It will be necessary to purify this enzyme and obtain monospecific antibodies before these mechanisms can be further investigated.

In summary, in the absence of exogenous choline, the ratio of PE to PC synthesis by the phosphoaminoalcohol pathway is most probably determined by the competition of the ethanolamine phosphate cytidylyltransferase and the AdoMet:ethanolamine phosphate N-methyltransferase for their common substrate, ethanolamine phosphate. When the cell has a ready supply of choline phosphate, the AdoMet:ethanolamine phosphate N-methyltransferase is repressed,[98,99] and control of flux into PC probably resides in the choline phosphate cytidylyltransferase activity.[47,57] Under these conditions, the control of flux into PE from ethanolamine is not known. *Lemna* plants also have a specific transport mechanism for the uptake of ethanolamine,[88] and free ethanolamine synthesis is repressed by exogenous ethanolamine in the media.[83] Presumably the site of regulation is the decarboxylation of serine, but there is no evidence for this to date.

V. EXCHANGE REACTIONS

There are enzyme activities in plant cells which catalyze the exchange of a head group of a preexisting phospholipid with a similar or different head group. The best-characterized of these reactions is the synthesis of PS by exchange of serine with the headgroup of PE. This is the only mechanism by which PS is made in animal cells,[4] but in plant cells the presence of this reaction appears to be tissue specific.[41] The reaction is catalyzed by the Ca^{2+}-dependent, microsomal enzyme PE:L-serine phosphatidyltransferase. The requirement for

Ca^{2+} by the animal enzyme can be circumvented by ATP,[4] but the equivalent experiments with plant cells have not been done. The reaction is reversible, and ethanolamine inhibits the forward reaction with L-serine.[11] Indeed, in a number of plant tissues, there appears to be a similar Ca^{2+}-dependent, base exchange enzyme which catalyzes the exchange of ethanolamine with the head group of a phospholipid to yield PE.[100] In castor bean endosperm, CDP-ethanolamine synthesis is located on the outer leaflet of endoplasmic reticulum (ER) membranes,[101] but both the ethanolamine exchange activity and the PE:L-serine phosphatidyltransferase are located on the luminal side of the ER.[101] In addition, both exchange enzymes have very similar properties.[11,101] It is possible, therefore, that ethanolamine exchange is the reverse of the L-serine exchange reaction, and is catalyzed by the PE:L-serine phosphatidyltransferase. In fact, L-serine does inhibit ethanolamine exchange,[100] albeit noncompetitively (which may simply indicate that ethanolamine and serine bind to separate sites on the same enzyme). Again, purification of the enzyme(s) responsible for L-serine and ethanolamine exchange is necessary to resolve this issue. However, even if the ethanolamine exchange activity is a reverse of the PE:L-serine phosphatidyltransferase reaction, it is not just an artifact of an *in vitro* assay, since ethanolamine exchange appears to operate in castor bean endosperms *in vivo*.[81] Despite a number of suggestions, such as modulating a rapid change in head group composition in response to an environmental change,[100] the function of the ethanolamine exchange reaction is far from being understood.

Another exchange reaction without an apparent function is the PI/inositol base exchange reaction reported to occur in castor bean endosperm.[11] This CMP-stimulated exchange of the inositol head group of PI with ambient inositol has also been demonstrated in many mammalian cells (see Reference 102 and references cited therein). Although purified yeast PI synthase has an inositol exchange activity,[103] the properties of both the castor bean and mammalian exchange activity appear to be significantly different from the reverse activity of PI synthase.[11,102]

VI. REGULATION OF PHOSPHOLIPID HEAD GROUPS

A number of environmental factors, such as temperature, salt stress, water stress, and light, have been shown to affect the phospholipid head group composition of plant membranes, particularly the major phospholipids PC and PE.[19] It is not entirely clear what physiological processes a change in the relative content of individual phospholipid types, such as PE or PC, would affect. Nevertheless, an experimentally induced change in the rate of synthesis of a particular head group can be useful in investigating the regulation of the synthesis of that phospholipid. Thus, in rye roots, an increase in PC content correlating with an increased rate of PC biosynthesis is observed in chilled roots when compared with warmer roots.[104] This temperature-induced increase can be linked to regulation of PC biosynthesis by the enzyme choline phosphate cytidylyltransferase.[47] In light of the above discussion, however, it now seems clear that a number of things must be established to better understand the change in amounts of PE and PC in response to an experimental stimulus. First, is the PE and PC in the tissue under investigation derived from CDP-DAG or DAG? If PE is derived from CDP-DAG, then regulation of its amount would be expected to reside in the rate-limiting step for its synthesis, either PS synthase or PS decarboxylase. If PC is derived from CDP-DAG, then regulation of its amount relative to PE would probably reside in the PE methylation reaction. Second, if PE and PC are derived from DAG, then what is the source of their aminoalcohol head groups? If both PE and PC are derived from ethanolamine, then control of their relative synthesis rates probably resides in AdoMet:ethanolamine phosphate N-methyltransferase. If there is a ready supply of choline or choline phosphate from exogenous sources, breakdown of choline compounds, cellular storage pools, or from other tissues, then control of PC biosynthesis probably resides in choline phosphate cytidylyltransferase. Control of PE biosynthesis under these conditions is not quite as obvious, but would probably reside in the ethanolamine phosphate cytidylyltransferase, since this is at the branch point of PE and PC biosynthesis in the phosphoaminoalcohol pathway. In addition, modification of individual phospholipids by head group exchange reactions must also be considered, particularly when changes in PE relative to PC are observed.

In summary, it is clear that a better understanding of the metabolism of phospholipid head groups in plants will be obtained once purified enzymes, monospecific antibodies, and cDNA clones are available. The recent appearance of plant cDNA libraries in yeast expression systems, the technique of colony autoradiography, and the availability of yeast mutants and gene disruptants should greatly assist in the cloning of plant phospholipid biosynthetic enzymes. It may also be possible to clone plant cDNAs, encoding enzymes which have an exact equivalent in yeast by the use of heterologous yeast probes. The purification of enzymes whose cDNAs cannot be isolated by these techniques will probably be assisted by the advent of better affinity columns.

REFERENCES

1. **Mazliak, P.,** Synthesis and turnover of plant membrane phospholipids, *Prog. Phytochem.,* 6, 49–102, 1980.
2. **Harwood, J. L.,** Plant acyl lipids: structure, distribution and analysis, in *The Biochemistry of Plants, Vol. 4, Lipids:Structure and Function,* Stumpf, P. K., Ed., Academic Press, New York, 1980, 1–55.
3. **Raetz, C. R. H. and Dowhan, W.,** Biosynthesis and function of phospholipids in *Escherichia coli, J. Biol. Chem.,* 265, 1235–1238, 1990.
4. **Vance, D. E.,** Phospholipid metabolism and cell signaling in eucaryotes, in *Biochemistry of Lipids, Lipoproteins and Membranes,* Vance, D. E. and Vance, J., Eds., Elsevier, Amsterdam, 1991, 205–240.

5. **Carman, G. M. and Henry, S. A.,** Phospholipid biosynthesis in yeast, *Annu. Rev. Biochem*, 58, 635–669, 1989.
6. **Kelley, M. J. and Carman, G. M.,** Purification and characterization of CDP-diacylglycerol synthase from *Saccharomyces cerevisiae*, *J. Biol. Chem.*, 262, 14563–14570, 1987.
7. **Klig, L. S., Homann, M. J., Kohlwein, S., Kelley, M. J., Henry, S. A., and Carman, G. M.,** *Saccharomyces cerevisiae* mutant with a partial defect in the synthesis of CDP-diacylglycerol and altered regulation of phospholipid biosynthesis, *J. Bacteriol.*, 170, 1878–1886, 1988.
8. **Homann, M. J., Bailis, A. M., Henry, S. A., and Carman, G. M.,** Coordinate regulation of phospholipid biosynthesis by serine in *Saccharomyces cerevisiae*, *J. Bacteriol.*, 169, 3276–3280, 1987.
9. **Homann, M. J., Henry, S. A., and Carman, G. M.,** Regulation of CDP-diacylglycerol synthase activity in *Saccharomyces cerevisiae*, *J. Bacteriol.*, 163, 1265–1266, 1985.
10. **Andrews, J. and Mudd, J. B.,** Phosphatidylglycerol synthesis in pea (*Pisum sativum* cultivar Laxton's Progress No. 9) chloroplasts. Pathway and localization, *Plant Physiol.*, 79, 259–265, 1985.
11. **Moore, T. S., Jr.,** Phospholipid biosynthesis, *Ann. Rev. Plant Physiol.*, 33, 235–259, 1982.
12. **Kleppinger-Sparace, K. F. and Moore, T. S., Jr.,** Biosynthesis of CDP-diacylglycerol in endoplasmic reticulum and mitochondria of castor bean (*Ricinus communis* cultivar Hale) endosperm, *Plant Physiol.*, 77, 12–15, 1985.
13. **Morlock, K. R., McLaughlin, J. J., Lin, Y.-P., and Carman, G. M.,** Phosphatidate phosphatase from *Saccharomyces cerevisiae*. Isolation of 45 and 104 KDa forms of the enzyme that are differentially regulated by inositol, *J. Biol. Chem.*, 266, 3586–3593, 1991.
14. **Block, M. A., Dorne, A.-J., Joyard, J., and Douce, R.,** The phosphatidic acid phosphatase of the chloroplast envelope is located on the inner envelope membrane, *FEBS Lett.*, 164, 111–115, 1983.
15. **Ichihara, K., Norikura, S., and Fujii, S.,** Microsomal phosphatidate phosphatase in maturing safflower seeds, *Plant Physiol.*, 90, 413–419, 1989.
16. **Ichihara, K.,** The action of phosphatidate phosphatase on the fatty acid composition of safflower triacylglycerol and spinach glycerolipids, *Planta*, 183, 353–358, 1991.
17. **Brindley, D. N.,** Intracellular translocation of phosphatidate phosphohydrolase and its possible role in the control of glycerolipid synthesis, *Prog. Lipid Res.*, 23, 115–133, 1984.
18. **Ichihara, K., Murota, N., and Fujii, S.,** Intracellular translocation of phosphatidate phosphatase in maturing safflower seeds, a possible mechanism of feedforward control of triacylglycerol synthesis by fatty acids, *Biochim. Biophys. Acta*, 1043, 227–234, 1990.
19. **Harwood, J.,** Lipid metabolism in plants, *Crit. Rev. Plant Sci.*, 8, 1–43, 1989.
20. **Browse, J. and Somerville, C.,** Glycerolipid synthesis: biochemistry and regulation, *Annu. Rev. Plant Mol. Biol.*, 42, 467–506, 1991.
21. **Kuchler, K., Daum, G., and Paltauf, F.,** Subcellular and submitochondrial localization of phospholipid synthesizing enzymes in *Saccharomyces cerevisiae*, *J. Bacteriol.*, 165, 901–911, 1986.
22. **Moore, T. S., Jr.,** Phosphatidylglycerol synthesis in castor bean endosperm. Kinetics, requirements and intracellular localization, *Plant Physiol.*, 54, 164–168, 1974.
23. **Mudd, J. B., and DeZacks, R.,** Synthesis of phosphatidylglycerol by chloroplasts from leaves of *Spinacia oleracea* L. (Spinach), *Arch. Biochem. Biophys.*, 209, 584–591, 1981.
24. **Belunis, C. J., Bae-Lee, M., Kelley, M. J., and Carman, G. M.,** Purification and characterization of phosphatidylinositol kinase from *Saccharomyces cerevisiae*, *J. Biol. Chem.*, 268, 18897–18903, 1988.
25. **Buxeda, R. J., Nickels, J. T., Belunis, C. J., and Carman, G. M.,** Phosphatidylinositol 4-kinase from *Saccharomyces cerevisiae*. Kinetic analysis using Triton X-100/phosphatidylinositol mixed micelles, *J. Biol. Chem.*, 266, 13859–13865, 1991.
26. **Berridge, M. J.,** Inositol lipids and cell proliferation, *Biochim. Biophys. Acta*, 907, 33–45, 1987.

27. **Uno, I., Fukami, K., Kato, H., Takenawa, T., and Ishikawa, T.,** Essential role for phosphatidylinositol 4,5-bisphosphate in yeast cell proliferation, *Nature*, 333, 188–190, 1988.

28. **Boss, W. F.,** Phosphoinositide metabolism and its relation to signal transduction in plants, in *Second Messengers in Plant Growth and Development*, Boss, W. F. and Morre, D. J., Eds., Alan R. Liss, New York, 1989, 29–56.

29. **Morse, M. J., Satter, R. L., Crain, R. C., and Cote, G. G.,** Signal transduction and phosphatidylinositol turnover in plants, *Physiol. Plant*, 76, 118–121, 1989.

30. **Einspahr, K. J. and Thompson, G. A.,** Transmembrane signaling via phosphatidylinositol 4,5-bisphosphate hydrolysis in plants, *Plant Physiol.*, 93, 361–366, 1990.

31. **Boss, W. F. and Massel, M. O.,** Polyphosphoinositides are present in plant tissue culture cells, *Biochem. Biophys. Res. Commun.*, 132, 1018–1023, 1985.

32. **Van Breeman, R. B., Wheeler, J. J., and Boss, W. F.,** Identification of carrot inositol phospholipids by fast atom bombardment mass spectrometry, *Lipids*, 25, 328–334, 1990.

33. **Sommarin, M. and Sandelius, A. S.,** Phosphatidylinositol and phosphatidylinositol phosphate kinases in plant plasma membranes, *Biochim. Biophys. Acta*, 958, 268–278, 1988.

34. **Chen, Q. and Boss, W. F.,** Short-term treatment with cell wall degrading enzymes increases the activity of the inositol phospholipid kinases and the vanadate–sensitive ATPase of carrot cells, *Plant Physiol.*, 94, 1820–1829, 1990.

35. **Memon, A. R. and Boss, W. F.,** Rapid light-induced changes in phosphoinositide kinases and H$^+$-ATPase in plasma membrane of sunflower hypocotyls, *J. Biol. Chem.*, 265, 14817–14821, 1990.

36. **Okpodu, C. M., Gross, W., and Boss, W. F.,** Characterization and partial purification of soluble phosphatidylinositol kinase from carrot suspension culture cells, *FASEB J.*, 4, A2079, 1990.

37. **Morse, M. J., Crain, R. C., and Satter, R. L.,** Light-stimulated inositolphospholipid turnover in *Samaea saman* leaf pulvini, *Proc. Natl. Acad. Sci. U.S.A.*, 84, 7075–7078, 1987.

38. **Morse, M. J., Crain, R. C., and Satter, R. L.,** Phosphatidylinositol cycle metabolites in *Samanea saman* pulvini, *Plant Physiol.*, 83, 640–644, 1987.

39. **Melin, P. M., Sommarin, M., Sandelius, A. S. and Jergil, B.,** Identification of Ca^{2+}-stimulated polyphosphoinositide phospholipase C in isolated plant plasma membranes, *FEBS Lett.*, 223, 87–91, 1987.

40. **Kinney, A. J., Bae-Lee, M., Panghaal, S. S., Kelley, M. J., Gaynor, P. M, and Carman, G. M.,** Regulation of phospholipid biosynthesis in *Saccharomyces cerevisiae* by cyclic AMP-dependent protein kinases, *J. Bacteriol.*, 172, 1133–1136, 1990.

41. **Zeringue, L. C. and Moore, T. S., Jr.,** Phosphatidylserine synthase distribution among tissues of castor bean, *Plant Physiol.*, 99(Suppl.), A452, 1992.

42. **McGraw, P. and Henry, S. A.,** Mutations in the *Saccharomyces cerevisiae* OPI 3 gene. Effects on phospholipid methylation, growth and cross-pathway regulation of inositol synthesis, *Genetics*, 122, 317–330, 1989.

43. **Singh, H. and Privett, O. S.,** Incorporation of ^{32}P in soybean phosphatides, *Biochim. Biophys. Acta*, 202, 200–202, 1970.

44. **Bell, R. M. and Coleman, R. A.,** Enzymes of glycerolipid synthesis in eukaryotes, *Ann. Rev. Biochem.*, 49, 459–487, 1980.

45. **Hosaka, K., Kodaki, T., and Yamashita, S.,** Cloning and characterization of the yeast *CK1* gene encoding choline kinase and its expression in *Escherichia coli*, *J. Biol. Chem.*, 264, 2053–2059, 1989.

46. **Brostrom, M. A. and Browning, E. T.,** Choline kinase from brewers yeast. Partial purification, properties and kinetic mechanism, *J. Biol. Chem.*, 248, 2364–2371, 1973.

47. **Kinney, A. J., Clarkson, D. T., and Loughman, B. C.,** The regulation of phosphatidylcholine biosynthesis in rye (*Secale cereale*) roots. Stimulation of the nucleotide pathway by low temperature, *Biochem. J.*, 242, 755–759, 1987.

48. **Kinney, A. J. and Moore, T. S., Jr.,** Phosphatidylcholine synthesis in castor bean endosperm, characteristics and reversibility of the choline kinase reaction, *Arch. Biochem. Biophys.*, 260, 102–108, 1988.

49. **Mellor, R. B., Christensen, T. M. I. E., and Werner, D.,** Choline kinase II is present only in nodules that synthesize stable peribacteroid membranes, *Proc. Natl. Acad. Sci. U.S.A.*, 83, 659–663, 1986.

50. **Weinhold, P. A., Rounsifer, M. E., and Feldman, D. A.,** The purification and characterization of CTP: phosphorylcholine cytidylyltransferase from rat liver, *J. Biol. Chem.*, 261, 5104–5110, 1986.

51. **Cornell, R. B.,** Chemical crosslinking reveals a dimeric structure for CTP, phosphocholine cytidylyltransferase, *J. Biol. Chem.*, 264, 9077–9082, 1989.

52. **Kalmar, G. B., Kay, R. J., LaChance, A., Aebersold, R., and Cornell, R. B.,** Cloning and expression of rat liver CTP: phosphocholine cytidylyltransferase, an amphipathic protein that controls phosphatidylcholine synthesis, *Proc. Natl. Acad. Sci. U.S.A.*, 87, 6029–6033, 1990.

53. **Tsukagoshi, Y., Nikawa, J., and Yamashita, S.,** Expression in *Escherichia coli* of the *Saccharomyces cerevisiae* CCT gene encoding cholinephosphate cytidylyltransferase, *Eur. J. Biochem.*, 169, 477–486, 1987.

54. **Wang, X. and Moore, T. S., Jr.,** Phosphatidylcholine biosynthesis in castor bean endosperm. Purification and properties of CTP: cholinephosphate cytidylyltransferase, *Plant Physiol.*, 93, 250–255, 1990.

55. **Wang, X. and Moore, T. S., Jr.,** Isolation and characterization of a complementary DNA clone encoding cholinephosphate cytidylyltransferase from castor bean endosperm, *Plant Physiol.*, 96(Suppl.), 126, 1991.

56. **Price-Jones, M. J. and Harwood, J. L.,** Purification of CTP: cholinephosphate cytidylyltransferase from pea stems, *Phytochemistry*, 24, 2523–2527, 1985.

57. **Price-Jones, M. J. and Harwood, J. L.,** Hormonal regulation of phosphatidylcholine biosynthesis in plants. The inhibition of cytidylyltransferase by indol-3-yl-acetic acid, *Biochem. J.*, 216, 627–631, 1983.

58. **Kinney, A. J. and Moore, T. S., Jr.,** Phosphatidylcholine biosynthesis in castor bean endosperm. Mechanisms of regulation during the immediate postgermination period, *Phytochemistry*, 28, 2635–2639, 1989.

59. **Rolph, C. E. and Goad, L. J.,** Phosphatidylcholine biosynthesis in celery cell suspension cultures with altered sterol compositions, *Physiol. Plant.*, 83, 605–610, 1991.

60. **Sauer, A. and Robinson, D. G.,** Subcellular localization of enzymes involved in lecithin biosynthesis in maize roots, *J. Exp. Bot.*, 36, 1257–1266, 1985.

61. **Kinney, A. J. and Moore, T. S., Jr.,** Phosphatidylcholine biosynthesis in castor bean endosperm: the localization and control of CTP: cholinephosphate cytidylyltransferase activity, *Arch. Biochem. Biophys.*, 259, 15–21, 1987.

62. **Wang, X. and Moore, T. S., Jr.,** Partial purification and characterization of CTP: cholinephosphate cytidylyltransferase from castor bean endosperm, *Arch. Biochem. Biophys.*, 274, 338–347, 1989.

63. **Price-Jones, M. J. and Harwood, J. L.,** The control of CTP:cholinephosphate cytidylyltransferase activity in pea, *Biochem. J.*, 240, 837–842, 1986.

64. **Tsukagoshi, Y., Nikawa, J., Hosak, K., and Yamashita, S.,** Expression in *Escherichia coli* of the *Saccharomyces cerevisiae* CCT gene encoding cholinephosphate cytidylyl-transferase, *J. Bacteriol.*, 173, 2134–2136, 1991.

65. **Harwood, J. L.,** The synthesis of acyl lipids in plant tissues, *Prog. Lipid Res.*, 18, 55–86, 1979.

66. **Wang, X. and Moore, T. S., Jr.,** Phosphatidylethanolamine synthesis by castor bean endosperm. Intracellular distribution and characteristics of CTP: ethanolaminephosphate cytidylyltransferase, *J. Biol. Chem.*, 266, 19981–19987, 1991.

67. **Kinney, A. J., Clarkson, D. T., and Loughman, B. C.,** The effect of temperature on phospholipid biosynthesis in rye roots, in *The Biochemistry and Metabolism of Plant Lipids,* Wintermans, J. F. G. M. and Kuiper, P. J. C., Eds., Elsevier Biomedical, Amsterdam, 1982, 437–440.

68. **Slack, C. R., Roughan, P. G., Browse, J. A., and Gardiner, S. E.,** Some properties of cholinephosphotransferase from developing safflower *Carthamus tinctorius* cotyledons, *Biochem. Biophys. Acta,* 833, 438–448, 1985.

69. **Cho, S. H. and Cheesbrough, T. M.,** Warm growth temperatures decrease soybean cholinephosphotransferase activity, *Plant Physiol.,* 93, 72–76, 1990.

70. **Zhuo, Z. C. and Moore, T. S., Jr.,** Some characteristics of a solubilized cholinephosphotransferase, *Plant Physiol.,* 80(Suppl.), 85, 1986.

71. **Lichtenberg, D., Robson, R. M., and Dennis, E. A.,** Solubilization of phopholipids by detergents. Structural and kinetic aspects, *Biochim. Biophys. Acta,* 737, 285–304, 1983.

72. **Harwood, J. L.,** Synthesis of molecular species of phosphatidylcholine and phosphatidylethanolamine by germinating soya bean, *Phytochemistry,* 15, 1459–1463, 1976.

73. **Justin, A. M., Demandre, C., Tremolieres, A., and Mazliak, P.,** No discrimination by choline and ethanolamine phosphotransferases from potato (*Solanum tuberosum*) tuber microsomes in molecular species of endogenous diacylglycerols, *Biochim. Biphys. Acta,* 836, 1–7, 1985.

74. **Justin, A. M., Demandre, C., and Mazliak, P.,** Choline- and ethanolaminephosphotransferases from pea leaf and soya beans discriminate 1-palmitoyl-2-linoleoyldiacylglycerol as a preferred substrate, *Biochim. Biophys. Acta,* 922, 364–371, 1987.

75. **Stymne, S. and Stobart, A. K.,** Triacylglycerol biosynthesis, in *The Biochemistry of Plants: A Comprehensive Treatise. Vol. 9, Lipids: Structure and Function,* Stumpf, P. K., Ed., Academic Press, Orlando, 1987, 175–214.

76. **Stymne, S., Bafor, M., Jonsson, L., Wiberg, E., and Stobart, K.,** Triacylglycerol assembly, in *Plant Lipid Biochemistry, Structure and Utilization,* Quinn, P. J. and Harwood, J. L., Eds., Portland Press, London, 1990, 191–197.

77. **Raetz, C. R. H.,** Isolation of *Escherichia coli* mutants defective in enzymes of membrane lipid synthesis, *Proc. Natl. Acad. Sci. U.S.A.,* 72, 2274–2278, 1975.

78. **Hjelmstad, R. H. and Bell, R. M.,** Mutants of *Saccharomyces cerevisiae* defective in sn-1,2-diacylglycerol choline phosphotransferase. Isolation, characterization and cloning of the CPT 1 gene, *J. Biol. Chem.,* 262, 3909–3917, 1987.

79. **Hjelmstad, R. H. and Bell, R. M.,** The sn-1,2-diacylglycerol ethanolaminephosphotransferase activity of *Saccharomyces cerevisiae.* Isolation of mutants and cloning of the EPT 1 gene, *J. Biol. Chem.,* 263, 19748–19757, 1988.

80. **Hjelmstad, R. H. and Bell, R. M.,** sn-1,2-Diacylglycerol choline- and ethanolaminephosphotransferases in *Saccharomyces cerevisiae.* Mixed micellar analysis of the CPT1 and EPT 1 gene products, *J. Biol. Chem.,* 266, 4357–4365, 1991.

81. **Kinney, A. J. and Moore, T. S., Jr.,** Phosphatidylcholine synthesis in castor bean endosperm. Metabolism of L-serine, *Plant Physiol.,* 84, 78–81, 1987.

82. **Mudd, J. B.,** Phospholipid biosynthesis, in *The Biochemistry of Plants: A Comprehensive Treatise. Vol. 4 Lipids: Structure and Function,* Stumpf, P. K., Ed., Academic Press, Orlando, 1980, 249–282.

83. **Mudd, S. H. and Datko, A. H.,** Synthesis of ethanolamine and its regulation in *Lemna paucicoctata, Plant Physiol.,* 91, 587–597, 1989.

84. **Tan, Z. and Boss, W. F.,** Association of phosphatidylinositol kinase, phosphatidylinositol monophosphate kinase and diacylglycerol kinase with the cytoskeleton and F-actin fractions of carrot cells grown in suspension culture, *Plant Physiol.,* 100, 2116–2120, 1992.

85. **Bygrave, F. L. and Dawson, R. M. C.,** Phosphatidylcholine biosynthesis and choline transport in the anaerobic protozoan *Entoninium caudatum, Biochem. J.,* 160, 481–490, 1976.

86. **Hitz, W. D., Rhodes, D., and Hanson, A. D.,** Radiotracer evidence implicating phosphoryl and phosphatidyl bases as intermediates in betaine synthesis by water-stressed barley leaves, *Plant Physiol.,* 68, 814–822, 1981.

87. **Hanson, A. D. and Rhodes, D.,** [14]C tracer evidence for synthesis of choline and betaine via phosphoryl base intermediates in salinized sugarbeet leaves, *Plant Physiol.,* 71, 692–700, 1983.

88. **Datko, A. H. and Mudd, S. H.,** Uptake of choline and ethanolamine by *Lemna paucicostata* Hegelm. 6746, *Plant Physiol.,* 81, 285–288, 1986.

89. **Che, F.-S., Cho, C., Hyeon, S.-B., Isogai, A., and Suzuki, A.,** Metabolism of choline chloride and its analogs in wheat seedlings, *Plant Cell Physiol.,* 31, 45–50, 1990.

90. **Mudd, S. H. and Datko, A. H.,** Phosphoethanolamine bases as intermediates in phosphatidylcholine synthesis by *Lemna, Plant Physiol.,* 82, 126–135, 1986.

91. **Bligny, R., Foray, M.-F., Roby, C., and Douce, R.,** Transport and phosphorylation of choline in higher plant cells. Phosphorus-31 nuclear magnetic resonance studies, *J. Biol. Chem.,* 264, 4888–4895, 1989.

92. **Gout, E., Bligny, R., Roby, C., and Douce, R.,** Transport of phosphocholine in higher plants. Phosphorus-31 NMR studies, *Proc. Natl. Acad. Sci. U.S.A.,* 87, 4280–4283, 1990.

93. **Martin, B. A. and Tolbert, N. E.,** Factors which affect the amount of inorganic phosphate, phosphorylcholine and phosphorylethanolamine in xylem exudate of tomato plants, *Plant Physiol.,* 73, 464–470, 1983.

94. **Moore, T. S., Jr., Price-Jones, M. J., and Harwood, J. L.,** Effect of IAA on phospholipid metabolism in pea (*Pisum sativum* cultivar Feltham-first stems), *Phytochemistry,* 22, 2421–2426, 1983.

95. **Mudd, S. H. and Datko, A. H.,** Methionine methyl group metabolism in *Lemna, Plant Physiol.,* 81, 103–114, 1986.

96. **Datko, A. H. and Mudd, S. H.,** Phosphatidylcholine synthesis. Differing patterns in soybean and carrot, *Plant Physiol.,* 88, 854–861, 1988.

97. **Datko, A. H. and Mudd, S. H.,** Enzymes of phosphatidylcholine synthesis in *Lemna,* soybean and carrot, *Plant Physiol.,* 88, 1338–1348, 1988.

98. **Mudd, S. H. and Datko, A. H.,** Synthesis of methylated ethanolamine moieties. Regulation by choline in *Lemna, Plant Physiol.,* 90, 296–305, 1989.

99. **Mudd, S. H. and Datko, A. H.,** Synthesis of methylated ethanolamine moieties. Regulation by choline in soybean and carrot, *Plant Physiol.,* 90, 306–310, 1989.

100. **Shin, S. and Moore, T. S., Jr.,** Phosphatidylethanolamine synthesis by castor bean endosperm. A base exchange reaction, *Plant Physiol.,* 93, 148–153, 1990.

101. **Shin, S. and Moore, T. S., Jr.,** Phosphatidylethanolamine synthesis by castor bean endosperm. Membrane bilayer distribution of phosphatidylethanolamine synthesized by the ethanolaminephosphotransferase and ethanolamine exchange reactions, *Plant Physiol.,* 93, 154–159, 1990.

102. **Cubitt, A. B. and Gershengorn, M. C.,** CMP activates reversal of phosphatidylinositol synthase and base exchange by distinct mechanisms in rat pituitary GH_3 cells, *Biochem. J.,* 272, 813–816, 1990.

103. **Fischl, A. S., Homann, M. J., Poole, M. A., and Carman, G. M.,** Phosphatidylinositol synthase from *Saccharomyces cerevisiae.* Reconstitution, characterization and regulation of activity, *J. Biol. Chem.,* 261, 3178–3183, 1986.

104. **Kinney, A. J., Clarkson, D. T., and Loughman, B. C.,** Phospholipid metabolism and plasma membrane morphology of warm and cool rye roots, *Plant Physiol. Biochem.,* 25, 769–774, 1987.

105. **Prud'homme, M. P. and Moore, T. S.,** Phosphatidylcholine synthesis in castor bean endosperm. Free bases as intermediates, *Plant Physiol.,* 100, 1527–1535, 1992.

106. **Prud'homme, M. P. and Moore, T. S.,** Phosphatidylcholine synthesis in castor bean endosperm. Occurrence of an S-adenosyl-L-methionine: ethanolamine N-methyltransferase. *Plant Physiol.,* 100, 1536–1540, 1992.

Chapter 9

SPHINGOLIPIDS

Daniel V. Lynch

TABLE OF CONTENTS

I. INTRODUCTION

Sphingolipids are structurally distinct from the more prevalent glycerolipids discussed in this book, and the respective biosynthetic pathways of these two acyl lipids apparently share no common intermediate other than palmitoyl coenzyme A (CoA). In contrast to glycerolipids that have fatty acids esterified to sn-glycerol-3-phosphate, sphingolipids consist of a long-chain base (amino alcohol) having a single amide-linked fatty acid. Complex sphingolipids are formed by the addition of polar groups such as phosphocholine or one or more sugar residues to the relatively nonpolar N-acyl long-chain base (ceramide). Sphingolipids are present in most, if not all, eukaryotic cells and a few prokaryotic organisms. They were discovered in brain tissue before the turn of the century, and the first reports of sphingolipids in plant tissues were those of Carter and co-workers during the 1950s, demonstrating the presence of complex glycophosphosphingolipids in seed tissue extracts. The reader is referred to the brief history of sphingolipid chemistry given in the excellent review by Hakomori.[1] Subsequent studies of plant sphingolipids have centered on the structural characterization of glucosylceramides (Figure 1), the predominant sphingolipid in plant tissues.[2-14]

A. OCCURRENCE OF SPHINGOLIPIDS IN PLANTS

Glucosylceramides and related sphingolipids are minor components in lipid extracts of plant tissues, typically accounting for less than 5% of the total lipid. Their low abundance in plant lipid extracts may have led some plant lipidologists to overlook their possible significance. Note that Hitchcock and Nichols[15] reviewed work on plant sphingolipids in 1971, but more recent treatises on plant lipid biochemistry[16-18] say little, if anything, about this group of lipids. With the advent of two-phase partitioning techniques to obtain highly purified plasma membrane from plant tissues, it was demonstrated that glucosylceramides are quantitatively important components of this membrane, comprising 7 to 26 mol% of the plasma membrane lipid, depending on the plant tissue.[19-22] Glucosylceramides have also been reported to be a quantitatively important component of tonoplast lipids.[19,23] These observations are consistent with those from mammalian systems where various sphingolipids (glucosyl- and galactosylceramides, sphingomyelin, sulfatides, and more complex glycosphingolipids) have been localized in the plasma membrane and related endomembranes.[24] It has also been reported that plasma membrane sphingolipids in erythrocytes[25] and brain tissue[26] are concentrated in the outer leaflet of the membrane exposed on the cell surface. It is not known if plant glucosylceramides adopt this asymmetric distribution.

B. SPHINGOLIPID STRUCTURE

Based on the studies cited above[2-14] and reviewed in Tables 1 and 2, the structural features typical of plant sphingolipids may be summarized. Long-chain bases common in plants are C_{18} amino alcohols (Figure 2) and include

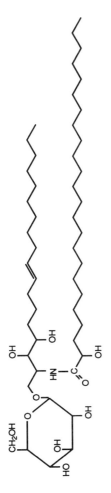

FIGURE 1. Structure of glucosylceramide having 4-hydroxy-8-sphingenine and 2-hydroxylignoceric (cerebronic) acid as the long-chain base and fatty acid component, respectively.

TABLE 1
Long-Chain Base Composition of Glucosylceramides from Plant Sources

	Long-chain base				
Plant tissue	d18:0	d18:1	d18:2	t18:0	t18:1
Wheat grain	9	72	15	1	3
Rice bran	—	5	83	1	11
Soy bean	—	2	78	3	17
Azuki bean	1	5	76	1	17
Pea seed	2	52	10	<1	36
Spinach leaf	2	2	41	<1	54
Rice leaf	—	1	39	3	57
Maize leaf	—	<1	57	—	41
Wheat leaf	—	4	14	—	80
Rye leaf	tr	2	17	3	77

Note: Values are expressed as weight percent and were obtained from published work.[5-14] Isomers differing in position and/or stereochemical configuration of double bonds are grouped together acording to the number of double bonds.

TABLE 2
Hydroxy Fatty Acid Composition of Glucosylceramides from Plant Sources

	Hydroxy fatty acid (HFA)								
Plant tissue	16:0	18:0	20:0	22:0	22:1	24:0	24:1	26:0	Other HFA
Wheat grain	39	8	44	3	—	3	1	<1	3
Rice bran	—	5	43	15	—	25	—	6	6
Soy bean	79	—	—	5	—	9	—	—	7
Azuki bean	41	1	1	14	—	29	—	2	11[a]
Pea seed	39	1	2	19	—	10	—	—	26[b]
Spinach leaf	42	—	—	12	—	32	—	6	7
Rice leaf	—	2	35	29	—	25	—	3	5
Maize leaf	3	4	28	33	—	30	—	1	1
Wheat leaf	7	1	6	7	—	24	24	3	19[c]
Rye leaf	3	<1	4	9	8	12	52	2	10[c]

Note: Values are expressed as weight percent and were obtained from published work.[5-14] For the plant sources shown below, 2-hydroxy fatty acids (α-hydroxy fatty acids) account for 90 to 99% of the total fatty acids of glucosylceramides.

[a] Hydroxy-14:0 accounts for 5%.
[b] Hydroxy-14:0 accounts for 21%.
[c] Odd chain saturated hydroxy fatty acids (C_{21}, C_{23}, C_{25}) constitute the majority of other hydroxy fatty acids.

the dihydroxy bases sphinganine (d18:0), 8-sphingenine (d18:1[8trans or cis]), and 4,8-sphingadienine (d18:2[4trans,8trans or 4trans,8cis]), and the trihydroxy bases 4-hydroxysphinganine (t18:0) and 4-hydroxy-8-sphingenine (t18:1[8trans or cis]). Other long-chain bases differing in chain length; number of hydroxyl groups; and number, position, and stereochemical configuration of double bonds occur as minor components in plant sphingolipids and parallel the diversity in long-chain base structure observed in other organisms.[27] The fatty acids of plant sphingolipids are almost exclusively 2-hydroxy fatty acids. Saturated C_{16}–C_{24} acyl chains are most abundant, although C_{14} and C_{26} acyl chains are present in some plants. Monounsaturated hydroxy fatty acids are common in some cereals. The fatty acid is linked to the long-chain base through an amide bond. The monosaccharide bound to ceramide of virtually all plant glycosphingolipids is glucose. The glucose moiety is glycosidically linked to the C-1 hydroxyl group of the long-chain base. Other chemical groups comprising the polar region of quantitatively minor sphingolipids (e.g., glycophosphosphingolipids) are also linked to the ceramide at C-1.

As shown in Tables 1 and 2, considerable structural diversity exists among the glucosylceramides from various plant tissues with respect to long-chain base and fatty acid composition. This is in contrast to the acyl chain compositions of individual glycerolipids of different plant tissues that are strikingly consistent, though not identical. Generally, glucosylceramides from seed tissues are enriched in long-chain (C_{16}–C_{20}) hydroxy fatty acids and dihydroxy long-chain bases. Glucosylceramides of dicot and cereal leaves are relatively enriched in trihydroxy long-chain bases. Very long-chain (>C_{20}) hydroxy fatty acids (saturated and monounsaturated) are common in glucosylceramides of cereal leaves, whereas those of spinach and other dicots such as squash and bean (not shown)[28] contain primarily hydroxypalmitic acid and hydroxylignoceric acid (cerebronic acid, h24:0). Glucosylceramides of root tissues are similar in composition to their leaf counterparts.[28]

C. SPHINGOLIPID FUNCTION

The ubiquity of sphingolipids in eukaryotic cells, their subcellular location, and the genotypic and tissue-specific differences in organisms suggest important biological roles for these lipids, and numerous functions have been ascribed to sphingolipids. Indeed, evidence of a vital function performed by sphingolipids is illustrated by a mutant strain of *Saccharomyces cerevisiae* that is deficient in long-chain base synthesis and requires exogenous sphinganine or phytosphingosine for growth.[29] Based on physical studies using model systems, glycosylceramides and related sphingolipids are thought to increase stability and decrease permeability of membranes as a consequence of intra- and intermolecular hydrogen bonding between amide and hydroxyl groups of the ceramide moiety[24,30-33] and have been implicated in regulating ion permeability.[32,33] The presence of sphingolipids having greater levels of hydroxylation in animal cells/tissues exposed to harsh chemical stresses (kidney tubules and intestine)[24] and in plant leaves and roots vs. seed tissue (see above)

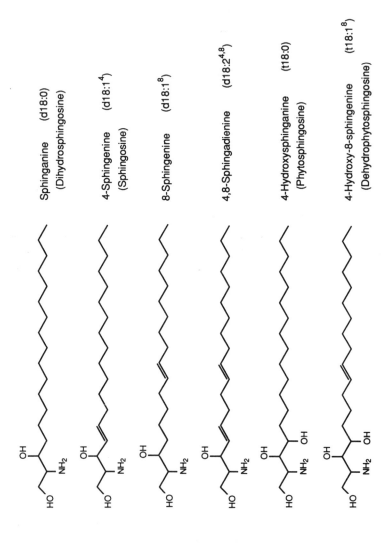

FIGURE 2. Structures of important long-chain bases. The name and shorthand designation described by Karlsson[27] is given for each. The first letter of the abbreviation designates the number of hydroxyl groups present on the free base ("d" for dihydroxy, "t" for trihydroxy), while the number before the colon refers to the number of carbon atoms in the alkyl chain, the number following the colon refers to the number of double bonds present, and the superscript designates the position of the double bond(s). The double bonds may be in the *trans* or *cis* configuration. The D-erythro isomer and the D-ribo isomer are the naturally occurring isomers of the dihydroxy and trihydroxy long-chain bases, respectively. For certain bases, the trivial name is also given in parentheses. Note that sphingosine (d18:1^{4trans}) is the predominant long-chain base in animal sphingolipids, whereas the positional isomer 8-sphingenine is more common in plants sphingolipids.

is consistent with a membrane-stabilizing function. The involvement of glucosylceramides in membrane-related phenomena associated with chilling sensitivity[13] and cold acclimation and freezing injury[34] has been suggested.

In addition to these functions as structural components in membranes, there is a vast literature on the role of glycosphingolipids in cell recognition and signal transduction[35] and the role of sphingolipids (especially sphingosine) in cellular regulation in animal and fungal systems.[36,37] In contrast to the tremendous volume of published research on these processes in animal systems, there is a paucity of published studies using plant systems. It is noteworthy, however, that sphingosine has been demonstrated to modulate plasma membrane ferricyanide reductase in response to blue light in oat mesophyll cells[38] and tonoplast pyrophosphatase activity in *Chenopodium rubrum* suspension cultures.[39] Certain molecular species of glucosylceramides from *Fusicoccum*, pea seeds, and wheat grain have been demonstrated to induce fruiting in the basidiomycete *Schizophyllum commune*.[40-43] Although a detailed exploration of the possible biological role(s) of plant sphingolipids is beyond the scope of this chapter, these examples suggest that studies of sphingolipid function should present exciting research opportunities in plant biology.

II. SPHINGOLIPID METABOLISM

A scheme for the proposed pathway for ceramide synthesis is shown in Figure 3. The steps shown are required for the synthesis of all complex sphingolipids, since the ceramide moiety is the common structural component of sphingolipids, just as diacylglycerol is common to glycerolipids. A brief review of our current understanding of each step is described below. Note that essentially all of the information reviewed here derives from studies of animal systems. Recent work from this laboratory on plant sphingolipid synthesis will also be described. The steps required for modification of long-chain bases (via hydroxylation and dehydrogenation) and of fatty acids (via hydroxylation) are not illustrated in Figure 3, but are discussed below. The reader should appreciate from the outset that many of the steps in sphingolipid metabolism are not well-defined, even in animal systems such as liver and brain that have been subject to intensive study over several decades.[44-47] Rather than present a comprehensive description of sphingolipid metabolism,[45,47] the goal of this review is to summarize our current understanding of sphingolipid synthesis, including studies of plant systems, and to highlight the more interesting and provocative questions remaining to be answered.

A. FORMATION OF LONG-CHAIN BASE (SPHINGANINE)

The biosynthetic pathway for complex sphingolipids begins with the synthesis of the long-chain base sphinganine. The first step in C_{18} long-chain base synthesis is the condensation of L-serine with palmitoyl-CoA, catalyzed by serine palmitoyltransferase. The product of this reaction, 3-ketosphinganine,

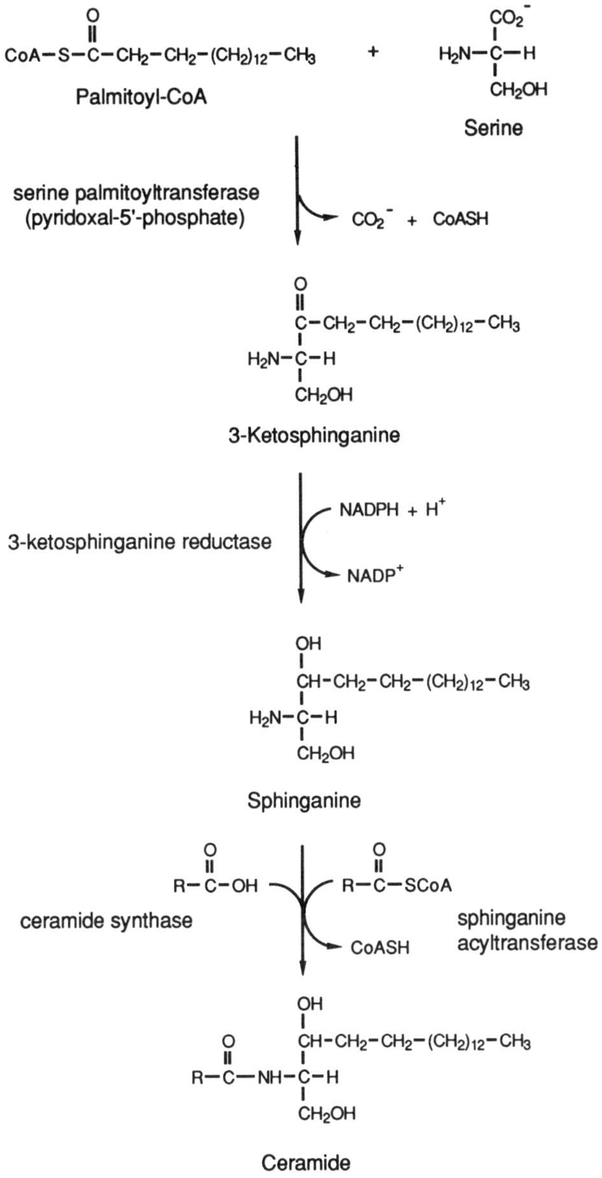

FIGURE 3. Tentative pathway for ceramide biosynthesis in plants. Only the major enzymatic steps for which there is evidence from plant systems are shown. Steps for modification of the long-chain base and fatty acid are not shown. Note that two mechanisms of long-chain base acylation (ceramide synthesis) have been found in squash membrane preparations, one utilizing free fatty acid and the other utilizing acyl-CoA.

is reduced to sphinganine by an NADPH-dependent reaction catalyzed by 3-ketosphinganine reductase.[48-54] Early studies of serine palmitoyltransferase by Snell and co-workers using a particulate fraction from the yeast *Hansenula cifferi*,[48-50] by Stoffel and co-workers using rat liver and yeast preparations,[51,52] and by others using brain preparations[53,54] demonstrated that serine and palmitoyl-CoA were substrates and pyridoxal 5'-phosphate was required for activity. The immediate product in the absence of NADPH was 3-ketosphinganine. A mechanism for this reaction was proposed requiring the loss of the α-hydrogen atom and the carboxyl group from serine, and their subsequent replacement by a proton from the medium and a palmitoyl group from the CoA thioester.[55] Serine palmitoyltransferase was partially purified (>100-fold) from the anaerobic bacterium *Bacterioides melaninogenicus* and was found to have properties similar to those of the eukaryotic enzyme.[56] The gene (*LCB1*) for serine palmitoyltransferase has been cloned from *Saccharomyces cerevisiae* and characterized.[57] DNA sequence analysis predicted a peptide of 558 amino acid residues having a globular, relatively hydrophobic structure with two potential membrane-spanning helical regions. Comparison of amino acid sequences demonstrated homology to 5-aminolevulinic acid synthase and to 2-amino-3-ketobutyrate CoA ligase, both pyridoxal 5'-phosphate-dependent enzymes.

Merrill and co-workers have characterized serine palmitoyltransferase activity in membrane fractions isolated from a range of mammalian tissues and cultured cell lines.[58-61] Activity was assayed by monitoring the incorporation of [³H]serine into the chloroform-soluble product, 3-ketosphinganine. The results obtained from these various tissues suggest that the activity of the enzyme (relative to glycerol 3-phosphate acyltransferase activity) is proportional to the sphingolipid content of the tissue. Holleran et al.[62] have reported high levels of serine palmitoyltransferase activity in microsomes from cultured human keratinocytes, consistent with the high content of sphingolipids in the stratum corneum, the epidermal barrier layer of mammals.

The activity of serine palmitoyltransferase has been characterized in summer squash fruit.[63-65] Enzyme activity in a microsomal membrane fraction from summer squash fruit was assayed using a procedure similar to that described above.[59] The apparent K_m for serine was 1.8 mM, and the enzyme exhibited a strong preference for palmitoyl-CoA, with other saturated acyl-CoA derivatives having an even number of carbon atoms exhibiting less than 10% maximal activity. This latter observation is consistent with the prevalence of C_{18} long-chain bases in sphingolipids of plant tissues. Pyridoxal 5'-phosphate was required for activity. Using marker enzyme assays of subcellular fractions obtained by differential centrifugation, serine palmitoyltransferase activity was localized to the endoplasmic reticulum. Two mechanism-based inhibitors of enzyme activity in animal and bacterial preparations, L-cycloserine[66] and β-chloro-L-alanine,[67] were effective inhibitors of enzyme activity in squash microsomes at concentrations yielding similar inhibition of the enzyme from other sources. Addition of NADPH (but not NADH) to the assay system promoted the conversion of 3-ketosphinganine to sphinganine. Although the properties of squash serine palmitoyltransferase appear similar to those of the animal, fungal, and bacterial enzymes in most respects, the specific activity of the squash enzyme was twofold to 20-fold higher than those previously reported for preparations from animal tissues.

Serine palmitoyltransferase activity has also been detected in microsomes isolated from leeks, bean hypocotyls, cucumber, apple, zucchini squash, potato tubers, and spinach leaves.[28] Attempts to demonstrate serine palmitoyltransferase activity in rye leaf microsomes were not successful; the preparations were found to contain an endogenous heat-labile component that inhibited liver or squash enzyme activity when rye microsomes were added to the assay.

The formation of 3-ketosphinganine is the first committed step in complex sphingolipid synthesis and has been proposed to be the rate-limiting step in animal tissues.[54,58,68,69] Based on the virtual absence of free long-chain bases and the low levels of ceramides in plant extracts, it is likely that this reaction is the rate-limiting step of sphingolipid biosynthesis in plants as well and thus may be involved in the regulation of this pathway. The specificity of the enzyme for acyl-CoA molecules of different chain lengths correlates with the composition of long-chain bases present in the tissues of animals[59,60,62] and plants (see above), indicating an important role for the enzyme in determining the species of long-chain base synthesized.

The regulation of serine palmitoyltransferase in animal tissue appears complex. Enzyme activity in cell cultures was dependent upon serine and palmitic acid concentrations in the medium, whereas other fatty acids appeared somewhat inhibitory.[69] Addition of lipoproteins or sphingolipids and long-chain bases decreased sphingolipid synthesis in cell cultures.[70,71] Note, however, that serine palmitoyltransferase activity was not inhibited by long-chain bases *in vitro*.[59,65]

Serine palmitoyltransferase activity in squash was found to vary with the size (age) of the fruit, with the smallest fruits tested having the highest specific activity (expressed per gram fresh weight). The specific activity decreased linearly with increasing fruit weight, although the total activity per fruit was greater in larger fruits and plateaued at a fruit weight of 250 to 300 g. Age-dependent decreases in enzyme-specific activity have been observed in elongating wax bean hypocotyls also.[28] These observations are taken as indirect evidence that sphingolipid synthesis is regulated, in part, by the level of serine palmitoyltransferase present, which in turn correlates with the relative growth rate of the tissue. As well, the apparent K_m of the enzyme for serine is similar to (or less than) the reported concentrations of this amino acid in plant tissues;[72] thus, substrate availability may influence sphingolipid biosynthesis in plant tissues. The potential effects of other small molecules or plant growth regulators on serine palmitoyltransferase activity and sphingolipid synthesis in plant tissues have not been investigated. Clearly, more needs to be done to understand the regulation of this enzyme, as well as its coordination with other enzymes involved in synthesizing phospholipids and sterols, the other lipid components of the plasma membrane and tonoplast.

B. CERAMIDE SYNTHESIS

Synthesis of ceramide involves formation of an amide bond via condensation of fatty acid with the amino group of long-chain base. Three mechanisms of ceramide synthesis in animal tissues have been proposed, differing with respect to the form of the fatty acid donor used in the reaction, and include (1) ceramide formation using free fatty acid via a reversal of ceramidase activity,[73-75] (2) ceramide formation utilizing acyl-CoA as an acyl donor and catalyzed by sphinganine (sphingosine) *N*-acyltransferase,[76-79] and (3) ceramide synthesis via acylation of the long-chain base with a fatty acid that is presumably activated by an as yet unidentified mechanism.[80-83] The latter two mechanisms are shown in Figure 3.

Reverse ceramidase activity was detected in the mitochondrial fraction of brain homogenates.[73-75] Ceramide synthesis activity via reverse ceramidase did not require ATP or CoA and was decreased when palmitoyl-CoA was substituted for palmitic acid. Optimum activity was observed at acid pH and was stimulated by detergent. Activity was not observed when very long-chain fatty acids were supplied. It is unlikely that this activity is involved in ceramide synthesis, since the reaction is thermodynamically unfavorable in the direction of synthesis, very long-chain fatty acids found in sphingolipids are not utilized as substrate, and the pH optimum and intracellular location are not consistent with synthetic activity.

Synthesis of ceramides using acyl-CoA as the donor was demonstrated in brain and liver homogenates[76] and microsomal preparations from brain.[76-79] Sribney[76] demonstrated a requirement for acyl-CoA. Palmitic acid and free CoA did not yield activity in the absence of ATP. Surprisingly, the acyltransferase was not specific for the natural *erythro* isomer of the long-chain base, being active with the *threo* isomer as well. However, Stoffel has demonstrated incorporation of various stereoisomers of long-chain bases into ceramides (but, interestingly, not into more complex sphingolipids) *in vivo*.[47] Morell and Radin[77] demonstrated that the specificity of the enzyme for CoA derivatives of stearic, lignoceric, palmitic, and oleic acids mirrored the ratio of these fatty acids found in brain sphingolipids, suggesting that the distribution of the fatty acids in sphingolipids is determined by the specificity of the acyltransferase(s). As well, sphinganine appeared to be a better substrate than sphingosine, at least in assays containing very long-chain acyl-CoA thioesters. Ullman and Radin[78] subsequently proposed the existence of four distinct acyltransferases differing in their specificities for long-chain and very long-chain fatty acids and their hydroxy analogs, based on competition experiments using different acyl-CoA molecules as substrates. Both the D- and L-hydroxyacyl-CoA isomers were utilized in the reaction, although the L isomer is not found in sphingolipids. This study also demonstrated the nonenzymatic formation of ceramide from long-chain base and acyl-CoA, particularly when the protein concentration in the assay was low. This observation was undoubtedly the impetus for Radin's admonition[46] that high reported activities for these types of reactions may be artifactual, since few reports include an appropriate control with heat-killed enzyme. Ceramide synthesis using lignoceroyl-CoA in brain microsomes was stimulated by the addition of a heat-stable factor from brain cytosol.[79] Ceramides containing both nonhydroxy fatty acids and hydroxy fatty acids were synthesized, with the formation of the latter presumably requiring NADPH that was included in the assay system (see below).

Another mechanism of ceramide synthesis has been proposed that involves the activation of the fatty acid by an as yet unidentified mechanism.[80-83] Activity was observed in preparations from rat brain in an assay system containing free fatty acid, heat-labile and heat-stable cytosolic factors, Mg^{2+}, and pyridine nucleotides (with NADPH stimulating activity the most). The reaction did not require exogenous CoA or ATP.[80-82] Note that although ceramide synthesis was observed using free fatty acids, this mechanism differed from the reverse ceramidase reaction in having different co-factor requirements, substrate specificities, sensitivity to inhibitors (including EDTA and respiratory chain inhibitors), and pH optimum.[80,81] This assay system was capable of producing labeled ceramides and glycosylceramides containing nonhydroxy and hydroxy fatty acids from [14C]lignoceric acid. The ratios of the products formed were altered by specific inhibitors.[83] Although the stimulation of enzyme activity by the addition of heat-labile and heat-stable cytosolic factors has not been clearly defined, it has been proposed that the factors activate the citric acid cycle and electron transport, providing ATP for the activation of fatty acid.[45,81] As evidence of this, it was demonstrated that respiratory electron transport chain inhibitors prevented NADPH-stimulated ceramide synthesis, but did not affect ATP-stimulated ceramide synthesis.[81] This proposed role for ATP is unclear, however, since exogenous ATP was stimulatory only at high concentrations, whereas pyridine nucleotide that may serve as an electron donor to the respiratory chain to produce ATP was much more effective than ATP itself and, surprisingly, NADPH was more effective at stimulating ceramide synthesis than NADH, the preferred electron donor for the mitochondrial electron transport chain. An alternative explanation may be suggested in which ceramide synthesis (or fatty acid activation) is somehow influenced by the redox state of some electron transport chain component of the endoplasmic reticulum.

Ceramide synthesis has been demonstrated in squash membranes.[64,84] The incubation of squash membranes or squash homogenate with [14C]palmitic acid and sphinganine resulted in the formation of a labeled product that co-chromatographed with authentic ceramide and was resistant to mild alkaline hydrolysis, but was degraded by strong acid hydrolysis. Formation of ceramide was linear over 60 min, exhibited a pH optimum of 6.8, and increased linearly with increasing protein concentration. Ceramide synthase activity in subcellular fractions obtained by differential centrifugation paralleled antimycin A-insensitive NADPH cytochrome c reductase activity, a marker enzyme for endoplasmic reticulum. Ceramide synthesis was proportional to the concentration of the naturally occurring *erythro* isomer of sphinganine, but the *threo* isomer was not utilized. Sphingosine was a poor substrate, and it did not affect ceramide synthesis when added with sphinganine. In contrast, phytosphingosine and stearylamine inhibited enzyme activity. Assays using labeled palmitate, stearate, or lignocerate exhibited similar rates of ceramide formation (at low fatty acid concentrations), and there was little evidence of competition between these fatty acids. Addition of unlabeled hydroxypalmitic acid did not alter incorporation of labeled palmitate into ceramides. Ceramide synthesis in squash membranes was lower in the presence of NADPH, but NADP, NAD, and NADH had no effect. ATP decreased ceramide formation, whereas other nucleoside triphosphates had little effect. Ceramide synthesis did not require CoA. In fact, the addition of CoA (in the presence of ATP) decreased the incorporation of labeled palmitic acid into ceramide, whereas the addition of unlabeled palmitoyl-CoA (in the presence of free CoA and ATP) increased the incorporation of labeled palmitate into ceramide. These observations are consistent with palmitic acid serving as substrate for ceramide synthesis, whereas palmitoyl-CoA is used for glycerolipid synthesis in the assay system. Thus, conditions that favor formation of acyl-CoA and so stimulate many other biosynthetic reactions actually diminished apparent ceramide synthase activity. Further evidence that thioester formation was not required for ceramide synthesis was suggested by experiments in which hydroxylamine added to the assay had no effect. Taken together, these results suggest that ceramide synthesis in squash membranes does not require acyl-CoA and does not reflect reverse ceramidase activity. The results are not entirely consistent with condensation of long-chain base and fatty acid via an unidentified mechanism as reported previously,[80,81] since the effects of ATP and NADPH differ from those demonstrated in brain preparations. Further studies are needed to more completely characterize enzyme activity, including the nature of the inhibition by phytosphingosine. It is noteworthy that fumonisins, compounds resembling phytosphingosine that are produced by *Fusarium moniliforme* and implicated as causative agents of various pathological conditions associated with consumption of grain contaminated by the fungus, were shown to inhibit ceramide synthesis *in vivo* and *in vitro*,[85] leading to an accumulation of free sphinganine in cells. Mycotoxins such as fumonisins may alter plant sphingolipid synthesis as well. In addition, the role of cytosolic components in activating ceramide synthesis in plants needs further exploration. We have evidence that addition of heat-killed squash homogenate to the assay increases ceramide formation in squash membranes.

In addition to ceramide synthesis by a mechanism utilizing free fatty acids, preliminary experiments using squash or bean hypocotyl microsomal membranes in an assay system containing [³H]sphinganine and palmitoyl-CoA have provided evidence of acyl-CoA-dependent ceramide synthesis in plant tissues. Of the results obtained to date, one of the more interesting aspects involves the apparent specificity for acyl-CoA thioesters of different chain lengths. Ceramide synthesis was greatest with palmitoyl-CoA, less active with behenoyl-CoA and lignoceroyl-CoA, and least active with stearoyl-CoA and arachidoyl-CoA. These relative activities parallel the distribution of (hydroxy) fatty acids in glucosylceramides from these sources. Further characterization of this reaction is in progress. Nevertheless, these studies demonstrating two routes for ceramide synthesis raise interesting questions concerning the respective roles of the acyl-CoA-dependent and -independent pathways. It may be that one is involved in *de novo* synthesis of ceramide, whereas the other may act as a retailoring mechanism or as a salvage pathway for fatty acids and long-chain bases, since both types of molecules are cytotoxic at even low concentrations.

C. FORMATION OF GLUCOSYLCERAMIDE

Ceramide serves as the precursor for all complex sphingolipids. The typical fate of plant ceramide is its conversion to glucosylceramide. Although galactosyl- and glucosyltransferase activities involved in glycosphingolipid formation have been characterized in brain and kidney homogenates and membrane fractions,[86,87] no such activity has been reported in plant preparations. Ceramide glucosyltransferase from animal tissues utilizes uridine diphosphate (UDP)-glucose as the glucose donor and nonhydroxyceramide as the preferred substrate (Figure 4), consistent with the finding that animal glycosphingolipids containing glucose typically contain nonhydroxy fatty acids.[87,88] In contrast, galactosyltransferase utilizes hydroxyceramides preferentially.[86,87] The glucosyltransferase reaction was found to be stimulated by Mg^{2+} or Mn^{2+}, and the mode of substrate (ceramide) delivery in detergent dispersion, on celite, or in liposomes affected apparent activity.[88-90] Inhibition of UDP-glucose pyrophosphatase activity by the addition of NAD has also been demonstrated to improve apparent enzyme activity.[88] UDP-glucose:ceramide glucosyltransferase in liver was localized to the cytosolic surface of Golgi membranes[91] and has been solubilized from Golgi membranes.[92]

In membrane preparations and homogenates of squash and bean tissue, the primary radiolabeled lipid product formed following incubation with [^{14}C]UDP-glucose was sterylglucoside,[64] as reported in other studies of plant glucosyltransferases.[93] No incorporation of radioactivity into glucosylceramides was observed, even in the presence of added ceramides. Various substrate (ceramide) delivery systems were investigated without success. We considered that long-chain base may be the substrate for glucosylation and the product (psychosine) subsequently acylated to form glucosylceramide. Glycosylation of long-chain base to form psychosine has been demonstrated in animal tissue preparations,[94] although this reaction is now thought to be of little physiological significance. However, glucosylation of long-chain base could not be demonstrated in plant preparations. To assess the possibility that something other than UDP-glucose may serve as the sugar donor, tritiated ceramides were used in the assay in conjunction with a number of potential sugar donors, including UDP, cytidine diphosphate (CDP), ADP, and guanosine diphosphate (GDP) derivatives of glucose, galactose, and mannose. However, none of the combinations of components promoted glucosylceramide synthesis. Thus, it appears that the assay conditions described for the enzymes of animal systems are not suitable for examining the activities in plant preparations. The possi-

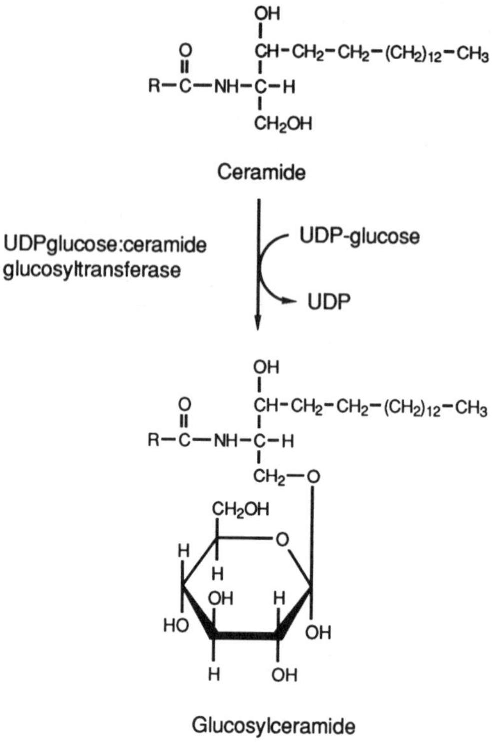

$$OH$$

Ceramide

UDPglucose:ceramide
glucosyltransferase

UDP-glucose

UDP

Glucosylceramide

FIGURE 4. Conversion of ceramide to glucosylceramide via UDP-glucose:ceramide glucosyl-transferase as found in animal tissues.

bility that plant tissues contain an endogenous inhibitor of the enzyme, as has been reported in brain preparations,[95] cannot be ruled out. However, given that UDP-glucose:sterol glucosyltransferase is active, one must also question the identity of the sugar donor for glucosylceramides.

D. MODIFICATION OF LONG-CHAIN BASE

Sphinganine is modified by dehydrogenation and/or hydroxylation to form the C_{18} bases commonly found in plants and animals (Figure 2). The modification of long-chain base apparently occurs in concert with complex sphingolipid synthesis. The enzymatic activities required for these modifications have not been demonstrated *in vitro*, and the true substrate(s) for these reactions (i.e., free long-chain base or ceramide) are not known with certainty.

Studies demonstrating the dehydrogenation of ketosphinganine,[96,97] sphinganine,[50] and ceramide (*N*-acylsphinganine)[68] have been published, but the most convincing evidence, from *in vivo*-labeling studies,[68,98] suggests that *N*-acylsphinganine is the true substrate for insertion of the double bond in the long-chain base, forming *N*-acylsphingosine. Indirect evidence supports this pathway for sphinganine modification in plant tissues as well.[28,84] Using an assay system containing flavin adenine dinucleotide (FAD) and other co-factors as employed in earlier studies,[96,97] we have found no evidence to date for dehydrogenation of labeled ketosphinganine or labeled sphinganine. Sphinganine served as substrate for ceramide synthesis (see above), whereas modified bases were ineffective as substrate or inhibited ceramide synthase. We have examined the effects of other long-chain bases in the assay system for acyl-CoA-dependent ceramide synthesis. Sphinganine was the preferred substrate for this reaction, as in animal preparations.[77] Analysis of squash tissue indicated that sphinganine is the predominant free long-chain base.[84] Taken together, these observations suggest that free long-chain base (or its keto derivative) is not the substrate for double bond insertion in plants. Future experiments will attempt to demonstrate dehydrogenation of *N*-acylsphinganine *in vitro*.

The hydroxylation of sphinganine has been demonstrated *in vivo*,[99,100] but to date an *in vitro* assay system has not been described. Nevertheless, studies of phytosphingosine formation in *H. cifferi* grown in the presence of $^{18}O_2$ or $H_2^{18}O$ demonstrated that molecular oxygen was the major source of oxygen of the hydroxyl group.[101] Note that this fungus produces and excretes large amounts of this long-chain base (as the tetra-acetyl derivative), thus its biosynthetic pathway may differ from that in other organisms that produce ceramides and complex sphingolipids. Studies demonstrating the conversion of sphinganine to phytosphingosine in rat[99,100] did not ascertain whether hydroxylation occurred on the free long-chain base, on the ceramide, or on a complex sphingolipid. Efforts are required to demonstrate hydroxylation *in vitro*, to characterize the identity of the substrate in the plant system, and to assess the possible role of electron transport chain components or other redox compounds in the reaction.

E. HYDROXYLATION OF FATTY ACID

One striking feature of many sphingolipids, particularly glucosylceramides of plant origin, is the presence of 2-hydroxy fatty acids as constituents of the molecules. Hydroxylation of fatty acids has been demonstrated in cell-free preparations from animal and plant tissues, typically as a reaction associated with fatty acid α-oxidation. However, the utilization of hydroxy fatty acids from α-oxidation for sphingolipid biosynthesis *in vivo* has not been unambiguously established.

The conversion of lignoceric acid to its 2-hydroxy analog, cerebronic acid, in rat brain preparations has been investigated by Kishimoto and co-work-ers.[45,102-105] The reaction was most active in the presence of O_2, Mg^{2+}, pyridine nucleotide (especially NADPH), and heat-stable and heat-labile cytosolic co-factors and appeared to be unique to brain, since preparations from extraneural tissues did not demonstrate activity.[102] Lignoceric acid hydroxylation was not stimulated by ATP or CoA, although lignoceroyl-CoA-also served as substrate for hydroxylation.[79,105] Mitochondrial electron transport inhibitors and EDTA inhibited the reaction. These conditions are the same as those described above for the acyl-CoA independent formation of ceramide involving an unidentified activated fatty acid.[79,80,83] In fact, labeled cerebronic acid produced in the assay was found only as a component of ceramide or cerebroside, not as a free acid or as an acyl-CoA thioester, and addition of sphingosine or psychosine stimulated cerebronic acid synthesis and ceramide formation. It was also demonstrated that added labeled cerebronic acid was not incorporated into sphingolipids. These results would suggest that hydroxylation occurs after ceramide formation, i.e., that ceramide is the actual substrate for fatty acid hydroxylation. However, neither labeled lignoceroyl-sphingosine (ceramide) nor galactosylceramide served as substrate in the assay. Based on these observations, Kishimoto proposed that activated lignoceric acid is hydroxylated and directly channeled into ceramide.[45]

It was subsequently reported that labeled lignoceric acid was oxidized by a mechanism unique to the brain, yielding labeled glutamate under the same assay conditions as those used to demonstrate hydroxylation of lignoceric acid.[106] As such, it is difficult to assess the physiological role of this reaction in sphingolipid synthesis. Moreover, the fact that rat brain preparation was not active with stearic acid (even though brain sphingolipids contain hydroxystearate) suggests some other role for this hydroxylation system. As with ceramide formation utilizing free fatty acid, the fatty acid activation mechanism is not known; although acyl-CoA has been ruled out as the immediate precursor, it would be informative to test whether respiratory inhibitors block hydroxylation and ceramide synthesis when lignoceroyl-CoA is provided in place of free fatty acid.

The proposed mechanism of lignoceric acid hydroxylation and channeling to ceramide contrasts with conclusions from other studies, in which CoA thioesters of hydroxy fatty acids were used as substrate for ceramide synthesis,[78] and from the results of elegant *in vivo* studies of sphingolipid fatty acid hydroxylation in thermally acclimating *Tetrahymena*.[107,108] In these latter studies, biosynthetically prelabeled complex sphingolipids containing nonhydroxy fatty acids were converted to hydroxy fatty acid-containing analogs with retention of radioactivity on the fatty acid moieties. Added hydroxy fatty acids were not incorporated into sphingolipids, but rather served as substrate for α-oxidation. These results indicate a direct conversion (hydroxylation) of the sphingolipid-linked fatty acid.

In vitro studies of α-oxidation have demonstrated the formation of 2-hydroxy fatty acids in plant preparations.[15,109,110] Hydroxypalmitic acid was formed from free palmitic acid as an intermediate of α-oxidation, particularly in the presence of glutathione and glutathione peroxidase that prevented the reactive intermediate, hydroperoxypalmitate, from undergoing decarboxylation. Experimental evidence suggested that a flavoprotein rather than cytochrome was required for hydroxylation. For the most part, these studies of plant systems examined free and esterified fatty acid products, but not amide-linked acids, i.e., ceramides. Future studies in plant systems will need to focus on hydroxylation of fatty acids destined for, or linked to, sphingolipids.

III. SUMMARY

Animal and fungal sphingolipid biosynthesis has been investigated intensively, but many of the enzymatic reactions have not been thoroughly characterized, and a definitive picture of the biosynthetic pathway(s) is lacking. Little is known about the enzymatic machinery and metabolic pathways for sphingolipid synthesis in plants. As we progress in characterizing the basic steps involved in glucosylceramide synthesis, we may learn new things not only about plant sphingolipid metabolism, but also about analogous metabolic steps in animal tissues that have yet to be characterized. For example, sphingolipids containing trihydroxy long-chain bases (common in plants) are prevalent in many organs such as kidney, but little is known about long-chain base hydroxylation in these tissues. In theory, plant preparations should be ideal systems for studies of sphingolipid biosynthetic enzymes (ignoring for the moment problems of enzyme degradation and interference or inhibition by secondary plant products), since many plant tissues are rapidly growing and require high rates of membrane lipid synthesis. Regulation of sphingolipid synthesis in plants and animals will differ considerably, and this fact emphasizes the need for further studies of sphingolipid metabolism in plants, particularly if improvement of crop plants may be achieved through manipulation of sphingolipid content. Given the evidence from other eukaryotic organisms for the role(s) of sphingolipids in cellular regulation, it is likely that these lipids play a critical function in the life of a plant. Based on this point alone, sphingolipid biochemistry in plants merits greater attention.

ACKNOWLEDGMENTS

The studies of plant sphingolipid metabolism described in this chapter were supported by a grant from the National Science Foundation (DCB 89-03816) to D. V. Lynch. The author gratefully acknowledges the assistance and contributions of research assistants and students at Williams College, Williamstown, MA.

REFERENCES

1. **Hakomori, S.,** Chemistry of glycosphingolipids, in *Handbook of Lipid Research, Vol. 3, Sphingolipid Biochemistry,* Kanfer, J. N. and Hakomori, S., Eds., Plenum Press, New York, 1983, 1.
2. **Carter, H. E., Hendry, R. A., Nojima, S., Stanacev, N. Z., and Ohno, K.,** Biochemistry of the sphingolipids XIII. Determination of the structure of cerebrosides from wheat flours, *J. Biol. Chem.,* 236, 1912, 1961.
3. **Sastry, P. S. and Kates, M.,** Lipid components of leaves. V. Galactolipids, cerebrosides, and lecithin of runner bean leaves, *Biochemistry,* 3, 1271, 1964.
4. **Ito, S. and Fujino, Y.,** Isolation of cerebroside from alfalfa leaves, *Can. J. Biochem.,* 51, 957, 1973.

5. **Fujino, Y. and Ohnishi, M.,** Constituents of ceramide and ceramide monohexoside in rice bran, *Chem. Phys. Lipids,* 17, 275, 1976.
6. **Ohnishi, M. and Fujino, Y.,** Chemical composition of ceramide and cerebroside in Azuki bean seeds, *Agric. Biol. Chem.,* 45, 1283, 1981.
7. **Ohnishi, M. and Fujino, Y.,** Sphingolipids in immature and mature soybeans, *Lipids,* 17, 803, 1982.
8. **Ohnishi, M., Ito, S., and Fujino, Y.,** Characterization of sphingolipids in spinach leaves, *Biochim. Biophys. Acta,* 752, 416, 1983.
9. **Fujino, Y. and Ohnishi, M.,** Sphingolipids in wheat grain, *J. Cereal Sci.,* 1, 159, 1983.
10. **Ito, S., Ohnishi, M., and Fujino, Y.,** Investigation of sphingolipids in pea seeds, *Agric. Biol. Chem.,* 49, 539, 1985.
11. **Fujino, Y., Ohnishi, M., and Ito, S.,** Further studies on sphingolipids in wheat grain, *Lipids,* 20, 337, 1985.
12. **Ohnishi, M., Ito, S., and Fujino, Y.,** Structural characterization of sphingolipids in leafy stems of rice, *Agric. Biol. Chem.,* 49, 3327, 1985.
13. **Ohnishi, M., Imai, H., Kojima, M., Yoshida, S., Murata, N., Fujino, Y., and Ito, S.,** Separation of cerebroside species in plants by reversed-phase HPLC and their phase transition temperature, *Proc. ISF-JOCS World Congr.,* 2, 930, 1988.
14. **Cahoon, E. B. and Lynch, D. V.,** Analysis of glucocerebrosides of rye (*Secale cereale* L. cv Puma) leaf and plasma membrane, *Plant Physiol.,* 95, 58, 1991.
15. **Hitchcock, C. and Nichols, B. W.,** *Plant Lipid Biochemistry,* Academic Press, New York, 1971.
16. **Stumpf, P. K.,** *The Biochemistry of Plants, Vol. 4, Lipids: Structure and Function,* Academic Press, Orlando, FL, 1980.
17. **Stumpf, P. K.,** *The Biochemistry of Plants, Vol. 9, Lipids: Structure and Function,* Academic Press, Orlando, FL, 1987.
18. **Harwood, J. L. and Bowyer, J. R.,** *Methods in Plant Biochemistry,* Vol. 4, Academic Press, San Diego, CA, 1990.
19. **Yoshida, S. and Uemura, M.,** Lipid composition of plasma membranes and tonoplasts isolated from etiolated seedlings of mung bean (*Vigna radiata* L.), *Plant Physiol.,* 82, 807, 1986.
20. **Lynch, D. V. and Steponkus, P. L.,** Plasma membrane lipid alterations associated with cold acclimation of winter rye seedlings (*Secale cereale* L. cv Puma), *Plant Physiol.,* 83, 761, 1987.
21. **Rochester, C. P., Kjellbom, P., Andersson, B., and Larsson, C.,** Lipid composition of plasma membranes isolated from light-grown barley (*Hordeum vulgare*) leaves: identification of cerebroside as a major component, *Arch. Biochem. Biophys.,* 255, 385, 1987.
22. **Sandstrom, R. P. and Cleland, R. E.,** Comparison of the lipid composition of oat root and coleoptile plasma membranes, *Plant Physiol.,* 90, 1207, 1989.
23. **Haschke, H.-P., Kaiser, G., Martinoia, E., Hammer, U., Teucher, T., Dorne, A. J., and Heinz, E.,** Lipid profiles of leaf tonoplasts from plants with different CO_2-fixation mechanisms, *Bot. Acta,* 103, 32, 1990.
24. **Karlsson, K.-A.,** Glycosphingolipids and surface membranes, in *Biological Membranes,* Vol. 4, Chapman, D., Ed., Plenum Press, New York, 1982, 1.
25. **Steck, T. L. and Dawson, G.,** Topographical distribution of complex carbohydrates in the erythrocyte membrane, *J. Biol. Chem.,* 249, 2135, 1974.
26. **Linington, C. and Rumsby, M. G.,** Accessibility of galactosyl ceramides to probe reagents in central nervous system myelin, *J. Neurochem.,* 35, 983, 1980.
27. **Karlsson, K.-A.,** Sphingolipid long chain bases, *Lipids,* 5, 878, 1970.
28. **Lynch, D. V.,** unpublished results, 1993.
29. **Wells, G. B. and Lester, R. L.,** The isolation and characterization of a mutant strain of *Saccharomyces cerevisiae* that requires a long chain base for growth and for synthesis of phosphosphingolipids, *J. Biol. Chem.,* 258, 10200, 1983.
30. **Abrahamsson, S., Pascher, I., Larsson, K., and Karlsson, K.-A.,** Molecular arrangements in glycosphingolipids, *Chem. Phys. Lipids,* 8, 152, 1972.

31. **Boggs, J. M.,** Lipid intermolecular hydrogen bonding: influence on structural organization and membrane function, *Biochim. Biophys. Acta*, 906, 353, 1987.
32. **Curatolo,W.,** Glycolipid function, *Biochim. Biophys. Acta*, 906, 137, 1987.
33. **Curatolo,W.,** The physical properties of glycolipids, *Biochim. Biophys. Acta*, 906, 111, 1987.
34. **Lynch, D. V., Caffrey, M., Hogan, J. L., and Steponkus, P. L.,** Calorimetric and x-ray diffraction studies of rye glucocerebroside mesomorphism, *Biophys. J.,* 61, 1289, 1992.
35. **Hakomori, S.,** Bifunctional role of glycosphingolipids: modulators for transmembrane signaling and mediators for cellular interactions, *J. Biol. Chem.,* 265, 18713, 1990.
36. **Hannun, Y. A. and Bell, R. M.,** Functions of sphingolipids and sphingolipid breakdown products in cellular regulation, *Science*, 243, 500, 1989.
37. **Merrill, A. H.,** Cell regulation by sphingosine and more complex sphingolipids, *J. Bioenerg. Biomembr.,* 23, 83, 1991.
38. **Dharmawardhane, S., Rubinstein, B., and Stern, A. I.,** Regulation of transplasmalemma electron transport in oat mesophyll cells by sphingoid bases and blue light, *Plant Physiol.,* 89, 1345, 1989.
39. **Bille, J., Weiser, T., and Bentrup, F.-W.,** The lysolipid sphingosine modulates pyrophosphatase activity in tonoplast vesicles and isolated vacuoles from a heterotrophic cell suspension culture of *Chenopodium rubrum, Physiol. Plant.,* 84, 250, 1992.
40. **Kawai, G., Ohnishi, M., Fujino, Y., and Ikeda, Y.,** Stimulatory effect of certain plant sphingolipids on fruiting of *Schizophyllum commune, J. Biol. Chem.,* 261, 779, 1986.
41. **Kawai, G. and Ikeda, Y.,** Fruiting-inducing activity of cerebrosides observed with *Schizophyllum commune, Biochim. Biophys. Acta,* 719, 612, 1982.
42. **Kawai, G. and Ikeda, Y.,** Structure of biologically active and inactive cerebrosides prepared from *Schizophyllum commune, J. Lipid Res.,* 26, 338, 1985.
43. **Kawai, G. and Ikeda, Y.,** Chemistry and functional moiety of a fruiting-inducing cerebroside in *Schizophyllum commune, Biochim. Biophys. Acta,* 754, 243, 1983.
44. **Merrill, A. H. and Jones, D. D.,** An update of the enzymology and regulation of sphingomyelin metabolism, *Biochim. Biophys. Acta,* 1044, 1, 1990.
45. **Kishimoto, Y.,** Sphingolipid formation, In *The Enzymes, Vol. 16, Lipid Enzymology,* 3rd ed., Boyer, P. D., Ed., Academic Press, New York, 1983, 358.
46. **Radin, N. S.,** Biosynthesis of the sphingoid bases: a provocation, *J. Lipid Res.,* 25, 1536, 1984.
47. **Kanfer, J. N.,** Sphingolipid metabolism, in *Handbook of Lipid Research, Vol. 3, Sphingolipid Biochemistry,* Kanfer, J. N., and Hakomori, S., Eds., Plenum Press, NY, 1983, 167.
48. **Braun, P. E. and Snell, E. E.,** Biosynthesis of sphingolipid bases, *J. Biol. Chem.,* 243, 3775, 1968.
49. **DiMari, S. J., Brady, R. N., and Snell, E. E.,** Biosynthesis of sphingolipid bases. IV. The biosynthetic origin of sphingosine in *Hansenula ciferri, Arch. Biochem. Biophys.,* 143, 553, 1971.
50. **Snell, E. E., DiMari, S. J., and Brady, R. N.,** Biosynthesis of sphingosine and dihydrosphingosine by cell-free systems from *Hansenula ciferri, Chem. Phys. Lipids,* 55, 116, 1970.
51. **Stoffel, W.,** Studies on the biosynthesis and degradation of sphingosine bases, *Chem. Phys. Lipids,* 55, 139, 1970.
52. **Stoffel, W., LeKim, D., and Sticht, G.,** Stereospecificity of the NADPH-dependent reduction of 3-oxodihydrosphingosine (2-amino-hydroxyoctadecane-3-one), *Hoppe-Seyler's Z. Physiol. Chem.,* 349, 1637, 1968.
53. **Kanfer, J. N. and Bates, S.,** Sphingolipid metabolism. II. The biosynthesis of 3-keto-dihydrosphingosine by a partially purified enzyme from rat brain, *Lipids,* 5, 781, 1970.
54. **Braun, P. E., Morell, P., and Radin, N. S.,** Synthesis of C_{18}- and C_{20}-dihydrosphingosines, ketodihydrosphingosines, and ceramides by microsomal preparations from mouse brain, *J. Biol. Chem.,* 245, 335, 1970.
55. **Krisnangkura, K. and Sweeley, C. C.,** Studies on the mechanism of 3-ketosphinganine synthetase, *J. Biol. Chem.,* 251, 1597, 1976.

56. **Lev, M. and Milford, A. F.,** The 3-ketodihydrosphingosine synthetase of *Bacteroides melaninogenicus*: partial purification and properties, *Arch. Biochem. Biophys.*, 212, 424, 1981.

57. **Buede, R., Rinker-Schaffer, C., Pinto, W. J., Lester, R. L., and Dickson, R. C.,** Cloning and characterization of *LCB1*, a *Saccharomyces* gene required for biosynthesis of the long-chain base component of sphingolipids, *J. Bacteriol.*, 173, 4325, 1991.

58. **Merrill, A. H.,** Characterization of serine palmitoyltransferase activity in Chinese hamster ovary cells, *Biochim. Biophys. Acta*, 745, 284, 1983.

59. **Williams, R. D., Wang, E., and Merrill, A. H.,** Enzymology of long-chain base synthesis by liver: characterization of serine palmitoyltransferase in rat liver microsomes, *Arch. Biochem. Biophys.*, 228, 282, 1984.

60. **Merrill, A. H. and Williams, R. D.,** Utilization of different fatty acyl-CoA thioesters by serine palmitoyltransferase from rat brain, *J. Lipid Res.*, 25, 185, 1984.

61. **Merrill, A. J., Nixon, D. W., and Williams, R. D.,** Activities of serine palmitoyltransferase (3-ketosphinganine synthase) in microsomes from different rat tissues, *J. Lipid Res.*, 26, 617, 1985.

62. **Holleran, W. M., Williams, M. L., Gao, W. N., and Elias, P. M.,** Serine-palmitoyltransferase activity in cultured human keratinocytes, *J. Lipid Res.*, 31, 1655, 1990.

63. **Lynch, D. V., Cahoon, E. B., Fairfield, S. R., and Tannishtha,** Glycosphingolipids of plant membranes, in *Plant Lipid Biochemistry, Structure and Utilization*, Quinn, P. J. and Harwood, J. L., Eds., Portland Press, London, 1990, 47.

64. **Lynch, D. V., Fairfield, S. R., Lapan, K. A., and Thomas, K. W.,** Characterization of enzymatic steps in glucosylceramide synthesis, *Plant Physiol.*, 96s, 135, 1991.

65. **Lynch, D. V. and Fairfield, S. R.,** Sphingolipid long-chain base synthesis in plants: characterization of serine palmitoyltransferase activity in squash fruit microsomes, 1993, in preparation.

66. **Sundaram, K. S. and Lev, M.,** Inhibition of sphingolipid synthesis by cycloserine in vitro and in vivo, *J. Neurochem.*, 42, 577, 1984.

67. **Medlock, K. A. and Merrill, A. H.,** Inhibition of serine palmitoyl-transferase in vitro and long-chain base biosynthesis in intact Chinese hamster ovary cells by β-chloroalanine, *Biochemistry*, 27, 7079, 1988.

68. **Merrill, A. H. and Wang, E.,** Biosynthesis of long-chain (sphingoid) bases from serine by LM cells. Evidence for introduction of the 4-trans-double bond after de novo biosynthesis of N-acylsphinganine(s), *J. Biol. Chem.*, 261, 3764, 1986.

69. **Merrill, A. H., Wang, E., and Mullins, R. E.,** Kinetics of long-chain (sphingoid) base biosynthesis in intact LM cells: effects of varying the extracellular concentrations of serine and fatty acid precursors of this pathway, *Biochemistry*, 27, 340, 1988.

70. **Verdery, R. B. and Theolis, R.,** Regulation of sphingomyelin long-chain base synthesis in human fibroblasts in culture. Role of lipoproteins and the low density lipoprotein receptor, *J. Biol. Chem.*, 257, 1412, 1982.

71. **van Echten, G., Birk, R., Brenner-Weis, G., Schmidt, R. R., and Sandhoff, K.,** Modulation of sphingolipid biosynthesis in primary cultured neurons by long chain bases, *J. Biol. Chem.*, 265, 9333, 1990.

72. **Lea, P. J. and Miflin, B. J.,** Transport and metabolism of asparagine and other nitrogen compounds within the plant, in *The Biochemistry of Plants*, Vol. 5, Miflin, B. J., Ed., Academic Press, Orlando, FL, 1980, 569.

73. **Gatt, S.,** Enzymatic hydrolysis and synthesis of ceramides, *J. Biol. Chem.*, 238, 3131, 1963.

74. **Gatt, S.,** Enzymatic hydrolysis of sphingolipids, *J. Biol. Chem.*, 241, 3721, 1966.

75. **Yavin, E. and Gatt, S.,** Enzymatic hydrolysis of sphingolipids VIII. Further purification and properties of rat brain ceramidase, *Biochemistry*, 8, 1692, 1969.

76. **Sribney, M.,** Enzymatic synthesis of ceramide, *Biochim. Biophys. Acta*, 125, 542, 1966.

77. **Morell, P. and Radin, N. S.,** Specificity in ceramide biosynthesis from long chain bases and various fatty acyl coenzyme A's by brain microsomes, *J. Biol. Chem.*, 245, 342, 1970.

78. **Ullman, M. D. and Radin, N. S.,** Enzymatic formation of hydroxy ceramides and comparison with enzymes forming nonhydroxy ceramides, *Arch. Biochem. Biophys.,* 152, 767, 1972.

79. **Akanuma, H. and Kishimoto, Y.,** Synthesis of ceramides and cerebrosides containing both α-hydroxy and nonhydroxy fatty acids from lignoceroyl-CoA by rat brain microsomes, *J. Biol. Chem.,* 254, 1050, 1979.

80. **Singh, I. and Kishimoto, Y.,** A novel synthesis of ceramide from lignoceric acid and sphingosine by rat brain preparation; the amide formation requires a pyridine nucleotide, *Biochem. Biophys. Res. Commun.,* 82, 1287, 1978.

81. **Singh, I.,** Ceramide synthesis from free fatty acids in rat brain: function of NADPH and substrate specificity, *J. Neurochem.,* 40, 1565, 1983.

82. **Mori, M., Shimeno, H., and Kishimoto, Y.,** Synthesis of ceramides and cerebrosides in rat brain: comparison with synthesis of lignoceroyl-coenzyme A, *Neurochem. Int.,* 7, 57, 1985.

83. **Singh, I., Kishimoto, Y., Vunnam, R. R., and Radin, N. S.,** Conversion by rat brain preparation of lignoceric acid to ceramides and cerebrosides containing both α-hydroxy and nonhydroxy fatty acids, *J. Biol. Chem.,* 254, 3840, 1979.

84. **Spence, R. A., Thomas, K. W., and Lynch, D. V.,** Ceramide formation in plant membranes by a reaction utilizing free fatty acid, 1993, in preparation.

85. **Wang, E., Norred, W. P., Bacon, C. W., Riley, R. T., and Merrill, A. H.,** Inhibition of sphingolipid biosynthesis by fumonisins, *J. Biol. Chem.,* 266, 14486, 1991.

86. **Morell, P. and Radin, N. S.,** Synthesis of cerebroside by brain from UDP-galactose and ceramide containing hydroxy fatty acid, *Biochemistry,* 8, 506, 1969.

87. **Brenkert, A. and Radin, N. S.,** Synthesis of galactosyl ceramide and glucosyl ceramide by rat brain: assay procedures and changes with age, *Brain Res.,* 36, 183, 1972.

88. **Shukla, G. S. and Radin N. S.,** Glucosylceramide synthase from mouse kidney: further characterization with an improved assay method, *Arch. Biochem. Biophys.,* 283, 372, 1990.

89. **Cestelli, A., White, F. V., and Constantino-Ceccarini, E.,** Use of liposomes as acceptors for the assay of lipid glycosyltransferases, *Biochim. Biophys. Acta,* 572, 283, 1979.

90. **Radin, N. S.,** Galactosyl ceramide synthetase from brain, *Methods Enzymol.,* 28, 488, 1972.

91. **Futerman, A. H. and Pagano, R. E.,** Determination of the intracellular sites and topology of glucosylceramide synthesis in rat liver, *Biochem. J.,* 280, 295, 1991.

92. **Durieux, I., Martel, M. B., and Got, R.,** Solubilization of UDPglucose-ceramide glucosyltransferase from the golgi apparatus, *Biochem. Biophys. Acta,* 1024, 263, 1990.

93. **Ullman, P., Bouvier-Nave, P., and Benveniste, P.,** Regulation by phospholipids and kinetic studies of plant membrane-bound UDP-glucose: sterol β-D-glucosyltransferase, *Plant Physiol.,* 85, 51, 1987.

94. **Cleland, W. W. and Kennedy, E. P.,** The enzymatic synthesis of psychosine, *J. Biol. Chem.,* 235, 45, 1960.

95. **Constantino-Ceccarini, E. and Suzuki, K.,** Isolation and partial characterization of an endogenous inhibitor of ceramide glucosyltransferases from rat brain, *J. Biol. Chem.,* 253, 340, 1978.

96. **Fujino, Y. and Nakano, M.,** Enzymatic conversion of labeled ketodihydrosphingosine to ketosphingosine in rat liver particulates, *Biochim. Biophys. Acta,* 239, 273, 1971.

97. **Hammond, R. K. and Sweeley, C. C.,** Biosynthesis of unsaturated sphingolipid bases by microsomal preparations from oysters, *J. Biol. Chem.,* 248, 632, 1973.

98. **Ong, O. E. and Brady, R. N.,** In vivo studies on the introduction of the 4-t-double bond of the sphingenine moiety of rat brain ceramides, *J. Biol. Chem.,* 248, 3884, 1973.

99. **Crossman, M. W. and Hirschberg, C. B.,** Biosynthesis of phytosphingosine by the rat, *J. Biol. Chem.,* 252, 5815, 1977.

100. **Crossman, M. W. and Hirschberg, C. B.,** Biosynthesis of 4D-hydroxysphinganine by the rat. En bloc incorporation of the sphinganine carbon backbone, *Biochim. Biophys. Acta,* 795, 411, 1984.

101. **Kulmacz, R. J. and Schroepfer, G. J., Jr.,** Sphingolipid base metabolism. Concerning the origin of the oxygen atom at carbon atom 4 of phytosphingosine, *J. Am. Chem. Soc.,* 100, 3963, 1978.
102. **Hoshi, M. and Kishimoto, Y.,** Synthesis of cerebronic acid from lignoceric acid by rat brain preparation. Some properties and distribution of the α-hydroxylation system, *J. Biol. Chem.,* 248, 4123, 1973.
103. **Murad, S., Strycharz, G. D., and Kishimoto, Y.,** α-Hydroxylation of lignoceric and nervonic acids in the brain, *J. Biol. Chem.,* 251, 5237, 1976.
104. **Murad, S. and Kishimoto, Y.,** α-Hydroxylation of fatty acid in brain. Effects of cerebroside components on the synthesis of cerebronic acid, *Biochim. Biophys. Acta,* 488, 102, 1977.
105. **Singh, I. and Kishimoto, Y.,** α-Hydroxylation of lignoceric acid in brain. Subcellular localization of α-hydroxylation and the requirement for heat-stable and heat-labile factors and sphingosine, *J. Biol. Chem.,* 254, 7698, 1979.
106. **Uda, M., Singh, I., and Kishimoto, Y.,** Glutamate formed from lignoceric acid by rat brain preparations in the presence of pyridine nucleotide and cytosolic factors: a brain-specific oxidation of very long chain fatty acids, *Biochemistry,* 20, 1295, 1981.
107. **Kaya, K., Ramesha, C. S., and Thompson, G. A.,** Temperature-induced changes in the hydroxy and non-hydroxy fatty acid-containing sphingolipids abundant in the surface membrane of *Tetrahymena pyrifirmis* NT-1, *J. Lipid Res.,* 25, 68, 1984.
108. **Kaya, K., Ramesha, C. S., and Thompson, G. A., Jr.,** On the formation of α-hydroxy fatty acids. Evidence for a direct hydroxylation of nonhydroxy fatty acid containing sphingolipids, *J. Biol. Chem.,* 259, 3548, 1984.
109. **Hitchcock, C. and James, A. T.,** The mechanism of α-oxidation in leaves, *Biochim. Biophys. Acta,* 116, 413, 1966.
110. **Shine, W. E. and Stumpf, P. K.,** Fat metabolism in higher plants. Recent studies on plant α-oxidation systems, *Arch. Biochem. Biophys.,* 162, 147, 1974.

Chapter 10

LIPID TRANSPORT IN PLANTS

Jean-Claude Kader

TABLE OF CONTENTS

0-8493-4907-9/93/$0.00+$.50
© 1993 by CRC Press Inc.

I. INTRODUCTION

The need for an intracellular transport of lipids in plants was suggested from various studies devoted to lipid metabolism. One example is given by the biosynthesis of a phospholipid, phosphatidylcholine (PC), which is a ubiquitous component of higher plant membranes. The synthesis of this lipid is mainly, or exclusively, located within endoplasmic reticulum and Golgi membranes.[1,2] A movement of this phospholipid is thus necessary to take into account its presence in mitochondrial membranes or other membranes lacking the PC-synthesizing enzymes.[3] *In vivo* experiments, involving a chase after incubation with labeled acetate, showed that a transfer between a microsomal fraction, rich in endoplasmic reticulum and mitochondria, is indeed operative.[4] Another example is given by the biosynthesis of galactolipids in chloroplasts, since it has been suggested that PC, synthesized in the endoplasmic reticulum, is transferred to chloroplasts where it is transformed into diacylglycerol (DAG) for galactolipid biosynthesis (see Chapter 7).[2,5] A third case is provided by the study of very long-chain fatty acids (VLCFA), which are actively synthesized by endoplasmic reticulum and not by plasma membranes, although they are important components of these membranes. An intracellular transport of these components is thus necessary.[6] These examples are sufficient to show that it is important to consider not only the enzymes involved in plant lipid metabolism, but also the intermembrane movement of lipids.

Although the main steps of lipid metabolism are now well-established, as shown in the other chapters of this book, the mechanisms of lipid transport are nowadays a matter of controversy, since different hypotheses are proposed, including first, transport of lipid monomers within the cytosolic phase; second, transport mediated by carrier proteins; and third, vesicular lipid transfer linked to membrane flow.[7,8] The aim of this review is to focus on the mechanisms which have been considered in higher plants.

II. VESICULAR LIPID TRANSFER

The involvement of a vesicular lipid transfer has been suggested particularly for the biogenesis of plasma membranes. The theory of the endomembrane flow, proposed by Morré,[9] involved the participation of a pathway linking the endoplasmic reticulum, the Golgi apparatus, and the plasma membrane for the intracellular trafficking of proteins and lipids. This pathway has been studied in plants by following *in vivo* and *in vitro* approaches.

A. *IN VIVO* STUDY OF INTRACELLULAR LIPID FLUX

The group of Cassagne has followed, in leek epidermal cells, the intracellular route of VLCFA having more than 18 carbons.[10-12] In leek cells, VLCFA are synthesized in endoplasmic reticulum and, to a lesser extent, in the Golgi apparatus. By contrast, plasma membrane, purified by a two-phase partition method, is unable to synthesize these lipids, although VLCFA are the most characteristic lipids of this membrane (up to 20% of the total fatty acids). It was suggested that the VLCFA are transferred to the plasma membrane. Intermembrane movements of VLCFA were studied *in vivo* by performing pulse-chase experiments followed by membrane fractionation of the microsomal pellets.[10] It was shown that when microsomal fractions, prepared from 7-day-old leek seedlings, labeled from [1-^{14}C]acetate, were incubated up to 2 h in labeled acetate-free medium and then fractionated by linear sucrose gradient centrifugation, changes in the specific radioactivities of the fractions were observed during the chase. The radioactivity of a light-density fraction (collected at a density of 1.08 g/cm^3) decreased, whereas an increase of the radioactivity was observed in a heavy fraction (corresponding probably to the plasma membrane). This result provided the first argument in favor of an *in vivo* transfer between endoplasmic reticulum and plasma membrane.

However, it remained to purify the putative plasma membrane fraction. This was recently done by a combination of sucrose gradient centrifugation and phase partitioning, leading to a pure plasma membrane which was identified both by electron microscopy and marker enzyme activities.[12] The characterization of these fractions revealed interesting features; the acyl elongase activities, responsible for the synthesis of VCLFA, are associated with the endoplasmic reticulum, the Golgi apparatus, and a light fraction corresponding to a density of 1.08 g/cm^3 which are involved in lipid transfer. No elongase activity was detected in the purified plasma membrane. This successful membrane fractionation allowed a study of the intracellular traffic of VLCFA by performing a chase of the membranes obtained from labeled leek seedlings. In the beginning of the experiments, the endoplasmic reticulum was the most labeled fraction; then after a chase of 30 min, the Golgi fraction exhibited the highest radioactivity (Figure 1). After 1 h of chase, the major part of the radioactive lipids was recovered in the plasma membrane. The transferred lipids were essentially the neutral lipids; the phospholipids [PC and PE (phosphatidylethanolamine)]; and the fatty acids, palmitic acid (C_{16}), stearic acid (C_{18}), and VLCFA. All of these observations are consistent with a transfer of VLCFA through membrane flow, from endoplamic reticulum to plasma membrane via the Golgi apparatus. These results are also in agreement with a vesicular lipid transfer as suggested by the observation of a lag phase of 30 min at the beginning of the chase.

In order to confirm the involvement of the Golgi apparatus in this process, monensin, a carboxylic ionophore which is known to disturb the intracellular transport of proteins at the level of the Golgi apparatus, was used. Monensin has been used in animal cells to establish the involvement of Golgi in the intracellular flux of proteins.[13] After a careful study of the uptake and the internal concentration of monensin, the effect of the drug on lipid biosynthesis was examined.[11] At a concentration lower than 10 mM and for an incubation time up to 30 min, monensin has no effect on the global incorporation of acetate in lipid. However, when the effect of monensin on the distribution of labeled

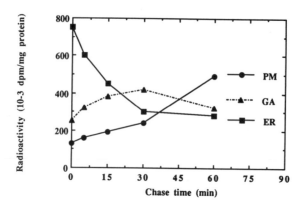

FIGURE 1. *In vivo* intermembrane movements of lipids in leek cells. Seven-day-old leek seedlings were incubated for 2 h with labeled acetate, then a chase in acetate-free medium was performed. The various membrane fractions (ER — endoplasmic reticulum; GA — Golgi; — PM — plasma membrane) were separated by centrifugation of the microsomal pellets on linear sucrose gradient. (From Bertho, P., et al., *Biochim. Biophys. Acta*, 1070, 127–134, 1991. With permission.)

lipids among membranes was studied, an accumulation of the radioactivity was noted in the Golgi-rich faction collected at a density of 1.13 g/cm^3. This result was confirmed recently by the characterization of membranes. The accumulation of newly synthesized lipids was noted in the Golgi, whereas plasma membranes contained less labeled lipids. When the lipids were analyzed, it was found that the changes concerned phospholipids and neutral lipids. An important result was found when the fatty acids were analyzed; in the plasma membrane, the decrease occurred much more in VLCFA (–67%) than the C_{16} and C_{18} fatty acids (–5 to 21%) (Figure 2).

These observations are of importance for several reasons. First, they indicate that monensin blocks preferentially the flux of VLCFA-containing phospholipids, suggesting a sorting process, allowing VLCFA to follow the endoplasmic reticulum → Golgi → plasma membrane pathway. It remains to be understood how this sorting process is controlled. In animal cells, it has been shown that the polar heads of phospholipids, lipid–lipid interaction, and lipid–protein interactions could be involved in sorting of lipids.[14] The data collected for VLCFA transfer are the first to suggest that a sorting of lipid in the intracellular flow could be controlled by the acyl moities of the lipids. The study of mechanisms of such acyl sorting is of high interest. It remains to be shown if transition vesicles derived from endoplasmic reticulum could be involved, as shown in animal cells, in the trafficking of lipids and proteins.[15] Second, the information given by the experiments with monensin led to the conclusion that, in addition to vesicular flux, other pathways are operative, since the *in vivo* movement of the lipid containing long-chain C_{16} and C_{18} is not affected by the drug. These data are thus in agreement with the hypothesis of the involvement of protein carriers in the intracellular lipid delivery.

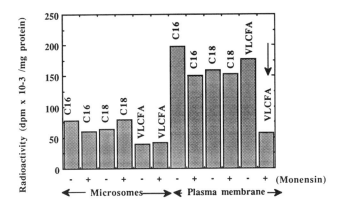

FIGURE 2. Effect of monensin on labeling of fatty acids of microsomes and plasma membrane of leek seedlings. Seedlings were preincubated with (+) or without (−) monensin, then incubated with labeled acetate for 30 min. The membrane fractions were then separated and their fatty acids were analyzed by radio gas liquid chromatography. A monensin-induced decrease was noted in VLCFA of the plasma membranes (arrow). (From Bertho, P., et al., *Biochim. Biophys. Acta*, 1070, 127–134, 1991. With permission.)

B. *IN VITRO* STUDY OF THE INTRACELLULAR LIPID FLUX

In addition to the *in vivo* approach, cell-free transfer of lipids have been studied in spinach[16] and in soybean.[17] These methods were based on the incubation of various membranes (Golgi, endoplasmic reticulum, mitochondria, tonoplast, chloroplast), some of them being used as donor labeled membranes, and others as unlabeled receptor membranes. The receptor membranes are immobilized on nitrocellulose strips; it is thus easy to recover the receptor membranes, bound on these strips, after incubation with radioactive donor membranes. This assay provides a convenient method for studying the effect of various combinations of membranes on the transfer of lipids, as well as the effect of ATP or temperature, which were shown to influence the transfer efficiency. ATP-dependent transfer was observed when endoplasmic reticulum is used as donor and Golgi as receptor membranes in the presence of cytosol. By contrast, the transfer between Golgi and plasma membrane was tempera- ture-, but not ATP-dependent. Only the combinations of membranes involved in vesicular transfer *in vivo* (endoplasmic reticulum/Golgi/plasma membrane/ nuclear membrane) allow a significant cell-free transfer of lipids *in vitro*. Again, a sorting of lipid seems operating, since phospholipids, but not triacylglycerols, for example, are transferred.

The same *in vitro* approach has been followed by the group of Sandelius,[17] which studied the intermembrane movement of phosphatidylinositol (PI) between labeled endoplasmic reticulum and unlabeled plasma membrane immobilized on nitrocellulose strips. A transfer of PI was observed followed by its phosphorylation to phosphatidylinositol-monophosphate and phospha- tidylinositol *bis*-phosphate.

All these cell-free reconstitution experiments are of interest for studying the mechanisms of intermembrane flux. However, the introduction of cytosolic proteins[16] or pure maize lipid transfer proteins[84] seems to enhance the *in vitro* transfers. These data are thus supportive of the involvement of protein mediators in lipid delivery.

III. LIPID TRANSFER PROTEINS

The membrane flux, although highly plausible in the endoplasmic reticulum → Golgi → plasma membrane pathway, seems inadequate for explaining the biogenesis of mitochondrial and plastidial membranes. It is why an alternative hypothesis assuming the involvement of protein shuttles carrying newly synthesized lipids from one membrane to another one has been proposed based on the finding of such proteins in animal tissues.[18]

A. DISCOVERY OF LIPID TRANSFER PROTEINS

This hypothesis has been reinforced by the finding, 17 years ago, that water-soluble proteins from potato tuber are able to enhance intermembrane phospholipid movements.[19] In these experiments, potato tuber homogenates, obtained after elimination of membranes by centrifugation and lowering the pH of the supernatant, were incubated with microsomal fractions (recovered by centrifugation at $100,000 \times g$ for 1 h) containing radioactive phospholipids in the presence of unlabeled mitchondria. It was found that mitochondria collected after incubation in the presence of tuber homogenate contained much more radioactive phospholipids than in the control experiments. It was then shown that the stimulating effect is due to a group of low molecular mass proteins (around 10 to 20 kDa). This first observation was then extended to several other plant organs including seeds,[20] and the improvement of purification procedures led to the preparation of pure proteins which were successively called phospholipid exchange proteins (PLEP), then phospholipid transfer proteins (PLTP), and, now, lipid transfer proteins (LTPs) or nonspecific lipid transfer proteins (nsLTPs).

B. HOW IS LIPID TRANSFER ACTIVITY ASSAYED?

Assaying LTPs is quite laborious and makes their monitoring in the purification steps somewhat difficult. However, improvements in these assays were recently obtained by using spectrophotometric techniques.

1. Discontinuous Methods

The basic method, followed in the first LTP discovery, is to incubate two types of membranes, one of which contained a radioactive lipid, and to separate them by centrifugation, such as incubation of microsomal fractions with mitochondria, for example.[19] However, one major problem is the fact that it is necessary to check for the absence of cross-contamination of a membrane fraction by another one. It is essential to determine if the enhancing effect of

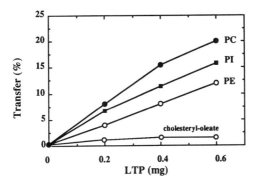

FIGURE 3. Lipid transfer assay. The activity of spinach LTP was determined by the liposome-mitochondria assay. Liposomes, made from ^{3}H-PC, ^{3}H-PI, ^{3}H-PE, and containing ^{14}C-cholesteryl oleate were incubated with mitochondria in the presence of increasing amounts of pure spinach LTP. The three phospholipids are transferred at various rates, whereas the transfer of cholesteryl-oleate is negligible.

LTP is due to a movement of lipids rather than to an increased sticking of labeled membrane vesicles on unlabeled mitochondria. This was accomplished by assays involving liposomes made from radioactive lipids and mitochondria (or other membranes). For example, these liposomes contained the phospholipid to be studied (^{3}H-labeled) and a lipid chosen as nontransferable by the protein, like ^{14}C-cholesteryl oleate (Figure 3). The transfer observed in the absence of LTP was low (up to 2% of the initial liposomal ^{3}H-radioactivity), whereas up to 30% transfer was found when LTP was present. In conclusion, the liposome-mitochondria assay clearly established that LTPs are acting by facilitating individual lipid movements. This type of assay which allows the study of transfer of different lipids, depending on which are radioactive, greatly helped the study of LTPs.

2. Continuous Assays

In order to facilitate the study of LTP, other types of assays which involve neither the preparation of mitochondria nor the separation of membranes after incubation have been recently designed. Liposomes were prepared from fluorescent phospholipids containing a fluorophore (NBD (nitrobenzoxadiazol) or pyrene).[21] These liposomes are quenched due to their high concentration or to the presence of a quencher. No fluorescence emission was detected in the absence of LTPs and liposomes containing normal lipids. When LTP is added, a fluorescence is observed due to the insertion of phospholipids into unlabeled membranes. This type of assay allowed a continuous monitoring of the LTP activity and was used to study the LTP from maize[21] and wheat.[22] Other types of assays were based on monitoring of the transfer of spin-labeled phospholipids[21] or galactolipid[23] by electron paramagnetic resonance. It is predicted that these types of assays will develop in the future. However,

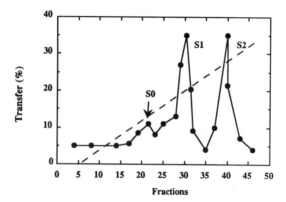

FIGURE 4. Purification of plant lipid transfer protein isoforms. Partially purified proteins from sunflower seedlings were separated on a carboxymethyl-Sepharose chromatography eluted by a phosphate gradient (- - -). Three peaks of PC-transfer activity (-●-) were detected, corresponding to isoforms of LTP.

difficulty is encountered due to the limited number of fluorescent or spin-labeled lipids which are commercially available.

C. SEVERAL LTP ISOFORMS, MAINLY BASIC, HAVE BEEN ISOLATED

The purification methods used for plant LTP purifications have been based on the assumption that these proteins are cytosolic and remain soluble at moderately low pH. For this reason, the supernatants obtained after centrifugation of mitochondria were treated at low pH to eliminate residual proteins and membranes. Remaining proteins were then precipitated from membrane-free supernatant by ammonium sulfate (saturated[19] or after an intermediate step at 40% to eliminate inactive protein[22]). The dialyzed proteins were then submitted to various chromatographic separations (gel filtration, cationic exchange columns, fast protein liquid chromatography [FPLC], and HPLC) (Figure 4). The purifications were facilitated by the fact that the plant LTPs are mainly basic, abundant (in general 20 to 100 mg of pure LTP from 500 g of seeds), and stable several months at +4°C. These properties greatly helped the preparation of highly purified LTPs from various plants.[24-29]

LTPs were purified from different plant organs as indicated in Table 1. In addition, LTPs have been detected, but not yet purified in tobacco, pea, and *Arabidopsis*.

It is remarkable that the molecular mass of the main isoform of LTPs, exhibiting the highest specific activity, is in the range 9 to 10 kDa as determined by sodium dodecyl sulfate (SDS)-electrophoresis. However, the migration pattern of LTPs depends on their oxido-reducing state.[85] This could explain the higher values (20 kDa) determined for potato tuber or maize in the first determinations.[19,24] However, it is important to note that several other peaks

TABLE 1
Properties of Lipid Transfer Proteins from Higher Plants

Protein	Source	Apparent M_r (kDa)	Number of residues	Calculated M_r (Da)	Isoelectric point	Substrate specificity	Ref.
LTP	Maize	9	93	9054	8.8	PC, PI, PE	25,44
LTP	Spinach	8.8	91	8833	9.3	PC, PI, PE	26,45,44
LTP-A	Castor bean	7	92	9313	High	PC, PE, MGDG	28
LTP-B	Castor bean	8	92	9593	High	PC, PE,MGDG	28
LTP-C	Castor bean	9	92	9847	High	PC, PE, MGDG	28
LTP-D	Castor bean	9	93	9326	High	PC, PE, MGDG	28
LTP	Wheat	9	90	9607	10	Only PC tested	22
LTP	Sunflower	9	—	—	9	PC, PI, PE	29
LTP-PAPI	Barley	10	92	—	9	Only PC tested	36
LTP-PAPI	Rice	9	91	8909	High	?	34
FABP	Oat	8.7	—	—	8.4	Acyl binding	39
GL-TP	Spinach chloroplast	11	—	—	High	MGDG	23
Acidic LTPs	Castor bean	11.1–69.2	—	—	5.4–6.6	PC or PI	30

of activity were detected in the last purification steps. These peaks could correspond to various isoforms differing in binding various lipids or other ligands. In connection with this observation, it is to be noted that the group of Yamada has detected and purified from castor bean seedlings several isoforms (nsLTP-A, -B, -C, -D) having M_r of 7, 8, and 9 kDa.[28] The four isoforms of castor bean LTP have the same specificity, since they transfer PC, PE, and monogalactosyldiacylglycerol (MGDG).

All plant LTPs purified until now are basic. However, partially purified and less characterized LTPs, having a low pI and specific for different phospholipids, have been obtained from castor bean seedlings.[30] Acidic LTPs have been also detected in maize seeds and in spinach leaves,[24,26] but have not yet been purified.

Plant LTPs are mainly nonspecific for phospholipids. When studied in liposome-mitochondria assays, they facilitate the transfer of PC, PE, or PI. This is why they are called nsLTP like the similar proteins present in mammalian tissues.[18] In addition to phospholipid transfer, LTPs from spinach leaves or castor bean are also able to transfer galactolipids, but not neutral lipids.[23,28] In connection with this observation, Nishida and Yamada[23] found that spinach leaf stroma contains a galactolipid transfer protein (GL-TP), able to transfer spin-labeled MGDG and having a M_r of about 28 kDa. In contrast, no PC-transfer activity was detected in the stroma of spinach chloroplasts.[31] It was not clearly established if plant LTPs are able to transfer sterols; the cholesterol ester is not transferred as indicated by liposome-mitochondria assays.

An additional complicating observation was recently resolved. Bernhard and Somerville[32] found sequence homologies between plant LTPs and a category of proteins called PAPI (putative amylase/protease inhibitors) purified from barley and rice,[33,34] but exhibiting no true inhibiting activity. These proteins have been designated PAPI, since they exhibit sequence homologies with a true amylase/protease inhibitor from indian finger millet.[35] Recent experiments by Breu et al.[36] established that in fact barley PAPI behaves like an LTP by transferring *in vitro* PC between membranes. The same also has been observed for rice LTP.[86]

D. LTPs BIND FATTY ACIDS

An unexpected observation on the specificity of LTPs was made several years ago when Spener's group was searching in plants for fatty acid binding proteins (FABPs) and acyl coenzyme A (CoA) binding proteins (ACBPs) similar to those studied in animal tissues.[37-39] A protein, purified from *Avena* seedlings, exhibited properties (molecular mass, pI) similar to those of plant LTPs.[39] It was found that an LTP from spinach was indeed able to bind several fatty acids and acyl-CoA esters,[40] whereas the *Avena* FABP was also able to transfer phospholipids.[87] This observation of a fatty acid binding property was recently extended to other plant LTPs, including those of castor bean,[41] sunflower,[29] and rapeseed.[88] In particular, castor bean nsLTP binds oleic acid as well as oleoyl-CoA, and the binding of each component is inhibited by the other, suggesting that the binding site is the same for the two compounds.[41] Moreover, castor bean nsLTP enhances the acyl-CoA oxidase activity by glyoxysomes. This important observation suggests a participation of nsLTP in fatty acid β-oxidation.[41]

Together, these data led to the concept that plant LTPs are bifunctional or multifunctional proteins.

E. STRUCTURAL PROPERTIES OF LTPs

The isolation of large amounts of LTPs allowed the determination of the complete amino acid sequence of these proteins in several plants.[42-44] These sequences were then confirmed, or corrected in the case of spinach LTP,[45] by the isolation and characterization of cDNAs. The comparisons of the sequences revealed remarkable similarities. First, the total number of amino acids is conserved (91 to 95); second, the number of cysteine residues is constant, and their positions are conserved; and third, positive charges are present in agreement with the high isoelectric point. The positive charges present in the central part of the LTP suggest that this domain of the protein might be an interactive site for the negatively charged polar head of the phospholipid.

It was suggested that the cysteine residues play an important role in the function of LTPs. In the case of castor bean LTP,[46] four disulfide bridges have been found between Cys 4/52, Cys 14/29, Cys 30/75, and Cys 50/89. In fact, the oxido-reducing conditions have a major effect on the activity of plant LTP, since when maize LTP is placed in high concentrations of dithiothreitol, the activity is suppressed.[85]

Other plant proteins such as thionins also contain high concentrations of cysteine.[47] It would be interesting to find common structural and functional features. Comparisons by alignment and a recently developed method, hydrophobic cluster analysis, have made possible the comparison between different domains of related proteins.[22,48] By these methods, it appeared that (1) plant LTPs possess homologous domains and (2) the presence or absence of a hydrophilic or hydrophobic domain in some proteins can be responsible for changes in their activity. In the case of a hydrophobic protein from soybean, with no functional activity determined until now, the polar domains of plant LTPs seem to be absent.[49]

It was tempting to deduce, from sequence determination, the secondary and tertiary structures of plant LTPs. Based on various computer programs, a model for plant LTPs was proposed, suggesting that the lipid molecule is embedded in a cavity with the polar moiety interacting with the central part of the LTP, whereas the acyl chains interact with two hydrophobic domains represented by several β-strands[44] (Figure 5). This model has been recently corrected by Madrid and von Wettstein[50] to take into account the determination in castor bean LTP of the position of disulfide bonds between the cysteines. According to this model, three amphiphilic regions consisting of two α-helices

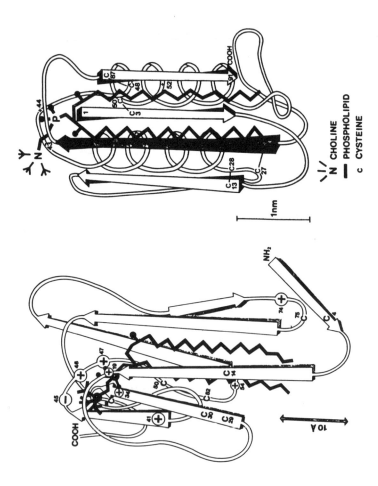

N CHOLINE
■ PHOSPHOLIPID
c CYSTEINE

FIGURE 5. Models of plant lipid transfer proteins. These models, based on the primary structures of several plant LTPs, have been proposed by Tchang et al.[44] (at left) and by Madrid and von Wettstein[50] (at right). These models differ by the higher proportion of helical structures and by the insertion of disulfide bridges in the latter model.

and one β-strand are present in LTPs. The amphiphilic structures could constitute a hydrophobic pocket for the binding of the lipid, while the surface of the LTP is hydrophilic, making the LTP-lipid complex soluble (Figure 5). Recently, structural studies have been performed on wheat LTP by using homonuclear two- and three-dimensional nuclear magnetic resonance (NMR) techniques.[22,51] Except for the C-terminal portion, the polypeptide chain appears to be organized mainly as helical fragments connected by disulfide bridges. The conformation of wheat LTP has also been studied by Raman and Fourier transform infrared spectroscopy.[22] In contrast to the predicted structures based on computer modeling, suggesting that LTP are primarily composed of β-sheets, infrared results showed that the wheat LTP contains 41% α-helix and 19% β-sheet (and 40% undefined or turns). By Raman spectroscopy, it was shown that among the four S-S, three adopt a *gauche-gauche-gauche* conformation, while the fourth one exhibits a *gauche-gauche-trans* conformation. When the disulfide bonds are cleaved by reducing agents, the conformation of the protein changes; the α-helix content decreases (from 25 to 40%), whereas the β-structures increase to 32%. Interestingly, the binding of a lyso-PC increases the α-helix content, suggesting that the binding of a lipid stabilizes the amphipathic helices, whereas the reduction of S-S bonds affects the stability of the N-terminal helix. This work suggests that the helical structure is essential for the lipid transfer activity. The length of the hydrophobic part of the amphipatic helix (23 Å) is sufficient to accommodate a long lipid hydrocarbon chain of about 20 Å.

These novel observations need to be completed by crystallographic studies.

F. MODE OF ACTION OF LTPs

These models can help to understand how LTPs can facilitate lipid movement between membranes. It is clear from the use of nontransferable tracer, like cholesteryl-oleate, that LTPs do not provoke a significant fusion of the acceptor and donor membranes. It was suggested several years ago that LTP can act by forming a reversible complex with a phospholipid molecule extracted from a membrane as recently confirmed by spectrofluorometric methods.[89] This phospholipid is then exchanged with a molecule originating from another membrane. This leads to an overall exchange process (Figure 6). As an example, when liposomes containing labeled PC were incubated with chloroplast envelope vesicles in the presence of spinach LTP, the specific radioactivities of PC decreased in the donor fraction and increased in the envelope membranes until an equilibrium was reached. As a consequence of this exchange, the fatty acid compositions of both membranes were modified due to the fact that new phospholipid molecules originating from liposomes were introduced into envelopes, with the reverse also being observed. This observation is of great interest, since it shows that LTPs could be used as "tools" to manipulate the lipid composition of membranes and, then, to modify some of their functions controlled by the lipid environment.

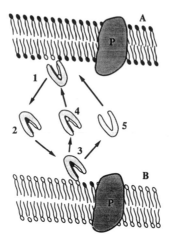

FIGURE 6. Scheme suggesting a mode of action of phospholipid transfer proteins from plants. A reversible protein-phospholipid complex is formed (1, 2). This complex interacts with membrane A and exchanges its phospholipid molecule with those of membrane B. This gives an overall exchange process (1, 4), implicating the outer layer of the membrane. This process can be controlled by the surface charges of the membranes, by the ionic composition of the medium, or by the nature of the transported lipid. The protein could also be released from the membrane without any lipid bound (5), leading to a net transfer process. The membrane proteins can also interfere with the exchange process.

Alternatively, the LTP can extract a lipid molecule from a membrane, carry it to another membrane, and leave this membrane without a bound lipid (Figure 6). This leads to a net transfer, i.e., an increase of lipids in acceptor membrane by incorporation of lipids provided by a donor membrane. This was observed with chloroplast envelope membranes incubated with liposomes and LTP, since an increase of 35% of the total amount of PC in envelope membrane was noted.[52]

However, one intriguing point is that in contrast to specific LTP from animal tissues, no one-for-one complex between lipid and LTP was found, with the ratio between LTP and bound lipid being very weak (around 1%). The same results were found for animal nsLTP, leading Nichols[53] to suggest that lipids can be transferred between membranes and LTP by a collision-dependent process; LTP must bind lipids from the membrane surface and carry them into the water phase. In the case of plant LTPs, their mechanism of action is still mysterious and has to be extensively studied by combining structural and functional approaches.

G. IMMUNOCHEMISTRY OF LTPs

The availability of specific polyclonal antibodies raised against maize[54] or castor bean[55] LTPs allowed several approaches for the study of the biosynthesis and the cellular localization of these proteins. In addition, other antibodies are available, either directed against barley LTP (PAPI)[33] or against a fusion protein comprising spinach LTP.[45]

The use of immunoblots led to establishing the presence of several cross-reacting isoforms in maize and castor bean seeds. A 7-kDa isoform was detected in maize,[56] whereas four isoforms (7, 8, and 9 kDa) were detected at various concentrations in the various parts of castor bean seedlings.[28] Immunochemical methods allowed determination of the changes in LTP levels during maturation and germination of seeds. These changes have been followed by Western blot or ELISA in maize[57] and by Western blot in castor bean,[55] in parallel with the determination of LTP activity. The augmentations in LTP levels were coherent with increases in lipid transfer activity.[55] Increases in LTP levels were observed during germination of both plants in the aerial parts of the seedlings and most intensely in the organs where membrane biogenesis is active. Maize endosperm, for example, which exhibits a low biosynthetic activity, contains low levels of LTP. Tsuboi et al.[55] found that the levels and the activity of basic LTPs are correlated with the growth of endosperm, whereas the changes in acidic LTPs are not correlated. The changes in the levels of the isoforms were specific for the different organs, e.g., in the cotyledons, the major one was the 9-kDa isoform, whereas the most abundant isoform in the axis was the 7-kDa form.

In castor bean, the biosynthesis of the isoforms was followed by immunoprecipitating the products of translation of poly (A$^+$)RNAs. Several immunoprecipitated products were detected, corresponding to precursors of LTP which were 3.5 to 4 kDa larger than the mature nsLTP. The levels of translatable mRNAs increased with the augmentations of the levels of nsLTP, indicating that LTPs are synthesized *de novo* by mRNAs formed during seed germination.

Immunochemical approaches were also useful in determining the cellular localization of LTPs. Plant LTPs have been isolated as water-soluble proteins and are generally considered to be cytosolic proteins. However, when immunoblots were carried out on carefully purified maize mitochondria, LTP was detected,[56] suggesting that maize LTPs may be partially membrane bound. This hypothesis was validated in maize by the use of immunocytochemical techniques at the electron microscope level, which revealed that the gold particles corresponding to LTP were located partly within the cytoplasm and partly associated with different membranes like endoplasmic reticulum and plasma membrane.[90] In the case of castor bean cotyledons, immunogold particles were located in the glyoxysomes.[4] However, a surprising observation was made in maize coleoptile,[58] castor bean,[41] carrot cells,[59] and recently in *Arabidopsis*;[91] some (or the major part in the case of *Arabidopsis*) of the LTP seem associated with the cell wall, suggesting an extracellular location for a portion of plant LTPs.

In connection with this unexpected location of LTP, it was found, in observations made at the photonic levels, that LTP is mainly located in the

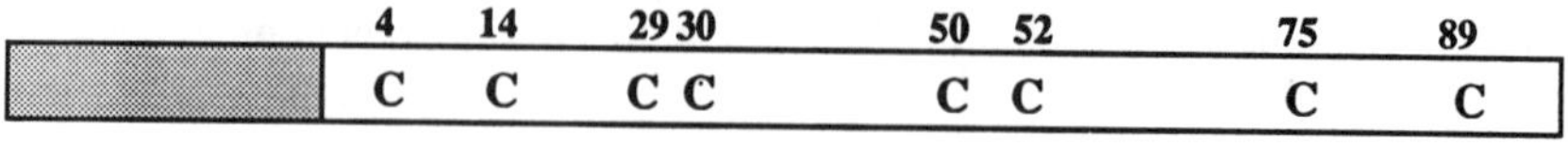

	4	14	29 30	50 52	75	89
	C	C	C C	C C	C	C

signal peptide number of residues	mature peptide number of residues	
27	93	maize cDNA (ref 44)
26	91	spinach cDNA (ref 45)
26	94	carrot cDNA (ref 59)
25	92	barley cDNA (ref 33)
24	92	castor bean C cDNA (ref 28)
21	92	castor bean C1 cDNA (ref 64)
23	92	castor bean C2 cDNA (ref 64)
24	95	tomato cDNA (ref 62)
23	91	tomato cDNA (ref 63)

FIGURE 7. Deduced sequences from cDNAs coding for plant lipid transfer proteins. The positions of the conserved cysteines are indicated in the mature protein.

superficial outer layer of maize coleoptile.[58] Recent data obtained in carrot[59] and in vine[92] established that LTPs are also secreted in the growth medium of somatic embryos. The physiological significance of these findings remains to be elucidated.

H. MOLECULAR BIOLOGY OF LTPs

The approaches of molecular biology have only recently been used to study plant LTPs.

1. Isolation and Characterization of cDNAs and Genes

The first cDNA coding for a plant LTP was obtained 4 years ago.[44] This cDNA, characterized as full length, was isolated by screening a cDNA library, prepared from mRNAs of maize coleoptile, with an antibody against maize LTP. The cDNA is 822 basepairs (bp) long with an open reading frame of 360 bp. The amino acid sequence deduced from the nucleotide sequence of this cDNA was identical to the amino acid sequence determined on the maize LTP after protease hydrolysis and microsequencing, except that an extra peptide of 27 amino acids was present at the N-terminal end (Figure 7).

The isolation of cDNA coding for a maize LTP allowed several investigations on the molecular biology of plant LTPs. Since this finding, a number of cDNAs coding for other plant LTPs have been isolated. Starting from the amino acid sequence determined for spinach, Bernhard et al.[45] isolated a cDNA clone encoding the spinach LTP by probing a library with synthetic oligonucleotides based on the amino acid sequence of the purified protein. This cDNA clone is full length and contains an open reading frame of 354 bp. An extra peptide of 26 amino acids was deduced from the cDNA sequence which is comprised of 900 bp.

Interestingly, a 25-amino acid leader sequence has been also deduced from a cDNA coding for the PAPI/LTP from barley.[33]

In addition to these three cDNAs, others have been recently identified by homology with maize LTP cDNA. This includes a partial cDNA from maize corresponding to an mRNA different from the first one,[60] a cDNA isolated from low temperature-treated barley,[61] a cDNA corresponding to a gene induced by drought stress in tomato,[62] and a cDNA corresponding to an LTP gene highly expressed in the stem of salt-treated tomato plants.[63] In castor bean, a cDNA clone, corresponding to the nsLTP-C, has been isolated and characterized,[28] whereas two cDNAs, one identical to the nsLTP-C and the other differing in 3 of the 92 amino acids of the predicted sequence, were recently identified.[64] Another cDNA corresponding to an LTP from tobacco has been used as a probe for studying the gene expression in tobacco flowers.[65] Other cDNAs coding for LTPs from rice, sorghum, or *Arabidopsis* have been characterized, but not yet published.[93]

The first isolation of genomic clones coding for barley[66,67] and tomato[62] LTPs have been very recently obtained. The deduced amino acid sequences are in perfect agreement with the amino acid sequences of purified proteins, as well as with the characterization of cDNAs.

2. Number of Genes

The isolation of cDNAs and genes provided probes to determine the number of LTP genes by Southern analysis. This method performed with genomic DNA, isolated from maize or spinach tissues, and cut by several restriction enzymes revealed only one major signal, suggesting the presence of only one gene for LTPs in maize or spinach genomes.[44,45] In barley, only one single LTP gene and one or two other genes with weak homology seem present.[66]

By using one of the six enzymes used to cut spinach DNA, two bands were detected suggesting the presence of at least one intron.[45] In the case of maize, the isolation of the two cDNAs differing in their coding region by a 74-bp insertion at the end of the coding region of one of them allowed the demonstration that LTP maize gene has at least one intron corresponding to the 74-bp insert.[60] The two mRNA species are assumed to arise from an alternatively spliced mRNA. This is of particular interest for the presence of isoforms of plant LTPs. In barley genes, the open reading frame is interrupted by one 133-bp intron.[66,67]

3. Gene Expression

Using cDNAs as probes, several laboratories have studied the expression of genes coding for plant LTPs. This expression was found to be organ specific and time regulated. In addition, a surprising cell specificity was noted in some plants. By probing RNA from various tissues of maize seedlings[44] or spinach plants[45] with corresponding cDNAs, it was found that the LTP mRNAs were accumulated in leaf or petiole of spinach or in endosperm and coleoptiles of maize. In contrast, only weak signals were detected in maize or spinach roots. This low abundance of LTP mRNA in roots raises the question of the presence and involvement of LTP in the physiology of this organ. The case of barley LTP is of interest, since the LTP genes are highly — or perhaps exclusively

— expressed in the aleurone layer cells.[66-68] The aleurone-specific expression was confirmed by following the expression of an LTP/GUS gene, delivered by the biolistic process, in developing barley seed tissues.[66] A high expression was detected histochemically in the aleurone layer. Analysis of the promoter region led to the identification of several putative regulatory motifs that could be linked to the control of LTP transcription in aleurone cells. In particular, a *cis*-acting sequence element, the *myb* protein recognition site, was detected.[67] An organ-specific gene expression was also noted for castor bean LTP. The gene coding for nsLTP-C is highly expressed in the cotyledons and weakly in the axis and endosperms,[28] whereas the cDNA corresponding to nsLTP-C mRNA was found neither in hypocotyl, roots, or endosperms of seedlings nor in leaves of castor bean plants; this mRNA was only found in the cotyledons.[64]

The LTP gene expression is time regulated. The levels of maize LTP mRNAs vary during the maturation and the germination of maize seeds.[58] An accumulation of LTP mRNA was noted in embryos and endosperms during seed maturation. After germination, the level of LTP mRNA in the coleoptile increased, whereas this level remained low in the scutellum and negligible in roots. These variations are coherent with a role of LTP in membrane biogenesis and lipid biosynthesis during seed maturation and germination.

In addition, a surprising cell specificity of the LTP gene expression was recently found by *in situ* hybridization performed in maize[58] and carrot.[59] The highest levels of LTP mRNAs were detected in the peripheral layers (epidermis layers) and vascular bundles of maize coleoptiles and embryos.[58] In carrot, the expression of the LTP gene was found in protoderm cells of somatic and zygotic embryos, as well as in epidermal cells of leaves and flower organs, whereas in maturing seeds the LTP gene is expressed in the outer epidermis of integument and in the pericarp epidermis.[59] As indicated earlier, the barley LTP gene is highly expressed in the aleurone layer.[66,67] In tobacco, some anther-specific mRNAs correspond to LTP and were found to be specifically expressed in tapetal cells.[65]

This peripheral localization of LTP gene expression is in agreement with several observations. First, immunogold methods showed a similar localization of the major LTP isoform; second, a signal peptide has been deduced from the isolation of several LTP cDNAs; and third, a secretion was indeed demonstrated in carrot cells,[59] in rice cells,[69] and in vine.[92]

4. Role of the Signal Peptide

The presence of an N-terminal leader peptide has been deduced from the analysis of several LTP cDNAs or genes; the length of the deduced sequence varies from 21 to 27 aminoacids (21, 23, and 24 for castor bean nsLTP-C;[28,64] 23 and 24 for tomato LTPs;[62,63] 25 for barley;[33] 26 for spinach[45] and carrot;[59] and 27 for maize[44]) (Figure 7). These findings strongly suggest that plant LTPs are synthesized as precursors containing the mature LTP and an *N*-terminal signal peptide which is cleaved after preprotein processing. The role of this *N*-terminal sequence was tested in spinach LTP by transcribing and translating the cDNA *in vitro* under conditions which lead to processing of signal peptide by the secretory pathway.[45] When transcripts from the spinach LTP cDNA were translated without addition of microsomes, the translation product was of a size expected from the open reading frame. By contrast, when microsomes are present during translation, a translation product is of the size expected for the mature LTP accumulated. In addition, the translation products were sensitive to proteinase K in the absence of microsomes and not in the presence of membranes, which indicates that the mature protein is inserted into the endoplasmic reticulum lumen and then becomes protected against the protease attack. This observation is consistent with the fact that plant LTPs (like maize LTP) are synthesized on membrane-bound polysomes[70] and indicates a translational insertion into lumen membranes. In similar studies, Madrid[71] showed that barley LTP is targeted into the lumen of the endoplasmic reticulum and has studied by radiosequencing the initiation site for pre-LTP translation. This correlates with the secretion of the barley PAPI/LTP from aleurone layers into incubation medium.[33] By contrast, no importation of castor bean LTP into microsomal vesicles was observed.[41] However, in agreement with the localization within glyoxysomes, when castor bean ^{35}S-pre-nsLTP, synthesized from nsLTP-C-specific mRNA, was incubated in the presence of castor bean glyoxysomes and proteinase K, a peptide corresponding to the mature nsLTP-C was observed.[41] This experiment clearly shows an import of an LTP into an organelle (glyoxysome). It is of interest to remember that an LTP seems to be present inside chloropasts.[23] Its insertion into this organelle remains to be studied.

I. POSTULATED ROLES OF LTPs

Based on the *in vitro* properties of LTPs, several *in vivo* roles for these proteins have been postulated. However, no direct evidence in favor of any of these roles has been provided until now.

1. Role in Membrane Biogenesis and Renewal

Since LTPs have been isolated as cytosolic proteins able to facilitate phospholipids between membranes, it has logically been suggested that plant LTPs are involved in transporting newly synthesized phospholipids from the sites of synthesis to membranes unable to actively synthesize their phospholipids like mitochondria or plasma membrane[7,8,72] (Figure 8). In favor of this hypothesis is the observation that the expression of genes coding for LTP, as well as the levels of the proteins or mRNAs coding for LTPs, vary in relation with membrane biogenesis and growth of the plants.

However, a major difficulty is the recent demonstration that the major LTP isoform does not behave as a housekeeping protein, since the gene coding for this isoform is expressed in a cell-specific manner and is not expressed in roots, for instance. Another difficulty is linked to the finding that in connection with the presence of a signal peptide the major LTP isoform seems to be located outside the cell, in the cell wall, for example, and is secreted into the growth medium of somatic embryos.[59] This unexpected extracellular location makes

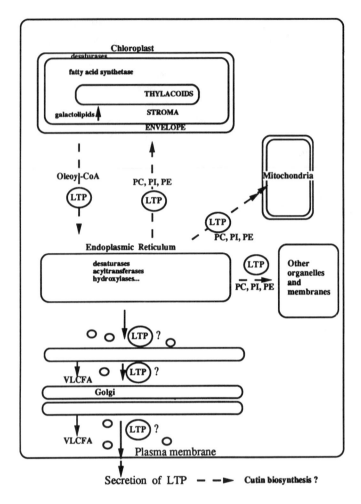

FIGURE 8. Postulated roles of lipid transfer proteins in membrane biogenesis and fatty acid metabolism. It is suggested that LTPs participate in membrane biogenesis and renewal of various membranes. Their ability to bind acyl-CoA esters makes LTPs candidates for the regulation of the intracellular pools of these compounds. In addition, the involvement of LTP in the secretory process through Golgi or to the building of cutin by secreted LTPs is indicated.

a direct role of the LTP unlikely. However, Madrid and von Wettstein[50] suggested that the LTP can act as a transporter of phospholipids inside the lumen of endoplasmic reticulum and then to the plasma membrane via the Golgi.

One possibility to preserve the validity of the hypothesis of LTP involvement in intracellular events is to suggest the existence of several LTP isoforms having different functions. The presence of several isoforms has been described.[7,8,72] Moreover, acidic and specific plant LTPs have been less explored than the basic ones. It is of interest to note that the first demonstration of an essential role for a PLTP in the cell was provided by studies in yeasts concerning an acidic lipid transfer protein which is specific for PI and PC.[73] The presence of such an LTP (called sec 14) remains to be demonstrated in higher plants.

Another reservation concerns the transfer of phospholipids into the internal membranes of mitochondria or chloroplasts. Since LTPs are assumed to act with the outer surface of the outer membranes of mitochondria or chloroplasts, it is not known how lipids are transported internally. Transmembrane movements of phospholipids could help provide such transport.[74] Another question concerns the intrachloroplastidial transfer of galactolipids which could be helped by LTPs from the stroma.[23]

2. Role in Phosphatidylinositol Cycle

The active transfer of PI by LTPs led to the hypothesis that they can participate to the functioning of the polyphosphoinositides cycle in plant cells. It is accepted that in response to various stimuli like hormone signals, phosphorylation and degradation of PI occurs, leading to the formation of second messengers (affecting the intracellular calcium level) like the inositol trisphosphate, via the synthesis of phosphatidylinsitol monophosphate and phosphatidylinositol bisphosphate.[17] The replacement of the hydrolyzed PI molecules seems necessary for the functioning of this cycle. Since PI is synthesized in the endoplasmic reticulum,[1,75] a transfer of PI to plasma membrane is needed and could be mediated by LTPs. Recent arguments in favor of this hypothesis have been provided by experiments incubating endoplasmic reticulum and plasma membranes; PI transported by maize LTP is metabolized into its phosphorylated derivatives[84.]

3. Role in Galactolipid and Fatty Acid Biosynthesis

One postulated role of LTPs, based on their *in vitro* multiple functions (ability to transfer phospholipids and galactolipids and to bind fatty acids and CoA esters), concerns the cooperative pathway (called "eukaryotic") operative for the biosynthesis of galactolipids containing linolenic acid[2,5] (Figure 8). The participation of LTPs in the cooperative pathway for galactolipid biosynthesis is suggested by the observation of an active transfer of PC toward chloroplasts can be mediated by LTP.[52] Moreover, reconstitution experiments have been performed by Ohnishi and Yamada[76] who incubated dilinoleoyl-PC liposomes with oat plastids and castor bean LTP and observed a synthesis of dilinolenoyl-MGDG. Oursel et al.[77] incubated spinach chloroplasts with PC liposomes in the presence of spinach LTP, then treated PC molecules transported to the chloroplasts by phospholipase C in order to form diacylglycerol; a synthesis of MGDG by galactosylation of diacylglycerol was observed. All these experiments provided some arguments in favor of the participation of LTPs in the cooperative "eukaryotic" scheme. LTPs could play a major role in regulating the intracellular level of oleoyl-CoA exported from plastids. In some experiments made by incubating together spinach chloroplasts, microsomes, and spinach LTP in the presence of [1-^{14}C]acetate, an increase in fatty acid labeling has been observed.[78]

LTPs in plant cells could play several roles related to the modulation of the intracellular levels of free fatty acids and oleoyl-CoA. It was recently shown that nsLTP from castor bean is able to activate glyoxysomal acyl oxidase.[41] This provides a strong argument in favor of the participation of these proteins in β-oxidation. It should be noted that the gene expression of other proteins, playing important roles in the accumulation of fatty acids in seeds like acyl carrier proteins[79] or associated with lipid storage bodies like oleosins,[80,81] have been studied during the maturation of seeds in relation with the accumulation of storage lipids. The developmental time course of LTP mRNA presents similar features with that of acyl carrier protein, for example.[79]

The secretion of LTPs, although making their roles in the intracellular lipid metabolism unlikely, could have an important physiological significance. The group of de Vries suggested that LTP secreted from carrot cells could be involved in the transport of cutin monomers through the extracellular matrix to sites of cutin synthesis.[59] Further evidence is necessary to confirm this interesting hypothesis.

4. Roles of LTPs in Reactions of Plants to Environmental Stress

Recent and unexpected observations shed light on the effects of environmental changes upon LTP gene expression. The group of Pintor-Toro has isolated and characterized, in tomato, a cDNA clone homologous to maize LTP by differential screening from a cDNA library synthesized from poly (A$^+$) RNAs isolated from seedlings treated with NaCl.[63] This LTP gene is highly expressed in the stem of plants treated by high concentrations of NaCl, whereas the expression is very weak in leaves and absent in roots. In tomato seedlings, the expression level is highly stimulated by salt and by a heat shock treatment at 37°C. Treatment by a hormone (abscisic acid) also enhances the LTP gene expression. In the same plant, the group of Bray has isolated a gene induced by drought stress, regulated by abscisic acid, and specifically expressed in the aerial photosynthetic tissues.[62] The gene is highly expressed in drought-stressed leaves, stems, and petioles. No signal is seen in roots or in nonstressed stems, petioles, and roots. The same result is noted, but to a lesser extent, when plants are treated by cold temperatures. Interestingly, a cDNA has been isolated from a library produced from shoot meristematic tissue mRNAs extracted from low temperature-treated plants of barley.[61] The expression of this gene is enhanced by both a low temperature (6°C) treatment and drought treatment, suggesting a relation with a dehydrative stress linked either to cold or to drought. By contrast, in the case of barley PAPI/LTP, no effect of abscisic acid has been found.[33]

All these observations are consistent with an involvement of LTPs in reactions of plants against external stress, possibly by participating in changes in the endomembrane system. In addition, the secretion of LTPs could also be involved. Masuta et al.[69] observed that an LTP-like protein is secreted in the incubation medium of cell suspension cultures of rice in the presence of salicylic acid, known to elicit pathogen defense processes. This result provides information about a possible participation of LTPs in defense reactions against pathogens.

IV. FUTURE DIRECTIONS

It is clear that what is missing now concerning LTPs is a direct demonstration of their *in vivo* role. Starting from a simple — simplistic? — idea of cytosolic proteins able to transport newly formed phospholipids from their sites of synthesis to sites of accumulation (membranes unable to form their own lipid constituents), extensive studies about these proteins have led to the discovery of several unexpected properties — secretion, cell specificity and organ specificity of their localization, broad specificity — which considerably complicated the first ideas about these proteins. However, their highly conserved structure in plants (all plants studied until now contain LTP, which have roughly the same structure), their ability to interact with various lipids, and their possible involvement in several developmental and environmental processes recently reinforced the interest of the studies of LTPs.

A direct demonstration of the *in vivo* role of plant LTPs could be provided by the strategy of using antisense RNA. This approach is being used for *Arabidopsis* LTP.[91] However, the presence of several LTP isoforms could complicate the results provided by such methods. Another possibility is to isolate a mutant deficient in LTP. Such a genetic approach, which has been successfully followed in *Saccharomyces cerevisiae*, could be adapted to *Arabidopsis*.[82] However, so far, no such mutant has been isolated.

In addition to these efforts to elucidate the *in vivo* role, studies are necessary on the regulation of the expression of LTP genes and the analysis of the promoters and regulatory elements acting on this expression. The characterization of other genes, coding, for example, for acidic LTPs, may reveal novel functional properties.

Further studies will consider the structure and the function of plant LTPs and, in particular, the regulation of the *in vitro* activity of these proteins in relation to the properties of the membranes or the composition of the medium. A careful examination of the structure, in particular by crystallography, could provide novel views of the mode of action of these fascinating proteins.

In parallel, studies performed *in vivo* should be continued in oder to reconcile the two main hypotheses of function, protein carriers, and membrane flow. The hypothesis of Madrid[71] that the LTPs carry phospholipids within the endoplasmic reticulum lumen and Golgi is one first possibility to reconcile both hypotheses. Another possibility, based on the recent discovery of an involvement of LTP in the secretory process in yeasts by affecting the phospholipid balance of the Golgi membranes,[83] should also be explored in the future.

ACKNOWLEDGMENTS

The author thanks Mitsuhiro Yamada (Japan), Friedrich Spener (Germany), Chris Somerville (U.S.A.), Anna Stina Sandelius (Sweden), Pedro Puigdomenech (Spain), Michel Delseny (France), and Claude Cassagne (France) for helpful discussions and communication of unpublished informations.

REFERENCES

1. **Moore, T. S.,** Regulation of phospholipid headgroup composition in castor bean endosperm, in *The Metabolism, Structure and Function of Plant Lipids*, Stumpf, P. K., Mudd, J. B., and Nes, W. D., Eds., Plenum Press, New York, 1987, 265–272.
2. **Browse, J. and Somerville, C.,** Glycerolipid synthesis: biochemistry and regulation, *Annu. Rev. Plant Physiol. Plant Mol. Biol.,* 42, 467–506, 1991.
3. **Kader, J. C., Douady, D., and Mazliak, P.,** Phospholipid transfer proteins, in *Phospholipids, A Comprehensive Treatise,* Hawthorne, J. N. and Ansell, G. B., Eds., Elsevier, Amsterdam, 1982, 279–311.
4. **Abdelkader, A. B. and Mazliak, P.,** Echanges de lipides entre mitochondries, microsomes et surnageant cytoplasmique de cellules de pomme de terre ou de chou-fleur, *Eur. J. Biochem.,* 15, 250–262, 1970.
5. **Harwood, J. L.,** Fatty acid metabolism, *Annu. Rev. Plant Physiol. Plant Mol. Biol.,* 39, 101–138, 1988.
6. **Moreau, P., Bertho, P., Juguelin, H., and Lessire, R.,** Intracellular transport of very long chain fatty acids in etiolated leek seedlings, *Plant Physiol. Biochem.,* 26, 173–178, 1988.
7. **Kader, J. C.,** Intracellular transfer of phospholipids, galactolipids and fatty acids in plant cells, in *Subcellular Biochemistry,* Vol. 16, Hilderson, H. J., Ed., Plenum Press, New York, 1990, 69–111.
8. **Arondel, V. and Kader, J. C.,** Lipid transfer in plants, *Experientia,* 46, 579–585, 1990.
9. **Morré, D. J.,** Membrane biogenesis, *Annu. Rev. Plant Physiol.,* 26, 441–481, 1975.
10. **Moreau, P., Juguelin, H., Lessire, R., and Cassagne, C.,** Intermembrane transfer of long chain fatty acid synthesized by etiolated leek seedlings, *Phytochemistry,* 25, 387–391, 1986.
11. **Bertho, P., Moreau, P., Juguelin, H., Gautier, M., and Cassagne C.,** Monensin-induced accumulation of neosynthesized lipids and fatty acids in a Golgi fraction prepared from etiolated leek seedlings, *Biochim. Biophys. Acta,* 978, 91–96, 1989.
12. **Bertho, P., Moreau, P., Morré, D. J., and Cassagne, C.,** Monensin blocks the transfer of very long chain fatty acid containing lipids to the plasma membrane of leek seedlings. Evidence for lipid sorting based on fatty acyl chain length, *Biochim. Biophys. Acta,* 1070, 127–134, 1991.
13. **Farqhar, M. G.,** Membrane biogenesis, *Annu. Rev. Cell Biol.,* 1, 447–488, 1985.
14. **Van Meer, G.,** Biosynthetic lipid traffic in animal eukaryotes, *Annu. Rev. Cell Biol.,* 5, 247–275, 1989.
15. **Moreau, P., Rodriguez, M., Cassagne, C., Morré, D. M., and Morré, D. J.,** Trafficking of lipids from the endoplasmic reticulum to the Golgi apparatus in a cell-free system from rat liver, *J. Biol. Chem.,* 266, 4322–4328, 1991.
16. **Morré, D. J., Penel, C., Morré, D. M., Sandelius, A. S., Moreau, P., and Andersson, B.,** Cell-free transfer and sorting of membrane lipids in spinach, *Protoplasma,* 160, 49–64, 1991.

17. **Harryson, P., Morré, D. J., and Sandelius, A. S.,** In vitro transfer of phosphatidylinositol from an endoplasmic reticulum-enriched fraction to a plasma membrane fraction, isolated from dark-grown soybean, in *Plant Lipid Biochemistry, Structure and Utilization*, Quinn, P. J. and Harwood, J. L., Eds., Portland Press, London, 1990, 260–262.

18. **Wirtz, K. W. A.,** Phospholipid transfer proteins, *Biochim. Biophys. Acta*, 60, 73–99, 1991.

19. **Kader, J. C.,** Proteins and the intracellular exchange of lipids. I. Stimulation of phospholipid exchange between mitochondria and microsomal fractions by proteins isolated from potato tuber, *Biochim. Biophys. Acta*, 380, 31–44, 1975.

20. **Yamada, M., Tanaka, T., Kader, J. C., and Mazliak, P.,** Transfer of phospholipids from microsomes to mitochondria in germinating castor bean endosperm, *Plant Cell Physiol.*, 19, 173–176, 1978.

21. **Geldwerth, D., de Kermel, A., Zachowski, A., Guerbette, F., Kader, J. C., Henry, J. P., and Devaux, P. F.,** Use of spin-labeled and fluorescent lipids to study the activity of the phospholipid transfer protein from maize seedlings, *Biochim. Biophys. Acta*, 1082, 255–264, 1991.

22. **Désormeaux, A., Blochet, J. E., Pézolet, M., and Marion, D.,** Amino acid sequence of a non-specific wheat phospholipid transfer protein and its conformation as revealed by infrared and Raman spectroscopy. Role of disulfide bridges and phospholipids in the stabilization of the α-helix structure, *Biochim. Biophys. Acta*, 1121, 137–152, 1992.

23. **Nishida, I. and Yamada, M.,** Semisynthesis of a spin-labeled monogalactosyl-diacylglycerol and its application in the assay for galactolipid transfer activity in spinach leaves, *Biochim. Biophys. Acta*, 813, 298–306,1986.

24. **Douady, D., Grosbois, M., Guerbette, F., and Kader, J. C.,** Purification of a basic phospholipid transfer protein from maize seedlings, *Biochim. Biophys. Acta*, 710, 143–153, 1982.

25. **Douady, D., Guerbette, F., and Kader, J. C.,** Purification of phospholipid transfer protein from maize seeds using a two-step chromatographic procedure, *Physiol. Veg.*, 23, 373–380, 1984.

26. **Kader, J. C., Julienne, M., and Vergnolle, C.,** Purification and characterization of a spinach leaf protein capable of transferring phospholipids from liposomes to mitochondria or chloroplasts, *Eur. J. Biochem.*, 139, 411–416, 1984.

27. **Watanabe, S. and Yamada, M.,** Purification and characterization of a non-specific lipid transfer protein from germinated castor bean endosperm which transfers phospholipids and galactolipids, *Biochim. Biophys. Acta*, 876, 116–123, 1986.

28. **Tsuboi, S., Suga, T., Takishima, K., Mamiya, G., Matsui, K., Ozeki, Y., and Yamada, M.,** Organ-specific occurrence and expression of the isoforms of nonspecific lipid transfer protein in castor bean seedlings and molecular cloning of a full-length cDNA for a cotyledon-specific isoform, *J. Biochem.* (Tokyo), 110, 823–831, 1991.

29. **Arondel, V., Vergnolle, C., Tchang, F., and Kader, J. C.,** Bifunctional lipid-transfer: fatty acid-binding proteins in plants, *Mol. Cell. Biochem.*, 98, 49–56, 1990.

30. **Tanaka, T. and Yamada, M.,** Phospholipid exchange proteins from castor bean seeds, *Plant Cell Physiol.*, 20, 533–542, 1979.

31. **Schwitzguebel, J. P. and Siegenthaler, P. A.,** Evidence for a lack of phospholipid transfer protein in the stroma of Spinach chloroplasts, *Plant Sci.*, 40, 167–171, 1985.

32. **Bernhard, W. R. and Somerville, C.,** Coidentity of putative amylase inhibitors from barley and finger millet with phospholipid transfer proteins inferred from amino acid sequence homology, *Arch. Biochem. Biophys.*, 269, 695–697, 1989.

33. **Mundy, J. and Rogers, J. C.,** Selective expression of a probable amylase/protease inhibitor in barley aleurone cells: comparison to the barley amylase/subtilisin inhibitor, *Planta*, 169, 51–63, 1986.

34. **Yu, Y. G., Chung, C. H., Fowler, A., and Suh, S. W.,** Amino acid sequence of a probable amylase/protease inhibitor from rice seeds, *Arch. Biochem. Biophys.*, 265, 466–475, 1988.

35. **Campos, F. A. P. and Richardson, M.**, The complete amino acid sequence of the α amylase inhibitor 1-2 from seeds of ragi (Indian finger millet, *Eleusine coracana* Gaertn.), *FEBS. Lett.*, 167, 221–225, 1984.

36. **Breu, V., Guerbette, F., Kader, J. C., Kannangara, C. G., Svensson, B., and von Wettstein-Knowles, P.**, A 10 kD barley basic protein transfer phosphatidylcholine from liposomes to mitochondria, *Carlsberg Res. Commun.*, 54, 81–84, 1989.

37. **Veerkamp, J. H., Peeters, R. A., and Maatman, R. G. H. J.**, Structural and functional features of different types of cytoplasmic fatty-acid binding proteins, *Biochim. Biophys. Acta*, 1081, 1–24, 1991.

38. **Knudsen, J.**, Acyl-CoA-binding protein (ACBP) and its relation to fatty-acid binding protein (FABP): an overview, *Mol. Cell. Biochem.*, 98, 217–224, 1990.

39. **Rickers, J., Spener, F., and Kader, J. C.**, A phospholipid transfer protein that binds fatty acids, *FEBS Lett.*, 180, 29–32, 1985.

40. **Rickers, J., Tober, I., and Spener, F.**, Purification and binding characteristics of a basic fatty acid binding protein from *Avena sativa* seedlings, *Biochim. Biophys. Acta*, 784, 313–319, 1984.

41. **Tsuboi, S., Suga, T., Takishima, K., Mamiya, G., Matsui, K., Ozeki, Y., and Yamada, M.**, Nonspecific lipid transfer protein in castor bean cotyledon cells. Subcellular localization and a possible role in β-oxidation, *J. Biochem.*, 1993, in press.

42. **Takishima, K., Watanabe, S., Yamada, M., and Mamiya, G.**, The amino-acid sequence of the nonspecific lipid transfer protein from germinated castor bean endosperms, *Biochim. Biophys. Acta*, 870, 248–255, 1986.

43. **Bouillon, P., Drischel, C., Vergnolle, C., Duranton, H., and Kader, J. C.**, The primary structure of spinach leaf phospholipid-transfer protein, *Eur. J. Biochem.*, 166, 387–391, 1987.

44. **Tchang, F., This, P., Stiefel, V., Arondel, V., Morch, M. D., Pages, M., Puigdomenech, P., Grellet, F., Delseny, M., Bouillon, P., Huet, J. C., Guerbette, F., Beauvais-Canté, F., Duranton, H., Pernollet, J. C., and Kader, J. C.**, Phospholipid transfer protein: full length cDNA and amino acid sequence in maize. Amino acid sequence homologies between plant phospholipid transfer proteins, *J. Biol. Chem.*, 263, 16489–16855, 1988.

45. **Bernhard, W. R., Thoma, S., Botella, J., and Somerville, C. R.**, Isolation of a cDNA clone for spinach lipid transfer protein and evidence that the protein is synthesized by the secretory pathway, *Plant Physiol.*, 95, 164–170, 1991.

46. **Takishima, K., Watanabe, S., Yamada, M., Suga, T., and Mamiya, G.**, Amino acid sequences of two nonspecific lipid-transfer proteins from germinated castor bean, *Eur. J. Biochem.*, 177, 241–249, 1988.

47. **Bohlmann, H. and Apel, K.**, Thionins, *Annu. Rev. Plant Physiol. Plant Mol. Biol.*, 42, 227–240, 1991.

48. **Henrissat, B., Popineau, Y., and Kader, J. C.**, Hydrophobic-cluster analysis of plant protein sequences, *Biochem. J.*, 255, 901–905, 1988.

49. **Odani, S., Koide, T., Ono, T., Seto, Y., and Tanaka, T.**, Soybean hydrophobic protein. Isolation, partial characterization and the complete primary structure, *Eur. J. Biochem.*, 162, 485–491, 1987.

50. **Madrid, S. M. and von Wettstein, D.**, Reconciling contradictory notions on lipid transfer proteins in higher plants, *Plant Physiol. Biochem.*, 29, 705–711, 1991.

51. **Simorre, J. P., Caille, A., Marion, D., Marion, D., and Ptak, M.**, Two- and three-dimensional ^{1}H NMR studies of a wheat phospholipid transfer protein: sequential resonance assignments and secondary structure, *Biochemistry*, 30, 11600–11608, 1991.

52. **Miquel, M., Block, M. A., Joyard, J., Dorne, A. J., Dubacq, J. P., Kader, J. C., and Douce, R.**, Protein mediated transfer of phosphatidylcholine from liposomes to spinach chloroplast envelope membranes, *Biochim. Biophys. Acta*, 937, 219–228,1987.

53. **Nichols, J. W.**, Kinetics of fluorescent-labeled phosphatidylcholine transfer between nonspecific lipid transfer protein and phospholipid vesicles, *Biochemistry*, 27, 1889–1896, 1988.

54. **Grosbois, M., Guerbette, F., Douady, D., and Kader, J. C.,** Enzyme immunoassay of a plant phospholipid transfer protein, *Biochim. Biophys. Acta,* 917, 162–168, 1987.

55. **Tsuboi, S., Watanabe, S. I., Ozeki, Y., and Yamada, M.,** Biosynthesis of nonpecific lipid transfer proteins in germinating castor bean seeds, *Plant Physiol.,* 90, 841–845, 1989.

56. **Douady, D., Grosbois, M., Guerbette, F., and Kader, J. C.,** Phospholipid transfer protein from maize seedlings is partly membrane-bound, *Plant Sci.,* 45, 151–156, 1986.

57. **Grosbois, M., Guerbette, F., and Kader, J. C.,** Changes in level and activity of phospholipid transfer protein during maturation and germination of maize seeds, *Plant Physiol.,* 90, 1560–1564, 1989.

58. **Sossountzov, L., Ruiz-Avial, L., Vignols, F., Jolliot, A., Arondel, V., Tchang, F., Grosbois, M., Guerbette, F., Miginiac, E., Delseny, M., Puigdomenech, P., and Kader, J. C.,** Spatial and temporal expression of a maize lipid tranfer protein gene, *Plant Cell,* 3, 923–933, 1991.

59. **Sterk, P., Booij, H., Scheleekens, G. A., Van Kammen, A., and de Vries, S. C.,** Cell-specific expression of the carrot EP2 lipid transfer protein gene, *Plant Cell,* 3, 907–921, 1991.

60. **Arondel, V., Tchang, F., Baillet, B., Vignols, F., Grellet, F., Delseny, M., Kader, J. C., and Puigdomenech, P.,** Multiple mRNA coding for phospholipid-transfer protein from *Zea mays* arise from alternative splicing, *Gene,* 99, 133–136, 1991.

61. **Dunn, M. A., Hughes, M. A., Zhang, L., Pearce, R. S., Quigley, A. S., and Jack, P. L.,** Nucleotide sequence and molecular analysis of the low temperature induced cereal gene, BLT4, *Mol. Gen. Genet.,* 229, 389–394, 1991.

62. **Plant, A. L., Cohen, A., Moses, M. S., and Bray, E. A.,** Nucleotide sequence and spatial expression pattern of a drought- and absissic acid-induced gene of tomato, *Plant Physiol.,* 97, 900–906, 1991.

63. **Torres-Schumann, S., Godoy, J. A., and Pintor-Toro, J. A.,** A probable lipid transfer protein gene is induced by NaCl in stems of tomato plants, *Plant Mol. Biol.,* 18, 749–757, 1992.

64. **Weig, A. and Komor, E.,** The lipid-transfer protein C of *Ricinus communis* L.: isolation of two cDNA sequences which are strongly and exclusively expressed in cotyledons after germination, *Planta,* 187, 367–371, 1992.

65. **Koltunow, A. M., Truettner, J., Cox, K. H., Wallroth, M., and Goldberg, R. B.,** Different temporal and spatial gene expression patterns occur during anther development, *Plant Cell,* 2, 1201–1224, 1990.

66. **Skriver, K., Leah, R., Müller-Uri, F., Olsen, F. L., and Mundy, J.,** Structure and expression of the barley lipid transfer protein gene Ltp1, *Plant Mol. Biol.,* 18, 585–589, 1992.

67. **Linnestad, C., Lönneborg, A., Kalla, E., and Alsen, O. A.,** Promoter of a lipid transfer protein gene expressed in barley aleurone cells contains similar myb and myc recognition sites as the maize Bz-Mcc allele, *Plant Physiol.,* 97, 841–843,1991.

68. **Olsen, O. A, Jkobsen, K. S., and Schmelzer, E.,** Development of barley aleurone cells: temporal and spatial patterns of accumulation of cell-specific mRNAs, *Planta,* 181, 462–466, 1990.

69. **Masuta, C., Van den Bulcke, M., Bauw, G., Van Montagu, M., and Caplan, A. B.,** Differential effects of elicitors on the viability of rice suspension cells, *Plant Physiol.,* 97, 619–629, 1991.

70. **Vergnolle, C., Arondel, V., Grosbois, M., Guerbette, F., Jolliot, A., and Kader, J. C.,** Synthesis of phospholipid transfer proteins from maize seedlings, *Biochem. Biophys. Res. Commun.,* 157, 37–41, 1988.

71. **Madrid, S. M.,** The barley lipid transfer protein is targeted into the lumen of the endoplasmic reticulum, *Plant Physiol. Biochem.,* 29, 695–703, 1991.

72. **Yamada, M.,** Lipid transfer proteins in plants and microorganisms, *Plant Cell Physiol.,* 33, 1–6, 1992.

73. **Bankaitis, V. A., Aitken, J. R., Cleves, A. E., and Dowhan, W.,** An essential role for a phospholipid transfer protien in yeast Golgi function, *Nature*, 347, 561–562, 1990.

74. **Zachowski, A. and Devaux, P. F.,** Transmembrane movements of lipids, *Experientia*, 46, 644–655, 1990.

75. **Moore, T. S., Jr.,** Enzymes of phospholipid synthesis, in *Methods in Plant Biochemistry*, Vol. 3, Lea, P. J., Ed., Academic Press, New York, 1990, 229–240.

76. **Ohnishi, J. L. and Yamada, M.,** Glycerolipid synthesis in *Avena* leaves during greening of etiolated seedlings, *Plant Cell Physiol.*, 23, 767–773, 1982.

77. **Oursel, A., Escoffier, A., Kader, J. C., Dubacq, J. P., and Trémolières, A.,** Last step in the cooperative pathway for galactolipid synthesis in spinach leaves: formation of monogalactosyldiacylglycerol with C18 polyunsaturated acyl groups at both carbon atoms of the glycerol, *FEBS Lett.*, 219, 393–399, 1987.

78. **Dubacq, J. P., Drapier, D., Trémolières, A., and Kader, J. C.,** Role of phospholipid transfer protein in the exchange of phospholipids between microsomes and chloroplasts, *Plant Cell Physiol.*, 25, 1197–1200, 1984.

79. **Hannapel, D. J. and Ohlrogge, J. B.,** Regulation of acyl carrier protein messenger RNA levels during seed and leaf development, *Plant Physiol.*, 86, 1174–1178, 1988.

80. **Qu, R. and Huang, A. H. C.,** Oleosin KD18 on the surface of oil bodies in maize. Genomic and cDNA sequences and the deduced protein structure, *J. Biol. Chem.*, 265, 2238–2243, 1990.

81. **Murphy, D. J., Keen, J. N., O'Sullivan, J. N., Au, D. M. Y., Edwards, E. W., Jackson, P. J., Cummins, I., Gibbons, T., Shaw, C. H., and Ryan, A. J.,** A class of amphipathic proteins associated with lipid storage bodies in plants. Possible similarities with animal serum lipoproteins, *Biochim. Biophys. Acta*, 1088, 86–94, 1991.

82. **Somerville, C. and Browse, J.,** Plant lipids: metabolism, mutants, and membranes, *Science*, 252, 80–87, 1991.

83. **Cleves, A., McGee, T., and Bankaitis, V. A.,** Phospholipid transfer proteins: a biological debut, *Trends Cell Biol.*, 1, 30–34, 1991.

84. **Sandelius, A. S. and Kader, J. C.,** unpublished data, 1992.

85. **Grosbois, M. and Kader, J. C.,** unpublished data, 1992.

86. **Vignols, F. and Kader, J. C.,** unpublished data, 1992.

87. **Spener, F.,** unpublished data, 1991.

88. **Ostergaard, J. and Kader, J. C.,** unpublished data, 1992.

89. **Moreau, F. and Kader, J. C.,** unpublished data, 1992.

90. **Sossountzov, L. and Kader, J. C.,** unpublished data, 1992.

91. **Thoma, S. and Somerville, C.,** unpublished data, 1992.

92. **Coutos-Thevenot, P.,** personal communication, 1992.

93. **Vignols, F., Delseny, M., Puigdomenach, P., and Kader, J. C.,** unpublished data, 1992.

SECTION III

BIOSYNTHESIS OF ISOPRENOID LIPIDS

Isoprenoid (C_5) base compounds are extremely diverse in plants, and many of these are lipids which have highly significant functions. Chapter 11 is concerned with the sterols, which function as membrane components and apparently also as hormones. The prenyllipids (in particular, carotenoids, chlorophylls, and quinones) functions and biosynthesis are described in Chapter 12. The role of such compounds in photosynthesis and respiration have long been established.

Chapter 10 begins the description of the isoprenoids with acetyl-CoA and describes the biosynthesis of the basic units. It also puts this area into perspective by describing some of the other isoprenoid classes and functions.

Chapter 11

TERPENOID BIOSYNTHESIS: THE BASIC PATHWAY AND FORMATION OF MONOTERPENES, SESQUITERPENES, AND DITERPENES

Jonathan Gershenzon and Rodney B. Croteau

TABLE OF CONTENTS

0-8493-4907-9/93/$0.00+$.50

I. INTRODUCTION

Plants produce a large variety of lipids that lack fatty acid moieties and are often called nonacyl lipids. The largest single category of nonacyl lipids in higher plants is undoubtedly the terpenoids, an extremely abundant class of natural products with a common biosynthetic origin. Terpenoids or terpenes are formed from the fusion of isopentenoid units derived from acetyl-coenzyme A (acetyl-CoA). Yet this common mode of biosynthesis belies the vast chemical diversity and complexity in the group. Hundreds of different terpenoid carbon skeletons are known, and a large assortment of oxygen-containing functions are often present. In addition, nitrogen-containing functional groups are not uncommon, and halogenated terpenoids are characteristic of many marine plants. The diversity of this class is increased even further by the fact that terpene moieties are frequently found joined to other molecules such as sugars, organic acids, or aromatic substances. Fortunately, for our purposes, terpenoid metabolism is not nearly as complex as this great structural variability might imply, and most steps of terpenoid biosynthesis can be accounted for by a few major reaction types.

A. TERPENOID STRUCTURE AND CLASSIFICATION

Terpenoids all share a common construction pattern: the linkage of five-carbon units having the branched carbon skeleton of isopentane (Figure 1). This structural unity has been appreciated for over 100 years, ever since the

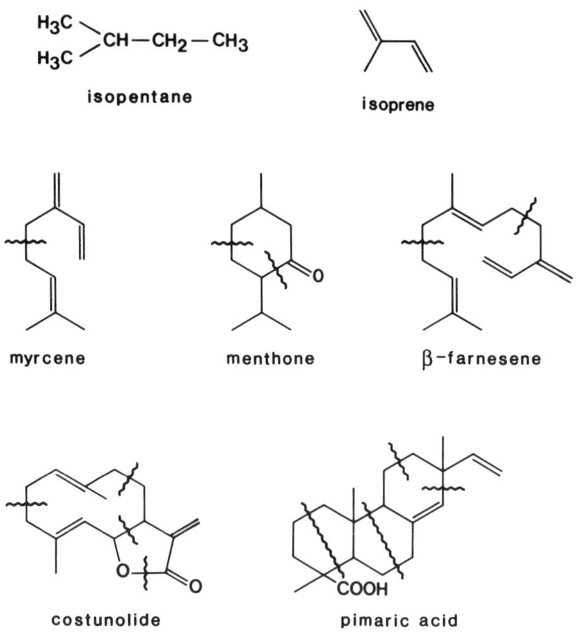

FIGURE 1. Terpenoids are formed from the fusion of five-carbon isopentanoid elements often referred to as isoprene units. The division of terpenoid structures into such units helps to visualize their biosynthetic assembly, although extensive metabolic rearrangements may complicate this task.

pioneering organic chemists of the 19th century discovered that many terpenoids could be pyrolyzed to give isoprene (2-methyl-1,3-butadiene, Figure 1), a diene with the isopentanoid skeleton. In 1887, the renowned German chemist, Otto Wallach, proposed that all terpenoids are derived from the ordered, head-to-tail condensation of isoprene units. This is the so-called "isoprene rule", which proved to be very useful in determining the structures of many unknown terpenoid substances. Leopold Ruzicka updated this concept in 1953[1] by formulating the biogenetic isoprene rule, which asserts that all terpenoids are biosynthesized from a biologically active isoprene unit, now known to be represented by isopentenyl pyrophosphate. In proposing the biogenetic isoprene rule, Ruzicka recognized that "regular" isoprenoid skeletons formed from the head-to-tail condensation of C_5 units can undergo substantial structural rearrangements. In such modified terpenoids, it may be quite difficult to discern the original organization of the isoprene units. Nevertheless, the formal dissection of terpenoid skeletons into isoprene units is often valuable for understanding their mode of biosynthesis (Figure 1). Within this historical context, it is not surprising that terpenoids are frequently referred to as isoprenoids.

Terpenoids are classified by the number of five-carbon isoprenoid units in their structures:

hemiterpenes	C_5	(1 isoprenoid unit)
monoterpenes	C_{10}	(2 isoprenoid units)
sesquiterpenes	C_{15}	(3 isoprenoid units)
diterpenes	C_{20}	(4 isoprenoid units)
triterpenes	C_{30}	(6 isoprenoid units)
tetraterpenes	C_{40}	(8 isoprenoid units)
polyterpenes	$(C_5)_n$	where $8 < n < 30,000$

This nomenclature is based on the fact that the term "terpene" was originally used to refer only to C_{10} compounds, once thought to be the smallest, naturally occurring representatives of the isoprenoid class. This later necessitated the naming of C_5 terpenoids as hemiterpenes ("half" terpenes), of C_{15} terpenes as sesquiterpenes ("one and one-half" terpenes), of C_{20} terpenes as diterpenes ("two" terpenes), and so on.

B. FUNCTIONS OF TERPENOIDS IN PLANTS

Terpenoids are found in all living organisms, but achieve their greatest structural and functional diversity in the plant kingdom. They constitute the largest group of organic compounds in plants, with over 15,000 representatives known[2] and thousands more undoubtedly remaining to be discovered. Terpenoids play a multitude of essential roles in plants. For instance, abscisic acid (C_{15}) and the gibberellins (C_{20}) are terpenoids that serve as plant hormones. Steroids (C_{21}–C_{30} triterpenoid derivatives) are important structural components of cell membranes. Carotenoids are red, orange, and yellow tetraterpenes (C_{40}) that function as accessory pigments in photosynthesis and protect photosynthetic tissues from deleterious photooxidative effects. Long-chain polyterpenoid alcohols called dolichols (C_{70}–C_{120}) act as membrane-bound sugar carriers in polysaccharide and glycoprotein assembly. Terpenoid side chains (C_5–C_{45}) are found in a variety of key plant metabolites, such as plastoquinone and ubiquinone (components of electron transport chains), chlorophyll, and cytokinins, and may facilitate anchoring to or movement within membranes. In fungal and animal cells, certain proteins have recently been reported to be posttranslationally modified by attachment of C_{15} and C_{20} terpenoid side chains,[3] a phenomenon likely to occur in plants as well.

The majority of plant terpenoids lack any apparent role in the basic processes of growth and development and are often designated as "secondary metabolites". These substances are scattered irregularly throughout the plant kingdom as constituents of essential oils, resins, latex, and waxes. Once thought of as metabolic waste products, they are now believed to function largely in an ecological context, serving as defenses against herbivores and pathogens, as attractants for pollinators and fruit-dispersing animals, or as agents of plant-plant competition.[4-6]

Numerous plant terpenoids have significant roles in human health and commerce. For example, many monoterpene (C_{10}) and sesquiterpene (C_{15})

constituents of essential oils are important flavoring and fragrance agents in foods, beverages, cosmetics, perfumes, soaps, and other products. Other plant terpenoids used in industry are turpentine (principally monoterpenoids, C_{10}), conifer rosin (principally diterpenoids, C_{20}), and rubber (a polyisoprenoid, $C_{5000-150,000}$). Turpentine and rosin are employed in the manufacture of solvents, adhesives, polymers, emulsifiers, coatings, and specialty chemicals. Terpenoids of nutritional or pharmacological significance include vitamins A, D, E, and K, as well as many plant steroids and diterpenes. For example, sterol glycosides isolated from foxglove (*Digitalis* sp.) are widely prescribed for the treatment of congestive heart disease, while the steroidal saponins of yam (*Dioscorea* sp.) are valuable starting materials in the synthesis of progesterone-like compounds for birth control pills. Taxol, a highly functionalized diterpene from yew (*Taxus* sp.), is a useful new anticancer drug.

C. OVERVIEW OF TERPENOID BIOSYNTHESIS

The pathway of terpenoid biosynthesis can be divided conveniently into several stages. The first stage (covered in Section II of this chapter) encompasses the synthesis of the biological C_5 isoprene unit, isopentenyl pyrophosphate (IPP), from three molecules of acetyl-CoA (Figure 2). This process occurs via a sequence of steps usually designated the mevalonic acid pathway, after its best known intermediate. In the second stage (Section III), IPP is isomerized to dimethylallyl pyrophosphate (DMAPP), and then these two C_5 compounds react to give geranyl pyrophosphate (C_{10}). Further condensations with additional IPP units generate successively larger acyclic prenyl pyrophosphates of C_{15}, C_{20}, etc. Finally, in the third stage (Section IV), the various prenyl pyrophosphates undergo a wide range of cyclizations, couplings, and rearrangements to produce the parent carbon skeletons of each class. Geranyl pyrophosphate (C_{10}) is converted to monoterpene skeletons, farnesyl pyrophosphate (C_{15}) to sesquiterpene skeletons, and geranylgeranyl pyrophosphate to diterpene skeletons (C_{20}). Farnesyl pyrophosphate can also dimerize in head-to-head fashion to form squalene (C_{30}), which leads to the formation of triterpenes, covered in Chapter 12. Similarly, geranylgeranyl pyrophosphate can dimerize to phytoene (C_{40}), the precursor of the tetraterpenoids, covered in Chapter 13. These major structural conversions may be followed by a variety of oxidations, reductions, isomerizations, hydrations, conjugations, and other transformations that eventually give rise to the many thousands of different terpenoid metabolites found in plants.

This chapter considers each of the three stages of terpenoid biosynthesis in turn (Sections II, III, and IV). We will describe the sequence of biosynthetic transformations of each stage and review what is known about the enzymology, mechanism, and regulation of various key steps of the pathway. The final section (Section V) will examine the roles of subcellular compartmentation, multienzyme complexes, assimilate partitioning, and morphological differentiation in the overall control of plant terpenoid biogenesis.

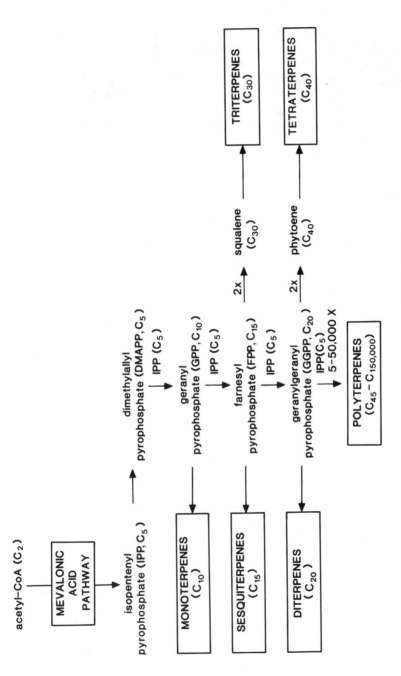

FIGURE 2.   Outline of terpenoid biosynthesis.

D. EXPERIMENTAL METHODS USED IN THE STUDY OF TERPENOID BIOSYNTHESIS

Investigation of a biosynthetic pathway necessarily begins with establishing the actual sequence of metabolic transformations. For terpenoids and other substances, such information was once obtained primarily by administering radiolabeled precursors to intact plant tissues, isolating the products, measuring their specific activities, and determining the site of labeling by chemical degradation. However, recent advances in the elucidation of terpenoid biosynthetic sequences have come largely from the use of cell-free extracts rather than intact tissue. Cell-free preparations allow more detailed examination of the immediate precursors and products of each step of the pathway and usually result in better incorporation of exogenous precursors because cellular and subcellular barriers to uptake have been removed. Another recent innovation has been the use of ^{13}C-labeled precursors for feeding studies. Nuclear magnetic resonance (NMR) spectroscopy of ^{13}C-labeled products permits the direct determination of isotopic enrichment and sites of labeling without the need for lengthy degradative procedures, although considerably more sample is required than when using radioactive precursors. Progress in the cell-free investigation of plant terpenoid biosynthesis nevertheless has been slow due to difficulties in extracting sufficient quantities of active enzyme for study. The enzymes of terpenoid biosynthesis are often present in extremely low amounts compared to other cellular proteins, and their activities in crude plant extracts are frequently quite labile.[7] In addition, certain steps of the pathway occur only in particular types of tissue at precise developmental stages. Thus, the choice of both plant species and tissue is often critical to the success of this work.

For the early stages of plant terpenoid biosynthesis (Sections II and III), the sequence of metabolic intermediates has been well-established, due in large measure to the detailed studies carried out on sterol formation in yeast and animals. Researchers have now begun to turn their attention to the individual enzymes catalyzing these early steps in plants, examining their properties (molecular weight, subunit architecture, co-factor requirements, kinetic parameters, etc.), catalytic mechanisms, subcellular locations, and possible regulatory significance. These investigations have generally employed standard enzymological and immunocytochemical procedures.

The methodology of biosynthetic research has been dramatically altered over the last few years by the explosive advances in molecular biology. For instance, the application of recombinant DNA techniques has now made it possible to readily isolate and manipulate genes encoding biosynthetic enzymes. Isolated genes can be used in prokaryotic and eukaryotic overexpression systems to generate large quantities of protein for detailed mechanistic studies. Site-directed mutagenesis can be employed to determine the key residues participating in catalysis. These powerful new techniques have just begun to be applied to the study of terpenoid biosynthesis in plants. Genes for three different enzymatic steps have recently been cloned,[8-15] providing new insights into genetic regulation, as well as aspects of enzyme structure and mechanism. Here again, the pace of research on plants has lagged behind that of research on animals and microorganisms. Yet studies on isoprenoid biosynthesis in nonplant systems have furnished valuable context and tangible assistance for plant work. For instance, the cloning of several plant terpenoid genes has been accomplished by screening cDNA libraries with heterologous oligonucleotide probes based on sequences encoding the corresponding enzymes from animals and fungi.[13-15] In the future, the widespread application of molecular biological methods to terpenoid biosynthesis will no doubt provide many exciting opportunities to study genomic organization, enzyme structure and function, and the regulation of metabolic flux through the pathway. However, despite the enormous potential of these new approaches, it is sobering to remember that the complete biosynthetic pathway to most terpenoids is still unknown, and little is understood about the regulation of plant terpenoid metabolism. Thus, much basic work on the elucidation of pathways and their control remains to be done.

II. THE MEVALONIC ACID PATHWAY: ACETYL-CoA TO IPP

The process of terpenoid biosynthesis begins with the mevalonic acid pathway. This well-characterized sequence starts with the central metabolic intermediate acetyl-CoA and results in the formation of IPP, the biological C_5 isoprene unit. The mevalonic acid pathway was first discovered through investigations of steroid biosynthesis in yeast and mammals, but all of the steps have now been demonstrated in plants.

A. CONVERSION OF ACETYL-CoA TO HYDROXYMETHYLGLUTARYL-CoA (HMG-CoA)

The initial reactions of the mevalonic acid pathway involve the stepwise fusion of three molecules of acetyl-CoA to produce the C_6 compound (3S)-3-hydroxy-3-methylglutaryl-CoA (HMG-CoA) (Figure 3). First, two units of acetyl-CoA condense to form acetoacetyl-CoA under catalysis by acetyl-CoA acetyltransferase (EC 2.3.1.9). Acetoacetyl-CoA then combines with the third unit of acetyl-CoA in a reaction catalyzed by HMG-CoA synthase (EC 4.1.3.5). Neither of these enzymes has been well-studied in plants. Acetyl-CoA acetyltransferase, under the name acetoacetyl-CoA thiolase, also catalyzes the reverse reaction, which is the final step in the β-oxidation of fatty acids. HMG-CoA synthase, on the other hand, has no analogy in fatty acid metabolism and is responsible for generating the branched-chain structure characteristic of terpenoid compounds. This branched assembly pattern of acetate units is at least as ancient in an evolutionary sense as the linear assembly pattern of acetate units seen in acyl lipid biosynthesis.[16]

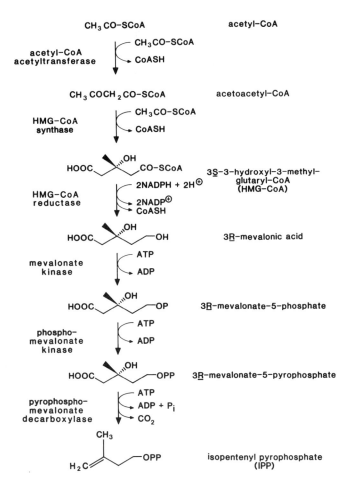

FIGURE 3. The mevalonic acid pathway.

Based on studies with animal and fungal extracts, the first two steps of terpenoid biosynthesis share a similar carbanionic mechanism.[17] The fusion of two acetyl-CoA molecules to give acetoacetyl-CoA is an enzyme-catalyzed Claisen condensation in which the intermediates are covalently bound to the enzyme itself (Figure 4). Initially, one acetyl-CoA attaches to the enzyme forming a bound intermediate linked via a cysteine thioester. Next, the other acetyl-CoA, acting as a nucleophilic α-carbanion, attacks the electron-deficient carbonyl carbon of the bound acetyl moiety displacing the sulfhydryl group of the cysteine residue and releasing acetoacetyl-CoA.

The reaction of acetoacetyl-CoA with the third acetyl-CoA unit is formally an aldol-type condensation (Figure 5). Again, catalysis is initiated with the covalent attachment of acetyl-CoA to the sulfhydryl group of an active site

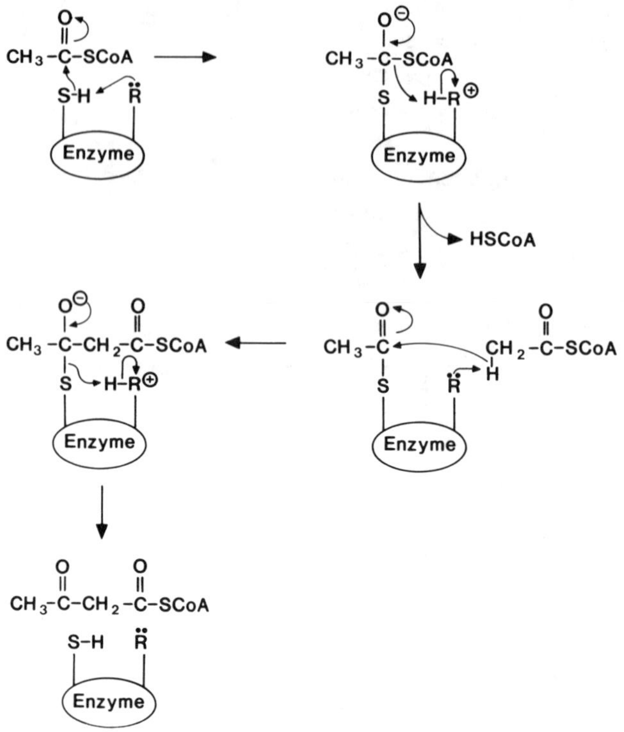

FIGURE 4. Proposed mechanism for conversion of two acetyl-CoA moieties to acetoacetyl-CoA by acetyl-CoA acetyltransferase.

cysteine,[18] the rate-limiting step of the overall transformation. But, in this case, the bound acetyl-CoA acts as the nucleophile in attacking the β-keto carbonyl function of acetoacetyl-CoA. The resulting enzyme-bound HMG-CoA is then released by hydrolysis.

The recent investigations of Bach and associates[8,19] have begun to shed some light on the enzymology of HMG-CoA formation in plants. In dark-grown radish seedlings, acetyl-CoA acetyltransferase and HMG-CoA synthase appear to form a tight, membrane-associated complex, since upon administration of acetyl-CoA, HMG-CoA accumulates in the incubation medium, but no free acetoacetyl-CoA is detectable. These two activities are inseparable through several chromatographic steps, and the two reactions could conceivably be catalyzed by a single enzyme, a proposal not inconsistent with their mechanistic similarities. The close association of these two activities would have a rational thermodynamic basis in that the equilibrium constant of acetyl-CoA acetyltransferase is very low, making the formation of acetoacetyl-CoA highly unfavorable.[20] However, if the two reactions are coupled physically as well as chemically, the overall formation of HMG-CoA is energetically favored.[19,20]

HMG-CoA in plants is formed not only in the process of terpenoid biosynthesis, but also as an intermediate in leucine catabolism. Thus, it has

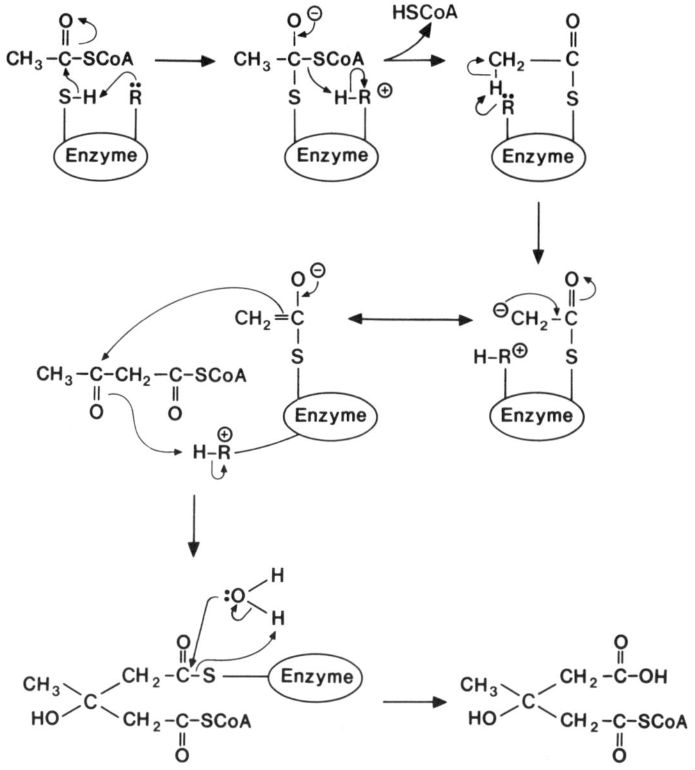

FIGURE 5. Proposed mechanism for conversion of acetoacetyl-CoA to HMG-CoA by HMG-CoA synthase.

been conjectured for many years that leucine could serve as a direct progenitor of terpenoids. However, recent experiments with *Andrographis paniculata* cell cultures have shown that isotopically labeled leucine is only incorporated into terpenoids after first being degraded to acetyl-CoA and acetoacetyl-CoA, suggesting that the pathway from leucine to HMG-CoA is not an important route to terpenoids in plants.[21] Similar results were obtained when leucine was fed to isolated spinach chloroplasts.[22]

B. HMG-CoA REDUCTASE

HMG-CoA is reduced to (3*R*)-mevalonic acid in a two-step, NADPH-dependent process catalyzed by HMG-CoA reductase (EC 1.1.1.34). In animals, reduction of HMG-CoA appears to be the critical, rate-controlling step in the biosynthesis of isoprenoids such as cholesterol. Therefore, given the relationship between serum cholesterol levels and cardiovascular disease, mammalian HMG-CoA reductase has been the focus of intense study. Interest has carried over to plants, and HMG-CoA reductase has now been investigated in over 20 different species of algae and higher plants.[19]

1. Enzymology

Plant HMG-CoA reductase is a membrane-bound enzyme.[19] But, beyond this fact, it is difficult to make any generalizations about the characteristics of this enzyme because of the wide disparities reported in the literature. This variation may be due to actual differences between plant species, to the occurrence of multiple isozymic forms of HMG-CoA reductase in a single tissue,[19] or to the presence of interfering substances typically found in crude membrane preparations.[23] The membranous and unstable nature of this enzyme has greatly hindered efforts at purification and characterization. However, HMG-CoA reductases have recently been solubilized and purified from radish seedlings by Bach's group[24] and from potato tubers by Kondo and Oba,[25] achievements that should inspire similar efforts in other plant species.

The most frequently reported subcellular location of HMG-CoA reductase is the endoplasmic reticulum.[19] However, this enzyme is also thought to occur in the membranes of mitochondria[26-28] and plastids,[27,29-32] based on subcellular fractionation studies and experiments with mevinolin, a specific inhibitor of HMG-CoA reductase.[33,34]

Several recent advances in our knowledge of the enzymology of plant HMG-CoA reductase have come from the application of molecular biological techniques. cDNA and genomic clones for this enzyme have been isolated from radish,[8] tomato,[15] pea,[14] rubber tree,[11] and *Arabidopsis thaliana*.[10,13,14] The radish, rubber tree, and *A. thaliana* genes have been shown to encode polypeptides of 63 kDa (approximately 590 amino acids), but the molecular mass and quaternary structure of the native enzymes are still uncertain.[8] The COOH-terminal domain (approximately 400 amino acid residues in length) contains the catalytic site and has an amino acid sequence that is highly conserved among all plant and other eukaryotic HMG-CoA reductases investigated so far. In contrast, the NH_2-terminal domain (about 160 amino acids), which contains the membrane-binding region, shows significantly more divergence. The portion of the protein linking these two domains contains a "PEST" sequence, a segment rich in proline (P), glutamic acid (E), serine (S), and threonine (T) residues that often marks proteins for rapid turnover. In all species examined, there are at least two genes for HMG-CoA reductase,[8,10,11,14,15,35] indicating that the regulation of this activity is probably complex.

2. Mechanism

The reduction of HMG-CoA to mevalonic acid is a two-step process that requires NADPH for both steps (Figure 6). First, HMG-CoA is reduced to a hemithioacetal, which by loss of CoA gives the corresponding aldehyde, mevaldic acid.[17] Next, the aldehyde function is reduced to an alcohol function, forming mevalonic acid. Both the hemithioacetal and the aldehyde are enzyme-bound intermediates that are not observed free in solution. However, the enzyme will accept the aldehyde intermediate as a substrate, reducing it to

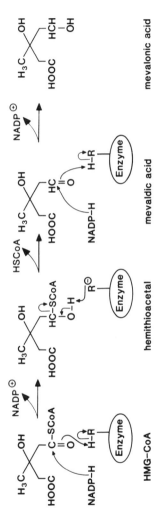

FIGURE 6. Proposed mechanism for conversion of HMG-CoA to mevalonic acid by HMG-CoA reductase.

mevalonic acid. Kinetic studies of the purified enzyme from radish seedlings show that substrate binding is ordered and sequential; HMG-CoA binds first, followed by NADPH. The oxidized co-factor $NADP^+$ must be released from the enzyme before the second NADPH can bind. The second reduction is believed to be rate limiting, with the release of $NADP^+$ being the slowest single step of the overall process.[17]

3. Regulatory Significance

The role of HMG-CoA reductase in regulating cholesterol biosynthesis in mammals has suggested that this enzyme also catalyzes a rate-controlling step in plant terpenoid formation. Indeed, there is often a strong correlation between the level of HMG-CoA reductase activity in plants and the rate of terpenoid formation. For example, elicitation of sesquiterpene phytoalexin production in potato,[36] sweet potato,[37] and tobacco[38] results in elevated levels of HMG-CoA reductase activity, and the peak of activity is closely associated with the onset of sesquiterpene accumulation. Similar data suggesting a regulatory role for this enzyme have been obtained from studies on the biosynthesis of sterols,[34] carotenoids,[39] and rubber.[40] In contrast, the formation of carotenoids in ripening tomato fruit occurs late in fruit development at a time when HMG-CoA reductase levels are less than 1% of what they were early in fruit development.[15] Thus, most of the available evidence is consistent with the notion that HMG-CoA reductase catalyzes a rate-limiting step in plant terpenoid biosynthesis. However, proof of this assertion requires further studies to compare the activity of HMG-CoA reductase with that of all the other pathway enzymes in a single tissue, while measuring the overall flux through the pathway and the actual concentrations of intermediates *in vivo*.

The regulatory importance of HMG-CoA reductase may be a consequence of the fact that the reduction of HMG-CoA to mevalonic acid is the first committed step of terpenoid biosynthesis in plants. The earlier intermediates of the mevalonate pathway, acetyl-CoA, acetoacetyl-CoA, and HMG-CoA, have alternate metabolic fates that can divert flux away from terpenoid biosynthesis. For example, HMG-CoA can be cleaved to acetoacetate and acetyl-CoA by the action of HMG-CoA lyase (EC 4.1.3.4). This activity has been detected in several plant species,[19,41,42] although its actual importance in the control of terpenoid formation is not known.

4. Mode of Regulation

The possibility that HMG-CoA reductase may control plant terpenoid biosynthesis leads to the question of how the level of HMG-CoA reductase activity itself is controlled. For animals, there is an abundance of information on this subject. Mammalian HMG-CoA reductase activity is subject to end product feedback inhibition by sterols that is elaborately mediated by (1) long-term changes in the rates of enzyme synthesis and degradation and (2) short-term changes in activity due to both allosteric effects and covalent modification involving reversible phosphorylation.[43-46]

There are preliminary indications that some of these control mechanisms may also operate in plants. In tomato fruit[15] and potato tubers,[35] increases in HMG-CoA reductase activity are associated with increased amounts of HMG-CoA reductase mRNA, suggesting that control occurs at the level of enzyme synthesis. Likewise, studies with sweet potato showed that the rise in HMG-CoA reductase activity, which accompanies the biosynthesis of sesquiterpene phytoalexins, is contingent upon *de novo* protein synthesis.[47,48] Scattered evidence also exists for other modes of HMG-CoA reductase regulation in plants, including reversible phosphorylation[23,33,49-51] and allosteric control.[24,31,50,52] However, much additional work is needed before the control of this activity in plants is understood to the same depth as that in mammals. Mammalian HMG-CoA reductases contain a much more extensive membrane-binding domain than do the plant enzymes studied thus far.[8,10,13] Since this membrane-binding region is implicated in certain regulatory phenomena in mammalian systems,[53] it would not be surprising if plant HMG-CoA reductases possess regulatory features very different from those of mammals.

C. CONVERSION OF MEVALONIC ACID TO IPP

Mevalonic acid, the product of HMG-CoA reductase, is converted to the C_5 intermediate IPP by a three-step biosynthetic sequence (Figure 3). First, mevalonate is pyrophosphorylated by the successive action of two enzymes, mevalonate kinase (EC 2.7.1.36) and phosphomevalonate kinase (EC 2.7.4.2). Then, the resulting mevalonate-5-pyrophosphate is subjected to an ATP-driven, decarboxylative elimination catalyzed by pyrophosphomevalonate decarboxylase (EC 4.1.1.33) to generate IPP. These three enzymes have only been characterized in a few plant species so far and have not been afforded the same attention as HMG-CoA reductase.

Mevalonate kinase and phosphomevalonate kinase both transfer the γ-phosphoryl moiety of ATP to their respective substrates. As a consequence, they have several properties in common with other kinases mediating similar types of reactions, including a need for thiols to stabilize activity in cell-free extracts and a requirement for a divalent metal ion (usually Mg^{2+}, occasionally Mn^{2+}) for catalysis.[54-57] The divalent cation forms a complex with ATP that neutralizes the negative charges on the phosphoryl oxygen atoms, thereby reducing electrostatic repulsion and facilitating nucleophilic attack on the phosphorus atom. Based on experiments with animal systems, both of these phosphorylations proceed by ordered, sequential mechanisms with mevalonate or mevalonate-5-phosphate binding first followed by Mg-ATP.[58,59] The two kinases are both reported to be localized primarily in the cytosol,[54-56,60] although there is some evidence that mevalonate kinase is also present in chloroplasts.[29]

The enzyme responsible for the third step of this sequence, pyrophosphomevalonate decarboxylase, also requires ATP and a divalent metal ion for activity.[56,57,61] The reaction catalyzed is essentially a concerted elimination in which an olefin is formed by the simultaneous loss of carbon dioxide and a hydroxyl function (Figure 7). Phosphorylation of the tertiary hydroxyl group

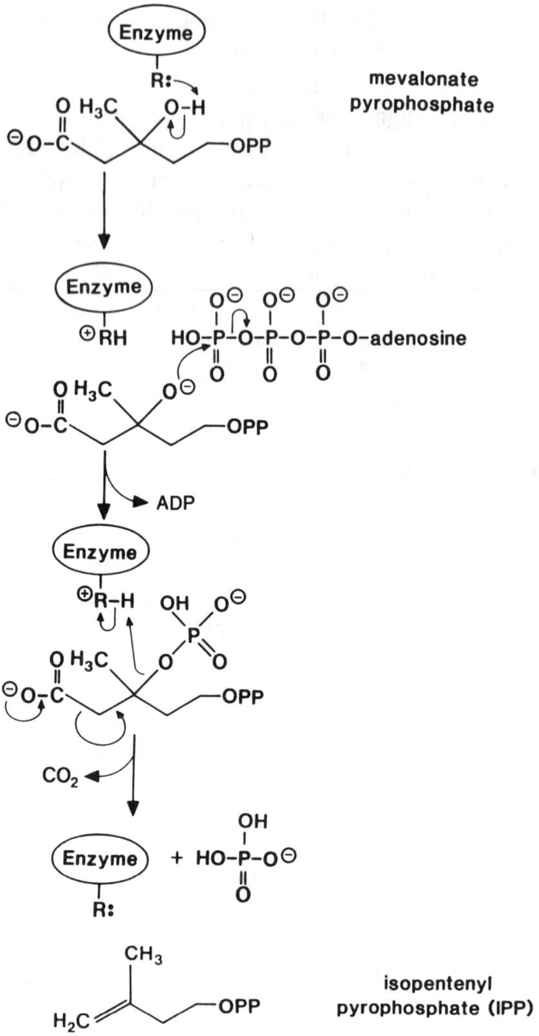

FIGURE 7. Proposed mechanism for conversion of mevalonate pyrophosphate to IPP by pyrophosphomevalonate decarboxylase.

makes it a better leaving group and so activates mevalonate-5-pyrophosphate for this elimination. The reaction is thought to be initiated by abstraction of a proton from the tertiary hydroxyl group, followed by phosphorylation at that site and then elimination.[17,62] However, the triphosphorylated intermediate has not yet been detected. The overall stereochemistry of the elimination is *anti*, and the mechanism is ordered and sequential with mevalonate-5-pyrophosphate binding first.[63]

The potential regulatory significance of either pyrophosphomevalonate decarboxylase or the two kinases is unclear. Neither of these three activities showed any change in rate during the induction of diterpene phytoalexin synthesis in castor bean[64] or the increased production of carotenoids and steroids in carrot cell suspensions.[39] However, the decarboxylase showed a rise in activity associated with the accumulation of sesquiterpenes in sweet potato roots,[48] and in extracts of peanut seedlings the decarboxylase had a lower activity than either of the two kinases or HMG-CoA reductase, suggesting that it might constitute a rate-limiting step at least in this tissue.[56]

All three enzymes involved in converting mevalonate to IPP require ATP. Thus, their activities might be influenced by the relative cellular concentrations of ATP, ADP, and AMP. Precedent for this type of control is seen in mammals, where these three enzymes are all stimulated by ATP at concentrations above that required for catalysis.[65] In plants, there are only scattered reports of such stimulation by "energy charge".[57,61] For example, the rate of biosynthesis of the diterpene kaurene in wild cucumber extracts was increased by added ATP and decreased by added ADP, and the site of control was demonstrated to be pyrophosphomevalonate decarboxylase.[61]

In mammals, the IPP formed by the mevalonic acid pathway is not irreversibly committed to terpenoid biosynthesis, but can be converted back to HMG-CoA and then to acetate by a series of transformations known as the mevalonate shunt.[66] Evidence for the operation of the mevalonate shunt in plants has recently been obtained in experiments with wheat.[67] This pathway could divert a significant fraction of IPP away from plant terpenoid biosynthesis, although its actual physiological importance has not been determined.

III. THE FORMATION OF PRENYL PYROPHOSPHATES

IPP, the product of the mevalonate pathway, is the biologically active C_5 isoprene unit that serves as the basic building block in terpenoid construction. The second phase of terpenoid biosynthesis begins with the isomerization of IPP to DMAPP. As the first allylic pyrophosphate in the pathway, DMAPP serves as a primer to which varying numbers of IPP units can be added in sequential chain elongation steps (Figure 8). Thus, IPP and DMAPP condense to form the allylic C_{10} compound geranyl pyrophosphate (GPP). Another molecule of IPP may then condense with GPP to generate the C_{15} allylic pyrophosphate, farnesyl pyrophosphate (FPP). Addition of a further IPP unit gives the C_{20} geranylgeranyl pyrophosphate (GGPP), and so on. This head-to-tail polymerization of isoprene units is catalyzed by enzymes called prenyltransferases. The process results in successively larger acyclic, allylic prenyl pyrophosphates, which ultimately serve (Section IV) as precursors of the major classes of terpenoids.

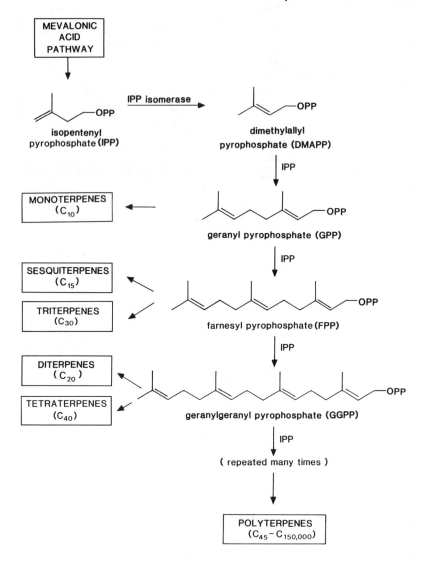

FIGURE 8. Pathway for the formation of prenyl pyrophosphates and the major groups of terpene end products derived from each.

A. IPP ISOMERASE

The allylic rearrangement of IPP to DMAPP is catalyzed by IPP isomerase (EC 5.3.3.2), an enzyme that has been well-characterized in several species of higher plants.[68-72] IPP isomerase typically possesses a single subunit of 30 to 40 kDa, requires a divalent metal ion (Mg^{2+} or Mn^{2+}) for catalysis, and is especially sensitive to sulfhydryl-directed reagents. The reaction is reversible, with equilibrium favoring DMAPP.[69] IPP isomerase is believed to be localized at multiple subcellular sites, including the mitochondria,[64] the plastids,[68,69,71] and the cytoplasm.[73]

Detailed investigations of the reaction mechanism, carried out with the enzyme from yeast and pig liver, have indicated that the isomerization of IPP proceeds via a carbocationic intermediate (or through a transition state with carbocationic character).[74,75] First, the C_3–C_4 double bond of IPP is protonated; then a proton is abstracted from C_2 to yield DMAPP (Figure 9). Proton removal and proton addition occur on opposite sides of the molecule, making the overall process antarafacial.[76] Thus, the enzyme would appear to require two separate bases for catalysis. At least one of these groups may be the sulfhydryl moiety of a cysteine residue.[75,77,78]

Nothing is known about the possible role of IPP isomerase in regulating terpenoid biosynthesis. However, by changing the relative levels of IPP and DMAPP available for subsequent biosynthetic steps, this enzyme could theoretically exert some general control over which classes of terpenoids are produced. For example, in monoterpene biosynthesis, one IPP moiety is needed to condense with each DMAPP cosubstrate to yield GPP (Figure 2). On the other hand, in polyisoprene biosynthesis, many IPP units are required for each DMAPP primer, so the rate of IPP isomerase activity could be considerably less than the rate of IPP production.

B. PRENYLTRANSFERASES

Both the substrate (IPP) and the product (DMAPP) of the IPP isomerase reaction participate in the next set of reactions of terpenoid biosynthesis. DMAPP is the primer and IPP the repeating unit for a series of chain elongation steps that form allylic pyrophosphates of 10, 15, 20, or more carbon atoms (Figure 8). These are the fundamental reactions that connect isoprene units to one another. They are catalyzed by prenyltransferases (EC 2.5.1.1), enzymes that mediate the condensation of IPP with an allylic pyrophosphate, such as DMAPP (C_5), GPP (C_{10}), or FPP (C_{15}), forming a new allylic pyrophosphate containing five carbon atoms more than the original.

1. Enzymology

A number of different types of prenyltransferases have been described that vary in substrate specificity, in the number of IPP units added, and in the geometry of the double bond(s) formed. Of the plant prenyltransferases characterized to date, most catalyze multistep reaction sequences, beginning with DMAPP and IPP and ending with a single product having an all *trans-*(*E*) configuration. For example, FPP synthases have been reported from several species that convert DMAPP and IPP to GPP and then condense the newly formed GPP with another IPP moiety to give FPP without appreciable accumulation of the intermediate GPP.[70,79,80] GGPP synthases of plant origin are also known that convert DMAPP and IPP to GGPP without accumulation of the intermediates GPP or FPP.[68,71,81,82]

Plant prenyltransferases possess a number of traits in common.[68,70,71,79-83] The majority of those investigated so far are homodimers with native M_r between 70,000 and 80,000. Each subunit has a single catalytic site.[76] Activity requires a divalent metal ion, either Mg^{2+} or Mn^{2+}, in the ratio of two molecules

358   *Lipid Metabolism in Plants*

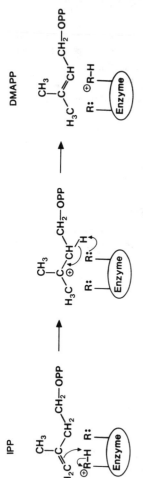

FIGURE 9. Proposed mechanism for the conversion of IPP to DMAPP by IPP isomerase.

of metal ion per catalytic site. Recently, Camara and colleagues in France isolated the first cDNA encoding a plant prenyltransferase.[12] The deduced amino acid sequence bears considerable homology with prenyltransferases from mammalian and fungal sources.[84-87]

Prenyltransferases are found in a variety of cellular and subcellular locations, depending on the nature of the reactions they catalyze.[73] For instance, a particular type of prenyltransferase from garden sage, that produces only GPP and does not catalyze further condensations with IPP, is restricted to the secretory cells involved in monoterpene biosynthesis.[88] GPP is the allylic pyrophosphate precursor of monoterpenes. GGPP synthases are found in the chromoplasts of tomato[71] and pepper,[12] organelles which are specialized for carotenoid production. Carotenoids are biosynthesized from GGPP via phytoene (Chapter 13). A polyprenyltransferase from guayule that catalyzes the addition of thousands of IPP units in the formation of rubber, a *cis*-(Z)-polyisoprenoid, is found only in rubber particles.[89]

Besides the head-to-tail condensation of isoprene units, prenyltransferases are involved in several other types of reactions in plants. A large class of these enzymes catalyzes condensations between allylic pyrophosphates and nonisoprenoid acceptors in which a dimethylallyl, geranyl, or larger terpenyl moiety is transferred to an aromatic nucleus, giving rise to products such as ubiquinone, furanocoumarins, and prenylated flavonoids. In addition, the dimerizations of FPP to presqualene pyrophosphate and of GGPP to prephytoene pyrophosphate may be regarded as prenyltransferase-catalyzed reactions in which the condensation of prenylated compounds occurs with a head-to-head orientation.

2. Mechanism

Prenyltransferases are one of the few known groups of enzymes whose mechanism, like that of IPP isomerase, involves carbocations as intermediates.[76] The reaction is initiated by the ionization of the allylic substrate through the loss of the pyrophosphate group (Figure 10). The resulting allylic cation adds to the double bond of IPP forming a new carbocation which is then stabilized by elimination of a proton.

Evidence for the involvement of carbocationic intermediates in this reaction has been obtained by Poulter, Rilling, and colleagues at the University of Utah from experiments with various fluorinated substrate analogs.[76,90] Fluorine-containing derivatives of the allylic pyrophosphate substrates slow the rate of the prenyltransferase reaction by several orders of magnitude. This is the expected result if the mechanism is carbocationic, since electron-withdrawing substituents, such as fluorine, should strongly retard the ionization of the allylic pyrophosphate by depleting electron density at the reaction site, but should have little effect on the rate of a direct nucleophilic displacement.

The divalent metal ion co-factor is thought to enhance the rate of catalysis by neutralizing the negative charges of the pyrophosphate group.[76,91] The complexation of the metal ion with the pyrophosphate moiety generates a

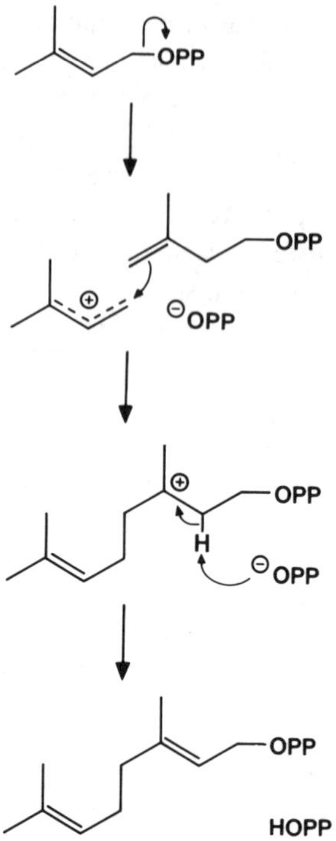

FIGURE 10. Proposed mechanism for the condensation of IPP and DMAPP to form GPP catalyzed by the prenyltransferase, GPP synthase.

better leaving group and thus promotes the ionization of the allylic pyrophosphate.

The stereochemistry of the prenyltransferase reaction was established in the 1960s by Cornforth, Popjak, and co-workers during their pioneering studies on cholesterol biogenesis in mammalian liver.[92,93] The net process is suprafacial, with addition to the double bond of IPP occurring on the same side of the molecule as proton loss. The departing pyrophosphate of the allylic substrate is thus in close proximity to the proton to be eliminated and may assist in proton removal (Figure 10).[76]

Nothing is currently known about the role of the functional groups of the enzyme itself in mediating prenyltransferase catalysis. However, the fact that all prenyltransferase amino acid sequences deduced so far contain aspartate-rich motifs[12,85,94,95] suggests that these regions could function in binding cationic intermediates or metal-substrate complexes. Another common feature is the presence of two adjacent arginine residues which may act to bind a pyrophosphate group.[12]

3. Regulatory Significance

Since prenyltransferases are situated at the primary branch points of the terpenoid pathway, the reactions that they catalyze could be important controlling steps in terpenoid biosynthesis. Metabolic pathways are often regulated at branch points, so that flux can be selectively directed among several possible routes. The rate of plant terpenoid formation *in vivo* does, in fact, show a close relationship with the level of prenyltransferase activity in many experimental systems,[40,64,89,96-100] suggestive of the regulatory importance of these enzymes. For example, in the biosynthesis of squalene from mevalonate in germinating peas, FPP synthase activity was more closely correlated with the rate of squalene formation than was any other enzyme of the pathway[97] and exhibited the most substantial increase in activity during germination.[98] Additional support for the regulatory role of prenyltransferases comes from studies indicating that the transformations mediated by these enzymes seem to be the slow steps of the pathway.[97] Furthermore, if significant amounts of IPP are diverted away from terpenoid biosynthesis by the mevalonate shunt (Section II.C), then prenyltransferases may actually catalyze the first committed steps of the pathway.

IV. FORMATION OF MAJOR TERPENOID CLASSES

Following formation of the prenyl pyrophosphates, the biosynthesis of most plant terpenoids can be divided into two parts: (1) cyclization of the prenyl pyrophosphates to one of a number of basic skeletal types and (2) secondary transformations of the initial cyclic products, including oxidation, reduction, rearrangement, conjugation, double bond isomerization, and hydration. In this chapter, there is only space to discuss the general characteristics of these biosynthetic steps. We will describe the broad outlines of monoterpene, sesquiterpene, and diterpene biosynthesis, and briefly consider the formation of several other important classes of terpenoids. Coverage of triterpene biosynthesis (Chapter 12) and tetraterpene biosynthesis (Chapter 13) can be found elsewhere in this book.

A. MONOTERPENES

The monoterpenes are the C_{10} representatives of the terpenoid family. Nearly all are derived from GPP and thus can be formally considered to arise from the head-to-tail fusion of two isoprene units. Almost 1000 different monoterpenes are known in plants (Figure 11), most of which are volatile compounds that accumulate in resin ducts, secretory cavities, and epidermal glands. These substances are responsible for the characteristic odors of many plants, including conifers, citrus, and mints.

1. Monoterpene Cyclases

The cyclization of GPP to the various skeletal types of monoterpenes is catalyzed by enzymes called monoterpene cyclases (or monoterpene synthases). Based on current evidence, GPP is believed to be the natural substrate for this enzyme class, since in preparations free of competing activities all cyclases investigated efficiently convert GPP to cyclic products without the formation

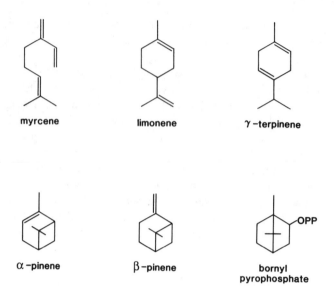

FIGURE 11.   Representative monoterpenes.

of free intermediates.[101] However, a glance at the structure of GPP (Figure 12) makes it clear that this intermediate cannot be cyclized directly to a six-membered ring because of the C_2–C_3 *trans* double bond. Thus, monoterpene cyclases must catalyze the isomerization of GPP to an intermediate capable of cyclization before mediating the actual cyclization process.[102] This cyclizable intermediate is thought to be the tertiary isomer of GPP, linalyl pyrophosphate (Figure 12).

During the last few years, a number of monoterpene cyclases have been partially purified, including those that synthesize limonene, α- and β-pinene, γ-terpinene, and bornyl pyrophosphate (Figure 11).[103-106] Most of these enzymes have similar properties, such as native molecular weight (50,000 to 100,000), the requirement for a divalent metal ion (either Mg^{2+} or Mn^{2+}) as a co-factor, and the presence of essential cysteine and histidine residues at the active site.[101,107,108] Monoterpene cyclases are operationally soluble in cell-free extracts, but may be associated *in vivo* with plastids[73,109] or with the endoplasmic reticulum.[110] Interestingly, individual enzymes of this class often catalyze the formation of multiple products. For example, a monoterpene cyclase recently purified from peppermint synthesizes small amounts of α-pinene, β-pinene, and myrcene in addition to its principal product, limonene.[104]

Given that the substrate GPP is unable to cyclize directly, the mechanism of monoterpene cyclization necessarily involves both isomerization and cyclization steps.[101] First, GPP ionizes with the assistance of the divalent metal ion (Figure 12), much like the first step of the prenyltransferase reaction. The resulting allylic cation then rearranges, via a double bond shift and the return of the pyrophosphate, to form the tertiary allylic isomer, linalyl pyrophosphate.

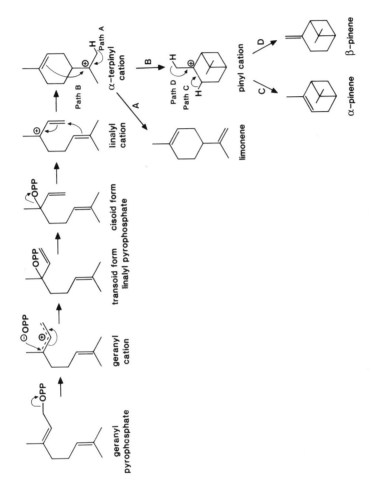

FIGURE 12. Proposed mechanism for the cyclization of GPP to limonene, α-pinene, and β-pinene catalyzed by monoterpene cyclases.

After rotation from the transoid to the cisoid conformer, linalyl pyrophosphate is itself ionized and then cyclized to the α-terpinyl cation via electrophilic addition to the C_6–C_7 double bond. From this point, the cyclization reaction may take one of several different paths involving further internal additions, hydride shifts, or rearrangements before the resulting cation is finally deprotonated to an olefin or captured by a nucleophile, such as water. For example, the α-terpinyl cation may be directly deprotonated to give limonene or instead undergo further cyclization via attack on the cyclohexene (C_1–C_2) double bond to afford α- and β-pinene following deprotonation (Figure 12). All monoterpene cyclases investigated to date are capable of catalyzing both the isomerization and cyclization reactions, and both steps are considered to take place at the same active site.

The mechanistic details of monoterpene cyclization have been deduced from numerous experiments with specifically modified substrates or reaction intermediates.[101,111] For instance, evidence for the cationic nature of both the isomerization and cyclization steps has been obtained through the use of fluorine-substituted analogs, as was done for the prenyltransferase reaction.[111,112] Experiments with ^{18}O-labeled and ^{32}P-labeled precursors suggest that the pyrophosphate anion remains tightly paired with the carbocationic intermediates formed during the course of the reaction.[111,113,114]

Based on similarities in mechanism, monoterpene cyclization appears to represent the intramolecular counterpart of the prenyltransferase reaction (Section III.B.2). Both of these key reactions of terpenoid biosynthesis involve the ionization of an allylic pyrophosphate, the attack of cationic intermediates on carbon-carbon double bonds, and the termination of the reaction by proton elimination. In the prenyltransferase reaction, the initial cationic center and the double bond attacked reside on separate molecules, while in cyclization they are part of the same molecule. The formation of carbon-carbon bonds by electrophilic addition is rather unusual in plant metabolism. The carbon-carbon bonds of most other substances, such as amino acids, carbohydrates, fatty acids, and nucleic acids, are normally constructed by nucleophilic condensation reactions involving carbonyl groups, like those of the first two steps of the mevalonate pathway (Section II.A).

From a regulatory standpoint, monoterpene cyclases and other terpene cyclases are good candidates for being rate-limiting enzymes in terpenoid biosynthesis because they catalyze the first steps leading specifically to monoterpenes, sesquiterpenes, or diterpenes. Monoterpene cyclase activity is, in fact, closely correlated with changes in the rate of monoterpene biosynthesis in several instances,[115-117] suggesting that these enzymes have regulatory importance. Similar trends have been noted for sesquiterpene and diterpene cyclases.[64,116,118,119] Interestingly, the elicitation of sesquiterpene phytoalexin biosynthesis in tobacco and potato cell cultures is accompanied not only by a rise in sesquiterpene cyclase activity, which converts FPP to sesquiterpenes, but also by a decline in the activity of squalene synthase, a competing enzyme

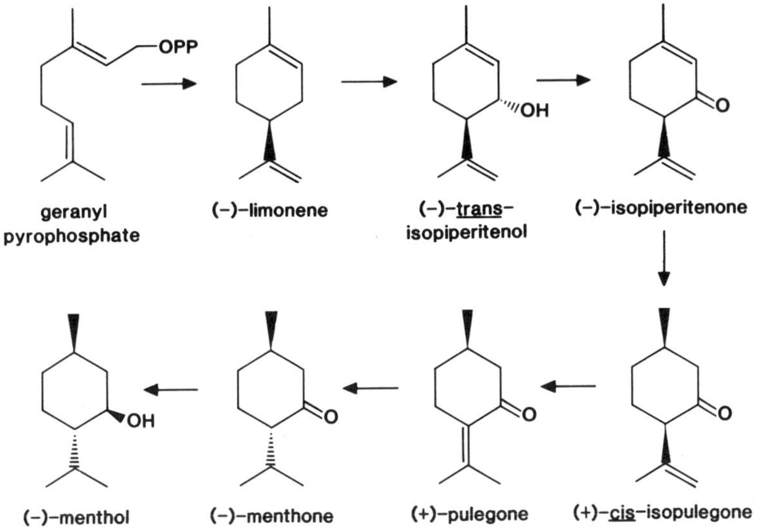

FIGURE 13. The pathway for menthol biosynthesis in peppermint. After GPP is cyclized to limonene, a series of oxidative and reductive secondary transformations convert limonene to menthol.

that catalyzes the dimerization of FPP to squalene, the first step in sterol and triterpene production.[118-120] Based on a few scattered reports, cyclase activity itself appears to be controlled at the level of transcription.[121,122]

2. Secondary Transformations

The cyclic monoterpenes initially formed from GPP are subject to an assortment of further enzymatic transformations, including oxidations, reductions, and isomerizations. These modifications produce the large number of monoterpene derivatives that naturally occur in plants. Unfortunately, very few secondary transformations have been well-studied, and there is little supporting evidence for most of the biosynthetic routes proposed. An exception is the pathway of menthol biosynthesis in peppermint (Figure 13). Limonene, the initial cyclization product of GPP, is hydroxylated by a cytochrome P-450-dependent monooxygenase to form isopiperitenol,[123] which is then converted by a series of largely redox transformations to menthol.[124]

The oxidation of monoterpene olefins by cytochrome P-450-dependent oxygenases is one of the few types of secondary transformations that has been investigated in any detail. These membrane-bound enzymes, probably localized in the endoplasmic reticulum, catalyze the position-specific hydroxylation of a number of monoterpene olefins utilizing molecular oxygen and NADPH.[123,125] Monoterpene hydroxylases exhibit remarkable substrate specificity, a trait displayed by many of the other enzymes involved in the secondary transformations of monoterpenes.[101]

3. Iridoid Monoterpenes

The iridoids are a group of more than 300 plant monoterpenes named for their structural similarity to compounds first isolated from the ant genus *Iridomyrmex*. Usually occurring as glycosides, iridoids possess a basic skeleton consisting of a cyclopentane ring fused to a pyran ring (Figure 14). These substances have attracted much interest because of their role in the formation of complex indole alkaloids. The early steps of iridoid biosynthesis have recently been investigated by Uesato and co-workers at Kyoto University in Japan (Figure 14).[126] The precursor is geraniol, presumably formed from the hydrolysis of GPP. Geraniol becomes hydroxylated at one of its terminal methyl groups by a cytochrome P-450 hydroxylase. The resulting diol is apparently oxidized in stepwise fashion to the corresponding dialdehyde which then cyclizes to form iridodial. The cyclizing enzyme has properties very different from those of the monoterpene cyclases previously discussed, such as a requirement for NADPH. Further oxidation and cyclization of iridodial leads to a wealth of different iridoid structures.[127,128]

4. Irregular Monoterpenes

A small group of monoterpenes, including artemisia ketone (Figure 15), are apparently assembled by head-to-head or head-to-middle linkage of isoprene units rather than by the typical head-to-tail condensations. These so-called irregular monoterpenes, which occur in sagebrush, chrysanthemum, and other members of the sunflower family, include such substances as the pyrethrins, a group of esters with potent insecticidal activities. Irregular monoterpenes have been postulated to be derived from chrysanthemyl pyrophosphate, rather than from GPP, by various cleavages of the cyclopropyl ring.[129] Chrysanthemyl pyrophosphate itself may arise from the condensation of two molecules of DMAPP in a manner analogous to the condensation of the two units of FPP that give rise to presqualene pyrophosphate.[130]

B. SESQUITERPENES

The sesquiterpenes are C_{15} terpenes, products of the head-to-tail linkage of three isoprene units. All sesquiterpenes are biosynthesized from the C_{15} prenyl pyrophosphate, FPP. The structural diversity of this class is greater than that of the monoterpenes due to the increased number of different cyclizations possible from a precursor with five additional carbon atoms (Figure 16). Almost 7000 sesquiterpenes have been isolated from plants,[2] nearly half of which are sesquiterpene lactones, highly oxygenated compounds found principally in members of the composite family. Sesquiterpenes, as a whole, are widely distributed in the plant kingdom. Many are volatile compounds, like monoterpenes, and are common constituents of secretory cavities, glands, and ducts.

1. Sesquiterpene Cyclization

Although most of our knowledge of sesquiterpene biosynthesis comes from studies with fungi rather than higher plants, the reactions involved in sesquit-

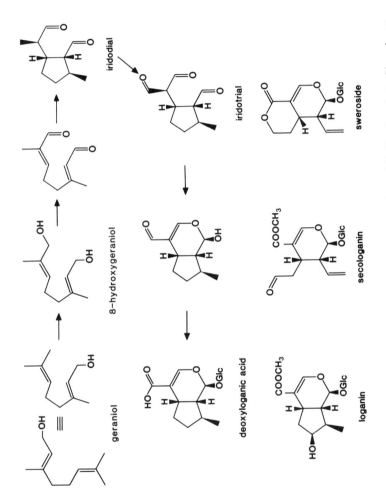

FIGURE 14. Early steps in iridoid biosynthesis and examples of some iridoid end products. Glc = glucose.

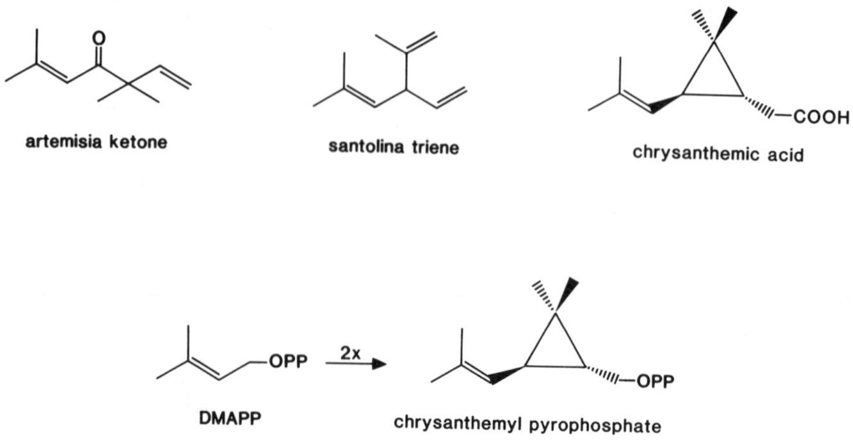

artemisia ketone   santolina triene   chrysanthemic acid

DMAPP   chrysanthemyl pyrophosphate

FIGURE 15. Examples of irregular monoterpenes and their possible biosynthetic precursor.

erpene formation in plants appear to be quite similar to those just discussed for monoterpene formation. The cyclization of FPP proceeds via carbocationic mechanisms directly analogous to those implicated in the cyclization of GPP.[131] However, after the initial ionization of FPP, electrophilic attack within the farnesyl cation may occur on either the central (C_6–C_7) or the distal double bond (C_{10}–C_{11}) (Figure 16). The leading researcher in this field, David Cane of Brown University, has demonstrated that attack on the central double bond necessitates the preliminary isomerization of FPP to an intermediate capable of cyclizing to a six- or seven-membered ring, in analogy with the isomerization of GPP to linalyl pyrophosphate occurring during monoterpene cyclization.[131-133] In this case, the intermediate is the C_{15} tertiary allylic isomer, nerolidyl pyrophosphate. However, attack on the distal double bond can proceed directly from the farnesyl cation without prior isomerization because there is no stereochemical impediment to formation of rings containing 10 or 11 carbon atoms.[134] Following initial cyclization, the resulting cation may undergo further internal additions, hydride shifts, or rearrangements before the reaction is terminated by deprotonation or capture of the cation by a nucleophile, in direct parallel to the processes occurring during the course of monoterpene cyclization. Sesquiterpene cyclases have been isolated from several species of plants.[135-137] These enzymes closely resemble the monoterpene cyclases in overall properties.

2. Secondary Transformations

Further transformations of sesquiterpenes are very common and may lead to extensive modification of the primary cyclic products. Unfortunately, almost nothing is known about the biosynthetic sequences involved in such secondary transformations. For example, in the formation of polycyclic sesquiterpenes, such as longifolene (Figure 16), it is unclear if multiple cyclization steps are

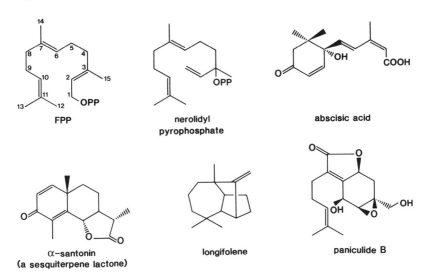

FIGURE 16. Representative sesquiterpenes and some intermediates of sesquiterpene biosynthesis. FPP is the precursor of all sesquiterpenes, while nerolidyl pyrophosphate is an intermediate in sesquiterpene cyclization processes that form six- or seven-membered rings. Abscisic acid is not a true sesquiterpene, but is formed via cleavage of a tetraterpene precursor.

catalyzed by a single enzyme or if there are multiple enzymes each acting on a free intermediate en route to the final product. For the many oxygenated sesquiterpenes, such as paniculide B, there is little evidence to indicate how oxygen is introduced into the carbon skeleton or the order of introduction of oxygen substituents.

One C_{15} terpenoid with several oxygen functions is the well-known plant hormone abscisic acid (Figure 16). Although the pathway of abscisic acid biosynthesis has not been fully elucidated, recent findings suggest that this substance is not derived directly from FPP, but instead arises from the cleavage of a C_{40} carotenoid precursor and thus is not in the strict sense a sesquiterpene.[138,139]

C. DITERPENES

After sesquiterpenes, the next size class is the diterpenes (C_{20}). These compounds are widespread and abundant in the plant kingdom, like the sesquiterpenes, and are often encountered in the resins of conifers, woody legumes, composites, and members of the Euphorbiaceae (the spurge family). Over 3000 different diterpene structures are known, all of which appear to be derived from the acyclic, C_{20} prenyl pyrophosphate, GGPP (Figure 8).

Although most plant diterpenes are cyclic, phytol (Figure 17), a derivative of geranylgeraniol in which three of the four double bonds are reduced, is an important acyclic diterpene. Phytol forms the lipophilic side chain of chlorophyll and, as a consequence, is nearly ubiquitous in higher plants. The biosynthesis of phytol proceeds by the attachment of GGPP to an incipient chlorophyll

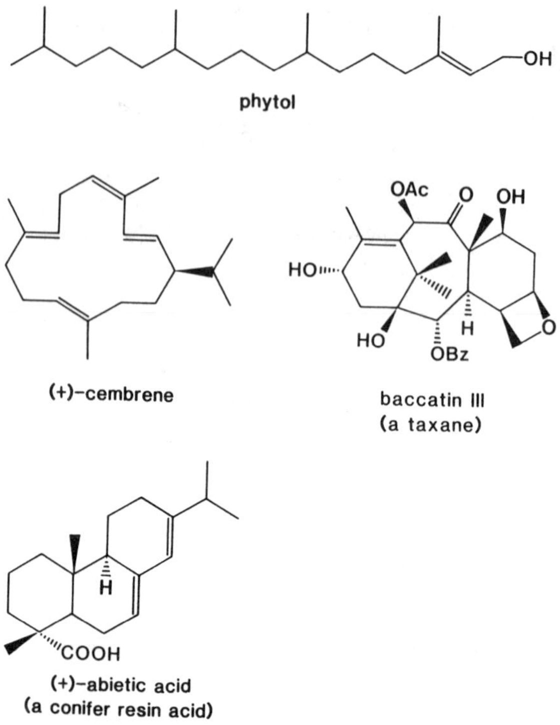

phytol

(+)-cembrene

OAc

OH

HO

HO

OBz

H

O

baccatin III
(a taxane)

H

COOH

(+)-abietic acid
(a conifer resin acid)

FIGURE 17.   Representative diterpenes.

molecule, forming a geranylgeranyl ester side chain that is then reduced *in situ* to give a phytol residue.[140]

1. Diterpene Cyclization

The origin of most structural types of diterpenes is not well-understood. There appear to be two major modes of cyclization in this class. The macrocyclic diterpenes, such as casbene and cembrene (Figures 17 and 18), are believed to be formed by cyclizations analogous to those of the monoterpene and sesquiterpene series.[140] Ionization of GGPP generates an allylic carbocationic center which adds intramolecularly to the terminal double bond of the geranylgeranyl chain (Figure 18). Proton elimination then gives a macrocyclic system with a 14- or 15-membered ring. The only reaction of this type that has been investigated in detail is the formation of casbene, a diterpene phytoalexin in castor bean seedlings. Robinson and West at UCLA have proposed the carbocationic mechanism in Figure 18 for the cyclization of GGPP to casbene[141] and have studied the enzymology of this conversion. Casbene cyclase is a monomer of M_r 59,000 that is localized in plastids, requires Mg^{2+} for activity, and possesses many other characteristics similar to those of monoterpene and sesquiterpene cyclases.[121,142,143] Several structural

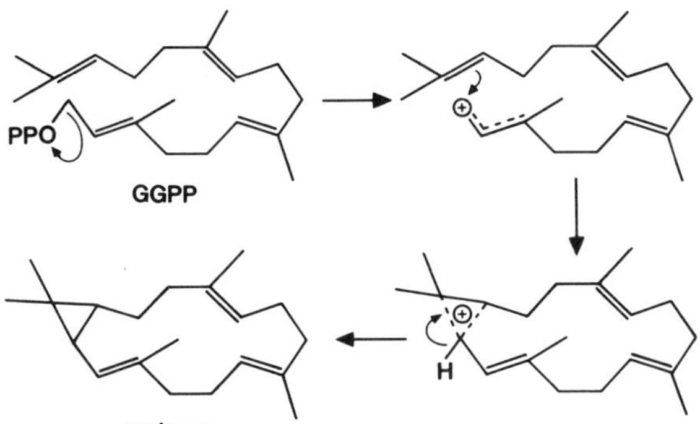

FIGURE 18. Proposed mechanism for the cyclization of GGPP to the diterpene phytoalexin casbene.

types of diterpenes with multiple rings, including the taxanes (Figure 17), are suspected to be derived from macrocyclic diterpenes via further cyclizations.[140,144]

The other mode of diterpene cyclization generates a bicyclic pyrophosphate, copalyl pyrophosphate, as the initial cyclic intermediate (Figure 19). This reaction is thought to begin by protonation of the terminal double bond of GGPP followed by two internal additions and proton elimination.[140] Copalyl pyrophosphate can then be converted to a large variety of bicyclic, tricyclic, and tetracyclic diterpene skeletal types via ionization of the pyrophosphate ester and subsequent internal additions, rearrangements, and eliminations. A curious feature of diterpene biosynthesis is that these reaction sequences proceed along two parallel routes, each leading to a different enantiomer of copalyl pyrophosphate. Many diterpenes, such as abietic acid of conifer resin (Figure 17), are formed from the (+)-antipode of copalyl pyrophosphate (Figure 19), while others, including the gibberellins (Figure 20), are derived from the (–)-antipode, sometimes called *ent*-copalyl pyrophosphate. Many plants produce compounds of both enantiomeric series.[140]

The enzymology of this second mode of diterpene cyclization has been examined by West's group in cell-free extracts from several plant species.[140,145,146] Enzyme activities catalyzing the cyclization of GGPP to (–)-copalyl pyrophosphate were found, but were difficult to resolve from activities catalyzing the immediately following biosynthetic steps. When separated activities were recombined, channeling effects were observed, suggesting that multienzyme complexes may be involved in the conversion of GGPP, via copalyl pyrophosphate, to polycyclic diterpenes.

2. Secondary Transformations

Plant diterpenes usually bear a variety of different oxygen-containing functional groups. Thus, the initial cyclic olefins must undergo a variety of sub-

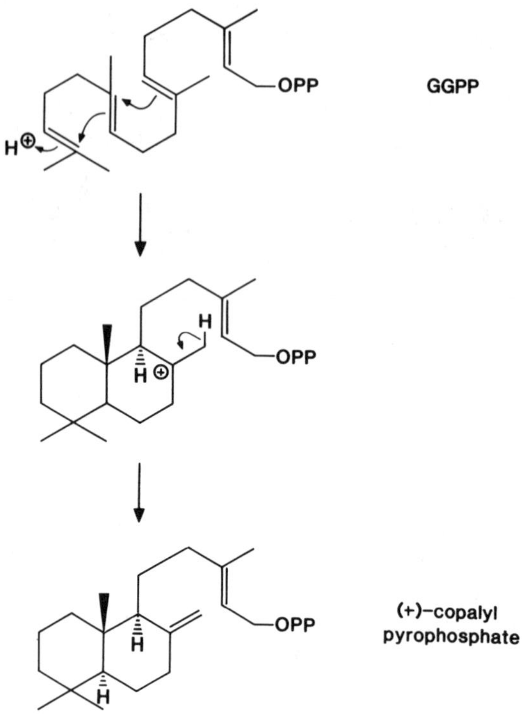

FIGURE 19. Proposed mechanism for the cyclization of GGPP to (+)-copalyl pyrophosphate.

sequent oxidative transformations. Most of these biosynthetic steps have not been studied, with the exception of those involved in gibberellin biosynthesis. The gibberellins, an important group of plant growth hormones, are highly functionalized diterpenes derived from (−)-copalyl pyrophosphate via the tetracyclic hydrocarbon *ent*-kaurene (Figure 20).[147,148] During gibberellin biosynthesis, one of the methyl groups at C_4 of *ent*-kaurene is oxidized to a carboxylic acid through a series of oxidations catalyzed by cytochrome P-450-dependent monooxygenases. Next, oxidation at C_7 and contraction of ring B from six to five members forms GA_{12} aldehyde, the precursor of the remaining 80 known gibberellins. From GA_{12} aldehyde, a variety of additional oxidative steps may occur, such as the loss of the methyl group at C_{10}, formation of a lactone bridge between C_{10} and C_{19}, or hydroxylation at various positions.

D. OTHER TERPENES

1. Hemiterpenes

Although individual isoprene units are common substituents of a variety of different plant metabolites, the only naturally occurring C_5 terpenoid is isoprene (Figures 1 and 21). Other five-carbon compounds with isopentenoid structures, such as valine, are not products of the terpenoid pathway.

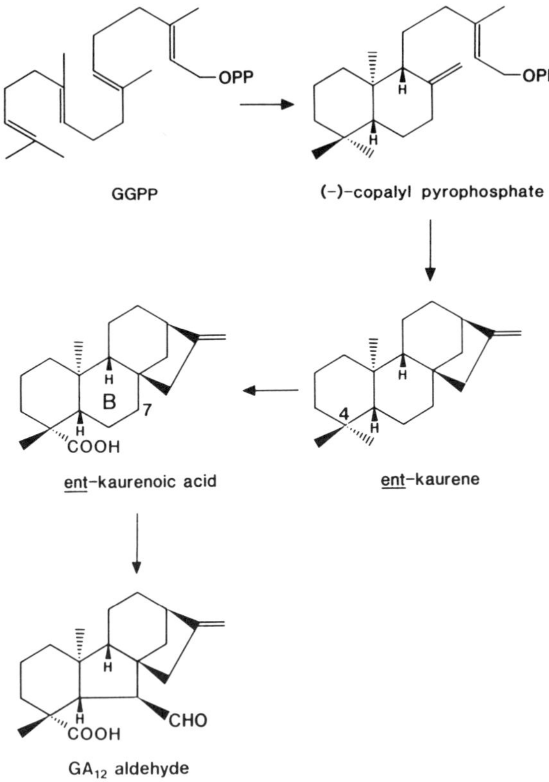

FIGURE 20. Pathway of gibberellin biosynthesis from GGPP to GA_{12} aldehyde.

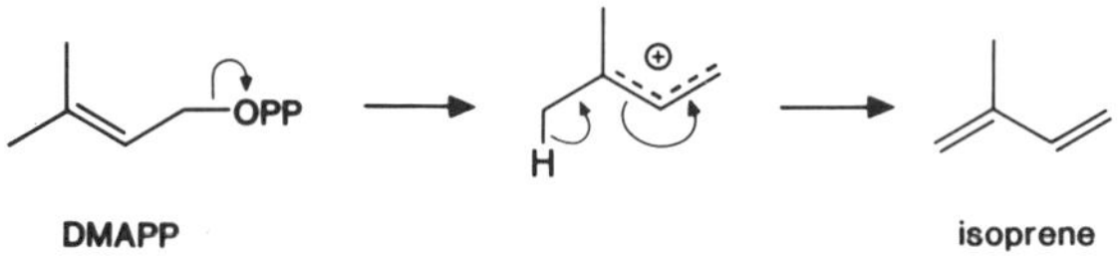

FIGURE 21. Proposed mechanism for the biosynthesis of isoprene from DMAPP.

Isoprene is a gas emitted from the photosynthesizing leaves of numerous species of plants.[149] Emission is not linked to photorespiration, as once believed,[150] and there is no connection between isoprene production and the accumulation of other terpenoids. Isoprene has received much attention recently because this substance is a major source of atmospheric hydrocarbon. However, from a plant perspective, the function of isoprene emission is unknown. Recently, isoprene was shown to be biosynthesized from DMAPP in aspen leaf extracts by a divalent metal ion-dependent activity.[151] The reaction is thought to proceed by ionization of the pyrophosphate moiety, followed by double bond migration and proton elimination (Figure 21).

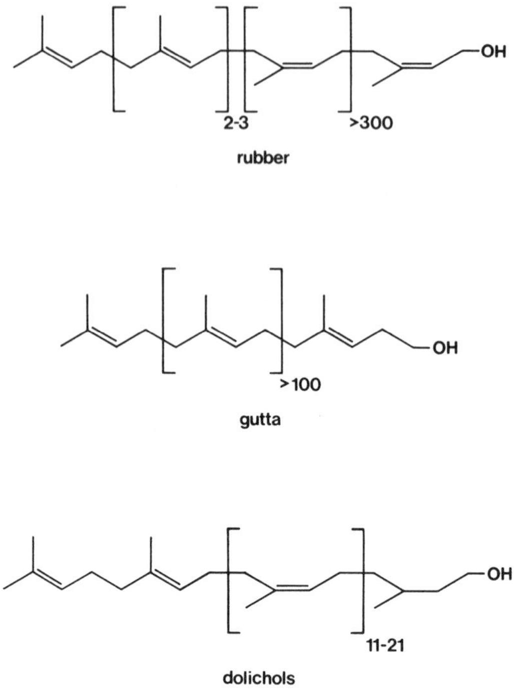

FIGURE 22.   Some plant polyterpenes.

2. Polyterpenes

Terpenoids containing more than eight isoprene units are known as polyterpenes. The largest of these substances are rubber, a linear hydrocarbon with nearly all *cis* double bonds that can have a molecular weight greater than 10^6, and gutta, a linear hydrocarbon of slightly lower molecular weight with *trans* double bonds (Figure 22).[152] Several kinds of long-chain polyterpene alcohols are found in plants, including the dolichols, compounds with 14 to 24 isoprene units that function as membrane-bound coenzymes in the glycosylation of proteins, lipids, and polysaccharides.[153] Nearly all the double bonds of dolichols have the *cis* configuration, except for the two isoprene residues farthest from the alcohol function (Figure 22).

The long carbon chains of polyterpenes are formed by specialized prenyltransferases (Section III.B) that catalyze the polymerization of many IPP units starting from a primer of DMAPP, GPP, FPP, or GGPP.[89,152,154] These enzymes specifically generate either *cis* or *trans* double bonds during the course of the reaction. The prenyltransferases involved in rubber biosynthesis in guayule and the rubber tree have been the subject of frequent investigation.[20,89,99,154-156]

3. Prenylated Compounds

A diverse assortment of plant metabolites contain a terpene moiety as a substituent on a nonterpenoid carbon skeleton, typically an aromatic ring

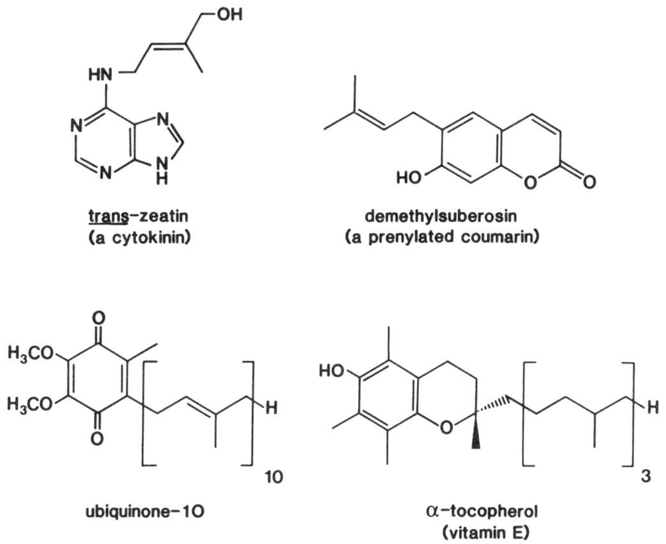

FIGURE 23. Examples of prenylated compounds found in plants.

system. These substances, referred to as prenylated compounds, include the cytokinins and the modified tRNA base isopentenyl adenosine (C_5 terpene substituents), chlorophyll (Section IV.C) and the tocopherols (C_{20} terpene substituents), and ubiquinone and plastoquinone (C_{45} or C_{50} terpene substituents) (Figure 23). The tocopherols are antioxidants that protect chloroplast membranes from photooxidative deterioration,[157] while ubiquinone and plastoquinone are electron transport components involved in phosphorylation in the mitochondrion and chloroplast, respectively.[158] The hydrophobic terpene side chains of these compounds facilitate attachment to or movement within membranes.

The biosynthesis of prenylated compounds involves the alkylation of a nonterpenoid intermediate by an allylic prenyl pyrophosphate. These reactions are catalyzed by specialized prenyltransferases that are thought to function in a manner similar to the prenyltransferases of the main terpenoid pathway (Section III.B.1) (Figure 10). Only a few enzymes of this type have been investigated in plants.[159-161] The properties of these are broadly similar to the properties of other prenyltransferases, except that prenyltransferases utilizing nonterpenoid substrates as prenyl acceptors are frequently found to be associated with membrane fractions in cell-free extracts rather than being operationally soluble.

V. REGULATION OF TERPENOID BIOSYNTHESIS

As this chapter attests, plants produce an enormous variety of terpenoid compounds. Some are found in all tissues (e.g., sterols, ubiquinone); some occur only in photosynthetic tissues (e.g., phytol, plastoquinone); and some are restricted solely to specialized cell types, such as secretory structures (e.g., most monoterpenes and sesquiterpenes). Therefore, regulatory mechanisms must exist to insure the selective operation of different branches of terpenoid biosynthesis in different parts of the plant at different stages of development. In this section, we describe a range of mechanisms that appear to regulate plant terpenoid formation at various levels of organization.

A. CHANGES IN ENZYME ACTIVITY

In discussing the individual reactions of terpenoid biosynthesis, we have already noted the possible regulatory importance of certain enzymes in the control of terpenoid formation. For example, HMG-CoA reductase shows changes in activity that are frequently well-correlated with changes in the overall rate of terpenoid biosynthesis (Section II.B.3), suggesting that this enzyme may catalyze the rate-limiting step of terpenoid biosynthesis in plants, as it does in animals. However, the prenyltransferases (Section III.B.3) and cyclases (Section IV.A.1), branch point enzymes situated later in the pathway, have also been implicated in the regulation of terpenoid formation. Different segments of the pathway could possess different regulatory features. Indeed, the branching of the terpenoid pathway in plants is more prolific than that in other organisms and may therefore necessitate more complex mechanisms of control. Determining the specific enzymes that regulate plant terpenoid biosynthesis will require much additional research.

B. SUBCELLULAR COMPARTMENTATION

Plant metabolism is extensively compartmented at the cellular level. Nearly all major reaction sequences contain segments that are known to occur in specific organelles.[162,163] Compartmentation has important consequences for the regulation of terpenoid biosynthesis, since it allows the independent control of different branches of the pathway at different sites in the cell. Within a compartment, the magnitude and direction of metabolic flux is dependent on the nature of the enzymes present and the permeability of intracellular membranes to precursors, intermediates, and products.

The intracellular compartmentation of terpenoid biosynthesis has been a frequent subject for investigation. Three separate compartments, the plastids, the mitochondria, and the cytoplasm, have been shown to be capable of converting IPP to various terpenoid metabolites.[73,164] Each compartment produces a unique range of products. For example, chloroplasts synthesize carotenoids, tocopherols, plastoquinone, and the phytol side chain of chlorophyll.[73,164] Other types of plastids are responsible for monoterpene and diterpene synthesis.[64,109,165] Elsewhere in the cell, the mitochondria are the apparent site of ubiquinone formation,[166] while the production of sesquiterpenes and sterols occurs in the cytoplasm and the endoplasmic reticulum.[73,110] Some of the secondary transformations of monoterpenes and diterpenes also seem to take place in the cytoplasm and endoplasmic reticulum.[73,167]

Although the incorporation of IPP into terpenoids has been demonstrated to take place at several distinct subcellular sites, the localization of the steps of terpenoid biosynthesis prior to the formation of IPP is still the subject of considerable debate. One point of view asserts that each site of synthesis (plastids, mitochondria, cytoplasm) has all of the enzymatic machinery of the mevalonate pathway starting from acetyl-CoA acetyltransferase.[22,168] An opposing hypothesis is that all the steps in terpenoid biogenesis prior to IPP are located solely in the cytoplasm. The IPP produced in the cytoplasm is then transported to various subcellular compartments for the formation of specific terpenoid metabolites.[164,169,170]

The results of a recent investigation on chloroplasts isolated from barley leaves may help to resolve this conflict.[171] Chloroplasts from young leaves were found to convert CO_2 efficiently to various terpenoid end products. Chloroplasts from mature leaves, on the other hand, were much less capable of terpenoid formation from CO_2, but were much more permeable to IPP than chloroplasts from young leaves. Mature chloroplasts thus appear to rely on an exogenous source of IPP for terpenoid biosynthesis. Therefore, both views on the intracellular organization of terpenoid biosynthesis may be substantially correct, each at a different stage of cell development.

C. MULTIENZYME COMPLEXES

The enzymes catalyzing sequential steps of a metabolic pathway are sometimes observed to form noncovalent aggregates called multienzyme complexes that serve to channel substrates through a reaction sequence without allowing intermediates to diffuse away.[172,173] Multienzyme complexes can be regarded as another form of compartmentation that could function in metabolic regulation by segregating competing pathways and by allowing the activities of a single complex to be controlled in unison. Unfortunately, multienzyme complexes are difficult to detect in higher plants, probably because they are easily disrupted during conventional extraction procedures.[173]

Several examples of multienzyme complexes are known in plant terpenoid biosynthesis. In radish seedling extracts, for instance, the first two enzymes of the mevalonate pathway, acetyl-CoA acetyltransferase and HMG-CoA synthase, bind together in a tight, membrane-associated complex (Section II.A).[8,19] In extracts of daffodil petals and tomato and pepper fruits, several enzymes of carotenoid biogenesis, including IPP isomerase, GGPP synthase (a prenyltransferase), and phytoene synthase, have been observed to form noncovalent aggregates.[68,69,174]

D. AVAILABILITY OF ACETYL-CoA, ATP, AND NADPH

The mevalonate pathway requires acetyl-CoA as its initial substrate and ATP and NADPH as co-factors (Figure 3). The availability of these metabolites could provide another mechanism for regulating terpenoid production, superimposed upon controls at the gene, enzyme, and subcellular levels. Acetyl-CoA, ATP, and NADPH are all ubiquitous cellular metabolites derived from the oxidation of stored carbohydrates or acyl lipids or arising directly from photosynthesis. One might therefore expect a close correlation between the rate of respiration (or the rate of photosynthesis) and the production of terpenoids. In fact, in developing lemongrass leaves, the enzyme activities involved in the generation of NADPH and the mobilization of starch and sucrose are highest during the period of maximal monoterpene formation and so could limit the overall rate of monoterpene production.[175,177] In a wide range of species, experimental manipulations that enhanced the rate of photosynthesis, such as increased light intensity, were also found to enhance the pace of terpenoid production.[178-184]

E. MORPHOLOGICAL DIFFERENTIATION

A large variety of plant monoterpenes, sesquiterpenes, and diterpenes are found in specialized secretory structures, such as epidermal hairs, resin ducts, resin cavities, and latex vessels.[185] Recent investigations indicate that these secretory structures are active sites of terpenoid synthesis. For example, the glandular hairs of several species of mints have been shown to have the capacity to produce monoterpenes.[186-188] For many plant monoterpenes, sesquiterpenes, and diterpenes, the presence of specialized secretory structures actually appears to be a prerequisite for synthesis.[189,190] Therefore, the differentiation of such structures may provide another form of control over terpene formation. Morphological differentiation is also of crucial importance in the synthesis of α-tocopherol, plastoquinone, the phytol side chain of chlorophyll, and other terpenoids by chloroplasts. These substances are not found in the undeveloped plastids of dark-grown, etiolated plants, but only appear in differentiated chloroplasts characteristic of green tissue.[143,191]

In this section, we have seen that terpenoid biosynthesis in plants is subject to a complex interplay of different controls acting at the enzyme, subcellular, cellular, and organ levels. Although, at present, there is only limited experimental support for the operation of most of these control mechanisms, the rapid advances in plant biochemistry and molecular biology (Section I.D) should lead to significant progress in the study of the regulation of terpene biosynthesis within the next few years.

ACKNOWLEDGMENTS

The work by the authors cited herein was supported by grants from the U.S. Department of Agriculture, National Science Foundation, Department of Energy, and National Institutes of Health. We thank Joyce Tamura-Brown for typing the manuscript.

REFERENCES

1. **Ruzicka, L.,** The isoprene rule and the biogenesis of terpenic compounds, *Experientia,* 9, 357, 1953.
2. **Connolly, J. D. and Hill, R. A.,** *Dictionary of Terpenoids,* Chapman and Hall, London, 1992.
3. **Sinensky, M. and Lutz, R. J.,** The prenylation of proteins, *BioEssays,* 14, 25, 1992.
4. **Gershenzon, J. and Croteau, R.,** Terpenoids, in *Herbivores: Their Interactions with Secondary Plant Metabolites,* Vol. 1, 2nd ed., Rosenthal, G. A. and Berenbaum, M. R., Eds., Academic Press, San Diego, CA, 1991, 165.
5. **Harborne, J. B.,** *Introduction to Ecological Biochemistry,* 3rd ed., Academic Press, London, 1988.
6. **Harborne, J. B.,** Recent advances in the ecological chemistry of plant terpenoids, in *Ecological Chemistry and Biochemistry of Plant Terpenoids,* Harborne, J. B. and Tomas-Barberan, F. A., Eds., Oxford University Press, Oxford, 1991, 399.
7. **Croteau, R. and Cane, D. E.,** Monoterpene and sesquiterpene cyclases, *Methods Enzymol.,* 110, 352, 1985.
8. **Bach, T. J., Boronat, A., Caelles, C., Ferrer, A., Weber, T., and Wettstein, A.,** Aspects related to mevalonate biosynthesis in plants, *Lipids,* 26, 637, 1991.
9. **Bartley, G. E., Viitanen, P. V., Bacot, K. O., and Scolnik, P. A.,** A tomato gene expressed during fruit ripening encodes an enzyme of the carotenoid biosynthesis pathway, *J. Biol. Chem.,* 267, 5036, 1992.
10. **Caelles, C., Ferrer, A., Balcells, L., Hegardt, F. G., and Boronat, A.,** Isolation and structural characterization of a cDNA encoding *Arabidopsis thaliana* 3-hydroxy-3-methylglutaryl coenzyme A reductase, *Plant Mol. Biol.,* 13, 627, 1989.
11. **Chye, M.-L., Tan, C.-T., and Chua, N.-H.,** Three genes encode 3-hydroxy-3-methylglutaryl-coenzyme A reductase in *Hevea brasiliensis: hmg1* and *hmg3* are differentially expressed, *Plant Mol. Biol.,* 19, 473, 1992.
12. **Kuntz, M., Romer, S., Suire, C., Hugueney, P., Weil, J. H., Schantz, R., and Camara, B.,** Identification of a cDNA for the plastid-located geranylgeranyl pyrophosphate synthase from *Capsicum annuum:* correlative increase in enzyme activity and transcript level during fruit ripening, *Plant J.,* 2, 25, 1992.
13. **Learned, R. M. and Fink, G. R.,** 3-Hydroxy-3-methylglutaryl-coenzyme A reductase from *Arabidopsis thaliana* is structurally distinct from the yeast and animal enzymes, *Proc. Natl. Acad. Sci. U.S.A.,* 86, 2779, 1989.
14. **Monfar, M., Caelles, C., Balcells, L., Ferrer, A., Hegardt, F. G., and Boronat, A.,** Molecular cloning and characterization of plant 3-hydroxy-3-methylglutaryl coenzyme A reductase, in *Biochemistry of the Mevalonic Acid Pathway to Terpenoids,* Towers, G. H. N. and Stafford, H. A., Eds., Plenum Press, New York, 1990, 83.
15. **Narita, J. O. and Gruissem, W.,** Tomato hydroxymethylglutaryl-CoA reductase is required early in fruit development but not during ripening, *Plant Cell,* 1, 181, 1989.
16. **Bu'lock, J. D., De Rosa, M., and Gambacorta, A.,** Isoprenoid biosynthesis in archaebacteria, in *Biosynthesis of Isoprenoid Compounds,* Vol. 2, Porter, J. W. and Spurgeon, S. L., Eds., John Wiley & Sons, New York, 1983, 159.
17. **Qureshi, N. and Porter, J. W.,** Conversion of acetyl-coenzyme A to isopentenyl pyrophosphate, in *Biosynthesis of Isoprenoid Compounds,* Vol. 1, Porter, J. W. and Spurgeon, S. L., Eds., John Wiley & Sons, New York, 1981, 47.
18. **Vollmer, S. H., Mende-Mueller, L. M., and Miziorko, H. M.,** Identification of the site of acetyl-S-enzyme formation on avian liver mitochondrial 3-hydroxy-3-methylglutaryl-CoA synthase, *Biochemistry,* 27, 4288, 1988.
19. **Bach, T. J., Weber, T., and Motel, A.,** Some properties of enzymes involved in the biosynthesis and metabolism of 3-hydroxy-3-methylglutaryl-CoA in plants, in *Biochemistry of the Mevalonic Acid Pathway to Terpenoids,* Towers, G. H. N. and Stafford, H. A., Eds., Plenum Press, New York, 1990, 1.

20. **Kekwick, R. G. O.**, The formation of isoprenoids in *Hevea* latex, in *Physiology of Rubber Tree Latex*, d'Auzac, J., Jacob, J.-L., and Chrestin, H., Eds., CRC Press, Boca Raton, FL, 1989, 145.

21. **Anastasis, P., Freer, I., Overton, K. H., Picken, D., Rycroft, D. S., and Singh, S. B.**, On the role of leucine in terpenoid metabolism, *Perkin Trans. I*, 2427, 1987.

22. **Schulze-Siebert, D. and Schultz, G.**, β-carotene synthesis in isolated spinach chloroplasts: its tight linkage to photosynthetic carbon metabolism, *Plant Physiol.*, 84, 1233, 1987.

23. **Russell, D. W.**, 3-Hydroxy-3-methylglutaryl-CoA reductases from pea seedlings, *Methods Enzymol.*, 110, 26, 1985.

24. **Bach, T. J., Rogers, D. H., and Rudney, H.**, Detergent-solubilization, purification, and characterization of membrane-bound 3-hydroxy-3-methylglutaryl-coenzyme A reductase from radish seedlings, *Eur. J. Biochem.*, 154, 103, 1986.

25. **Kondo, K. and Oba, K.**, Purification and characterization of 3-hydroxy-3-methylglutaryl CoA reductase from potato tubers, *J. Biochem.*, 100, 967, 1986.

26. **Boll, M., Kardinal, A., and Berndt, J.**, Properties and regulation of 3-hydroxy-3-methylglutaryl coenzyme A reductase from spruce (*Picea abies*), *Biol. Chem. Hoppe-Seyler*, 368, 1024, 1987.

27. **Brooker, J. D. and Russell, D. W.**, Subcellular localization of 3-hydroxy-3-methylglutaryl coenzyme A reductase in *Pisum sativum* seedlings, *Arch. Biochem. Biophys.*, 167, 730, 1975.

28. **Suzuki, H. and Uritani, I.**, Subcellular localization of 3-hydroxy-3-methylglutaryl coenzyme A reductase and other membrane-bound enzymes in sweet potato roots, *Plant Cell Physiol.*, 17, 691, 1976.

29. **Arebalo, R. E. and Mitchell, E. D.**, Cellular distribution of 3-hydroxy-3-methylglutaryl coenzyme A reductase and mevalonate kinase in leaves of *Nepeta cataria*, *Phytochemistry*, 23, 13, 1984.

30. **Camara, B., Bardat, F., Dogbo, O., Brangeon, J., and Moneger, R.**, Terpenoid metabolism in plastids: isolation and biochemical characteristics of *Capsicum annuum* chromoplasts, *Plant Physiol.*, 73, 94, 1983.

31. **Ramachandra Reddy, A. and Rama Das, V. S.**, Partial purification and characterization of 3-hydroxy-3-methylglutaryl coenzyme A reductase from the leaves of guayule (*Parthenium argentatum*), *Phytochemistry*, 25, 2471, 1986.

32. **Wong, R. J., McCormack, D. K., and Russell, D. W.**, Plastid 3-hydroxy-3-methylglutaryl coenzyme A reductase has distinctive kinetic and regulatory features: properties of the enzyme and positive phytochrome control of activity in pea seedlings, *Arch. Biochem. Biophys.*, 216, 631, 1982.

33. **Bach, T. J.**, Synthesis and metabolism of mevalonic acid in plants, *Plant Physiol. Biochem.*, 25, 163, 1987.

34. **Bach, T. J.**, Hydroxymethylglutaryl-CoA reductase, a key enzyme in phytosterol synthesis?, *Lipids*, 21, 82, 1986.

35. **Yang, Z., Park, H., Lacy, G. H., and Cramer, C. L.**, Differential activation of potato 3-hydroxy-3-methylglutaryl coenzyme A reductase genes by wounding and pathogen challenge, *Plant Cell*, 3, 397, 1991.

36. **Stermer, B. A. and Bostock, R. M.**, Involvement of 3-hydroxy-3-methylglutaryl coenzyme A reductase in the regulation of sesquiterpenoid phytoalexin synthesis in potato, *Plant Physiol.*, 84, 404, 1987.

37. **Oba, K. and Uritani, I.**, Biosynthesis of furanoterpenes by sweet potato cell culture, *Plant Cell Physiol.*, 20, 819, 1979.

38. **Chappell, J., VonLanken, C., and Vogeli, U.**, Elicitor-inducible 3-hydroxy-3-methylglutaryl coenzyme A reductase activity is required for sesquiterpene accumulation in tobacco cell suspension cultures, *Plant Physiol.*, 97, 693, 1991.

39. **Nishi, A. and Tsuritani, I.**, Effect of auxin on the metabolism of mevalonic acid in suspension-cultured carrot cells, *Phytochemistry*, 22, 399, 1983.

40. **Ramachandra Reddy, A., Suhasini, M., and Rama Das, V. S.,** Enhanced photorespiration and rubber yield in Cycocel-treated guayule plants (*Parthenium argentatum* Gray), *Acta Physiol. Plant.*, 12, 193, 1990.

41. **Skrukrud, C. L., Taylor, S. E., Hawkins, D. R., Nemethy, E. K., and Calvin, M.,** Subcellular fractionation of triterpenoid biosynthesis in *Euphorbia lathyris* latex, *Physiol. Plant.*, 74, 306, 1988.

42. **Hepper, C. M. and Audley, B. G.,** The biosynthesis of rubber from β-hydroxy-β-methylglutaryl-coenzyme A in *Hevea brasiliensis* latex, *Biochem. J.*, 114, 379, 1969.

43. **Dugan, R. E.,** Regulation of HMG-CoA reductase, in *Biosynthesis of Isoprenoid Compounds*, Vol. 1, Porter, J. W. and Spurgeon, S. L., Eds., John Wiley & Sons, New York, 1981, 95.

44. **Nakanishi, M., Goldstein, J. L., and Brown, M. S.,** Multivalent control of 3-hydroxy-3-methylglutaryl coenzyme A reductase: mevalonate-derived product inhibits translation of mRNA and accelerates degradation of enzyme, *J. Biol. Chem.*, 263, 8929, 1988.

45. **Sabine, J. R.,** *3-Hydroxy-3-Methylglutaryl Coenzyme A Reductase*, CRC Press, Boca Raton, FL, 1983.

46. **Zammit, V. A. and Easom, R. A.,** Regulation of hepatic HMG-CoA reductase *in vivo* by reversible phosphorylation, *Biochim. Biophys. Acta*, 927, 223, 1987.

47. **Ito, R., Oba, K., and Uritani, I.,** Mechanism for the induction of 3-hydroxy-3-methylglutaryl coenzyme A reductase in $HgCl_2$-treated sweet potato root tissue, *Plant Cell Physiol.*, 20, 867, 1979.

48. **Oba, K., Tatematsu, H., Yamashita, K., and Uritani, I.,** Induction of furano-terpene production and formation of the enzyme system from mevalonate to isopentenyl pyrophosphate in sweet potato root tissue injured by *Ceratocystis fimbriata* and by toxic chemicals, *Plant Physiol.*, 58, 51, 1976.

49. **Budde, R. J. A. and Chollet, R.,** Regulation of enzyme activity in plants by reversible phosphorylation, *Physiol. Plant.*, 72, 435, 1988.

50. **Sipat, A. B.,** Hydroxymethylglutaryl CoA reductase (NADPH) in the latex of *Hevea brasiliensis*, *Phytochemistry*, 21, 2613, 1982.

51. **Wititsuwannakul, R., Wititsuwannakul, D., and Dumkong, S.,** *Hevea* calmodulin: regulation of the activity of latex 3-hydroxy-3-methylglutaryl coenzyme A reductase, *Phytochemistry*, 29, 1755, 1990.

52. **Brooker, J. D. and Russell, D. W.,** Properties of microsomal 3-hydroxy-3-methylglutaryl coenzyme A reductase from *Pisum sativum* seedlings, *Arch. Biochem. Biophys.*, 167, 723, 1975.

53. **Chun, K. T. and Simoni, R. D.,** The role of the membrane domain in the regulated degradation of 3-hydroxy-3-methylglutaryl coenzyme A reductase, *J. Biol. Chem.*, 267, 4236, 1992.

54. **Gray, J. C. and Kekwick, R. G. O.,** Mevalonate kinase in green leaves and etiolated cotyledons of the French bean *Phaseolus vulgaris*, *Biochem. J.*, 133, 335, 1973.

55. **Lalitha, R. and Ramasarma, T.,** Mevalonate phosphorylation in lemon grass leaves, *Indian J. Biochem. Biophys.*, 23, 249, 1986.

56. **Lalitha, R., George, R., and Ramasarma, T.,** Mevalonate-metabolizing enzymes in *Arachis hypogaea*, *Mol. Cell. Biochem.*, 87, 161, 1989.

57. **Skilleter, D. N. and Kekwick, R. G. O.,** The enzymes forming isopentenyl pyrophosphate from 5-phosphomevalonate (mevalonate 5-phosphate) in the latex of *Hevea brasiliensis*, *Biochem. J.*, 124, 407, 1971.

58. **Eyzaguirre, J. and Bazaes, S.,** Phosphomevalonate kinase from pig liver, *Methods Enzymol.*, 110, 78, 1985.

59. **Porter, J. W.,** Mevalonate kinase, *Methods Enzymol.*, 110, 71, 1985.

60. **Kreuz, K. and Kleinig, H.,** On the compartmentation of isopentenyl diphosphate synthesis and utilization in plant cells, *Planta*, 153, 578, 1981.

61. **Knotz, J., Coolbaugh, R. C., and West, C. A.,** Regulation of the biosynthesis of *ent*-kaurene from mevalonate in the endosperm of immature *Marah macrocarpus* seeds by adenylate energy charge, *Plant Physiol.*, 60, 81, 1977.

62. **Jabalquinto, A. M., Alvear, M., and Cardemil, E.,** Physiological aspects and mechanism of action of mevalonate 5-diphosphate decarboxylase, *Comp. Biochem. Physiol.*, 90B, 671, 1988.

63. **Jabalquinto, A. M. and Cardemil, E.,** Substrate binding order in mevalonate 5-diphosphate decarboxylase from chicken liver, *Biochim. Biophys. Acta*, 996, 257, 1989.

64. **Dudley, M. W., Dueber, M. T., and West, C. A.,** Biosynthesis of the macrocyclic diterpene casbene in castor bean (*Ricinus communis* L.) seedlings: changes in enzyme levels induced by fungal infection and intracellular localization of the pathway, *Plant Physiol.*, 81, 335, 1986.

65. **Chiew, Y. E., O'Sullivan, W. J., and Lee, C. S.,** Studies on pig liver mevalonate-5-diphosphate decarboxylase, *Biochim. Biophys. Acta*, 916, 271, 1987.

66. **Landau, B. R. and Brunengraber, H.,** Shunt pathway of mevalonate metabolism, *Methods Enzymol.*, 110, 100, 1985.

67. **Nes, W. D. and Bach, T. J.,** Evidence for a mevalonate shunt in a tracheophyte, *Proc. R. Soc. London*, B225, 425, 1985.

68. **Dogbo, O. and Camara, B.,** Purification of isopentenyl pyrophosphate isomerase and geranylgeranyl pyrophosphate synthase from *Capsicum* chromoplasts by affinity chromatography, *Biochim. Biophys. Acta*, 920, 140, 1987.

69. **Lutzow, M. and Beyer, P.,** The isopentenyl-diphosphate Δ−isomerase and its relation to the phytene synthase complex in daffodil chromoplasts, *Biochim. Biophys. Acta*, 959, 118, 1988.

70. **Ogura, K., Nishino, T., and Seto, S.,** The purification of prenyltransferase and isopentenyl pyrophosphate isomerase of pumpkin fruit and their some properties, *J. Biochem.*, 64, 197, 1968.

71. **Spurgeon, S. L., Sathyamoorthy, N., and Porter, J. W.,** Isopentenyl pyrophosphate isomerase and prenyltransferase from tomato fruit plastids, *Arch. Biochem. Biophys.*, 230, 446, 1984.

72. **Widmaier, R., Howe, J., and Heinstein, P.,** Prenyltransferase from *Gossypium hirsutum*, *Arch. Biochem. Biophys.*, 200, 609, 1980.

73. **Kleinig, H.,** The role of plastids in isoprenoid biosynthesis, *Annu. Rev. Plant Physiol. Plant Mol. Biol.*, 40, 39, 1989.

74. **Muehlbacher, M. and Poulter, C. D.,** Isopentenyl-diphosphate isomerase: inactivation of the enzyme with active-site-directed irreversible inhibitors and transition-state analogues, *Biochemistry*, 27, 7315, 1988.

75. **Reardon, J. E. and Abeles, R. H.,** Mechanism of action of isopentenyl pyrophosphate isomerase: evidence for a carbonium ion intermediate, *Biochemistry*, 25, 5609, 1986.

76. **Poulter, C. D. and Rilling, H. C.,** Prenyl transferases and isomerase, in *Biosynthesis of Isoprenoid Compounds*, Vol. 1, Porter, J. W. and Spurgeon, S. L., Eds., John Wiley & Sons, New York, 1981, 161.

77. **Street, I. P. and Poulter, C. D.,** Isopentenyldiphosphate:dimethylallyldiphosphate isomerase: construction of a high-level heterologous expression system for the gene from *Saccharomyces cerevisiae* and identification of an active-site nucleophile, *Biochemistry*, 29, 7531, 1990.

78. **Street, I. P., Coffman, H. R., and Poulter, C. D.,** Isopentenyl diphosphate isomerase. Site-directed mutagenesis of cys139 using "counter" PCR amplification of an expression plasmid, *Tetrahedron*, 47, 5919, 1991.

79. **Hugueney, P. and Camara, B.,** Purification and characterization of farnesyl pyrophosphate synthase from *Capsicum annuum*, *FEBS Lett.*, 273, 235, 1990.

80. **Perez, L. M., Lozada, R., and Cori, O.,** Biosynthesis of allylic isoprenoid pyrophosphates by an enzyme preparation from the flavedo of *Citrus paradisii*, *Phytochemistry*, 22, 431, 1983.

81. **Laferriere, A. and Beyer, P.,** Purification of geranylgeranyl diphosphate synthase from *Sinapis alba* etioplasts, *Biochim. Biophys. Acta*, 1077, 167, 1991.

82. **Ogura, K., Shinka, T., and Seto, S.,** The purification and properties of geranylgeranyl pyrophosphate synthetase from pumpkin fruit, *J. Biochem.*, 72, 1101, 1972.

83. **Dudley, M. W., Green, T. R., and West, C. A.,** Biosynthesis of the macrocyclic diterpene casbene in castor bean (*Ricinus communis* L.) seedlings: the purification and properties of farnesyl transferase from elicited seedlings, *Plant Physiol.*, 81, 343, 1986.

84. **Anderson, M. S., Yarger, J. G., Burck, C. L., and Poulter, C. D.,** Farnesyl diphosphate synthetase: molecular cloning, sequence, and expression of an essential gene from *Saccharomyces cerevisiae, J. Biol. Chem.*, 264, 19176, 1989.

85. **Carattoli, A., Romano, N., Ballario, P., Morelli, G., and Macino, G.,** The *Neurospora crassa* carotenoid biosynthetic gene (albino 3) reveals highly conserved regions among prenyltransferases, *J. Biol. Chem.*, 266, 5854, 1991.

86. **Clarke, C. R., Tanaka, R. D., Svenson, K., Wamsley, M., Fogelman, A. M., and Edwards, P. A.,** Molecular cloning and sequence of a cholesterol-repressible enzyme related to prenyltransferase in the isoprene biosynthetic pathway, *Mol. Cell. Biol.*, 7, 3138, 1987.

87. **Sheares, B. T., White, S. S., Molowa, D. T., Chan, K., Ding, V. D.-H., Kroon, P. A., Bostedor, R. G., and Karkas, J. D.,** Cloning, analysis, and bacterial expression of human farnesyl pyrophosphate synthetase and its regulation in Hep G2 cells, *Biochemistry*, 28, 8129, 1989.

88. **Croteau, R. and Purkett, P. T.,** Geranyl pyrophosphate synthase: characterization of the enzyme and evidence that this chain-length specific prenyltransferase is associated with monoterpene biosynthesis in sage (*Salvia officinalis*), *Arch. Biochem. Biophys.*, 271, 524, 1989.

89. **Cornish, K. and Backhaus, R. A.,** Rubber transferase activity in rubber particles of guayule, *Phytochemistry*, 29, 3809, 1990.

90. **Poulter, C. D. and Rilling, H. C.,** The prenyl transfer reaction. Enzymatic and mechanistic studies of the 1'-4 coupling reaction in the terpene biosynthetic pathway, *Acc. Chem. Res.*, 11, 307, 1978.

91. **Brems, D. N. and Rilling, H. C.,** On the mechanism of the prenyltransferase reaction. Metal ion-dependent solvolysis of an allylic pyrophosphate, *J. Am. Chem. Soc.*, 99, 8351, 1977.

92. **Cornforth, J. W., Cornforth, R. H., Popjak, G., and Yengoyan, L.,** Studies on the biosynthesis of cholesterol. XX. Steric course of decarboxylation of 5-pyrophosphomevalonate and of the carbon to carbon bond formation in the biosynthesis of farnesyl pyrophosphate, *J. Biol. Chem.*, 241, 3970, 1966.

93. **Popjak, G. and Cornforth, J. W.,** Substrate stereochemistry in squalene biosynthesis, *Biochem. J.*, 101, 553, 1966.

94. **Ashby, M. N. and Edwards, P. A.,** Elucidation of the deficiency in two yeast coenzyme Q mutants: characterization of the structural gene encoding hexaprenyl pyrophosphate synthetase, *J. Biol. Chem.*, 265, 13157, 1990.

95. **Ashby, M. N., Spear, D. H., and Edwards, P. A.,** Prenyltransferases: from yeast to man, in *Molecular Biology of Atherosclerosis*, Attie, A. D., Ed., Elsevier, Amsterdam, 1990, 27.

96. **Banthorpe, D. V., Long, D. R. S., and Pink, C. R.,** Biosynthesis of geraniol and related monoterpenes in *Pelargonium graveolens*, *Phytochemistry*, 22, 2459, 1983.

97. **Green, T. R. and Baisted, D. J.,** Development of the squalene-synthesizing system during early stages of pea seed germination, *Biochem. J.*, 125, 1145, 1971.

98. **Green, T. R. and Baisted, D. J.,** Development of the activities of enzymes of the isoprenoid pathway during early stages of pea-seed germination, *Biochem. J.*, 130, 983, 1972.

99. **Madhavan, S., Greenblatt, G. A., Foster, M. A., and Benedict, C. R.,** Stimulation of isopentenyl pyrophosphate incorporation into polyisoprene in extracts from guayule plants (*Parthenium argentatum* Gray) by low temperature and 2-(3,4-dichlorophenoxy)-triethylamine, *Plant Physiol.*, 89, 506, 1989.

100. **Ramachandra Reddy, A. and Rama Das, V. S.,** Enhanced rubber accumulation and rubber transferase activity in guayule under water stress, *J. Plant Physiol.*, 133, 152, 1988.

101. **Croteau, R.,** Biosynthesis and catabolism of monoterpenoids, *Chem. Rev.*, 87, 929, 1987.

102. **Croteau, R.,** A stereochemical model for monoterpene cyclization, in *Models in Plant Physiology and Biochemistry*, Vol. 2, Newman, D. W. and Wilson, K. G., Eds., CRC Press, Boca Raton, FL, 1987, 55.

103. **Alonso, W. R. and Croteau, R.,** Purification and characterization of the monoterpene cyclase γ-terpinene synthase from *Thymus vulgaris*, *Arch. Biochem. Biophys.*, 286, 511, 1991.

104. **Alonso, W. R., Rajaonarivony, J. I. M., Gershenzon, J., and Croteau, R.,** Purification of 4S-limonene synthase, a monoterpene cyclase from the glandular trichomes of peppermint (*Mentha × piperita*) and spearmint (*Mentha spicata*), *J. Biol. Chem.*, 267, 7582, 1992.

105. **Croteau, R. and Shaskus, J.,** Biosynthesis of monoterpenes: demonstration of a geranyl pyrophosphate:(−)-bornyl pyrophosphate cyclase in soluble enzyme preparations from tansy (*Tanacetum vulgare*), *Arch. Biochem. Biophys.*, 236, 535, 1985.

106. **Lewinsohn, E., Gijzen, M., and Croteau, R.,** Wound-inducible pinene cyclase from grand fir: purification, characterization, and renaturation after SDS-PAGE, *Arch. Biochem. Biophys.*, 293, 167, 1992.

107. **Rajaonarivony, J. I. M., Gershenzon, J., and Croteau, R.,** Characterization and mechanism of (4S)-limonene synthase, a monoterpene cyclase from the glandular trichomes of peppermint (*Mentha × piperita*), *Arch. Biochem. Biophys.*, 296, 49, 1992.

108. **Rajaonarivony, J. I. M., Gershenzon, J., Miyazaki, J., and Croteau, R.,** Evidence for an essential histidine residue in 4S-limonene synthase and other terpene cyclases, *Arch. Biochem. Biophys.*, 299, 77, 1992.

109. **Gleizes, M., Pauly, G., Carde, J.-P., Marpeau, A., and Bernard-Dagan, C.,** Monoterpene hydrocarbon biosynthesis by isolated leucoplasts of *Citrofortunella mitis*, *Planta*, 159, 373, 1983.

110. **Belingheri, L., Pauly, G., Gleizes, M., and Marpeau, A.,** Isolation by an aqueous two-polymer phase system and identification of endomembranes from *Citrofortunella mitis* fruits for sesquiterpene hydrocarbon synthesis, *J. Plant Physiol.*, 132, 80, 1988.

111. **Croteau, R., Miyazaki, J. H., and Wheeler, C. J.,** Monoterpene biosynthesis: mechanistic evaluation of the geranyl pyrophosphate:(−)endo-fenchol cyclase from fennel (*Foeniculum vulgare*), *Arch. Biochem. Biophys.*, 269, 507, 1989.

112. **Croteau, R.,** Evidence for the ionization steps in monoterpene cyclization reactions using 2-fluorogeranyl and 2-fluorolinalyl pyrophosphates as substrates, *Arch. Biochem. Biophys.*, 251, 777, 1986.

113. **Cane, D. E., Saito, A., Croteau, R., Shaskus, J., and Felton, M.,** Enzymatic cyclization of geranyl pyrophosphate to bornyl pyrophosphate. Role of the pyrophosphate moiety, *J. Am. Chem. Soc.*, 104, 5831, 1982.

114. **Croteau, R., Wheeler, C. J., Aksela, R., and Oehlschlager, A. C.,** Inhibition of monoterpene cyclases by sulfonium analogs of presumptive carbocationic intermediates of the cyclization reaction, *J. Biol. Chem.*, 261, 7257, 1986.

115. **Croteau, R., Felton, M., Karp, F., and Kjonaas, R.,** Relationship of camphor biosynthesis to leaf development in sage (*Salvia officinalis*), *Plant Physiol.*, 67, 820, 1981.

116. **Croteau, R., Gurkewitz, S., Johnson, M. A., and Fisk, H. J.,** Biochemistry of oleoresinosis: monoterpene and diterpene biosynthesis in lodgepole pine saplings infected with *Ceratocystis clavigera* or treated with carbohydrate elicitors, *Plant Physiol.*, 85, 1123, 1987.

117. **Gijzen, M., Lewinsohn, E., and Croteau, R.,** Characterization of the constitutive and wound-inducible monoterpene cyclases of grand fir (*Abies grandis*), *Arch. Biochem. Biophys.*, 289, 267, 1991.

118. **Threlfall, D. R. and Whitehead, I. M.,** Co-ordinated inhibition of squalene synthetase and induction of enzymes of sesquiterpenoid phytoalexin biosynthesis in cultures of *Nicotiana tabacum*, *Phytochemistry*, 27, 2567, 1988.

119. **Vogeli, U. and Chappell, J.,** Induction of sesquiterpene cyclase and suppression of squalene synthetase activities in plant cell cultures treated with fungal elicitor, *Plant Physiol.*, 88, 1291, 1988.

120. **Brindle, P. A., Kuhn, P. J., and Threlfall, D. R.,** Biosynthesis and metabolism of sesquiterpenoid phytoalexins and triterpenoids in potato cell suspension cultures, *Phytochemistry*, 27, 133, 1988.

121. **Lois, A. F. and West, C. A.,** Regulation of expression of the casbene synthetase gene during elicitation of castor bean seedlings with pectic fragments, *Arch. Biochem. Biophys.*, 276, 270, 1990.

122. **Vogeli, U. and Chappell, J.,** Regulation of a sesquiterpene cyclase in cellulase-treated tobacco cell suspension cultures, *Plant Physiol.*, 94, 1860, 1990.

123. **Karp, F., Mihaliak, C. A., Harris, J. L., and Croteau, R.,** Monoterpene biosynthesis: specificity of the hydroxylations of (–)-limonene by enzyme preparations from peppermint (*Mentha piperita*), spearmint (*Mentha spicata*), and perilla (*Perilla frutescens*) leaves, *Arch. Biochem. Biophys.*, 276, 219, 1990.

124. **Croteau, R. and Venkatachalam, K. V.,** Metabolism of monoterpenes: demonstration that (+)-*cis*-isopulegone, not piperitenone, is the key intermediate in the conversion of (–)-isopiperitenone to (+)-pulegone in peppermint (*Mentha piperita*), *Arch. Biochem. Biophys.*, 249, 306, 1986.

125. **Karp, F., Harris, J. L., and Croteau, R.,** Metabolism of monoterpenes: demonstration of the hydroxylation of (+)-sabinene to (+)-*cis*-sabinol by an enzyme preparation from sage (*Salvia officinalis*) leaves, *Arch. Biochem. Biophys.*, 256, 179, 1987.

126. **Uesato, S., Ikeda, H., Fujita, T., Inouye, H., and Zenk, M. H.,** Elucidation of iridodial formation mechanism — partial purification and characterization of the novel monoterpene cyclase from *Rauwolfia serpentina* cell suspension cultures, *Tetrahedron Lett.*, 28, 4431, 1987.

127. **Inouye, H. and Uesato, S.,** Biosynthesis of iridoids and secoiridoids, *Prog. Chem. Org. Nat. Prod.*, 50, 169, 1986.

128. **Jensen, S. R.,** Plant iridoids, their biosynthesis and distribution in angiosperms, in *Ecological Chemistry and Biochemistry of Plant Terpenoids*, Harborne, J. B. and Tomas-Barberan, F. A., Eds., Oxford University Press, Oxford, 1991, 133.

129. **Epstein, W. W. and Poulter, C. D.,** A survey of some irregular monoterpenes and their biogenetic analogies to presqualene alcohol, *Phytochemistry*, 12, 737, 1973.

130. **Poulter, C. D.,** Biosynthesis of non-head-to-tail terpenes. Formation of 1′-1 and 1′-3 linkages, *Acc. Chem. Res.*, 23, 70, 1990.

131. **Cane, D. E.,** Enzymatic formation of sesquiterpenes, *Chem. Rev.*, 90, 1089, 1990.

132. **Cane, D. E., Iyengar, R., and Shiao, M.-S.,** Cyclonerodiol biosynthesis and the enzymatic conversion of farnesyl to nerolidyl pyrophosphate, *J. Am. Chem. Soc.*, 103, 914, 1981.

133. **Cane, D. E. and Ha, H.-J.,** Trichodiene biosynthesis and the role of nerolidyl pyrophosphate in the enzymatic cyclization of farnesyl pyrophosphate, *J. Am. Chem. Soc.*, 110, 6865, 1988.

134. **Harrison, P. H. M., Oliver, J. S., and Cane, D. E.,** Pentalenene biosynthesis and the enzymatic cyclization of farnesyl pyrophosphate. Inversion at C-1 during 11-membered-ring formation, *J. Am. Chem. Soc.*, 110, 5922, 1988.

135. **Dehal, S. S. and Croteau, R.,** Partial purification and characterization of two sesquiterpene cyclases from sage (*Salvia officinalis*) which catalyze the respective conversion of farnesyl pyrophosphate to humulene and caryophyllene, *Arch. Biochem. Biophys.*, 261, 346, 1988.

136. **Munck, S. L. and Croteau, R.,** Purification and characterization of the sesquiterpene cyclase patchoulol synthase from *Pogostemon cablin, Arch. Biochem. Biophys.*, 282, 58, 1990.

137. **Vogeli, U., Freeman, J. W., and Chappell, J.,** Purification and characterization of an inducible sesquiterpene cyclase from elicitor-treated tobacco cell suspension cultures, *Plant Physiol.*, 93, 182, 1990.

138. **Parry, A. D. and Horgan, R.,** Carotenoids and abscisic acid (ABA) biosynthesis in higher plants, *Physiol. Plant.*, 82, 320, 1991.

139. **Zeevart, J. A. D. and Creelman, R. A.,** Metabolism and physiology of abscisic acid, *Annu. Rev. Plant Physiol. Plant Mol. Biol.*, 39, 439, 1988.

140. **West, C. A.,** Biosynthesis of diterpenes, in *Biosynthesis of Isoprenoid Compounds*, Vol. 1, Porter, J. W. and Spurgeon, S. L., Eds., John Wiley & Sons, New York, 1981, 375.

141. **Robinson, D. R. and West, C. A.,** Biosynthesis of cyclic diterpenes in extracts from seedlings of *Ricinus communis* L. II. Conversion of geranylgeranyl pyrophosphate into diterpene hydrocarbons and partial purification of the cyclization enzymes, *Biochemistry*, 9, 80, 1970.

142. **Moesta, P. and West, C. A.,** Casbene synthetase: regulation of phytoalexin biosynthesis in *Ricinus communis* L. seedlings, *Arch. Biochem. Biophys.*, 238, 325, 1985.

143. **Dueber, M. T., Adolf, W., and West, C. A.,** Biosynthesis of the diterpene phytoalexin casbene: partial purification and characterization of casbene synthetase from *Ricinus communis*, *Plant Physiol.*, 62, 598, 1978.

144. **Adolf, W. and Hecker, E.,** Diterpenoid irritants and cocarcinogens in Euphorbiaceae and Thymelaeaceae: structural relationships in view of their biogenesis, *Isr. J. Chem.*, 16, 75, 1977.

145. **Duncan, J. D. and West, C. A.,** Properties of kaurene synthetase from *Marah macrocarpus* endosperm: evidence for the participation of separate but interacting enzymes, *Plant Physiol.*, 68, 1128, 1981.

146. **Railton, I. D., Fellows, B., and West, C. A.,** *ent*-Kaurene synthesis in chloroplasts from higher plants, *Phytochemistry*, 23, 1261, 1984.

147. **Phinney, B. O. and Spray, C. R.,** Plant hormones and the biosynthesis of gibberellins: the early-13-hydroxylation pathway leading to GA$_1$, in *Biochemistry of the Mevalonic Acid Pathway to Terpenoids*, Towers, G. H. N. and Stafford, H. A., Eds., Plenum Press, New York, 1990, 203.

148. **Graebe, J. E.,** Gibberellin biosynthesis and control, *Annu. Rev. Plant Physiol.*, 38, 419, 1987.

149. **Sharkey, T. D., Loreto, F., and Delwiche, C. F.,** The biochemistry of isoprene emission from leaves during photosynthesis, in *Trace Gas Emissions by Plants*, Sharkey, T. D., Holland, E. A., and Mooney, H. A., Eds., Academic Press, San Diego, CA, 1991, 153.

150. **Hewitt, C. N., Monson, R. K., and Fall, R.,** Isoprene emissions from the grass *Arundo donax* L. are not linked to photorespiration, *Plant Sci.*, 66, 139, 1990.

151. **Silver, G. M. and Fall, R.,** Enzymatic synthesis of isoprene from dimethylallyl diphosphate in aspen leaf extracts, *Plant Physiol.*, 97, 1588, 1991.

152. **Tanaka, Y.,** Structure and biosynthesis mechanism of natural polyisoprene, *Prog. Polym. Sci.*, 14, 339, 1989.

153. **Hemming, F. W.,** Biosynthesis of dolichols and related compounds, in *Biosynthesis of Isoprenoid Compounds*, Vol. 2., Porter, J. W. and Spurgeon, S. L., Eds., John Wiley & Sons, New York, 1983, 305.

154. **Archer, B. L. and Audley, B. G.,** New aspects of rubber biosynthesis, *Bot. J. Linn. Soc.*, 94, 181, 1987.

155. **Cornish, K.,** Natural rubber biosynthesis: a branch of the isoprenoid pathway in plants, in *Regulation of Isopentenoid Metabolism*, Nes, W. D., Parish, E. J., and Trzaskos, J. M., Eds., American Chemical Society, Washington, D.C., 1992, 18.

156. **Paterson-Jones, J. C., Gilliland, M. G., and van Staden, J.,** The biosynthesis of natural rubber, *J. Plant Physiol.*, 136, 257, 1990.

157. **Fryer, M. J.,** The antioxidant effects of thylakoid vitamin E (α-tocopherol), *Plant Cell Environ.*, 15, 381, 1992.

158. **Pennock, J. F. and Threlfall, D. R.,** Biosynthesis of ubiquinone and related compounds, in *Biosynthesis of Isoprenoid Compounds*, Vol. 2, Porter, J. W. and Spurgeon, S. L., Eds., John Wiley & Sons, New York, 1983, 191.

159. **Hamerski, D., Schmitt, D., and Matern, U.,** Induction of two prenyltransferases for the accumulation of coumarin phytoalexins in elicitor-treated *Ammi majus* cell suspension cultures, *Phytochemistry*, 29, 1131, 1990.

160. **Heide, L. and Tabata, M.,** Geranylpyrophosphate:*p*-hydroxybenzoate geranyltransferase activity in extracts of *Lithospermum erythrorhizon* cell cultures, *Phytochemistry*, 26, 1651, 1987.

161. **Welle, R. and Grisebach, H.,** Properties and solubilization of the prenyltransferase of isoflavonoid phytoalexin biosynthesis in soybean, *Phytochemistry*, 30, 479, 1991.

162. **ap Rees, T.,** Compartmentation of plant metabolism, in *The Biochemistry of Plants: A Comprehensive Treatise*, Vol. 12, Stumpf, P. K. and Conn, E. E., Eds., Academic Press, San Diego, CA, 1987, 87.

163. **Hrazdina, G. and Jensen, R. A.,** Spatial organization of enzymes in plant metabolic pathways, *Annu. Rev. Plant Physiol. Plant Mol. Biol.*, 43, 241, 1992.

164. **Gray, J. C.,** Control of isoprenoid biosynthesis in higher plants, *Adv. Bot. Res.*, 14, 25, 1987.

165. **Mettal, U., Boland, W., Beyer, P., and Kleinig, H.,** Biosynthesis of monoterpene hydrocarbons by isolated chromoplasts from daffodil flowers, *Eur. J. Biochem.*, 170, 613, 1988.

166. **Lutke-Brinkhaus, F., Liedvogel, B., and Kleinig, H.,** On the biosynthesis of ubiquinones in plant mitochondria, *Eur. J. Biochem.*, 141, 537, 1984.

167. **Karp, F. and Croteau, R.,** Role of hydroxylases in monoterpene biosynthesis, in *Bioflavour '87*, Schreier, P., Ed., Walter de Gruyter, Berlin, 1988, 173.

168. **Rogers, L. J., Shah, S. P. J., and Goodwin, T. W.,** Compartmentation of biosynthesis of terpenoids in green plants, *Photosynthetica*, 2, 184, 1968.

169. **Kreuz, K. and Kleinig, H.,** Synthesis of prenyl lipids in cells of spinach leaf: compartmentation of enzymes for formation of isopentenyl diphosphate, *Eur. J. Biochem.*, 141, 531, 1984.

170. **Lutke-Brinkhaus, F. and Kleinig, H.,** Formation of isopentenyl diphosphate via mevalonate does not occur within etioplasts and etiochloroplasts of mustard (*Sinapis alba* L.) seedlings, *Planta*, 171, 406, 1987.

171. **Heintze, A., Gorlach, J., Leuschner, C., Hoppe, P., Hagelstein, P., Schulze-Siebert, D., and Schultz, G.,** Plastidic isoprenoid synthesis during chloroplast development: change from metabolic autonomy to a division-of-labor stage, *Plant Physiol.*, 93, 1121, 1990.

172. **Srere, P. A.,** Complexes of sequential metabolic enzymes, *Annu. Rev. Biochem.*, 56, 89, 1987.

173. **Stafford, H. A.,** Compartmentation in natural product biosynthesis by multienzyme complexes, in *The Biochemistry of Plants: A Comprehensive Treatise*, Vol. 7, Stumpf, P. K. and Conn, E. E., Eds., Academic Press, New York, 1981, 117.

174. **Maudinas, B., Bucholtz, M. L., Papastephanou, C., Katiyar, S. S., Briedis, A. V., and Porter, J. W.,** The partial purification and properties of a phytoene synthesizing enzyme system, *Arch. Biochem. Biophys.*, 180, 354, 1977.

175. **Singh, N. and Luthra, R.,** Sucrose metabolism and essential oil accumulation during lemongrass (*Cymbopogon flexuosus* Stapf.) leaf development, *Plant Sci.*, 57, 127, 1987.

176. **Singh, N., Luthra, R., and Sangwan, R. S.,** Oxidative pathways and essential oil biosynthesis in the developing *Cymbopogon flexuosus* leaf, *Plant Physiol. Biochem.*, 28, 703, 1990.

177. **Singh, N., Luthra, R., and Sangwan, R. S.,** Mobilization of starch and essential oil biogenesis during leaf ontogeny of lemongrass (*Cymbopogon flexuosus* Stapf.), *Plant Cell Physiol.*, 32, 803, 1991.

178. **Burbott, A. J. and Loomis, W. D.,** Effects of light and temperature on the monoterpenes of peppermint, *Plant Physiol.*, 42, 20, 1967.

179. **Firmage, D. H.,** Environmental influences on the monoterpene variation in *Hedeoma drummondii*, *Biochem. Syst. Ecol.*, 9, 53, 1981.

180. **Flesch, V., Jacques, M., Cosson, L., Teng, B. P., Petiard, V., and Balz, J. P.,** Relative importance of growth and light level on terpene content of *Ginkgo biloba*, *Phytochemistry*, 31, 1941, 1992.

181. **Gleizes, M., Pauly, G., Bernard-Dagan, C., and Jacques, R.,** Effects of light on terpene hydrocarbon synthesis in *Pinus pinaster*, *Physiol. Plant.*, 50, 16, 1980.

182. **Gref, R. and Tenow, O.,** Resin acid variation in sun and shade needles of Scots pine (*Pinus sylvestris* L.), *Can. J. For. Res.*, 17, 346, 1987.

183. **Metivier, J. and Viana, A. M.,** The effect of long and short day length upon the growth of whole plants and the level of soluble proteins, sugars, and stevioside in leaves of *Stevia rebaudiana* Bert., *J. Exp. Bot.*, 30, 1211, 1979.

184. **Yamaura, T., Tanaka, S., and Tabata, M.,** Light-dependent formation of glandular trichomes and monoterpenes in thyme seedlings, *Phytochemistry*, 28, 741, 1989.

185. **Fahn, A.,** *Secretory Tissues in Plants*, Academic Press, London, 1979.

186. **Gershenzon, J., Maffei, M., and Croteau, R.,** Biochemical and histochemical localization of monoterpene biosynthesis in the glandular trichomes of spearmint (*Mentha spicata*), *Plant Physiol.*, 89, 1351, 1989.

187. **McCaskill, D., Gershenzon, J., and Croteau, R.,** Morphology and monoterpene biosynthetic capabilities of secretory cell clusters isolated from glandular trichomes of peppermint (*Mentha piperita* L.), *Planta*, 187, 445, 1992.

188. **Yamaura, T., Tanaka, S., and Tabata, M.,** Localization of the biosynthesis and accumulation of monoterpenoids in glandular trichomes of thyme, *Planta Med.*, 58, 153, 1992.

189. **Gershenzon, J. and Croteau, R.,** Regulation of monoterpene biosynthesis in higher plants, in *Biochemistry of the Mevalonic Acid Pathway to Terpenoids*, Towers, G. H. N. and Stafford, H. A., Eds., Plenum Press, New York, 1990, 99.

190. **Keene, C. K. and Wagner, G. J.,** Direct demonstration of duvatrienediol biosynthesis in glandular heads of tobacco trichomes, *Plant Physiol.*, 79, 1026, 1985.

191. **Spurgeon, S. L. and Porter, J. W.,** Carotenoids, in *The Biochemistry of Plants: A Comprehensive Treatise*, Vol. 4, Stumpf, P. K. and Conn, E. E., Eds., Academic Press, New York, 1980, 419.

Chapter 12

REGULATION OF PHYTOSTEROL BIOSYNTHESIS

**W. David Nes, Stephen R. Parker, Farrist G. Crumley,
and Samir A. Ross**

TABLE OF CONTENTS

I. INTRODUCTION

Crop plants are an essential source of energy and raw materials that are used in the food chain. In the past, research was directed toward optimizing crop yields by genetic improvement, application of fertilizers, and the use of chemicals to control insects, nematodes, microorganisms, and weeds. For example, in America in 1950, one farmer produced food and fiber for 27 people; in 1990, the production was for 128 people.[1] This increased efficiency, however, has not come without costs. For instance, the U.S. Environmental Protection Agency has identified agriculture as the largest nonpoint source of water pollution with many of the polluting substances being pesticides and nitrates from fertilizers.[2] The question for the 1990s then is how we may improve the plant's health, so that it may annually produce both economically and profitably without endangering the environment in the process of improvement.

One recent approach to increasing the efficiency of crop production is to induce host defense mechanisms that stimulate chemical protection by switching carbon flow from sterol to terpene and triterpene production.[3] A second approach is to use sterol biosynthesis inhibitors (SBIs) that control disease development.[4,5] A third approach, which is still is in the experimental stage of development, is to spray plants with brassinosteroids.[6] Interest, however, in the sterol biochemistry of plants has lacked the attention devoted to other biomolecules produced by plants because their biological properties were poorly understood until recently. For many years, phytosterols (and mycosterols) were regarded as secondary waste metabolites of interest only to natural product chemists, for the study of chemotaxonomy, and industrial chemists who required certain sterols, e.g., stigmasterol and ergosterol, as starting materials in the synthesis of steroid pharmaceuticals ("the Pill" for birth control, cortisone, and vitamin D). Phytosterols are now known to play multiple physiological roles (structural and hormonal), the expression of which controls plant growth, development, and reproduction.[7] Therefore, the development of control measures that exploit some fundamental difference between the sterol biochemistry of the host and its pathogen may now be realized as promising to agriculture.

Phytosterols may be defined as compounds produced by plants. However, sterols synthesized by plants are not unique in origin and may also be synthesized by microorganisms and animals.[7,8] The sterol which predominates in tracheopytes is sitosterol (24α-ethyl cholesterol), whereas ergosterol (24β-methyl cholesta-5,7,22-trienol), poriferasterol (24β-ethyl cholesta-5,22-dienol), and cholesterol are typically the major sterols which occur in fungi, algae, and animals, respectively.[8] Cycloartenol may be considered the first compound in plants to enter the tetracyclic transformation process, a set of transformations we operationally refer to as the sterol pathway. The sterol pathway may be distinguished from other isopentenoid (isoprenoid:terpenoid) pathways through mechanistic enzymology. The isopentenoid pathway involves enzymes that catalyze the polymerization of C_5 units to squalene oxide, whereas the sterol pathway involves enzymes that act on the tetracyclic structure. Not all tetracycles will necessarily pass through the sterol pathway because some of them lack the necessary features that permit binding and transformation by enzymes that act on the sterol substrate. This biochemical determinant becomes functionally significant as squalene oxide is known to cyclize to at least five isomeric forms of cycloartenol,[7] each of which could effectively compete with cycloartenol in their abilities to be processed through the sterol pathway (in the absence of enzyme specificity) and replace sterols as membrane inserts.

II. STEROL STRUCTURE AND STEREOCHEMISTRY

Sterols may be defined as any chiral tetracyclic isopentenoid which may be formed by the cyclization of squalene oxide through the transition state possessing stereochemistry similar to the *trans-syn-trans-anti-trans-anti* configuration, i.e., the protosteroid cation,[8,9] and retains a polar group at C-3 (hydroxyl or keto), an all *trans-anti* stereochemistry in the ring system, and a side chain (20*R*) configuration. Molecules that fail to incorporate the steroidal stereochemistry, such as the biogenetically related tetracyclic and pentacyclic triperpenoids, may be considered sterol-like compounds.[9] The enzymes that compose the sterol pathway are specifically mated for tetracycles that possess the (20*R*) stereochemistry, and through that mating the sterol pathway differs from other isopentenoid pathways. Thus, squalene oxide is not steroidal, although it is an isopentenoid. The systematic nomenclature of sterols is based on the tetracyclic skeleton of cholestane, which has an acyclic isooctyl side chain, as shown in Figure 1. The cholestane skeleton is composed of carbons C-1 to C-27. The carbons numbered C-28, C-29, C-30, C-31, and C-32 are additions to the skeleton. The α/β-notation of the nucleus has a different meaning from the stereochemical notation used in the side chain. For a deeper discussion of nomenclature rules for the systematic naming of sterols that are used by sterol phytochemists relative to those introduced by IUPAC, the reader should consult Reference 9.

III. PATHWAY OF CARBON FLUX TO STEROL

Sterols are ultimately produced in plants from the photosynthetic fixation of carbon dioxide to sugar. The sugar formed in plastids is metabolized to acetyl coenzyme A (CoA), the basic C_2 building block of cellular biosynthesis. Whereas the cyclization of squalene oxide to cycloartenol is associated with organisms having a photosynthetic lineage,[10-13] it is not necessary for plants to be operating in the phytosynthetic mode to produce sterols. Nevertheless, sterol biosynthesis is stimulated by light.[13] Acetyl-CoA may be directed into the isopentenoid pathway or accumulate as a pool of compartmentalized carbon reserve. The acetyl-CoA pool may be fed by sugar degradation and β-oxidation of acyl lipids (fatty acids). The pool may also be increased by metabolism of isopentenoids and amino acids through the so-called mevalonic acid (MVA) shunt.[14] The MVA shunt is responsible for channeling amino acids into sterols and triterpenoids,[14-17] so that carbon flux between protein degradation and polycyclic isopentenoids may bypass the sugar/acyl lipid-acetyl-CoA circuit.

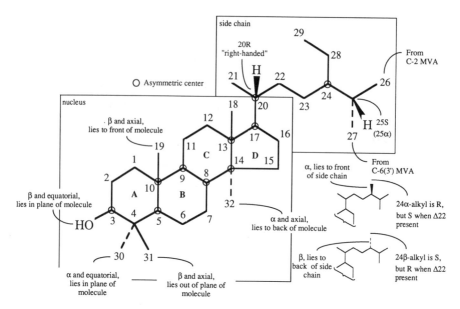

FIGURE 1. Nomenclature and structural features of the sterol skeleton.

Recent studies have indicated the MVA shunt is responsible for 10% of the carbon flux into sterols and triterpenes in germination of some plants.[17]

Because Δ^5 sterols are primary metabolites formed in active cell growth,[9] we suggested the formation of the Δ^5 bond represents completion of the sterol pathway.[7,9] However, Δ^5 sterols may serve as precursors for saturated sterols (stanols) and highly functionalized steroids which play different roles in plant physiology than Δ^5 sterols.[18-20] The metabolism of Δ^5 sterols to other compounds generally occurs at or after the onset of growth arrest. Δ^5 sterols and intermediates in the sterol pathway may also be conjugated with acyl lipids and sugars.[21] An exciting new area of research in animal physiology is the demonstration of prenylated proteins, which through their covalent interactions affect disease states.[22] Undoubtedly, similar observations are forthcoming in plant biochemistry.

IV. PHYLOGENETIC DISTRIBUTION OF STEROLS

The current view in the field is that plants, fungi, and animals operate different sterol pathways, and bacteria fail to produce sterols. The sterol pathway is thought to have developed after the advent of atmospheric oxygen, and lanosterol and cycloartenol are considered the first two tetracyclic products of squalene oxide cyclization (reviewed in Reference 7). A cycloartenol-based pathway is associated with a photosynthetic lineage, whereas a lanosterol-based pathway is associated with a nonphotosynthetic lineage.[10] The biosynthetic bifurcation is sometimes viewed, incorrectly, as representing a plant-animal division. Cycloartenol is assumed to be cyclized into the bent conformation[23-27] because the resultant sterol possesses a *syn-cis* relationship at the B/C-ring junction formed by the 9β,19-cyclopropyl bridgehead, and 8β-H. Lanosterol, with an 8,9-double bond, assumes the flat conformation.[28]

Additional phylogenetic determinants in sterol biochemistry are the ability of an organism to alternatively alkylate or reduce the 24,25-double bond, the resulting stereochemistry of the alkyl group introduced at C-24, and the configuration at C-25 following saturation of the 24,25-double bond.[29,30] Fungi, particularly the more advanced fungi, are thought to be unable to produce cholesterol (sterol with a saturated side chain) and to produce a configuration at the two chiral centers, C-24 and C-25, which are opposite to that formed by vascular plants (Figure 2). Animals are thought to be unable to produce 24-alkyl sterols, but insects may dealkylate phytosterols to produce cholesterol.[8] The alkylating enzyme is referred to as the C-24 S-adenosylmethyltransferase (SMT).

Plants produce another specific enzyme, the 9β,19-cyclopropyl to Δ^8 isomerase, also referred to as the cycloeucalenol to obtusifoliol isomerase (COI) (Figure 2). The COI acts with a high degree of substrate specificity on the bent cyclopropyl sterol substrate.[31] Plants and fungi contain a C-4 demethylase enzyme (C-4 DEM). The C-4 DEM operates similar reaction courses in plants and fungi.[8]

Pentacyclic analogs of sterols, which are sterol-like functionally,[25] are thought to precede sterol production in biochemical evolution.[26] For instance, tetrahymanol, a saturated pentacyclic triterpenoid, is formed by the anaerobic cyclization of squalene, and this event occurs in bacteria.[32,33] Pentacyclic triterpenoids also occur widely in crop plants. Their biosynthesis in plants is developmentally delayed relative to sterol production, and they occur mainly as part of the surface wax.[7]

The apparent differences in the sterol conformation and side chain stereochemistry assumed by sterols in fungi and plants suggest the enzymes and reaction courses involved in catalysis and the genetic sequence coding for the respective enzymes might similarly be different in the two taxa, i.e., independent evolution. With these considerations in mind, sterol biosynthesis inhibitors have been rationally designed to interfere with the steps in the sterol pathway that distinguishes plants from fungi.[34]

V. ANATOMIC DISTRIBUTION OF STEROLS

With recent advances in sterol analysis, the sterol composition of plants and fungi have been found to be much more similar than once thought. Moreover, the complex sterol patterns observed in marine invertebrates, which are often considered to be unique to this group of less advanced animals,[35] are now known to be produced to varying degrees by bacteria,[7,8] plants,[36] and fungi.[37] The major differences in the sterol compositions in nature are associated with variations in the side chain structure. Many vascular plants produce primarily sitosterol in the leaves.[30] However, other sterols with a variety of modified side

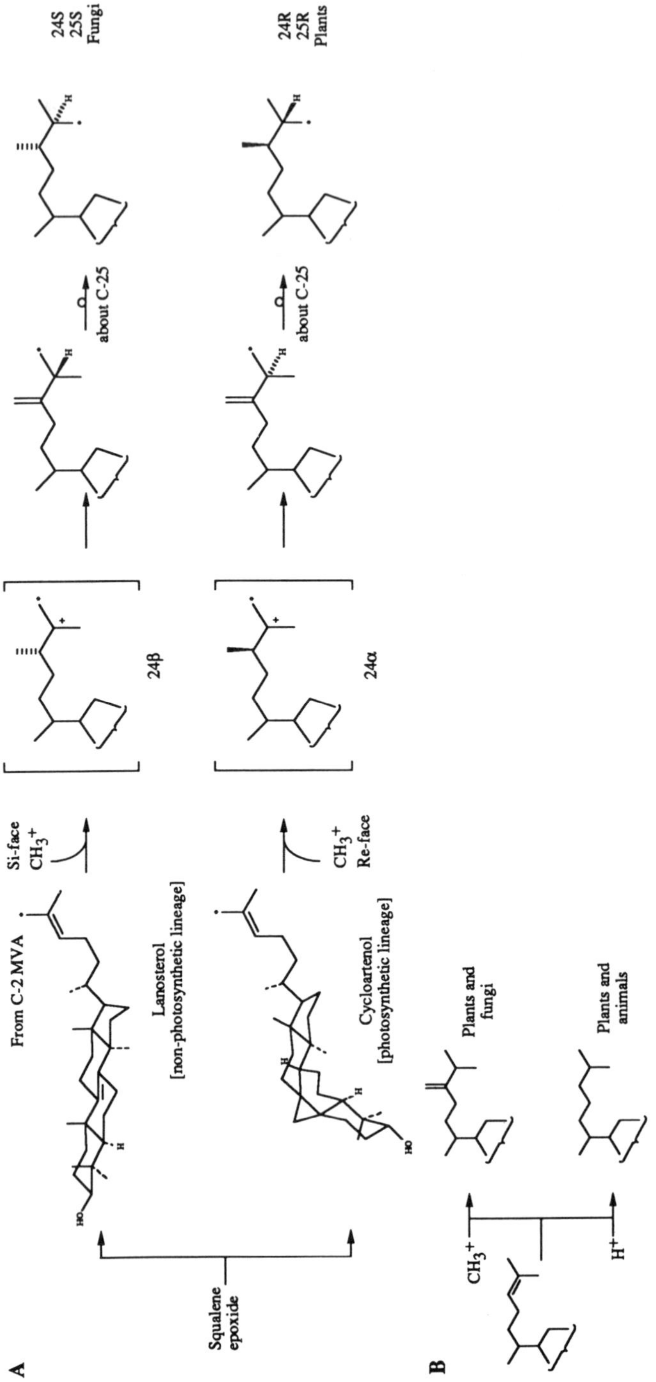

A
B
From C-2 MVA
Si-face
CH3+
Re-face
Lanosterol
[non-photosynthetic lineage]
Cycloartenol
[photosynthetic lineage]
Squalene epoxide
24β
24α
about C-25
about C-25
24S
25S
Fungi
24R
25R
Plants
CH3+
H+
Plants and fungi
Plants and animals

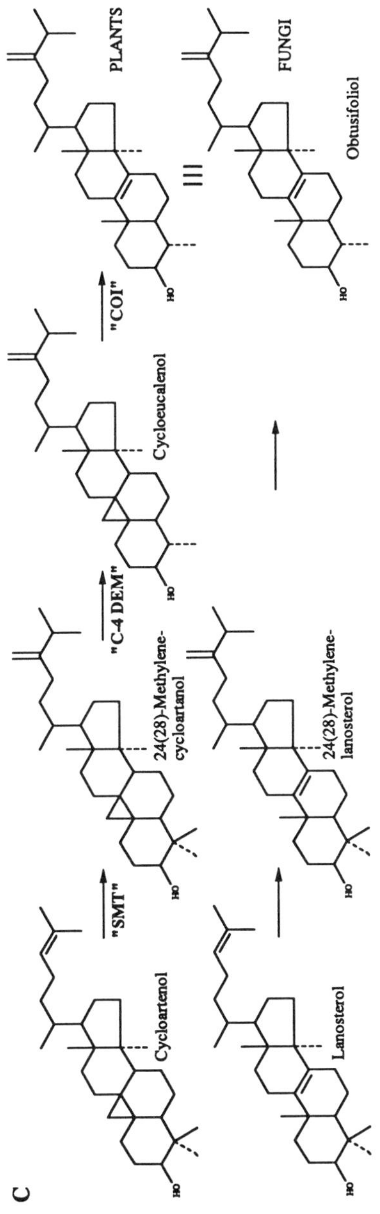

FIGURE 2. Origins of the phylogenetic bifurcations in sterol biosynthesis.

TABLE 1
Level of Sterol in Specific Plant Anatomy and Isolated Cell Types from Sunflower

Tissue/cell type	Total sterol[a]	End product/ intermediate ratio[b]	% Cholesterol of total sterol
Seed	83 μg/seed	8:1	<1
Sprout (4 day; 2.7 cm)[c]	15 μg/sprout	8:2	<1
Cultured cells (from embryo tissue)	33,000 fg/cell	97:3	2
Pollen grain (hand collected)	585,000 fg/grain	6:4	<1
Primary leaf (22 day)[c]	15–20 μg/block		
Whole leaf (28 day)[c]	80–100 μg/blade	9:1	10
Leaf wax (28 day)[c]	0.5–2.2 μg/leaf	100:tr	100
Cotyledon (22 day)[c]	65 μg/blade		
Stem trichome (40 day)[c]	None	—	—
Immature green flower bud	35 μg/bud	250:1	2
Whole ligule flower wax (72 day)[c]	<0.1 μg/flower	—	—
Whole ligule flower (72 day)[c]	200 μg/flower[d]	10:1	<1
Ray flower mesophyll cells	800–1,200 fg/cell	4:1	3–5
Leaf mesophyll cells (30 day)[c]	5,000–8,000[e] fg/cell	4:1	15–35
Disk flowers (72 day)	11.2 μg/flower	8:2	2

[a] Free sterols liberated by alkaline saponification of sample.

[b] Ratio of 4,4-desmethyl sterols (end product) to 4-mono and 4,4-dimethyl sterols recovered from TLC plate. TLC pure fractions were eluted from a reversed-phase HPLC column and analyzed by gas chromatography–mass spectroscopy (GC-MS).

[c] Age of plant following germination.

[d] Significant pentacyclic triterpenoids present in the cells.

[e] No significant pentacyclic triterpenoids present in the cells.

chains may occur in leaves or they may occur in specific plant parts (seeds) or cell types.[36] In sunflower, we recently observed that trichomes do not possess sterol, whereas cholesterol may be the major sterol of mesophyll cells and the surface wax (Table 1). Moreover, the chirality at C-24 may shift from C-24α-ethyl to 24β-ethyl with maturation of *Kalanchoe*[38] and the cucurbits,[39] leading to epimerically different sterol compositions in the plant anatomy.[35-37]

Plants die and wither away, leaving their organic matter to be used by microorganisms. The resultant naked 24-alkyl sterols deposited into the sediments are degraded by soil bacteria and fungi as an energy source, whereas other fungi such as the oomycetous fungus, *Phytophthora cactorum*, may use the sterol to regulate its growth and reproduction.[40,41]

VI. THE KINETICALLY FAVORED PATHWAY IN PHYTOSTEROL GENESIS

One school of thought regarding C-24 alkylation consists of four points. First, the mechanism of C-24 methylation is thought to proceed by *re*-face attack (α-face attack leading to the 24-methyl chirality of campesterol) with the 1,2-hydride shift of H-24 to C-25 proceeding along the *si*-face to produce the 25-*R* configuration. Here, C-26 becomes the pro-R methyl group.[42-44] Second, the C-24 alkyl group and its stereochemistry is functionally unimportant during cell proliferation, since model membrane systems and certain microorganisms auxotrophic for sterol use ergosterol, sitosterol, and cholesterol equally well.[27] Third, side chain modification, as shown in Figure 3, may either precede ring A/B-modification or be partly independent of the A/B-system because phytosterol genesis operates within a metabolic grid with enzymes of low substrate specificity.[45,46] Fourth, by associating hydroxymethylglutaryl coenzyme A (HMG-CoA) reductase (HMGR) activity with C-24 alkyl sterol production,[47] sitosterol genesis is thought to be controlled by regulation at the level HMGR activity similar to cholesterol and ergosterol production in animals and fungi, respectively, which means that no step in the pathway should be significantly rate limiting under physiological conditions, which is consistent with the previous point.

An alternative position is that taxa-specific, kinetically favored sterol pathways operate in phytosterol production. The C-24 methylation reaction is the rate-limiting step, and the biosynthetic inclusion of the 24-alkyl group and its stereochemistry are physiologically significant in cell proliferation and membrane structure and function.[9,28,29,48]

For an enzyme to be rate limiting of phytosterol genesis, i.e., involving only those enzymes that act on the sterol substrate, the following rules should be considered: (1) the activity of the enzymes should be controlled by the substrate specificity and attendant co-factor requirements, both of which are determined by the mechanism of the reaction sequence rather than simply the level of sterol (regardless of its structure) that passes through the pathway; (2) the regulatory step should involve an enzymatic catalysis in which the concentrations of substrates and products involved in the catalysis are far from thermodynamic equilibrium; (3) the maximal velocity of the regulatory enzyme, as measured in cell-free extracts under optimal conditions, should be one of the slower enzymes in the pathway; and (4) the concentration of the natural substrate for the regulatory enzyme can be caused to accumulate in the cell after treatment with a sterol biosynthesis inhibitor or by changing the co-factor availability.[9] It is expected that the effect of accumulating the 4,4,14-trimethyl sterol intermediate cycloartenol, at the expense of forming a 24-alkyl Δ^5 sterol end product(s), sitosterol, is to upset the natural distribution of sterol in the cell, which then should interfere with cell proliferation because of an imbalance in sterol homeostasis.[48]

The kinetically favored sterol pathway in crop plants, such as in sunflower and corn, that likely gives rise to sitosterol is the linear sequence shown in Figure 4. The K_m values for the specific substrates illustrated in Figure 4 are those obtained by ourselves[28,48] and by the Benveniste group[31,49-51] using microsomal precipitates. Note the kinetic evidence indicates the first methylation is a, if not the, critical slow step in the 24-alkyl sterol pathway. Benveniste

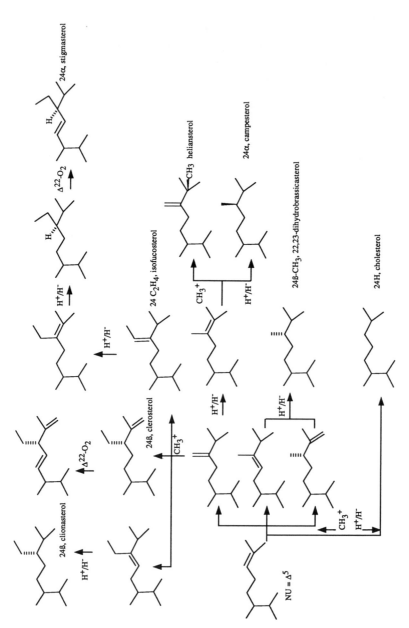

FIGURE 3.   Alternative C-24 alkylation pathways in tracheophytes.

and co-workers have examined steps 2 and 7 in Figure 4, and they also appear to possess a high degree of substrate specificity and to be slow reactions from the preliminary data reported.[52,53] It is important to recognize that alternative outcomes are possible (branch points along the pathway).

The sitosterol pathway, involving the stereospecific transfer of two methyl groups to C-24, may be favored during active cell proliferation. Alternatively, the formation of the other end products typically found with sitosterol is favored at, or after, the onset of growth arrest or plant maturation (flowering). A relationship between sterol composition and content is known to exist in sorghum in providing the necessary critical mass of free sterol to promote flowering.[7,54] The shift in apparent methylation outcomes (degree and location of methylation and the proportion of C-24 epimers) may result from kinetic and thermodynamic control principles discussed elsewhere,[9] rather than resulting from developmentally delayed expression of novel methyl transferase and 24,25-reductase enzymes, our first thought.[28] It follows that we should find that heliansterol, campesterol, 24-epicampesterol (24β-methyl cholesterol), and cholesterol (see Figure 3 for a key to the structures) are formed by the same side chain modifying enzyme that operates to produce sitosterol. The new sterols should be formed secondarily to sitosterol production, for which we have supporting data in sunflower (Nes, W. D., Kalinowska, M., Norton, R. A., and Parker, S. R., unpublished data, 1992). As reviewed in References 7 and 8, 24-methyl sterols are never the major sterols in the sterol composition of tracheophytes. Stigmasterol is another sterol that we expect should be formed after sitosterol production. A determination of the synthesis of stigmasterol relative to that of the other sterols is complicated by the possibility that multiple sterol pools exist.[55,56] A precursor-product relationship of sitosterol and stigmasterol has been implicated by studies in this laboratory[54] and that of Goad and co-workers.[55]

VII. UNIFORMITY IN THE STEROL TRANSFORMATION PROCESS

The fungal sterol pathway is often associated with the transformation of lanosterol to ergosterol.[27] The ergosterol pathway is thought to operate within a biosynthetic cube,[57] similar in nature to the metabolic grid operative in plants.[45] A primary control point in the fungal pathway is the C-32 demethylation step. The C-32 demethylation reaction has attracted much attention as a site in the pathway to block. In fact, the mode of action of several antifungal agents are designed to block this step.[58] The removal of the C-32 angular methyl will only proceed with a suitable $\Delta^{8(9)}$ sterol substrate, such as with lanosterol. A requirement for a double bond specifically located at the 8(9)-position was shown by *in vivo*[59] and enzymological[50,60,61] studies which indicated that the $\Delta^{8(9)}$ bond might stabilize the *pi*-electronic system through conjugation and thus facilitate C-32 decarbonylation.[8] Consequently, a rationale has been provided

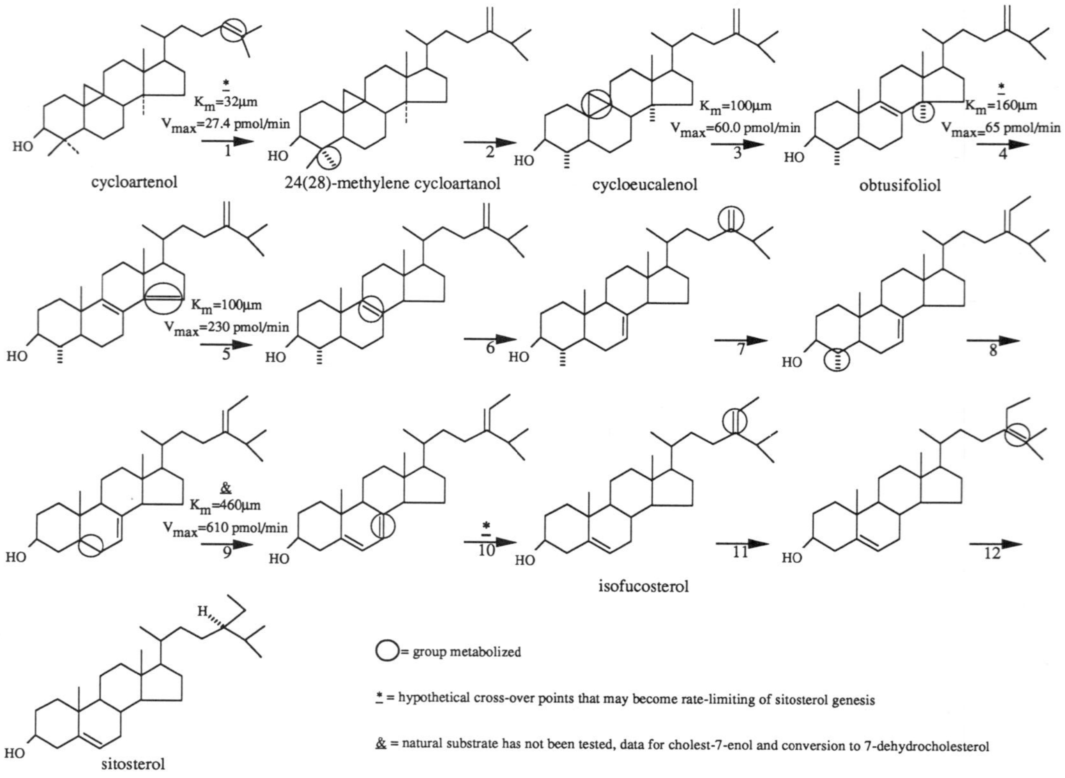

FIGURE 4. Hypothetical kinetically favored pathway in phytosterol biosynthesis.

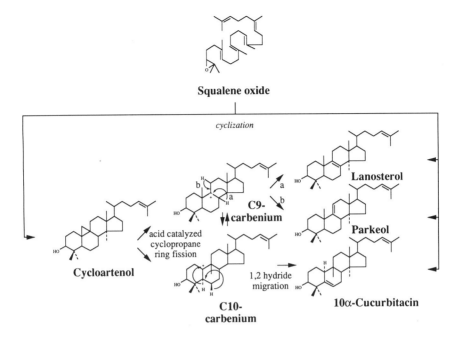

FIGURE 5. Cyclization of squalene oxide to tetracyclic products and proposed conversions.

why plants seemingly open the cyclopropyl ring and stereospecifically form the $\Delta^{8(9)}$ bond prior to C-32 demethylation.[8,13,45]

However, plants and some marine organisms[36,62] are now known to form parkeol, the $\Delta^{9(11)}$ isomer of cycloartenol (Figure 5), which suggests that the $\Delta^{9(11)}$ bond may be introduced into the tetracyclic structure either by the direct cyclization of squalene oxide or by the conversion of cycloartenol into parkeol. Parkeol may be transformed into stigmast-9(11)-enol by plants by a route that bypasses the intermediacy of the 8(9)-bond,[63] or it may undergo transformation to a Δ^5 sterol[64] by a route that presumably involves isomerization of the 9(11)-bond into the 8(9)-bond. These results raise the possibility that C-32 demethylation may proceed with a sterol substrate other than with one that contains the 8(9)-bond. Recent data from this laboratory on cycloartenol supplementation to the yeast mutant *GL7* supports this view.[65] Thus, we discovered that *GL7*, after the cultures were sterol adapted[27] and given sparing ergosterol (0.5 ppm), transforms cycloartenol to ergosterol via a 9(11)-sterol (Figure 6). C-32 Demethylation may proceed either to form the $\Delta^{14(15)}$ or $\Delta^{8(14)}$ bond. We suggest that the precise C-32 demethylated outcome is dependent on substrate specificity and sterol allosterism (the competitive binding of more than one sterol to a polymeric protein operating cooperatively). Parkeol is also metabolized to ergosterol by *GL7*, and 24,25-dihydroparkeol is converted to cholesta-5,7,22-trienol by the cells. The affect of demethylating C-32 of, e.g., cycloartenol, to form an 8(14)-bond is to induce a shift in the conformation

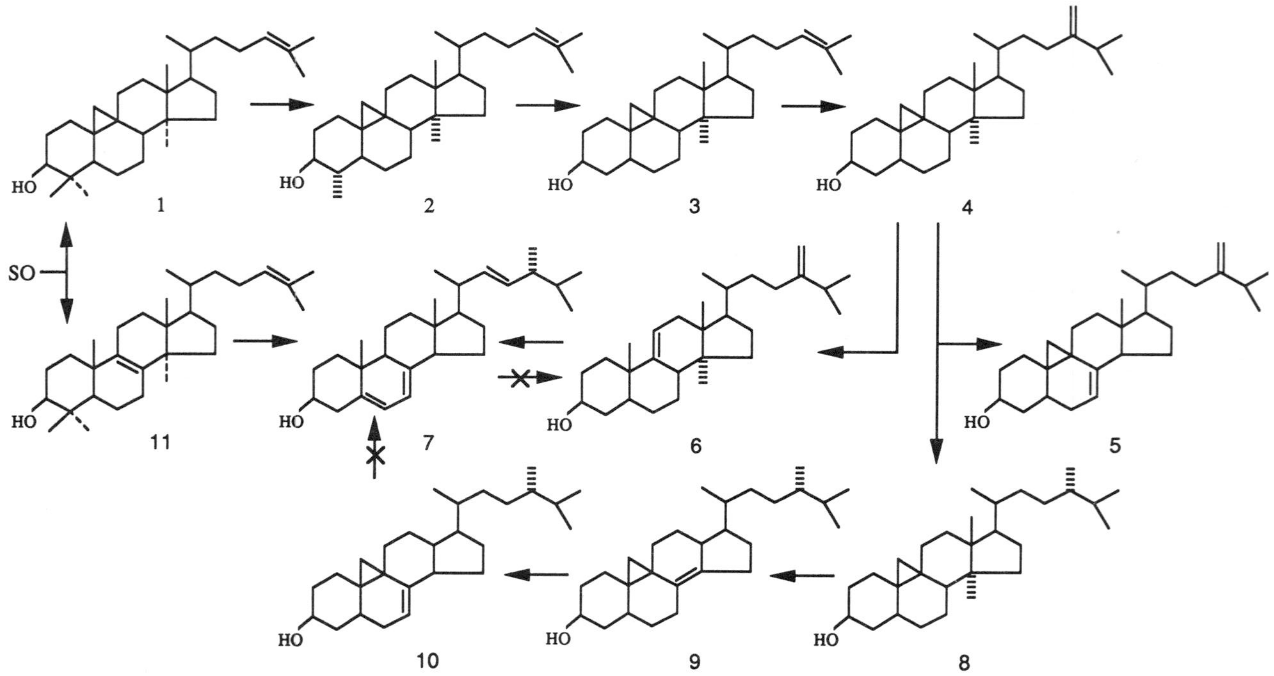

FIGURE 6. Pathway of cycloartenol metabolism by *Saccharomyces cerevisiae* strain *GL7*. Lanosterol and cycloartenol were each fed to the *GL7*. Lanosterol was transformed to ergosterol with no significant accumulation of other sterols. Cycloartenol was converted to ergosterol with the accumulation of sterols indicated in the figure. Proof of structure of cycloartenol metabolites was obtained by MS and ¹H-NMR.

of cyclopropyl sterols from flat to bent (unpublished), whereas there is no change in the relative planar shape of other sterols that participate in the reaction, e.g., $\Delta^{7(8)}$ or $\Delta^{8(9)}$ sterols. Flat sterols will support growth, whereas bent sterols and sterol-like molecules (10α-cucurbitacin) will not support normal growth of yeast auxotrophic for sterol, nor will yeast transform bent cyclopropyl sterols to ergosterol.[65]

Hence, plants, fungi, and animals (including marine invertebrates and insects which dealkylate the C-24 alkyl group of phytosterols) may operate the same basic sterol pathway, which implies a common origin of genetic material. Moreover, one gene may code for an enzyme, such as the C-24 methyl transferase, capable of performing multiple reactions on sterols with appropriate functionality. As indicated in the "steric-electric plug" model shown in Figure 7, the proposed juxtaposition in the active site of S-adenosyl-L-methionine (SAM, CH_3^+) and NAD(P)H (H–) relative to the sterol (requiring the existence of multiple binding sites on a polymeric protein), the substrate/co-factor binding order (multiple domains) and type, and amount of bound sterol (operationally referred to as sterol allosterism) are biochemical determinants that act to regulate the reaction kinetics and position of internal equilibria mediating the stereochemistry of methylation and reduction of the 24,25-double bond. It follows that a single enzyme (or family of closely related enzymes) may produce the diversity of sterols in nature, where the variation in occurrence lies in the structure of the sterol side chain. Similarly, the enzyme that is considered to be unique to plants, the COI, may be found in fungi (and for that matter plants and animals) as the enzyme that is used to isomerize the double bonds from the 9(11)- and 8(9)-positions to the 8(9)- and 7(8)-positions. Each of these reaction courses involve acid-induced isomerizations. A possible reason for why yeast have been found unable to convert cycloartenol to ergosterol[27,66] is that the cell-free preparations used to study the transformation contained endogenous lanosterol, which competed with the cycloartenol for binding/catalysis. In the case of the mutant cells, they most likely were partially heme amplified (sterol adapted), so that the oxidative removal of the methyl groups at C-14 was impaired.

VIII. RATIONALLY DESIGNED STEROL BIOSYNTHESIS INHIBITORS

Studies of stable analogs of the substrate portion of the activated complex (AC) and high energy intermediate (HEI) forms, which are of lower free energy in the transition state progress than the AC, can reveal what substrate bond-making and bond-breaking steps occur during turnover. Competitive and noncompetitive inhibitors have been used to probe the transition state progress in C-24 methylation.[44,48,66,67]

Synthetic inhibitors prepared to inhibit C-24 methylation are designed as transition state analogs (TSA). They were first designed to interfere with the carbocation intermediate (the HEI) that forms during the normal reaction

Binding here (domain B) of potential substrate, product or stable isoelectric / isosteric analog of reaction intermediaes will influence the catalytic character, quantitatively and qualitatively, of a second communicating active site (domain A) This is illustrated schematically in schemes A, B and C below for catalysis in the absence or presence (at limiting and saturating concentrations), respectively, of such an analog.

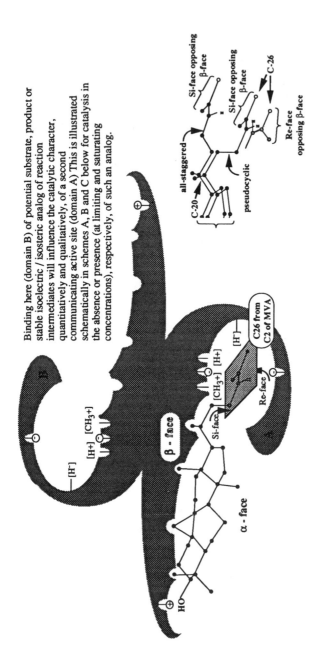

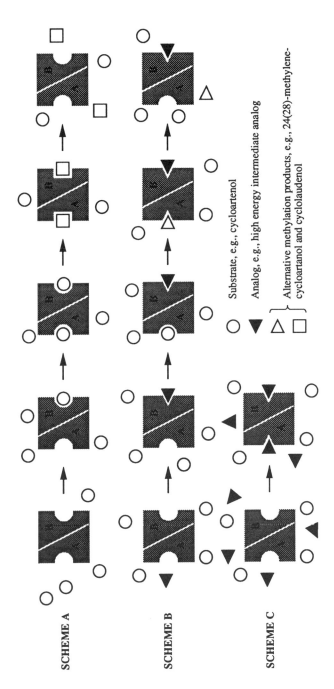

FIGURE 7. Schematic representation of the "steric-electric plug" model as an allosteric protein with communicating active sites. Drawn to illustrate how both the initial disposition of the reactive moiety of the substrate (Δ^{24} bond) in the active site and binding of an "allosteric modulator" (e.g., high energy intermediate analog) may influence the character of catalysis.

course that involves SAM and the 24,25-double bond of the sterol.[44] We have taken a new approach and that is to consider the reaction course in its entirety. We have developed a series of novel inhibitors that interfere with the intermolecular nucleophilic catalysis necessary to promote formation of the dative bond similar to the transition state that is established between SAM and the C-24=C-25 nucleophile and a second set of inhibitors that interfere with the electrostatic interactions formed between charged species that facilitate substrate binding and product outcome. We have also followed the transfer of tritium at C-24 to C-25 and determined the number of deuterium atoms in the sterol side chain after sunflower microsomes were fed 24-tritiolanosterol and deuterated SAM.

Based on our recent efforts using the new inhibitors of the C-24 methylation reaction and radio and stable isotopes introduced into the sterol side chain (Reference 68 and Nes, W. D., Kalinowska, M., Norton, R. A., and Parker, S. R., unpublished data, 1992), we conclude the methylation reaction proceeds by an S_n1 reaction in two discrete stages. The different transition states use substrates of different conformation and electrical stability. A single enzyme mediates the first and second C_1 transfers and the stereochemically varied methylation outcomes, viz., to produce a 24(28)-methylene, 24 β-methyl, and a $\Delta^{23(24)}$ methyl sterol and isomerization of the 24(28)-bond and reduction of the 24,25-bond. The first stage is postulated to involve an S_n2 nucleophilic attack of the *si*-face 24,25-double bond of the sulfonium methyl group of SAM, and this may be considered the rate-limiting step for methylation and hence of phytosterol genesis. Here, two nucleophiles compete for interaction with SAM: one presumably comes from the active site itself and the other comes from the incoming double bond on the sterol side chain (cf., Figure 7, the "steric-electric plug" model). The mechanism (S_n2 displacement) of methylation predicts inversion of configuration of the SAM molecule when substitution occurs at the chiral carbon. There is experimental support for this mechanism in the biosynthesis of fungal ergosterol.[69] The second stage of the methylation reaction is considered to center around the electrostatic interactions that develop between the intermediate C-24 and C-25 carbenium ions and the nucleophile in the active site which otherwise should be interacting with SAM as part of an electrostatic bridge.

A hypothetical reaction progress for the C-24 methylation events in sunflower is given in Figure 8. The formation of the dative bond requires a higher energy barrier to overcome than other reactions and for chemical reasons is the rate-limiting step. However, it is through the stabilization of the C-25 carbenium ion, which is of a lower free energy in the reaction coordinate, that the stereochemistry of the methylation product(s) is determined, i.e., kinetic control. Stabilization of the C-25 carbenium ion may proceed from the 25(27)-carbon or from a 1,2-hydride shift of the 24-hydrogen along the *re*-face of the original double bond to C-25. Whereas proton to proton migration in the symmetrical bridged intermediate could lead to epimerization at C-24 and C-25 or produce a side chain with either a saturated or unsaturated system, e.g., with a $\Delta^{22(23)}$

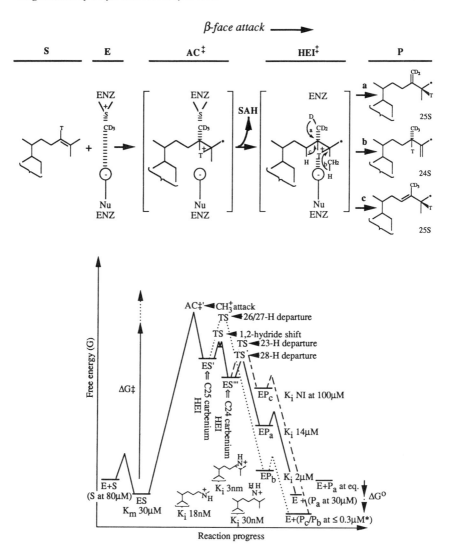

FIGURE 8. Hypothetical free-energy diagram for the alternative outcomes of C-24 methylation. A first approximation of the relative energies of the reaction course intermediates are shown: ES — enzyme-substrate complex; AC — activated complex; TS — transition state; and EP — enzyme-product complex. Relative energies of the intermediate species were deduced from studies of product inhibition and the use of compounds designed as high energy intermediate (HEI) analogs.[49,58,64] NI — not inhibitory. Endogenous microsomal concentrations are as indicated on the figure: S — substrate (cycloartenol), P_a — product (24(28)-methylenecycloartanol), P_b — (cyclolaudenol), and P_c — (cyclosadol), where * — 0.3 μM represents our limit of detection by GC/HPLC. Microsomal preparations contained circa 30 μM endogenous cycloartenol. A typical assay was performed with 50 μM cycloartenol, 50 μM SAM, and Tween 80 for 45 min at 30°C. Enzyme assays from Reference 104.

bond, this seems unlikely at the present time. A C-24 carbenium ion is formed after the hydrogen shift from C-24 to C-25, and this structure leads to a $\Delta^{23(24)}$-24 methyl sterol. From the kinetic data reported in Figure 8, cyclolaudenol is the most effective sterol inhibitor of C-24 methyl transferase activity, which is consistent with *si*-face methyl attack on the 24,25-double bond.[29]

In sunflower, the kinetically favored methylation outcome appears to be a 24(28)-sterol. The formation of the 24(28)-methylene sterol as the favored methylation outcome of the first C_1 transfer may proceed in most, if not all, tracheophytes.[70,71] Certain algae produce 25(27)-sterols as the preferred C-24 methylation outcome. These algae often produce minor amounts of 24-ethyl sterols,[72,73] suggesting kinetic control is favored here to produce the 25(27)-sterols. The tacit assumption that 25(27)-sterols, such as cyclolaudenol, may enter into the sterol pathway of vascular plants and be transformed by a 25(27)-reductase to 24β-methyl sterols[30,70,71] now appears unlikely in view of recent studies by Uomori et al., who demonstrated that 24β-methyl cholesterol was formed by reduction of the 24,25-bond in cultured plant cells,[74] a route we found earlier to operate in fungi.[75] Bladocha and Benveniste also observed a similar reaction course after their plants were treated with tridemorph.[34] The ability of 25(27)-sterols to proceed to 24-methyl or 24-ethyl sterols is likely due to the isomerization of the 25(27)-bond to the 24(25)-bond.[76]

The second methylation may not be regiospecific, but catalytic competence should be mediated by substrate specificity and sterol allosterism. Thus, in terms of methylation kinetics, either a 24(28)-methylene or $\Delta^{24(25)}$ C-24 methyl sterol may be methylated.[77] As shown in the mechanisms proposed in Figure 9, the phytosterols indicated in Figures 5 and 6 may be formed by the same alkylating enzyme. The reduction of the 24,25-double bond, initiated by a H$^+$ species in the active site, may be regiospecific and proceed on the same enzyme surface where methylation occurs as shown in Figure 9. Another mechanism for reduction, based on the tacit assumption that the C-24 methyl transferase and Δ^{24} reductase are different enzymes, has been proposed by Benveniste and co-workers.[34,78]

There are several types of TSAs which may be rationally designed to interfere with the transition state progress which characterizes the different methylated products distributed in nature. The TSAs may also be used to study sterol allosterism of SMT activity. As shown in Figure 10, we have prepared three classes of TSAs. First, are the mechanism-based inhibitors, which may be either monosubstrate (directed to the nucleophile binding site) or bisubstrate (directed to the enzyme nucleophile and SAM binding sites). Second are the HEI analogs (directed to the nucleophile binding site). Third, are the ground-state mimics (as isosteres they are directed to the binding site to quantitatively replace the preferred substrate, a competitive inhibitor).

In the design of a TSA of C-24 methylation that leads to properties that characterize a mechanism-based inhibitor, the modified sterol should undergo covalent catalysis by forming an irreversible covalent bond with SAM and/ or a nucleophile in the active site. The TSA should be a stable compound that

Second C₁-Transfer

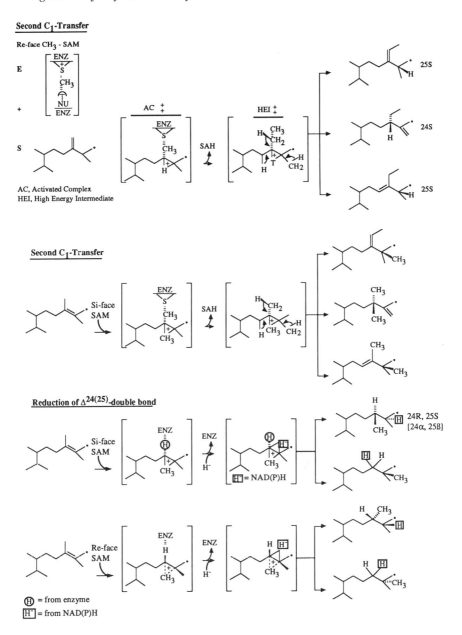

AC, Activated Complex
HEI, High Energy Intermediate

Second C₁-Transfer

Reduction of Δ²⁴⁽²⁵⁾-double bond

Ⓗ = from enzyme
H̄ = from NAD(P)H

FIGURE 9. Alternative mechanisms for C-24 methylation and Δ²⁴ reduction.

possesses a geometry (three-dimensional shape) comparable to the natural substrate (which in sunflower is cycloartenol) before it is bound to the enzyme, i.e., in the ground state which is known to be flat with the side chain oriented to the right,[28] and resembles the conformation of the substrate at a postulated

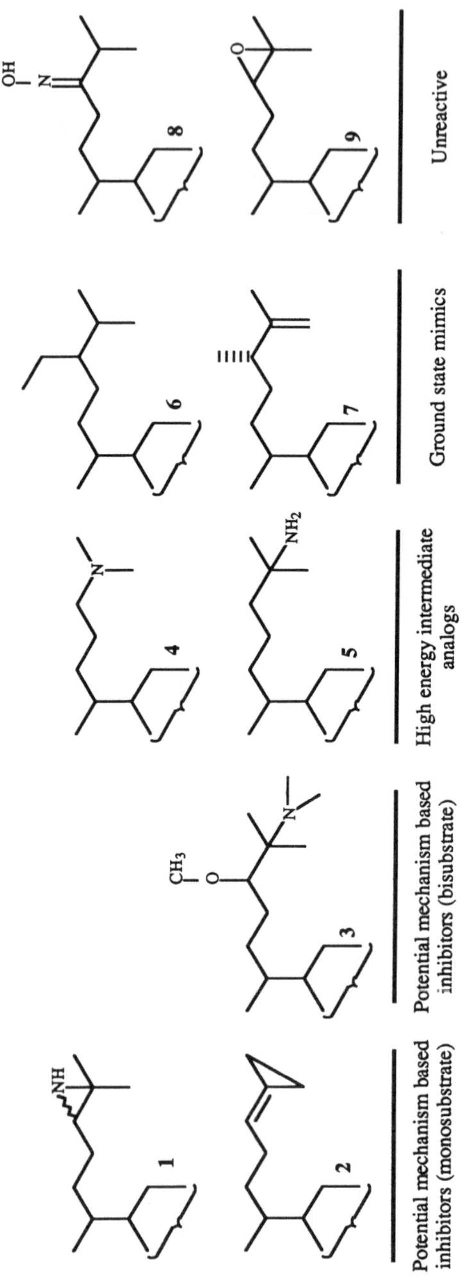

FIGURE 10. Rationally designed analogs of sterol methyl transfer reaction progress intermediates.

transition state of the reaction in which the bound conformer is assumed to be the one with the side chain in the pseudocyclic conformation. Here, the side chain and nuclear conformations are equally important to binding and catalysis. In contrast to HEI analogs, which possess a strategically located heteroatom in the side chain, giving it charge, mechanism-based inhibitors may be prepared without adding a heteroatom into the side chain (Figure 10).

For HEI analogs, the global three-dimensional shape is of less significance than in the case of mechanism-based or ground-state mimic inhibitors, since the HEI analogs bind with a different set of ionic parameters from those which establish the activated complex and from those involved in sterol allosterism. Examples of some HEI analogs which form a reversible complex with the enzyme are sterols that possess an ammonium or a related heteroatom in the terminal segment of the sterol side chain, such as 25-azalanosterol and 25-aminolanosterol.[44,48] If the *N*-sterol binds as the ammonium salt (supposedly protonated at physiological pH), the only mechanism available for the quaternary amine is electrostatic facilitation. In order to facilitate disassociation, the ammonium salt needs to be deprotonated. As it so happens, the active site is known to possess a deprotonating agent, e.g., as demonstrated in the conversion of the 24-methyl sterol carbocation intermediate to a 24(28)-methylene sterol product. If the *N*-sterol is to disassociate from the enzyme by using the same deprotonating agent that is used in the normal course of the methylation of the 24,25-double bond, then the side chain conformation of the bound *N*-sterol should be essential to activity, as it is for the natural substrate. The key features which distinguish whether an ammonium compound blocks methylation by a mechanism-based or some other type of interference is to assess the chemical reactivity of the nitrogen grouping (i.e., whether it is transformed to another compound) and to study the kinetics of inhibition.

With these assumptions in mind, we studied the inhibitory potency of several ammonium containing sterols (modified *N*-sterols) in growth support of plant cells,[48] cancer cells,[79] and pathogenic fungi.[80,81] Several other groups have adopted this approach.[44,82-84] In each system that we studied, the 24(*R,S*),25-epiminolanosterol (IL) was a potent inhibitor of cell proliferation. IL inhibited growth at very low concentrations, I_{50} of 200 n*M*, in cultured sunflower cells. An order of magnitude greater concentration of IL was necessary to inhibit the fungal and animal systems.[48,79-81] To our knowledge, IL is the most potent sterol biosynthesis inhibitor in plants.

The mechanism of action of IL and the other *N*-sterols was not entirely clear, since the ammonium compound may exist either in the neutral or salt form (Figure 11). Depending on the state of protonation, the nitrogen can mimic either an electrophilic or nucleophilic center. In the case of the isosteric interchange of the 24,25-double bond with the aziridine ring (cf., IL), the *N*-sterol was observed to be a potent inhibitor of the C-24 methylation reaction with a K_i of 3 n*M* (Table 2). The nitrenium ion bears a positive charge at physiological pH and therefore should interfere with methylation by preferentially binding to the nucleophilic site, which normally interacts with

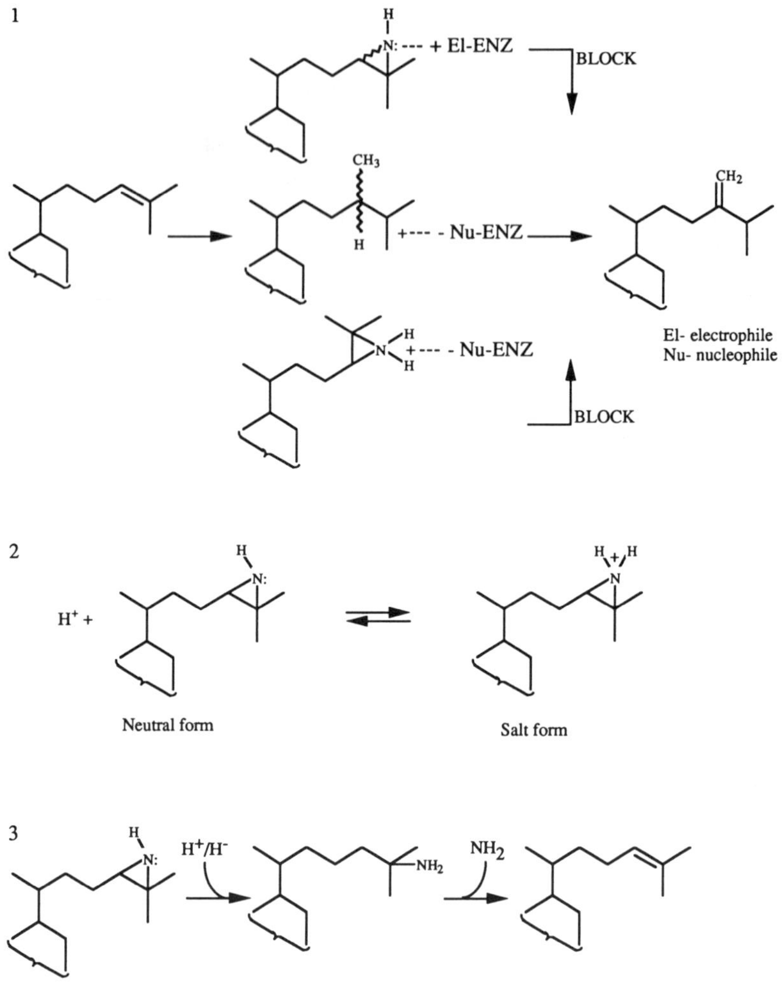

FIGURE 11. Consideration of the basis for the inhibitory activity of 24,25-epiminosterol. 1. Aziridine moiety interacting with either an electrophile or nucleophile, depending on its state of protonation. 2. Equilibrium between the protonated (salt) and deprotonated (neutral) forms of 24, 25-epiminosterol. 3. Reductive deamination of the 24,25-epiminosterol.

the C-25 carbenium ion generated by the natural substrate during the course of methylation. There was no indication [thin layer chromatography (TLC) or gas chromatography (GC)] that the aziridine was methylated by the enzyme, although the kinetics of IL inhibition was a noncompetitive type of inhibition and it demonstrated a time-dependent inactivation (Reference 68 and Nes, W. D., Kalinowska, M., Norton, R. A., and Parker, S. R., unpublished data, 1992). These data appear to be at variance, but they are not. The aziridine most likely entered the active site as the protonated ammonium salt. Therefore, it

Substrate	K_i	Substrate	K_i
1	3 nM	10	17 nM
2	30 nM	11	45 nM
3	33 nM	12	600 nM
4	52 nM	13	55 nM
5	120 nM	14	250 nM
6	24 nM	15	37 nM
7	18 nM	16	500 nM
8	NI	17	NI
9	NI	18	25 μM

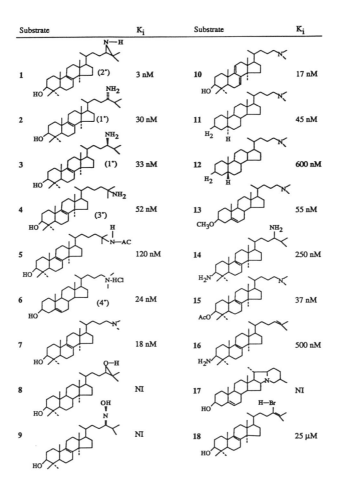

TABLE 2. Inhibition of C-24 methyl transferase activity by sterols with ammonium and related heteratoms at C-3, C-24, and C-25.

was not readily available to be methylated by SAM. The aziridine may be released from the enzyme by the same process that other *N*-sterols are likely released, that is by the action of a deprotonating agent present in the site for methylation. Because the configuration of the nitrogen in the aziridine ring is chiral, inversion of the nitrogen center may proceed after deprotonation of the bound ammonium salt form. The *N*-hydrogen then will be toward the nucleophile and the lone pair of electrons away from the interaction with the nucleophile in the active site. The newly formed interactions that result from deprotonation are expected to cause the side chain to swing from the pseudocyclic into the all-staggered conformation. However, if the aziridine is trapped by a H⁺ species before it dissociates from the enzyme, then it may undergo

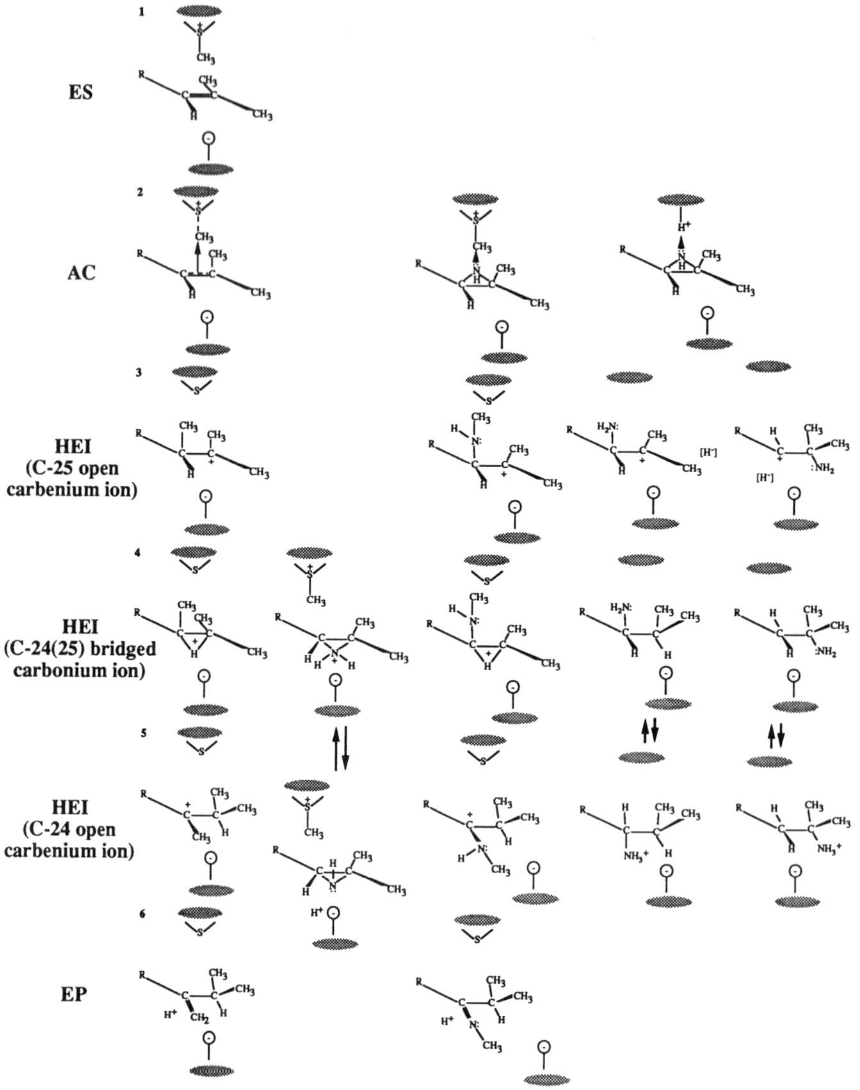

FIGURE 12. Comparison of the proposed C-24 methylation reaction progress and possible modes of interaction between the catalytic site and various *N*-steroid species designed as HEI analogs of the sterol methyl transfer.

stereospecific ring opening, as we recently demonstrated in the fungal metabolism of IL to 25-aminolanosterol and lanosterol.[81] Our speculative scheme for the reaction course for IL interaction and transformation by the SMT is shown in Figure 12.

Rahier et al. outlined the molecular parameters for inhibition of C-24 methyl transferase activity in plants.[44] According to these authors, for an HEI analog to be an effective inhibitor, the compound should possess a steroid skeleton.

However, any molecular species that enters the active site and exploits a favorable interaction with the catalytic site may function as a competitive inhibitor. The suggestion by Rahier et al. may well form the basis for rational design of HEI analogs, but the "postulate" should not exclude the use of "conformationally adaptive" species, e.g., triparanol,[67] which may exhibit more effective electrostatic interactions despite the obvious lack of structural complimentarity. We are now investigating the possibility that sterol biosynthesis inhibitors which interfere with C-24 alkylation may once again be prepared for commercial use — tailored as fungicides, algicides, or herbicides because they may be designed for a specific transition state coordinate and undergo a time-dependent biodegradation (i.e., the basis for them to be environmentally safe). Alternatively, because of the proposed allosteric nature of sterolic enzymes, it may be possible to influence the qualitative character of catalysis, e.g., methylation vs. reduction, through the selective use of mechanism-based inhibitors.

IX. STATUS AND FUNCTION OF PHYTOSTEROLS

Phytosterols play three main roles: as architectural components of membranes, as precursors to hormones (e.g., brassinosteroids and ecdysteroids) and phytoalexins (e.g., steroidal alkaloids), and as trigger molecules of cell proliferation.[7,18,85] In the conversion of cycloartenol to stigmasterol (and sitosterol), there is an underlying assumption that pathway sequencing is not random, and a purpose is served by the conversion of the intermediate sterol into one or more structurally similar end products.

Examination of Dreiding models of cycloartenol and stigmasterol show the two molecules possess several similar features; that is, they are amphiphilic, flat, and the side chain is free to orient to the right, so that the length of the molecule may approximate the 20 Å length of the monolayer of the lipid leaflet. However, the transformation process affects other features, such as the nucleophilicity and ability of the molecule to assume one or another three-dimensional shapes. The sterol structural features, which are modified by transformation, are shown in Figure 13. Of potential functional significance is the removal of the C-4 geminal methyl groups, which is expected to strengthen the OH-bonding capability at C-3; the introduction of one or more methyl groups at C-24, which should increase the molecular volume, as the added bulk should allow for the side chain to sweep out a larger cone than a side chain that contains double bonds either at Δ^{22} or Δ^{24}. The terminally located double bonds should be expected to restrict rotation of the side chain. Removal of the 14-methyl group should establish a planar α-face, and the opening of the cyclopropane ring to a $\Delta^{8(9)}$ sterol and the subsequent conversion of the 8(9)-bond to a 5(6)-bond should affect the global flexibility of the molecule.

Cycloartenol is now known to be flat in solution,[86] the solid state,[87] in membranes,[88] and when bound to the C-24 methyl transferase from sunflower.[28] Thus, we surmised that the different activities that result from incu-

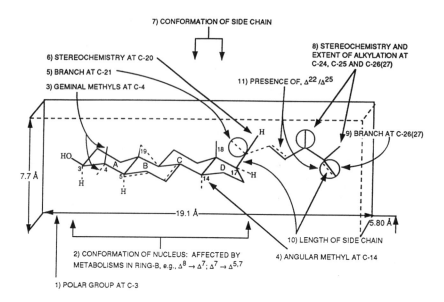

FIGURE 13. Functionally significant structural features of sterols altered by metabolism.

bation with cycloartenol and lanosterol are not due to a difference in molecular shape, bent vs. flat, as indicated by others,[27,89-91] but due to conformational affects on the tilt and H-bonding strength of the OH group.[91] Therefore, fungicides may not be designed to exploit differences in the conformation of cyclopropyl sterols produced by plants. That plants may accumulate certain cyclopropyl sterols after treatment with sterol biosynthesis inhibitors is often interpreted to imply that all naturally occurring cyclopropyl sterols are suitable sitosterol surrogates.[89,90,93] Interestingly, the cyclopropyl sterols which accumulate in the plant plasma membrane lack the C-4 geminal groups.[90-92] The inhibited plants are not accumulating cycloartenol. We recently demonstrated that cycloartenol is not a sitosterol surrogate in cultured sunflower cells.[48]

In previous discussions of sterol functioning as a membrane insert, the sterol was viewed as "cone-shaped"[94] and that it induced an increased condensing effect on the acyl chain order of lipids in the liquid crystalline state and by its liquefying effect decreased the chain order of lipids in the gel state.[95] Thus, sterol-containing membranes maintained the lamellar phase (bilayer) in an intermediate state of fluidity.[94-96] The most persistent and currently prevalent models for sterol-lipid interactions in the lipid leaflet propose complex formation from 22 to 50 mol% of the sterol.[94-96] The maintenance of the lamellar phase structures in the intermediate gel states by sterol is the so-called bulk role.[27]

The general view is that the structural requirements of the sterol to function in the bulk role are rather broad:[27] simply that it possess a free 3β-OH group, a planar tetracyclic nucleus, and an intact side chain of eight to ten carbon atoms (meaning the side chain could be alkylated at C-24). No sterol is superior to cholesterol in mediating membrane fluidity, and pentacylic triterpenoids may replace sterol as a membrane insert.[26,27] The Rothman-Engelman model suggests the interactions between sterol (cholesterol) and the phospholipid are primarily steric in nature, and the dramatic effects of sterol on the motional state of the lamellar phase resides in the relative cross-sectional area of the nucleus compared with that of the aliphatic tail.[97] Thus, this is the basis for sterol-like molecules to replace sterols as membrane inserts. As the polar head of the sterol is much smaller than the nucleus, sterols are viewed as cone shaped.[94] Recent biophysical data indicate the amount of sterol influences the state of molecular motion of the sterol side chain, and the ability of the 3β-OH of the sterol to engage in hydrogen bonding with the carbonyl oxygen of the fatty acid groups of the phospholipid are equally important to the functioning of sterol in the membrane.[98-100] Sterol as low as 5 mol% sterol to lipid may promote bilayer to hexagonal H_{II} phase transitions. Because sterol may concentrate in discrete regions of the bilayer,[88,94] the sterol may promote the evolution of intramembrane structures which cause cell fusion or other forms of morphological change. Hence, the sterol role in plant membrane function may extend beyond simply affecting permeability.

In an effort to assess the structural requirements of sterols to play the bulk role, which is assumed to be the same in fungi and plants, we cultured two phylogenetically different fungi that are auxotrophic for sterol for membrane biogenesis. The two sterol auxotrophs are *P. cactorum* (an Oomycete) (reviewed in Reference 7) and *Saccharomyces cerevisiae* strain GL7 (an Ascomycete) cultured aerobically[65] and sterol adapted to amplify heme-competence. For growth support of the GL7 (a measure of sterol membrane competence), the 3β-OH group could not be replaced by H, OMe, OBu, NH_2, NHOH, OAc, keto, or 3α-OH (Table 3). In some cases, the aberrant sterols supported growth of GL7, but they were recovered from the cells as transformed sterols. When yeast cells are cultured under strict anaerobiosis, many of the sterols which supported growth of the GL7 failed to support growth of the anaerobic cultures.[8,101] The OH group is therefore essential for activity. We further interpret the results to imply that the amphiphilic character necessary for sterols to function in the lipid leaflet must have the polar group at C-3 act as a H donor in the hydrogen bonding with the ester carbonyl moiety of the fatty acid, which is associated

TABLE 3
Growth Support and Metabolism of Sterols and Sterol-Like Molecules by the Sterol Auxotroph Yeast Strain GL7

Compound fed[a,b]	Cell no. (×10⁶ cells/ml)	Wet wt. (g/l)	Visual OBS[c]	Total sterol fg/cell	Major cellular sterol(s)
Wild-type strain A1842	240	14.0	N	11	Ergosterol
No sterol	0	0	N	0	—
Ergosterol1	228	13.6	N	10	Substrate
9(11)-Dehydroergosterol2	171	10.5	N	6.1	Substrate
Ergosta-7,9(11)22-trienol3	194	13.8	N	4.3	Ergosta-5,7,9(11),27-tetraenol
Ergosterol endoperoxide (+ sp. E)4	0 (1.0)	–(–)	–(–)	–(–)	–(–)
14α-Methyl ergost-9-enol (+ sp. E)5	201 (200)	13.8 (12.2)	N	6.6 (10)	Ergosterol/14α-methylergost-9-enol
14α-Methyl stigmast-9-enol (+ sp. E)6	214 (212)	13.4 (14.8)	N	6.7 (8.3)	Substrate
Sitosterol7	330	19.9	N	11.0	Substrate
Cholesterol8	238	14.2	N	10	Substrate
Cholest-8(14)-enol (+ sp. E)9	85	9.5	N	4.9	Substrate
Cholest-14(15)-enol (+ sp. E)10	127	12.3	N	14.5	Substrate
Cholesta-5,7-dienol11	200	10.5	N	8	Cholesta-5,7,22-trienol
Cholest-4-enol (+ sp. E)12	0.5 (39)	–(8.2)	(N)	–(2)	(Cholesterol)
Cholestanol13	256	14	N	8	Substrate
3-Epicholestanol (+ sp. E)14	35(96)	–(12.7)	A (N)	–(8.5)	(Cholestanol)
Coprostanol (+ sp. E)15	0 (0)	–(–)	–(–)	–(–)	–(–)
5α-Cholestane (+ sp. E)16	0.0 (0.0)	–(–)	–(–)	–(–)	–(–)
E-17(20)-Dehydrocholesterol17	220	13.0	N	10	Substrate
20-Epicholesterol (+ sp. E)19	0 (0)	–(–)	N	–(–)	–(–)
Wingsterol20	1.0 (1.0)	–(–)	–(–)	–(–)	–(–)
Pregn-5-en-3β-ol (+ sp. E)24	0 (0)	–(–)	–(–)	–(–)	–(–)
(20R)-N-Butyl pregn-5-en-3β-ol (+ sp. E)25	185 (173)	13.6 (11.5)	N	10 (11)	Substrate (substrate)
(20R)-N-Heptyl pregn-5-en-3β-ol (+ sp. E)26	0 (0)	–(–)	–(–)	–(–)	–(–)
(20R)-N-Nonyl pregn-5-en-3β-ol (+ sp. E)27	0 (0)	–(–)	–(–)	–(–)	–(–)

24,25-Dehydropollinastanol28	240	14.5	N	-(5)	Ergosterol/mixture of 24-alkyl-pollinastanols
Cholest-4-en-3-one (+ sp. E)29	1.3 (75)	-(9.0)	(N)	-(9.0)	(Cholesterol)
Cholest-5-en-3-one (+ sp. E)30	1.4 (32)	-(5.3)	(N)	-(4.6)	(Cholesterol)
3-Epicholesterol (+ sp. E)31	0.7 (80)	-(8.7)	A(N)	-(5.8)	(Cholestanol)
4,4-Dimethyl cholesterol (+ sp. E)32	1.0 (1.0)	-(-)	-(-)	-(-)	-(-)
Cholesteryl methyl ether (+ sp. E)33	228 (232)	16.9 (19.6)	(N)	16 (18)	Cholesterol (cholesterol)
Cholesteryl buty ether (+ sp. E)34	1 (11)	-(-)	-(-)	-(-)	-(-)
Cholesteryl 3-OAc (+ sp. E)35	107 (179)	13.4 (12.2)	N (N)	4.3 (9.8)	Cholesterol (cholesterol)
3-Aminocholesterol (+ sp. E)36	27 (82)	-(-)	A (A)	-(-)	-(-)
3-Oximocholesterol (+ sp. E)ψ37	110 (113)	9.7 (9.8)	A (A)	6.7 (7.9)	Cholestanol (cholestanol)
Lanosterol (+ sp. E)38	111 (150)	9.3 (10.9)	N (N)	4 (14)	Ergosterol (ergosterol)
Lanosteryl methyl ether (+ sp. E)39	1 (1.1)	-(-)	-(-)	-(-)	-(-)
3-Aminolanosterol (+ sp. E)40	1.8 (36)	-(-)	A	-(-)	-(-)
3-Oximolanosterol (+ sp. E)41	10 (76)	-(5.6)	A	-(-)	-(-)
Lanosterol-3-one (+ sp. E)42	83 (172)	-(12.9)	A	-(3.75)	-(Ergosterol)
24,25-Dihydrolanosterol (+ sp. E)43	23 (151)	-(10.6)	N	4 (14)	(Cholest-7-enol/cholest-5-enol cholest-5,7,22-trienol/cholesta-5,7,9(11),27-tetraenol)
Lanostanol (+ sp. E)44	1.1 (1.0)	-(-)	-(-)	-(-)	-(-)
Parkeol (+ sp. E)45	0 (139)	-(10.6)	N	-(4)	Ergosterol/substrate/others
24,25-Dihydroparkeol (+ sp. E)46	23 (151)	-(10.6)	N	-(6)	(Substrate/31-nor-24, 25-dihydroparkeol/cholesta-5,7,22-trienol/others)
Cycloartenol47	1.0 (106)	-(7.4)	N	-(4)	(Ergosterol/substrate/31 nor-cycloartenol/mixture of 4-desmethyl-24-alkylcyclopropyl sterols)
Euphol (+ sp. E)48	1.0 (1.0)	-(-)	-(-)	-(-)	-(-)
Tircucallol (+ sp. E)49	1.0 (1.0)	-(-)	-(-)	-(-)	-(-)
10α-Cucurbitadienol (+ sp. E)50	1.0 (1.0)	-(-)	-(-)	-(-)	-(-)
Dammaradienol (+ sp. E)51	0 (0)	-(-)	-(-)	-(-)	-(-)
Tetrahymanol (+ sp. E)52	1.0 (1.0)	-(-)	-(-)	-(-)	-(-)
Isoarborinol (+ sp. E)53	0 (2.0)	-(-)	-(-)	-(-)	-(-)

TABLE 3 (continued)
Growth Support and Metabolism of Sterols and Sterol-Like Molecules by the Sterol Auxotroph Yeast Strain GL7

Compound fed[a,b]	Cell no. (×10⁶ cells/ml)	Wet wt. (g/l)	Visual OBS[c]	Total sterol fg/cell	Major cellular sterols(s)
Motiol (+ sp. E)54	1.0 (1.0)	-(-)	-(-)	-(-)	-(-)
α-Amyrin (+ sp. E)55	1.0 (1.0)	-(-)	-(-)	-(-)	-(-)
Epifriedelinol (+ sp. E)56	1.0 (52)	-(3.6)	A	-(31)	(Substrate)

Note: 26-Homocholesterol (21), Halosterol (22), and 21-Norcholesterol (23) supported growth to the same extent as ergosterol, but the cells were not examined for sterol content. Z-17(20)-dehydrocholesterol (18) failed to support growth, but due to insufficient sample, we could not evaluate efficacy with sparing ergosterol.

[a] Sterols were fed to adapted *S. cerevisiae* cells at 5 μg/ml. Sparing (sp.) levels of ergosterol (0.5 μg/ml) were added to select sterol treatments as shown above. Dash line indicates not determined. Experiments were performed in triplicate, and the data reported did not vary by more than 5%. No sterol treatment contained Tween 80. In the absence of Tween 80, the cultures underwent several cell doublings in 75 h.

[b] Structures of sterol supplements are given in Figure 1.

[c] N — normal cells; A — aberrant cells which were very small or large or in clusters; cell clusters were counted as one cell.

[d] Mixture of 3-*E* and Z in a 40:60 ratio as deduced by ¹H-NMR.

with the phospholipid head group, either directly or by way of a water bridge. Alternatively, the electron pair on the oxygen is significant for the sterol to bind to enzymes and other kinds of proteins that act or interact with the sterol substrate (the bridge model[102]). In the yeast mutant, there is a lack of architectural parity between sterols and sterol-like molecules. A methyl group at C-14 was neither deleterious nor essential for activity. A narrow window for side chain length was observed, so that growth support required a sterol with the longest methylene segment extending from C-20 not to exceed six contiguous C atoms and the stereochemistry to be C-(20*R*). No significance could be attributed to branching at C-20 (i.e., to C-21), C-24 (when alkylated), or to C-25 (regarding the isopropyl group). Sometimes, the addition of trace (sparing) levels of dietary ergosterol (0.5 μg/ml) was necessary to promote growth and transformation of the bulk sterol to a membrane competent insert, a function we refer to specifically as ergosterol allosterism. The latter function is similar to the sitosterol allosterism that we propose operates to downregulate C-24 methyl transferase activity in plant cells. Similar observations have been reported for the structural requirements of sterols to affect oospore production (a measure of sterol promoting oosphere membrane genesis).[7,40]

The detailed structural requirements of sterol to play the trigger function are not known, but a 24-alkyl group seems to be important for activity.[85]

In summary, in the consideration of the structural features of membrane inserts that satisfy membrane function, one needs to consider that these requirements, as regards to the three-dimensional shape of the insert, may vary with the dynamic activity of the membrane. Hence, the amount of insert and the structural features required to promote/inhibit phase structure transitions and maintain bilayer order may differ according to the lipid composition of the leaflet. The superiority of a molecule to function as a membrane insert is related to the "lipid background" and temperature. To date, the considerations of an evolutionary hierarchy of sterols, namely the ability of one sterol or sterol-like molecule to replace another, has been limited by the supposition that sterols have uniform function and shape. There now are cogent reasons to believe that sterols play multiple functions, the satisfaction of which is not necessarily met by a single population of sterol existing in the cone shape.

REFERENCES

1. **Hess, C. E.,** New challenges in agriculture, *Phytopathology*, 82, 40, 1992.
2. **Pesek, J.,** *Alternative Agriculture*, National Academy Press, Washington, D.C., 1989.
3. **Brindle, P. A., Kuhn, P. J., and Threlfall, D. R.,** Biosynthesis and metabolism of sesquiterpenoid phytoalexins and triterpenoids in potato cell suspension cultures, *Phytochemistry*, 27, 133, 1988.
4. **Mercer, E. I.,** Sterol biosynthesis inhibitors in plants, fungi, *INFORM*, 1, 904, 1990.
5. **Burden, R. S., Cooke, D. T., and Carter, G. A.,** Inhibitors of sterol biosynthesis and growth in plants and fungi, *Phytochemistry*, 28, 1791, 1989.

6. **Ikekawa, N. and Zhao, Y.-J.,** Application of 24-epibrassinolide in agriculture, *ACS Symp. Ser.,* 474, 280, 1991.

7. **Nes, W. D.,** Control of sterol biosynthesis and its importance to developmental regulation and evolution, *Rec. Adv. Phytochem.,* 24, 283, 1990.

8. **Nes, W. D. and McKean, M. L.,** *Biochemistry of steroids and other isopentenoids,* University Park Press, Baltimore, MD, 1977, 412.

9. **Parker, S. R. and Nes, W. D.,** Regulation of sterol biosynthesis and its phylogenetic implications, *ACS Symp. Ser.,* 497, 110, 1992.

10. **Nes, W. R. and Nes, W. D.,** *Lipids in Evolution,* Plenum Press, New York, 1980, 157.

11. **Giner, J.-L., Wünsche, L., Andersen, R. A., and Djerassi, C.,** Dinoflagellates cyclize squalene oxide to lanosterol, *Biochem. System. Ecol.,* 19, 143, 1991.

12. **Nes, W. D., Norton, R. A., Crumley, F. G., Madigan, S. J., and Katz, E. R.,** Sterol phylogenesis and algal evolution, *Proc. Natl. Acad. Sci. U.S.A.,* 87, 7565, 1990.

13. **Goodwin, T. W.,** Biosynthesis of plant sterols, in *Sterols and Bile Acids,* Danielsson, H. and Sjovall, J., Eds., Elsevier Science Publishers, New York, 1985, 175.

14. **Nes, W. D. and Bach, T. J.,** Evidence for a mevalonate shunt in a tracheophyte, *Proc. R. Soc. London B,* 225, 425, 1985.

15. **Goad, L. J.,** How is sterol synthesis regulated in higher plants?, *Biochem. Soc. Trans.,* 11, 548, 1983.

16. **Koops, A. J. and Groeneveld, H. W.,** Triterpenoid biosynthesis in the etiolated seedling of *Euphorbia lathyris* L.: developmental changes and the regulation of local triterpenoid production, *J. Plant Physiol.,* 138, 142, 1991.

17. **Koops, A. J., Italiaander, E., and Groeneveld, H. W.,** Triterpenoid biosynthesis in the seedling of *Euphorbia lathyris* L. from sucrose and amino acids, *Plant Sci.,* 74, 193, 1991.

18. **Roddick, J. G.,** Antifungal activity of plant steroids, *ACS Symp. Ser., 325,* 286, 1987.

19. **Heftmann, E.,** Metabolism of cholesterol in plants, in *Isopentenoids in Plants: Biochemistry and Function,* Nes, W. D., Fuller, G., and Tsai, L.-S., Eds., Marcel Dekker, New York, 1984, 487.

20. **Cutler, H. G., Yokota, T., and Günter, A.,** *Brassinosteroids: Chemistry, Bioactivity and Applications,* Vol. 474, American Chemical Society Press, Washington, D.C., 1991.

21. **Wojceichowski, Z. A.,** Biochemistry of phytosterol conjugates, in *Physiology and Biochemistry of Sterols,* Patterson, G. W. and Nes, W. D., Eds., American Oil Chemists Society Press, Champaign, IL, 1991, 361.

22. **Rilling, H. C., Leining, L. M., Bruenger, E., Lever, D., and Epstein, W. W.,** Prenylated amino acid compositions of tissues, in *Regulation of Isopentenoid Metabolism,* Nes, W. D., Parish, E. J., and Trzaskos, J. M., Eds., American Chemists Society Press, Washington, D.C., 1992, in press.

23. **Rees, H. H., Goad, L. J., and Goodwin, T. W.,** Studies in phytosterol biosynthesis, *Biochem. J.,* 107, 417, 1968.

24. **Heintz, R. and Benveniste, P.,** Plant sterol metabolism, *J. Biol. Chem.,* 249, 4267, 1974.

25. **Nes, W. R.,** Role of sterols in membranes, *Lipids,* 9, 596, 1974.

26. **Ourisson, G. and Rohmer, M.,** Prokaryotic polyterpenes: phylogenetic precursors of sterols, *Curr. Top. Membr. Transp.,* 17, 153, 1982.

27. **Bloch, K. E.,** Sterol structure and membrane function, *CRC Crit. Rev. Biochem.,* 14, 47, 1983.

28. **Nes, W. D., Janssen, G. G., and Bergenstrahle, A.,** Structural requirements for transformation of substrates by the (S)-adenosyl-L-methionine: $\Delta^{24(25)}$-sterol methyl transferase, *J. Biol. Chem.,* 266, 15202, 1991.

29. **Nes, W. D., Norton, R. A., and Benson, M.,** Carbon-13 NMR studies on sitosterol biosynthesized from [^{13}C] mevalonates, *Phytochemistry,* 31, 805, 1992.

30. **Nes, W. R., Krevitz, K., and Patterson, G. W.,** The phylogenetic distribution of sterols in tracheophytes, *Lipids,* 12, 511, 1977.

31. **Rahier, A., Taton, M., and Benveniste, P.,** Cycloeucalenol-obtusifoliol isomerase: structural requirements for transformation on binding of substrates and inhibitors, *Eur. J. Biochem.* 181, 615, 1989.

32. **Caspi, E., Zander, J. M., Greig, J. B., Mallory, F. B., Conner, R. L., and Landrey, J. R.,** Evidence for a nonoxidative cyclization of squalene in the biosynthesis of tetrahymanol, *J. Am. Chem. Soc.,* 90, 3563, 1968.

33. **Ochs, D., Tappe, C. H., Gartner, P., Kellner, R., and Poralla, K.,** Properties of purified squalene-hopene cyclase from *Bacillus acidocadarius, Eur. J. Biochem.,* 194, 75, 1990.

34. **Bladocha, M. and Benveniste, P.,** Stereochemical aspects of the biosynthesis of the side chain of 9 β,19-cyclopropyl sterols in maize seedlings treated with tridemorph, *Plant Physiol.,* 79, 1098, 1985.

35. **Djerassi, C. and Silva, C. J.,** Sponge sterols: origin and biosynthesis, *Acc. Chem. Res.,* 24, 371, 1991.

36. **Akihisa, T., Kokke, W. C. M. C., and Tamura, T.,** Naturally occurring sterols and related compounds from plants, in *Physiology and Biochemistry of Sterols,* Patterson, G. W. and Nes, W. D., Eds., American Oil Chemists Society Press, Champaign, IL, 1991, 172.

37. **Shimizu, S., Kawashima, H., Wada, M., and Yamada, H.,** Occurrence of a novel sterol, 24,25-methylenecholest-5-en-3β-ol, in *Mortierella alpina* 1S-4, *Lipids,* 27, 481, 1992.

38. **Kalinowska, M., Nes, W. R., Crumley, F. G., and Nes, W. D.,** Stereochemical differences in the anatomical distribution of C-24 alkylated sterols in *Kalanchoe diargremontiana, Phytochemistry,* 29, 3427, 1990.

39. **Fenner, G. P. and Patterson, G. W.,** 24-Ethyl orientation of the 24-ethyl sterols during the life cycle of the squash, *Phytochemistry,* 31, 73, 1992.

40. **Nes, W. D., Saunders, G. A., and Heftmann, E.,** The role of steroids and triterpenoids in the growth and reproduction of *Phytophthora cactorum, Lipids,* 17, 178, 1982.

41. **Brassell, S. C., Eglinton, G., and Maxwell, J. R.,** The geochemistry of terpenoids and steroids, *Biochem. Soc. Trans.,* 11, 595, 1983.

42. **Nicotra, F., Ronchetti, F., Russo, G., Lugaro, G., and Castellato, M.,** Stereochemistry of hydrogen migration from C-24 to C-25 during isofucosterol biosynthesis in *Pinus pinea, J. Chem. Soc. Chem. Commun.,* 889, 1977.

43. **Seo, S., Uomori, A., Yoshimura, Y., and Takeda, K.,** Stereospecificity in the biosynthesis of phytosterol side chains: ^{13}C NMR signals of C-26 and C-27, *J. Am. Chem. Soc.* 105, 6343, 1983.

44. **Rahier, A., Génot, J.-C., Schuber, F., Benveniste, P., and Narula, A. S.,** Inhibition of S-adenosyl-*L*-methionine sterol-C-24-methyltransferase by analogues of a carbocationic high-energy intermediate, *J. Biol. Chem.,* 259, 15215, 1984.

45. **Benveniste, P.,** Sterol biosynthesis, *Annu. Rev. Plant Physiol.,* 37, 275, 1986.

46. **Gill, H. K., Smith, R. W., and Whiting, D. A.,** Biosynthesis of the nicadrenoids: stages in the oxidative elaboration of the side chain and the fate of the diastereotopic 25-methyl groups of 24-methylene cholesterol, *J. Chem. Soc. Chem. Commun.,* 1459, 1986.

47. **Bach, T. J.,** Hydroxymethylglutaryl-CoA reductase, a key enzyme in phytosterol synthesis?, *Lipids,* 21, 82, 1986.

48. **Nes, W. D., Janssen, G. G., Norton, R. A., Kalinowska, M., Crumley, F. G., Tal, B., Bergenstrahle, A., and Jonsson, L.,** Regulation of sterol biosynthesis in sunflower by 24(R,S),25-epiminolanosterol, a novel C-24 methyl transferase inhibitor, *Biochem. Biophys. Res. Commun.,* 177, 566, 1991.

49. **Taton, M. and Rahier, A.,** Properties and structural requirements for substrate specificity of cytochrome P-450-dependent obtusifoliol 14α-demethylase from maize (*Zea mays*) seedlings, *Biochem. J.,* 277, 483, 1991.

50. **Taton, M., Benveniste, P., and Rahier, A.,** Microsomal $\Delta^{8,14}$-sterol Δ^{14}-reductase in higher plants, *Eur. J. Biochem.,* 185, 605, 1989.

51. **Pascal, S., Taton, M., and Rahier, A.,** Oxidative C-4 demethylation of 24-methylene cycloartanol by a cyanide-sensitive enzymatic system from higher plant microsomes, *Biochem. Biophys. Res. Commun.,* 172, 98, 1990.

52. **Cattel, L., Delprino, L., Benveniste, P., and Rahier, A.,** Effect of the configuration of the methyl group at C-4 on the capacity of 4-methyl-9β-cyclosteroids to be substrates of a cyclopropyl cleavage enzyme from maize, *J. Am. Oil Chem. Soc.,* 56, 6, 1979.

53. **Rahier, A., Taton, M., Bouvier-Nave, P., Schmitt, P., Benveniste, P., Schuber, F., Narula, A. S., Cattel, L., Anding, C., and Place, P.,** Design of high energy intermediate analogues to study sterol biosynthesis in higher plants, *Lipids,* 21, 52, 1986.

54. **Heupel, R. C., Sauvaire, X., Le, P. H., Parish, E. J., and Nes, W. D.,** Sterol composition and biosynthesis in sorghum: importance to developmental regulation, *Lipids,* 21, 69, 1986.

55. **Rendell, N., Misso, N. L. A., and Goad, L. J.,** Biosynthesis of 24-methycholest-5-en-3β-ol and 24-ethylcholest-5-3n-3β-ol in *Zea mays, Lipids,* 21, 63, 1986.

56. **Huang, L.-S. and Grunwald, C.,** Mevalonate incorporation into alfalfa sterols, *Phytochemistry,* 28, 465, 1989.

57. **Weete, J. D.,** *Lipid Biochemistry of Fungi and Other Organisms,* Plenum Press, New York, 1980, 261.

58. **Vanden Bossche, H.,** Ergosterol biosynthesis inhibitors, in *Candida Albicans,* Prasad, R., Ed., Springer-Verlag, Berlin, 1991, 239.

59. **Nes, W. D., Norton, R. A., Parish, E. J., Meenan, A., and Popják, G.,** Concerning the role of 24, 25-dihydrolanosterol and lanostanol in sterol biosynthesis by cultured cells, *Steroids,* 53, 461, 1989.

60. **Sharpless, K. R., Snyder, T. E, Spencer, T. A., Makeshwari, K. K., Guhn, G., and Clayton, R. B.,** Biological demethylation of 4,4-dimethyl sterols. Initial removal of the 4α-methyl group, *J. Am. Chem. Soc.,* 90, 6874, 1968.

61. **Aoyama, Y., Yoshida, Y., Sonada, Y., and Sato, Y.,** Role of the 8(9)-double bond of lanosterol in the enzyme-substrate interaction of cytochrome P-450$_{14DM}$, *Biochim. Biophys. Acta,* 1001, 196, 1989.

62. **Cordeiro, M. L., Kerr, R. G., and Djerassi, C.,** Biosynthetic studies of marine lipids 15. Conversion of parkeol (lanosta-9(11)-24-dien-3β-ol) to 14 α-methylcholest-9(11)-en-3β-ol in the sea cucumber *Holothuria arenicola, Tetrahedron Lett.,* 29, 2159, 1988.

63. **Akhila, A., Gupta, M. M., and Thakur, R. S.,** Direct cyclization of squalene to 5 α-stigmast-9(11)-en-3β-ol via $\Delta^{9(11)}$-lanosterol in *Costus speciosus:* a unique finding in sterol biosynthesis, *Tetrahedron Lett.,* 28, 4085, 1987.

64. **Palmer, M. A., Goad, L. J., Goodwin, T. W., Copsey, D. B., and Boar, R. B.,** The conversion of 5 α-lanost-24-ene-3β,9α-diol and parkeol into poriferasterol by the alga *Ochromonas malhamensis, Phytochemistry,* 17, 1577, 1978.

65. **Nes, W. D., Janssen, G. G., Crumley, F. G., Kalinowska, M., and Akihisa, T.,** The structural requirements of sterols for membrane function in *Saccharomyces cerevisiae, Arch. Biochem. Biophys.,* 300, 724, 1992.

65a. **Nes, W. D., Janssen, G. G., Crumley, F. G., Kalinowska, M., and Akihisa, T.,** The structural requirements of sterols for membrane function in *Saccharomyces cerevisiae, FASEB J.,* 5(A), 423, 1991.

66. **Anding, C., Parks, L. W., and Ourisson, G.,** Enzymic modification of cyclopropane sterols in yeast cell-free systems, *Eur. J. Biochem.,* 43, 459, 1974.

67. **Malhotra, H. C. and Nes, W. R.,** The mechanisms of introduction of alkyl groups at C-24 of sterols: IV. Inhibition by triparanol, *J. Biol. Chem.,* 246, 4934, 1971.

68. **Janssen, G. G., Kalinowska, M., Norton, R. A., and Nes, W. D.,** (S)-adenosyl-*L*-methionine: Δ^{24}-sterol methyl transferase: mechanism, enzymology, inhibitors and physiological importance, in *Physiology and Biochemistry of Sterols,* Patterson, G. W. and Nes, W. D., Eds., American Oil Chemists Society Press, Champaign, IL, 1991, 83.

69. **Arigoni, D.,** Stereochemical studies of enzymic C-methylations, *Ciba Found. Symp.,* 60, 243, 1978.

70. **Misso, N. L. A. and Goad, L. J.,** The synthesis of 24-methylene cycloartanol, cyclosadol and cyclolaudenol by a cell-free preparation from *Zea mays* shoots, *Phytochemistry,* 22, 2473, 1983.

71. **Scheid, F., Rohmer, M., and Benveniste, P.,** Biosynthesis of $\Delta^{5,23}$-sterols in etiolated coleoptiles from *Zea Mays, Phytochemistry,* 21, 1959, 1982.

72. **Goad, L. J., Lenton, J. R., Knapp, F. F., and Goodwin, T. W.,** Phytosterol side chain biosynthesis, *Lipids,* 9, 582, 1974.

73. **Patterson, G. W.,** Sterols of algae, in *Physiology and Biochemistry of Sterols*, Patterson, G. W. and Nes, W. D., Eds., American Oil Chemists Society Press, Champaign, IL, 1991, 118.

74. **Uomori, A., Nakaguwa, Y., Yoshimatsu, S., Seo, S., Sankawa, U., and Takeda, K.,** Biosynthesis of campesterol and dihydrobrassicasterol in cultured cells of *Amsonia elliptica*, *Phytochemistry*, 31, 1569, 1992.

75. **Nes, W. D. and Le, P. H.,** Evidence for separate intermediates in the biosynthesis of multiple 24β-methylsterol end products by *Gibberella fujikuroi*, *Biochim. Biophys. Acta*, 1042, 119, 1990.

76. **Balliano, G., Caputo, O., Viola, F., Delprina, L., and Cattel, L.,** Conversion of cyclolaudenol to 24α- and 24β-ethyl sterols in the cucurbitaceae, *Lipids*, 18, 302, 1983.

77. **Giner, J.-L. and Djerassi, C.,** Biosynthesis of 24-methylene-25-methylcholesterol in *Phaseolus vulgaris*, *Phytochemistry*, 30, 811, 1991.

78. **Costet-Corio, M.-F. and Benveniste, P.,** Sterol metabolism in wheat treated by *N*-substituted morpholines, *Pestic. Sci.*, 22, 343, 1988.

79. **Popják, G., Meenan, A., Parish, E. J., and Nes, W. D.,** Inhibitors of cholesterol synthesis and cell growth by 24(R,S),25-epiminolanosterol and triparanol in cultured rat hepatoma cells, *J. Biol. Chem.*, 264, 6230, 1989.

80. **Nes, W. D. and Le, P. H.,** Regulation of sterol biosynthesis in *Saprolegnia ferax* by 25-azacholesterol, *Pestic. Biochem. Physiol.*, 30, 87, 1988.

81. **Nes, W. D., Xu, S., and Parish, E. J.,** Metabolism of 24(R,S),25-epiminolanosterol to 25-aminolanosterol and lanosterol by *Gibberella fujikuroi*, *Arch. Biochem. Biophys.*, 272, 323, 1989.

82. **Ator, M. A., Schmidt, S. J., Adams, J. L., and Dolle, R. E.,** Mechanism and inhibition of Δ^{24}-sterol methyl transferase from *Candida albicans* and *Candida tropicalis*, *Biochemistry*, 28, 9633, 1989.

83. **Giner, J.-L. and Djerassi, C.,** Biosynthetic studies of marine lipids. No. 31. Evidence for a protonated cyclopropyl intermediate in the biosynthesis of 24-propylidenecholesterol, *J. Am. Chem. Soc.*, 113, 1386, 1991.

84. **Oehlschlager, A. C., Angus, R. H., Pierce, A. M., Pierce, H. D., Jr., and Srinivasan, R.,** Azasterol inhibition of Δ^{24}-sterol methyl transferase in *Saccharomyces cerevisiae*, *Biochemistry*, 23, 3582, 1984.

85. **Haughan, P. A., Lenton, J. R., and Goad, L. J.,** Sterol requirements and palclobutrazol inhibition of celery cell culture, *Phytochemistry*, 27, 2491, 1988.

86. **Nes, W. D., Benson, M., Lundin, R. E., and Le, P. H.,** Conformational analysis of 9β,19-cyclopropyl sterols: detection of the pseudoplanar conformer by nuclear Overhauser effects and its functional implications, *Proc. Natl. Acad. Sci. U.S.A.*, 85, 5759, 1988.

87. **Yoshida, K., Hirose, Y., Imai, Y., and Kondo, T.,** Conformational analysis of cycloartenol, 24-methylene cycloartanol and their derivatives, *Agric. Biol. Chem.*, 53, 1901, 1989.

88. **Collins, J. M., Nes, W. D., Quinn, P. J., Wolfe, D. H., Cunningham, B. A., Kucuk, O., Westerman, M. P., and Lis, L. J.,** Inference of sterol pseudoplanar conformations using lipid phase relations, *FASEB J.*, 6(A), 241, 1992.

89. **Schuler, I., Duportail, G., Glasser, N., Benveniste, P., and Hartmann, M.-A.,** Soybean phosphatidylcholine vesicles containing plant sterols: a fluorescence anisotropy study, *Biochim. Biophys. Acta*, 1028, 82, 1990.

90. **Milhaud, J., Bolard, J., Benveniste, P., and Hartmann, M. A.,** Interaction of the polyene antibiotic filipin with model and natural membranes containing plant sterols, *Biochim. Biophys. Acta*, 943, 315, 1985.

91. **Roederstorff, D. and Rohmer, M.,** Polyterpenoids as cholesterol and tetrahymanol surrogates in the ciliate *Tetrahymena pyriformis*, *Biochim. Biophys. Acta*, 960, 190, 1988.

92. **Xu, S., Norton, R. A., Crumley, F. G., and Nes, W. D.,** Comparison of the chromatographic properties of sterols, select additional steroids, and triterpenoids: gravity-flow column liquid chromatography, thin-layer chromatography, gas-liquid chromatography and high-performance liquid chromatography, *J. Chromatogr.*, 452, 377, 1988.

93. **Schular, I., Milon, A., Nakatani, Y., Ourisson, G., Albrecht, A.-M., Benveniste, P., and Hartmann, M.-H.,** Differential effects of plant sterols on water permeability and on acyl chain ordering of soybean phosphatidyl choline bilayers, *Proc. Natl. Acad. Sci. U.S.A.*, 88, 6922, 1991.

94. **Gallay, J. and DeKruijff, B.,** Correlation between molecular shape and hexagonal H_{II} phase promoting ability of sterols, *FEBS Lett.*, 143, 133, 1982.

95. **Demel, R. A. and DeKruyff, B.,** The function of sterols in membranes, *Biochim. Biophys. Acta*, 457, 109, 1976.

96. **McKersie, B. D. and Thompson, J. E.,** Influence of plant sterols on the phase properties of phospholipial bilayers, *Plant Physiol.*, 63, 802, 1979.

97. **Rothman, J. E. and Engelmen, D. E.,** Molecular mechanism for the interaction of phospholipid with cholesterol, *Nature New Biol.*, 237, 42, 1972.

98. **Kroon, P. A., Kainosho, M., and Chan, S. I.,** State of molecular motion of cholesterol in lecithin bilayers, *Nature*, 256, 582, 1975.

99. **Huang, C.,** A structural model for the cholesterol phosphatidylcholine complex in bilayer membranes, *Lipids*, 12, 348, 1977.

100. **Martin, R. B. and Yeagle, P. L.,** Models for lipid organization in cholesterol-phospholipid bilayers including cholesterol dimer formation, *Lipids*, 13, 594, 1978.

101. **Nes, W. R., Sekula, B. C., Nes, W. D., and Adler, J. H.,** The functional importance of structural features of ergosterol in yeast, *J. Biol. Chem.*, 253, 6218, 1978.

102. **Nes, W. D. and Heftmann, E.,** A comparison of triterpenoids with steroids as membrane components, *J. Nat. Prod.*, 44, 377, 1981.

103. **Nes, W. D., Kalinowska, M., and Norton, R. A.,** unpublished data, 1992.

104. **Janssen, G. G. and Nes, W. D.,** Structural requirements for tranformation of substrates by the S-Adrenosyl-L-methionine: $\Delta^{24(25)}$-sterol methyl transferase: II. Inhibition by analogs of the transition state coordinate, *J. Biol. Chem.*, 267, 25856, 1992.

Chapter 13

THE PLANT PRENYLLIPIDS, INCLUDING CAROTENOIDS, CHLOROPHYLLS, AND PRENYLQUINONES

Hartmut K. Lichtenthaler

TABLE OF CONTENTS

0-8493-4907-9/93/$0.00+$.50

427

I. INTRODUCTION

The group of isoprenoid or terpenoid plant lipids consists of various classes of lipid-soluble plant constituents which differ in their chemical structures, but share in common that they are biosynthetically constructed either totally or in part from C_5 isoprene building units with isopentenyl pyrophosphate as their precursor. The major prenyllipid classes of higher plants and algae are the sterols, carotenoids, chlorophylls, prenylquinones, and the prenols, all of which are regular plant constituents. Among these prenyllipids, one can differentiate between (1) the pure prenyllipids, the carbon skeleton of which is made up exclusively of isoprene units such as sterols, prenols (isoprenoid alcohols), and carotenoids; and (2) the mixed prenyllipids (e.g., chlorophylls, prenylquinones, prenylchromanols) which consist of an isoprenoid side chain (prenyl chain) bound to an aromatic nucleus, e.g., a porphyrin ring in the case of chlorophylls a and b; a benzoquinone ring in the case of the prenylquinones plastoquinone-9, α-tocoquinone, as well as the ubiquinones-9 or -10; and a naphthoquinone ring in the case of phylloquinone (vitamin K_1). To these mixed prenyllipids also belong isoprenoid chromanols (tocopherols and tocotrienols), which represent the cyclic forms of reduced tocoquinones (quinols, hydroquinones) with α-tocopherol (vitamin E) as a major component. Some of the plant prenyllipids, which are synthesized in the plant cell and accumulate in chloroplasts, represent vitamins such as β-carotene (provitamin A), α-tocopherol (vitamin E), and phylloquinone (vitamin K_1) and are then referred to as prenylvitamins.[1] The chemical structure of typical representatives of the different prenyllipid classes of higher plants is shown in Figure 1.

A typical feature of the plant prenyllipids group is the fact that they represent regular lipid components of green plants which often have particular functions in the cellular biomembranes. Other terpenoid substances, such as monoterpene, sesquiterpene, and diterpene derivatives, e.g., in fragrant oils or particular isoprenoid phytoalexins which only show up in particular plant families, are not considered as plant prenyllipids, even though some of them are fat soluble and their biosynthesis is connected to the isoprenoid pathway.

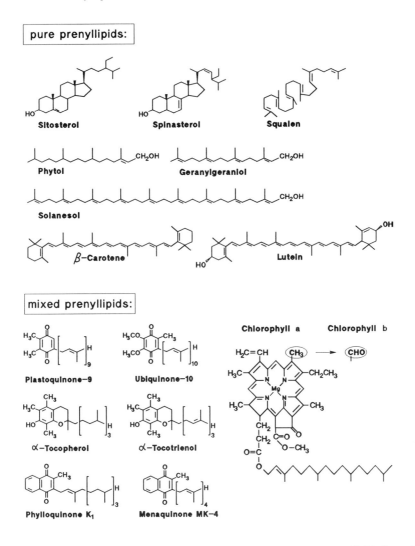

FIGURE 1. Chemical structure of representatives of each plant prenyllipid class of higher plants.

This chapter will give some basic general information on the biosynthesis, localization, and physiological function of individual plant prenyllipids with special emphasis on carotenoids and chlorophylls, as well as on mitochondrial (ubiquinones) and plastidic prenylquinones (plastoquinone-9, phylloquinone K_1, α-tocoquinone, and α-tocopherol). Sterols, which represent a major class of plant prenyllipids, are treated in Chapter 12. Since not all aspects of plant prenyllipids can be treated here, access is given to older original literature and reviews, where further details can be looked up.

Basic introductions to the prenyllipids of higher plants are given by Goodwin[2] and Lichtenthaler.[3-5] The various chromatographic methods of different authors for the separation of the different classes of plant prenyllipids by paper, column, thin layer chromatography (TLC), and gas liquid chromatography (GLC), as well as high performance liquid chromatography (HPLC) and TLC on silver nitrate-impregnated plates, are found in the comprehensive review of Lichtenthaler.[6] A major achievement in prenylpigment determination is the simultaneous determination of the chlorophylls a and b together with total carotenoids (x + c) in the same extract solution (acetone, methanol, ethanol, diethylether) using the new and redetermined absorption coefficients.[7] An elegant though expensive method is the determination of all individual carotenoids, including zeaxanthin and antheraxanthin, together with chlorophylls by reversed-phase HPLC.[8-10] Details of the quantitative determination of prenylquinones and α-tocopherol have been provided for separation by TLC[3,11,12] and by HPLC.[13,14] Separation and determination of prenols (e.g., phytol, geranylgeraniol, solanesol) have been reported.[6,14,15] The separation of such prenylquinones, prenylvitamins (tocopherols, tocotrienols), and prenols, which only differ in one or two double bonds, is easily achieved using silver nitrate-coated plates[16] or by reversed-phase HPLC.[13,14] A combination of adsorption HPLC followed by reversed-phase HPLC allows successive separation and quantification of the various ubiquinone homologs Q-6, Q-7, Q-8, Q-9, Q-10, and Q-11 in plants,[17,18] and the application of this method has settled much of the controversy on the nature of the ubiquinone homologs in a particular plant.

II. BIOSYNTHESIS OF PRENYLLIPIDS

A. GENERAL PATTERN OF PRENYLLIPID FORMATION

The biogenetic relationship between the different classes of plant prenyllipids is shown in Figure 2. Biosynthesis starts from acetyl-CoA via mevalonic acid (mevalonate) to yield the physiologically active isoprene unit isopentenyl pyrophosphate (IPP). The latter is isomerized by an IPP isomerase to its isomer dimethyllallyl pyrophosphate (DMAPP). The condensation of IPP with DMAPP by an IPP-prenyl transferase in a head-to-tail addition yields a C_{10} prenyl chain, the monoterpene geranyl pyrophosphate. Further sequential additions of IPP units in a head-to-tail manner results in the consecutive formation of the sesquiterpene farnesyl pyrophosphate (a C_{15} prenyl chain), the diterpene geranylgeranyl pyrophosphate (a C_{20} prenyl chain), and eventually the C_{45} prenyl chain solanesyl pyrophosphate and the C_{50} decaprenyl pyrophosphate. Inorganic pyrophosphate is released in each step of the prenyl chain elongation. In some plants and plant families, longer prenyl chains are formed, the intermediate and the long polyprenols. The intermediate polyprenols, such as castaprenols, are found in horse chestnut.[15,19] Long polyprenols are present in the milky sap (latex) of Euphrobiaceae and Compositae such as caoutchouc and guttapercha. However, these isoprenoid products do not represent regular functional plant prenyllipids.

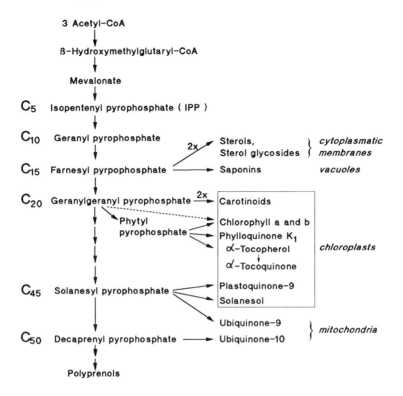

FIGURE 2. Biogenetic relationship of plant prenyllipids starting from acetyl-CoA via mevalonate and isopentenyl pyrophosphate (IPP) to the individual plant prenyllipids which accumulate in different compartments and membranes of the plant cell.

The pure plant prenyllipids such as sterols (triterpene derivatives) and carotenoids (tetraterpenes) are formed by dimerization, i.e., the condensation of two prenyl chain units in a tail-to-tail manner. From two farnesyl pyrophosphates, the triterpene squalene (see Figure 1) is formed, which is the precursor for sterols,[20] steryl glycosides,[21] and saponins.[22,23] If the two diterpenes (i.e., two geranylgeranyl pyrophosphates) are joined tail-to-tail, the result is the still colorless C_{40} precursors of the different plant carotenoids such as phytoene and phytofluene.[24,25]

In the case of mixed prenyllipids, the prenyl side chains of different length are bound to a nonisoprenoid nucleus, which is a porphyrin ring in the case of chlorophylls a and b; a benzoquinone ring in the case of the plastidic plastoquinone-9, α-tocoquinone, as well as its chromanol α-tocopherol and the mitochondrial ubiquinones Q-9 and Q-10; and a naphthoquinone nucleus in the example of the plastidic phylloquinone (vitamin K_1).

In the case of the ubiquinones and plastoquinone-9 the prenyl side chain is unsaturated and exhibits a *trans* double bond in each isoprene unit of the C_{45} and C_{50} chain (Figure 3). There also exist derivatives of plastoquinone with

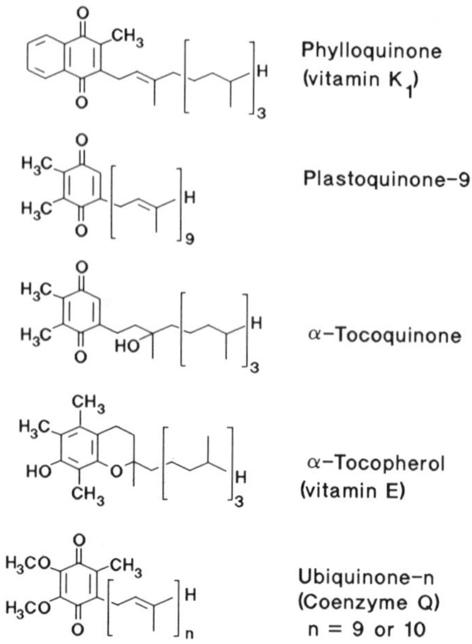

FIGURE 3. Chemical structure of the major prenylquinones of higher plants and algae.

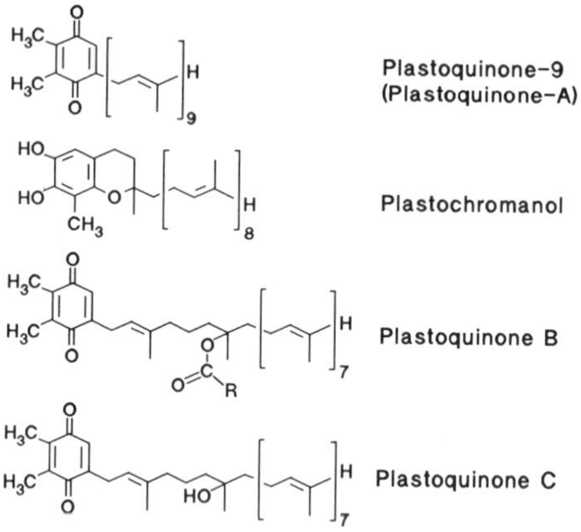

FIGURE 4. Chemical structure of plastoquinone-9 (=PQ-A) and its derivatives PQ-Bs and PQ-Cs, which can occur in older and senescent plant tissues up to 10 or 20% of the total plastoquinone-9 content. The position of the hydroxy group with a particular side chain of the hydroxy plastoquinones (PQ-Cs) and their fatty acid esters (PQ-Bs), as given here for the second isoprene unit, seems to be variable for the individual members of the PQ-B and PQ-C series.

a hydroxy prenyl side chain (PQ-C series) and their esterified acylated forms (PQ-B series) (Figure 4). The chlorophylls and phylloquinone, in turn, contain a C_{20} phytyl chain in which only the first isoprene unit contains a double bond. Its biosynthetic precursor geranylgeranyl pyrophosphate with four double

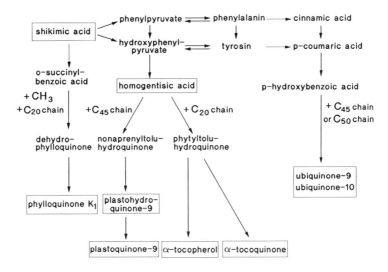

FIGURE 5. Scheme for the biosynthesis of the nonisoprenoid nucleus of plant prenylquinones of chloroplasts (plastoquinone-9, α-tocoquinone, α-tocopherol, and phylloquinone) and of mitochondria (ubiquinone homologs Q-9 and Q-10) from intermediates of the shikimic pathway. (Adapted from Lichtenthaler, H. K., *Lipids and Lipid Polymers in Higher Plants*, Tevini, M. and Lichtenthaler, H. K., Eds., Springer-Verlag, Berlin, 1977, 231.)

bonds (one in each isoprene unit) is thought to be transformed by a prenyl chain saturase to phytyl pyrophosphate, which is then connected by a prenyltransferase to the nonisoprenoid nucleus. There is still discussion whether — at least in some cases — the geranylgeranyl pyrophosphate is directly used for the prenylation (see Section II.B.).The phytyl chain is also found in the chromanol α-tocopherol and α-tocoquinone.

1. Origin of the Quinone Ring Moiety
In the biosynthesis of α-tocopherol and α-tocoquinone, the C_{20} phytyl chain is bound to homogentisic acid to finally yield phytyl toluhydroquinone, the common precursor of α-tocopherol and α-tocoquinone (Figure 4).[26,27] The transfer of the C_{45} solanesyl chain to homogentisic acid yields nonaprenyl toluhydroquinone, the precursor of plastoquinone-9. The acetic acid side chain of homogentisate is converted into a chiral methyl group of α-tocopherol, α-tocoquinone, and plastoquinone-9 by stereospecific decarboxylation.[28] The naphthoquinone nucleus of phylloquinone is formed from shikimic acid via succinylbenzoic acid. The benzoquinone ring of ubiquinones also is derived from the shikimic acid pathway with cinnamic acids and p-hydroxybenzoic acid as intermediates (Figure 5).

2. Formation of the Porphyrin Structure
The porphyrin ring of chlorophylls of higher plants and algae is synthesized not from glycine and succinyl, as, e.g., in the case of animals, but by conversion

of glutamate, perhaps via glutaryl-tRNA, to δ-aminolaevulinic acid (ALA).[29,30] Dimerization of ALA yields porphobilinogen, and subsequent condensation of four porphobilinogens results in the formation of the first tetrapyrrole uroporphyrinogen III. The porphyrin ring of protochlorophyllide (PChlide) is formed via Mg-protoporphyrin IX,[30] which accumulates in dark-grown seedlings and becomes esterified in higher plants with phytol (possibly with phytyl pyrophosphate as endogenous donor) to yield the chlorophyll precursor protochlorophyll (PChl). Dark-grown etiolated plants usually contain both PChlide and its phytolester PChl,[31] both of which are bound in particular ways to the PChl(ide) holochrome of etioplasts.[32] Upon illumination with white light, one double bond in ring D of the porphyrin is reduced (with NADPH as donor) by the enzyme NADPH-protochlorophyllide oxidoreductase to yield either chlorophyll a (from PChl) or chlorophyllide a (from PChlide), which then is esterified by the enzyme chlorophyll synthase with phytol or geranylgeraniol (see Section II.B.). Access to further details and to original literature concerning porphyrin ring formation in chlorophyll biosynthesis can be found in recent reviews.[30,33]

3. Chlorophyll b Formation

The question of how chlorophyll b, which bears an aldehyde group in place of the B-ring methyl substituent of the tetrapyrrole of chlorophyll a (Figure 1), is formed still seems to be open. Several possibilities are under discussion. Upon illumination of etiolated seedlings, at first, only chlorophyll a is formed by photoreduction of PChl/PChlide, whereas chlorophyll b accumulates later with a time delay of several minutes. This results in very high values for the ratio chlorophyll a to b of 100:30 in the first 3 to 30 min of illumination, whereas the chlorophyll a to b ratio of normal green leaves and chloroplasts amounts to values around 3 (with values of 2.6 to 2.8 in the shade and 3.0 to 3.5 in sun-exposed leaves). From these results with greening etiolated plant leaves, it was generally accepted that chlorophyll b could be derived by a direct oxidation of chlorophyll a (possibly by a monooxygenase reaction). This view is supported by tracer studies of Shlyk[34] and Akoyunoglou.[35] Other authors believe that chlorophyll b may derive from chlorophyllide a, which would require the formation of chlorophyllide b and its esterification with a C_{20} prenol, or geranylgeranyl pyrophosphate with subsequent hydrogenation of the side chain to phytyl chlorophyll b.[36] A particular protochlorophyllide b, however, has not been found and can be excluded as a precursor of chlorophyll b.

B. THE PHYTYLATION STEP OF PRENYLLIPIDS
1. Chlorophylls

The prenylalcohol found in the mature chlorophylls a and b is the diterpene phytol. The C_{20} prenol initially bound to the tetrapyrrole protochlorophyllide must not always be phytol (one double bond), it can also be its biosynthetic precursor geranylgeraniol (four double bonds; one in each of the four isoprene units).[37] Etiolated tissue also contains a small pool of free phytol, besides phytyl protochlorophyllide and unesterified protochlorophyllide (PCHlde),[31] and also small soluble and membrane-bound pools of phytyl pyrophosphate and geranylgeranyl pyrophosphate.[37]

Under conditions of herbicide treatment, additional chlorophyll forms appear, besides the normal phytyl chlorophyll (phChl), which are esterified with geranylgeraniol chlorophyll (ggChl).[38] Upon illumination of dark-grown etiolated oak and bean seedlings, which contain small pools of PChlide and some esterified PChl, the latter accumulate geranylgeranyl chlorophyll a (ggChl), which is then hydrogenated in three steps to phChl.[39,40] Esterification of chlorophyllide by geranylgeranyl pyrophosphate (ggPP) could also be shown in a cell-free system of maize shoots.[41] The chlorophyll synthase from oat etioplasts more readily accepts ggPP than phytyl pyrophosphate (phPP).[42] However, the situation in green plants is quite different. The chlorophyll synthase of broken spinach chloroplasts more readily accepts phPP than ggPP.[43] It was also shown that phPP is formed from ggPP in the envelope of chloroplasts.[43,44] In green tobacco-cell cultures, exogenously applied labeled ggPP may first become transformed to phPP before it is bound to chlorophyllide a, since its radiolabel is found primarily in the form of phChl a.[45] From these results, it appears that in freshly illuminated etiolated leaf tissue chlorophyll a formation proceeds via ggChl a, which is successively hydrogenated to phChl, whereas in chloroplasts phChl a is directly formed from chlorophyllide (Chlide) a and phPP.[37] In chloroplasts, the chlorophyll synthase is bound to the thylakoids and in etioplasts to the prothylakoid/prolamellar body fraction, but not to the envelope.[37] Since phPP is the only substrate in green barley and is only formed in the envelope, but not in thylakoids or the stroma, it must be transported from the envelope to the thylakoids.[37] Prenyl chain formation from IPP including ggPP proceeds in the plastid stroma, which indicates that the formation of chlorophyll a must be a coordinated cooperation of stroma, envelope, and thylakoids. Though the phytylation step of chlorophyll a formation is established,[33,37] that of chlorophyll b still remains open. Some observations would favor the existence of a ggChl b, which is successively hydrogenated to phChl b.[37] A direct formation from phChl a or the phytylation of a Chlide b cannot hitherto be excluded. Several possible alternative pathways in the final steps of chlorophyll a and b biosynthesis are summarized in Figure 6.

2. α-Tocopherols and α-Tocoquinones

The major tocopherol and tocoquinone of functional chloroplasts are the α-forms with three methyl groups as substituents in the aromatic nucleus.[46-48] There also exist β-, γ-, and δ-forms in trace amounts in chloroplasts,[49] which contain a lower number of methyl groups (β-T and γ-T two methyl groups and δ-T one methyl group). In senescent tissue and in nongreen plant tissue, e.g., seeds, fruits, or young etiolated seedlings,[50] and in the latex of Euphorbiaceae, additional tocopherol species show up, of which the free prenyl chain consisting of three isoprene units is unsaturated. These are the corresponding α-, β-, γ-, and δ-tocotrienols (Figure 1). Correspondingly, as in chlorophyll biosynthesis,

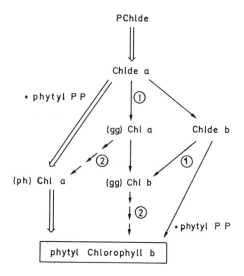

FIGURE 6. Alternative pathways leading to the formation of chlorophyll b from chlide a (gg-Chl — geranylgeraniol chlorophyll; phChl — phytyl chlorophyll). Step 1 would be catalyzed by geranylgeranyl prenyltransferases and step 2 by saturases eliminating three of the four double bonds in the C_{20} prenyl chain.

alternative pathways in α-tocopherol and α-tocoquinone biosynthesis may exist using either ggPP or phPP for prenylation, as indicated in Figure 7. The detection in etiolated tissue of monodehydroforms of α-tocopherol and α-tocoquinone with one double bond in the prenyl side chain[51] would favor the assumption that geranylgeranyl toluquinone, not phytyl toluquinone, is the biosynthetic intermediate in α-tocopherol and α-tocoquinone biosynthesis. There still remains, however, the possibility that the geranylgeranyl forms, the tocotrienols, represent side products and only accumulate when the main biosynthetic pathway of these thylakoid prenyllipids is disturbed or blocked, e.g., by darkness or the application of inhibitors.

3. Phylloquinone

In the case of the 2-dimethyl-3-phytyl-naphthoquinone vitamin K_1, a geranylgeranyl derivative (=menaquinone-4 as found in some bacteria) hitherto has not been detected in green algae or higher plants. Etiolated leaves contain, besides phylloquinone K_1, a second naphthoquinone K′, the pool of which declines during the light-induced enhanced phylloquinone accumulation.[12] This was shown to be a monohydrophylloquinone.[51] This may be an indication that phylloquinone biosynthesis proceeds or can proceed under certain conditions via the geranylgeranyl derivative.

The general question, however, whether geranylgeranyl or phytyl species are the real intermediates of the different phytyl prenyllipids can only be decided at the enzyme level, when the particular prenyltransferases of tocopherol and phylloquinone biosynthesis are isolated and their substrate specificity will be determined. It may turn out that depending on the developmental and

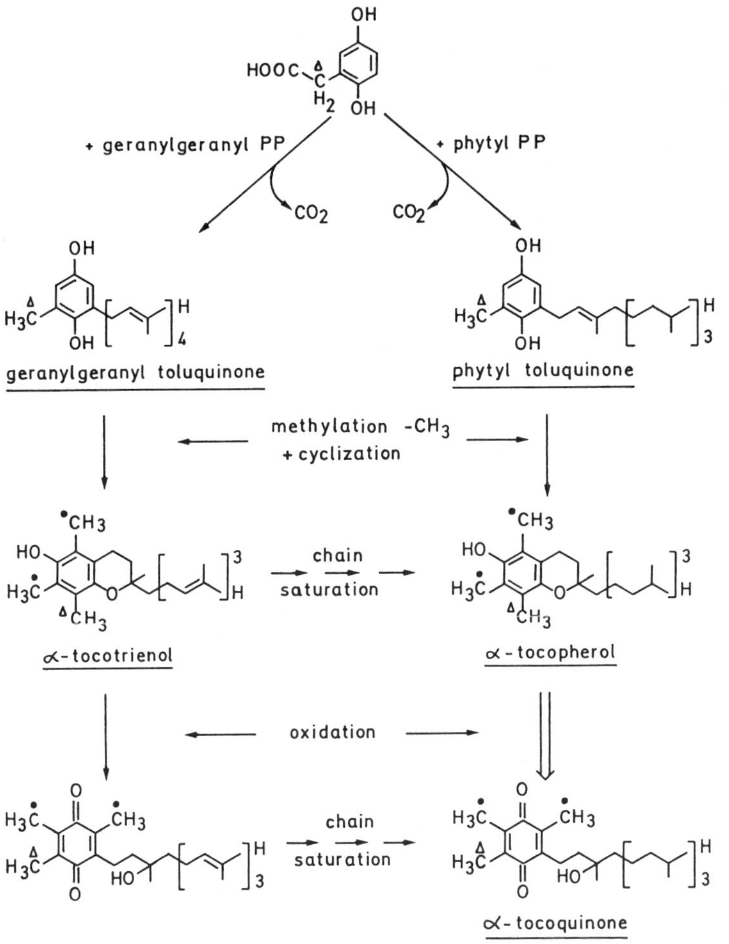

FIGURE 7. Alternative pathways leading to the formation of α-tocopherol and α-tocoquinone from either geranylgeranyl or phytyl toluquinone. (Adapted from Lichtenthaler, H. K., in *Advances in the biochemistry and physiology of plant lipids*, Appleqvist, L.-A. and Liljenberg, C., Eds., Elsevier, Amsterdam, 1979, 57.)

physiological stages both pathways can exist. Similar to the final stage of chlorophyll biosynthesis, it may be possible that in etioplasts and briefly illuminated tissue, ggPP is used for tocopherol and phylloquinone biosynthesis, whereas in chloroplasts phPP is the endogenous precursor.

C. INTRACELLULAR LOCALIZATION AND SITE OF BIOSYNTHESIS OF PLANT PRENYLLIPIDS

Sterols are synthesized from two farnesyl pyrophosphates via squalene and successive ring closure (dimethyl sterols) and desmethylation (methyl and desmethyl sterols),[20,23] which apparently proceeds in the cytoplasm (microsomes, endoplasmic reticulum). Sterols, their glycoside, and acylglycoside derivatives[21] are bound in particular ways to the different cytoplasmatic membranes, whereas other triterpenoid compounds and saponins can be stored in the vacuole.[22,23] The cytoplasm apparently possesses a complete biosynthetic machinery of the prenyl chain pathway from acetyl coenzyme A (CoA) via IPP to farnesyl pyrophosphate, including its dimerization to squalene and further modifications to the final sterols. This biosynthetic prenyl chain sequence can easily be fed not only by exogenously applied labeled acetate, but also by mevalonate.[23,24,46]

Two essential key enzymes of the cytoplasmic isoprenoid pathway, the hydroxymethylglutaryl-CoA synthase (HMGS)[52] and the hydroxymethylglutaryl-CoA reductase (HMGR),[53,54] have recently been isolated and defined, which supports the existence of a complete biosynthetic sequence in the cytoplasm. In contrast to animal tissue, the plant HMGS enzyme seems to be the acetoacetyl-CoA thiolase (AACT), which condenses two acetyl-CoA to acetoacetyl-CoA.[52] By application of the antibiotic mevinolin, a specific inhibitor of the HMG-CoA reductase,[55] the formation of mevalonate and consequently that of sterols is blocked in different plants and plant tissues.[56-60]

Chloroplasts and their photochemically active thylakoids do not possess sterols, except for some minor amounts [21] which may be bound to the chloroplast envelope. Chloroplasts, however, exhibit several prenyllipids in three localization sites. The envelope contains carotenoids, with violaxanthin as a major compound; the prenylquinones, plastoquinone-9, phylloquinone, α-tocoquinone; and its chromanol α-tocopherol (Table 1).[61] The photochemically active thylakoids, which perform the photosynthetic light reactions, possess chlorophylls and carotenoids bound to several particular chlorophyll-carotenoid proteins and also the three prenylquinones and α-tocopherol.[47,48] Though the same carotenoids and prenylquinones occur in envelope and thylakoids, the percentage composition of carotenoids and prenylquinones is, however, quite different. In thylakoids, lutein and β-carotene are predominant among the carotenoids and plastoquinone-9 among the prenylquinones,[48,49,62] whereas in the envelope violaxanthin and α-tocopherol are the dominant substances of these two prenyllipid classes (Table 1). The osmophilic plastoglobuli,[63] which function as lipid stores for excess plastid lipids, mainly contain plastoquinone-9 (PQ-9) and its reduced form plastohydroquinone-9 (PQ-9·H_2) together with α-tocopherol and minor amounts of α-tocoquinone.[64,65] Phylloquinone K_1 and primary carotenoids are found in plastoglobuli of functional chloroplasts only in trace amounts. Chlorophylls are restricted to thylakoids and are never found in plastoglobuli or the envelope.

Only in senescent leaves and during chromoplast development, which is associated with thylakoid and chlorophyll breakdown, does one find carotenoids[66] and secondary carotenoids[67] in plastoglobuli. In cases where plastoglobuli become very large, they may also contain triacylglycerols.[68]

1. Chloroplast Site

That chloroplasts possess their own biosynthetic prenyl chain pathway is assumed by many authors. The fact that the antibiotic mevinolin, the specific

TABLE 1
Percentage Composition (Weight %) of the Carotenoids and Prenylquinones of the Chloroplast Envelope and Isolated Thylakoids of Spinach Chloroplasts

	Envelope	Thylakoids
β-Carotene	9.4	33.8
Zeaxanthin	12.5	2.6
Antheraxanthin	4.7	1.0
Violaxanthin	40.6	15.9
Lutein	28.1	40.0
Neoxanthin	4.7	6.7
Plastoquinone-9 (PQ-9)	28	70
Phylloquinone (K_1)	3	6
α-Tocoquinone (α-TQ)	5	4
α-Tocopherol (α-T)	64	20

Note: The total amounts in milligram per milligram protein of envelope and thylakoids were 6.4 and 19.5 for the sum of carotenoids and 4.3 and 5.5 for the sum of prenylquinones, respectively.

inhibitor of mevalonate formation (Figure 8), only blocks sterol biosynthesis in the cytoplasm, whereas the prenyllipid formation in chloroplasts (chlorophylls, carotenoids, prenylquinones) is unaffected,[56] is strong evidence of an independent biosynthetic prenyl chain system in chloroplasts. Chloroplasts seem to possess the capacity of mevalonate formation from applied acetate. Isolated chloroplasts were found to exhibit some activity of the key enzyme HMG-CoA reductase. In several plants, two or more genes for the HMG-CoA reductase have been found,[54] which indicates that (1) two or three different HMG-CoA reductases exist in plants, and (2) there may be two or three different sites of mevalonate formation within the plant cell. In view of the endosymbiotic origin of chloroplasts, it seems possible that chloroplasts may have preserved their own isoprenoid biosynthesis machinery during evolution. The manner in which chloroplasts may make use of cytoplasmic IPP is under discussion. Exogenously applied mevalonate can be used for prenyllipid biosynthesis in chloroplasts.[46]

The steps in prenyl chain synthesis from IPP are apparently present in chloroplasts. The synthesis of ggPP and geranylgeraniol from IPP by a combined system of chloroplast envelope and stroma fraction has been demonstrated.[69] One contribution proposed a compartmentation of IPP formation,[70] whereas others have evidence of full autonomy of isoprenoid synthesis in chloroplasts.[71-73] The plastidic prenyl chain synthesis in young chloroplasts seems to be fully autonomous and synthesizes IPP from photosynthetically fixed CO_2, whereas at later stages in development IPP is imported from the cytoplasm.[74] The multienzymes ggPP synthase, as well as phytoene synthase, were recently localized using polyclonal antibodies against both enzymes. The latter were found to be restricted to the plastid compartment, and to be localized in the plastid stroma.[75] The site of synthesis of the final steps of the chloroplast

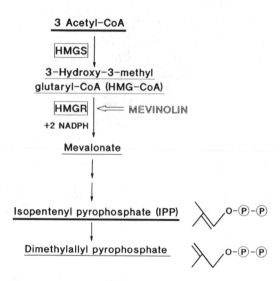

FIGURE 8. Formation of mevalonate and IPP from acetyl-CoA showing the key enzymes HMGS and HMGR. The inhibition of HMGR by the antibiotic mevinolin is indicated.

prenylquinones are the chloroplasts, as has been demonstrated by several publications.[77-79] In these, it was shown that the chloroplast envelope plays a major role in prenylquinone formation.[80] It is the site of prenylation[79] and methylation[78] in α-tocopherol biosynthesis, whereas the thylakoid membrane is also involved in the prenylation and methylation sequence of PQ-9 biosynthesis.[79] It was also shown that phytol, but not geranylgeraniol was the precursor of the side chain of α-tocopherol in spinach chloroplasts. The site of prenylation in phylloquinone K_1 biosynthesis is the chloroplast envelope,[81] whereas the methylation takes place in a thylakoid plus stroma fraction.[82] Plastids are also the site of synthesis of the tetraterpene structure of carotenes, as has been shown for chloroplasts[72,83] and chromoplasts.[71,75,84] Except for the phytoene synthase, all other enzyme steps in carotenoid biosynthesis (desaturation, cyclization, and hydroxylation) seem to proceed on membrane-bound enzymes (compare Section II.E and Figure 14).

2. Mitochondrial Site

Whether mitochondria possess the capacity to synthesize mevalonate and IPP and the C_{45} and C_{50} prenyl side chains of their ubiquinones Q-9 and Q-10 is not known. By growing plant tissue cultures in the presence of mevinolin, which inhibits cytoplasmic mevalonate biosynthesis, ubiquinone homologs with shorter lengths of their prenyl chain, e.g., Q-6, Q-7, and Q-8, also show up.[17,18,59] This is evidence that the formation of the ubiquinones in the mitochondria, with respect to the biosynthesis of the prenyl side chain, is dependent in some way on a functional mevalonate synthesis in the cytoplasm.

From all these observations, it appears that the regulation of the isoprenoid pathway within the plant cell occurs on the level of the HMG-CoA reductase and mevalonate formation. The question whether two or three parallel pathways can exist in a cell in different compartments (cytoplasm, chloroplasts, and mitochondria) can be solved through regulation studies of the activity of the different genes for HMG-CoA reductase, a topic which is under investigation in several laboratories.[54] Further studies must prove whether there are exchange mechanisms for mevalonate and IPP between the three different plant compartments, cytoplasm, chloroplasts, and mitochondria, in such a way that a block in one compartment can be compensated for by synthesis and import from another compartment. This is apparently the case in older chloroplasts.[74] A substantial export of mevalonate or IPP from chloroplasts to the cytoplasm, however, seems not to occur, since in the case of an inhibition of cytoplasmic mevalonate formation by mevinolin prenyllipid synthesis proceeds in chloroplasts, whereas sterol formation in the cytoplasm is stopped.[56,58,59]

D. BIOSYNTHESIS OF CAROTENOIDS

The formation of the C_{20} geranylgeranyl pyrophosphate (GGPP) proceeds by head-to-tail condensation of the biosynthetic isoprene units. IPP and the longer prenylpyrophosphates all exhibit all-*trans* configuration of the one double bond. The first tetraterpene is, however, formed by tail-to-tail condensation of two all-*trans* GGPP, yielding the colorless C_{40} compound phytoene[24,25] (with three conjugated double bonds), which already exhibits the typical feature of the colored carotenoids of three absorption maxima/shoulders in the absorption spectrum, yet these lie in the ultraviolet (UV) region. The biosynthesis of phytoene, with prephytoene pyrophosphate as intermediate,[85] is catalyzed by the two enzymes prophytoene pyrophosphate synthase and phytoene synthase.[86] In contrast to the all-*trans* colored carotenoids such as lycopene, β-carotene, and xanthophylls, the phytoene formed in plastids is a C_{15}-*cis* tetraterpene (Figure 9).

1. Desaturation of Phytoene

The further steps in carotenoid biosynthesis consist of a series of desaturation reactions, whereby additional conjugated double bonds are introduced into the tetraterpene structure which gives rise to the colored carotene lycopene with 11 conjugated double bonds (Figure 9) and a shift of the absorption maxima/shoulder of the absorption spectrum from the UV to the visible region. The *cis-trans* isomerization to the all-*trans* tetraterpene structure is supposed to take place at the transformation of phytoene via phytofluene to ζ-carotene as catalyzed by the enzyme phytoene desaturase (Figure 10). The ζ-carotene desaturase also catalyzes two desaturation steps and introduces two additional double bonds to form lycopene with neurosporene as an intermediate (Figure 9).

2. Formation of Cyclic Carotenoids

The next step in carotenoid biosynthesis is the cyclization of lycopene at both ends to form either α-carotene or β-carotene with its derivatives zeaxanthin,

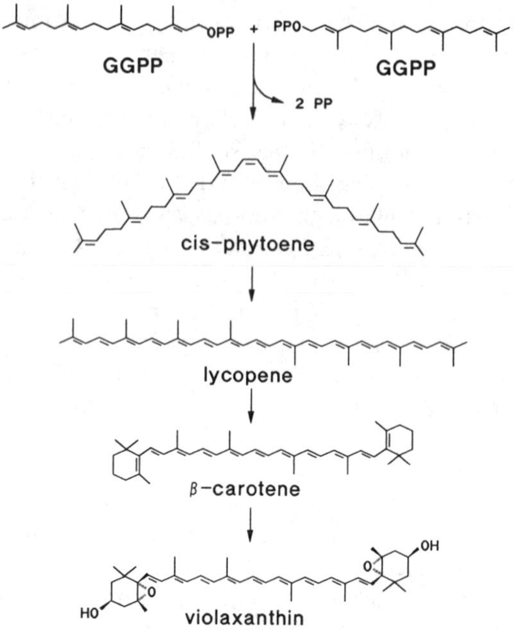

FIGURE 9. Chemical structure of some intermediates (C_{15}-*cis* phytoene and all-*trans* lycopene) in the biosynthesis of the cyclic carotenoids. GGPP = geranylgeranyl pyrophosphate.

antheraxanthin, and violaxanthin. The lycopene cyclase forms two β-ionone rings, one at each end of the lycopene molecule, in which the double bond of the β-ionone rings are conjugated to the other double bonds of the linear part of the tetraterpene structure (see β-carotene in Figure 9). In the case of α-carotene, only one β-ionone ring is formed and an ε-ionone ring in which the double bond is not conjugated with other double bonds (as seen in α-carotene and lutein). It is assumed that two different cyclases (see Figure 10) act with γ-carotene (one β-ionone ring) as an intermediate in β-carotene formation and with δ-carotene (with one ε-ionone ring) in the synthesis of α-carotene.[87,88]

3. Formation of Xanthophylls

The next steps in the biosynthesis of the different primary carotenoids is the introduction of two hydroxyl groups at position 3 and 3′ of the ionone rings to yield the xanthophylls lutein in the α-carotene series (α: 3-OH; 3′-OH) and zeaxanthin in the β-carotene series (β: 3-OH; 3′-OH). The formation of zeaxanthin proceeds in a two-step hydroxylation with β-cryptoxanthin (β: 3-OH) as the intermediate. The latter is thought to be formed from β-carotene by the action of a NADPH-dependent mixed functional oxygenase. The hydroxyl group derives from molecular oxygen.[25] Whether the same oxygenase can hydroxylate α-carotene and possibly also introduce the second hydroxyl

Carotenoid precursors:

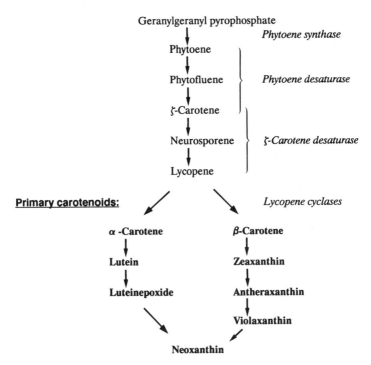

FIGURE 10. Biosynthetic sequence leading to the synthesis of the primary carotenoids of the α- and β-carotene series from common carotenoid precursors.

group is not known. Further modifications of the xanthophyll structure are the introduction of 5,6-epoxy groups. Zeaxanthin is epoxidized to violaxanthin (5,6- and 5′,6′-epoxy groups) with antheraxanthin as an intermediate (one 5,6-epoxy group), see also Section V. In the α-carotene series, lutein epoxide (α: 3-OH, 3′-OH, and 5,6-epoxy group) is found as a minor component of the chloroplast carotenoids; the corresponding xanthophyll with two epoxy groups does not occur. Further oxidation of violaxanthin opens the epoxy ring yielding neoxanthin (β: 3-OH, 3′-OH, 6-OH, and 5′,6′ epoxy group), which exhibits an unusual cumulated double bond. In the α-carotene series, one could also arrive at neoxanthin by oxidation of lutein, whereby an epoxy group is introduced into the β-ionone ring and an OH-group at position 6 of the ε-ionone ring. There is a general agreement that the final steps in carotenoid biosynthesis proceed within the chloroplasts (plastids) and are membrane bound. This particularly applies to all steps from the desaturase of lycopene and its successive transformation to colored cyclic carotenes and xanthophylls. Once the first tetraterpene is formed, it becomes channeled in a quasilinear sequence into the final chloroplast carotenoids.

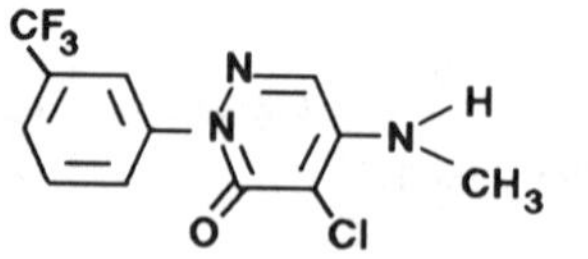

Norfluorazone (SAN 9789)

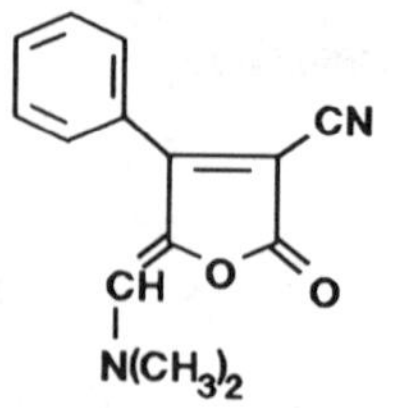

Difunone (EMD-TT 5914)

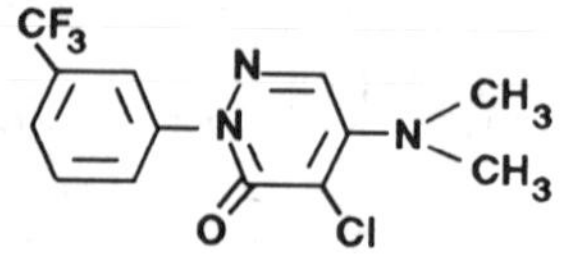

Metflurazone (SAN 6706)

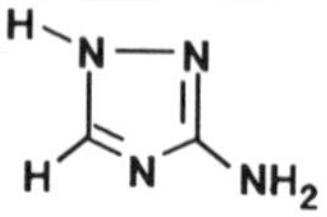

Amitrol

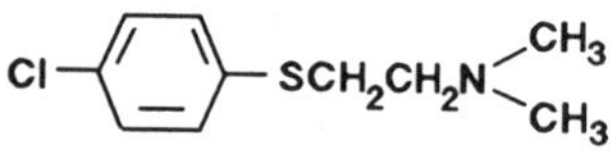

CPTA

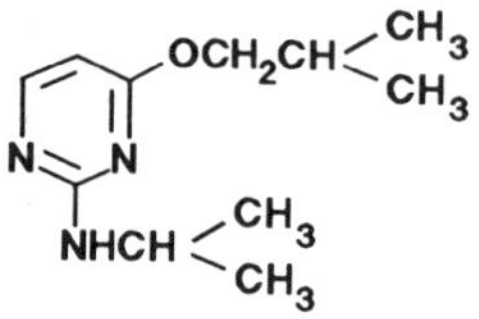

J 852

FIGURE 11. Chemical structure of some inhibitors of carotenoid biosynthesis which block different enzymes in the biosynthetic sequence.

E. CAROTENOID BIOSYNTHESIS AS TARGET SITE OF HERBICIDES

Much research has been done in the last decade to find and develop new herbicides which block or interfere with carotenoid biosynthesis. This gave rise to a whole group of inhibitors and herbicides of quite a different chemical structure which efficiently block carotenoid synthesis in higher plants and algae. This group has been termed "bleaching herbicides" and also "chlorosis-inducing herbicides".[86,89,90] The chemical structure of some of these inhibitors is shown in Figure 11. These inhibitors gave new insight into the organization and regulation of carotenoid biosynthesis. The first of these components was the pyridazinone derivative SAN 6706, known as methfluorazone.[89] Further compounds are norfluorazone, dichlormate, amitrol, pyrichlor, difunone, and fluridone. The bleaching herbicides interfere at early steps of carotenoid biosynthesis, whereby different colorless carotenoid precursors are accumulated, such as phytoene, phytofluene, ζ-carotene, neurosporene, and lycopene (Figure 10). It depends upon the plant and the development stage whether mainly phytoene and phytofluene are accumulated or also ζ-carotene and lycopene. The pool size of these noncyclic carotenoid precursors is too small to be detected under normal biosynthetic conditions, which proceed in a membrane-bound ordered sequence. The overall effect of these inhibitors is the formation of white seedlings and leaves which are free of chlorophylls and carotenoids, but exhibit the regular morphological growth response.[90,91]

It turned out that only the biosynthesis of cyclic carotenoids is blocked and that at very low light intensities, i.e., in dim light, the herbicide-treated plants are green, since the chlorophyll formation and accumulation is unaffected. At medium light and higher light levels, the chlorophylls are, however, destroyed ("bleached"), since the cyclic carotenoids with their protective function against photooxidation of the chlorophylls, e.g., by singlet oxygen $^1O_2$, are missing. The accumulation of the chloroplast prenylquinones is also blocked by these inhibitors of carotenoid biosynthesis.[90] Whether the prenylquinones are bleached and undergo photooxidative destruction like the chlorophylls and apparently all other thylakoid components or whether their biosynthesis is directly affected has yet to be determined.

The point of interaction of norfluorazone, metfluorazone, fluridone, and difunone is the membrane-bound enzyme phytoene desaturase, which catalyzes the synthesis of ζ-carotene with phytofluene as the intermediate.[86] The direct target site for amitrol may not to be phytoene desaturase, but another biosynthetic step.[86] The consecutive enzyme steps in carotenoid biosynthesis are also membrane bound. These are ζ-carotene desaturase, which desaturates ζ-carotene via neurosporene to lycopene, and the lycopene cyclases. The ζ-carotene desaturase can be inhibited by dihydropyranones and pyrimidine analogs (e.g., J 852) and the lycopene cyclase by substituted trialkylamines such as chlorophenylthiotriethylamine (CTPA) (for chemical structure see Figure 11). This thylakoid-bound enzymic sequence of carotenoid biosynthesis from the colorless C_{40} precursors to the cyclic carotenoids β-carotene and xanthophylls, which are integrated into the photosynthetic reaction centers and light-harvesting protein complex respectively, is shown in Figure 12. The interaction of inhibitors with the different steps of carotenoid formation is indicated. The formation of xanthophylls can be blocked by the inhibitor tetracyclasis.[86]

Present research in this interesting field is devoted to finding the reasons why particular plants are resistant to the carotenoid biosynthesis inhibitors. The concept is that in resistant plants the corresponding enzymes responsible for a particular step in carotenoid biosynthesis exhibit a different structure and amino acid sequence, so that the inhibitor can no longer bind. This stimulated the search for the genes of the different enzymes of carotenogenesis. This work is in progress in several laboratories at the level of photosynthetic bacteria, algae, and higher plants; cloned genes are listed in Table 2. The future goal will be to transfer the genes of resistant weeds into crop plants so that these become resistant to bleaching herbicides.

III. FUNCTION OF THE PHOTOSYNTHETIC PRENYLPIGMENTS

A. ORGANIZATION OF THE PRENYLPIGMENTS

1. The Chlorophylls

Chlorophylls are bound together with carotenoids in specific ways to the different chlorophyll-carotenoid proteins of the photochemically active

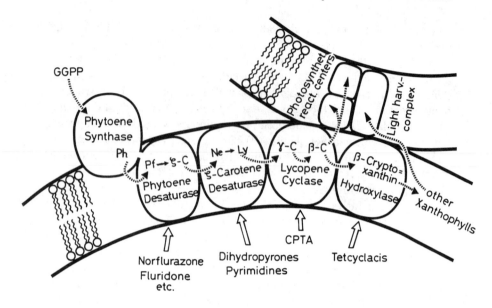

FIGURE 12.   Scheme of the membrane-bound biosynthesis of carotenoids in chloroplasts from geranylgeranyl pyrophosphate (GGPP) via the enzyme sequence system phytoene synthase, phytoene desaturase and ζ-carotene desaturase, lycopene cyclase, and β-cryptoxanthin hydroxylase with indication of the point of interference of particular herbicides and inhibitors. (Ph — phytoene; Pf — phytofluene; ζ-C — ζ-carotene; Ne — neurosporene; Ly — lycopene; γ-C and β-C — γ- and β-carotene. (Courtesy of G. Sandmann, Konstanz 1992).

thylakoids of functional chloroplasts. The chlorophylls of higher plants and green algae contain chlorophyll a as a major component and chlorophyll b as an accessory light-absorbing pigment in an chlorophyll a to b ratio of about 3.[107] Minor amounts of an additional chlorophyll, termed a′, are found in older literature.[108] It is a C-10 epimer of chlorophyll a and is considered to be identical in the form of a dimer with the reaction center chlorophyll P700 of photosystem I.[109]

2. The Carotenoids

The group of primary carotenoids, which are bound to the photosynthetic thylakoid membrane, consist of the oxygen-free carotenes and of the xanthophylls, which contain oxygen in the form of hydroxy or epoxy groups. At each oxidation level, there exist two isomers which differ in the position of the double bond in one of the two ionone rings: α-carotene and its derivatives and β-carotene and its derivatives. Derivatives of α-carotene are the xanthophylls lutein and luteinepoxide. β-carotene, in turn, is characterized by two β-ionone rings; its corresponding xanthophylls are zeaxanthin, antheraxanthin, and violaxanthin. The xanthophyll neoxanthin with three hydroxy groups, which no longer contains a double bond in its one ionone ring, can be attributed to both carotenoid groups.

TABLE 2

**Cloned Genes of Enzymes of the Carotenoid Biosynthesis Pathway
from Various Organisms Including Higher Plants
(*Capsicum, Glycine, Solanum*)**

Enzyme	Gene	Organism	Ref.
Geranylgeranyl	*al-3*	*Neurospora crassa*	92
pyrophosphate synthase		*Capsicum annuum*	93
Prephytoene pyro-	*crtB*[a]	*Rhodobacter capsulatus*	94
phosphate synthase	*crtB*[a]	*Erwinia uredovora*	95
	crtB[a]	*E. Herbicola*	96
			97
Phytoene synthase	*crtE*[a]	*R. capsulatus*	94
	crtE[a]	*E. uredovora*	95
	crtE[a]	*E. herbicola*	96
			97
	crtE[a]	*Cyanophora paradoxa*	98
	pys	*Synechococcus* PCC7942	99
	Psyl	*Solanum lycopersicum*	100
Phytoene desaturase	*crtI*	*R. capsulatus*	94
			101
	crtI	*E. uredovora*	95
	crtI	*E. herbicola*	96
	crtI	*Aphanocapsa* PCC6714	102
	pds	*Synechococcus* PCC7942	103
	al-1	*N. crassa*	104
	pds1	*Glycine max*	105
		Solanum lycopersicum	103
Lycopene cyclase	*crtY*	*E. uredovora*	95
	crtZ	*E. herbicola*	97
β-Carotene	*crtZ*	*E. uredovora*	95
hydroxylase	*crtH*	*E. herbicola*	97

[a] Recent analysis revealed for the *crtB* and *crtE* gene product of *E. uredovora* phytoene synthase activity and GGPP synthase activity, respectively.[106]

In higher plants, the major carotene is β-carotene, with only traces or minor amounts of α-carotene, whereas in certain green algae α- and β-carotene can occur in almost equal concentrations. The predominant xanthophyll of thylakoids is lutein, followed by violaxanthin, neoxanthin, and the other xanthophylls in minor amounts.[7,47,62] In photosynthetically active chloroplasts, the percentage composition of carotenoids (weight percent) can vary within the ranges: β-carotene 25 to 40%, lutein 40 to 57%, violaxanthin 9 to 20%, and neoxanthin 5 to 13%. The other xanthophylls usually contribute only up to 5 to 10% of the total leaf carotenoids. Zeaxanthin only shows up under high light conditions and is formed by deepoxidation of violaxanthin with antheraxanthin as an intermediate. These three carotenoids participate in the xanthophyll cycle, the possible function of which is treated in more detail in Section V. All these carotenoids are regarded as primary carotenoids which are needed for photosynthetic function, whereas particular carotenoids and xanthophyll ester with often quite different chemical structures, e.g., of red fruits and flower petals, are classified as secondary carotenoids, which are not treated here.

B. THE CHLOROPHYLL-CAROTENOID PROTEINS

Small amounts of violaxanthin and some minor amounts of lutein and β-carotene are found in the chloroplast envelope (Table 2).[61] Within the chloroplasts, all carotenoids and chlorophylls are bound to the thylakoids.[47,48,62] There the pigments are associated with several chlorophyll-carotenoid proteins[7,110] which possess particular chlorophyll and carotenoid compositions and which can be separated by polyacrylamide gel electrophoresis (PAGE) of digitonin or sodium dodecyl sulfate (SDS)-treated chloroplasts or thylakoid preparations.[111,112] These are (1) the two P700-containing chlorophyll a/β-carotene proteins CPI and CPIa of photosystem I; (2) CPa, a chlorophyll a/β-carotene protein, with the reaction center of photosystem II; and (3) several forms of light-harvesting pigment proteins LHCPs (LHCP$_1$, LHCP$_2$, LHCP$_3$, LHCP$_y$), which represent chlorophyll a + b/lutein/neoxanthin proteins.[111] The carotenoid composition of these different pigment proteins has been determined in a comparative way only for radish chloroplasts. As compared to whole chloroplasts, β-carotene is enriched in the pigment proteins CPIa and CPa, which contain the reaction centers of photosystem I and II, respectively, whereas lutein and neoxanthin are enriched in the LHCPs (see Table 3). The LHCPs contain chlorophyll a and b in a ratio of 1.1 to 1.3. In contrast, the proteins CPIa, CPI, and CPa exhibit higher values for the ratio chlorophyll a/b and also low values for the ratio chlorophyll a/β-carotene (a/c). Chlorophyll a/β-carotene proteins can be further purified from smaller antenna pigment proteins (LHCPs) to yield fairly pure chlorophyll core complexes of the reaction centers of photosystem I and II. In these preparations, β-carotene always sticks to chlorophyll a.[113]

Violaxanthin and zeaxanthin are not enriched in any of the analyzed pigment proteins (Table 3). In contrast, in the case of radish chloroplasts, which at the stage of investigation contained only violaxanthin and no zeaxanthin, the major position of violaxanthin was found in the free pigment fraction (FP) even under very mild disintegration and gel electrophoresis conditions (Table 3). In cases where both violaxanthin and zeaxanthin were present in the leaf, both carotenoids, which can be interconverted, were primarily found in the free pigment fraction. From these results, we assume that *in vivo* violaxanthin and zeaxanthin, including the intermediate xanthophyll antheraxanthin, are not bound to a particular pigment protein, but are freely organized in the bilayer of the photosynthetic membrane in such a way that they are accessible to interconversion by the enzymes epoxidase and deepoxidase (see xanthophyll cycle in Section V). A nonprotein-bound organization of violaxanthin within the photosynthetic membrane could be the reason that violaxanthin can be

TABLE 3
Ratio of Prenylpigments and Percentage Composition of Carotenoids in Different Pigment Proteins of Thylakoids[a]

Protein	Chloroplasts	CPI[a]	CPI	CP[a]	LHCP$_{1-3}$	FP[a]
Pigment ratios						
Chlorophyll a/b	3.2	4–7	9–21	3–8	1.1–1.3	2.6
(a + b)/(x + c)[b]	4	5–6	7–9	3–5	3–5	3
a/c[b]	12	12	9	4–8	60–180	14
Percentage carotenoid composition[c]						
β-Carotene	**30**	**56**	**70**	**75**	1–2	24
Lutein	**45**	24	15	13	**56–65**	37
Neoxanthin	6	9	4	5	**21–29**	10
Violaxanthin	**11**	5	6	4	4–5	**20**
Others	8	6	5	3	6	9

Note: Adapted from Lichtenthaler, H . K., *Methods in Enzymology*, Vol. 148, Douce, R. and Packer, L., Eds., Academic Press, New York, 1987, 350.)

[a] Isolated by SDS-PAGE from radish chloroplasts. FP — free pigments.
[b] (a + b)/(x + c) — weight ratio chlorophylls to carotenoids; a/c — weight ratio chlorophyll a to β-carotene.
[c] The enrichment of particular carotenoids in the pigment protein as compared to whole chloroplasts is indicated in boldface.

found in a low content in all pigment proteins (Table 3), apparently resulting in nonspecific binding to all proteins when the thylakoid membrane is disintegrated by treatment with detergents. Similar results of carotenoid distribution between pigment proteins to those shown here for radish chloroplasts were obtained by varying techniques of pigment protein isolation (see review in Reference 113).

The individual levels of the light-harvesting proteins LHCPs can vary depending on the developmental stage and upon light conditions. In the first hours of greening of etiolated seedlings, the chlorophyll a/β-carotene proteins (CPI and CPa) are primarily accumulated, whereas the accumulation of the LHCPs is retarded.[112] Consequently, the ratio of chlorophyll a/b in greening etiolated seedlings is initially very high and then declines to values of 3 with increasing levels of LHCPs, which possess low values for the chlorophyll a/b ratio of about 1.1 to 1.3. The percentage contribution of β-carotene to the total leaf carotenoid content is also higher (30 to 35%) when the level of LHCPs is fairly low. On the other hand, the weight percentage of lutein is higher, e.g., in chloroplasts with higher amounts of LHCPs. Chloroplasts can adapt to high light and low light growth conditions, which results in the formation of sun-type and shade-type chloroplasts.[114] Under low light growth conditions (shade), chloroplasts exhibit a higher proportion of LHCPs with reference to CPI and CPa than under high light growth conditions (sun).[115,116] The particular prenylpigment ratios such as chlorophyll a/b, total chlorophylls to total caro-tenoids (a + b)/(x + c), and chlorophyll a to β-carotene (a/c) thus reflect the relative pigment composition of leaves under a particular light regime.

These pigment ratios are very useful indicators and reflect changes and damage of the photosynthetic pigment apparatus under natural and anthropogenic stress conditions (photoinhibition, photooxidation, herbicide treatment and photobleaching, air pollution, ozone exposition).[117,118] The ratio of chlorophylls to carotenoids (a + b)/(x + c) is a general stress indicator and quickly declines from normal values of 5 to 6 to values around 4 under the influence of various stressors. In spruces with progressive forest decline symptoms, the ratio (a + b)/(x + c) can drop to values between 2 and 3. In some cases, the LHCPs are primarily degraded (high light stress), which results in higher a/b values (at a lower total chlorophyll content), while under other stress constraints the photosystems may be the first point of damage, which then results in lower values of 2.3 for the a/b ratio. Corresponding changes can be found in the ratio a/c and the ratio x/c (xanthophylls to β-carotene).

C. FUNCTIONAL ROLE OF CAROTENOIDS AND CHLOROPHYLLS

1. Chlorophylls and Lutein-Neoxanthin

According to their differential binding to the various chlorophyll-carotenoid proteins of thylakoids, the particular chlorophylls and carotenoids possess different functions in photosynthesis. Chlorophyll a and b in the LHCPs exhibit their main function in the absorption of light energy. In the LHCPs, which show particular light absorption properties different from those of other pigment proteins (CPI and CPa), chlorophyll b and the xanthophylls lutein and neoxanthin function as accessory pigments (Table 4). They also absorb light quanta in wavelength regions (blue green and orange red) which are not absorbed by the chlorophyll a present in the same LHCP. The main functions of the xanthophylls lutein and neoxanthin are absorption of light and transfer of the excited states to neighboring chlorophyll b and chlorophyll a molecules. The transfer of excited states from chlorophyll b to chlorophyll a proceeds to 100%, since *in vivo* in thylakoids or in isolated LHCPs one only observes chlorophyll a fluorescence, but not that of chlorophyll b. In contrast, isolated chlorophyll b exhibits red fluorescence.[91]

2. β-Carotene

The absorbed light energy is transferred via excited states (exciton migration) from chlorophyll a forms in the LHCPs to chlorophyll a forms in the chlorophyll a/β-carotene proteins CPI and CPa and within these finally to the reaction center molecules P700 and P680 of photosystem I and II, respectively. The function of β-carotene in photosystems I and II seems not to be that of an accessory light-harvesting pigment, though it might also absorb incident light quanta. The major physiological and photosynthetic function of carotenes

TABLE 4
Possible Function of the Major Leaf and Thylakoid Carotenoids in Photosynthesis

Lutein	Accessory pigment in LHCPs	Light absorption
Neoxanthin	Accessory pigment in LHCPs	Light absorption
β-Carotene	Reaction center pigment of photosystem I and II (CPI and CPa)	Dissipation of excited states,[a] e.g., ^{3}Chl or $^1O_2$
Violaxanthin	Component of the xanthophyll cycle	Oxidation product of zeaxanthin
Zeaxanthin	Component of the xanthophyll cycle	Photoprotective function, dissipation of highly reactive oxygen species? Destruction of epoxy states of acyllipids?

[a] Chl — triplet chlorophyll a; $^1O_2$— singlet oxygen.

seems to lie in the protection of the chlorophylls in the two photosystem reaction centers against photooxidative degradation by taking over the triplet state of chlorophyll a (^{3}Chl) yielding triplet carotenes (^{3}Car) which are dissipated by heat emissions to carotene in the ground state (Car). It is also discussed that carotenes may react with singlet oxygen ($^1O_2$) which also yields triplet carotene (^{3}Car), the energy of which is dissipated by heat emission. Though this functional roles of carotenoids in photoprotection of chlorophylls is generally considered for the total carotenoids of the chloroplast without any differentiation,[119-123] we believe that in view of the specific association of β-carotene with the two photosystems I and II, this photoprotective function is associated with β-carotene (Table 4). In any case, the protective function of carotenoids in photosynthesis is evident with the observation that in plants treated with bleaching herbicides blocking the biosynthesis of carotenoids,[86,90,91] the chlorophylls still accumulate at very dim light intensities, but become photodegraded at medium and higher light intensities, resulting in the formation of fully white leaves.

One of the open problems of this photoprotective function of β-carotene is the question of whether β-carotene in the triplet state is degraded or restored. How often can β-carotene be excited to the triplet state before it is destroyed — 5, 10, or 50 times? What are the decomposition products of β-carotene in this functional role? Do oxidized derivatives and possibly other carotenoids and xanthophylls of the β-ionone structure show up? This field of the functional role of β-carotene in the reaction centers of the two photosynthetic photosystems and the fate of the triplet carotene need much more attention in future research. That the β-carotene pool of thylakoids is not stable, but undergoes a rapid turnover under saturating light conditions is evidenced by the observations that it exhibits, together with phylloquinone K_1, the highest degree of labeling of all chloroplast prenyllipids.[124,125]

IV. FUNCTION OF PRENYLQUINONES

A. PLASTOQUINONE-9

PQ-9 always occurs in the thylakoids together with its quinol form plastohydroquinone-9 (PQ-9·H$_2$) forming a lipophilic redox system. In older chloroplasts of light-exposed leaves, the major portion of the PQ-9 is found outside the thylakoids in the osmophilic plastoglobuli of the plastid stroma. Within the thylakoids, PQ-9 + PQ-9·H$_2$ are associated with the "heavy particle fraction" obtained by digitonin fractionation of thylakoid membranes, which primarily consist of photosystem II + antenna complex.[62,126] PQ-9, which is present in thylakoids at a five to ten times higher concentration than other prenylquinones and electron carriers,[49,62] plays an essential role in the transfer of electrons between the two photosynthetic photosystems I and II. Its concentration amounts to 4 to 8 mol/100 mol of total chlorophyll, depending on the growth conditions and the plant.

In the form of a semiquinone, which is a one-electron carrier, PQ-9 is identical to the primary chlorophyll fluorescence quencher Q$_A$. The latter transfers its electron to the secondary quencher, the two-electron acceptor Q$_B$, which is also supposed to be a PQ-9. Q$_A$ and Q$_B$ are bound in specific ways to the reaction center proteins. Reduced Q$_B$ can apparently freely move away from the 32-kDa Q$_B$ protein, take up two protons deriving from the stroma site, and transfer its two hydrogens (electrons and protons) to the large plastoquinone pool. Other oxidized mobile plastoquinone molecules then dock at the 32-kDa protein, become reduced, and shuttle electrons and protons to the plastoquinone pool. Many herbicides such as diuron (DCMU), bentazon, and atrazine can bind instead of PQ-9 to the 32-kDa Q$_B$ protein and thus block photosynthetic electron flow.[127]

The plastoquinone pool transfers its electrons via the Rieske center and the cytochrome b$_6$/f complex to photosystem I. Since cytochromes as redox carriers can only transfer electrons, the two protons are released to the thylakoid lumen, giving rise to the proton gradient across the thylakoid membrane, which is a prerequisite for the light-induced ATP formation. In fact, the combination of a prenylquinone, which can carry electrons and protons, together with a cytochrome, which only carries electrons, is the basic element not only of the vectorial photosynthetic electron transport reactions, but also in mitochondrial electron flow and in bacterial respiratory systems. This combination of a prenylquinone with a cytochrome in a biomembrane yielding a pH gradient across the membrane, which can be used by an ATP synthase system to form ATP, is one of the essential requirements and basic features of life.

Besides plastoquinone A (PQ-9) in older green and senescent leaves and fruits, hydroxy derivatives (PQ-Cs) and acylated form (PQ-Bs) are found (Figure 4); these, however, possess no physiological function in photosynthesis.

B. α-TOCOQUINONE AND α-TOCOPHEROL

α-Tocoquinone together with its chromanol α-tocopherol, the cyclic version of its reduced form, α-tocohydroquinone, represent another lipophilic redox system of the photochemically active thylakoids. The exact physiological function of both compounds is hitherto not known. Partition studies of both substances between the two photosystems I and II (heavy and light particle fractions of digitonin-treated chloroplasts) did not give a preferential enrichment in either photosystem.[62,126]

α-Tocoquinone is a potential redox carrier which could have a function in photosynthetic electron transport at a different site than PQ-9. PQ-9, in the form of Q_A and Q_B, is the acceptor of electrons of photosystem II and is responsible for the reduction of the large plastoquinone pool. α-Tocoquinone, in turn, might control the oxidation of the large plastoquinone pool in the electron flow from plastohydroquinone to the cytochrome b_6/f complex pool and the transfer of protons to the thylakoid lumen. In such a function, it could block possible uncontrolled backflow of electrons and protons and direct the electron flow toward photosystem I. This presumed function has been postulated,[128] but not yet established. The fact that the halogenized benzoquinone derivative DBMIB (dibromo methyl isopropyl benzoquinone) blocks the oxidation of the plastoquinone pool at low concentration (at the presumed α-tocoquinone site?), but is required at a higher concentration for the reduction of PQ-9 at the Q_B site would be in agreement with such an assumed function of α-tocoquinone. In any case, α-tocoquinone is a genuine constituent of photosynthetically active thylakoids and is present in a fairly constant and similar concentration as other photosynthetic electron carriers such as cytochromes, phylloquinone, or Q_A.[62] Its concentration in thylakoids amounts to 0.3 to 0.5 mol/100 mol of total chlorophyll. Though α-tocoquinone may also arise as an oxidation product of α-tocopherol, one should see its function primarily as a potential electron carrier which has a physiological function apart from that of α-tocopherol.

Within chloroplasts, the major part of α-tocopherol is bound as an excess lipid to the osmiophilic plastoglobuli;[64,68] only a small portion is found in thylakoids. The level of α-tocopherol with respect to total chlorophylls is not constant and can vary in plastoglobuli-free thylakoids from 1 to 4 mol/100 mol of chlorophyll. In fact, during greening of etiolated seedlings,[12] when the young chloroplasts are practically free of plastoglobuli (the lipids of which are used up for thylakoid formation), the level of α-tocopherol in thylakoids is very low and only a little higher than that of α-tocoquinone. There are two proposals concerning the physiological function of α-tocopherol in the photosynthetic membrane: (1) It is a lipophilic antioxidant and protects the thylakoid lipids against oxidation by peroxides and highly oxidizing excited oxygen species and radicals[120,121,123,129] which are formed in the photosynthetic membrane under high light[129] and other stress conditions. It is regarded as a very efficient scavenger of free radicals[120,123,129] which has been proven in isolated test systems. (2) α-Tocopherol may possess a structural role in the photosynthetic membrane. This is based on the observation that in liposomes it specifically interacts with polyunsaturated phospholipids.[131] α-Tocopherol may, in the sterol-free thylakoids, play a similar structural function as the sterols possess in the other cellular biomembranes.[132]

Whether α-tocopherol in its function as a lipid antioxidant of the photo-synthetic membrane is oxidized to α-tocoquinone or other oxidation products is not known. It is also doubtful that the fairly stable pool of α-tocoquinone in the photosynthetic membrane can, after reduction to the α-tocohydroquinone, be readily transformed back to the chromanol α-tocopherol. A regeneration of α-tocopherol in the membrane via reduction of its oxidation products has hitherto not yet been proven. In any case, α-tocopherol is a lipid antioxidant, the oxidation of which occurs, however, relatively late under photooxidative conditions. Under various stress conditions of either isolated chloroplasts or intact leaves, we found that at first the PQ-9·H_2 pool is oxidized, then the level of β-carotene begins to decline, and only thereafter could we see a loss in the level of α-tocopherol. In addition, in adult chloroplasts of sun-exposed plants, the major part of α-tocopherol is localized in the plastoglobuli,[64-66] i.e., outside the photosynthetic membrane; therefore, the lipid antioxidant function of α-tocopherol in the photosynthetic membrane cannot be as essential as usually anticipated.

C. PHYLLOQUINONE K_1

Vitamin K_1, a 2-methyl-3-phytyl-1,4-naphthoquinone, which like chlorophyll a contains a phytyl side chain, predominantly occurs in green photosynthetically active plant tissues,[49] whereas chlorophyll-free tissue contains phylloquinone in only trace amounts.[11,50] Within the plant cell, phylloquinone K_1 was shown to be quantitatively bound to chloroplasts and thylakoid fragments.[47] Further studies by Lichtenthaler and co-workers on the distribution of phylloquinone between the two photosystems showed that it is associated with photosystem I in an estimated ratio of two vitamin K_1 per one P700.[62,126] Additional experiments in 1981 with digitonin-fragmented chloroplasts and subsequent SDS gel electrophoresis proved that K_1 is tightly bound to the photosystem I containing chlorophyll a protein CPI in a stable level of 1 mol K_1 per 100 mol of chlorophyll.[133] These studies also demonstrated that there exists a second relatively loosely bound pool of K_1 within photosystem I which is lost during gel electrophoresis. Based on these results, the physiological function of the naphthoquinone K_1 as an electron carrier in the electron flow of photosystem I was evident. This function and the level of two K_1 per one P700 were later reconfirmed in studies with isolated reaction centers of photosystem I.[134,135]

According to our present knowledge, the plastidic naphthoquinone phylloquinone K_1 seems to play a defined functional role in the photosynthetic electron transport within photosystem I. The latter was considered as a light-driven plastocyanin:ferredoxin oxidoreductase of higher plants, green algae, and cyanobacteria.[136] Upon light excitation in photosystem I, charge separation takes place, and P700 donates an electron to A_0. This electron is further transferred from A_0 via the intermediate acceptor A_1 (one of the two K_1 molecules in the reaction center) and the iron sulfur center F_X to the iron sulfur cluster F_B/F_A (see Figure 13). Upon UV irradiation leading to the destruction

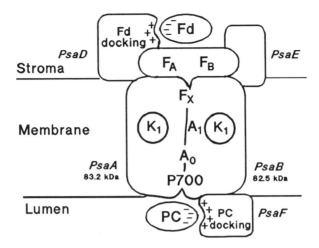

FIGURE 13. Simplified scheme of the photosystem I reaction center showing the two molecules of phylloquinone K_1 equally distributed between the two reaction center proteins PsaA and PsaB. P700 — reaction center chlorophyll dimer and primary electron donor; A_0 — a primary acceptor; A_1 — an intermediate quinone acceptor (K_1?); F_A, F_B, F_X — three iron sulfur centers; F_d — ferredoxin; PC — plastocyanin. Some of the integral proteins PsaD, PsaE, and PsaF are indicated. (Modified from Golbeck, J. H., *Annu. Rev. Plant Physiol. Plant Mol. Biol.*, 43, 293, 1992.)

of phylloquinone and also upon extraction of K_1 by hexane, the electron flow within photosystem I does not proceed, as seen from particular optical-flash spectroscopic signatures, but can be restored after addition of phylloquinone (for original literature see the review in Reference 136). One of the two phylloquinone K_1 molecules is tightly bound and apparently essential for the photosystem I electron flow (K_1 at position A_1), whereas the more "loosely bound" second molecule K_1 seems to be inactive.[136] Further research is, however, required to fully define the exact role of both K_1 molecules in the reaction center of photosystem I. From the observation that synthetic naphthoquinones can stimulate cyclic electron flow around photosystem I, it was concluded that phylloquinone represents the endogenous co-factor of cyclic electron flow.[137] Whether the loosely bound K_1 can also be reduced in photosystem I and then take over such a role needs to be investigated.

The endogenous naphthoquinone phylloquinone K_1 may have an additional function in photosynthesis which is different from its role as an electron carrier mentioned above. Halogenated naphthoquinones are excellent quenchers of the variable chlorophyll fluorescence of chloroplast preparations.[138] It is therefore quite feasible that the endogenous phylloquinone K_1 exhibits a double function (1) as a quencher of excitation energy and (2) as an electron carrier. The fact that at room temperature the chlorophyll fluorescence of intact chloroplasts almost exclusively derives from photosystem II and the light-harvesting antenna proteins, whereas photosystem I contribute little or none to the chlorophyll fluorescence (for details see reviews in References 139 and 140), could be due to its K_1 content and a presumed quencher function of K_1 in its oxidized or

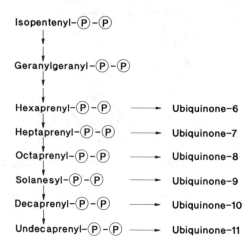

FIGURE 14. Biosynthesis of the different ubiquinone homologs Q-9 to Q-11 occurring in higher plants from IPP.

reduced form in photosystem I. A mobile part of K_1 could thus function as quencher in the regulation of the two photosynthetic photosystems I and II and control spillover processes from photosystem II to I. It is of interest in this respect that in the green alga *Chlorella*, of all the plastidic prenyllipids, it is phylloquinone (together with β-carotene) which exhibits the highest labeling degree from $^{14}CO_2$.[124,125] This points to a function of K_1 with a high turnover of K_1. For the quencher of excitation energy, a high turnover rate could be anticipated.

D. UBIQUINONES Q-9 AND Q-10

Ubiquinones with different lengths of the isoprenoid side chain are genuine constituents of mitochondria and function as electron and proton carriers in the electron flow of the mitochondrial membranes. The ubiquinone homolog generally occurring in higher plants and ferns is in most cases Q-10, with 10 isoprene units in the prenyl side chain, except for membranes of the Poaceae, which possess the Q-9 homolog.[17,18] There are, however, dicot families in which either Q-10 or Q-9 may occur. Also, within one plant species, e.g., radish, one variety may contain Q-10, while another exhibits Q-9. One may speculate that the Q-10 is the usually ubiquinone homolog of higher plants, which is then replaced by Q-9 with a C_{45} prenyl chain (also needed for PQ-9 formation) when the ability to form the C_{50} prenyl chain is lost. With regard to the function of ubiquinones in the mitochondrial membrane, a chain length of 6 to 10 isoprene units (homologs Q-6 to Q-10) is required to bind in the membrane.

Most plants also contain, besides the major ubiquinone homolog Q-9 or Q-10, the shorter prenyl chain homologs, though in much lower concentrations of 1 to 10% of the major ubiquinone homolog.[17,18] These shorter-chain ubiquinone homologs could be explained as biosynthetic pool sizes, which may show up when biosynthesis declines. The other possibility would be that the prenyltransferase which transfers the C_{45} or C_{50} prenyl chain to the p-hydroxybenzoic acid moiety (see Figure 5) is not specific to the C_{45} or C_{50} chain and can also use shorter prenyl chains, though with a lower efficiency. This biosynthesis of ubiquinone homologs with different lengths of the prenyl side chain is shown in Figure 14. In red pepper (*Capsicum annuum* L.), an even longer homolog Q-11 is found besides Q-9. Since the major and minor ubiquinone homologs are found in the intact plant tissue and in isolated mitochondria in the same relative concentrations, it is assumed that they are all bound to mitochondria and integrated in the respiratory electron transport chain. Most authors assume that within the plant cell the ubiquinones are exclusively bound to the mitochondria. The possibility that a small pool of the cellular ubiquinones may be associated in plants with other cell membranes, in particular with the endoplasmic reticulum (ER) or Golgi vesicles as in animal tissue,[141] is still a matter of discussion and could also be due to degradation of mitochondria. Q-9 and Q-10 function in the mitochondrial membrane and respiratory electron flow as electron and proton carriers in a way similar to that of PQ-9 in thylakoids. Ubiquinone, e.g., Q-9, becomes reduced through NADH to its quinol $Q-9 \cdot H_2$ in the first step of the mitochondrial electron flow and passes its electrons over to the cytochrome b, which follow as the next electron carrier in the respiratory electron transport chain. The protons, in turn, are transferred to the inner mitochondrial membrane lumen, thus yielding a pH gradient across the mitochondrial membrane which can be utilized by the mitochondrial ATP synthase to form ATP. With its functional position in the first step of the respiratory chain the ubiquinone functions in the cyanide-sensitive normal respiratory chain, as well as in the cyanide-resistant respiration. The concentration of the ubiquinone homologs range in general between 3 to 5 nmol/mg of mitochondrial protein.[18] The ratio of ubiquinones (Q) to cytochromes (Cyt) amounts to values of 6 to 20 for Q to Cyt b + c and 15 to 40 for the ratio Q to Cyt a + a_3.[18] In contrast to plastidic PQ-9, which also accumulates in the osmiophilic plastoglobuli, the mitochondrial ubiquinones seem to be exclusively bound to the mitochondrial membrane and not to occur in excess amounts.

V. FUNCTION OF THE XANTHOPHYLL CYCLE IN PHOTOSYNTHESIS

Violaxanthin (v), together with antheraxanthin (a) and zeaxanthin (z) form the xanthophyll cycle carotenoids, which are interconvertible (for chemical structure see Figure 15). The operation of the high light-induced xanthophyll cycle and zeaxanthin formation, as detected by Sapozhnikov,[143] has been established by several research groups.[144-147] The forward reaction, i.e., the deepoxidation of violaxanthin (two epoxy, two hydroxy groups), to zeaxanthin (two hydroxy groups) proceeds under high light conditions, whereas in dark-

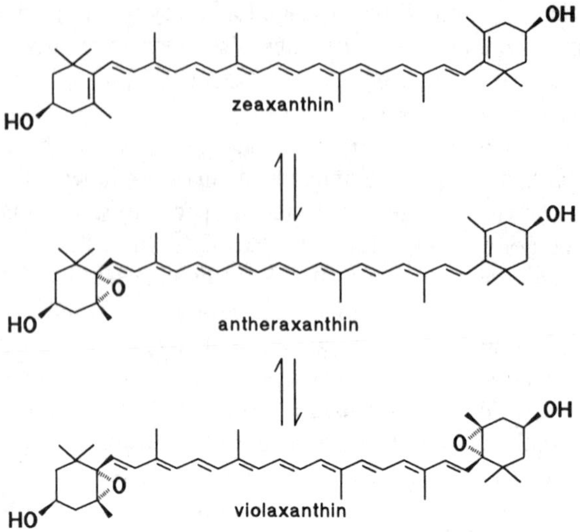

FIGURE 15. Chemical structure of the xanthophyll cycle carotenoids zeaxanthin, antheraxanthin, and violaxanthin which can be interconverted by epoxidation and deepoxidation.

ness or at low light conditions the back reaction takes place with the epoxidation of zeaxanthin to violaxanthin. Antheraxanthin is an intermediate in both the forward and the back reaction, but its pool size remains low as compared to zeaxanthin and violaxanthin. The basic conditions of these enzymic carotenoid transformations, which proceed in the thylakoids, are established. The deepoxidation requires a reducing milieu (NADPH, reduced ascorbate) and a pH in the thylakoid lumen of about 5 which activates the deepoxidase.[145,146] In contrast, the back reaction consumes NADPH and oxygen and proceeds at pH 7.5.

The percentage of the xanthophyll cycle carotenoids $v + a + z$ of the total carotenoid $(x + c)$ content of chloroplasts varies from plant to plant and seems to be genetically controlled, but also varies depending on the growth conditions. In normal green plants, the percentage $v + a + z$ lies in the range of 14 to 24%, and usually 60 to 80% of the violaxanthin is transformable under high light conditions to zeaxanthin (see Table 5). In an aurea variety of tobacco, the xanthophyll cycle carotenoids $v + a + z$ make up 30 to 33% of the total carotenoids, and 90% of violaxanthin can be phototransformed to zeaxanthin. That percentage of violaxanthin which can be transformed to zeaxanthin apparently represents that violaxanthin pool which is bound to thylakoids, whereas the remainder may sit in the chloroplast envelope and not be available to the enzyme deepoxidase. In aurea tobacco, the percentage proportion of $v + a + z$ can be increased by *de novo* accumulation in the younger leaves under high light conditions, whereas the level of chlorophylls and the other carotenoids such as β-carotene, lutein, and neoxanthin does not change. In fact, the zeaxanthin content doubled within 5 h of continuous high light exposure

TABLE 5
Percentage of Xanthophyll Cycle Carotenoids v + a + z of the Total Carotenoid Content (x + c), as Well as Maximum Percentage of the Violaxanthin to Zeaxanthin Transformation after 5 to 10 min of High Light Irradiation (2200 mmol/m²/s) in Different Plants

	% v + a + z of total x + c	% v → z (forward reaction)
Phaseolus vulgaris	14–15	60 ± 3
Raphanus sativus	15–16	80 ± 2
Acer platanus	21–24	81 ± 2
Nicotiana tabacum		
Green variety su/su	15–16	67 ± 2
Aurea variety Su/su	30–33	90 ± 3
Su/su after 5 h high light	41	90 ± 4

from 7.5 to 15 mg/cm² leaf area and that of total v + a + z from 9.5 to 19 mg/cm² leaf area.[9] In outdoor plants, the level of zeaxanthin increases during the day and reflects the incident photon flux density on the upper leaf surface.[148] It was also shown that shade leaves possess a lower content of v + a + z than sun-exposed leaves.[148]

Though much information is available on the operation and occurrence of the xanthophyll cycle and the transformation of xanthophylls, very little is known of the physiological function of the light-driven xanthophyll cycle and zeaxanthin accumulation. Most authors assume a photoprotective function of the xanthophyll cycle operation and the accumulated zeaxanthin. The tobacco aurea mutant, which exhibits a much higher v + a + z content than the normal/green variety (Table 5) and also doubles the zeaxanthin level by *de novo* biosynthesis within 5 h of high light exposure, is much less sensitive to photoinhibition than the green variety. This, as well as the higher proportion of v + a + z carotenoids in sun than shade leaves, points to a photoprotective role of zeaxanthin and the xanthophyll cycle. However, the exact mechanism of how this should be achieved is not yet known.

More recently, a function of the high light accumulated zeaxanthin in the nonradiative energy dissipation of excited chlorophyll states was proposed,[149,150] which would cause a decline in the variable chlorophyll fluorescence Fv as seen in a decrease of the fluorescence ratio Fv to Fm. In kinetic measurements of maple leaves and tobacco aurea leaves, we could, however, demonstrate that zeaxanthin accumulation and the decline in Fv and Fv to Fm under high photon flux densities are not linearly correlated.[9,10] Zeaxanthin formation at high photon fluxes proceeds quickly within 2 to 5 min with little or no change in the variable chlorophyll fluorescence, Fv, and the various fluorescence ratio (Fv to Fm, Fv to Fo, or in Rfd values), whereas a decrease in variable fluorescence occurs much later. The back reaction of zeaxanthin to violaxanthin in the dark is also a two-step procedure which could not be correlated with the regeneration of variable fluorescence. Furthermore, at a medium light intensity (250 to 500 $mmol/m^2/s$) below the photosynthetic light saturation point (650 $mmol/m^2/s$), we found a decrease in variable fluorescence upon continuous illumination, even though the initially formed pool of zeaxanthin had already been epoxidized back to violaxanthin. From these results, we concluded that zeaxanthin formation and the decrease in variable fluorescence are two high light-induced processes which proceed independently of each other and are not causally related. This is also assumed in a recent paper by Havaux et al.[151] who could demonstrate an increased energy dissipation at high light conditions even though zeaxanthin formation was blocked. We also found that under high light stress the back reaction from zeaxanthin to violaxanthin is increasingly retarded and needs many hours in darkness for regeneration. From this, it appears that under high light stress the xanthophyll cycle can only work in the forward reaction, whereas the epoxidase is blocked. Does there exist another reaction which readily produces violaxanthin from zeaxanthin at high light conditions?

In view of these results, a new hypothesis concerning the mechanism of zeaxanthin and the operation of the xanthophyll cycle was established.[152] It is well-known that under high photon flux densities above the light saturation point of photosynthetic CO_2 fixation not only photoinhibition occurs,[130,139,140] but that various reactive oxygen species are formed such as singlet oxygen ($^1O_2$), the superoxide radical anion (O_2^{*-}), the hydroxyl radical (OH^*), and hydrogen peroxide (H_2O_2).[120,123] These reactive oxygen species are formed under high light stress conditions with an overreduction of the photosynthetic electron transport chain and closed reaction centers and lead to a photooxidative degradation of pigments and lipids. Such oxygen species can be formed, e.g., from triplet chlorophyll (^3Chl) and reduced ferredoxin (Fd_{red}):

$$^3Chl + O_2 \rightarrow Chl + {}^1O_2 \quad \text{(singlet oxygen)}$$

$$^3Chl + O_2 \rightarrow Chl + O_2^{*-} \quad \text{(superoxide)}$$

$$Fd_{red} + O_2 \rightarrow Fdox + O_2^{*-} \quad \text{(superoxide)}$$

We propose the working hypothesis[152] that zeaxanthin either responds directly with such highly reactive oxygen species and becomes epoxidized to violaxanthin or may take over the epoxy structure from membrane acyl lipids which had been epoxidized by the reactive oxygen species (see Figure 16). The function of the xanthophyll cycle would then be to continuously restore the zeaxanthin — the acceptor substance for the dissipation of the reactive oxygen species — by the light-induced deepoxidation of violaxanthin. With such a physiological role of zeaxanthin and the xanthophyll cycle, one could explain the basic findings about the xanthophyll cycle carotenoids and, in particular, the photoprotective function of zeaxanthin and the necessity of a higher level of the xanthophyll cycle operation in sun than shade leaves.

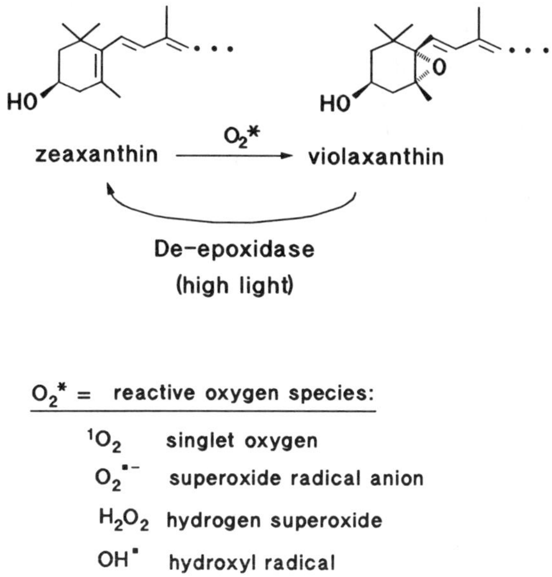

FIGURE 16. Postulated function of the xanthophyll cycle and of zeaxanthin in photochemically active thylakoids in the dissipation of highly reactive oxygen species $O_2^{\cdot}$.

VI. CONCLUSION

Many aspects of biosynthesis, localization, and functional role of plant prenyllipids are well-established. This particularly applies to the carotenoids (including chlorophylls) and the various plant prenylquinones. At first view, it appears that most scientific questions and problems associated with plant prenyllipids have been solved. On a second, more detailed view, one realizes that there are still many open questions, some of which have been pointed out in this review. Certain steps in the biosynthesis and, in particular, the regulation of the biosynthesis of the different prenyllipid classes and the possible exchange of common substrates between different pathways and compartments still require a detailed investigation. We know very little about the metabolism and the breakdown pathway of carotenoids and prenylquinones and their physiological control. Details of the physiological function, e.g., of phylloquinone K_1 and zeaxanthin, and the significance of the xanthophyll cycle are still under investigation. The function of α-tocoquinone is not known. The mode of action and fate of β-carotene in its photoprotective function is not yet understood. Though basic features are known in all fields of plant prenyllipids, there also exist many hypotheses and assumptions which have to be proven by well-designed biochemical and physiological experiments. Progress in science comes from the application of new methods. Plant molecular biology represents such a new method, which provides the tools to solve many of the intriguing existing questions and controversial observations.

The isolation of genes of the enzymes of the different prenyllipid pathways, studies on the regulation of their expression, the application of antibodies to particular enzymes for the study of their intracellular localization and compartmentation, the transfer of genes from resistant to sensitive plants, and the manipulation of cell metabolism through transformation by suitable vectors for over- and underexpression of genes will give new insight into the regulation and organization of the plants prenyllipid and cell metabolism. In this respect, the area of plant prenyllipids, carotenoids, and prenylquinones represents an exciting field with many challenges to students and young scientists who can here prove their scientific merits.

ACKNOWLEDGMENTS

Part of the work described here was supported by the BMFT Bonn within the EUREKA research project No. 380 LASFLEUR, which is gratefully acknowledged. Thanks are due to my Ph.D. students Andrea Golz, Michael Lang, Christiane Schindler, and Kai Vollack, as well as to Ms. Edith Höchst for excellent assistance during the preparation of the manuscript.

REFERENCES

1. **Lichtenthaler, H. K.**, Prenyl vitamins of vascular plants and algae, in *CRC Handbook of Biosolar Resources, Vol. 1, (Part 1), Basic Principles*, Mitsui, A. and Black, C. C., Eds., CRC Press, Boca Raton, FL, 1982, 451.
2. **Goodwin, T. W.**, The prenyllipids of the membranes of higher plants, in *Lipids and Lipid Polymers in Higher Plants*, Tevini, M. and Lichtenthaler, H. K., Eds., Springer-Verlag, Berlin, 1977, 29.
3. **Lichtenthaler, H. K.**, Regulation of prenylquinone synthesis in higher plants, in *Lipids and Lipid Polymers in Higher Plants*, Tevini, M. and Lichtenthaler, H. K., Eds., Springer-Verlag, Berlin, 1977, 231.
4. **Lichtenthaler, H. K.**, Synthesis of prenyllipids in vascular plants (including chlorophylls, carotenoids, prenylquinones), in *CRC Handbook of Biosolar Resources, Vol. 1, (Part 1), Basic Principles*, Mitsui, A. and Black, C. C., Eds., CRC Press, Boca Raton, FL, 1982, 405.
5. **Lichtenthaler, H. K.**, Occurrence and function of prenyllipids in the photosynthetic membrane, in *Advances in the Biochemistry and Physiology of Plant Lipids*, Appelqvist, L.-A. and Liljenberg, C., Eds., Elsevier/North-Holland Biomedical Press, Amsterdam, 1979, 57.
6. **Lichtenthaler, H. K.**, Chromatography of prenyllipids including chlorophylls, carotenoids, prenylquinones and fat-soluble vitamins, in *Handbook of Chromatography: Lipids*, Vol. 2, Mangold, H. K., Ed., CRC Press, Boca Raton, FL, 1984, 115.
7. **Lichtenthaler, H. K.**, Chlorophylls and carotenoids, the pigments of photosynthetic biomembranes, in *Methods in Enzymology*, Vol. 148, Douce, R. and Packer, L., Eds., Academic Press, New York, 1987, 350.
8. **Siefermann-Harms, D.**, High performance liquid chromatography of chloroplast pigments. One step separation of carotene and xanthophyll isomers, chlorophylls and phaeophytins, *J. Chromatogr.*, 20, 411, 1988.

9. **Schindler, C. and Lichtenthaler, H. K.**, Accumulation of zeaxanthin in aurea tobacco at high light conditions: a biphasic process, *Plant Physiol.*, 96(Suppl.) (Abstr. 854), 143, 1992.

10. **Lichtenthaler, H. K., Burkart, S., Schindler, C., and Stober, F.**, Changes in photosynthetic pigments and in vivo chlorophyll fluorescence parameters under photoinhibitory growth conditions, *Photosynthetica*, 27, 1992, in press.

11. **Lichtenthaler, H. K., Karunen, P., and Grumbach, K. H.**, Determination of prenylquinones in green photosynthetically active moss and liver moss tissues, *Physiol. Plant.*, 40, 105, 1977.

12. **Lichtenthaler, H. K.**, Light stimulated synthesis of plastid quinones and pigments in etiolated barley seedlings, *Biochim. Biophys. Acta*, 184, 164, 1969.

13. **Lichtenthaler, H. K. and Prenzel, U.**, High-performance liquid chromatography of natural prenylquinones, *J. Chromatogr.*, 135, 493, 1977.

14. **Prenzel, U. and Lichtenthaler, H. K.**, High performance liquid chromatography of prenylquinones, prenyl-vitamins and prenols, *J. Chromatogr.*, 242, 9, 1982.

15. **Hemming, F. W.**, Polyisoprenoid alcohols (prenols), in *Terpenoids in Plants*, Pridham, J. B., Ed., Academic Press, London, 1967, 223.

16. **Lichtenthaler, H. K., Börner, K., and Liljenberg, C.**, Separation of prenylquinones, prenyl-vitamins and prenols on thin-layer plates impregnated with silver nitrate, *J. Chromatogr.*, 242, 196, 1982.

17. **Schindler, S. and Lichtenthaler, H. K.**, Distribution of levels of ubiquinone homologs in higher plants, in *Biochemistry and Metabolism of Plant Lipids*, Wintermans, J. F. G. M. and Kuiper, P. J. C., Eds., Elsevier/North Holland Biomedical Press, Amsterdam, 1982, 545.

18. **Schindler, S., Lichtenthaler, H. K., Dizengremel, P., Rustin, P., and Lance, C.**, Distribution and significance of different ubiquinone homologues in purified mitochondria and in intact plant tissue, in *Structure, function and metabolism of plant lipids*, Siegenthaler, P.-A. and Eichenberger, W., Eds., Elsevier Science Publishers, Amsterdam, 1984, 267.

19. **Wellburn, A. R. and Hemming, F. W.**, The subcellular distribution and biosynthesis of castaprenols and plastoquinone in the leaves of *Aesculus hippocastanum*, *Biochem. J.*, 104, 173, 1967.

20. **Goad, J.**, The biosynthesis of plant sterols, in *Lipid and Lipid Polymers in Higher Plants*, Tevini, M. and Lichtenthaler, H. K., Eds., Springer-Verlag, Berlin, 1977, 147.

21. **Eichenberger, W.**, Steryl glycosides and acylated steryl glycosides in *Lipids and Lipid Polymers in Higher Plants*, Tevini, M. and Lichtenthaler, H. K., Eds., Springer-Verlag, Berlin, 1977, 169.

22. **Fässler, L., Lichtenthaler, H. K., Gruiz, K., and Biacs, P. A.**, Accumulation of saponins and sterols in *Avena* seedlings under high-light and low-light growth conditions, in *Structure, Function and Metabolism of Plant Lipids*, Siegenthaler, P.-A. and Eichenberger W., Eds., Elsevier Science Publishers, Amsterdam, 1984, 225.

23. **Wellburn, A. R., Gounaris, J., Fässler, L., and Lichtenthaler, H. K.**, Changes in plastid ultrastructure and fluctuations of cellular isoprenoid and carbohydrate compounds during continued etiolation of dark-grown oat seedlings, *Physiol. Plant*, 59, 347, 1983.

24. **Davies, B. H.**, Carotenoids in higher plants, in *Lipids and Lipid Polymers in Higher Plants*, Tevini, M. and Lichtenthaler, H. K., Eds., Springer-Verlag, Berlin, 1977, 199.

25. **Britton, G.**, Carotenoid biosynthesis in higher plants, *Physiol. Veg.*, 20, 735, 1982.

26. **Threlfall, D. R. and Whistance, G. R.**, Biosynthesis of isoprenoid quinones and chromanols, in *Aspects of Terpenoid Chemistry and Biochemistry*, Goodwin, T. W., Ed., Academic Press, London, 1971, 357.

27. **Pennock, J. F. and Threlfall, D. R.**, Biosynthesis of ubiquinone and related compounds, in *Biosynthesis of Isoprenoid Compounds*, Vol. II, Porter, J.W. and Spurgeon, S.L., Eds., John Wiley & Sons, New York, 1982, 191.

28. **Krügel, R., Grumbach, K. H., Lichtenthaler, H. K., and Retey, J.**, Biosynthesis of vitamin E and of the plastoquinones in chloroplasts, steric course of the decarboxylation, *Bioorg. Chem.*, 13, 187, 1985.

29. **Beale, S. I. and Castelfranco, P. A.**, The biosynthesis of δ-aminolevulinic acid in higher plants, *Plant Physiol.*, 53, 297, 1974.

30. **Redlinger, T. E. and Beale, S. I.**, Chlorophylls, in *Photoreceptor Evolution and Function*, Holmes, M. G., Ed., Academic Press, London, 1991, 151.

31. **Liljenberg, C.**, Chlorophyll formation: the phytylation step, in *Lipids and Lipid Polymers in Higher Plants*, Tevini, M. and Lichtenthaler, H. K., Eds., Springer-Verlag, Berlin, 1977, 259.

32. **Lichtenthaler, H. K. and Sundqvist, C.**, Association of plastoquinone-9 and phylloquinone K_1 with photoconvertible protochlorophyllide holochrome, *Physiol. Plant*, 45, 381, 1979.

33. **Rüdiger, W. and Schoch, S.**, Chlorophylls, in *Plant Pigments*, Goodwin, T. W., Ed., Academic Press, London, 1988.

34. **Shlyk, A. A.**, Biosynthesis of chlorophyll b, *Annu. Rev. Plant Physiol.*, 22, 169, 1971.

35. **Akoyunoglou, G.**, Chlorophyll a as a precursor of chlorophyll b synthesis in barley leaves, *Chem. Chron. A*, 32 5, 1967.

36. **Oelze-Karow, H., Kasemir, H., and Mohr, H.**, Control of chlorophyll b formation by phytochrome and a threshold level of chlorophyllide A, in *Chloroplast Development*, Akoyunoglou G. and Agyroudy-Akoyunoglou, J. H., Eds, Elsevier, Amsterdam, 1978, 787.

37. **Rüdiger, W.**, Chlorophyllsynthase and its implication for regulation of chlorophyll biosynthesis, in *Progress in Photosynthesis Research*, Vol. 4, Biggins, J., Ed., Martinus Nijhoff, Boston, 1987, 461.

38. **Rüdiger, W., Benz, J., Lempart, U., Schoch, S., and Steffens, D.**, Hemmung der Phytol-Akkumulation mit Herbiziden. Geranylgeraniol- und Dihydrogeranylgeraniol-haltiges Chlorophyll aus Weizenkeimlingen, *Z. Pflanzenphysiol.*, 80, 131, 1976.

39. **Schoch, S., Lampert, U., and Rüdiger, W.**, Über die letzten Stufen der Chlorophyll biosynthese. Zwischenprodukte zwischen Chlorophyllid und phytolhaltigem Chlorophyll, *Z. Pflanzenphysiol.*, 83, 427, 1977.

40. **Schoch, S.**, The esterification of chlorophyllide a in greening bean leaves, *Z. Naturforsch.*, 33c, 712, 1978.

41. **Rüdiger, W., Hedden, P., Köst, H.-P., and Chapman, D. J.**, Esterification of chlorophyllide by geranylgeranyl pyrophosphate in a cell free system from maize shoots, *Biochem. Biophys. Res. Commun.*, 74, 1268, 1977.

42. **Rüdiger, W., Benz, J., and Guthoff, C.**, Detection and partial characterization of activity of chlorophyll synthetase in etioplast membranes, *Eur. J. Biochem.*, 109, 193, 1980.

43. **Soll, J., Schultz, G., Rüdiger, W., and Benz, J.**, Hydrogenation of geranylgeraniol — pathways exist in spinach chloroplast, *Physiol. Plant.*, 71, 849, 1983.

44. **Soll, J. and Schultz, G.**, Phytol synthesis from geranylgeraniol in spinach chloroplasts, *Biochem. Biophys. Res. Commun.*, 99, 907, 1981.

45. **Benz, J., Lempert, U., and Rüdiger, W.**, Incorporation of phytol precursor into chlorophylls of tobacco cell cultures, *Planta*, 162, 215, 1984.

46. **Lichtenthaler, H. K.**, Functional orientation of carotenoids and prenylquinones in the photosynthetic membrane, in *The Metabolism, Structure and Function of Plant Lipids*, Stumpf, P. K., Mudd, J. B., and Nes, W. D., Eds., Plenum Press, New York, 1987, 63.

47. **Lichtenthaler, H. K. and Calvin, M.**, Quinone and pigment composition of chloroplasts and quantasome aggregates from *Spinacea oleracea*, *Biochim. Biophys. Acta*, 79, 30, 1964.

48. **Lichtenthaler, H. K. and Park, R. B.**, Chemical composition of chloroplast lamellae form spinach, *Nature*, 198, 1070, 1963.

49. **Lichtenthaler, H. K.**, Verbreitung und relative Konzentration der lipophilen Plastidenchinone in grünen Pflanzen, *Planta*, 81, 140, 1968.

50. **Lichtenthaler, H. K.**, Die Verbreitung der lipophilen Pastidenchinone in nicht-grünen Pflanzengeweben, *Z. Pflanzenphysiol.*, 59, 195, 1968.

51. **Threlfall, D. R. and Whistance, G. R.**, Dehydrophylloquinone, α-dehydrotoco-pheroquinone and dehydrotocopherols from etiolated maize and barley shoots, *Phytochemistry*, 16, 1903, 1977.

52. **Weber, T.**, Studies on the synthesis and metabolism of 3-hydroxy-3-methylglutaryl-coenzyme A in *Raphanus sativus, Karlsr. Contribut. Plant Physiol.*, 24, 1, 1992.
53. **Bach, T. J.**, Current trends in the development of sterol biosynthesis inhibitors: early aspects of the pathway, *J. Am. Oil Chem. Soc.*, 65, 591, 1988.
54. **Bach, T. J., Wettstein, A., Boronat, A., Ferrer, H., Enjuto, M., Gruissem, W., and Narita, J. O.**, Properties and molecular cloning of plant HMG-CoA reductase, in *Physiology and Biochemistry of Sterols*, Patterson, G. W. and Nes, W. D., Eds., American Oil Chemists Society, Champaign, IL, 1991, 29.
55. **Bach, T. and Lichtenthaler H. K.**, Mevinolin: a highly specific inhibitor of microsomal 3-hydroxy-3-methylglutaryl-coenzyme A reductase of radish plants, *Z. Naturforsch.*, 37c, 46–49, 1982.
56. **Bach, T. J. and Lichtenthaler, H. K.**, Inhibition by mevinolin of plant growth, sterol formation and pigment accumulation, *Physiol. Plant.*, 59, 50, 1983.
57. **Bach, T. J. and Lichtenthaler, H. K.**, Inhibition of mevalonate biosynthesis and of plant growth by fungal metabolite mevinolin, in *Biochemistry and Metabolism of Plant Lipids*, Wintermans, J. F. G. M. and Kuiper, P. J. C., Eds., Elsevier/North Holland Biomedical Press, Amsterdam, 1982, 515.
58. **Bach, T. J. and Lichtenthaler, H. K.**, Plant growth regulation by mevinolin and other sterol biosynthesis inhibitors, in *Ecology and Metabolism of Plant Lipids*, Fuller, G. and Nes, W.D., Eds., ACS Symposium Series No. 325, American Chemical Society, Washington, DC, 1987, 109.
59. **Döll, M., Schindler, S., Lichtenthaler, H. K., and Bach, T. J.**, Differential inhibition by mevinolin of prenyllipid accumulation in cell suspension cultures of *Sibybum marianum* L., in *Structure, Function and Metabolism of Plant Lipids*, Siegenthaler, P.-A. and Eichenberger, W., Eds., Elsevier Science Publishers, Amsterdam, 1984, 273.
60. **Schindler, S., Bach, T .J., and Lichtenthaler, H. K.**, Differential inhibition by mevinolin of prenyllipid accumulation in radish seedlings, *Z. Naturforsch.*, 40c, 208, 1985.
61. **Lichtenthaler, H. K., Prenzel, U., Douce, R., and Joyard, J.**, Localisation of prenylquinones in the envelope of spinach chloroplasts, *Biochim. Biophys. Acta*, 641, 99, 1981.
62. **Lichtenthaler, H. K.**, Localisation and functional concentrations of lipoquinones in chloroplasts, in *Progress in Photosynthesis Research*, Vol. 1, Metzner, H., Ed., H. Laupp, Tübingen, 1969, 304.
63. **Lichtenthaler, H. K.**, Plastoglobuli and the fine structure of plastids, *Endeavour*, 27, 144, 1968.
64. **Lichtenthaler, H. K. and Sprey, B.**, Über die osmiophilen globulären Lipideinschlüsse der Chloroplasten, *Z. Naturforsch.* 21b, 690. 1966.
65. **Lichtenthaler, H. K. and Weinert, H.**, Die Beziehungen zwischen Lipochinonsynthese und Plastoglobulibildung in den Chloroplasten von *Ficus elastica*, *Z. Naturforsch.*, 25b, 619, 1970.
66. **Lichtenthaler, H. K.**, Die Plastoglobuli von Spinat, ihre Größe und Zusammensetzung während der Chloroplastendegeneration, *Protoplasma*, 68, 315, 1969.
67. **Lichtenthaler, H. K.**, Lokalisation der Plastidenchinone und Carotinoide in den Chromoplasten der Petalen von *Sarothamnus scoparius*, *Planta*, 90, 142, 1970.
68. **Tevini, M. and Steinmüller, D.**, Composition and function of plastoglobuli. II. Lipid composition of leaves and plastoglobuli during beech leaf senescence, *Planta*, 163, 91, 1985.
69. **Block, M. A., Joyard, J., and Douce, R.**, Site of synthesis geranyl geraniol derivatives in intact spinach chloroplasts, *Biochim. Biophys. Acta*, 631, 210, 1980.
70. **Kreuz, K. and Kleinig, H.**, Prenyllipid synthesis in spinach leaf cells. Compartmentation of enzymes of isopentenyl pyrophosphate formation, *Eur. J. Biochem.*, 141, 531, 1984.
71. **Camara, B.**, Carotene synthesis in *Capsicum* chromoplasts, *Methods Enzymol.*, 110, 244, 1985.

72. **Reddy, A. R. and Das, V. S. R.**, Chloroplast autonomy for the biosynthesis of isopentyl diphosphate in guayule C *Parthenium argentatum* Gray I., *New Phytol.*, 106, 457, 1987.

73. **Schulze-Siebert, D. and Schultz, G.**, Full autonomy in isoprenoid synthesis in spinach chloroplasts, *Plant Physiol. Biochem.*, 25, 145, 1987.

74. **Henitze, A., Görlach, J., Leuschner, C., Hoppe, P., Hagelsctein, P., Schulze-Siebert, D., and Schultz, G.**, Plastidic isoprenoid synthesis during chloroplast development. Change from metabolic autonomy to division of labor stage, *Plant Physiol.*, 93, 1121, 1990.

75. **Camara, B., Bousquet, J., Cheniclet, C., Carde, J. P., Kuntz, M., Evrad, J.-L., and Weil, J.-H.**, Enzymology of isoprenoid biosynthesis and expression of plastid and nuclear genes during chromoplast differentiation in pepper fruits (*Capsicum annuum*), in *Physiology and Biochemistry of Nongreen Plastids,* Boyer, C. D., Shannon, J. C., and Hardison, R. C., Eds., The American Society of Plant Physiologists, Rockville, MD, 1989, 141.

76. **Soll, J. and Schultz, G.**, 2-Methyl-6-phytylquinol and 2,3-dimethyl-5-phytylquinol as precursors of tocopherol synthesis in spinach chloroplasts, *Phytochemistry*, 19, 215, 1980.

77. **Soll, J., Douce, R., and Schultz, G.**, Site of biosynthesis of α–tocopherol in spinach chloroplasts, *FEBS Lett.*, 112, 243, 1980.

78. **Schultz, G., Soll, J., Fiedler, E., and Schulze-Siebert, D.**, Synthesis of prenylquinones in chloroplasts, *Physiol. Plant.*, 64, 123, 1985.

79. **Soll, J., Kemmerling, M., and Schultz, G.**, Tocopherol and plastoquinone synthesis in spinach chloroplast subfractions, *Arch. Biochem. Biophys.*, 204, 544, 1980.

80. **Soll, J., Schultz, G., Joyard, J., and Block, M. A.**, Localization and synthesis of prenylquinones in isolated outer and inner envelope membranes from spinach chloroplast, *Arch. Biochem. Biophys.*, 238, 290, 1985.

81. **Schultz, G., Ellerbrock, B. H., and Soll, J.**, Site of prenylation reaction in synthesis of phylloquinone (vitamin K_1), *Eur. J. Biochem.*, 117, 329, 1981.

82. **Kaiping, S., Soll, J., and Schultz, G.**, Site of methylation of 2-phytyl-1,4-naphtoquinol in phylloquinone (vitamin K_1) synthesis in spinach chloroplasts, *Phytochemistry*, 23, 89, 1984.

83. **Schulze-Siebert, D. and Schultz, G.**, β–carotene synthesis in isolated spinach chloroplasts. Its tight linkage to photosynthetic carbon metabolism, *Plant Physiol.*, 84, 1233, 1987.

84. **Beyer, P.**, Carotene biosynthesis in daffodil chromoplasts: on the membrane-integral desaturation and cyclization reactions, in *Physiology and Biochemistry Genetics of Nongreen Plastids*, Boyer, C. D., Shannon, J. C., and Hardison, R. C., Eds., The American Society of Plant Physiologists, Rockville, MD, 1989, 157.

85. **Bramley, P. M.**, The *in vitro* biosynthesis of carotenoids, *Adv. Lipid Res.*, 21, 243, 1985.

86. **Sandmann, G. and Böger, P.**, Inhibition of carotenoid biosynthesis by herbicides, in *Target Sites of Herbicide Action*, Böger, P. and Sandmann, G., Eds., CRC Press, Boca Raton, FL, 1989, 25.

87. **Goodwin, T. W.**, *The Biochemistry of Carotenoids*, Vol. 1, Chapmann and Hall, London, 1980, chap. 2.

88. **Sandmann, G. and Bramley, P. M.**, The *in vitro* biosynthesis of β-cryptoxanthin and related xanthophylls with *Aphanaocapsa* membranes, *Biochim. Biophys. Acta*, 843, 73, 1985.

89. **Bartels, P. G. and McCullough, C.**, A new inhibitor of carotenoid synthesis in higher plants: 4-chloro-5(dimethylamino)-2-α,α,α(trifluoro-*m*-tolyl)3(2H)-pyridazinone(Sandoz 6706), *Biochem. Biophys. Res. Commun.*, 48, 16, 1972.

90. **Lichtenthaler, H. K. and Kleudgen, H. K.**, Effect of the herbicide San 6706 on bio-synthesis of photosynthetic pigments and prenylquinones in *Raphanus* and in *Hordeum* seedlings, *Z. Naturforsch.*, 32c, 236, 1977.

91. **Lichtenthaler, H. K. and Pfister, K.**, *Praktikum der Photosynthese*, Quelle & Meyer, Heidelberg, Germany, 1978.

92. **Carattoli, A., Romano, N., Ballario, P., Morelli, G., and Macino, G.**, The *Neurospora crassa* carotenoid biosynthesis gene (albino 3) reveals highly conserved regions among prenyltransferases, *J. Biol. Chem.*, 266, 5854, 1991.

93. **Kuntz, M., Romer, S., Suire, C., Hugueney, P., Weil, J. H., Schantz, R., and Camara, B.**, Identification of a cDNA for the plastid-located geranygeranyl pyrophosphate synthase from *Capsicum annuum* — correlative increase in enzyme-activity and transcript level during fruit ripening, *Plant J.*, 2, 25, 1992.

94. **Armstrong, G. A., Alberti, M., Leach, F., and Hearst, J. E.**, Nucleotide sequence, organization, and nature of the protein products of the carotenoid biosynthesis gene cluster of *Rhodobacter capsulatus*, *Mol. Gen. Genet.*, 216, 254, 1989.

95. **Misawa, N., Nakagawa, M., Kobayashi, K., Yamano, S., Izawa, Y., Nakamura, K., and Harashima, K.**, Elucidation of *Erwinia uredovora* carotenoid biosynthetic pathway by functional analysis of gene products expressed in *Escherichia coli*, *J. Bacteriol.*, 172, 6704, 1990.

96. **Armstrong, G. A., Alberti, M., and Hearst, J. E.**, Conserved enzymes mediate the early reactions of carotenoid biosynthesis in nonphotosynthetic and photosynthetic prokaryotes, *Proc. Natl. Acad. Sci. U.S.A.*, 87, 9975, 1990.

97. **Schnurr, G., Schmidt, A., and Sandmann, G.**, Mapping of a carotenogenic gene cluster from *Erwinia herbicola* and functional identification of six genes, *FEMS Microbiol. Lett.*, 78, 157, 1991.

98. **Michalowski, C. B., Loffelhardt, W., and Bohnert, H. J.**, An ORF 323 with homology to crtE, specifying prephytoene pyrophosphate dehydrogenase, is encoded by cyanelle DNA in the eukaryotic alga *Cyanophora paradoxa*, *J. Biol. Chem.*, 266, 11866, 1991.

99. **Chamovitz, D., Misawa, N., Sandmann, G., and Hirschberg, J.**, Molecular cloning and expression in *Escherichia coli* of a cyanobacterial gene coding for phytoene synthase, a carotenoid biosynthesis enzyme, *FEBS Lett.*, 296, 305, 1992.

100. **Bartley, G., Viitanen, P. V., Bacot, K. O., and Scolnik, P. A.**, A tomato gene expressed during fruit ripening encodes an enzyme of the carotenoid biosynthesis pathway, *J. Biol. Chem.*, 267, 5036, 1992.

101. **Bartley, G. E. and Scolnik, P. A.**, Carotenoid biosynthesis in photosynthetic bacteria, *J. Biol. Chem.*, 264, 13109, 1989.

102. **Schmidt, A. and Sandmann, G.**, Cloning and nucleotide sequence of the crtI gene encoding phytoene dehydrogenase from the cyanobacterium *Aphanocapsa* PCC6714, *Gene*, 91, 113, 1990.

103. **Chamovitz, D., Pecker, I., and Hirschberg, J.**, The molecular-basis of resistance to the herbicide norflurazon, *Plant Mol. Biol.*, 16, 967, 1991.

104. **Schmidhauser, T. J., Lauter, F. R., Russo, V. E. A., and Yanofsky C.**, Cloning, sequence, and photoregulation of al-1, a carotenoid biosynthetic gene of *Neurospora crassa*, *Mol. Cell., Biol.* 10, 5064, 1990.

105. **Bartley, G. E., Viitanen, P. V., Pecker, I., Chamovitz, D., Hirschberg, J., and Scolnik, P. A.**, Molecular cloning and expression in photosynthetic bacteria of a soybean cDNA coding for phytoene desaturase, an enzyme of the carotenoid biosynthesis pathway, *Proc. Natl. Acad. Sci. U.S.A.*, 88, 6532, 1991.

106. **Sandmann, G. and Misawa, N.**, New functional assignment of the carotenogenic genes crtB and crtE with constructs of these genes from *Erwinia* species, *FEMS Microbiol. Lett.*, 90, 253, 1992.

107. **Willstätter, R. and Stoll, A.**, *Untersuchungen über Chlorophyll*, Springer-Verlag, Berlin, 1913.

108. **Strain, H. H., Cope, B. T., and Svec, W. A.**, Analytical procedures for the isolation, identification, estimation and investigation of chlorophyll b, *Methods Enzymol.*, 23, 452, 1971.

109. **Watanabe, T., Kobayashi, M., Hongu, A., Nakazoto, M., Hiyama, T., and Murata, N.**, Evidence that a chlorophyll a dimer constitutes the photochemical reaction centre I (P700) in photosynthetic apparatus, *FEBS Lett.*, 191, 252, 1985.

110. **Thornber, J. P.**, Chlorophyll proteins: light-harvesting and reaction center components of plants, *Annu. Rev. Plant Physiol.*, 26, 127, 1975.

111. **Lichtenthaler, H. K., Prenzel, U., and Kuhn, G.**, Carotenoid composition of chlorophyll-carotenoid-proteins from radish chloroplasts, *Z. Naturforsch.*, 37c, 10, 1982.
112. **Lichtenthaler, H. K., Burkard, G., Kuhn, G., and Prenzel, U.**, Light induced accumulation and stability of chlorophylls and carotenoid proteins during chloroplast development in radish seedlings, *Z. Naturforsch.*, 36c, 421, 1981.
113. **Siefermann-Harms, D.**, Carotenoids in photosynthesis. I. Location in photosynthetic membranes and light-harvesting function, *Biochim. Biophys. Acta*, 811, 325, 1985.
114. **Lichtenthaler, H. K., Buschmann, C., Döll, M., Fietz, H.-J., Bach, T., Kozel, U., Meier, D., and Rahmsdorf, U.**, Photosynthetic activity, chloroplast ultrastructure, and leaf characteristics of high-light and low-light plants and of sun and shade leaves, *Photosynth. Res.*, 2, 115, 1981.
115. **Lichtenthaler, H. K., Kuhn, G., Prenzel, U., Buschmann, C., and Meier D.**, Adaptation of chloroplast-ultrastructures and of chlorophyll-protein levels to high-light and low-light growth conditions, *Z. Naturforsch.*, 37c, 464, 1982.
116. **Lichtenthaler, H. K., Kuhn, G., Prenzel, U., and Meier, D.**, Chlorophyll-protein levels and stacking degree of thylakoids in radish chloroplasts from high-light, low-light and bentazon-treated plants, *Physiol. Plant.*, 56, 183, 1982.
117. **Lichtenthaler, H.K. and Buschmann, C.**, Photooxidative changes in pigment composition and photosynthetic activity of air-polluted spruce needles (*Picea abies* L.) in *Advances in Photosynthesis Research*, Vol 4., Sybesma, C., Ed., Martinus Nijhoff/Dr. W. Junk Publisher, The Hague, 1984, 245.
118. **Lichtenthaler, H. K., Burgstahler, R., Buschmann, C., Meier, D., Prenzel, U., and Schöntal, A.**, Effect of high-light and high light stress on composition function and structure of the photosynthetic apparatus, in *Effects of Stress on Photosynthesis*, Marcelle, R., Ed., Dr. W. Junk Publisher, The Hague, 1982, 353.
119. **Siefermann-Harms, D.**, The light-harvesting and protective functions of carotenoids in photosynthetic membranes, *Physiol. Plant.*, 69, 561, 1987.
120. **Asada, K. and Takahashi, M.**, Production and scavenging of active oxygen in photosynthesis, in *Photoinhibition*, Kyle, D. J., Osmond, C. B., and Arntzen, C. J., Eds., Elsevier Science Publishers, Amsterdam, 1987, 227.
121. **Goodwin, T. W.**, Carotenoids, in *Photoreceptor Evolution and Function*, Holmes, M. G., Ed., Academic Press, London, 1991, 125.
122. **Mathis, P. and Schenk, C. C.**, The functions of carotenoids in photosynthesis, in *Carotenoid Chemistry and Biochemistry*, Britton, G. and Goodwin, T. W., Eds., Pregamon Press, Oxford, 1982, 339.
123. **Elstner, E. F.**, *Der Sauerstoff (Biochemie, Biologie, Medizin)*, BI-Wissenschaftsverlag, Mannheim, 1990.
124. **Grumbach, K. H. and Lichtenthaler, H. K.**, Incorporation of $^{14}CO_2$ in prenylquinones of *Chlorella pyrenoidosa*, *Planta*, 141, 253, 1978.
125. **Grumbach, K. H., Lichtenthaler, H. K., and Erismann, K. H.**, Incorporation of $^{14}CO_2$ in photosynthetic pigments of *Chlorella pyrenoidosa*, *Planta*, 140, 37, 1978.
126. **Tevini, M. and Lichtenthaler, H. K.**, Untersuchungen über die Pigment- und Lipochinonausstattung der zwei photosynthetischen Pigmentsysteme, *Z. Pflanzenphysiol.*, 62, 17, 1970.
127. **Mets, L. and Thiel, A.**, Biochemistry and genetic control of the photosystem II herbicide target site, in *Target Sites of Herbicide Action*, Böger, P. and Sandmann, G., Eds., CRC Press, Boca Raton, FL, 1989, 1.
128. **Lichtenthaler, H. K.**, Prenylquinones in plant leaves, in *Biogenesis and Function of Plant Lipids*, Mazliak, P., Benveniste, P., Costes, C., and Douce, R., Eds., Elsevier/North Holland, Amsterdam, 1980, 299.
129. **Kunert, K. J. and Dodge, A. D.**, Herbicide induced radical damage and antioxidative systems, in *Target Sites of Herbicide Action*, Böger, P. and Sandmann, G., Eds., CRC Press, Boca Raton, FL, 1989, 45.

130. **Voss, J., Styring, S., Hundal, T., Koivuniemi, A., Aro, E.-M., and Anderson, B.,** Reversible and irreversible intermediates during photoinhibition of photosystem II: stable reduced Q_A species promote chlorophyll triplet formation, *Proc. Natl. Acad. Sci. U.S.A.,* 89, 1408, 1992.

131. **Lucy, J. A.,** in *Tocopherol, Oxygen and Biomembranes,* de Duve, C. and Hayaishi, H., Eds., Elsevier, Amsterdam, 1978, 109.

132. **Huang, C.,** Configuration of fatty acyl chains in egg phosphatidylcholine-cholesterol mixed bilayers, *Chem. Phys. Lipids,* 19, 150, 1977.

133. **Interschick-Niebler, E. and Lichtenthaler, H. K.,** Partition of phylloquinone K_1 between digitonin particles and chlorophyll-proteins of chloroplast membranes from *Nicotiana tabacum, Z. Naturforsch.,* 36c, 276, 1981.

134. **Malkin, R.,** On the function of two vitamin K_1 molecules in the PSI electron acceptor complex, *FEBS Lett.,* 208, 343, 1986.

135. **Schoeder, H.-U. and Lockau, W.,** Phylloquinone copurifies with the large subunit of photosystem I, *FEBS Lett.,* 199, 23, 1986.

136. **Golbeck, J. H.,** Structure and function of photosystem I, *Annu. Rev. Plant Physiol. Plant Mol. Biol.,* 43, 293, 1992.

137. **Arnon, D. I.,** Role of vitamin K and other quinones in photosynthesis, *Fed. Proc.,* 20, 1012, 1961.

138. **Pfister, K., Lichtenthaler, H. K., Burger G., Musso, H., and Zahn, M.,** The inhibition of photosynthetic light reactions by halogenated naphthoquinones, *Z. Naturforsch.,* 36c, 645, 1981.

139. **Krause, G. H. and Weiss, E.,** Chlorophyll fluorescence and photosynthesis: the basics, *Annu. Rev. Plant Physiol. Plant Mol. Biol.,* 42, 313, 1991.

140. **Lichtenthaler, H. K. and Rinderle, U.,** The role of chlorophyll fluorescence in the detection of stress conditions in plants, *CRC Crit. Rev. Analyt. Chem.,* 19(Suppl. I), S29, 1988.

141. **Crane, F. L. and Morré, D. L.,** Evidence for coenzyme Q function in Golgi membranes, in *Biomedical and Clinical Aspects of Coenzyme Q,* Vol. 1, Folkers, K. and Yamamura, Y., Eds., Elsevier/North Holland Biomedical Press, Amsterdam, 1973, 3.

142. **Day, D.A., Arron, G.P., and Laties, G.G.,** Nature and control of respiratory pathways in plants: the interaction of cyanide-resistant respiration with the cyanide-sensitive pathway, in *The Biochemistry of Plants,* Vol. 2, Davies, D. D., Ed., Academic Press, New York, 1980, 197.

143. **Sapozhnikov, D. I.,** Transformation of xanthophylls in chloroplasts, in *Progress in Photosynthesis Research,* Vol. 2, Metzner, H., Ed., H. Laupp, Tübingen, 1968, 694.

144. **Yamamoto, H. Y.,** Biochemistry of violaxanthin cycle in higher plants, *Pure Appl. Chem.,* 51, 639, 1979.

145. **Hager, A.,** Die reversiblen, lichtabhängingen Xanthophyllumwandlungen in Chloroplasten, *Ber. Dtsch. Bot. Ges.,* 88, 27, 1975.

146. **Hager, A.,** The reversible, light-induced conversions of xanthophylls in the chloroplast, in *Pigments in Plants,* Czygan, F.-C., Ed., G. Fischer-Verlag, Stuttgart, 1980, 57.

147. **Siefermann-Harms, D.,** The xanthophyll cycle in higher plants, in *Lipids and Lipid Polymers in Higher Plants,* Tevini, M. and Lichtenthaler, H. K., Eds., Springer Verlag, Berlin, 1977, 218.

148. **Demmig-Adams, B. and Adams W. W., III,** Photoprotection and other responses of plants to high light stress, *Annu. Rev. Plant Physiol. Plant Mol. Biol.,* 43, 599, 1992.

149. **Demmig, B., Winter, K., Krüger, A. and Czygan, F.-C.,** Photoinhibition and zeaxanthin formation in intact leaves. A possible role of the xanthophyll cycle in the dissipation of excess light energy, *Plant Physiol.,* 84, 218, 1987.

150. **Demming-Adams, B., Winter, K., Winkelmann, E., Krüger, A., and Czygan, F.-C.,** Photosynthetic characteristics and the ratios of chlorophyll, β-carotene, and the components of the xanthophyll cycle upon a sudden increase in growth light regime in several plant species, *Bot. Acta,* 102, 261, 1989.

151. **Havaux, M., Gruszecki, W. I., Dupont, I., and Leblanc, R. M.**, Increased heat emission and its relationship to the xanthophyll cycle in pea leaves exposed to strong light stress, *J. Photochem. Photobiol. B: Biol.*, 8, 361, 1991.

152. **Lichtenthaler, H. K. and Schindler, C.**, Studies on the photoprotective function of zeaxanthin at high-light conditions, in *Research in Photosynthesis*, Murata, N., Ed., Kluwer Academic Publishers, Dordrecht, 1992, 517.

SECTION IV

ACYL LIPID CATABOLISM

The preceding chapters have been devoted to studies on lipid biosynthesis, but lipids do turn over or are mobilized for their stored carbon and energy. Research in this area has been uneven, with the most attention being devoted to breakdown and mobilization of stored lipids. Much research has been performed on fatty acid degradation and the association of the major pathway with the glyoxysomes. This is described in Chapter 16. The reactions which remove the acyl units from the backbone, the lipases, are described in Chapter 14.

Research on the phospholipases was very active several years ago, but has received only limited attention recently. Chapter 15 reports on some of the recent data, but particularly the renewed interest in these enzymes as potential mediators in signal transduction.

Chapter 14

LIPASES

Anthony H. C. Huang

TABLE OF CONTENTS

0-8493-4907-9/93/$0.00+$.50
© 1993 by CRC Press Inc.

I. INTRODUCTION

A true lipase catalyzes the hydrolysis of triacylglycerols (TAG). In plants, the enzyme is present in organs in which stored TAG are being mobilized. These organs include seeds, pollens, and some roots and leaves of special species. Only the lipase from seeds, which can be obtained in abundance, has been studied. The fruit mesocarp of some species, such as oil palm, avocado, and olive, also contain high amounts of TAG. However, fruit mesocarp TAG are not for mobilization by the plants, and it is uncertain if internal lipases are present during any stage of fruit development, including ripening. In seeds, lipase controls the first step in TAG mobilization, and thus it may control the rate of germination.

A group of related lipid hydrolyzing enzymes, termed fatty acyl hydrolase, are present in high activities in many plant organs. The enzymes catalyze the hydrolysis of the acyl groups from most acyl lipids except TAG. They have been studied to some extent because of their extreme high activities and because they may be involved in membrane turnover and the production of rancidity in agricultural products.

Although plant lipases and fatty acyl hydrolases have been studied for many years, our current knowledge of them is still quite limited. The nomenclature for plant lipolytic enzymes, especially with reference to substrate specificity, is a common source of confusion. This confusion, together with technical difficulties in the assay of lipase activities, has discouraged researchers. However, many of the problems have now been recognized and overcome.

Little progress on plant lipase and acyl hydrolase studies has been made in the past several years since the last major review on the subject was published in 1987.[1] New, as well as the established, lipid laboratories focus on or switch to studying lipid synthesis rather than degradation, since the knowledge can be directly applied, via genetic engineering or breeding, to the modification of seed oil for different commercial purposes, as well as to the production of plant varieties that are resistant to extreme temperatures. In direct contrast, tremendous progress has been made in the past several years on the studies of mammalian and microbial lipases due to the importance of the enzymes in health and industrial applications. The complete amino acid sequences of more than ten of these lipases, as deduced from their encoded DNA, have been reported.[2,3] Several of them have been crystallized and their three-dimensional structures visualized. The active site has been revealed to be composed of a triad of serine, histidine, and aspartate.[4,5] The mechanism of several human genetic defects of lipases have been pinpointed.[6] In the industrial utilization of lipases,[7] the microbial enzymes are employed as detergent additives. They are also used for the production of fatty acids and glycerols via mild hydrolysis, the randomization of the positional specificity of TAG in order to generate special oils, and many other processes.

This chapter is an up-date of the latest 1987 review article. In this chapter, two topics will be described. The first topic deals with true lipases (E.C. 3.1.1.3) in oilseeds. The other topic deals with a group of apparently similar enzymes occurring in diverse tissues that possess the combined activities of phospholipases A_1 (E.C. 3.1.1.32), A_2 (E.C. 3.1.1.4), B (E.C. 3.1.1.5), glycolipase, sulfolipase, and monoacylglycerol lipase. Reviews on plant lipases include those on cereal lipases,[8] plant lipases,[1,9,10] and oil bodies with details on lipase assays.[11,12] A multiauthored book on lipolytic enzymes is also available.[13]

II. LIPASES (E.C. 3.1.1.3)

A. GLUCONEOGENESIS IN OILSEEDS

True lipases, which attack the fatty acyl linkage of water-insoluble TAG, are known to occur in oilseeds and cereals. Oilseeds generally contain 20 to 50% of their dry weight as storage TAG. In postgerminative growth, the oil reserve is rapidly mobilized to provide energy and carbon skeletons for the growth of the embryo.[14] Highly active lipases are present to catalyze the hydrolysis of reserve TAG. The TAG are localized in subcellular organelles called oil bodies (oleosomes, lipid bodies, spherosomes) which are bounded by a half-unit membrane.[15,16] Depending on the plant species, the lipase may be localized on the membrane of oil bodies or in other subcellular compartments. Lipase activity is absent in ungerminated seeds and increases rapidly in postgermination.[10] The fatty acids released by lipase activity are further metabolized in the glyoxysomes.[14] The oil bodies and glyoxysomes *in vivo* are in close proximity or direct physical contact with one another;[17] presumably, this proximity could facilitate transport of fatty acids from the oil bodies to the glyoxysomes.

Naturally, the studies of lipases in plants center on those in oilseeds, especially in postgermination when lipolysis is active. In fact, high activities of true lipases in tissues or organs other than seeds have not been well-documented.

B. TECHNICAL DIFFICULTIES IN THE ASSAY OF LIPASE ACTIVITIES

In the study of lipases, technical difficulties in the assay of enzyme activities need to be overcome. Some of the difficulties are not unusual and are common among enzyme assays dealing with water-insoluble substrates. In the assay of lipase activity, the substrate emulsions are generally prepared by sonicating TAG in the presence of an emulsifying agent such as Gum Acacia (gum arabic), which is a carbohydrate, or a detergent such as Triton X-100. The fatty acid released is quantitated by a pH stat continuously or at time intervals by colorimetry or radioactive analysis. The assay is time consuming. The substrate emulsion is unstable, and the sizes of the emulsion are nonuniform and are not totally reproducible from experiment to experiment. Therefore, some researchers use artificial TAG containing shorter acyl moieties, so that the emulsion is much more stable and uniform. Some other researchers measure the esterase activity of the lipase by using an artificial substrate which upon enzymatic hydrolysis generates a fluorescent product. As will be explained, these shortcuts actually generate more problems than those they are intended to solve. An account of the technical problems in measuring lipase activities is itemized as follows, and some of them are especially critical when dealing with plant tissues:

1. As will be described in Section II.C, a lipase from a certain plant species is likely to be relatively specific on the native storage TAG or artificial TAG containing the major acyl components of the storage TAG in that same species. Thus, when assaying seed lipase activity, a suitable TAG should be used. For example, based on information described in Section II.C, it is inappropriate to use tristearin or tripalmitin to study maize lipase or to use triolein to study elm lipase.

2. Highly active nonspecific fatty acyl hydrolases are present in many plant organs (Section III). These hydrolases are active on various acyl lipids other than TAG. Under ideal assay conditions, their activities are several orders of magnitude higher than the lipase activity in the extract from the same organs. Thus, using an artificial ester substrate (generally, monoester of a fatty acid and a fluorescent moiety) to study seed lipases generates uncertain results. The problem may be relatively less important in the study of lipases in other organisms, but is very critical in plant organs due to the presence of the ubiquitous and highly active acyl hydrolases.

3. The problem described in item 2 is compounded by the general unawareness of the purity of commercial TAG preparations used as substrates. Impurity of a small amount of monoacylglycerols or other acyl lipids would lead to the assay of the dominating acyl hydrolases mentioned above. It has been reported that commercially available tributyrin preparation contains monobutyrin as a contaminant,[18] and trilinolein preparation, labeled as 99% pure, contains a small amount of monoacylglycerols (from one company) or fatty acids (from another company).[19] A simple preparative thin layer chromatography has been described to remove fatty acid, di-, and monoacylglycerols from TAG preparations.[11] Alternatively, a one-step silicic acid column chromatography using hexane-diethyl ether as the eluting solvent also works well.[20]

4. The optimal pH for lipase activity on the native TAG may be quite different from that on an artificial substrate. For example, rapeseed lipase has an optimal activity at pH 8.5 on *N*-methylindoxylmyristate, but at pH 6.5 on internal lipid (autolipolysis of the oil bodies), trilinolein, or trierucin.[19] It is inappropriate to assay the optimal pH for activity using an artificial substrate (or impure substrate) and then use this pH to study substrate specificity and other enzyme properties.

5. Lipase inhibitors are present in the seeds of some plant species before and after germination.[21-24] These inhibitors are not the classical enzyme inhibitors which bind to or act on the enzyme molecule. Instead, they are proteins that bind to the surface of the substrate emulsions in an *in vitro* enzyme assay. The binding prevents the normal functioning of the

lipase which acts on the interfacial area between the aqueous medium and the emulsion surface. If the seed extract does not contain an overwhelming amount of the protein inhibitors, as is the case of peanut cotyledons, lipase activity can be detected and measured by simply adding more substrate emulsions to the assay system.[24] However, in other seed extracts, such as those of soybean cotyledons, the amount of protein inhibitors is in great excess, and a simple increase in the substrate emulsions is not sufficient to overcome the inhibition. Methods should be designed to remove these proteins prior to assay of lipase activities. The methods of removal could take advantage of the fact that these protein inhibitors are of amphipathic nature.

In an overall assessment, items 1 to 4 are minor technical difficulties that can be overcome easily. The substrate TAG should be the native TAG extracted from the ungerminated seed or artificial TAG containing the major storage acyl moieties. The release of fatty acid is monitored by a pH stat continuously or at time intervals by colorimetry or radioactive assay (if the TAG is available in radioactive form). The TAG obtained commercially, irrespective of the purity claimed by the chemical company, should be checked for purity and, if necessary, purified by thin layer or column chromatography. Artificial substrates such as those used in fluorescence measurement could be employed in routine enzyme assays such as in enzyme purification, but should be used only after the lipase has been fairly well-studied and separated from the nonspecific acyl hydrolases. Overcoming the lipase inhibitors described in item 5 is more difficult, and procedures to remove them will have to be designed and are likely to be species specific. So far, no such procedure has been published.

C. SUBSTRATE SPECIFICITY

A distinct feature of plant lipases is their substrate specificity. Seed lipase from a certain plant species is relatively specific for the native TAG or TAG containing the major acyl moieties of the storage TAG of that same species (Table 1). For example, maize lipase is most active on TAG containing linoleoyl and oleoyl moieties, which are the major acyl constituents of maize oil.[25] Rapeseed and mustard seed (storage TAG possessing substantial amount of erucoyl moiety) lipases are more active on TAG containing crucoyl than stearoyl or behenoyl moiety.[19] The elm seed (containing a high percent of caproyl moiety in TAG) lipase[26] and palm kernel (containing a high percent of lauroyl moiety in TAG) lipase[27] are more active on tricaprin than on trilaurin, tripalmitin, or trilinolein. The above substrate specificity has been observed in direct comparisons of different seed lipases using individual TAG as well as mixed TAG preparations. The pattern of fatty acyl specificity can also be observed on diacylglycerols, monoacylglycerols, and fatty acyl 4-methylumbelliferone, although the pattern becomes less distinct.[26] This gradual loss in the acyl specificity of the seed lipases from tri- to di- to monoacylglycerols may be of physiological significance. Each storage TAG molecule generally

TABLE 1
Hydrolysis of Various Triacylglycerols by Lipases from Various Sources[a]

Relative activity

Substrate	Porcine pancreas[b]	Castor bean[b]	Maize[b]	Rapeseed[b]	Erucic acid-free rapeseed[b]	Elm[b]	Mustard[c]	Palm[d]
Tricaprin	207	43	127	89	—	100[e]	—	100[e]
Trilaurin	92	60	0	31	—	4	—	60
Trimyristin	50	26	0	92	—	3	—	15
Tripalmitin	5	46	0	27	51	0	39	35
Tristearin	0	62	0	36	89	0	40	—
Triolein	97	55	38	44	138	4	96	—
Trilinolein	100[e]	57	100[e]	89	116	6	89	—
Triricinolein	53	100[e]	0	83	—	0	—	—
Tribehenin	—	—	0	16	—	0	—	—
Trierucin	92	36	45	100[e]	100[e]	0	100[e]	—

[a] Glycoxysomal lipases from soybean[48] and castor bean[52] are also very active on trilinolein and triricinolein, respectively.

[b] From Lin et al.[26] Similar results with castor bean lipase were reported earlier.[31]

[c] Lin and Huang.[19]

[d] Oo and Stumpf.[27]

[e] The standard (100%) used to calculate the relative activity of the lipase on various substrates.

is not composed of only the major acyl moieties, but rather of the major as well as other acyl moieties in the same molecule.[28] Thus, after the lipase has hydrolyzed the major acyl moieties from a TAG molecule, it should still have high capacity to hydrolyze the remaining diacylglycerol and monoacylglycerol containing the other acyl moieties.

Very few lipases from mammalian and microbial sources are known to have high acyl specificity.[29] A notable exception is the lipase from *Geotrichum candidum*, which is partially specific for fatty acyl moieties with *cis*-9 configuration.[30] In this and in other reported cases, the physiological relevance of acyl specificity is unknown. On the contrary, the physiological significance of acyl specificity in seed lipases is obvious. The plant species can afford to produce more specific lipases for a higher efficient catalysis because the acyl compositions of the storage TAG also are well-defined and largely inherited. This acyl specificity of plant lipases could be exploited in lipid biotechnology.

D. DIFFERENCES IN PROPERTIES OF SEED LIPASES AMONG SPECIES

A common feature among seed lipases from diverse species is that the enzyme activities are absent in ungerminated seeds and increase in postgermination.[10] Other than this similarity, oilseed lipases from diverse species exhibit differences in their properties. These differences include acyl specificity, pH for optimal activity, reactivity toward sulfhydryl reagents, hydrophobicity of the molecule, and subcellular location. In the last aspect, depending on the species, three subcellular locations of lipases have been reported: namely, the oil bodies, the glyoxysomes, and the cytosol. The lipases that are associated with oil bodies have been studied most extensively. Oil bodies have been reviewed recently[16] and will not be described in detail in this chapter.

E. OIL BODY LIPASES

Lipases are present in the seed oil bodies and, specifically, on the surrounding membrane of the organelles of several plant species, including castor, maize, rape, mustard, and jojoba. During subcellular fractionations of the seed extracts, about 50 to 80% of the lipase activity could be recovered in the oil body fraction. Isolated oil bodies can be subjected to TAG removal by ether extraction, and the remaining fraction contains the membranes of the organelles. Because the membrane is composed of a monolayer of phospholipids,[15,16] the oil body ghosts tend to adhere to one another. The membrane contains the lipase, presumably as a minor protein, and the major structural proteins called oleosins.[16]

1. Castor Lipase

The classic work of Ory and his colleagues[31,32] on castor bean acid lipase has been reviewed,[9,33] and will not be discussed here. Among lipases from different plant species, the castor lipase is unique in several aspects, including the acidic pH (pH 5) for optimal activity; its presence in an active form in the ungerminated seeds; its lack of substrate specificity on diverse TAG; and its being equally active on tri-, di-, and monoacylglycerols (Table 1).[26,32] During seed maturation, lipase activity increases concomitant with the accumulation of storage TAG. In postgermination, it decreases concomitant with the disappearance of TAG.[34]

In postgermination of castor bean, the activity of a "neutral" lipase, which is apparently different from the acid lipase, increases concomitant with lipolysis.[35,36] The activity is associated with the oil bodies and is dependent on Ca^{2+}. An activity vs. pH curve shows that this lipase is inactive at pH 4 to 5, in which the acid lipase is most active. Its activity increases from pH 6 to 8; its activity beyond pH 8 is unknown. The *in vitro* activity of the lipase at pH 7 is sufficient to account for the *in vivo* rate of lipolysis. At the peak of the appearance of enzyme activity in postgermination, the oil bodies still contain acid lipase activity that is several times higher than the Ca^{2+} dependent activity at pH 7. It is unknown if this neutral lipase is a protein newly synthesized in postgermination or a modified form of the acid lipase. It is possible that Ca^{2+} modifies the acid lipase or the substrate emulsions such that the acid lipase can work at both pH 5 and 7. The neutral lipase activity is inhibited by acyl coenzyme A (CoA).[37] Whether this inhibition is due to the detergent effect of the acyl-CoA or represents an *in vivo* feedback mechanism to control lipolysis remains to be seen.

2. Maize Lipase

Lipase from the scutella of maize seedlings has been extensively studied.[25,26,38] Similar to the castor lipase, the maize enzyme is tightly associated with the oil body membrane and resists solubilization by repeated washing with dilute buffer or salt solution. Lipase activity is absent in ungerminated seed and appears in postgermination. Oil bodies isolated from the seedlings undergo autolipolysis, releasing fatty acids. The enzyme has an optimal activity at pH 7.5 in the autolipolysis of oil bodies or on trilinolein or *N*-methylindoxyl-myristate. The activity is not greatly affected by salt or pretreatment of the enzyme with *p*-chloromercuribenzoate or mersalyl.

Among the various TAG examined, maize lipase is active only on those containing linoleoyl and oleoyl moieties (and the shorter-chain acyl moieties) which are the major fatty acyl constituents of maize oil (Table 1). The enzyme is inactive on lecithins. It is more active on tri- than di- or monolinolein. It releases linoleic acid from both primary and secondary positions, although it exerts some preference in releasing fatty acid from the primary rather than the secondary position of a TAG.[26] At the primary positions, it is more active on oleoyl than stearoyl moiety.

Maize lipase apparently catalyzes acyl transfer.[26] During its catalysis on the hydrolysis of 1,2-dilinolein, trilinolein is produced in addition to linoleic acid and monolinolein. No trilinolein is produced when 1,3-dilinolein is the substrate.

Presumably, the esterification of linoleic acid is much more active on the primary position of 1,2-dilinolein than on the secondary position of 1,3-dilinolein. Similar catalysis of acyl transfer has been observed in pancreatic lipase and other lipases.[18]

Maize lipase has been purified 272-fold to apparent homogeneity as evidenced by sodium dodecyl sulfate polyacrylamide gel electrophoresis (SDS-PAGE) and double immunodiffusion.[38] As estimated by SDS-PAGE, the enzyme has an approximate molecular mass of 65 kDa. This molecular mass is not reduced by pretreatment of the enzyme with 2-mercaptoethanol, and the molecule contains no cysteine. The amino acid composition as well as a biphasic partition using Triton X-114 reveal the enzyme to be a hydrophobic protein. This hydrophobicity is in accord with the enzyme being tightly associated with the membrane.

3. Rape and Mustard Lipase

Rape and mustard belong to the family Cruciferae, whose species generally contain a high amount of erucoyl moieties in the seed storage TAG. In rapeseed, lipase activity is absent in the ungerminated seed and increases in postgermination.[19,39] About 50% of the activity can be recovered in the oil bodies after subcellular fractionation. However, unlike the castor and maize lipases, rape lipase can be washed away easily from the oil bodies by dilute buffer. The enzyme detached from the oil body fraction (which contains all the TAG and floats during centrifugation) is still attached to membrane fragments. These fragments were proposed to be appendices of the oil bodies and the location for the synthesis of nascent lipase.[40] However, the fragments seem more likely to represent the ghosts, or part of the ghosts, of the oil bodies after the TAG have been consumed.[41] Rape lipase has an optimal activity at pH 6.5 on the native TAG, trilinolein, or trierucin, but at pH 8.5 on the artificial substrate N-methylindoxylmyristate. Among the acylglycerols examined, the enzyme is most active on trierucin and trilinolein (Table 1), and is less active on the corresponding di- and monoacylglycerols.[19,26] This substrate specificity is meaningful, since rapeseed contains roughly 40% erucoyl and 20% linoleoyl moieties in the storage TAG. Depending on how the substrate is prepared, the enzymatic activity may be inhibited by erucic acid or promoted by NaCl and mild detergents.[19,39]

Rape lipase has been solubilized from the membrane by detergents.[42] The solubilized enzyme in detergent has a molecular mass of 250 kDa as estimated by gel filtration. Earlier, it was reported that a rapeseed protein, identified by antiserum raised against a preparation of castor bean glyoxysomal lipase (to be described), had a molecular mass of 62 kDa.[35] This protein was subsequently shown to be malate synthase which was identified due to the antiserum containing strong antibodies against contaminating malate synthase in the original castor bean lipase preparation.[43,44]

Many varieties of rape containing no erucoyl moiety in the seed TAG have been produced through breeding. Lipase obtained from one of these varieties

(Tower) possesses properties very similar to those of the normal rape lipase,[19] including its relatively higher activity on trierucin and trilinolein (Table 1). This high activity on trierucin is not unexpected, since erucic acid-free varieties were obtained by breeding through the alterations of the fatty acid elongation enzymes.[45] Apparently the lipase gene was not affected.

Lipase from the seed of mustard possesses properties very similar to those of the rapeseed lipase.[19] These similarities include the pH for optimal activity on the native substrates and *N*-methylindoxylmyristate, the substrate specificity (highest on trierucin and trilinolein) (Table 1), subcellular location, and the appearance of activity in the seed only in postgermination. The similarities are quite expected in view of the close relationship between the two species.

4. Oil Palm Lipase

In the fruit of oil palm (*Elaeis guneensis*), massive TAG are present in the kernel (endosperm and embryo) and the mesocarp. Lipase activity was not detected in the extracts of the endosperm and haustorium of the kernel in postgermination, but was detected in the shoot extract.[27] The shoot lipase is most active on tridecanoin and trilaurin (Table 1). In the fruit mesocarp, lipase is present in the oil bodies.[46] The partially purified enzyme has an optimal activity at pH 4.5 and is active on its native substrate (palm oil, which contains mostly oleoyl and palmitoyl moieties), triolein, and tripalmitin. Sodium cyanide, resorcinal, cholesterol, lecithin, and glycylglycine strongly inhibit, whereas phenol, *L*-cysteine, and EDTA enhance its enzyme activities. The physiological role of lipase in the fruit mesocarp is unclear, since the TAG are not to be mobilized for internal use by the plant. Perhaps the fatty acids produced somehow hasten the ripening process. The lipase activity is of some economic importance because the mesocarp oils deteriorate rapidly after the fruit has been picked.

5. Jojoba Lipase

Jojoba seed is unique among oilseeds as it contains intracellular wax esters instead of TAG as food storage. The activity of a wax ester hydrolase is absent in ungerminated seed and increases in postgermination.[47] The enzyme is associated with the membrane of the oil bodies and has an optimal activity at pH 8.5 (on *N*-methylindoxylmyristate). It has the highest activities on monoacylglycerols, wax esters, and jojoba wax ester and the lowest activities on dipalmitin, tripalmitin, and triolein. Since the enzyme hydrolyzes TAG, it can be called a true lipase (E.C. 3.1.1.3), although its native substrates are monoesters. The enzyme is inactivated by *p*-chloromercuribenzoate, and the inactivation is reversed by subsequent addition of dithiothreitol.

6. Soybean Lipase (Acyl Hydrolase)

Soybean oil bodies isolated from ungerminated or germinated seed do not undergo autolipolysis.[48] However, the membranes of oil bodies isolated from ungerminated seed contain a monoacylglycerol hydrolase. This hydrolase has an optimal activity at pH 6.5 (on *N*-methylindoxylmyristate) and hydrolyzes monolinolein, but not trilinolein. In postgermination, the enzyme activity declines before the total TAG. The above observations suggest that the enzyme is not involved in TAG hydrolysis in postgermination. Of all the oilseeds examined, only soybean possesses oil bodies having the activity of this monoacylglycerol hydrolase. Its physiological role is unknown. Whether it is a nonspecific acyl hydrolase as described in Section III requires further studies.

F. GLYOXYSOMAL LIPASES

A lipase/acyl hydrolase is present in the glyoxysomes of germinated seeds of diverse species.[49-51] The enzyme is tightly bound to the membrane of the organelles and resists solubilization by buffer or high salt solutions. Treatment of intact glyoxysomes with trypsin strongly diminishes the enzyme activity, but does not affect the glyoxysomal matrix enzyme activities.[52] The enzyme can be solubilized by deoxycholate and KCl and has been purified to apparent homogeneity. The molecular mass of the purified enzyme, as determined by SDS-PAGE, is 62 kDa.

Originally, the glyoxysomal lipase activity was assayed as a fatty acyl hydrolase using *N*-methylindoxylmyristate in a fluorescence assay.[49] The enzyme has an optimal activity at alkaline pH on this artificial substrate. It is also active on monoacylglycerols, but is relatively inactive on TAG. Nevertheless, it does hydrolyze TAG, especially those containing the dominant acyl moieties of the storage TAG.[48,52] Even so, the enzyme from soybean hydrolyzes trilinolein at a rate of only about 10% of that on monolinolein. Since the enzyme can hydrolyze TAG, it can be termed a true lipase. Its role in TAG hydrolysis in postgermination remains unclear in view of its activity being much higher on monoacylglycerol than on TAG, its physical separation from the TAG in the oil bodies, and its low activity in comparison with the true lipase in the same cell.

G. SOLUBLE LIPASES

In oilseeds of some species, lipase activity is present mostly in the soluble fraction after subcellular fractionations. The enzyme is truly soluble in the sense that it is not associated with small membrane fragments. The soluble nature of these lipases is probably not due to a harsh fractionation procedure for the following two reasons. First, under the same fractionation procedure carried out in the same laboratory, seeds of other species yield lipase activity that is associated with oil bodies.[26] Second, in the latter species, the enzymes cannot be solubilized easily. These lipases are either tightly associated with the oil bodies and not easily removed (e.g., castor, maize, and jojoba) or still bound to membrane fragments after their relatively easy removal from the oil bodies (e.g., rape and mustard). If the soluble lipases are indeed those involved in lipolysis *in vivo*, they should be amphipathic proteins that can associate with the membrane of the oil bodies in order to carry out lipolysis. None of these soluble lipases, to be described in the following paragraphs, has been purified or even partially purified.

In elm seed, lipase activity is present in germinated, but not ungerminated seed.[26] The lipase has an optimal activity at pH 9.0 on tricaprin or trilinolein. It is highly specific on tricaprin and hydrolyzes trilaurin, trilinolein, and triolein at a rate less than 10% of that on tricaprin (Table 1). After subcellular fractionation, more than 90% of the lipase activity can be recovered in the 100,000-*g*, 90-min supernatant, whereas the oil body fraction has no detectable activity.

In cotton seed, lipase activity is present only in germinated seed, and most of the activity appears in the soluble fraction after subcellular fractionation.[10] About 10 to 15% of the lipase activity is associated with the oil body fraction, and it can be removed from the fraction easily by washing the fraction with dilute buffer.

In Douglas fir, two lipase activities, measured at pH 5.1 and 7.1, are present in the storage gametophytes of seed.[53] The activities increase in postgermination. With native neutral lipids as the substrate for enzyme assays, about 80% of the lipase activity is present in the soluble fraction, and the rest are distributed among oil body, mitochondrial, and protein body fractions. The lack of lipase activity in fir oil bodies is similar to that in pine,[10] in which the oil bodies of ungerminated and germinated seed are unable to undergo autolipolysis.

H. OTHER OILSEED LIPASES

Lipase activities, besides those described above, have been reported in oilseeds of other species, but their subcellular location are unknown. In many cases, acetone powders of the seeds were used as enzyme sources.

In peanut, a lipase was partially purified from an acetone powder of maturing seeds, using tributyrin as the substrate in the enzyme assay.[54] The enzyme activity increases during seed maturation. It has higher activity on tributyrin than on maize oil. Similar findings were observed from germinated peanut, although the data were not presented. It is unknown whether this partially purified lipase is the same as the peanut glyoxysomal lipase;[51] they both share the same alkaline pH for optimal activity and increase in activity in postgermination. Even though the glyoxysomal alkaline "lipase" activity on *N*-methylindoxylmyristate is dominant over activities at pH 5 or 7 throughout postgermination, its involvement in storage TAG hydrolysis in postgermination is still uncertain. An earlier report shows that the maximal lipolytic (autolytic) activity in crude extracts of germinated peanut occurs at pH 4 to 5.[55]

A lipase was partially purified from ungerminated seed of *Veronica anthelmintica*,[56] an oilseed in which the major storage oil is trivernolin (vernolic acid is *cis*-12,13-epoxy-*cis*-9-octadecenoic acid). The enzyme has a molecular mass of >200 kDa and an optimal activity at pH 7.5 to 8.0. It hydrolyzes both primary and secondary ester bonds of TAG and is equally active on saturated (palmitate, stearate) and unsaturated (oleate) TAG. Its activity on trivernolin was not examined.

An acid lipase (pH 5) is present in the acetone powder of germinated seed of *Cucumeriopsis edulis*.[57] The enzyme activity increases in postgermination. Since the enzyme indiscriminately releases fatty acids from *rac*-glyceryl-1-palmitate-2-oleate-3-stearate or *rac*-glyceryl-1-stearate-2-palmitate-3-oleate, its activity is quite nonspecific.

In acetone powder of apple seed, two lipase activities are present, with optimal activities at pH 5 and 7.5.[58] The activities are low in ungerminated or dormant seed and increase in postgermination.

I. LIPASES IN CEREAL GRAINS

Cereal grains are not considered oilseeds because of their low content of oils which are not utilized commercially (except maize oil and rice bran oil). They generally contain 2 to 10% oils, depending on the species and varieties. Most of the oils are TAG, of which 80 to 90% of the acyl moieties are linoleoyl and oleoyl moieties.[28] The oils are usually located in the embryo (germ) and the aleurone layer (the bran, which also includes the pericarp, testa, and some endosperm).[59,60] In the aleurone layer of wheat[61] and barley,[62] the oils are present in oil bodies. In maize, the oils are present in the oil bodies of scutella and embryonic axis and are mobilized in postgermination (see Section II.E).

Lipolytic activities are present in different parts of cereal grains. Studies of these enzymes have been carried out largely because of their potential in causing rancidity during grain or bran storage. The physiological role of these enzymes is unclear, and in some cases their exact tissue location is still unknown. In addition, the lipolytic activities might have been contributed by contaminating microbes.[8] These microbes proliferate upon grain storage, especially under humid conditions, and utilize the grain oils and other food reserves for growth.

It has been suggested that in barley aleurone layer treated with gibberellic acid, the lipase originally associated with the protein bodies is transferred to the oil bodies by membrane flow via a contact point between the two organelles.[63] The suggestion is based mostly on electron microscopic observations; apparently, the limitation in the amount of tissues and the low lipase activities prevent extensive biochemical analyses.

Lipases in cereal grains have been reviewed by Galliard,[9] and there has been little progress since. They will not be discussed further in this chapter. Other review articles covering cereal lipases include those by Galliard[8] and Huang.[10]

J. LIPASES IN NONSEED TISSUES

True lipase in nonseed organs has been reported occasionally. In maize roots, lipase activity with an optimal activity at pH 5 on olive oil occurs mostly in the zone of elongation.[64] A lipase with a molecular mass of 77 kDa has been separated from other acyl hydrolases in potato tuber extract.[65] The potato lipase releases fatty acids preferentially from triolein instead of diolein or monoolein and has no activity on phospholipids or galactolipids. It has an optimal activity at pH 7.5 to 8.0 on *N*-methylindoxyl esters and preferentially hydrolyzes *N*-methylindoxyl esters of longer chain length.

K. INTRACELLULAR TRANSPORT OF LIPOLYTIC PRODUCTS

In seeds, TAG are localized in the oil bodies, and the hydrolyzed fatty acids are activated and β-oxidized in the glyoxysomes.[18] The other product, glycerol, is metabolized initially in the cytosol. Fatty acid is water insoluble, and its removal from the oil bodies is desirable and probably essential for the continuation of lipolysis. Little fatty acid accumulates in seeds in post-germination.[55] The mode of tranport of fatty acid from the oil bodies to the glyoxysomes is unknown.

There are several possible mechanisms for the above-mentioned fatty acid transport. Phospholipid transfer proteins are present in mammalian and plant systems (Chapter 12), but their ability to transfer free fatty acid in oilseeds has not been established. In those seeds where soluble instead of oil body-bound lipases are present, the enzymes may carry out lipolysis as well as transport of fatty acid; however, there is no information to support this notion. As observed *in situ* under the electron microscope, oil bodies and glyoxysomes occasionally are in physical contact with each other.[17] Even in those electron micrographs where the two organelles are observed to be in close proximity but not in direct contact, the organelles may still have a point of contact if the three-dimensional nature of the organelles is taken into account. If so, it is possible that fatty acid is transported from the oil body to the glyoxysome by membrane flow. Fatty acid released by lipase on the oil body membrane would flow along the membrane by simple diffusion, due to a concentration gradient, through the contact junction of the two organelles, to the membrane of the glyoxysome. Such a flow of fatty acid along the membranes of different organelles (and cells) has been proposed to occur in mammalian systems.[66] Another possibility is that the glyoxysomal membrane acyl hydrolase, which does not have a known metabolic role in lipolysis, is somehow involved in the transport of fatty acid.

L. BIOSYNTHESIS OF SEED LIPASES

Lipase activity is absent in ungerminated seed and appears in postgermination. The enzyme may be synthesized as an inactive precursor during seed maturation or as an active form in postgermination. The mode of lipase synthesis has been studied only with maize, mainly because the maize lipase is the only seed lipase that has been purified to homogeneity and its antibodies are available. As analyzed with an *in vitro* protein synthesis system, the mRNA for lipase is present in germinated, but not maturing seeds.[67] The *in vitro* and *in vivo*-synthesized lipase exhibit the same molecular mass of 65 kDa by SDS-PAGE, and thus there is no apparent co- or posttranslational processing of the protein. The enzyme is synthesized by mRNA extracted from free and not bound polyribosomes. Apparently, after its synthesis, the enzyme will attach itself specifically to the membrane of the oil bodies and not other cell organelles. The specific recognition signals present on both the enzyme and the oil bodies are unknown. It has been speculated that the recognition signal on the surface of the oil bodies is contributed by oleosins, since the proteins are unique to the organelles.[16] Whether or not the above scheme of lipase synthesis in postgermination in maize is the same in other species remains to be seen.

The maize lines, Illinois High Oil and Illinois Low Oil, containing about 18 and 0.5%, respectively, of kernel lipids, have been obtained through continuous breeding for oil quantity.[68] The lipase activities in the kernels of the above two lines, as well as the F1 generation, in postgermination are proportional to the lipid contents.[69] On the contrary, the activities of glyoxysomal enzymes are the same in the three maize lines irrespective of the lipid content. Thus, there is a difference in the genetic control of lipase and the other gluconeogenic enzymes and a co-selection for high lipid content and high lipase activity through breeding. The mechanism of coordinate selection for both high lipid and high lipase activity is unknown. There are about 50 genes for the expression of high lipid content in Illinois High Oil,[68] and apparently they are expressed only in seed maturation and not in seedling growth. Yet lipase activity is absent in maturing and ungerminated seeds and appears only after germination. It is unlikely that the genes for high lipid content are tightly linked to the lipase gene(s). Since lipase is *de novo* synthesized in postgermination, it is likely that the enzyme is synthesized or degraded in proportion to the availability of specific binding surface on the oil bodies or substrate and thus metabolic need.

M. CONTROL OF LIPASE ACTIVITY

Since storage TAG are used for gluconeogenesis during germination and in postgerminative growth of seeds, lipase activity which represents the first reaction of the gluconeogenic pathway should be a possible target for metabolic control. In most seeds, lipase activity increases during germination and in postgermination. There are a few studies where hormones applied externally to the seeds or to excised seed parts enhance lipase activity and lipid consumption. However, when such a stimulation occurs, it is unclear whether the hormone effect is specific on the increase in lipase activity preceding other developmental processes or whether it is merely a general influence on cell development or differentiation. In the seeds of many species, lipolysis or lipase activity is correlated with the germination process or rate and is affected by dormancy breaking[70] or salt stress.[71]

In castor bean, excised endosperm provided with water appears to undergo the normal postgerminative development, including gluconeogenesis from lipids, independent of the embryonic axis or cotyledons.[72] Gibberellin added externally enhances lipolysis slightly.[73] Thus, if any hormonal influence on lipolysis is involved, the hormone must have been present in the endosperm.

In the embryo of dormant apple seeds, two lipase activities measured at pH 5 and 7.5 are present.[58] In excised embryo, light acting through the phytochrome system enhances the internal gibberellin content, as well as the alkaline lipase activity.[74] Externally applied gibberellin also exerts similar effects. These two treatments allow the embryo, originally from dormant seeds, to undergo germination, though slowly. Cold treatment has a similar effect on enhancing internal gibberellin content and lipase activity, and it also initiates normal germination. AMO-1618, a gibberellin biosynthetic inhibitor, blocks the effect of light or gibberellin application. In contrast, light, gibberellin, or AMO-1618 application generally causes a decrease in the acid lipase activity. It has been proposed that the alkaline lipase is involved in lipolysis during germination and that light and gibberellin are part of the control mechanism by which low temperature breaks dormancy.

In wheat grain, most of the lipase activities are present in the endosperm and bran, and they increase rapidly in postgermination.[75] The activities in excised endosperm or bran do not increase unless embryo diffusate or growth regulators are supplied. Application of hydroxylamine or glutamine at 1 mM induces the activity increase in the endosperm, but not in the bran. Lipase activity in the excised bran increases only after the addition of both hydroxylamine and indoleacetic acid. In excised aleurone layers of barley, application of gibberellin causes no change in the amount of neutral lipids in the first 12 h, even though the development of other processes is induced.[62] Neutral lipid consumption is only evident after 24 h of incubation. Apparently, lipolysis is not one of the very first events in the hormone-induced development of the aleurone layer.

As mentioned in Section II.B, lipase inhibitors of protein nature are present in some oilseeds. They are present in the soluble fraction after subcellular fractionation. They bind to the substrate emulsions in *in vitro* enzyme assays, thereby preventing the access of substrate to the enzyme. Such an inhibitory action is different from that of the classical enzyme inhibitors which act on the enzyme molecules directly. It is doubtful that the lipase inhibitors in oilseeds exert a regulatory role. *In vivo*, the TAG are packed inside the oil bodies, which are surrounded by a monolayer of phospholipids. Thus, the substrate TAG are not accessible to the inhibitors. It is inconceivable that a lipase regulatory mechanism is built on a massive encasement of the TAG by proteins. It is more likely that some proteins, such as special storage proteins in massive amount, are amphipathic and fortuitously bind to the TAG emulsions in an *in vitro* lipase assay and thus prevent lipolysis.

III. LIPID ACYL HYDROLASES

A. INTRODUCTION

An enzyme that hydrolyzes acyl groups from several classes of lipids, including glycolipids, phospholipids, sulfolipids, and mono- and diacylglycerols, but is inactive on TAG, is present in many plant organs. The acyl hydrolase releases both fatty acids from diacyl glycerolipids, and in many cases there is no preference for either the 1- or 2-position of the ester linkage. Thus, the enzyme possesses a combined catalytic capacity of phospholipase A_1, A_2, and B, as well as glycolipase, sulfolipase, and monoacylglycerol lipase. Similarities of the enzymes from various organs include the following: (1) they exert a similar pattern of substrate specificity as described above; (2) they occur as isozymes in each organ, and the isozymes have fairly similar patterns of substrate specificity; (3) they have optimal activities on a particular substrate at a similar pH, which shifts to a more alkaline value in the presence of detergent; and (4) they catalyze acyltransferase reactions.

The hydrolytic activities on various classes of acyl lipids are apparently carried out by a single protein. The following evidence with occasional exceptions has been obtained using purified or partially purified enzymes on, usually but not exclusively, galactolipids and phosphatidylcholine.

1. The activity ratio of the enzyme preparation on galactolipid and phospholipid remains fairly constant throughout an enzyme purification procedure.
2. The activity ratio is also similar after treatment of the enzyme with high or low temperature and chemical-modifying reagents.
3. The enzyme carries out acyltransferase reactions with each of the substrates.
4. Each substrate at a similar high concentration inhibits the activity.
5. The optimal activity is at an acidic pH and shifts to a high pH in the presence of detergent.
6. Competition between the two substrates for enzyme activity exists, suggesting that the activities reside not only in a single protein, but also within the same active site.
7. Potato tuber of a special variety in which the acyl hydrolase activity is very low contains a proportional reduction of activity on each of the substrates.[9]

Historically, a galactolipase was first observed in bean leaves in 1964,[76] and it was subsequently purified to homogeneity.[77] In 1971, a partially purified enzyme from potato tuber was shown to possess glycolipase, phospholipase, and other acyl hydrolytic activities,[78] and this enzyme was purified to homogeneity in 1975.[79] Reexamination of the leaf galactolipase revealed that the enzyme also catalyzes the removal of acyl moiety from phospholipids and other acyl lipids.[80] The enzyme from rice bran was also purified.[81] In the following description, the properties of the potato tuber enzyme will be given in some detail. Then, the enzymes from leaves and rice bran will be mentioned, with emphasis on the major similarities and differences between them and the potato tuber enzyme. The enzymes in other tissues known to hydrolyze at least one class of the acyl lipids will be reviewed briefly. Finally, the physiological role of the enzyme will be discussed.

B. POTATO TUBER ACYL HYDROLASE

The potato tuber enzyme is the best-studied nonspecific acyl hydrolase. There were some interesting uncertainties in earlier studies, and they have recently been clarified. Galliard[78] partially purified (five-fold) an acyl hydro-lase from potato tuber. He was unable to attain a higher fold of purification by various procedures; the enzyme apparently was co-purified with the bulk of the soluble proteins. Hirayama et al.[79] achieved a 350-fold purification of an acyl hydrolase from potato tuber to apparent homogeneity. Even though the fold of purification of the enzymes by Galliard[78] and by Hirayama et al.[79] were enormously different, the specific activities of the two enzyme prepa-rations were quite similar. Since the potato tuber acyl hydrolase exists in isozyme forms,[82] it is likely that Galliard's enzyme (13% yield) represents a mixture of the isozymes, whereas Hirayama et al.'s enzyme (1% yield) rep-resents one of the isozymes. Presumably, Hirayama et al.'s procedure selec-tively excluded or inactivated some of the isozymes. Nevertheless, the two enzyme preparations were remarkably similar in their substrate specificity (described later in this section).

The above discrepancies have been clarified in recent years. The potato tuber acyl hydrolases are identified as the vegetative storage glycoproteins termed patatin.[83] Patatins are a group of closely related proteins of molecular mass of 40 kDa and are present in the cell vacuoles.[84] They can be separated into 6 to 15 closely associated protein bands by isoelectric focusing, and all of these protein bands possess acyl hydrolase activities. They are immunologi-cally indistinguishable and are almost identical in at least the 22 *N*-terminal amino acids.[85] The fact that patatins represent 20 to 40% of the total soluble proteins of the tuber explains the unsuccessful attempt by Galliard[78] to purify the enzyme to more than five-fold. Patatins in a potato plant are subgrouped into class I and class II. Class I patatins are abundant in tubers, whereas class II patatins are present in low amounts in the leaves, roots, stems, and flowers. Class I patatins in the latter tissues can be induced by the removal of the metabolic sink (tuber) from the whole plant or the application of excess sucrose. Patatins are encoded by 10 to 15 genes per haploid genome. The reason for the vegetative storage proteins to possess acyl hydrolase activities is unknown. It has been proposed that the acyl hydrolases are for defense against predators; upon chewing injury, the released enzymes hydrolyze the membrane of the invader and/or synthesize wax to cover the wound. However, upon wounding of potato, a new set of "wound-inducing" genes is activated, whereas the patatin genes are turned off.[86] This finding suggests that the patatins are more likely for normal growth rather than for defense or wound healing, unless the proteins are synthesized earlier during normal growth as a reserve for defense. Because of their abundance and the importance of potato tuber in human nutrition, patatins have been subjected to intensive studies. These studies include mapping the genes on the chromosomes,[87] using the gene promoters for expression of foreign genes in the tubers,[88] charting the targeting signals on the proteins involved in the intracellular passage of the newly synthesized protein from the endoplasmic reticulum to the vacuoles,[89] and using the expression of the patatin genes as a signal to detect tuber induction.[90]

A description of these intensive studies is beyond the scope of this chapter in a lipid book. The following is an account of the enzymatic properties of the acyl hydrolase.

The molecular mass of patatins is about 40 kDa, which is comparable with the subunit molecular mass of the studied acyl hydrolase.[78,79] Chemical modifications and photooxidation of the protein show that histidine, serine, and tyrosine residues are important for enzyme activity. The enzyme hydrolyzes fatty acyl groups from various glycolipids, phospholipids, sulfolipids, monoolein, and diolein.[78,79] It has lower activities on short-chain esters and does not act on triolein, tristearin, and wax esters. Table 2 shows the relative activities of the enzyme obtained from two different laboratories on various lipids. It should be emphasized that there are many factors affecting the activity on each individual substrate, and a comparison of activities toward different substrates under one particular assay condition is only a rough approximation. These factors include the pH of the assay, the substrate concentration and method of preparation, the presence of an activator such as Triton X-100, and the purity of the enzyme preparation. Nevertheless, the data show a similar pattern of substrate specificity reported by two different laboratories, with the exception of acylated steryl glycoside. In addition to the substrates listed in Table 2, the enzyme also hydrolyzes methyl esters and *p*-nitrophenyl esters of long-chain fatty acids.[78] In the hydrolysis of monogalactodiacylglycerol, fatty acid is released from the 1-position first, producing a monogalacto-monoacylglycerol as an intermediate, and further hydrolysis generates another fatty acid and a galactoglycerol.[78,79] In the hydrolysis of phosphatidylcholine, digalactodiacyl-glycerol, or diolein, a monoacyl intermediate is not detectable, presumably because the first acyl group removal is much slower than the second or because the intermediate does not leave the active site. The enzyme also hydrolyzes potato tuber mitochondrial membranes.[91] The pH for optimal activity of the enzyme varies with individual substrates. For example, the optimal activity is at pH 5 in citrate buffer for monogalactodiacylglycerol and pH 8.5 in (Tris(hydroxymethyl)aminomethane) buffer for phosphatidylcholine.[79] Generally speaking, the pH for optimal activity is around 5 in the absence of detergent and shifts to a more alkaline value in the presence of Triton X-100.[78]

The activity on phosphatidylcholine is stimulated by Ca^{2+} at pH higher than 7.5 and inhibited by Ca^{2+} at lower pH values.[91] This calcium stimulation does not appear to be mediated by calmodulin,[92] unlike the potato leaf enzyme (Section III.C). Triton X-100 enhances the activity on phosphatidylcholine, but has relatively little effect on glycolipids.[78] The activity toward membrane phospholipids is enhanced by detergents, especially unsaturated fatty acids. In fact, the totally purified enzyme is inactive in the absence of a detergent. Thus, an "autocatalysis" was suggested to occur in tissue extracts[9] because the fatty acids released by the enzyme alter the membrane structure to allow access of the enzyme to more substrates. Deoxycholate at low concentrations (10 to 50

TABLE 2
Comparison of Substrate Specificities of Lipid Acyl Hydrolases Purified from Tubers of Potato, Leaves of Two *Phaseolus* Species and Potato, and Bran of Rice

	Relative activity(%)[a]					
			Leaves			
	Potato tubers		*P.* *multiflora*	*P.* *vulgaris*	Potato	Rice bran
Substrate	(78)[b]	(79)	(80)	(98)	(97)	(81)
Monogalactodiacylglycerol	100	100	100	100	100	100
Monogalactomonoacylglycerol	—	68	—	132	394	110
Digalactodiacylglycerol	56	62	60	39	44	169
Digalactomonoacylglycerol	—	26	—	49	146	—
Phosphatidylcholine	42	100	23	31	14	27
Lysophosphatidylcholine	233	304	142	89	26	33
Phosphatidylglycerol	40	151	—	47	77	—
Lysophosphatidylglycerol	—	20	—	93	89	—
Phosphatidylethanolamine	35	33	—	11	1	Trace
Lysophosphatidylethanolamine	—	71	—	32	6	—
Phosphatidylinositol	—	36	—	—	—	—
Phosphatidic acid	40	16	—	7	—	—
Monoolein	325	—	161	—	—	—
Diolein	68	—	—	—	—	—
Triacylglycerol	< 1	—	< 1	< 0.1	< 0.1	Trace
Tributyrin	< 2	—	—	—	—	—
Sulfoquinovosyldiacylglycerol	—	52	13	14	43	—
Acylated steryl glucoside	< 2	500	—	—	—	—

[a] The activities are relative to that on monogalactodiacylglycerol. There are many factors (e.g., pH, detergent inclusion) affecting the activity on each individual substrate, and a comparison of activities toward different substrates under one particular assay condition is only a rough approximation.
[b] Numbers in parentheses refer to the references.

μM) inhibits the rate of autolysis of phosphatidylcholine in tuber homogenates, whereas at higher concentrations (100 to 1000 μM) it stimulates the autolysis.[92] Three common calmodulin inhibitors (dibucaine, chlorpromazine, and trifluoperazine) exhibit similar effects. The effects of the inhibitors are probably due to their detergent-like properties rather than their interaction with calmodulin.

Several isozymes of the enzyme are present in potato tuber.[62,82,93] There are slight differences in the substrate specificities of the various isozymes. The isozyme pattern also varies, depending on the variety of the species. In potato tuber, besides the acyl hydrolase, there is a true lipase (Section II.J) and a low molecular mass (23 kDa) esterase that hydrolyzes monoolein, phosphatidylcholine, and *N*-methylindoxylbutyrate, but does not hydrolyze triolein, diolein, glycolipids, and *N*-methylindoxylmyristate.[65]

The enzyme also catalyzes an acyl transfer reaction:[94]

$$RCOOX + YOH \leftrightarrow RCOOY + XOH$$

When Y = H the reaction is a hydrolysis, and when Y = CH_3 a fatty acyl methyl ester is produced. The affinity of the enzyme for methanol is about ten times that for water.

The acyl hydrolase is always present in the soluble fraction in potato tuber extract and does not appear to be associated with any cytoplasmic organelles. However, in view of its acid pH for optimal activity and its apparent lack of activity *in vivo*, it may be localized in the cell vacuoles, which are generally broken in cell fractionation. This suggestion is validated by the identification of the acyl hydrolase to be the storage proteins, patatin, in the vacuoles. The enzyme has also been suggested to be present in the plastids (amyloplasts), which are also extremely fragile in cell fractionation. This latter possibility is raised in view of the suggested chloroplastic location of the leaf enzyme (Section III.C). Such a possibility has been discounted based on the lack of enzyme activity in isolated amyloplasts, although information on the integrity of the amyloplasts was not presented.[95]

C. LEAF ACYL HYDROLASE

The earliest reference to this enzyme appeared in 1964 when Sastry and Kates[76] reported a "galactolipase" in the leaves of *Phaseolus multiflorus*. The enzyme is present in the leaves of diverse species.[9] It has been purified to homogeneity from the leaves of two *Phaseolus* species[77,96] and potato.[97] An important purification step is the use of palmitoylated gauze chromatography, which is supposed to stabilize the enzyme activity.[97] The enzymes from *P. multiflorus*[77] and potato[97] have a molecular mass of 110 kDa, and the *P. vulgaris* enzyme has a molecular mass of 90 kDa.[97] Whereas about 50% of the *P. multiflorus* enzyme (galactolipase) activity is inhibited by 5 to 10 m*M* 2-mercaptoethanol and completely by 1 m*M* cysteine,[80] the *P. vulgaris* enzyme (galactolipase) activity is enhanced some 50% by the two sulfhydryl reagents at 1 to 10 m*M*. Chemical-modification studies of the *P. vulgaris* enzyme suggest that histidine and tryptophan residues are important for activity.

The leaf enzymes from three different species[80,88,97] exhibit a pattern of substrate specificity fairly similar to the potato tuber acyl hydrolase (Table 2). In addition, the potato tuber enzyme and the leaf enzymes generally have an optimal activity at acidic pH, and in the presence of detergent the pH for optimal activity shifts to a higher value. Acyltransferase activity has been observed with the *P. vulgaris* enzyme.[98] Unlike the potato tuber enzyme, the *P. vulgaris*[98] and potato[98] leaf enzymes do not produce detectable amounts of monoacyl lipid intermediates from either monogalactodiacylglycerol or phosphatidylcholine.

Like the potato tuber enzyme, leaf acyl hydrolase occurs as isozymes of fairly similar substrate specificity.[97] Two hydrolases from *P. multiflorus* leaves have been partially separated.[99,100] One hydrolase is active on monoolein and phosphatidylcholine and the other hydrolase is active on monoolein and galactolipid. Whereas competition between monoolein and monogalactodiacylglycerol for activity occurs with the "glycolipase", little competition is present between phosphatidylcholine and digalactodiacylglycerol for the enzyme activity. The relationship among the two enzymes and the acyl hydrolase isozyme systems in the leaves of other species and in potato tuber and rice bran is unknown.

Whereas the potato tuber acyl hydrolase is a "soluble" enzyme, the leaf enzyme has been reported to be present in the chloroplasts of *P. multiflorus*,[79] *P. vulgaris*,[101] and potato.[97] Washing and sonication are ineffective in releasing the enzyme from the chloroplasts; the release can be achieved by acetone extraction. Subsequent reports show that the enzyme activity in potato leaf extracts is much higher than those reported earlier and that the enzyme activity is present largely in the soluble fraction.[102] Apparently, the enzyme activity is regulated by calcium and calmodulin[103] and by reversible protein phosphorylation.[104]

D. RICE BRAN ACYL HYDROLASE

The enzyme from rice bran has been purified to apparent homogeneity using a purification procedure that includes palmitoylated gauze chromatography.[81] The enzyme has a molecular mass of 40 kDa, which is similar to that of the potato tuber enzyme. Serine and cystine residues are important for enzyme activity. The enzyme exhibits a pattern of substrate specificity fairly similar to the potato tuber and leaf enzymes (Table 2). Like the leaf enzyme, the rice enzyme produces no detectable monoacyl lipid intermediate from either monogalactodiacylglycerol or phosphatidylcholine.

E. OTHER ACYL HYDROLASES

Washing of isolated potato tuber mitochondria with $CaCl_2$ induces enzymatic hydrolysis of internal membrane phospholipids or externally supplied phospholipids with the release of free fatty acids.[105] As a result, the mitochondrial oxidative and phosphorylative properties are damaged. The enzyme has been identified as a membrane-bound acyl hydrolase which is unmasked by $CaCl_2$. The potato tuber nonspecific acyl hydrolase described in Section III.B is a soluble enzyme and is activated by Ca^{2+}. Whether this potato tuber enzyme is related to the mitochondrial enzyme is unknown.

Several forms of lysophospholipase have been purified from postgerminative barley endosperm.[106] The enzyme has a polypeptide of 36 kDa and an extra mass of 10 to 12% carbohydrates. It can release fatty acids from different lysophospholipids, with the highest activity on lysophosphatidylcholine containing palmitic acid. It has no activity on *p*-nitrophenyl palmitate, no phospholipase A activity on phosphatidylcholine, and no transacylation activity on two lysophospholipids. The metabolic role of the enzyme appears to be in the hydrolysis of lysophospholipid starch in postgermination.

Acyl hydrolases are present in other plant tissues. Since their substrate specificity on different classes of lipids has not been probed, it is unknown if they are related to the nonspecific acyl hydrolases of potato tuber, leaves, and rice bran. Hydrolase activity that releases fatty acids from phosphatidylcholine and lysophosphatidylcholine is present in barley grains.[107] The enzyme activity increases and decreases in accordance with the metabolic activity of the seed during maturation and germination. An acyl hydrolase that releases fatty acid from sulfolipids has been detected in alfalfa leaves and roots and in maize roots.[108]

F. PHYSIOLOGICAL ROLE

The activities of the nonspecific acyl hydrolases in many organs are extremely high. For example, in potato tuber, if all the internal phospholipids and glycolipids were available to the enzyme under the most suitable condition for activity, all the lipids would be deacylated within 1 s at 25°C.[9] Despite this high activity, the physiological role of the enzyme is still unknown. The situation is equivalent to phospholipase D in various plant organs in which the enzyme is extremely active *in vitro*, but its physiological role is also uncertain.[9] Recent studies clearly show that the potato tuber enzymes are the abundant vegetative storage proteins in the vacuoles; the reason for the storage proteins having high hydrolase activities is unknown.

Speculation on the role of the nonspecific acyl hydrolases has been centered on their involvement in the turnover of membrane lipids. In green leaves, galactolipids[109] and sulfolipids[110] are the dominant lipid components of the chloroplasts which are the intracellular sites of acyl hydrolases. Thus, the enzyme may be involved in the turnover or rearrangement of the chloroplast membrane system, especially during leaf greening and senescence. In leaf greening, the internal membrane system of a chloroplast undergoes a vigorous structural alteration from a prolamellar body to the final thylakoid membrane system. During leaf senescence, the membranes are broken down and some of the components are translocated to other parts of the plant for reutilization. In senescing rose petal, there is a decrease in phospholipids and an increase in acyl hydrolase activity.[111] In pea leaves, senescence is promoted by the activation effect of calmodulin on acyl hydrolase.[112] Since acyl hydrolase also catalyzes acyl transferring reactions, it could participate in the synthetic pathway of acyl lipids.[113] In organs other than leaves, alterations of membrane lipids occur in development and differentiation, such as during seed maturation and germination and fruit ripening. The free fatty acids released from the enzymatic reaction may undergo oxidation catalyzed by several known oxidases to produce nonvolatile and volatile metabolites that may be of hormonal nature.[33] In injured potato tuber cells, the hydrolytic and acyl transferring activities of acyl hydrolase, together with lipoxygenase, may release cytotoxic, oxidized fatty acid derivatives and water-insoluble waxes that inhibit microbial invasion.[83,114] In cereal grains in postgermination, the acyl hydrolase may be involved in the mobilization of starch-bound lysophosphatidylcholine.[115] All the above-mentioned physiological processes may require the involvement of acyl hydrolase activities. However, there is no sufficient evidence for these involvements.

Some evidence is available on the involvement of acyl hydrolases in the hydrolysis of internal lipids under pathological conditions. Mechanical or freezing injury disrupts subcellular compartmentation. Acyl hydrolases, which otherwise would be physically separated from the substrates, will be in direct contact with various cytoplasmic organelles such that deacylation can occur readily. The fatty acids released could inhibit the activities of cellular organelles; this inhibition has been observed *in vitro* on chloroplasts,[116] mitochondria,[91] and microsomes.[117] Alternatively, the fatty acids released could be oxidized to volatile and nonvolatile components which may be the sources of rancidity;[33] the involvement of acyl hydrolase in the formation of rancidity in rice bran during storage has been demonstrated.[118] During the "aging" of potato tuber slices, acyl hydrolase is thought to be involved in a major turnover of the internal membrane lipids.[91] In the formation of leaf galls of oak, there is a decrease in phospholipids and an enhancement of acyl hydrolase activity.[119] In fungal infection, the fungus may simply secrete a few enzymes to disrupt the subcellular compartments, and all subsequent reactions of cell disintegration can be accomplished by internal acyl hydrolases and other enzymes.[120] For example, during the infection of potato tuber by *Botrytis cinerea*, the fungus produces a phospholipase which does not attack isolated protoplasts, but is able to enter the cells and hydrolyze the tonoplast.

IV. PERSPECTIVE

Research on oilseed lipases has not progressed rapidly, partly due to various experimental difficulties described earlier in this chapter. These difficulties are now recognized, and the potential of future research to yield valuable information is tremendous.

Most of the oilseed lipases that have been studied in some detail are associated with the membranes of the oil bodies. Further studies on these membrane-associated lipases should be carried out not only with purified enzymes, but also with enzymes still attached to the membrane. The latter aspect is especially important in that the substrate TAG have to interact with the membrane-bound lipase. The effect of association of the enzyme with the membrane on the catalytic mechanism should be analyzed. The positional specificity of the enzyme with reference to the types of fatty acyl moieties attached to different carbon atoms of the glycerol and the mechanisms of progressive hydrolysis of one TAG to three fatty acids are important information. As far as the purified maize oil body lipase (or castor glyoxysomal lipase) is concerned, one single enzyme can catalyze the removal of all the three fatty acids. How the product fatty acids move from the oil bodies to the glyoxysomes, where fatty acid activation and β-oxidation occur, remains to be elucidated.

The storage tissues do not accumulate free fatty acids in postgermination, even though the lipase activity as measured *in vitro* is much higher than the rate of *in vivo* lipolysis. The control of lipase activity by internal factors is apparent and requires elucidation.

In the seeds of some species, the lipases do not associate with the oil bodies, but remain in the soluble fraction during subcellular fractionation. The properties of these soluble lipases are unknown. To play a role in TAG hydrolysis, the enzyme must come into contact with the oil bodies. It is likely that the enzyme is an amphipathic protein and is actually associated with the oil body membrane permanently, but loosely *in vivo*. Alternatively, the enzyme may be associated with the membrane only transiently during catalysis. If so, it may also be involved in the transport of fatty acids from the oil bodies to the glyoxysomes (unless the oil bodies and the glyoxysomes are in direct contact). There are plenty of intriguing questions waiting to be answered.

The time is ripe to study the lipase structure by molecular biology techniques. It should be relatively straightforward to isolate the lipase gene, obtain the deduced amino acid sequence of the protein, and analyze the structure of the enzyme using methodology that have been employed successfully to the mammalian and microbial enzymes. Modification of the structure of the lipase protein as induced by its association with the oil body surface should be elucidated. The control of lipase gene expression should also be studied. The synthesis of massive amounts of plant lipase by microbes that have been transformed with the plant lipase gene may be of interest to the lipid industry, since the plant lipases have unique substrate specificities that are not found in microbial lipases.

The physiological role of the glyoxysomal lipase remains a puzzle. Its physical separation from the substrate TAG and its low activity on TAG relative to monoacylglycerols argue against its role in TAG hydrolysis. In addition, there is always a highly active true TAG-hydrolyzing lipase present in the oil bodies or cytosol. Since the glyoxysomal lipase is facing the cytosol, there is a possibility that its real physiological role is to aid the reception of fatty acids from the cytosol or from the contact point between the glyoxysomes and oil bodies.

The activity of nonspecific acyl hydrolase is extremely high in many organs. Although its action in causing rancidity in stored agricultural products and in damaged or infected organs has been quite well-documented, its *in vivo* physiological role is still unknown. Circumstantial evidence has suggested that the enzyme participates in membrane lipid turnover or net synthesis; however, direct evidence is lacking. Investigations should be carried out with organs that have a high rate of *in vivo* turnover, net synthesis, or degradation of acyl lipids. Examples are senescing leaves and greening leaves in which the prolamellar bodies in the plastids are actively being converted to the thylakoids. The subcellular location of the enzyme and its isozymes should be studied. In view of its hydrolytic nature and its optimal activity being at an acidic pH, the possible localization of the enzyme, or a fraction thereof, in the vacuoles should be examined. At least part of the extractable enzyme activity is associated with the chloroplasts; information on its exact subchloroplast location, especially in relation to the acyl lipids, may give hints to its physiological role. The catalytic mechanism of the enzyme, which can act on so many different types of acyl lipids, but not TAG, should be studied.

The recent identification of the potato tuber acyl hydrolase to be the abundant vegetative storage proteins termed patatins opens up exciting new areas for investigation. The investigation will be facilitated substantially by the available genes encoding the proteins. The three-dimensional structure of the enzyme can be probed by analyzing the deduced amino acid sequence and by X-ray crystallography. In potato, the enzymes in the tuber have been identified as the class I patatins, but whether the enzymes in the leaves and other organs are actually the related class II patatins remains to be elucidated. Investigation should be made to see if the well-studied enzymes from other organs of other species, including the *Phaseolus* leaves and rice bran, are also vegetative storage proteins similar to the potato patatins. If these acyl hydrolases are indeed the vegetative storage proteins localized in the cell vacuoles, the physiological role of the enzymes in membrane turnover or synthesis should be reevaluated.

REFERENCES

1. **Huang, A. H. C.**, Lipases, in *The Biochemistry of Plants*, Vol. 9, Stumpf, P. K. and Conn, E. E., Eds., Academic Press, New York, 1987, 91.
2. **Boel, E., Huge-Jensen, B., Christensen, M., Thim, L., and Fiil, N. P.**, *Rhizomucot miehei* triglyceride lipase is synthesized as a precursor, *Lipids*, 23, 701, 1988.
3. **Kirchgessner, T. G., Chuat, J.-C., Heinzmann, C., Etienne, J., Guilhot, S., Svenson, K., Ameis, D., Pilon, C., d'Auriol, L., Andalibi, A., Schotz, M. C., Galibert, F., and Lusis, A. J.**, Organization of the human lipoprotein lipase gene and evolution of the lipase gene family, *Proc. Natl. Acad. Sci, U.S.A.*, 86, 9647, 1989.
4. **Winkler, F. K., D'Arcy, A., and Hunziker, W.**, Structure of human pancreatic lipase, *Nature*, 343, 771, 1990.
5. **Brady, L., Brzozowski, A. M., Derewenda, Z. S., Dodson, E., Dodson, G., Tolley, S., Turkenburg, J. P., Christiansen, L., Huge-Jensen, B., Norskov, L., Thim, L., and Menge, U.**, A serine protease triad forms the catalytic centre of a triacylglycerol lipase, *Nature*, 343, 767, 1990.
6. **Beg, O. U., Meng, M. S., Skarlatos, S. I., Previato, L., Brunzell, J. D., Brewer, H. B., Jr., and Fojo, S. S.**, Lipoprotein lipase$_{Bethesda}$: a single amino acid substitution (Ala-176→Thr) leads to abnormal heparin binding and loss of enzymic activity, *Proc. Natl. Acad. Sci. U.S.A.*, 87, 3474, 1990.
7. **Bjorkling, F., Godtfredsen, S. E., and Kirk, O.**, The future impact of industrial lipases, *Trends Biotechnol.*, 9, 360, 1991.
8. **Galliard, T.**, Enzymic degradation of cereal lipids, in *Lipids in Cereal Technology*, Barnes, P. J., Ed., Academic Press, New York, 1983, 111.

9. **Galliard, T.**, Degradation of acyl lipids — hydrolytic and oxidative enzymes, in *The Biochemistry of Plants*, Vol. 4, Stumpf, P. K. and Conn, E. E., Eds., Academic Press, New York, 1980, 85.

10. **Huang, A. H. C.**, Plant lipases, in *Lipases*, Borgström, B. and Brockman, H. L., Eds., Elsevier, Amsterdam, 1984, 419.

11. **Huang, A. H. C.**, Lipid bodies, in *Modern Methods of Plant Analyses*, Jackson, J. F. and Linskens, H. F., Eds., Springer-Verlag, Berlin, 1985, 145.

12. **Huang, A. H. C.**, Enzymes of lipid degradation, in *Methods in Plant Biochemistry*, Dey, P. M. and Harborne, J. B., Eds., Academic Press, London, 1990, 219.

13. **Borgström, B. and Brockman, H. L.**, *Lipases*, Elsevier, Amsterdam, 1984.

14. **Beevers, H.**, The role of the glyoxylate cycle, in *The Biochemistry of Plants*, Vol. 4, Stumpf, P. K. and Conn, E. E., Eds., Academic Press, New York, 1980, 117.

15. **Yatsu, L. Y. and Jacks, T. J.**, Spherosome membranes. Half unit membrane, *Plant Physiol.*, 49, 937, 1972.

16. **Huang A. H. C.**, Oil bodies and oleosins in seeds, *Annu. Rev. Plant Physiol. Plant Mol. Biol.*, 43, 177, 1992.

17. **Huang, A. H. C., Trelease, R. N., and Moore, T. S.**, *Plant Peroxisomes*, Academic Press, New York, 1983.

18. **Brockerhoff, H. and Jensen, R. G.**, *Lipolytic Enzymes*, Academic Press, New York, 1974.

19. **Lin, Y. H. and Huang, A. H. C.**, Lipase in lipid bodies of cotyledons of rape and mustard seedlings, *Arch. Biochem. Biophys.*, 225, 360, 1983.

20. **Christie, W. W.**, *Lipid Analysis*, Pergamon Press, Oxford, 1973, 158–161.

21. **Widner, F.**, Pancreatic lipase effectors extracted from soybean meal, *J. Agric. Food Chem.*, 25, 1142, 1977.

22. **Satouchi, K. and Matsushita, S.**, Purification and properties of a lipase-inhibiting protein from soybean cotyledons, *Agric. Biol. Chem.*, 40, 889, 1976.

23. **Gargouri, Y., Julien, R., Pieroni, G., Verger, R., and Sarda, L.**, Studies on the inhibition of pancreatic and microbial lipases by soybean proteins, *J. Lipid Res.*, 25, 1214, 1984.

24. **Wang, S. M. and Huang, A. H. C.**, Inhibitors of lipase activities in soybean and other oil seeds, *Plant Physiol.*, 76, 929, 1984.

25. **Lin, Y. H., Wimer, L. T., and Huang, A. H. C.**, Lipase of lipid bodies of corn scutella in seedling growth, *Plant Physiol.*, 73, 460, 1983.

26. **Lin, Y. H., Yu, C., and Huang, A. H. C.**, Substrate specificities of lipases from corn and other seeds, *Arch. Biochem. Biophys.*, 244, 346, 1986.

27. **Oo, K. C. and Stumpf, P. K.**, Some enzymic activities in the germinating oil palm (*Elaeis guineensis*) seedling, *Plant Physiol.*, 73, 1028, 1983.

28. **Hilditch, T. P. and Williams, P. N.**, *The Chemical Constitution of Natural Fats*, Wiley, New York, 1964.

29. **Jensen, R. G., De Jong, F. A., and Clark, R. M.**, Determination of lipase specificity, *Lipids*, 18, 239, 1983.

30. **Jensen, R. G.**, Characteristics of the lipase from the mold, *Geotrichum candidum*: a review, *Lipids*, 9, 149, 1974.

31. **Ory, R. L.**, Acid lipase of the castor bean, *Lipids*, 4, 177, 1969.

32. **Ory, R. L., St. Angelo, A. J., and Altschul, A. M.**, The acid lipase of the castor bean: properties and substrate specificity, *J. Lipid Res.*, 3, 99, 1962.

33. **Galliard, T. and Chan, H. W. S.**, Lipoxygenase, in *The Biochemistry of Plants*, Vol. 4, Stumpf, P. K. and Conn, E. E., Eds., Academic Press, New York, 1980, 131.

34. **Moreau, R. A., Liu, K. D. F., and Huang, A. H. C.**, Sperosomes in castor bean endosperm: membrane components, formation, and degradation, *Plant Physiol.*, 65, 1176, 1980.

35. **Hills, M. J. and Beevers, H.**, An antibody to the castor bean glyoxysomal lipase (62 kD) also binds to a 62 kD protein in extracts from many young oilseed plants, *Plant Physiol.*, 85, 1084, 1987.

36. **Hills, M. J. and Beevers, H.**, Ca^{++}-stimulated neutral lipase activity in castor bean lipid bodies, *Plant Physiol.*, 84, 272, 1989.

37. **Hills, M. J., Murphy, D. J., and Beevers, H.**, Inhibition of neutral lipase from castor bean lipid bodies by coenzyme A (CoA) and Oleoyl CoA, *Plant Physiol.*, 89, 1006, 1989.

38. **Lin, Y. H. and Huang, A. H. C.**, Purification and initial characterization of lipase from the scutella of corn seedlings, *Plant Physiol.*, 76, 719, 1984.

39. **Rosnitschek, I. and Theimer, R. R.**, Properties of a membrane-bound triglyceride lipase of rapeseed (*Brassica napus* L.) cotyledons, *Planta*, 148, 193, 1980.

40. **Wanner, G. and Theimer, R. R.**, Membranous appendices of spherosomes (oleosomes): possible role in fat utilization in germinating oilseeds, *Planta*, 140, 163, 1978.

41. **Bergfeld, B., Hong, Y. N., Kuhnl, T., and Schopfer, P.**, Formation of oleosomes (storage lipid bodies) during embryogenesis and their breakdown during seedling development in cotyledons of *Sinapsis alba* L., *Planta*, 143, 297, 1978.

42. **Weselake, R. J., Thomson, L. W., Tenaschuk, D., and MacKenzie, S. L.**, Properties of solubilized microsomal lipase from germinating *Brassica napus*, *Plant Physiol.*, 91, 1303, 1989.

43. **O'Sullivan, J. N., Hills, M. J., and Murphy, D. J.**, Purification and properties of lipase from oilseed rape (*Brassica rapa* L.), in *Plant Lipid Biochemistry, Structure, and Utilization*, Quinn, P. J. and Harwood, J. L., Eds., Portland Press, London, 1990, 313.

44. **Hoppe, A. and Theimer, R. R.**, Separation of lipase and malate synthase from cotyledons of rape (*Brassica napus* L.) seedlings, *Biol. Chem. Hoppe-Seyler*, 370, 798, 1989.

45. **Downey, R. K. and Craig, B. M.**, Genetic control of fatty acid biosynthesis in rapeseed (*Brassica napus* L.), *J. Am. Oil Chem. Soc.*, 41, 475, 1964.

46. **Abigor, D. R., Opute, F. I., Opoku, A. R., and Osagie, A. U.**, Partial purification and some properties of the lipase present in oil palm (*Elaeis guineensis*) mesocarp, *J. Sci. Food Agric.*, 36, 599, 1985.

47. **Moreau, R. A. and Huang, A. H. C.**, Enzymes of wax catabolism in jojoba seedlings, in *Methods in Enzymology*, Vol. 71 (Part C), Lowenstein, J. M., Ed., Academic Press, New York, 1981, 804.

48. **Lin, Y. H., Moreau, R. A., and Huang, A. H. C.**, Involvement of glyoxysomal lipase in the hydrolysis of storage triacylglycerols in the cotyledons of soybean seedlings, *Plant Physiol.*, 70, 108, 1982.

49. **Muto, S. and Beevers, H.**, Lipase activities in castor bean endosperm during germination, *Plant Physiol.*, 54, 23, 1974.

50. **Huang, A. H. C.**, Comparative studies on glyoxysomes from different fatty seedlings, *Plant Physiol.*, 55, 555, 1975.

51. **Huang, A. H. C. and Moreau, R. A.**, Lipase in the storage tissues of peanut and other oilseeds during germination, *Planta*, 141, 111, 1978.

52. **Maeshima, M. and Beevers, H.**, Purification and properties of glyoxysomal lipase from castor bean, *Plant Physiol.*, 79, 489, 1985.

53. **Ching, T. M.**, Intracellular distribution of lipolytic activity in the female gametophyte of germinating Douglas fir seeds, *Lipids*, 3, 482, 1968.

54. **Sanders, T. H. and Pattee, H. E.**, Peanut alkaline lipase, *Lipids*, 10, 50, 1975.

55. **St. Angelo, A. J. and Altschul, A. M.**, Lipolysis and the free fatty acid pool in seedlings, *Plant Physiol.*, 39, 880, 1964.

56. **Olney, C. E., Jensen, R. G., Sampugna, J., and Quinn, J. G.**, The purification and specificity of a lipase from *Vernonia anthelmintica* seed, *Lipids*, 3, 498, 1968.

57. **Opute, F. I.**, Lipase activity in germinating seedlings of *Cucumeropsis edulis*, *J. Exp. Bot.*, 26, 379, 1975.

58. **Somolenska, G. and Lewak, S.**, The role of lipases in the germination of dormant apple embryo, *Planta*, 116, 361, 1974.

59. **Morrison, W. R.**, Acyl lipids in cereals, in *Lipids in Cereal Technology*, Barnes, P. J., Ed., Academic Press, New York, 1983, 11.

60. **Hammond, E. G.**, Oat lipids, in *Lipids in Cereal Technology*, Barnes, P. J., Ed., Academic Press, New York, 1983, 331.

61. **Jelsema, C., Morre, D. J., Ruddat, M., and Turner, C.** Isolation and characterization of the lipid reserve bodies, spherosomes, from aleurone layers of wheat, *Bot. Gaz.*, 38, 138, 1977.

62. **Firn, R. D. and Kende, H.**, Some effects of applied gibberellin acid on the synthesis and degradation of lipids in isolated barley aleurone layers, *Plant Physiol.*, 54, 911, 1974.

63. **Fernandez, D. E. and Staehelin L. A.**, Does gibberellic acid induce the transfer of lipase from protein bodies to lipid bodies in barley aleurone cells?, *Plant Physiol.*, 85, 487, 1987.

64. **Heimann-Matile, J. and Pilet, P. E.**, Lipase activity in growing roots of *Zea mays*, *Plant Sci. Lett.*, 9, 247, 1977.

65. **Hasson, E. P. and Laties, G. G.**, Separation and characterization of potato lipid acylhydrolases, *Plant Physiol.*, 57, 142, 1976a.

66. **Scow, R. O., Desnuelle, P., and Verger, R.**, Lipolysis and lipid movement in a membrane model: action of lipoprotein lipase, *J. Biol. Chem.*, 254, 6456, 1979.

67. **Wang, S. M. and Huang, A. H. C.**, Biosynthesis of lipase in the scutellum of maize kernel, *J. Biol. Chem.*, 262, 2270, 1987.

68. **Dudley, J. W., Lambert, F. J., and Alexander, D. E.**, in *Seventy Generations of Selection for Oil and Protein in Maize*, Dudley, J. W., Ed., Crop Science Society of America, Madison, WI, 1974, 181.

69. **Wang, S. M., Lin, Y. H., and Huang, A. H. C.**, Lipase activities in scutella of maize lines having diverse lipid content, *Plant Physiol.*, 76, 837, 1984.

70. **Li, L. and Ross, J. D.**, Lipid mobilization during dormancy breakage in oilseed of *Corylus avellana*, *Ann. Bot.*, 66, 501, 1990.

71. **Younis, M. E., Hasaneen, M. N. A., and Nemet-Alla, M. M.**, Plant growth, metabolism and adaptation in relation to stress conditions. IV. Effects of salinity on certain factors associated with the germination of three different seeds high in fats, *Ann. Bot.*, 60, 337, 1987.

72. **Huang, A. H. C. and Beevers, H.**, (57) Developmental changes in endosperm of germinating castor bean independent of embryonic axis, *Plant Physiol.*, 54, 277, 1974.

73. **Marriott, K. M. and Northcote, D. H.**, The induction of enzyme activity in the endosperm of germinating castor bean, *Biochem. J.*, 152, 65, 1975.

74. **Zarsla-Maciejewska, B., Sinska, I., Witkowska, E., and Lewak, S.**, Low temperature, gibberellin and acid lipase activity in removal of apple seed dormancy, *Physiol. Plant.*, 48, 532, 1980.

75. **Clarke, N. A., Wilkinson, M. C., and Laidman, D. L.**, Lipid metabolism in germinating cereals, in *Lipids in Cereal Technology*, Barnes, P. J., Ed., Academic Press, London, 1983, 57.

76. **Sastry, P. S. and Kates, M.**, Hydrolysis of monogalactosyl and digalactosyl diglycerides by specific enzymes from runner-bean leaves, *Biochemistry*, 3, 1280, 1964.

77. **Helmsing, P. J.**, Purification and properties of galactolipase, *Biochim. Biophys. Acta*, 178, 519, 1969.

78. **Galliard, T.**, The enzymic deacylation of phospholipids and galactolipids in plants, *Biochem. J.*, 121, 379, 1971.

79. **Hirayama, O., Matsuda, H., Takeda, H., Maenaka, K., and Takatsuka, H.**, Purification and properties of a lipid acyl-hydrolase from potato tubers, *Biochim. Biophys. Acta*, 384, 127, 1975.

80. **Burns, D. D., Galliard, T., and Harwood, J. L.**, Catabolism of sulpholipid by an enzyme from the leaves of *Phaseolus multiflorus*, *Biochem. Soc. Trans.*, 5, 1302, 1977.

81. **Matsuda, H. and Hirayama, O.**, Purification of a galactolipase from rice bran by affinity chromatography on palmitoylated gauze, *Agric. Biol. Chem.*, 43, 463, 1979a.

82. **Galliard, T. and Dennis, S.**, Isoenzymes of lipolytic acyl hydrolase and esterase in potato tuber, *Phytochemistry*, 13, 2463, 1974b.

83. **Racusen, D.**, Lipid acyl hydrolase of patatin, *Can. J. Bot.* 62, 1640, 1984.
84. **Andrew, D. L., Beames, B., Summers, M. D., and Park, W. D.**, Characterization of the lipid acyl hydrolase activity of the major potato (*Solanum tuberosum*) tuber protein, patatin, by cloning and abundant expression in a baculovirus vetcor, *Biochem. J.*, 252, 199, 1988.
85. **Park, W. D., Blackwood, C., Mignery, G. A., Hermodson, M. A., and Lister, R.**, Analysis of the heterogeneity of the 40,000 molecular weight tuber glycoprotein of potatoes by immunological methods and by NH_2- terminal sequence analysis, *Plant Physiol.*, 71, 156, 1983.
86. **Logemann, J., Mayer, J. E., Schell, J., and Willmitzer, L.**, Differential expression of genes in potato tubers after wounding, *Proc. Natl. Acad. Sci. (U.S.A.)*, 85, 1136, 1988.
87. **Ganal, M. W., Bonierbale, M. W., Roeder, M. S., Park, W. D., and Tanksley, S. D.**, Genetic and physical mapping of the patatin genes in potato and tomato, *Mol. Gen. Genet.*, 225, 501, 1991.
88. **Rocha-Sosa, M., Sonnewald, U., Frommer, W., Stratmann, M., Schell, J., and Willmitzer, L.**, Both developmental and metabolic signals activate the promoter of a class I patatin gene, *EMBO J.*, 8, 23, 1989.
89. **Sonnewald, U., von Schaewen, A., and Willmitzer, L.**, Expression of mutant patatin protein in transgenic tobacco plants: role of glycans and intracellular location, *Plant Cell*, 2, 345, 1990.
90. **Hannapel, D. J.**, Differential expression of potato tuber protein genes, *Plant Physiol.*, 94, 919, 1990.
91. **Hasson, E. P. and Laties, G. G.**, Purification and characterization of an A type phospholipase from potato and its effect on potato mitochondria, *Plant Physiol.*, 57, 148, 1976b.
92. **Moreau, R. A., Isett, T. F., and Piazza, G. J.**, Dibucaine, chlorpromazine, and detergents mediate membrane breakdown in potato tuber homogenates, *Phytochemistry*, 24, 2555, 1985.
93. **Shepard, D. V. and Pitt, D.**, Purification and physiological properties of two lipolytic enzymes of Solanum tuberosum, *Phytochemistry*, 15, 1471, 1976b.
94. **Galliard, T. and Dennis, S.**, Phospholipase, galactolipase, and acyl transferase activities of a lipolytic enzyme from potato, *Phytochemistry*, 13, 1731, 1974a.
95. **Wardale, D. A.**, Lipid-degrading enzymes from potato tubers, *Phytochemistry*, 19, 173, 1980.
96. **Matsuda, H., Tanaka, G., Morita, K., and Hirayama, O.**, Purification of a lipolytic acyl hydrolase from *Phaseolus vulgaris* leaves by affinity chromatography on palmitoylated gauze and its properties, *Agric. Biol. Chem.*, 43, 563, 1979.
97. **Matsuda, H. and Hirayama, O.**, Purification and properties of a lipolytic acyl-hydrolase from potato leaves, *Biochim. Biophys. Acta*, 573, 155, 1979b.
98. **Matsuda, H., Morita, K., and Hirayama, O.**, Further properties of a lipolytic acyl-hydrolase from *Phaseolus vulgaris* leaves, *Agric. Biol. Chem.*, 44, 783, 1980.
99. **Burns, D. D., Galliard, T., and Harwood, J. L.**, Purification of acyl hydrolase enzymes from the leaves of *Phaseolus multiflorus*, *Phytochemistry*, 18, 1793, 1979.
100. **Burns, D. D., Galliard, T., and Harwood, J. L.**, Properties of acyl hydrolase enzymes from *Phaseolus multiflorus* leaves, *Phytochemistry*, 19, 2281, 1980.
101. **Anderson, M. M., McCarty, R. E., and Zimmer, E. A.** The role of galactolipids in spinach chloroplast lamella membranes, *Plant Physiol.*, 53, 699, 1974.
102. **Moreau, R. A.**, Membrane-degrading enzymes in the leaves of *Solanum tuberosum*, *Phytochemistry*, 24, 411, 1985.
103. **Moreau, R. A. and Isett, T. F.**, Autolysis of membrane lipids in potato leaf homogenates: effects of calmodulin and calmodulin antagonists, *Plant Sci.*, 40, 95, 1985.
104. **Moreau, R. A.**, Regulation of phospholipase activity in potato leaves by calmodulin and protein phosphorylation-dephosphorylation, *Plant Sci.*, 47, 1, 1986.
105. **Bligny, R. and Douce, R.**, Calcium-dependent lipolytic acyl-hydrolase activity in purified plant mitochondria, *Biochim. Biophys. Acta*, 529, 419, 1978.

106. **Fujikura, Y. and Baisted, D.**, Purification and characterization of a basic lysophospholipase in germinating barley, *Arch. Biochem. Biophys.*, 234, 570, 1985.

107. **Von Rebmann, H. and Acker, L.**, Über die Aktivitat der Phospholipase B der Gerste im Laufe der Entwicklung des Kornes, *Fette Seifen Anstrichm.*, 75, 409, 1973.

108. **Yagi, R. and Benson, A. A.**, Plant sulfolipid V. Lysosulfolipid formation, *Biochim. Biophys. Acta*, 57, 601, 1962.

109. **Douce, R. and Joyard, J.**, Plant galactolipids, in *The Biochemistry of Plants*, Vol. 4, Stumpf, P. K. and Conn, E. E., Eds., Academic Press, New York, 1980, 321.

110. **Harwood, J. L.**, Sulfolipids, in *The Biochemistry of Plants*, Vol. 4, Stumpf, P. K. and Conn, E. E., Eds., Academic Press, New York, 1980, 301.

111. **Borochov, A., Halevy, A. H., and Shinitzky, M.**, Senescence and fluidity of rose petal membranes, *Plant Physiol.*, 69, 296, 1982.

112. **Leshem, Y. Y., Sridhara, S., and Thompson, J. E.**, Involvement of calcium and calmodulin in membrane deterioration during senescence of pea foliage, *Plant Physiol.*, 75, 329, 1984.

113. **Harwood, J. L.**, The synthesis of acyl lipids in plant tissues, *Prog. Lipid Res.*, 18, 55, 1979.

114. **Davis, D. A., Currier, W. W., and Racusen, D.**, Release of esterase from wounded potato tubers and evidence for its presence on the plasma membrane, *Can. J. Bot.*, 67, 1009, 1989.

115. **Fujikura, Y. and Baisted, D.**, Changes in starch-bound lysophospholipids, and lysophospholipase in germinating Glacier and Hi Amylose glacier barley varieties, *Phytochemistry*, 22, 865, 1983.

116. **McCarty, R. E. and Jagendorf, A. T.**, Chloroplast damage due to enzymatic hydrolysis of endogenous lipids, *Plant Physiol.*, 40, 725. 1965.

117. **Abdelkader, B., Cherif, A., Demandre, C., and Mazliak, P.**, The oleyl-coenzyme-A desaturase of potato tubers. Enzymatic properties, intracellular localization and induction during aging of tuber slices, *Eur. J. Biochem.*, 32, 155, 1973.

118. **Hirayama, O. and Matsuda, H.**, Purification and characterization of lipolytic acyl hydrolases from rice bran, *Nippon Neogei Kagaku Kaishi*, 49, 569, 1975.

119. **Bayer, M. H.**, Phospholipids and lipid acyl hydrolase (phospholipase) in leaf galls (Hymenoptera: Cynipidae of black oak [*Quercus robor* L.]), *Plant Physiol.*, 73, 179, 1983.

120. **Shepard, D. V. and Pitt, D.**, Purification of a phospholipase from Botrytis and ants effects on plant tissues, *Phytochemistry*, 15, 1465, 1976a.

Chapter 15

PHOSPHOLIPASES

Xuemin Wang

TABLE OF CONTENTS

0-8493-4907-9/93/$0.00+$.50
© 1993 by CRC Press Inc.

I. INTRODUCTION

Phospholipases are a group of enzymes which hydrolyze phospholipids. There are five basic types of phospholipases, A_1 (EC 3.1.1.32), A_2 (EC 3.1.1.4), B (EC 3.1.1.5), C (EC 3.1.4.10), and D (EC 3.1.4.4), which are classified according to the sites of their cleavage on phospholipid substrates (Figure 1). Phospholipase C (PLC) and phospholipase D (PLD) are phosphodiesterases; the former acts on the glycerophosphate ester linkage, whereas the later cleaves the head group-phosphate bond. Phospholipase A_1 (PLA$_1$), phospholipase A_2 (PLA$_2$), and phospholipase B (PLB) are acyl hydrolases; the former two hydrolyze at the *sn*-1 and *sn*-2 acyl ester linkages, respectively, while the latter can act on both acyl positions. The common reaction products of the acyl hydrolases are fatty acids and lysophospholipids. Lysophospholipids may be further degraded by lysophospholipases. The action of PLB can remove both acyl chains and generate a glycerol-phospho-head group.

The occurrence of the various types of phospholipases has been presumed in plants. PLD and inositol phospholipid-specific PLC have been unambiguously documented. PLD has been purified from several plant species,[1-4] and its reaction mechanisms have been investigated in some detail.[5] The polyphospho-inositide-hydrolyzing PLC has received some attention recently because of its possible role in generating second messengers in plants. The PLC activity has been partially characterized in terms of its subcellular location and catalytic properties.[6-12] Partial purification of PLC has been reported from plasma membranes.[13] Reports have also emerged describing acyl-hydrolyzing phospholipase activities in connection with the involvement of their reaction products, fatty acids and lysophospholipids, in mediating cell function,[14-16] but precise identification of PLA$_1$, PLA$_2$, and PLB is yet to be obtained.

Current understanding of the regulation and cellular function of phospholipases in plants is rather limited, particularly when compared with the enzymes in animals. It is now evident that phospholipases in animals may, in addition to their involvement in lipid catabolism, play an important role in regulating various cell functions, particularly, in generating second messengers in the transduction of agonist-dependent signals (Figure 2).[17] There is growing evidence for a role of PLC in plants in transducing signals similar to the role already described in animal systems.[18-20] It has also been proposed that plant PLA may be involved in regulating a number of cellular processes, which are either unique to plants or analogous to animal systems.[21]

This chapter is intended to provide an overview on the current status of research on plant phospholipases. Some comparisons will be made between the plant and animal enzymes, particularly in the areas where a clear grasp of information on the plant enzymes is lacking.

II. PHOSPHOLIPASE D

Some enzymatic properties of PLD were first described with extracts of carrots in 1947,[22] which was prompted by an observation that a very low

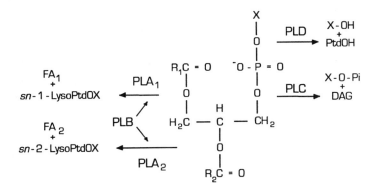

FIGURE 1. Action sites of phospholipases (PL) A_1, A_2, B, C, and D on phospholipid and their reaction products. DAG — diacylglycerol; PtdOH — phosphatidic acid; PtdOX — phospholipid.

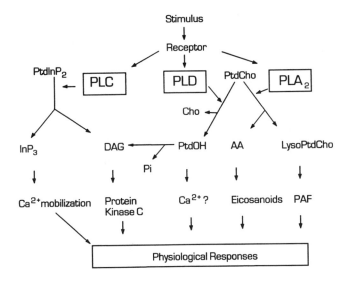

FIGURE 2. A simplified scheme showing the role of phospholipases in animals in generating second messengers during signal transducing processes. These phospholipases are not necessarily coupled with the same receptor. The hydrolysis of phosphatidylinositol 4,5-bisphosphate (PtdInP$_2$) by PLC is coupled with G protein upon stimulus perception. The mediators between signals and PLD activation are not well understood, and they are thought to be G protein, protein kinase C, or Ca^{2+}. PLA$_2$ action on phosphatidylcholine (PtdCho) generates arachidonic acid (AA) and lyso-phosphatidylcholine (LysoPtdCho). The later can be acetylated to form platelet-activating factor (PAF) which is a potent metabolic regulator. Cho — choline; InP$_3$ — inositol 1,4,5,-trisphosphate. See Figure 1 for other abbreviations.

amount of choline or lipid nitrogen content was present in phospholipid fractions from carrot slices unless the enzymes were heat inactivated.[23] PLD has since been purified from several plant species (Table 1). More has been learned about PLD than any other plant phospholipases, and the development of this work until the end of the 1970s was thoroughly reviewed.[5,24] Unfortunately,

TABLE 1
Properties of Purified PLD from Plant Sources

Plant source	Mol. Wt. (kDa)		K_m, mM PTCho	pH optimum	pI	Ca^{2+} mM	Specific activity[e]	N-terminal amino acid	Ref.
	Native	Denatured							
Peanut seed	200	22	2.2	5.6	4.7	nd[a]	234	Gly	2
Cabbage leaf	116.6	112.5	2.7[b]	6.3[c]	5.0	0.21[d]	8.8	nd	1
Citrus callus	90.5	90.0	nd	6.5	nd	50	464	nd	3
Rice bran	75.0	nd	1.7	5.9	nd	100	0.4	nd	4
Castor bean	nd	92	2.0	6.5	4.2	20	302	Gln	[f]

Note: nd — not determined

[a] Optimal Ca^{2+} not determined, but Ca^{2+} is required.
[b] Dihexanoyl PtdCho used as substrate.
[c] pH optimum varied with Ca^{2+} concentrations; optimal pH 6.3 and 7.3, respectively, at 50 and 0.5 mM Ca^{2+}.
[d] K_m value for Ca^{2+}.
[e] Specific activity of final purified enzymes, which was defined as micromole substrate hydrolyzed per minute per milligram of protein.
[f] Unpublished data from author's laboratory.

little progress has been made during the last decade, particularly with regard to understanding the *in vivo* control and cellular function of this enzyme.

A. OCCURRENCE

PLD activity is widely distributed in the plant kingdom.[25] Investigations on the intracellular distribution of PLD activity has produced mixed results. PLD has been found in both soluble and membrane-associated forms, with the majority of the enzyme being cytosolic. The enzyme activity in germinating mung bean cotyledons was reported to be associated largely with protein bodies.[26] It appears that PLD does not occur in plastids and mitochondria.[27] Most of the activity in developing peanut seeds was found in membrane-associated fractions, whereas it is completely soluble in dry seeds.[5] The structure and function relationship between soluble and membrane-associated forms of the enzymes is not known. In germinating castor bean endosperm, antibodies raised against purified soluble PLD cross-react with both cytosolic and membrane-associated PLD.[81]

By way of comparison, mammalian PLD was originally thought to be primarily membrane bound.[28] A recent study, however, has shown that the majority of the enzyme activity occurs in the cytosol.[29] The soluble and membrane-associated forms differ in their substrate specificity, their dependence on substrate concentration and divalent cations, and their detergent effect, as well as their chromatographic mobilities on anion exchange and gel filtration columns. An intriguing possibility has been suggested that this enzyme may translocate between the membranous and cytosolic fractions. The intracellular translocation appears to be an important regulatory mechanism which is operative for a number of phospholipid-involving enzymes in animals such as PLA_2,[30] phosphatidylinositol-hydrolyzing PLC,[31] protein kinase C,[32] and cholinephosphate cytidylyltransferase.[33]

B. GENERAL REACTIONS

PLD is a hydrolytic enzyme which releases an aminoalcohol head group such as choline from phospholipids. The hydrolysis products are phosphatidic acid and a water-soluble free head group (Figure 1). However, under special conditions, PLD also catalyzes a transphosphatidylation reaction in which water is replaced by primary alcohols such as methanol and ethanol.[34] The end lipid product is phosphatidylalcohol rather than phosphatidic acid. This reaction is generally higher than its hydrolytic one at alcohol concentrations below a denaturing level. In cabbage leaves, about 80% of the PLD product was identified as phosphatidylethanol in the reaction mixture containing ethanol.[34] The formation of transphosphatidylation product has been used as an indicator of the PLD activity in recent investigations on the distinction between PLC- and PLD-mediated hydrolysis of phospholipids in animal tissues.[28] Phosphatidylalcohol is not further metabolized, while the hydrolytic product, phosphatidic acid, is readily converted to other lipids. The unique transphosphatidylation reaction has also been exploited to generate unusual phospholipids and analogs such as dimethylethanolamine plasmologen, phosphatidylhomoserine, phosphatidylnucleoside, phosphatidylmethanol, and phosphatidylethanol.[35,36]

C. PURIFICATION

The soluble form of PLD has been purified to apparent homogeneity from several plant species (Table 1). Although the purification procedure for one source may not be directly applied to the others, similarities among the procedures used have been noted. A general purification sequence includes (A) high speed centrifugation (20,000 to 100,000 × *g*); (B) bulk fractionation with ammonium sulfate and/or acetone precipitation or heat coagulation; (C) column chromatography with gel filtration, anion exchange, and/or hydrophobic interaction; and finally, native polyacrylamide gel electrophoresis and electro-elution. Nondenaturing gel electrophoresis followed by electro-elution appears to be a crucial step for obtaining highly purified enzyme of peanut seed,[2] citrus callus,[3] and castor bean endosperm.[81]

The reported specific activity of the apparent homogenous PLD varied substantially from the different sources (Table 1). The PLD activity from citrus callus is more than 1000 times higher than that from rice bran. Factors contributing to the differences are not clear, and they may, in addition to the enzyme sources, include the degree of enzyme purity, assay conditions, as well as activity instability.

The molecular structure of PLD remains to be studied. The enzyme from cabbage leaves, citrus callus, and rice bran appears to be composed of one polypeptide, whereas oligomeric polypeptides were reported for the peanut seed enzyme (Table 1). The M_r reported for PLD ranged from 75 to 200 KDa. It is not clear how these proteins from the various sources are related and whether the variability arises from differences in species or specific tissues used for the purification. The membrane-associated PLD has not been purified.

D. CATALYTIC PROPERTIES

The purified PLD from various plant sources shares considerable catalytic similarities (Table 1). It requires divalent cations, particularly Ca^{2+}, for activity, and the optimal concentration is usually high, ranging from 20 to 100 m*M*. PLD activity is stimulated by detergents or organic solvents. Sodium dodecyl sulfate (SDS) is a powerful stimulator of the enzyme, and the optimal concentration varies from 1 to 3 m*M*. K_ms of the purified enzyme from various sources are around 2 m*M* when assayed in the presence of Ca^{2+}. The enzyme had a slightly acidic pH optimum when it was assayed in millimolar concentrations of Ca^{2+}. A shift of pH optimum of cabbage PLD activity from 6.3 to 7.3 was observed when the Ca^{2+} level was reduced from 50 to 0.5 m*M*.

PLD generally exhibits a broad substrate specificity, hydrolyzing a variety of phospholipids.[5] The selectivity of substrates by the enzyme is strongly influenced by the physical states of substrate and by the presence of different *in vitro* modulators. Early studies using crude cell extracts or partially purified enzyme showed that the enzyme hydrolyzed phosphatidylethanolamine with a much higher rate than phosphatidylcholine in the absence of stimulators such as SDS or diethyl ether.[5] On the contrary, with the stimulators, phosphatidylcholine was the much preferred substrate. However, detailed studies on the substrate specificity have not been described with the highly purified PLD. Whether or not there exist distinct forms of PLD with varied substrate specificity is open to question. Different substrate specificity was reported for the soluble and membrane-associated PLD activities of cabbage; the former degraded ethanolamine plasmologen, whereas the later acted on choline plasmologen.[37,38]

E. PHYSIOLOGICAL ROLE

Despite the wide occurrence and high activity of PLD in plants, there has been no agreement in literature on the physiological function of this enzyme. Speculation was originally centered on the involvement of PLD in the degradation of storage and membrane phospholipids. A high activity of this enzyme has been found in storage tissues.[25] During seed germination and seedling growth, changes in PLD activity have been correlated with a decrease in phospholipid content.[26,39] This enzyme in germinating mung bean cotyledons was found in protein bodies together with a number of other hydrolases.[26] It is possible that the change in PLD activity is related to both phospholipid catabolism and changes in membrane lipid composition. Active membrane proliferation is essential in seed germination. The reaction product phosphatidic acid is a central intermediate for synthesis of various glycerolipids.[40]

Changes in PLD activity associated with membrane phospholipid content have been described in relation to freeze injury,[41,42] water stress,[43] pathogen infection,[44] and senescence.[45] In cold- and freezing-injured tissues, a loss of membrane phospholipids was attributed to an elevated activity of this enzyme which was thought to result from the breakdown of compartmentation that allows PLD to come in contact with membranes.[41,42] Studies suggest that phosphatidylcholine and phosphatidylinositol in senescent membranes are attacked initially by PLD, and the resulting phosphatidic acid is further hydrolyzed to diacylglycerol by phosphatidic acid phosphatase.[45] The PLD activity of microsomal membranes was activated by physiological concentrations of Ca^{2+}.[46] It is, however, difficult to distinguish whether the increase in PLD activity is an onset of the stress-related membrane deterioration, a consequence of release of PLD from its sequestered or compartmentalized form, or due to the availability of Ca^{2+}. In any case, the activity of this enzyme may manifest the stress injury by further deteriorating cellular membranes. On the other hand, PLD activity which prefers phosphatidylcholine with oxidized acyl groups has recently been observed in microsomal membranes of castor bean, safflower, rapeseed, wheat, and avocado.[47] One of its possible roles may be to protect membranes from oxidative changes.

Recent studies in animal systems suggest that PLD plays an important role in signal transduction pathways (Figure 2; for review see Billah and Anthes,[28] and references therein). PLD-catalyzed breakdown of phosphatidylcholine has been observed within seconds or minutes in response to a variety of agonists and other factors including hormones, neurotransmitters, growth factors, chemotactic compounds, phorbol esters, and hypotonic shock. Phosphatidic acid generated as a result of PLD has been shown to be a potential cellular mediator, and it can also be converted by a phosphatidate phosphohydrolase to diacylglycerol. The later is a known activator of protein kinase C, which phosphorylates a variety of cellular proteins and thereby modulates a variety of physiological responses. The participation of G protein (guanosine triphosphate [GTP] binding regulatory proteins), protein kinase C, and Ca^{2+} has been recently shown to regulate PLD activity.[48-50] In some cell types, the D-type hydrolysis of phosphatidylcholine followed by a phosphatidate phosphohydrolase appears to be the major source for diacylglycerol generation. The increase in diacylglycerol level in many animal systems occurs in a biphasic manner. The early phase coincides with phosphatidylinositol 4,5-bisphosphate hydrolysis which is catalyzed by a C-type phospholipase (Figure 2; discussed below), whereas diacylglycerol in the later phase comes from the D-type cleavage of phosphatidylcholine. These results suggest that the D-type hydrolysis of phosphatidylcholine acts in concert with the C-type breakdown of polyphosphoinositide.

The involvement of the PLD-catalyzed hydrolysis of phosphatidylcholine in transmembrane signaling is yet to be studied in plants. Scattered reports, however, exist on the changes in PLD activity and phosphatidylcholine hydrolysis in response to various stimuli. Alteration in the PLD activity was reported in response to initial imbibition temperature during pea seed germination.[51] Inoculation of wheat leaves with brown rust pathogen resulted in an increase in the enzyme activity followed by a decrease of this activity.[44] Auxin-stimulated rapid turnover of phospholipids (effect seen in 5 min) was reported for soybean hypocotyls and isolated microsomal membranes.[21,52] Although the basis for these effects remains to be established, a D-type hydrolysis was assumed. In wheat aleurone layers, gibberellin stimulated the turnover of phosphatidylcholine.[53] An immediate rise (within 1 min) in the hydrolysis of phosphatidylcholine together with that of phosphatidylinositol has been recently described in hypoosmotic shock of green alga *Dunaliella salina*[54,55] and in irradiation of etiolated *Brassica* hypocotyls.[56] However, it should be emphasized that the enzymatic basis for the rapid breakdown of phosphatidylcholine described above is not clear. Further studies are needed to establish whether the PLD-mediated hydrolysis of phosphatidylcholine in plants plays a role analogous to animal tissues in transmembrane signaling.

III. PHOSPHOLIPASE C

A. OCCURRENCE

Enzymatic evidence for the presence of PLC in plants was first reported in 1955.[57] The release of water-soluble organic phosphate from egg lecithin was observed with particulate preparations of carrots, spinach, sugar beet, and cabbage. The PLC activity that hydrolyzes phosphatidylcholine has since been reported largely in cytosol of several plant species including germinating pea,[51] rice bran,[58] *Lilium* pollen,[11] and spinach leaves.[59] The spinach PLC has been partially purified.[56] However, catalytic properties and substrate specificity of the phosphatidylcholine-hydrolyzing PLC have not been described. Recently, a group of PLC activities which are specific for inositol phospholipids has been found to be primarily associated with plasma membranes (Table 2).[6-13]

According to the substrate specificity, PLC may be grouped into several subclasses. The best-studied groups of PLCs in animals and plants are specific for inositol phospholipids. Another group of PLC generally shows a broad substrate specificity, but does not hydrolyze polyphosphoinositides. In animals, the other subclasses include a lysosomal PLC which acts on preferentially peroxidized phospholipids and an activity that is specific for 1-*O*-alkylphosphatidylcholine.[60,61] Diacylglycerol generated by phosphatidylinositol-specific enzyme stimulates protein kinase C, while 1-*O*-alkyl, 2-acyl-glycerol from 1-*O*-alkylphosphatidylcholine-specific PLC inhibits protein kinase C.[61] The different specificity in animals is postulated to play an important role in a complex regulation of protein kinase C.

B. CONSIDERATIONS IN THE ACTIVITY ASSAY

A main concern for the reported plant PLC activities is whether or not the assay procedures were specific to PLC, since the reaction products of PLC and PLD are interconvertible via the activity of kinases and phosphatases (Figure 3). The various methods used for assaying phosphatidylcholine-hydrolyzing PLC in plants include measuring inorganic phosphate released by a coupled alkaline phosphatase assay;[57] using an artificial substrate, p-nitrophenolphosphocholine;[58] and analyzing formation of diacylglycerol.[59] Given phosphatidylcholine as the substrate, the formation of phosphocholine can come from either C-type cleavage or D-type linked with a choline kinase reaction (Figure 3). The production of diacylglycerol can result from either C-type action or D-type followed by a phosphohydrolase. The activity of PLD is generally much higher than that of PLC, and specific or nonspecific phosphatases and kinases are present in most plant tissues. Thus, caution should be exercised, particularly when the PLC activity is assayed. Vigorous verification of all other possible enzymatic reactions is deemed necessary to identify the actual phospholipases involved. For example, when a tissue without much background on the various enzyme activities is utilized, a useful starting approach would be to analyze both water-soluble and lipid-phase reaction products and compare kinetics of their release such as choline, phosphocholine, phosphatidic acid, and diacylglycerol. The criteria for the PLC action will be that (1) diacylglycerol appears before phosphatidic acid and (2) there is stoichiometric production of diacylglycerol and phosphocholine. A recently developed method in animals involves the use of double labeling of cellular phosphatidylcholine with ^{32}P and ^{3}H (acyl group).[28] In the absence of endogenous [^{32}P]-ATP, the formation of [^{32}P]phosphocholine and [^{32}P]phosphatidic acid

TABLE 2
Occurrence and Properties of Inositol Phospholipid-Hydrolyzing PLC in Plants

Plant source	Subcellular fraction	Substrate specificity	Activation	pH	Ref.
Celery stalk	Particulate[a]	PtdIn, PtdInP$_2$	Ca^{2+}, 1 μM; deoxycholate, 0.15%	7.2	7
Wheat root and shoot	Soluble	PtdIn	Ca^{2+}, 5 mM		6
	Plasma	PtdInP, PtdInP$_2$	Ca^{2+}, 10 μM	5.5–6, PtdInP$_2$; 5.5–7.5, PtdInP	8
	Soluble	PtdIn			
Wheat root plasma	Partially purified	PtdInP, PtdInP$_2$	Ca^{2+}, 5–10 μM; Mg^{2+}, 5 mM;[b] deoxycholate, 0.015%	6–6.5, PtdInP$_2$; 6–7, PtdInP	13
Soybean hypocotyl	Plasma	PtdIn	Ca^{2+}, 50 μM	6.6	9
	Soluble	PtdIn	Ca^{2+}, 50 μM	6.6	
Oat root	Plasma	PtdInP$_2$ > PtdInP > PtdIn	Ca^{2+}, 10 μM	nd[c]	10
Lilium pollen	Particulate	PtdIn	Ca^{2+}	nd[c]	11
	Soluble	PtdIn, PtdEtn, PtdCho	Ca^{2+}		
Green alga	Plasma	PtdInP$_2$	Ca^{2+}, 10 μM; GTP	nd[c]	12

Note: PtdCho — phosphatidylcholine; PtdIn — phosphatidylinositol.

[a] 100,000 g precipitates of post-600 g supernatant.
[b] Stimulation requires the presence of Ca^{2+}.
[c] nd — not determined, but pH 6.5 was used for activity assay.

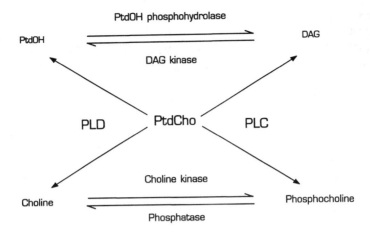

FIGURE 3. Interconversion of reaction products of PLC and PLD via phosphatase and kinase. See Figures 1 and 2 for abbreviations.

must be derived form the attack of PLC and PLD, respectively. Moreover, by comparison of ^{32}P to ^{3}H ratios between phosphatidic acid and phosphatidylcholine, the contribution of PLC and diacylglycerol kinase reactions to phosphatidic acid formation can be estimated. While [^{32}P]phosphatidic acid may come exclusively from PLD action, [^{3}H]phosphatidic acid can be produced by both PLD and PLC, followed by diacylglycerol kinase reaction.

C. INOSITOL PHOSPHOLIPID-SPECIFIC PHOSPHOLIPASE C (PLC)

The occurrence of PLC activity specific for inositol phospholipids has been documented in a number of plant species (Table 2). The PLC activity has been found in both membrane-bound (particularly with plasma membranes) and soluble fractions. In general, the membranous PLC hydrolyzes phosphatidylinositol phosphates, whereas the soluble form acts on either phosphatidylinositol only or a wide range of phospholipids such as phosphatidylcholine, phosphatidylethanolamine, and phosphatidylglycerol, depending on the enzyme sources. The PLC activity for most plants exhibits a neutral pH optimum. Ca^{2+} is required for the enzyme activity. The maximal activation of the membrane-associated PLC is obtained by micromolar Ca^{2+}, whereas millimolar ranges of Ca^{2+} are needed for the soluble PLC.[10] The PLC was solubilized with deoxycholate from wheat plasma membrane and subsequent purification of the enzyme resulted in a 25-fold enhancement of the specific activity.[13] The partially purified enzyme was found to be stimulated five-fold by Mg^{2+} in the presence of Ca^{2+}.

Variation in the PLC properties has also been reported among different plants. A high level of Ca^{2+} was reported to be inhibitory for the enzyme from celery,[6] soybean,[9] and wheat,[8] but a lack of inhibition was also observed for the enzyme from oat.[10] Deoxycholate stimulated the partially purified enzyme activity at a low concentration (0.015%) and severely inhibited its activity at high concentrations.[13] In contrast, the PLC activity in particulate fractions of celery was enhanced by the high level of deoxycholate (0.15%).[7] The hydrolysis rate was reported to be stimulated by GTP in green alga[12] and a crude membrane fraction from cultured cells of *Daucus carota*.[63] However, attempts on other plants failed to observe the GTP-stimulated hydrolysis of inositol phospholipids.[8]

The inositol phospholipid-specific PLC has been purified and its gene has been cloned from a number of mammalian tissues.[31] Comparison of molecular sizes, deduced amino acid sequences, and immunological cross-reactivity of the PLC has revealed that mammalian PLC is a heterogeneous group which can be divided into four types PLC-α, PLC-β, PLC-γ, and PLC-σ. Each type contains several distinct family members. A putative gene encoding a plant endomembrane protein which resembles vertebrate protein disulfide isomerase and PLC-α has been cloned recently, using a human protein disulfide isomerase as heterologous probe.[64] Expression of the cloned gene in *Escherichia coli* showed elevated protein disulfide isomerase activity, but its identity for a plant PLC gene is not established. It is not known, therefore, whether the plant PLC is a member of the PLC types already classified in animal tissues. Whether or not plant researchers will be able to capitalize on the available information on the mammalian PLC genes and proteins for the plant PLC investigation remains to be seen. Recent success in solubilization and partial purification of the PLC from wheat plasma membranes shows promise for obtaining highly purified PLC which may lead subsequently to cloning of the plant PLC genes.

D. PHYSIOLOGICAL ROLE

Much of the interest in the plant PLC stemmed from the animal model that the PLC-catalyzed hydrolysis of polyphosphoinositides plays an important role in transmembrane signaling (Figure 2). The agonist-stimulated hydrolysis of inositol phospholipids by PLC produces at least two second messengers, inositol 1,4,5-trisphosphates and diacylglycerol.[65] The former mobilizes Ca^{2+} from intracellular and extracellular stores. The free Ca^{2+} in cytosol interacts with various proteins such as protein kinase and calmodulin, which initiate and modulate a variety of physiological responses. Diacylglycerol activates phospholipid-dependent protein kinase C, which controls a number of cell functions.

It appears that most of the elements involved in the PLC-polyphosphoinositide signaling pathway as described in animals are also present in plants, and this work has been a subject of several recent reviews.[18-20,66] In addition to the discussed PLC activity, the occurrence of phosphatidylinositol bisphosphate and its hydrolysis products, lipid stimulated protein kinase (but not a typical protein kinase C whose activation requires both diacylglycerol and Ca^{2+}), Ca^{2+} mobilization, and G protein have been described. Changes in inositol 1,4,5-trisphosphate level have been observed in response to a number of hormonal and environmental stimuli.[67-72] Inositol trisphosphate has been shown to stimulate

Ca^{2+} mobilization in various plant membrane systems.[66] An immediate rise in diacylglycerol concentration in plasma membrane has been demonstrated within 1 min by hypoosmotic shock of a green alga.[54,55] However, the cellular processes that are mediated by the changes in diacylglycerol level are not clear in plants. A model plant system is not yet available in which the various components function in concert to achieve end physiological responses.

IV. PHOSPHOLIPASES A_1, A_2, AND B

A. OCCURRENCE

Evidence for the presence of PLAs and PLB in plants is very sparse. PLA activity which hydrolyzes phosphatidylcholine to lysophosphatidylcholine and a free fatty acid has been observed with a plasma membrane fraction of oat roots.[14] PLA activity of protoplast plasma membrane has been reported based on *in vitro* assays measuring the release of fluorescent fatty acid from 1 acyl-2(*N*-4-nitrobenzo-2-oxa-1,3-diazole) aminocaproyl phosphatidylcholine.[16] An increase in lysophospholipid formation has been associated with the application of biological active auxin, but not inactive analogs, to zucchini membranes of hypocotyls, presumably as a result of A-type phospholipase.[15] Other indirect evidence also exists for the occurrence of the acyl-hydrolyzing phospholipases.[73] The precise enzyme basis of the acyl hydrolysis (i.e., A_1-, A_2-, or B-type) remains to be established.

Very little is known about substrate specificity and activity modulation of these reported activities. A PLA_2 specific to ricinoleoyl-phosphatidylcholine has been reported in microsomal membranes of developing castor bean endosperm.[74] PLAs preferring oxygenated acyl groups have also been observed in microsomes of several other plant species.[47] A soluble PLA has been purified to apparent homogeneity from suspension carrot cells.[75] Interestingly, substrate specificity of the isolated PLA is influenced by Ca^{2+} concentration; it exhibits PLA_2 activity in the presence of millimoles of Ca^{2+}, whereas it shows predominantly PLA_1 activity in the presence of micromoles of Ca^{2+}.

It should be pointed out that plant tissues contain a highly active acyl hydrolase which removes fatty acids from glycerol backbone of various glycerolipids including glycolipids, phospholipids, sulfolipids, and mono- and diacylglycerols (see Chapter 13 for more details). In terms of end products, this enzyme could be misidentified as phospholipase A_1, A_2, B, galactolipase, sulfolipase, and monoacyl- or diacyllipase. Whenever the use of crude extract as enzyme sources is involved in assaying PLAs, careful verification is needed to distinguish the possibility of contribution from nonspecific acyl hydrolases.

B. PHYSIOLOGICAL ROLE

The most completely studied group of phospholipases in animals is the abundant pancreatic PLA_2 and its apparent role is involved in food digestion. Animals also possess distinct intracellular PLA_2.[17,30] The PLA_2 activity has been documented to be a key enzyme in stimulus-response signaling, and its

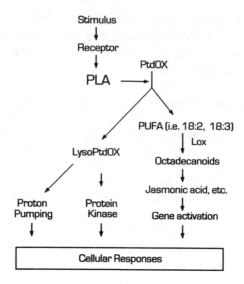

FIGURE 4. A proposed scheme for the functional significance of phospholipase A in plant cellular metabolism. PUFA — polyunsaturated fatty acid; Lox — lipoxygenase. See Figure 1 for other abbreviations.

activity results in the release of arachidonic acid (Figure 2). Arachidonic acid is an immediate precursor for the synthesis of eicosanoids, a group of 20-carbon, potent metabolic regulators. The function of PLA_1 and PLB in animals are less well-defined. In addition to their possible function as digestive enzymes, PLA_1 and PLB may play an important role in membrane remodeling, involving a deacylation-reacylation cycle and transacylation.[76,77]

PLAs in plants have been proposed to play an important role in cell activation and regulation (Figure 4). One line of information comes from the investigations on the effects of PLA reaction products, lysophospholipids and fatty acids, on various cellular processes. Lysophospholipids and fatty acids have been shown to activate proton pumping and affect the activities of a number of membrane enzymes in plants (see review by Morré[21] and references therein). These include stimulation of the activities of auxin-elevated NADH-oxidase, 1,3-β-D-glucan synthetase, phytase, protein kinase, and vanadate-sensitive ATPase. The other line of information derives from the observations of the cellular processes which lead to the A-type hydrolysis of phospholipids. For example, application of auxin to zucchini hypocotyl membranes was found to stimulate the formation of lysophospholipids.[15] The release of lysophospholipids in the membrane fractions was coupled with auxin reception on plasma membrane, since blockage of auxin receptors with antibodies also inhibited the hydrolysis of phospholipids. Thus, the PLA activity has also been suggested as an integral step signaling auxin-mediated events in isolated zucchini plasma membrane.

The recent discovery of PLAs that prefer oxygenated acyl groups in plant membranes suggests further physiological functions for PLAs.[47,74] In developing castor bean endosperm, the PLA activity may be involved in seed triacylglycerol synthesis by supplying free ricinoleic acid from phosphatidylcholine. The removal of oxidized acyl groups from membrane phospholipids may play a role in protecting membranes from oxidative damages, releasing precursors for hormone synthesis, and generating second messengers.

The acyl-hydrolyzing phospholipase activities in plants have been implicated in octadecanoid signaling cascades (Figure 4). Octadecanoids refer to the oxygenated 18-carbon fatty acids which are plant equivalent of the animal 20-carbon eicosanoids.[78] These compounds can be the precursor of metabolic regulatory substances such as jasmonic acid. A recent study has shown that the octadecanoid precursors of jasmonic acid, hydroperoxy-linolenic acid, phytodienoic acid, and linolenic acid act as signals for induction of proteinase inhibitor in tomato leaves.[79] In the proposed fatty acid-based signaling scheme, a lipase on plasma membrane is implicated to couple with the stimulus perception to initiate the signaling cascade. The precise nature of the lipase (phospholipases, nonspecific acyl hydrolases, galatolipases) remains to be defined. Senescence in plants has been hypothesized to be mediated by a PLA_2 fatty acid cascade.[73] In the proposed model, senescence is triggered by release of Ca^{2+} into cytoplasm, which in turn activates PLA_2. This activation results in release of linoleic and linolenic acids, which is then oxidized by lipoxygenase to generate a series of potent substances including oxy-free radicals, ethylene, and jasmonic acid. The cycling of this process would eventually result in senescing membranes. However, there is lack of direct evidence on the activation of PLA activities during these physiological processes (for more information on jasmonic acid, see Chapter 5).

V. PERSPECTIVE

Phospholipases are ubiquitous enzymes present in all tissues of plants examined, and their functional significance in plants as a whole may be considered in three general aspects: (1) lipid catabolism, (2) changing membrane lipid composition, and (3) cell regulation. The suggestive evidence for the catabolic role is the changes of phospholipase activities in stress-damaged tissues and during seed germination. Phospholipids are essential constituents of cell membranes and serve to build the hydrophobic barrier between the internal and external environments. The breakdown of phospholipids by phospholipases will inevitably alter membrane integrity, which is crucial for cell viability and normal function.

The participation of phospholipases in changing membrane lipid composition remains an open question in plants. The reaction products of one phospholipid by phospholipases are potential substrates for synthesis of other lipids. Lysophospholipids from PLA actions may be reacylated, and the deacylation-reacylation cycle is important in animal membrane remodeling. Diacylglycerol and phosphatidic acid from the PLC and PLD activities are central intermediates for the synthesis of various phospholipids, triacylglycerols, and glycolipids.

It was speculated that the transphosphatidylation by PLD might contribute to phosphatidylglycerol formation.[34] Phospholipases with specificities to both polar and acyl groups have been suggested to occur in senescent membranes.[80] Activation of such phospholipases may result in a selective decline in certain populations of phospholipids, hence altering membrane composition. The change of phospholipid composition may play an important role in various cellular processes, since many cell functions depend on appropriate membrane lipid composition.

In terms of cell regulation, a central and challenging issue is to identify the processes that are regulated by phospholipases and their products. The exciting development in current investigations of phospholipases deals with the involvement of phospholipases in signal transduction. The occurrence of inositol phospholipid-specific PLC in plasma membranes of plants has been used as one of the indicators that the hydrolysis of polyphosphoinositides is involved in transmembrane signaling analogous to animals. PLAs have been suggested as an alternative route in mediating plant responses to various signals.[21] Hypothesis and methodology in this area have been heavily influenced by those used in mammalian systems. Although the animal models have provided useful parallel to plant studies, properties unique to plant systems have also been encountered,[19,21] and more can be expected.

However, progress in understanding the physiological roles of phospholipases in plants will be hindered unless better knowledge on the enzymology and regulation of these enzymes is obtained. Measurement of PLD activity *in vitro* shows that this enzyme is highly active and destructive. Hence this enzyme *in vivo* must be properly compartmentalized or sequestered from its substrates. It is also likely that multiple forms of the enzyme occur in plants and they may reside and be regulated differently. Further investigations on PLD should capitalize on the availability of purified enzyme from several plant species by obtaining antibodies and genes for this enzyme. Studies using antibodies and molecular approaches should enable us to address some important questions, including the enzyme's structure, intracellular distribution, cell- and tissue-specific expression, developmental changes, control at translation and transcription level, and multiple forms and genes encoding the enzyme. Answers to these questions would provide valuable insights into the regulation and function of this enzyme in plant metabolism.

One of the directions on the studies of polyphosphoinositide-specific PLC in the near future will be to obtain highly purified enzyme. Indeed, progress on solubilization and partial purification has been reported.[13] Isolation of this enzyme will allow detailed studies of the enzyme substrate specificity, activity modulation, and catalytic kinetics. Knowledge of the mechanisms by which the PLC activity is regulated is essential for establishing a direct link of the enzyme activity with stimulus-signaling processes.

Despite the attractiveness of the role proposed for PLAs in cell regulation and signal transduction in plants, A-type phospholipases are the least understood group of plant phospholipases. There seem now to be fewer doubts than before on the presence of the enzyme, but a challenge is to distinguish it from the highly active, but nonspecific acyl hydrolases. Systematic studies are needed on the distribution, reaction mechanisms, substrate specificity, and activity modulation of the enzymes.

In summary, phospholipases are a diverse group of enzymes. Their roles, which by no means are the same, and importance in plant cellular metabolism, growth, and development are being gradually recognized. There are growing research interests in the question of whether or not the hydrolysis of membrane phospholipids in plants constitute one of the mechanisms to coordinate plant response to hormonal and external stimuli. The search of the physiological functions of lipid-derived metabolites will stimulate investigations on the mechanism and control of phospholipases. With the advent of molecular biology and advanced analytical techniques, probing rapid changes in cellular metabolites, ion flow, and gene expression, it is hoped that rapid progress will be made in the future for a better understanding of the regulation and cell function of phospholipases in plants.

ACKNOWLEDGMENTS

The author is grateful to Mrs. Hallie Fulmer for typing the manuscript and to Mr. Dave Manning for his assistance in the illustrations. This work was in connection with a project of the Kansas State Agricultural Experimental Station (No. 93-67-B) and is published with approval of the director.

REFERENCES

1. **Allgyer, T. T. and Wells, M. A.**, Phospholipase D from savoy cabbage: purification and preliminary kinetic characterization, *Biochemistry*, 24, 5348, 1979.
2. **Heller, M., Mozes, N., Peri, I., and Maes, E.**, Phospholipase D from peanut seeds. IV. Final purification and some properties of the enzyme, *Biochim. Biophys. Acta*, 369, 397, 1974.
3. **Witt, W., Yelenosky, G., and Mayer, R. T.**, Purification of phospholipase D from citrus callus tissue, *Arch. Biochem. Biophys.*, 259, 164, 1987.
4. **Lee, M. H.**, Phospholipase D of rice bran. I. Purification and characterization, *Plant Sci.*, 59, 25, 1989.
5. **Heller, M.**, Phospholipase D, *Adv. Lipid Res.*, 16, 267, 1978.
6. **Irvine, R. F., Letcher, A. J., and Dawson, R.M.C.**, Phosphatidylinositol phosphodiesterase in higher plants, *Biochem. J.*, 192, 279, 1980.
7. **McMurray, W. C. and Irvine, R. F.**, Phosphatidylinositol 4,5-bisphosphate phosphodiesterase in higher plants, *Biochem. J.*, 249, 877, 1988.
8. **Melin, P. M., Sommarin, M., Sandelius, A. S., and Jergil, B.**, Identification of Ca^{2+} stimulated polyphosphoinositide phospholipase C in isolated plant plasma membranes, *FEBS Lett.*, 223, 87, 91, 1987.
9. **Pfaffman, H., Hartmann, E., Brightman, A., and Morré, D. J.**, Phosphatidylinositol specific phospholipase C of plant stems: membrane associated activity concentrated in plasma membranes, *Plant Physiol.*, 85, 1151, 1987.

10. **Tate, B. F., Schaller, G. E., Sussman, M. R., and Crain, R. C.**, Characterization of a polyphopshoinositide phospholipase C from plasma membrane of *Avena sativa, Plant Physiol.*, 91, 1275, 1989.

11. **Helsper, P.F.G., de Groot, P.F.M., Linskens, H. F., and Jackson, J. F.**, Phosphatidyl-inositol phospholipase C activity in pollen of *Lilium longiflorum, Phytochemistry*, 25, 2053, 1986.

12. **Einspahr, K. J., Peeler, T. C., and Thompson, G. A., Jr.**, Phosphatidyl 4,5-bisphosphate phospholipase C and phosphomonoesterase in *Danaliella salina* membranes, *Plant Physiol.*, 90, 1115, 1989.

13. **Melin, P.-M., Pical, C., Jergal, B., and Sommariu, M.**, Polyphosphoinositide phospholipase C in wheat root plasma membranes, *Biochim. Biophys. Acta*, 1123, 163, 1992.

14. **Palmgren, M. G., Sommarin, M., Ulvskov, P., and Jorgensen, P. L.**, Modulation of plasma membrane H$^+$-ATPase from oat roots by lysophosphatidylcholine, free fatty acids and phospholipase A$_2$, *Physiol. Plant.*, 74, 11, 1988.

15. **Andre, B. and Sherer, G.F.E.**, Stimulation by auxin of phospholipase A in membrane vesicles from an auxin-sensitive tissue is mediated by an auxin receptor, *Planta*, 185, 209, 1991.

16. **Dengler, L. A., Rincon, M., and Boss, W. F.**, NBD-PC: a tool to study endosytosis and phospholipase activity in plant protoplasts, in *Cell-Free Analysis of Membrane Traffic*, Morre, D. F., Howell, K., Cook, G.M.W., and Evans, W. H., Eds., Alan R. Liss, New York, 1988, 291.

17. **Dennis, E. A., Rhee, S. G., Billah, M. M., and Hannun, Y. A.**, Role of phospholipases in generating lipid second messengers in signal transduction, *FASEB J.*, 5, 2068, 1991.

18. **Einspahr, K. J. and Thompson, G. A.**, Transmembrane signaling via phosphatidylinositol 4,5-bisphosphate hydrolysis in plants, *Plant Physiol.*, 93, 361, 1990.

19. **Lehle, L.**, Phosphatidyl inositol metabolism and its role in signal transduction in growing plants, *Plant Mol. Biol.*, 15, 647, 1990.

20. **Sandelus, A. S. and Sommarin, M.**, Membrane-localized reactions involved in polyphosphoinositide turnover in plants, in *Inositol Metabolism in Plants*, Morré, D. J., Boss, D. F., and Loewus, F. A., Eds., Wiley-Liss Press, New York, 1990, 139.

21. **Morré, D. J.**, Activation of phospholipase A: an alternative mechanism for signal transduction, in *Inositol Metabolism in Plants*, Morré, D. J., Boss, D. F., and Loewus, F. A., Eds., Wiley-Liss Press, New York, 1990, 227.

22. **Hanahan, D. J. and Chaikoff, I. L.**, A new phospholipid splitting enzyme specific for an ester linkage between the nitrogenous base and the phosphoric acid group, *J. Biol. Chem.*, 169, 699, 1947.

23. **Hanshan, D. J. and Chaikoff, I. L.**, The phosphorus-containing lipids of the carrot, *J. Biol. Chem.*, 168, 233, 1946.

24. **Galliard, T.**, Degradation of acyl lipids: hydrolytic and oxidative enzymes, in *The Biochemistry of Plants*, Stumpf, P. K., Ed., Academic Press, New York, 1980, 85.

25. **Quarles, R. H. and Dawson, R. M. C.**, Distribution of phospholipase D in developing and mature plants, *Biochem. J.*, 112, 787, 1969.

26. **Herman, E. M. and Chrispeels, M. J.**, Characteristics and subcellular localization of phospholipase D and phosphatidic acid phosphatase in mung bean cytoledons, *Plant Physiol.*, 66, 1001, 1980.

27. **Clermont, H. and Douce, R.**, Localization of phospholipase activity in plant tissues. 1. Absence of phospholipase D activity in isolated mitochondria and plastids, *FEBS Lett.*, 9, 284, 1970.

28. **Billah, M. M. and Anthes, J. C.**, The regulation and cellular functions of phosphatidyl-choline hydrolysis, *Biochem. J.*, 269, 281, 1990.

29. **Wang, P., Athens, J. C., Siegel, M. I., Egan, R. W., and Billah, M. M.**, Existence of cytosolic phospholipase D. Identification and comparison with membrane-bound enzyme, *J. Biol. Chem.*, 266, 14877, 1991.

30. **Channon, J. Y. and Leslie, C. C.**, A calcium-dependent mechanism for associating a soluble arachidonyl-hydrolysing phospholipase A_2 with membrane in the macrophage cell line RAW 264.7, *J. Biol. Chem.*, 265, 5409, 1990.

31. **Rhee, S. G., Suh, P. G., Ryu, S. H., and Lee, S. Y.**, Studies of inositol phospholipid-specific phospholipase C, *Science*, 244, 546, 1989.

32. **Kraft, A. S. and Anderson, W. B.**, Phorbol esters increase the amount of Ca^{2+}, phospholipid-dependent protein kinase associated with plasma membrane, *Nature*, 301, 621, 1983.

33. **Pelech, S. L. and Vance, D. E.**, Regulation of phosphatidylcholine biosynthesis, *Biochim. Biophys. Acta*, 779, 219, 1984.

34. **Yang, S. F., Freer, S., and Benson, A. A.**, Transphosphatidylation by phospholipase D, *J. Biol. Chem.*, 242, 477, 1967.

35. **Shuto, S., Imamura, S., Fukakukawa, K., Sakakibara, H., and Murese, J.**, A facile one-step synthesis of phosphatidylhomoserine by phospholipase D catalysed trans-phosphatidylation, *Chem. Pharm. Bull.*, 35, 447, 1987.

36. **Eibl, H. and Kovatchev, S.**, Preparation of phospholipids and their analogs by phospholipase D, *Methods Enzymol.*, 72, 632, 1981.

37. **Lands, W. E. M. and Hart, P.**, Metabolism of plasmologen III. Relative reactivities of acyl and alkenyl derivatives of glycerol-3-phosphorylcholine, *Biochim. Biophys. Acta*, 98, 532, 1965.

38. **Slotboom, A. J., de Haas, G. H., and van Deenen, L.L.M.**, On the synthesis of plasmalogens, *Chem. Phys. Lipids*, 1, 192, 1967.

39. **Lee, M. H.**, Phospholipase of rice bran. II. The effects of the enzyme inhibitors and activators on the germination and growth of root and seedling of rice, *Plant Sci.*, 59, 35, 1989.

40. **Moore, T. S.**, Phospholipid biosynthesis, *Annu. Rev. Plant Physiol.*, 33, 235, 1982.

41. **Willemot, C.**, Rapid degradation of polar lipids in frost damaged winter wheat crown and root tissue, *Phytochemistry*, 22, 861, 1983.

42. **Yoshida, S.**, Freezing injury and phospholipid degradation *in vivo* in woody plant cells, *Plant Physiol.*, 64, 241, 1979.

43. **Chetal, S., Wagle, D. S., and Nainawatee, H. S.**, Phospholipase D activity in leaves of water-stresses wheat and barley, *Biochem. Physiol. Pflanzen*, 177, 92, 1982.

44. **Saini, R. S., Chawla, H.K.L., and Wagle, D. S.**, Catalytic activity of two phosphoric diester hydrolases in wheat leaves inoculated with brown rust, *Puccininia recondita*, *Biol. Plant.*, 32(4), 313, 1990.

45. **Paliyath, G. and Thompson, J. E.**, Calcium- and calmodulin-regulated breakdown of phospholipid by microsomal membranes from bean cotyledons, *Plant Physiol.*, 83, 63, 1987.

46. **Paliyath, G., Lynch, D. V., and Thompson, J. E.**, Regulation of membrane phospholipid catabolism in senescing carnation flowers, *Physiol. Plant.*, 71, 503, 1987.

47. **Banas, A., Johasson, I., and Stymne, S.**, Plant microsomal phospholipases exhibit preference for phosphatidylcholine with oxygenated acyl groups, *Plant Sci.*, 84, 137, 1992.

48. **Hurst, K. K., Hughes, B. P., and Barritt, G. J.**, The roles of phospholipase D and a GTP-binding protein in guanosine 5'-[r-thio]triphosphate-stimulated hydrolysis of phosphatidylcholine in rat liver plasma membranes, *Biochem. J.*, 272, 749, 1990.

49. **Pai, J. K., Pachter, J. A., Weinstein, J. B., and Bishop, W. R.**, Overexpression of protein kinase C B1 enhance phospholipase D activity and diacylglycerol formation in phorbol ester-stimulated rat fibroblasts, *Proc. Natl. Acad. Sci. U.S.A.*, 88, 598, 1991.

50. **Huang, R., Kucera, G. L., and Rittenhouse, S. E.**, Elevated cytosolic Ca^{2+} activates phospholipase D in human platelets, *J. Biol. Chem.*, 266, 1652, 1991.

51. **Di Nola, L. and Mayer, A. M.**, Effect of temperature on glycerol metabolism in membranes and on phospholipases C and D of germinating pea embryos, *Phytochemistry*, 25, 2255, 1986.

52. **Morré, D. J., Gripshwer, B., Monroe, A., and Morre, J. J.**, Phosphatidylinositol turn-over in isolated soybean membranes stimulated by the synthetic growth hormone 2,4-dichlorophenoxyacetic acid, *J. Biol. Chem.*, 259, 15364, 1984.

53. **Hetherington, P. R. and Laidman, D. L.**, Influence of gibberellic acid and the Rht3 gene on choline and phospholipid metabolism in wheat aleurone tissue, *J. Exp. Bot.*, 42, 1357, 1991.

54. **Ha, K. S. and Thompson, G. A.**, Diacylglycerol metabolism in *Dunaliella salina* and its possible role in transmembrane signalling, *Plant Physiol.*, 96S, 43, 1991.

55. **Ha, K. S. and Thompson, G. A.**, Biophasic changes in the level and composition of *Dunallella-Salina* plasma membrane diacylglycerols following hypoosmotic shock, *Biochemistry*, 31, 596, 1992.

56. **Acharya, M. K., Dureja-Munjal, I., and Guha-Mukherjee, S.**, Light-induced rapid changes in inositolphospholipids and phosphatidylcholine in *Brassica* seedlings, *Phytochemistry*, 30, 2895, 1991.

57. **Kates, M.**, Hydrolysis of lecithin by plant plastid enzymes, *Can. J. Biochem. Physiol.*, 33, 575, 1955.

58. **Chrastil, J. and Parrish, F. W.**, Phospholipases C and D in rice grains, *J. Agric. Food Chem.*, 35, 624, 1987.

59. **Oursel, A., Grenier, G., and Tremolieres, A.**, Isolation and assays of purification of one soluble phospholipase C from spinach leaves, in *Plant Lipid Biochemistry, Structure and Utilization*, Quinn, P. J. and Harwood, J. L., Eds., Portland Press, London, 1990, 316.

60. **Gamache, D. A., Fawzy, A. A., and Fanson, R. C.**, Preferential hydrolysis of peroxidized phospholipid by lysosomal phospholipase C, *Biochim. Biophys. Acta*, 958, 116, 1988.

61. **Nishihira, J. and Ishibashi, T.**, A phospholipase C with a high specificity for platelet activating factor in rabbit liver light mitochondria, *Lipids*, 21, 780, 1986.

62. **Daniel, L. W., Small, G. W., and Schmitt, J. D.**, Alkyl-linked diglycerides inhibit protein kinase C activation by diacylglycerides, *Biochem. Biophys. Res. Commun.*, 151, 291, 1988.

63. **Zbell, B., Walter-Back, C., and Bucher, H.**, Evidence of an auxin-mediated phosphoinositide turnover and an inositol 1,4,5-trisphosphate effect on isolated membranes of *Daucus carota* L., *J. Cell Biochem.*, 40, 1, 1989.

64. **Shorrosh, B. and Dixon, R. A.**, Molecular cloning of a putative plant endomembrane protein resembling vertebrate protein disulfide-isomerase and a phosphatidylinositol-specific phospholipase C, *Proc. Natl. Acad. Sci. U.S.A.*, 88, 10941, 1991.

65. **Berridge, M. J.**, Inositol triphosphate and diacylglycerol as second messengers, *Biochem. J.*, 220, 345, 1984.

66. **Ricon, M. and Boss, W. F.**, Second-messenger role of phosphoinositides, in *Inositol Metabolism in Plants*, Morré, D. J., Boss, D. F., and Loewus, F. A., Eds., Wiley-Liss Press, New York, 1990, 173.

67. **Morse, M. J., Crain, R. C., and Satter, R. L.**, Light-stimulated phosphatidyl inositol turnover in *Samanea saman* pulvini, *Proc. Natl. Acad. Sci. U.S.A.*, 84, 7075, 1987.

68. **Zbell, B. and Walter-Back, C.**, Signal transduction of auxin on isolated plant cell membranes: indications for a rapid polyphosphoinositide response stimulated by indoleacetic acid, *J. Plant Physiol.*, 133, 353, 1988.

69. **Murthy, P. P. N., Renders, J. M., and Keranenl, L. M.**, Phosphoinositides in barley aleurone layers and gibberellic acid-induced changes in metabolism, *Plant Physiol.*, 91, 1266, 1989.

70. **Ettlinger, C. and Lehle, L.**, Auxin induces rapid changes in phosphatidylinositol metabolism, *Nature*, 331, 176, 1988.

71. **Einspahr, K. J., Peeler, T. C., and Thompson, G. A., Jr.**, Rapid changes in polyphosphoinositide metabolism associated with the response of *Dunaliella salina* to hypoosmotic shock, *J. Biol. Chem.*, 263, 5775, 1988.

72. **Srivastava, V. and Guha-Mukherjee, S.**, Phosphatidylinositol turnover in crown gall tumour, *Phytochemistry*, 31, 773, 1992.

73. **Leshem, Y. Y.**, Membrane phospholipid catabolism and Ca^{+2} activity in control of senescence, *Physiol. Plant*, 69, 551, 1987.

74. **Bafor, B., Smith, M. A., Johansson, L., Stobart, A.K., and Stymne, S.**, Ricinoleic acid biosynthesis and triacylglycerol assembly in microsomal preparations from developing castor bean (*Ricinus commuis*) endosperm, *Biochem. J.*, 280, 597, 1991.

75. **Brglez, I., Kuralt, M.-C., and Boss, W. F.**, Purification and characterization of phospholipase A isolated from culture medium, *Plant Physiol.*, 99S, 717, 1992.

76. **Kuwae, T., Schmid, P. C., and Schmid, H.H.O.**, Assessment of phospholipid deacylation-reacylation cycles by a stable isotope technique, *Biochem. Biophys. Res. Commun.*, 142, 86, 1987.

77. **Waite, M. and Sisson, P.**, Studies on the substrate specificity of the phospholipase A_1 of the plasma membrane of rat liver, *J. Biol. Chem.*, 249, 6401, 1974.

78. **Vick, B. A. and Zimmerman, D. C.**, Oxidative system for modification of fatty acid: the lipoxygenase pathway, in *The Biochemistry of Plants: Lipids*, Vol. 9, Stumpf, P. K. and Conn, E. E., Academic Press, New York, 1987, 53–90.

79. **Farmer, E. F. and Ryan, C. A.**, Octadecanoid precursors of jasmonic acid activate the synthesis of wound-inducible proteinase inhibitors, *Plant Cell*, 4, 129, 1992.

80. **Brown, J. H., Chamber, J. A., and Thompson, J. E.**, Distinguishable patterns of phospholipid susceptibility to catabolism in senescing carnation petals, *Phytochemistry*, 30, 2537, 1991.

81. **Zheng, L., Dyer, J. H., Zhang, Y., and Wang, X.**, Characterization of phospholipase D purified from castor bean, *Plant Physiol.*, 99S, 447, 1992.

Chapter 16

CATABOLISM OF FATTY ACIDS
(α- AND β-OXIDATION)

Bernt Gerhardt

TABLE OF CONTENTS

I. INTRODUCTION

A capacity for fatty acid degradation determined by *in vitro* studies has been demonstrated in all higher plant tissues examined to date. *In vivo* experiments using a variety of plant tissues and techniques presented evidence that fatty acid degradation happens in higher plants. A massive carbon flux in consequence of fatty acid degradation occurs only in lipid-storing nutrient tissues of seeds when the lipid reserves are mobilized (and utilized for sucrose synthesis) during germination,[1,2] but in the majority of plant tissues, fatty acid degradation is surely not a major metabolic process. Nevertheless, fatty acid degradation should be important in every cell, considering the harmful properties of free fatty acids, which may result from both protein and membrane lipid turnover in mature, nonlipid-storing tissues. Utilization of fatty acids or those amino acids which are catabolized via acyl coenzyme A (CoA) intermediates as respiratory substrate in higher plants has been demonstrated[3,4] or suggested[4a,4b] in very few cases. Fatty acid degradation causes, however, initial wound respiration which may reflect breakdown of cellular membranes.[5]

Catabolism of fatty acids — in the strict sense of the term — occurs by the oxidative process known as β-oxidation. It proceeds by oxidation in the β-position of the fatty acid carbon chain to the carboxyl group and sequential removal of C_2 units (acetyl-CoA). Where the molecular structure of fatty acids causes barriers to the sequential removal of the C_2 units, the basic reaction sequence of the β-oxidation pathway is flanked by additional reactions allowing degradation of the fatty acids by the pathway in principle. In higher plants, metabolism of fatty acids by basic, supplemented or modified β-oxidation results in complete degradation of fatty acids to their constituent acetyl units.

Besides β-oxidation, there are other oxidative processes acting on fatty acids: α-oxidation, ω-oxidation, in-chain oxidation, and the lipoxygenase pathway. With respect to fatty acid catabolism, the importance of α-oxidation resulting in the removal of a C_1 unit (CO_2) from fatty acids is presently not well-defined. However, this chapter will include current knowledge on α-oxidation. In-chain oxidation and ω-oxidation of fatty acids generate hydroxyl, oxo, and epoxy derivatives and are involved in the formation of polyfunctional fatty acids, constituents of surface lipid polymers (cutin, suberin). In-chain oxidation and ω-oxidation as biosynthetic rather than catabolic processes are outside the scope of this chapter. The lipoxygenase pathway and its possible functions are covered in Chapter 5.

The review is strictly related to and traces only current knowledge of fatty acid catabolism in higher plants. At any level, the aspect of comparing fatty acid catabolism in higher plants with that in other organisms (including algae) is generally not covered in order to achieve perspicuity of the central topic. Reviews on β-oxidation in mammalian cells have been published very recently.[6,7]

II. β-OXIDATION

In 1956, Stumpf and Barber,[8] studying fatty acid degradation by *in vitro* systems from lipid-mobilizing tissues of germinating oilseeds, suggested that plant cells degrade fatty acids by the β-oxidation pathway known for mammalian and microbial cells at that time. Subsequent studies, mainly from Stumpf's laboratory, established the existence of the β-oxidation pathway in lipid-mobilizing tissues of germinating seeds. The pathway was ascribed to mitochondria based on the results of plant cell fractionation studies. Later studies on lipid-mobilizing tissues of germinating seeds showed that the β-oxidation activity was associated primarily with the soluble protein fraction.[9,10] In 1969, Cooper and Beevers,[11] working with organelles separated from castor bean (*Ricinus communis* L.) endosperm on sucrose density gradients, discovered and established that the β-oxidation pathway is located in this tissue within fragile organelles, referred to as glyoxysomes.[12] Following the discovery by Cooper and Beevers,[11] it was shown for lipid-mobilizing tissues of germinating seeds of various species that β-oxidation is located in glyoxysomes, organelles characteristic of such tissues. Concerning nonlipid-storing tissues, i.e., the majority of higher plant tissues, fatty acid degradation (β-oxidation) was still considered to be a mitochondrial process. But there was little, if any, experimental support for this assumption.

A. SUBCELLULAR LOCATION

Advances in cell fractionation led to the discovery of glyoxysomes,[12] organelles characteristic of lipid-mobilizing tissues of germinating seeds. According to ultrastructure and marker enzyme (catalase, H_2O_2-generating oxidase) composition, glyoxysomes represent a subcategory of peroxisomes which are common organelles of higher plant cells.[13,14] Among plant peroxisomes, glyoxysomes are characterized by housing the glyoxylate cycle required for the conversion of stored lipid to sucrose during germination of oilseeds. Directly following their discovery, glyoxysomes were also recognized to contain the β-oxidation pathway.[11]

From 1981 to 1983, the first reports were published which presented evidence that the ability to carry out β-oxidation is not restricted to glyoxysomes among plant peroxisomes.[15-18] When examined since, β-oxidation enzymes and/or β-oxidation activity have been demonstrated in peroxisomes isolated from photosynthetic tissue, roots, and other plant tissues/organs which are devoid of storage lipids. Moreover, recent studies which will be reviewed in the next sections established that peroxisomes are able to metabolize physiologically relevant fatty acids of different molecular structure and to degrade completely the carbon chain of these fatty acids to acetyl-CoA. Thus, current experimental evidence strongly supports the concept[19] that the enzymes of fatty acid catabolism are common constituents of the higher plant peroxisome and that the peroxisomes of higher plant cells, independent of tissue-specific functions (e.g., conversion of acetyl-CoA to succinate via the glyoxylate cycle in tissues mobilizing reserve lipids; involvement in photorespiration in photosynthetic tissues), generally represent a compartment competent for fatty acid catabolism which follows in principle the β-oxidation pathway.[20]

Fatty acid catabolism by β-oxidation is located in and a general and basic function of the higher plant peroxisome which, on the other hand, is a regular

constituent of the higher plant cell. There are also reports on fatty acid degrading activity (β-oxidation) located in plant mitochondria. They will be
treated in Section H. Up to that section, all information presented is related
to the peroxisomal system of fatty acid catabolism.

B. FATTY ACID ACTIVATION

Prior to degradation by β-oxidation, fatty acids have to be activated to an
acyl-CoA. Two mechanisms by which fatty acids are activated in higher plant
peroxisomes are known: activation by acyl-CoA synthetase and activation by
oxidative decarboxylation.

1. Acyl-CoA Synthetase

Activation of straight-chain fatty acids occurs in a reaction requiring ATP,
Mg^{2+}, and CoA-SH[21,22] and resulting in the formation of acyl-CoA, AMP, and
pyrophosphate in a 1:1:1 stoichiometry.[22] This observation demonstrates that
straight-chain fatty acids are activated by an acyl-CoA synthetase (EC 6.2.1.3)
reaction. The acyl-CoA synthetase has not yet been purified, and the following
kinetic properties of the enzyme are based on studies using broken peroxisomes
as the enzyme source.

Straight-chain fatty acids of different chain length are activated by acyl-
CoA synthetases with different rates.[21-23] Optimum activity of the enzymes has
mainly been observed with C_{12}, C_{14}, or C_{16} saturated fatty acids or C_{18} unsaturated
fatty acids. Short-chain fatty acids (including propionic acid) are rather poor
substrates for acyl-CoA synthetases (reaction rates $\leq$ 30% of the optimum
activity observed). Half maximum rates of activation in most cases were
obtained at substrate concentrations in the range 3 to 30 μM. Activities of some
acyl-CoA synthetases are listed in Table 1.

As long as flux rates through fatty acid degrading pathways are unknown,
physiological aspects on fatty acid activating systems may pose the question
as to whether the molecular structure of the fatty acids which are, or are thought
to be, substrates for fatty acid degradation within a tissue is mirrored by the
substrate specificity of the fatty acid activating system. Ricinoleic acid which
comprises more than 80% of the fatty acids esterified in the storage
triacylglycerols of castor bean endosperm is activated in glyoxysomes from
this tissue at rates somewhat lower than those obtained with palmitate as
substrate (which in this case is a rather unphysiological substrate).[21,24] Based
on data reported by Cooper,[21] the rate of ricinoleate activation accounts for
approximately 40% of the rate required to support the calculated carbon flux
from fat to sucrose in the castor bean endosperm. Acyl-CoA synthetases from
tissues which contain appreciable amounts of oleate esterified in their storage
triacylglycerols showed, in some cases, low or no activity with oleate as
substrate.[23] The acyl-CoA synthetase of glyoxysomes isolated from rape
(*Brassica napus* L.) cotyledons did not activate the erucic acid predominant
in the fatty acid components of the reserve lipids of the rape variety studied.[23]

TABLE 1
Relative Activities of β-Oxidation Enzymes in Peroxisomal Fractions Isolated from Lipid-Mobilizing (A) and Other (B) Tissues

			Multifunctional protein		
		Activity related to acyl-CoA oxidase activity			
Peroxisomes isolated from:	Acyl-CoA synthetase	Acyl-CoA oxidase	Enoyl-CoA hydratase	3-OH-acyl-CoA-DH	Thiolase
A. Cucumis sativus L., cotyledons		1	27	0.7	0.03
Ricinus communis L., endosperm	0.5	450[a]	9167[a]	217[a]	67[a]
Helianthus annuus L., cotyledons	0.8	1 (3.3)[b]	15	15	1.0
B. Spinacia oleracea L., leaves	0.2	1 (7.7)	64	8.4	0.3
Pisum sativum L., leaves	0.3	1 (0.6)	8.3	0.5	0.5
Vigna radiata L., hypocotyls	1.0	1 (0.7)	9.6	0.7	0.3
Zea mays L., seedling roots	1.3	1 (1.0)	37	2.5	2.2
Solanum tuberosum L., tubers	0.1	1 (0.4)	40	1.8	0.5
P. sativum L., cotyledons		1 (1.4)	16	1.5	0.2
		1 (0.4)	2.8	2.5	0.3

Note: Relative activities were calculated from data compiled by Gerhardt,[19] except those concerning cucumber cotyledons; these were calculated from data reported by Kindl.[26a] Acyl-CoA synthetase and acyl-CoA oxidase activities assayed with C_{16} substrates; activities of multifunctional protein and thiolase assayed with C_4 substrates.

a   Specific activities (nkat mg^{-1}) of the purified enzymes.
b   The values in parentheses give the absolute activities (nkat (mg organelle protein)$^{-1}$) of acyl-CoA oxidase.

Thus, current data on the *in vitro* substrate specificity of the peroxisomal acyl-CoA synthetase from lipid-mobilizing tissue of oilseeds are only partly in accord with the *in vivo* demand on the enzyme.

The peroxisomal acyl-CoA synthetase from mung bean (*Vigna radiata* L.) hypocotyls, in which free fatty acids may result from membrane lipid turnover, activates oleate, linoleate, and linolenate with at least twice the rate obtained with palmitate as substrate.[22] However, the enzyme also activates ricinoleate at a relative rate comparable to that obtained with the acyl-CoA synthetase of glyoxysomes from the castor bean endosperm.[24] This indicates that the substrate specificity of an acyl-CoA synthetase is not necessarily related to physiological requirements.

2. Oxidative Decarboxylation

Intermediates of the catabolism of the branched-chain amino acids leucine, isoleucine, and valine are branched-chain 2-oxo acids which are further catabolized by β-oxidation or a β-oxidation system including additional or modified reactions (Section II.E.4). Therefore, the branched-chain 2-oxo acids have to be activated prior to degradation. In the presence of branched-chain 2-oxo acid, CoA-SH, and NAD, peroxisomes from mung bean hypocotyls catalyze the formation of CO_2 and NADH and acyl-CoA.[25] The acyl-CoA contains one carbon atom less than the branched-chain 2-oxo acid used as substrate. CO_2 release, NADH, and acyl-CoA formation occur in a 1:1:1 stoichiometry. According to these data, the branched-chain 2-oxo acids are activated by oxidative decarboxylation. The underlying branched-chain 2-oxo acid dehydrogenase complex has not yet been purified or characterized. Using mung bean hypocotyl peroxisomes as an enzyme source, rates of 50 to 100 pkat per mg of peroxisomal protein have been reported for the activation of 2-oxoisocaproate, 2-oxo-3-methylvalerate, and 2-oxoisovalerate, the branched-chain 2-oxo acids formed by transamination from leucine, isoleucine, and valine, respectively.

Hydroxyl or oxo groups located at an even-numbered carbon atom of a straight-chain fatty acid form a barrier to the degradation of the fatty acid by continuous β-oxidation. Fatty acids carrying such a functional group occur in some storage triacylglycerols of oilseeds, as, for example, ricinoleic acid (D-12-hydroxy-9-*cis*-octadecenoic acid) in the castor bean oil or licanic acid (4-oxo-9,11,13-all *cis*-octadecatrienoic acid) in the oiticica oil (oil of *Licania regida* Benth.). It has been demonstrated that the catabolism of ricinoleic acid initiated by the acyl-CoA synthetase reaction and subsequent β-oxidation leads to the formation of intermediates which are not esterified to CoA-SH when the degradation of the carbon chain reaches the region of the hydroxyl group (Section II.E.2).[24] Following oxidation of the hydroxyl group, the intermediate 2-oxooctanoate is formed. It is metabolized next to heptanoyl-CoA. Co-factor requirements (CoA-SH, NAD) of this reaction indicate that the 2-oxooctanoate is activated by oxidative decarboxylation.

By analogy to the catabolism of ricinoleic acid, it may be supposed that the catabolism of straight-chain fatty acids containing a hydroxyl or oxo group at an even-numbered carbon atom involves two activation reactions: (1) activation of the parent fatty acid by acyl-CoA synthetase and (2) activation of a 2-oxo acid intermediate by oxidative decarboxylation. The latter activation reaction acting at the intermediate level where the functional group forms a barrier to further β-oxidation allows return of the catabolic process to the acyl-CoA track of β-oxidation.

C. β-OXIDATION REACTION SEQUENCE

Following activation of a fatty acid, the acyl-CoA formed can be catabolized. In the simplest case represented by the catabolism of straight-chain, saturated, common fatty acids, the acyl-CoA next undergoes a series of four reactions (β-oxidation reaction sequence). The intermediates (acyl-CoA thioesters) of the β-oxidation reaction sequence in higher plant peroxisomes appear to be identical with those established for the β-oxidation reaction sequence in mammalian mitochondria and in bacteria. However, this statement is based predominantly on the type of the individual enzymatic reactions involved in β-oxidation instead of identification of the intermediates themselves or the products of the individual enzymatic reactions.

The β-oxidation reaction sequence (Figure 1) starts by oxidation of acyl-CoA to 2-*trans*-enoyl-CoA. Subsequent *trans* addition of water leads to the formation of L-3-hydroxyacyl-CoA, which is then oxidized to 3-oxoacyl-CoA. The 3-oxoacyl-CoA is cleaved by a thiolytic step with CoA-SH to acetyl-CoA and an acyl-CoA containing two carbon atoms less than the parent acyl-CoA.

Except for the thiolytic cleavage step, which has not yet been studied closely, the individual steps of the β-oxidation reaction sequence are catalyzed in peroxisomes by enzymes quite different from those catalyzing the corresponding steps in mammalian mitochondria and in bacteria. At present, the enzymes of the β-oxidation reaction sequence of higher plant peroxisomes have only been purified from glyoxysomes of cucumber (*Cucumis sativur* L.) cotyledons. Catalytic properties of the enzymes have however been studied using not only the purified proteins, but using also broken peroxisomes as the enzyme source. Data on the activity of the enzymes catalyzing the β-oxidation reaction sequence in higher plant peroxisomes are given in Table 1.

1. First Oxidation Step

The first β-oxidation step of the peroxisomal β-oxidation reaction sequence leading from acyl-CoA to 2-*trans*-enoyl-CoA is an oxygen-dependent reaction. Reaction products are H_2O_2[11,16] and enoyl-CoA.[16] Oxygen uptake and H_2O_2 and enoyl-CoA formation occur in a 1:1:1 stoichiometry.[16] Thus, the two electrons resulting from acyl-CoA oxidation are transferred to oxygen yielding the reaction product H_2O_2. The electron transfer does not require an electron carrier other than that bound to the enzyme as the prosthetic group (FAD).

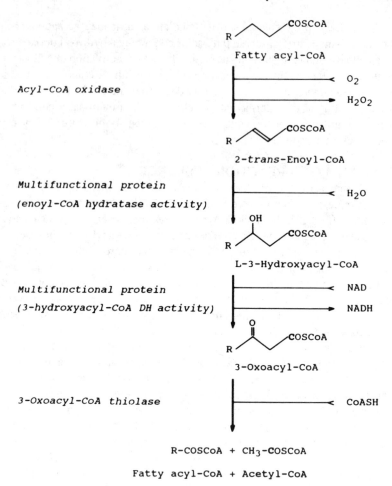

FIGURE 1. Peroxisomal β-oxidation reaction sequence.

Therefore, the enzyme catalyzing the first oxidation step of the β-oxidation reaction sequence is an oxidase, named acyl-CoA oxidase (EC 1.1.3.-).

The plant acyl-CoA oxidase, a homodimer (M_r of 150 kDa),[26] is believed to be a FAD-containing protein like the acyl-CoA oxidase of mammalian[27] and yeast[28] peroxisomes. The purified enzyme of glyoxysomes from cucumber cotyledons preferentially acted on long-chain acyl-CoAs; the activity toward linoleoyl-CoA and linolenoyl-CoA was, however, low in comparison to that obtained with other long-chain fatty acids.[26] The K_m values observed with palmitoyl-CoA, stearoyl-CoA, and oleoyl-CoA as substrates were in the range of 20 μM; for butyryl-CoA, the K_m value was five times higher. Similar results on the substrate specificity of acyl-CoA oxidase were obtained with spinach leaf (*Spinacia oleracea* L.) peroxisomes.[29,30] The enzyme exhibited both highest activity and highest affinity toward lauroyl-CoA and myristoyl-CoA (half maximum rates at approximately 10 μM substrate concentration) and low activity and affinity toward short-chain acyl-CoAs (half maximum rate with butyryl-CoA at approximately 30 μM substrate concentration). Activity and affinity of the enzyme toward unsaturated C_{18} fatty acids were intermediate with respect to those observed with short-chain and medium-chain saturated fatty acids.

The acyl-CoA oxidase of mung bean hypocotyl peroxisomes, which in other respects exhibited kinetic properties similar to those outlined above for the enzyme from other sources, showed an additional, pronounced optimum of activity with butyryl-CoA as substrate.[29,30] The affinity of the enzyme toward butyryl-CoA (half maximum rate at 60 μM substrate concentration) was, however, low in comparison to that observed with medium-chain and long-chain acyl-CoAs. Activity with butyryl-CoA considerably higher than that obtained with palmitoyl-CoA has also been reported for the acyl-CoA oxidase of peroxisomes from Jerusalem artichoke (*Helianthus tuberosus* L.) tubers and pea (*Pisum sativum* L.) cotyledons (no data are available on the affinity of the enzymes toward substrates of different chain length).[17,18] The acyl-CoA oxidases of peroxisomes from rape cotyledons and maize (*Zea mays* L.) scutellum showed a very strong preference for C_{12} and/or C_{14} acyl-CoA with respect to their activity;[23] in contrast, the affinity of these enzymes to these acyl-CoAs was exceptionally low (half maximum rates at 100 to 200 μM substrate concentration) in comparison to their affinities to long-chain acyl-CoAs. Thus, the data currently available on the kinetic properties of plant acyl-CoA oxidases indicate that activity and affinity of the enzyme toward acyl-CoAs of increasing chain length do not always change in parallel.

2. Hydration Step and Second Oxidation Step

The reactions leading from 2-*trans*-enoyl-CoA to L-3-hydroxyacyl-CoA and from L-3-hydroxyacyl-CoA to 3-oxoacyl-CoA are catalyzed by one and the same protein[31] which in addition to its enoyl-CoA hydratase (EC 4.2.1.17) and L-3-hydroxyacyl-CoA dehydrogenase (EC 1.1.1.35) activity exhibits still a third enzymic activity involved in fatty acid degradation (Section II.E.1). The multifunctional protein uses NAD as acceptor of the electrons resulting from the oxidation of L-3-hydroxyacyl-CoA. The occurrence of NAD in peroxisomes has been demonstrated for glyoxysomes from castor bean endosperm.[32] Different mechanisms leading to reoxidation of the NADH formed have been proposed (Section II.G.3).

In glyoxysomes from cucumber cotyledons (and peroxisomes from green leaves of *Lens culinaris* L.), two isoforms of the multifunctional protein have been demonstrated.[33] The two monomeric isoforms (M_r of 74 and 76.5 kDa) differ in their molecular structure and kinetic properties. The ratio of 2-enoyl-CoA hydratase activity to L-3-hydroxyacyl-CoA dehydrogenase activity, both activities determined with the corresponding C_{10} substrate, was for the 76.5-

kDa protein three times higher (approximately 10.0) than for the 74-kDa protein. The K_m value of both 2-enoyl-CoA hydratase activities for crotonyl-CoA was approximately 150 μM, but the 76.5-kDa protein showed a K_m value (30 μM) for hexenoyl-CoA distinctly lower than that of the 74-kDa protein. The 2-enoyl-CoA hydratase activity of glyoxysomes from cotton (*Gossypium hirsutum* L.) cotyledons has been reported to decrease (100-fold) with increasing chain length of the enoyl-CoA (C_4 to C_{16}).[34] The K_m value (in the micromolar range) for the enoyl-CoAs of chain length from C_4 to C_{16} increased approximately tenfold over the chain length range studied.

Data on kinetic properties of the L-3-hydroxyacyl-CoA dehydrogenase activity of the multifunctional protein have not been reported up to now.

3. Thiolytic Cleavage Step

So far, there are no reports concerning kinetic properties of the 3-oxoacyl-CoA thiolase (EC 2.3.1.16), a homodimeric protein (M_r of 90 kDa),[35] which catalyzes the final step of the β-oxidation reaction sequence.

D. β-OXIDATION PATHWAY

The β-oxidation reaction sequence acting completely on an acyl-CoA generates, besides acetyl-CoA, an acyl-CoA which contains two carbon atoms less than the parent acyl-CoA. If the acyl-CoA formed by the β-oxidation reaction sequence undergoes another round of the β-oxidation reaction sequence, it is shortened again by a C_2 unit. Thus, at each passage through the β-oxidation reaction sequence, an acetyl-CoA is removed from the acyl-CoA which entered the first round of the β-oxidation reaction sequence. Successive removal of acetyl-CoA by continuous repetition of the β-oxidation reaction sequence finally leads to complete degradation of the parent fatty acid. With a few exceptions however (Section II.E.4), straight-chain, saturated, common fatty acids (carbon chain length >3) can only be catabolized completely by continuous repetition of the β-oxidation reaction sequence, the β-oxidation pathway in the strict sense.

It is well-established that peroxisomes isolated from different tissues of higher plants and from various species and when provided with a straight-chain, saturated, common fatty acid (plus ATP, CoA, and NAD) or acyl-CoA (plus CoA and NAD) catalyze the formation of NADH (3-hydroxyacyl-CoA dehydrogenase reaction) as well as the formation of acetyl-CoA. This indicates that the enzyme activities catalyzing the individual steps of the β-oxidation reaction sequence in the higher plant peroxisome (Section II.C) are linked to form a capacity for complete β-oxidation. The stoichiometry of oxygen uptake (acyl-CoA oxidase reaction) and NADH and acetyl-CoA formation during palmitoyl-CoA oxidation has been determined to be 1:1:1,[11] corresponding to the stoichiometry of the β-oxidation reaction sequence.

Peroxisomal β-oxidation assays are preferentially performed with palmitate, or palmitoyl-CoA, as a substrate. However, only a few data have been reported on the dependency of the β-oxidation rate on the chain length of the straight-chain, saturated acyl-CoA used as substrate. At low substrate concentration (about 10 μM; below the critical micellar concentration of the substrate), the rate increased up to about fourfold with an increase in chain length of the acyl-CoA (C_4 to C_{16}).[11,36] No dependency on the chain length was observed at high acyl-CoA concentrations (200 μM).[37]

With respect to physiological aspects, it is also important to know whether straight-chain, saturated, common fatty acids (acyl-CoAs) of long chain length are degraded completely by the β-oxidation system of the higher plant peroxisome, since an alternate (mitochondrial) β-oxidation system has not been demonstrated unequivocally in higher plant cells (Section II.H). Complete degradation of the fatty acid type considered here has been demonstrated for a cell-free system prepared from peanut (*Arachis hypogaea* L.) cotyledons.[8] Labeled CO_2 was evolved by the cell-free system when it was provided with (15-^{14}C)palmitate and additions required to allow reactivity of the tricarboxylic acid cycle. Capability of the β-oxidation system of higher plant peroxisomes for complete degradation of straight-chain, saturated fatty acids of long chain length is supported by the facts that medium- and short-chain acyl-CoAs, intermediates of long-chain acyl-CoA catabolism, are metabolized by the enzymes of the β-oxidation reaction sequence (Section II.C) and butyryl-CoA is degraded to acetyl-CoA by peroxisomes.[36] Finally, direct evidence for complete palmitate catabolism in assay mixtures containing peroxisomes as the enzyme source has been obtained recently. Using substrate concentrations <10 μM and following product formation until it ceased, it was shown that 7 nmol of NADH per 1 nmol of palmitoyl-CoA were formed[38] and that each nanomole of [U-^{14}C]palmitate disappearing from the reaction mixture gave rise to 8 nmol of [^{14}C]acetyl-CoA.[39] A few temporarily accumulating intermediates were observed during [U-^{14}C]palmitate catabolism (Section F).

E. MODIFICATIONS OF THE β-OXIDATION PATHWAY

Catabolism of straight-chain, saturated, common fatty acids (carbon chain length >3) is accomplished by the β-oxidation pathway. Catabolism of physiologically relevant fatty acids possessing other molecular structures is, in general, more complex. Double bonds of naturally occurring unsaturated fatty acids are in the *cis* configuration which forms a barrier to continuous β-oxidation. Hydroxyl or oxo groups at even-numbered carbon atoms of a fatty acid are another barrier to continuous passages of the fatty acid through the β-oxidation reaction sequence. Branching of fatty acids at certain positions of the carbon chain also interferes with steps of the β-oxidation reaction sequence. Reactions in addition to those of the β-oxidation reaction sequence allow circumvention of the barriers when the degradation of the fatty acid by the β-oxidation pathway reaches the intermediate level where the barrier prevents further degradation by the β-oxidation reaction sequence. Following circumvention of the barrier, the fatty acid chain may be degraded further by the β-oxidation pathway. The reactions by which barriers are circumvented involve intermediates esterified to CoA-SH in the case where the barrier is formed by a double bond, but where intermediates not esterified to CoA-SH are formed following circumvention of the barrier the catabolic pathway returns to the acyl-CoA track of β-oxidation by a new activation reaction (oxidative decarboxylation; Section II.B.2). Intermediates not esterified to CoA-SH also occur at the catabolism of propionate to acetyl-CoA and CO_2.

1. Modifications Due to Double Bonds

A rather high percentage of the fatty acids contained in storage triacylglycerols and structural lipids are (poly)unsaturated. Naturally occurring unsaturated fatty acids usually have the double bond(s) in the *cis* configuration. This configuration forms a barrier to the degradation of unsaturated, straight-chain fatty acids by repetitive passages through the β-oxidation reaction sequence. Degradation by the β-oxidation pathway up to the point of the double bond yields an intermediate which cannot be metabolized by an enzyme of the β-oxidation reaction sequence. The barrier is circumvented by enzymes additional to those of the β-oxidation reaction sequence. Different auxiliary enzymes and acyl-CoA intermediates are involved, depending on whether the double bond extends from an odd-numbered or even-numbered carbon atom of the fatty acid carbon chain. Following circumvention of the barrier, degradation by the β-oxidation pathway continues (at least up to the next barrier caused by a *cis* double bond). The acyl-CoA track of β-oxidation is never left at catabolism of unsaturated, straight-chain fatty acids.

Degradation by the β-oxidation pathway of a straight-chain, unsaturated fatty acid with a double bond extending from an odd-numbered carbon atom (e.g., 9-*cis*-octadecenoic acid, oleic acid) yields a 3-*cis*-enoyl-CoA intermediate (3-*cis*-dodecenoyl-CoA) when the position of the double bond has been reached (cf., Figure 2). The enoyl-CoA formed by acyl-CoA oxidase and functioning as intermediate in the β-oxidation reaction sequence has, however, the 2-*trans* configuration. The 3-*cis*-enoyl-CoA intermediate is converted into the 2-*trans*-enoyl-CoA by action of a Δ²,Δ³-enoyl-CoA isomerase. Occurrence of this isomerase in plant tissue was first reported by Hutton and Stumpf[40] who observed [^{14}C]acetyl-CoA formation from [8-^{14}C]-3-*cis*-dodecenoic acid by a crude particle preparation from castor bean endosperm, as well as formation of [^{14}C]acetyl-CoA from [10-^{14}C]oleic acid by glyoxysomes from the same tissue. Using 3-*cis*-dodecenoyl-CoA as substrate, the intermediate formed at the point of the double bound at oleic acid degradation, isomerase activity was demonstrated in glyoxysomes from cotton cotyledons.[37]

The Δ²,Δ³-enoyl-CoA isomerase (EC 5.3.3.8) of glyoxysomes from cucumber cotyledons was purified (specific activity: 0.7 μkat/mg) and characterized very recently.[41] The enzyme, a homodimer (M_r of 50 kDa), catalyzes the conversion of 3-*cis*-enoyl-CoA into 2-*trans*-enoyl-CoA. The reaction is reversible. Neither 2-*cis*-enoyl-CoA, which is also a substrate for the enoyl-CoA hydratase ac-

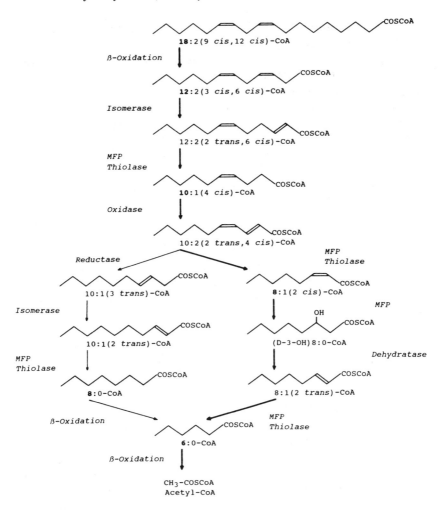

FIGURE 2. Scheme of peroxisomal linoleoyl-CoA catabolism including the alternative pathways from 2-*trans*, 4-*cis*-decadienoyl-CoA to hexanoyl-CoA. MFP — multifunctional protein.

tivity of the multifunctional protein (Section II.C.2), 4-*cis*-enoyl-CoA, nor 2-*trans*,4-*cis*-dienoyl-CoA, substrate of the 2,4-dienoyl-CoA reductase (see below), are accepted as substrates by the isomerase.

Besides 3-*cis*-enoyl-CoA, 3-*trans*-enoyl-CoA is also converted to 2-*trans*-enoyl-CoA by the isomerase.[41] When the activity of the enzyme toward 3-*cis*- and 3-*trans*-hexenoyl-CoA was compared, the activity with the *cis* isomer was approximately 30 times higher than that observed with the *trans* isomer. Increase of the chain length of 3-*trans*-enoyl-CoA from C_6 to C_{12} increased the relative activity of the enzyme, reaching a rate with 3-*trans*-dodecenoyl-CoA comparable to that observed with 3-*cis*-hexenoyl-CoA (see below). Activity of the purified enzyme toward 3-*cis*-dodecenoyl-CoA, intermediate at oleate catabolism, was not assayed.

The Δ^2,Δ^3-enoyl-CoA isomerase exhibited considerable substrate inhibition with all CoA esters used as substrates.[41] A K_m value of 170 μM was estimated for the substrate 3-*cis*-hexenoyl-CoA.

Degradation by the β-oxidation pathway of an unsaturated, straight-chain fatty acid with the *cis* double bond(s) extending from an even-numbered carbon atom (e.g., 6-*cis*-octadecenoic acid, petroselenic acid) yields a 2-*cis*-enoyl-CoA intermediate (2-*cis*-tetradecenoic acid) when the position of the double bond has been reached (cf., Figure 2). As the 2-enoyl-CoA hydratase activity of the multifunctional protein does not show specificity for the geometrical configuration of the double bond at position 2, hydration of the 2-*cis*-enoyl-CoA intermediate is catalyzed by the multifunctional protein (as is the hydration of the 2-*trans*-enoyl-CoA intermediate formed by the activity of the acyl-CoA oxidase). However, hydration of 2-*cis*- and 2-*trans*-enoyl-CoA yields isomers of the 3-hydroxyacyl-CoA formed. Hydration of 2-*cis*-enoyl-CoA by the enoyl-CoA hydratase reaction results in the formation of D-3-hydroxyacyl-CoA, which is not a substrate for the subsequent 3-hydroxyacyl-CoA dehydrogenase reaction. The D-3-hydroxyacyl-CoA can be epimerized to its L isomer by a 3-hydroxyacyl-CoA epimerase (EC 5.1.2.3) activity. Using D-3-hydroxydecanoyl-CoA as substrate, 3-hydroxyacyl-CoA epimerase activity has been demonstrated in and partially purified from glyoxysomes of cucumber cotyledons.[42] The activity has been attributed to three distinct proteins. Two of them were the two isoforms of the multifunctional protein (Section II.C.2). The third, monofunctional protein (M_r of 50 kDa) carried the predominant proportion of the overall epimerase activity and showed immunological relationship to both isoforms of the multifunctional protein.

Following the demonstration of 3-hydroxyacyl-CoA epimerase activity in plant peroxisomes,[42] it has been found in rat liver peroxisomes that the epimerization of 3-hydroxyacyl-CoAs consists of two reaction steps and that a distinct 3-hydroxyacyl-CoA epimerase does not exist.[43] The "epimerase activity" results from dehydration of D-3-hydroxyacyl-CoA to 2-*trans*-enoyl-CoA by a novel D-3-hydroxyacyl-CoA dehydratase and subsequent hydration of the 2-*trans*-enoyl-CoA to L-3-hydroxyacyl-CoA by the enoyl-CoA hydratase activity of the multifunctional protein.

Two homodimeric isoforms (M_r of 65 kDa) of the novel D-3-hydroxyacyl-CoA dehydratase which exhibit very similar kinetic and molecular properties have been very recently purified from glyoxysomes of cucumber cotyledons.[44] It is assumed that the two isoforms represent the earlier described monofunctional "epimerase" protein which carried the predominant proportion of the overall epimerase activity in cucumber cotyledons (see above). Both D-3-hydroxyacyl-CoA dehydratases reversibly catalyzed the conversion of D-3-hydroxydecanoyl-CoA to 2-*trans*-decenoyl-CoA, as demonstrated by product identification, and were inactive toward L-3-hydroxydecanoyl-CoA or 2-*cis*-decenoyl-CoA. The purified isoforms exhibited a specific activity of 4 μkat/mg with 3-hydroxydecenoyl-CoA as substrate. The activity/affinity of the isoforms with/ to 2-*trans*-decenoyl-CoA (K_m = 9.5 μ*M*) was approximately 5 to 10 times higher than that with/to 2-*trans*-butenoyl-CoA.

Activity of the D-3-hydroxyacyl-CoA dehydratase allows circumvention of the barrier, resulting in the degradation of unsaturated, straight-chain fatty acids from a *cis* double bond extending from an even-numbered carbon atom. According to the enzymatic composition of peroxisomes, however, there exists the potential possibility to surmount this barrier by another reaction. β-Oxidation of an unsaturated, straight-chain fatty acid containing a *cis* double bond extending from an even-numbered carbon atom leads at one point to the formation of a 4-*cis*-enoyl-CoA intermediate. Oxidation of this intermediate by the acyl-CoA oxidase yields a 2-*trans*,4-*cis*-dienoyl-CoA. Besides the multifunctional protein, there is another enzyme known which uses 2-*trans*,4-*cis*-dienoyl-CoA as substrate (Figure 2). The enzyme, 2,4-dienoyl-CoA reductase (EC 1.3.1.-), which has been purified from mammalian mitochondria[45] and peroxisomes[46] and characterized reduces 2-*trans*,4-*cis*- as well as 2-*trans*,4-*trans*-dienoyl-CoA to 3-*trans*-enoyl-CoA in an NADPH-dependent reaction. In order to return to the β-oxidation pathway, the 3-*trans*-enoyl-CoA formed by the 2,4-dienoyl-CoA reductase must subsequently be isomerized to 2-*trans*-enoyl-CoA by the isomerase (see above). Using 2-*trans*,4-*trans*-decadienoyl-CoA (cf., Figure 2) as substrate, 2,4-dienoyl-CoA reductase activity has been demonstrated in glyoxysomes from cucumber cotyledons.[42] Occurrence of 2,4-dienoyl-CoA reductase activity (assayed with sorbyl-CoA as substrate) has also been reported for peroxisomal fractions from pineapple (*Ananas comosus* Merr.) fruit tissue.[47]

Enzymes in addition to those of the β-oxidation reaction sequence are required to achieve catabolism of unsaturated, straight-chain fatty acids containing *cis* double bonds. *Cis* double bonds form a barrier which prevents continuous β-oxidation and must be surmounted. Enzymes which fulfill this task have been demonstrated in higher plant peroxisomes and purified from cucumber cotyledons (see above). The most common unsaturated, straight-chain fatty acids are oleic, linoleic, and linolenic acid. Figure 2 represents the catabolic pathway of linoleoyl-CoA as summarized from current knowledge. The catabolic pathways of oleic and linolenic acid, as well as of other unsaturated, straight-chain fatty acids, can easily be deduced from the pathway shown in Figure 2.

Completeness of oleic acid (18:1, 9-*cis*) degradation in higher plant peroxisomes has been demonstrated using glyoxysomes from sunflower (*H. annuus* L.) cotyledons or peroxisomes from potato (*Solanum tuberosum* L.) tubers as the enzyme source and 1 nmol of [18-^{14}C]oleate as substrate.[39] After a lag phase, continuous formation of [^{14}C]acetyl-CoA (identified as such by different methods) from the substrate was observed until the substrate was nearly completely consumed. Temporary accumulation of a few intermediates occurred during [18-^{14}C]oleate catabolism (Section F). Activity of the isomerase toward 3-*cis*-dodecenoyl-CoA, intermediate during oleate catabolism (cf. Figure 1), has been demonstrated.[37]

Complete degradation of linoleic acid (18:2, 9-*cis*, 12-*cis*) by higher plant peroxisomes is indicated by the following results. In addition to [^{14}C]acetyl-CoA, other products finally did not accumulate when peroxisomes metabolized [U-^{14}C]linoleate (1 μ*M*) (Section F).[39] The amount of [^{14}C]acetyl-CoA formed after nearly complete consumption of the linoleate corresponded to that calculated for complete degradation of the linoleate consumed. Under steady state conditions of [U-^{14}C]linoleate degradation, accumulation was not observed at the C_{12}, C_{10}, or C_8 intermediate level[42] where the barriers to continuous passages through the β-oxidation reaction sequence have to be surmounted (Figure 2).

Considering the question as to whether linoleate is prevalently catabolized via the pathway involving D-3-hydroxyacyl-CoA dehydratase or the pathway involving 2,4-dienoyl-CoA reductase (Figure 2), there are strong arguments, with respect to higher plant peroxisomes, in favor of the former pathway. (1) Complete degradation of linoleate as well as the rate of linoleate catabolism were uneffected by NADPH (or NADH) required for participation of the 2,4-dienoyl-CoA reductase in linoleate degradation.[39] (2) The activity of the D-3-hydroxyacyl-CoA dehydratase in glyoxysomes from cucumber cotyledons was found to be 100 times higher than the 2,4-dienoyl-CoA reductase activity, which also amounted to only one tenth of the activity of thiolase,[44] which appears to be the rate-limiting enzyme of the β-oxidation reaction sequence *in vitro* (Table 1). (3) The low 2,4-dienoyl-CoA reductase activity should lead to intermediate accumulation at or above the C_{10} level; this was not observed (see above).[42] (4) The hydration of 2-*trans*,4-*cis*-decadienoyl-CoA to 3-hydroxy,4-*cis*-decenoyl-CoA (Figure 2), which has been considered to be a thermodynamically unfavorable reaction,[48] proceeded to an equilibrium with K approximately 1 in glyoxysomes from cucumber cotyledons,[44] allowing entrance into the D-3-hydroxyacyl-CoA dehydratase pathway.

Of the different reactions required in addition to the β-oxidation reaction sequence at linoleate catabolism following the D-3-hydroxyacyl-CoA dehydratase pathway (Figure 2), it has been demonstrated by *in vitro* reconstitution experiments using purified enzymes that 4-*cis*-decenoyl-CoA is metabolized to 2-*cis*-octenoyl-CoA. Activity of the isomerase and D-3-hydroxyacyl-CoA dehydratase has not been assayed using the purified enzymes with substrates corresponding exactly to the intermediates upon which the enzymes should act at linoleate catabolism (Figure 2). But the data on the kinetic properties of the enzymes (see above) strongly support the assumption that the enzymes catalyze the reactions required to circumvent the *cis* double bond barriers at linoleate catabolism according to the scheme presented in Figure 2.

Catabolism of linolenic acid (18:3, 9-*cis*, 12-*cis*, 15-*cis*) is thought to proceed by the pathway(s) of linoleate catabolism, except that a second isomerase reaction is involved at the C_6 intermediate level in order to surmount the barrier caused by the parent 15-*cis* double bond. It has been demonstrated that 3-*cis*-hexenoyl-CoA formed at the C_6 intermediate level is converted to 2-*trans*-hexenoyl-CoA by purified isomerase.[41]

2. Modifications Due to In-Chain Hydroxyl Groups

Straight-chain fatty acids containing an in-chain L-hydroxyl (or oxo) group located at an odd-numbered carbon atom should be degraded by the β-oxidation pathway without problems (no data on the catabolism of such a fatty acid are currently available). Degradation by β-oxidation up to the point of the functional group results in the formation of an L-3-hydroxyacyl-CoA (or 3-oxoacyl-CoA) intermediate which can be metabolized by the multifunctional protein (and thiolase, respectively). If the in-chain hydroxyl group of the parent fatty acid has D-configuration, a D-3-hydroxyacyl-CoA intermediate is formed, which is not a substrate for the multifunctional protein. However, the recently discovered D-3-hydroxyacyl-CoA dehydratase (Section II.E.1) can convert the D-3-hydroxyacyl-CoA into 2-*trans*-enoyl-CoA, an intermediate of the β-oxidation reaction sequence.

An in-chain hydroxyl (or oxo) group located at an even-numbered carbon atom of a straight-chain fatty acid forms a barrier to continuous degradation by the β-oxidation pathway. The barrier is encountered at the latest at the intermediate level where a 2-hydroxyacyl-CoA intermediate results in the course of degradation. Enzymes acting on 2-hydroxyacyl-CoAs are unknown at present. Catabolism of fatty acids possessing an in-chain hydroxyl group located at an even-numbered carbon atom has only been studied in the case of ricinoleic acid (D-12-hydroxy-9-*cis*-octadecenoic acid, D-12-hydroxyoleic acid).[24,40] Ricinoleic acid comprises more than 80% of the fatty acyl moiety in the triacylglycerols stored in the castor bean endosperm and mobilized during germination. Ricinoleic acid is, however, catabolized not only by the peroxisomal β-oxidation system of the castor bean endosperm, but also, and without qualitative differences, by the peroxisomal β-oxidation system of tissues which do not contain this uncommon fatty acid.[24,40]

The catabolic pathway of ricinoleic acid was first studied and found to be located in peroxisomes by Hutton and Stumpf.[40] Based on experimental evidence, the pathway could not be resolved conclusively at the C_8 intermediate level where the hydroxyl group of ricinoleic acid forms a barrier to further degradation of the ricinoleic acid by the β-oxidation reaction sequence. Of the intermediates detected at ricinoleate catabolism, two belonged to the C_8 intermediate level and were identified as 2-hydroxyoctanoate and 2-oxooctanoate (for methodological reasons, the studies did not distinguish between intermediates esterified or not esterified to CoA-SH). It was also demonstrated that 2-oxooctanoate was metabolized beyond the C_8 level and that C-3 of this compound became C-2 of acetate. Based on these data, it was tentatively suggested that an α-oxidation (Section III) step was involved at the C_8 intermediate level of ricinoleate catabolism in order to circumvent the β-oxidation barrier caused by the hydroxyl group. The α-oxidation would yield CO_2 and heptanoate which can be degraded further by β-oxidation.

Involvement of α-oxidation in ricinoleate catabolism would have some consequences. (1) Since α-oxidation acts on substrates (fatty acids) not esterified to CoA-SH, the acyl-CoA track of β-oxidation has to be left at the C_8 intermediate level. (2) The heptanoate formed by α-oxidation has to be newly activated by acyl-CoA synthetase, consuming ATP, in order to allow return to the acyl-CoA track of β-oxidation. (3) α-Oxidation, which is considered to be a process located at the endoplasmic reticulum (Section III), must (also) be located in peroxisomes, since all reactions of ricinoleate catabolism appeared to be associated with the glyoxysomes.

Recent studies aimed at resolving the pathway which surmounts the hydroxyl group barrier at ricinoleate catabolism in peroxisomes and differentiating between acyl-CoA thioester intermediates and intermediates not esterified to CoA-SH led to the following results.[24] As already reported by Hutton and Stumpf,[40] only a few intermediates of ricinoleate catabolism could be detected. Among them were acyl-CoA thioesters as well as two intermediates not esterified to CoA-SH. These were 2-hydroxyoctanoate and 2-oxooctanoate. Heptanoate was not detected. Peroxisomes oxidized 2-hydroxyoctanoate to 2-oxooctanoate in an H_2O_2-generating reaction showing a stoichiometry of H_2O_2 and 2-oxooctanoate formation of 1:1, indicating oxidation of 2-hydroxyoctanoate by an 2-hydroxy acid oxidase. 2-Hydroxyacid oxidases known from higher plants act preferentially on L-2-hydroxy acids, while the 2-hydroxyoctanoate formed at ricinoleate catabolism should have the D-configuration. However, both isomers of racemic 2-hydroxyoctanoate used as substrate were oxidized by the peroxisomes. 2-Oxooctanoate was metabolized to heptanoyl-CoA (and degraded further to propionyl-CoA and acetyl-CoA). The heptanoyl-CoA formation could be due to either α-oxidation of 2-oxooctanoate followed by activation of the reaction product heptanoate by acyl-CoA synthetase or an oxidative decarboxylation (Section II.B.2) of 2-oxooctanoate releasing directly heptanoyl-CoA.

There are several arguments in favor of an oxidative decarboxylation of 2-oxooctanoate at ricinoleate catabolism:[24] (1) heptanoyl-CoA was formed in an ATP-independent, NAD-dependent reaction; (2) heptanoate was not detected as an intermediate when 2-oxooctanoate was metabolized to heptanoyl-CoA (the metabolism of which was prevented), although the concomitant NADH formation (due to either α-oxidation or oxidative decarboxylation) was two times higher than the acyl-CoA synthetase activity assayed with heptanoate as substrate; (3) 2-oxooctanoate metabolism was inhibited by arsenite, an inhibitor of oxidative decarboxylations, but was insensitive to imidazole, an inhibitor of α-oxidation.

A hydroxyl group located at an even-numbered carbon atom of a straight-chain fatty acid prevents repetitive passages through the β-oxidation reaction sequence at the intermediate level where a 2-hydroxyacyl-CoA is formed. Based on the results of the studies on ricinoleate catabolism, it can be proposed that this hydroxyl group barrier is surmounted by the following reaction sequence acting at the 2-hydroxyacyl-CoA intermediate level: hydrolysis of the 2-hydroxyacyl-CoA by a hydrolase (an enzyme type involved repeatedly in fatty acid catabolism; Section II.E.3, 4) yields the 2-hydroxy acid and thereby the acyl-CoA track of β-oxidation is left; oxidation of the 2-hydroxy acid to the 2-oxo acid by a 2-hydroxy acid oxidase (glycolate oxidase?); and oxidative decarboxylation of the 2-oxo acid to the acyl-CoA containing one carbon atom less than the 2-oxo acid. The oxidative decarboxylation allows return, without the requirement of ATP for the renewed activation process, to the acyl-CoA track for further degradation of the parent fatty acid by β-oxidation. Enzymes catalyzing the proposed additional reactions at catabolism of hydroxy fatty acids have not been purified from plant tissue up to now.

Figure 3 represents the pathway of ricinoleate catabolism as currently proposed.[24] Following activation of ricinoleic acid by acyl-CoA synthetase (Section II.B.1) and subsequent degradation of ricinoleoyl-CoA by the β-oxidation pathway, the *cis* double bond of the molecule located at an odd-numbered carbon atom causes the first barrier to degradation by repetitive passages through the β-oxidation reaction sequence. At the C_{12} intermediate level, D-6-hydroxy-3-*cis*-dodecenoyl-CoA should be formed, and D-6-hydroxy-3-*cis*-dodecenoic acid has been identified as an intermediate in ricinoleate catabolism.[40] Conversion of D-6-hydroxy-3-*cis*-dodecenoyl-CoA into D-6-hydroxy-2-*trans*-dodecenoyl-CoA can be envisaged on the basis that Δ^2,Δ^3-enoyl-CoA isomerase has been demonstrated in peroxisomes (Section II.E.1). The further steps of the pathway up to heptanoyl-CoA have been discussed above. The alternative reaction sequence leading from D-4-hydroxydecanoyl-CoA via 4-oxodecanoyl-CoA to 2-oxooctanoate has been proposed due to detection of 4-oxodecanoate at ricinoleate catabolism.[40] In-chain oxidation of hydroxyl groups to oxo groups is known at the biosynthesis of wax components (a process not located in peroxisomes).[49] Catabolism of heptanoyl-CoA as well as ricinoleoyl-CoA to propionyl-CoA and acetyl-CoA has been demonstrated.[24] Since propionyl-CoA is metabolized to acetyl-CoA in peroxisomes (Section II.E.3), complete degradation of ricinoleic acid to acetyl-CoA (and CO_2) by the proposed peroxisomal pathway (Figure 3) appears to be possible. Following NADH formation during degradation of 1 nmol of ricinoleate, a stoichiometry of 9:1 was observed when the NADH formation had ceased.[24] This stoichiometry corresponds to that expected for complete degradation of 1 nmol of ricinoleate to acetyl-CoA.

3. Propionate Catabolism

Catabolism of odd-numbered, straight-chain fatty acids by the β-oxidation pathway results in the formation of propionyl-CoA. But such fatty acids are rarely found in plant tissues as constituents of storage triacylglycerols or membrane lipids. Propionyl-CoA is, however, generated at certain catabolic processes such as the catabolism of ricinoleic acid (Section II.E.2) and branched-chain 2-oxo acids (Section II.E.4).

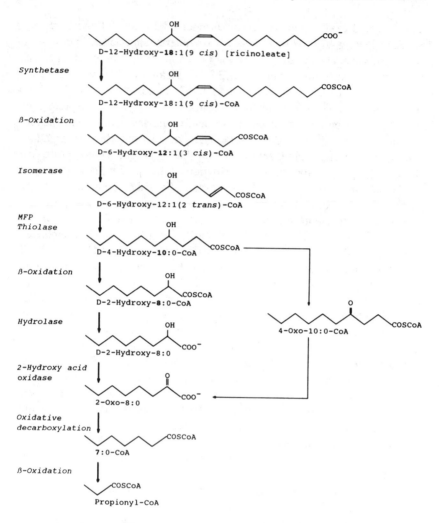

FIGURE 3. Proposed peroxisomal pathway for degradation of ricinoleate. (Adapted from Gerbling, H. and Gerhardt, B., *Bot. Acta*, 104, 233, 1991; and Hutton, D. and Stumpf, P. K., *Arch. Biochem. Biophys.*, 142, 48, 1971.)

Stumpf and co-workers[50,51] studied propionate metabolism in plant tissues with respect to the fact that coenzyme B_{12}, required for propionyl-CoA catabolism in mammalian mitochondria, had not been detected in plant tissue. In mammalian mitochondria, propionyl-CoA is carboxylated to methylmalonyl-CoA in a biotin-dependent reaction, and the methylmalonyl-CoA is subsequently rearranged to succinyl-CoA in a coenzyme B_{12}-dependent mutase reaction. Employing intact tissues from different plant species as well as cell-free systems, Stumpf and co-workers[50,51] demonstrated that (1) the C-1 of propionate was metabolized to CO_2, and the C-2 and C-3 of propionate became C-2 and C-1, respectively, of acetyl-CoA; (2) CO_2 was not required for the oxidative degradation of propionate; and (3) 3-hydroxypropionate was an intermediate of propionate catabolism. On the basis of these results, the pathway of propionate catabolism shown in Figure 4 was proposed.[51,52] Of the pathway called "modified β-oxidation", only the intermediate 3-hydroxy-propionate had been identified directly. Catabolism of propionate via 3-hydroxypropionate to acetate has also been reported recently.[53]

Catabolism of propionyl-CoA to acetyl-CoA via the proposed pathway of modified β-oxidation (Figure 4) has now been confirmed by kinetic experiments using peroxisomes from mung bean hypocotyls as the enzyme source.[54,55] The acyl-CoA thioester intermediates, the intermediates not esterified to CoA-SH, and the end product acetyl-CoA were demonstrated by HPLC. Succinyl-CoA, a product of propionyl-CoA metabolism in mammalian mitochondria, was not detected. Oxidation of propionyl-CoA to acrylyl-CoA resulted in concomitant H_2O_2 formation, indicating that the reaction was catalyzed by acyl-CoA oxidase. The subsequent hydration of acrylyl-CoA to 3-hydroxypropionyl-CoA may be catalyzed by the multifunctional protein which, however, has never been assayed with acrylyl-CoA as substrate (Section II.C.2). The subsequent reactions of the modified β-oxidation are catalyzed by enzymes not yet characterized. At the second oxidation step of the modified β-oxidation, the electrons are transferred to NAD. In the absence of NAD, 3-hydroxypropionate accumulated as an end product of propionyl-CoA degradation. Malonic semialdehyde accumulated when CoA-SH was omitted from the reaction mixture.

The data presented demonstrate that propionyl-CoA is catabolized to acetyl-CoA plus CO_2 by modified β-oxidation in plant peroxisomes. This pathway of propionyl-CoA catabolism avoids both a biotin-dependent carboxylation and a coenzyme B_{12}-dependent mutase reaction involved in propionyl-CoA catabolism in mammalian mitochondria. Whether the pathway of modified β-oxidation is a peculiarity of the plant cell or characteristic of the peroxisomal compartment, which exists also in animal cells, is unknown at present, since propionyl-CoA degradation by animal peroxisomes has not been studied up to now.

Propionate/propionyl-CoA catabolism by modified β-oxidation and its localization in peroxisomes has been demonstrated in plant tissue. However, there are a few reports suggesting that propionate may additionally be metabolized in potato tubers by a pathway similar to that occurring in mammalian mitochondria (see above). This suggestion is based on results of labeling experiments performed with potato slices[56] and the detection of vitamin B_{12}-stimulated methylmalonyl-CoA mutase activity in potato tuber extracts.[57] If the additional pathway of propionate catabolism occurs in potato tuber (and other plant organs/tissues), its subcellular localization is unknown at present.

4. Modifications Due to Methyl Branching

To ease comparisons between the catabolism of branched-chain 2-oxo acids/acyl-CoAs and that of straight-chain fatty acids/acyl-CoAs, the systematic names of the branched-chain 2-oxo acids/acyl-CoAs are used in the following

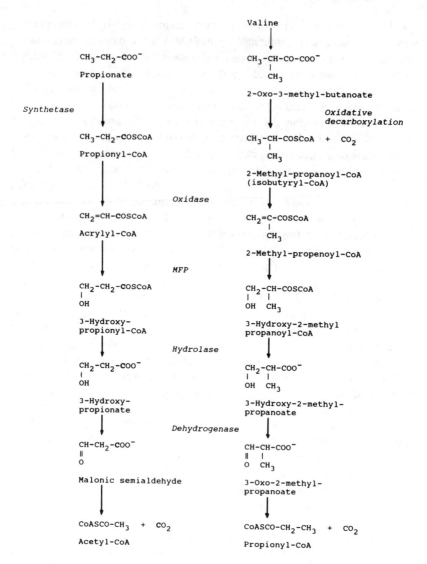

FIGURE 4. Peroxisomal catabolism of propionate and 2-oxo-3-methyl-butanoate (2-oxo-isovalerate) by modified β-oxidation. MFP — multifunctional protein. (Adapted from Hatch, M. D. and Stumpf, P. K., *Arch. Biochem. Biophys.*, 96, 193, 1962; Gerbling, H. and Gerhardt, B., *Biological Role of Plant Lipids*, Biacs, P. A., Gruiz, K., and Kremmer, T., Eds., Plenum Press, New York, 1989, 21; Gerbling, H. and Gehardt, B., *Eur. J. Cell Biol.*, 54 (Suppl. 32), 74, 1991; and Gerbling, H. and Gerhardt, B., *Plant Physiol.*, 91, 1387, 1988.)

instead of the trivial names. The systematic names give a better impression of the molecular structure to readers who are not so familiar with fatty acids.

Branched-chain 2-oxo acids form a group of fatty acids which are generated by the catabolism of the branched-chain amino acids. Catabolism of leucine, isoleucine, and valine is initiated by an aminotransferase reaction yielding the branched-chain 2-oxo acids 2-oxo-4-methylpentanoic acid (2-oxoisocaproic acid), 2-oxo-3-methyl-pentanoic acid (2-oxo-3-methyl-valeric acid), and 2-oxo-3-methyl-butanoic acid (2-oxoisovaleric acid), respectively. Catabolism of the branched-chain 2-oxo acids, i.e., catabolism of the branched-chain amino acids in plant tissues, has received very little attention and been only very recently studied.[25,55,58,59] It has been demonstrated that the branched-chain 2-oxo acids are activated by oxidative decarboxylation (Section II.B.2) and then catabolized to propionyl-CoA and subsequently to acetyl-CoA in peroxisomes from mung bean hypocotyls. Based on both identification of intermediates and the results of kinetic experiments, pathways of the catabolism of the branched-chain 2-oxo acids were proposed (Figures 4 and 5).[55,58,59] Individual reactions of the proposed pathways have been directly assayed only in a few cases. They include the oxidative decarboxylation and the subsequent reaction leading from acyl-CoA to 2-enoyl-CoA. Oxidation of the branched-chain acyl-CoAs to the corresponding 2-enoyl-CoAs resulted in concomitant H_2O_2 formation, indicating that the reaction is catalyzed by acyl-CoA oxidase.[25] It is unknown at present whether steps of the β-oxidation reaction sequence involved in the catabolism of branched-chain acyl-CoAs are catalyzed by proteins identical with those acting on straight-chain acyl-CoAs.

Oxidative decarboxylation of 2-oxo-4-methyl-pentanoic acid, 2-oxo-3-methyl-pentanoic acid, and 2-oxo-3-methyl-butanoic acid results in the formation of 3-methyl-butanoyl-CoA, 2-methyl-butanoyl-CoA, and 2-methyl-propanoyl-CoA, respectively. In the following, the proposed catabolic pathways of these branched-chain acyl-CoAs are considered only up to the formation of propionyl-CoA, which is traceably catabolized further to acetyl-CoA.[58] The pathways differ considerably (Figures 4 and 5). 2-Methyl-butanoyl-CoA is catabolized to propionyl-CoA plus acetyl-CoA by one complete passage through the β-oxidation reaction sequence (Figure 5). A reaction sequence corresponding to the modified β-oxidation by which propionyl-CoA is metabolized to acetyl-CoA appears to degrade 2-methyl-propanoyl-CoA to propionyl-CoA plus CO_2 (Figure 4). The proposed conversion of C-1 of 2-methyl-propanoyl-CoA to CO_2 has not been demonstrated. Omission of NAD from a reaction mixture containing 2-methyl-propanoyl-CoA as substrate resulted in an accumulation of 3-hydroxy-2-methyl-propanoate, indicating NAD-dependent oxidation of 3-hydroxy-2-methyl-propanoate to 3-oxo-2-methyl-propanoate.[55] The latter intermediate accumulated when CoA-SH was omitted from the reaction mixture. These results correspond to those obtained in studies on propionyl-CoA catabolism (Section II.E.3).

The proposed pathway for the catabolism of 3-methyl-butanoyl-CoA in plant peroxisomes (Figure 5) differs totally from that established in mammalian mitochondria. In the latter organelles, 3-methyl-butanoyl-CoA is metabolized to 3-hydroxy-3-methyl-pentanedioyl-mono-CoA (HMG-CoA) by a reaction sequence which includes a biotin-dependent carboxylation. HMG-CoA was not detected in reaction mixtures containing peroxisomes from mung bean

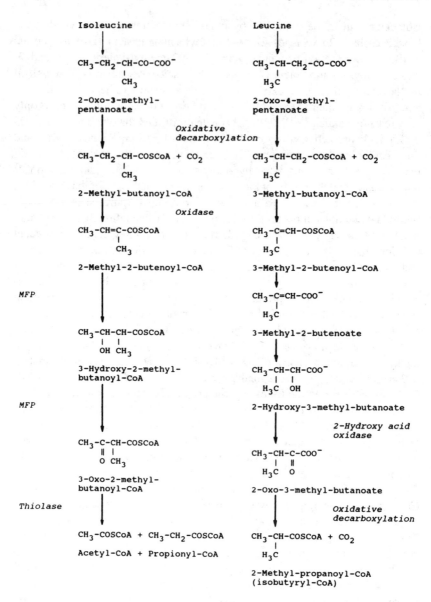

FIGURE 5. Peroxisomal catabolism of 2-oxo-3-methyl-pentanoate and proposed pathway for peroxisomal degradation of 2-oxo-4-methyl-pentanoate (2-oxo-isocaproate) to 2-methyl-propanoyl-CoA (isobutyryl-CoA). MFP — multifunctional protein. (Adapted from Gerbling, H. and Gerhardt, B., *Plant Physiol.*, 91, 1387, 1988; and *Biol. Chem. Hoppe-Seyler*, 372, 530, 1991.)

hypocotyls as the enzyme source and 3-methyl-butanoyl-CoA as the substrate.[58] On the contrary, the substrate was metabolized via 3-methyl-2-butenoyl-CoA formed by acyl-CoA oxidase activity (see above) to 2-methyl-propanoyl-CoA (isobutyryl-CoA). This overall reaction implies the loss of one carbon atom (CO_2). As in the case of propionyl-CoA catabolism in plant peroxisomes, 3-methyl-butanoyl-CoA was metabolized by a pathway which apparently does not involve a biotin-dependent carboxylation. Plant peroxisomes are apparently not able to carry out biotin-dependent carboxylations. There is some experimental evidence in support of the pathway leading from 3-methyl-2-butenoyl-CoA to 2-methyl-propanoyl-CoA as shown in Figure 4: detection of the specified intermediates at peroxisomal catabolism of 2-oxo-4-methyl-pentanoate and metabolism of these intermediates, when used as substrates, to 2-methyl-propanoyl-CoA.[59] The reaction(s) introducing the hydroxyl group at C-2 of 3-methyl-2-butenoate is (are) unknown. Oxidation of 2-hydroxy-3-methyl-butanoate to 2-oxo-3-methyl-butanoate in an H_2O_2-generating reaction and subsequent oxidative decarboxylation of 2-oxo-3-methyl-butanoate to 2-methyl-propanoyl-CoA has been demonstrated.[60] The reaction sequence corresponds to that involved in ricinoleate catabolism at the C_8 intermediate level (Section II.E.2). The 2-methyl-propanoyl-CoA formed as intermediate at the catabolism of 3-methyl-butanoyl-CoA is metabolized to propionyl-CoA by the reaction sequence (modified β-oxidation) catabolizing the 2-methyl-propanoyl-CoA resulting from oxidative decarboxylation of 2-oxo-3-methyl-butanoate (see above).

F. REGULATION

Information on the regulation of both fatty acid catabolism and the enzymes involved is scarce regarding plant cells. According to the data presented in Table 1, thiolase appears mainly to be the rate limiting enzyme of the β-oxidation reaction sequence.

Under nonsteady state conditions and using peroxisomes from sunflower cotyledons, transient accumulation of a few intermediates was observed at [U-^{14}C]palmitate, [18-^{14}C]oleate and [U-^{14}C]linoleate catabolism (Sections II.D and II.E.1).[39] In all cases, temporary intermediate accumulation occurred at the short-chain (C_4) intermediate level. In contrast to complete degradation of palmitate and oleate, complete degradation of linoleate required removal of acetyl-CoA from the reaction mixture or else a medium-chain intermediate accumulated as an end product along with acetyl-CoA. Under steady state conditions, linoleate degradation by peroxisomes from cucumber cotyledons led to intermediate accumulation at the C_4 level (acetyl-CoA removed from the assay mixture).[42] Thus, fatty acid catabolism generally appears to slow down, at least at the C_4 intermediate level. This is in accordance with, but cannot solely be explained by, the known kinetic properties of acyl-CoA oxidase (Section II.C.1) and the results on the dependence of fatty acid degradation on the fatty acid chain length (Section II.D).

During germination of oilseeds, the β-oxidation enzyme activities rise and fall in the lipid-mobilizing tissue of the seeds.[37,61,62] An increasing amount of data demonstrates that peroxisomal enzyme activities rise during germination due to increased transcriptional activity, although processes operating at other levels can also be involved.[63-65] It seems likely then that the rise in β-oxidation enzyme activities during germination (and that reported for senescing leaves)[66] is primarily due to increased expression of the genes encoding β-oxidation enzyme proteins. As all peroxisomal proteins studied under the aspect of biosynthesis have been shown to be synthesized on free polyribosomes and to be inserted posttranslationally into the organelle, β-oxidation enzyme proteins are certainly synthesized in the cytosol. Detection of pulse-labeled multifunctional protein in the cytosol prior to appearance of the protein in glyoxysomes has been reported.[67]

Loss in β-oxidation enzyme activities at later stages of oilseed (cucumber) germination is due both to decreases in the amount of encoding RNAs and protein degradation (turnover).[68]

The stimuli causing changes in transcriptional activity with respect to β-oxidation enzyme proteins are virtually unknown.

G. INTEGRATION INTO CELL METABOLISM

The location of fatty acid catabolism in a membrane-bound compartment (peroxisome) raises the questions (1) as to how fatty acids and required cofactors enter the organelle and (2) what happens to the products of fatty acid degradation which cannot, and do not, accumulate in the peroxisome. It is absolutely unknown how fatty acids reach the peroxisome from their point of origin. Involvement of lipid transfer/fatty acid binding proteins or mediation by membrane contact between oil bodies and peroxisomes in lipid-mobilizing tissues are, at best, speculative, and the former is rather unlikely.[69]

1. Aspects Related to Acyl-CoA Synthetase Reaction

When examined, enzymes of fatty acid catabolism in higher plant peroxisomes appear to be either soluble matrix proteins or loosely associated with the organelle membrane,[22,31,35,37,42] except acyl-CoA synthetase which is tightly bound to the membrane.[22] Following membrane fractionation, the acyl-CoA synthetase activity was recovered in the fraction containing very hydrophobic proteins. The acyl-CoA synthetase activity of peroxisomes from mung bean hypocotyls,[22] pea leaves,[22] and sunflower cotyledons[70] showed latency and resistance to protease (thermolysin) treatment, indicating location of the enzyme, or at least essential domain(s) of the enzyme, on the matrix face of the peroxisome membrane.

The latency of acyl-CoA synthetase activity implies that the peroxisome membrane has, at least, restricted permeability for the substrate or one of the cofactors of the reaction. However, transport processes across the membrane of higher plant peroxisomes have not been thoroughly studied and, in particular, not at all with respect to β-oxidation. A carnitine transfer system mediating entry of fatty acids into the peroxisomes does not appear to exist. Both carnitine acyltransferase and carnitine acetyltransferase activity could not be demonstrated in plant peroxisomes.[37,71] In addition, acyl-carnitines are not oxidized by plant peroxisomes.[11,36,37] Unknown at present are how the peroxisomes are provided with ATP required for the acyl-CoA synthetase reaction, which evidence indicates is certainly not synthesized in the organelles, and whether plant peroxisomes possess their own CoA-SH pool.

Hydrolysis of the pyrophosphate formed in the acyl-CoA synthetase reaction is essential for pulling the activation process in the direction of acyl-CoA formation. An inorganic pyrophosphatase, however, has not been detected in peroxisomes,[72,73] although acyl-CoA synthetase appears to be located inside the organelle (see above). This problem has not yet been resolved. Unknown also is the fate of the AMP formed in the acyl-CoA synthetase reaction.

2. Fate of Acetyl-CoA

With respect to carbon, catabolism of physiologically relevant fatty acids in higher plant cells leads to acetyl-CoA as the sole end product, apart from CO_2 evolved at oxidative decarboxylations. The fate of this acetyl-CoA depends on the enzymatic composition of the fatty acid degrading compartment (peroxisome), which itself depends on the physiological and/or developmental status of the cell.[20] Metabolism of the acetyl-CoA appears to be confined to the organelle where the acetyl-CoA is formed, and according to the fate of acetyl-CoA, two subgroups of peroxisomes may be distinguished: (1) glyoxysomes which contain the enzyme activities comprising the glyoxylate cycle/pathway[2,74] which occur in lipid-mobilizing tissues of germinating seeds[2] and also appear to develop in senescing tissue,[66,75] and (2) nonglyoxysomal peroxisomes which lack activity of isocitrate lyase and malate synthase, key enzymes of the glyoxylate cycle, and occur in the majority of plant tissues.

Metabolism of acetyl-CoA via the glyoxylate cycle/pathway in glyoxysomes results in the formation of C_4 dicarboxylic acid(s) and enables the cell to direct carbon of fatty acids into glyconeogenesis.[2] The reactions involved in glyoxysomal acetyl-CoA metabolism can be arranged diagrammatically in two different ways, resulting in the glyoxylate cycle[2] or the glyoxylate pathway.[74] If glyoxysomes perform the glyoxylate cycle, 2 mol of acetyl-CoA are converted to 1 mol of succinate by a cyclic process, allowing a self-sustained acetyl-CoA metabolism in the organelles. Succinate, itself not an intermediate of the glyoxylate cycle, is released from the glyoxysomes. If glyoxysomes perform the glyoxylate pathway, both succinate and malate are end products of noncyclic metabolism of 2 mol acetyl-CoA and are released from the organelles. In this case, the glyoxysomes need a constant input of carbon skeleton (oxaloacetate, aspartate) in order to balance the loss of carbon not provided by acetyl-CoA. Whether the glyoxylate cycle or glyoxylate pathway reflect the reality has yet to be established.

A pathway for acetyl-CoA metabolism in nonglyoxysomal peroxisomes, which lack acetyl-CoA hydrolase activity, has been suggested.[20,76] It is based on enzyme activities which were detected in purified preparations of potato tuber peroxisomes and were demonstrated to belong to peroxisomal enzyme isoforms. The proposed pathway includes formation of citrate, conversion of citrate to isocitrate, and subsequent decarboxylation of isocitrate by NADP-dependent isocitrate dehydrogenase. Oxaloacetate required for citrate synthesis can be generated by an aminotransferase acting on the 2-oxoglutarate formed within the organelles and aspartate imported into the organelles in exchange for the glutamate formed in the aminotransferase reaction from 2-oxoglutarate. According to the proposed pathway, the carbon of acetyl-CoA becomes part of the exported glutamate.

As envisaged at present, acetyl-CoA metabolism in both glyoxysomes and nonglyoxysomal peroxisomes involves aconitase. Occurrence of this enzyme in peroxisomes has been questioned very recently.[77] If peroxisomes lack aconitase, the carbon of acetyl-CoA generated at fatty acid catabolism within the organelles would be expected to be exported from the organelles into the cytosol following citrate formation for conversion to isocitrate (isocitrate must then, however, return to the glyoxysomes, the site of isocitrate lyase and malate synthase).

3. Fate of Electrons

As an oxidative process, fatty acid catabolism proceeds only in the presence of electron acceptors. There are two electron acceptors, molecular oxygen and NAD, involved in fatty acid degradation. Oxygen serves as the electron acceptor at the first oxidation step of the β-oxidation reaction sequence (Section II.C.1) and at the oxidation of 2-hydroxy acid intermediates formed at the catabolism of certain fatty acids (Sections II.E.2 and 4). The (flavin-containing) enzymes involved, acyl-CoA oxidase and 2-hydroxy acid oxidase(s), transfer two electrons to oxygen, reducing it to H_2O_2. The H_2O_2 is decomposed by catalase, a marker enzyme of peroxisomes, without energy conservation.

The multifunctional protein catalyzing the second oxidation step of the β-oxidation reaction sequence (Section II.C.2), the dehydrogenase(s) involved in oxidation of the ω-hydroxyl group during modified β-oxidation (Sections II.E.3 and 4), and oxidative decarboxylations (Section II.A.2) depend on NAD as an electron acceptor. NAD (as well as NADP) has been demonstrated in castor bean endosperm glyoxysomes (0.2 to 0.6 nmol NAD per milligram of organelle protein) isolated by aqueous methods.[32,74] Due to the catalytic amount of NAD in the organelles, continuous β-oxidation (as well as any other NADH-generating process in peroxisomes) requires sustained regeneration of NAD, i.e., concomitant reoxidation of the NADH formed.

Two different mechanisms, a malate-aspartate shuttle transferring reducing equivalents from glyoxysomes to mitochondria[11,74] and an electron transfer system located in the glyoxysomal membrane,[78] have been proposed for reoxidation of the NADH formed in glyoxysomes. The concept of a malate-aspartate shuttle is mainly based on the fact that both glyoxysomes and mitochondria contain highly active malate dehydrogenase and glutamate:oxalacetate aminotransferase. There are also several lines of indirect experimental evidence in support of the concept. A malate-oxalacetate/aspartate shuttle has also been proposed[79] and indirectly been shown to operate[80,81] between leaf peroxisomes and an NADH source. In this case, the shuttle functions in transferring reducing equivalents into leaf peroxisomes in order to sustain photorespiration. As the glycerate pathway, the directly NADH-consuming part of photorespiration, is reversible,[79] the shuttle should also function in the reverse direction. The concept of a malate-aspartate shuttle as a system transferring reducing equivalents between peroxisomes and their surroundings, however, may not be applicable to all peroxisomes, since, for example, potato tuber peroxisomes which perform β-oxidation (Table 1) have been reported to lack malate dehydrogenase activity.[82]

Electron transport components have been demonstrated in the membrane of castor bean endosperm glyoxysomes. They include a flavoprotein NADH reductase and cytochrome b_5.[78] The orientation of the reductase in the membrane allows electron uptake from the organelle matrix, that of the cytochrome b_5 allows electron discharge to the cytosol.[83] The physiologically final electron acceptor of the supposed electron transport system is yet unknown (and probably located in the cytosol). Oxidation of palmitoyl-CoA (and other NAD(P)H-generating glyoxysomal processes) have been coupled to the reduction of artificial electron acceptors accepted by the electron transport components.[84] The capacity of the membrane redox system assayed with NADH and artificial electron acceptors surpassed the actually measured β-oxidation activity of the glyoxysomes.

Since the functioning of both malate-aspartate shuttle and membrane located redox system is supported by experimental data, electron flow through the two routes may operate in parallel or alternatively to achieve NADH oxidation in castor bean endosperm glyoxysomes at least. NADH oxidation by the shuttle system can link peroxisomal fatty acid catabolism to mitochondrial ATP formation by oxidative phosphorylation, resulting in energy conservation. Whether the membrane redox system leads eventually to energy conservation depends on the nature of the external (cytosolic) physiological electron acceptor of the system which has yet to be identified.

H. MITOCHONDRIAL β-OXIDATION?

Catabolism of physiologically relevant fatty acids is located in the peroxisome, a regular compartment of the higher plant cell. This finding poses the question as to whether higher plant cells possess an additional, mitochondrial β-oxidation system. Such dual subcellular localization of β-oxidation activity is known for mammalian cells, and peroxisomal and mitochondrial β-oxidation systems differ in their properties,[6] the latter representing the classic β-oxidation system. On the other hand, the peroxisomal β-oxidation systems of mammalian and higher plant cells are similar with respect to many of their properties. The two peroxisomal β-oxidation systems differ, however, in their ability to catabolize long-chain fatty acids. Complete degradation of long-chain fatty acids has been demonstrated only in plant peroxisomes. An additional β-oxidation system, therefore, appears not to be a necessity to accomplish fatty acid catabolism in higher plant cells. Results related to this question are outlined in the following.

The inner mitochondrial membrane is impermeable for acyl-CoA thioesters which are formed at the outer membrane of animal mitochondria. Transfer of the fatty acyl group across the inner membrane of animal mitochondria occurs in the form of acyl-carnitine which is generated from acyl-CoA and L-carnitine by a carnitine acyltransferase. A second carnitine acyltransferase reconverts the acyl-carnitine to the corresponding acyl-CoA inside the mitochondria. L-Carnitine has been demonstrated in plant tissues[85,86] and carnitine acyltransferase activity in plant mitochondria.[71,87-90] Carnitine acyltransferase activity has been detected using long-, medium-, and short-chain acyl-CoAs as substrate. The carnitine long-chain acyltransferase activity has also been reported to be associated with the outer face of the inner membrane of pea cotyledon mitochondria, as well as to be localized intramitochondrially.[90] The carnitine acyltransferase activities of pea cotyledon mitochondria, which are detectable with long- or short-chain acyl-CoAs, have optionally been ascribed to two different enzymes with different functions.[88-91] The enzyme thought to react with long-chain acyl-CoAs is thought to be involved in mitochondrial β-oxidation.[89,90] Studies on the mitochondrial carnitine acyltransferase activity of mung bean hypocotyls suggested that the activity may be due to a protein resembling a short-chain rather than a long-chain carnitine acyltransferase.[71,92] Solution of the problem must await purification of the plant mitochondrial carnitine acyltransferase(s), and only then may it be decided whether the demonstrated carnitine long-chain acyltransferase activity is a valid argument in favor of the occurrence of a mitochondrial β-oxidation system in plant cells.

Oxidation of fatty acids by the peroxisomal β-oxidation system is KCN insensitive in the presence of NAD.[11,16] This is due to the fact that the first oxidation step of the peroxisomal β-oxidation reaction sequence (Section II.C.1) is catalyzed by an acyl-CoA oxidase transferring electrons directly from the substrate to oxygen. In contrast, mitochondrial β-oxidation in animal cells is KCN sensitive, since an acyl-CoA dehydrogenase tightly coupled to the respiratory chain catalyzes the first oxidation step. KCN sensitive,[90] palmitoyl-carnitine (or palmitoyl-CoA plus L-carnitine)-dependent oxygen uptake by pea cotyledon mitochondria has repeatedly been observed in the presence of sparker malate.[90,93-96] The KCN sensitivity of the reaction excludes peroxisomes contaminating the mitochondrial fraction and thereby providing the site of the palmitoyl-carnitine-dependent oxygen uptake. Moreover, peroxisomes do not oxidize palmitoyl-carnitine (Section G.1).

NADH formation due to oxidation of sparker malate, malate, or other substrates oxidized in an NAD-dependent reaction in mitochondria from avocado (*Persea americana* L.) mesocarp or potato tubers was greatly stimulated by palmitoyl-carnitine at micromolar concentrations.[97] Palmitoyl-L-carnitine and palmitoyl-DL-carnitine were equally effective while L-carnitine was ineffective.

Mitochondrial reactions depending on the mitochondrial electron transport chain were stimulated at low, but inhibited at higher palmitoyl-carnitine concentrations (micromolar range). The stimulation of mitochondrial reactions by palmitoyl-carnitine was reduced or abolished in the presence of bovine serum albumin, depending on the ratio of palmitoyl-carnitine to bovine serum albumin. At certain ratios, palmitoyl-carnitine-dependent oxygen uptake was observed with potato tuber mitochondria, provided that sparker malate was present in the assay mixture.[97] The preliminary data suggest that palmitoyl-carnitine has a disintegrating effect on mitochondrial membranes at micromolar concentrations, resulting in facilitated access of external substrates to intramitochondrial reaction sites. Thus, palmitoylcarnitine-dependent oxygen uptake by mitochondria in the presence of sparker malate must be interpreted with caution as evidence for mitochondrial β-oxidation activity.

When potato tuber mitochondria were provided with palmitoyl-L-carnitine labeled at the C-1 position of the fatty acid moiety, neither formation of [^{14}C]acetyl-CoA nor labeled citrate, in the presence of sparker malate, nor loss of radioactivity from the substrate were detected.[97]

Activities of β-oxidation enzymes are normally detected in mitochondrial fractions isolated from plant tissues on density gradients.[15-18,95,98] If these activities are not due to peroxisomal contamination, they can represent mitochondrial constituents, with the exception of acyl-CoA oxidase activity. So far, acyl-CoA oxidase is considered to be a peroxisomal marker enzyme. The first oxidation step at mitochondrial β-oxidation in animal cells is catalyzed by acyl-CoA dehydrogenase. All attempts to demonstrate acyl-CoA dehydrogenase activity in higher plant mitochondria using long-, medium-, and short-chain acyl-CoAs as substrates have been unsuccessful.[16-18,26,29,37] Spectrophotometric assays employed had detection limits of approximate 20 pkat per mg of mitochondrial fraction protein.[29] Inactivation of acyl-CoA dehydrogenase does not seem to occur during mitochondria isolation, as acyl-CoA dehydrogenase of insect mitochondria did not become inactivated when insect mitochondria were added to a homogenate from mung bean hypocotyls and were reisolated with the plant mitochondria.[29]

When the ratio of enoyl-CoA hydratase, 3-hydroxyacyl-CoA dehydrogenase, or thiolase activity to the activity of the peroxisomal marker catalase was calculated, a statistically significant difference between the ratios obtained for mitochondrial and peroxisomal fractions did not result, except in one case concerning thiolase (see below).[15,16] Therefore, the β-oxidation enzyme activities in mitochondrial fractions were attributed to contaminating peroxisomes rather than to mitochondrial constituents. However, values of the above-mentioned ratios higher in mitochondrial fractions than in peroxisomal fractions have also been reported and were interpreted to demonstrate true mitochondrial β-oxidation activity.[95,98] Interpretations of such ratios, however, may be problematic, as indicated by two examples. (1) Mitochondrial fractions from potato tubers exhibited a ratio of thiolase to catalase activity higher than the peroxisomal fractions.[16] However, thiolase appeared to leak out very easily from its housing organelle; the thiolase activity exhibited hyperbolic distribution on the gradient, showing a pronounced peak nowhere.[99] (2) As an intact inner membrane of mitochondria is not penetrable for acyl-CoAs, Thomas and co-workers[95,98] used a special procedure to obtain ruptured mitochondria for enzyme assays. By applying this procedure to mitochondrial fractions from sunflower cotyledons, the specific activity of acyl-CoA oxidase as well as the ratio of acyl-CoA oxidase to catalase activity rose in the treated mitochondrial fraction in comparison to the values obtained for the untreated mitochondrial fraction.[73]

Enoyl-CoA hydratase and 3-hydroxyacyl-CoA dehydrogenase activity of the peroxisomal β-oxidation system are activities of the multifunctional protein. In contrast, these activities belong to individual proteins in animal mitochondria. A protein exhibiting only enoyl-CoA hydratase activity and also differing from the peroxisomal multifunctional protein in other properties has very recently been isolated from pea cotyledon mitochondria.[100] In addition, antibodies raised against rat liver mitochondrial enoyl-CoA hydratase gave a positive signal in Western blots of total pea mitochondrial proteins; no signal was observed when blots of total pea peroxisomal proteins were probed.

In summing up current data and their interpretations on the question of whether a mitochondrial β-oxidation system exists in higher plant cells, so far, the answer to this question remains equivocal. Direct demonstration of acetyl-CoA formation from palmitoyl-carnitine by mitochondria and/or of the first oxidation step of mitochondrial β-oxidation are primary lines of evidence needed to establish the existence of a mitochondrial β-oxidation system in higher plant cells.

III. α-OXIDATION

Catabolism of straight-chain, common fatty acids by the β-oxidation pathway results in successive removal of C_2 units (acetyl-CoA) from the parent fatty acid. There is another pathway known by which straight, long-chain, common fatty acids may be degraded. It proceeds by oxidation in the α-position of the fatty acid carbon chain and removal of a C_1 unit (CO_2). Unlike β-oxidation, this α-oxidation process does not involve acyl-CoA thioester intermediates, but acts on free fatty acids and is inhibited by imidazole.[101]

The α-oxidation system is generally thought to be membrane bound and probably located in the endoplasmic reticulum. However, preparations reported to contain the α-oxidation activity differ considerably in their behavior at centrifugation. The α-oxidation activity of pea leaves has been found, on the one hand, to be associated with a particulate fraction pelleting between 700 to 7000 g (30 min centrifugation time)[102,103] and, on the other hand, to not pellet at 105,000 g (3 h).[104] The α-oxidation activity of peanut cotyledons was predominantly associated with a fraction which pelleted between 34,000 to

105,000 g (3 h).[104] A fraction pelleting at similar centrifugation forces from pea leaf preparations has also been reported to contain α-oxidation activity.[105]

Studies on the substrate specificity of the α-oxidation sytem showed that the C_{14} and C_{16} saturated fatty acids and C_{18} unsaturated fatty acids ($C_{18:1}$, $C_{18:2}$, $C_{18:3}$) were readily attacked.[56,101,106] Short-chain fatty acids, lauric acid ($C_{12:0}$), and stearic acid ($C_{18:0}$) were very poor or unsuitable substrates. These results suggest that a long-chain fatty acid cannot be degraded beyond the C_{12} carbon chain length by repeated α-oxidation.

The α-oxidation of a C_n fatty acid (n > 12) requires molecular oxygen and a reductant.[104] Besides the products CO_2 and C_{n-1} fatty acid, 2-hydroxy-C_n fatty acid and C_{n-1} fatty aldehyde were detected in assay mixtures. The aldehyde is finally oxidized to the C_{n-1} fatty acid by an NAD-dependent aldehyde dehydrogenase.[101] The 2-hydroxy-C_n fatty acid accumulating in reaction mixtures of α-oxidation has been identified to be the D-isomer;[104,107] the L-isomer was detected only in very low amounts[107] or not at all.[104] When used as substrates for α-oxidation, neither the D-isomer nor the L-isomer of 2-hydroxy-C_{16} fatty acid fulfilled the requirement of an intermediate, as the rate of CO_2 evolution from both isomers was less than that from the C_{16} fatty acid.[104] A considerable decrease in CO_2 formation from C_{16} fatty acid and a concomitant increase in the formation of 2-hydroxy-C_{16} fatty acid were observed when α-oxidation was performed in the presence of a system reducing alkyl hydroperoxides to alkyl alcohols,[104] indicating involvement of an alkyl hydroperoxide in α-oxidation.

Based on such results, Shine and Stumpf[104] proposed a pathway of α-oxidation which involves D-2-hydroperoxyl-C_n fatty acid as a central intermediate (Figure 6). The proposal differed markedly from that reported earlier by Hitchcock and co-workers[102,103] who suggested that both D- and L-isomers of 2-hydroxy-C_n fatty acid are intermediates of α-oxidation, but metabolized to the C_{n-1} fatty acid by different routes involving different intermediates and intermediate channeling. The α-oxidation pathway proposed by Shine and Stumpf[104] is consistent with the results reported on α-oxidation of fatty acids. According to the proposal, the D-2-hydroperoxyl-C_n intermediate is preferentially decarboxylated to the C_{n-1} fatty aldehyde, but to some degree reduced to D-2-hydroxy-C_n fatty acid, which is considered to be a dead end product. The D-configuration of the postulated, and indirectly demonstrated, alkyl hydroperoxide intermediate has been deduced from the observations (1) that D-2-hydroxy-C_n fatty acid accumulates in reaction mixtures of α-oxidation and (2) that the label of L-[2-³H]C_{16} fatty acid was retained in the C_{15} fatty aldehyde formed by α-oxidation, but that the label of D-[2-³H]C_{16} fatty acid was lost.[103]

The electron transfer system required for α-oxidation appears to involve a flavoprotein in some cases[104] and a metal-requiring protein in other cases.[106] Since the work by Shine and Stumpf,[104] the mechanism of α-oxidation in plants has not been specifically studied.

The physiological functions of α-oxidation are not well-understood. The α-oxidation process has been demonstrated in lipid mobilizing as well as other

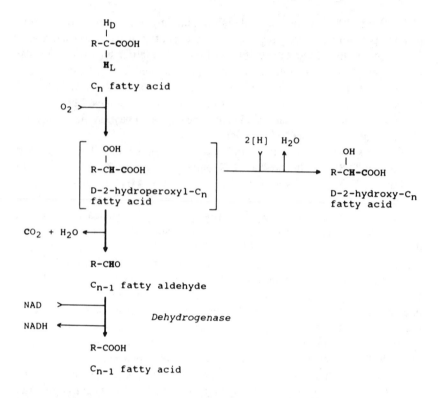

FIGURE 6. Proposed mechanism of α-oxidation. (Adapted from Shine, W. E. and Stumpf, P. K., *Arch. Biochem. Biophys.*, 162, 147, 1974.)

tissues. Initial wound respiration in certain bulky storage organs (e.g., potato) is mainly due to α-oxidation of fatty acids which arise probably from membrane degradation initiated by wounding.[56,108,109] However, complete degradation of the fatty acids attacked by α-oxidation requires subsequent β-oxidation due to the restriction of α-oxidation to long-chain fatty acids. Long-chain fatty aldehydes which can be synthesized by α-oxidation are components of volatile products which are formed by plants and serve different functions. The synthesis of odd chain alkanes, components of plant surface waxes, from very long-chain, even-numbered fatty acids does not include an α-oxidation. The alkanes are synthesized from acyl-CoAs which are reduced to the corresponding aldehydes by an acyl-CoA reductase; the aldehydes subsequently undergo decarbonylation to generate the alkanes and CO.[105,110] Odd-numbered fatty acids are rarely found in plant tissues. They may be formed from even-numered fatty acids by α-oxidation and degraded subsequently by β-oxidation, yielding propionyl-CoA which in turn is catabolized by modified β-oxidation (Section II.E.3). An intermediate of the latter pathway is malonic semialdehyde, a precursor of β-alanine.[51] Thus, α-oxidation may be indirectly involved in the synthesis of a component of CoA-SH and acyl carrier protein.

REFERENCES

1. **Beevers, H.,** Metabolic production of sucrose from fat, *Nature,* 191, 433, 1961.
2. **Beevers, H.,** The role of the glyoxylate cycle, in *The Biochemistry of Plants,* Vol. 4, Stumpf, P. K., Ed., Academic Press, New York, 1980, 117.
3. **Walker, D. B., Gysi, J., Sternberg, L., and DeNiro, M. J.,** Direct respiration of lipids during heat production in the inflorescence of *Philodendron selloum, Science,* 220, 419, 1983.
4. **Dunford, R., Kirk, D., and ap Rees, T.,** Respiration of valine by higher plants, *Phytochemistry,* 29, 41, 1990.
4a. **Salon, C., Raymond, P., and Pradet, A.,** Quantification of carbon fluxes through the tricarboxylic acid cycle in early germinating lettuce embryos, *J. Biol. Chem.,* 263, 12278, 1988.
4b. **Raymond, P., Spiteri, A., Dieuaide, M., Gerhardt, B., and Pradet, A.,** Peroxisomal β-oxidation of fatty acids and citrate formation by a particulate fraction from early germinating sunflower seeds, *Plant Physiol. Biochem.,* 30, 153, 1992.
5. **Jacobson, B. S., Smith, B. N., Epstein, S., and Laties, G. G.,** The prevalence of carbon-13 in respiratory carbon dioxide as indicator of the type of endogenous substrate, *J. Gen. Physiol.,* 55, 1, 1970.
6. **Schulz, H.,** Beta oxidation of fatty acids, *Biochim. Biophys. Acta,* 1081, 109, 1991.
7. **Osmundson, H., Bremer, J., and Pedersen, J. J.,** Metabolic aspects of peroxisomal β-oxidation, *Biochim. Biophys. Acta,* 1085, 141, 1991.
8. **Stumpf, P. K. and Barber, G. A.,** Fat metabolism in higher plants. VII. β-Oxidation of fatty acids by peanut mitochondria, *Plant Physiol.,* 31, 304, 1956.
9. **Rebeiz, C. A. and Castelfranco, P.,** An extramitochondrial enzyme system from peanuts catalyzing the β-oxidation of fatty acids, *Plant Physiol.,* 39, 932, 1964.
10. **Yamada, M. and Stumpf, P. K.,** Fat metabolism in higher plants. XXIV. A soluble β-oxidation system from germinating seeds of *Ricinus communis, Plant Physiol.,* 40, 653, 1965.
11. **Cooper, T. G. and Beevers, H.,** β-Oxidation in glyoxysomes from castor bean endosperm, *J. Biol. Chem.,* 244, 3514, 1969.
12. **Breidenbach, R. W. and Beevers, H.,** Association of glyoxylate cycle enzymes in a novel subcellular particle from castor bean endosperm, *Biochem. Biophys. Res. Commun.,* 27, 462, 1967.
13. **Gerhardt, B.,** *Microbodies/Peroxisomen pflanzlicher Zellen,* Springer-Verlag, Vienna, 1978.
14. **Huang, A. H. C., Trelease, R. N., and Moore, T. S., Jr.,** *Plant Peroxisomes,* Academic Press, New York, 1983.
15. **Gerhardt, B.,** Enzyme activities of the β-oxidation pathway in spinach leaf peroxisomes, *FEBS Lett.,* 126, 71, 1981.
16. **Gerhardt, B.,** Localization of β-oxidation enzymes in peroxisomes isolated from nonfatty plant tissues, *Planta,* 159, 238, 1983.
17. **Macey, M. and Stumpf, P. K.,** β-Oxidation enzymes in microbodies from tubers of *Helianthus tuberosus, Plant Sci. Lett.,* 28, 207, 1982.
18. **Macey, M.,** β-Oxidation and associated enzyme activities in microbodies from germinating peas, *Plant Sci. Lett.,* 30, 53, 1983.
19. **Gerhardt, B.,** Basic metabolic function of the higher plant peroxisome, *Physiol. Vég.,* 24, 397, 1986.
20. **Gerhardt, B.,** Peroxisomal catabolism of fatty and amino acids and its integration into cell metabolism, in *Molecular Approaches to Compartmentation and Metabolic Regulation,* Huang, A. H. C. and Taiz, L., Eds., American Society of Plant Physiologists, Rockville, MD, 1991, 121.
21. **Cooper, T. C.,** The activation of fatty acids in castor bean endosperm, *J. Biol. Chem.,* 246, 3451, 1971.

22. **Gerbling, H. and Gerhardt, B.,** Activation of fatty acids by nonglyoxysomal peroxisomes, *Planta,* 171, 386, 1987.
23. **Olsen, J. A. and Huang, A. H. C.,** Glyoxysomal acyl-CoA synthetase and oxidase from germinating elm, rape and maize seed, *Phytochemistry,* 27, 1601, 1988.
24. **Gerbling, H. and Gerhardt, B.,** Ricinoleic acid catabolism in peroxisomes, *Bot. Acta,* 104, 233, 1991.
25. **Gerbling, H. and Gerhardt, B.,** Oxidative decarboxylation of branched-chain 2-oxo fatty acids by higher plant peroxisomes, *Plant Physiol.,* 88, 13, 1988.
26. **Kirsch, T., Löffler, H.-G., and Kindl, H.,** Plant acyl-CoA oxidase. Purification, characterization, and monomeric apoprotein, *J. Biol. Chem.,* 261, 8570, 1986.
26a. **Kindl, H.,** β-Oxidation of fatty acids by specific organelles, in *The Biochemistry of Plants,* Vol. 9, Stumpf, P. K., Ed., Academic Press, New York, 1987, 31.
27. **Osumi, T., Hashimoto, T., and Ui, N.,** Purification and properties of acyl-CoA oxidase from rat liver, *J. Biochem.,* 87, 1735, 1980.
28. **Coudron, P. E., Frerman, F. E., and Schowalter, D. B.,** Chemical and catalytic properties of the peroxisomal acyl-coenzyme A oxidase from *Candida tropicalis, Arch. Biochem. Biophys.,* 226, 324, 1983.
29. **Gerhardt, B.,** Peroxisomes — site of β-oxidation in plant cells, in *Structure, Function and Metabolism of Plant Lipids,* Siegenthaler, P. A. and Eichenberger, W., Eds., Elsevier Science Publishers, Amsterdam, 1984, 189.
30. **Gerhardt, B.,** Substrate specificity of peroxisomal acyl-CoA oxidases, *Phytochemistry,* 24, 351, 1985.
31. **Frevert, J. and Kindl, H.,** A bifunctional enzyme from glyoxysomes. Purification of a protein possessing enoyl-CoA hydratase and 3-hydroxyacyl-CoA dehydrogenase activities, *Eur. J. Biochem.,* 107, 79, 1980.
32. **Donaldson, R. P.,** Nicotinamide cofactors (NAD and NADP) in glyoxysomes, mitochondria, and plastids isolated from castor bean endosperm, *Arch. Biochem. Biophys.,* 215, 274, 1982
33. **Behrends, W., Engeland, K., and Kindl, H.,** Characterization of two forms of the multifunctional protein acting in fatty acid β-oxidation, *Arch. Biochem. Biophys.,* 263, 161, 1988.
34. **Miernyk, J. A. and Trelease, R. N.,** Substrate specificity of cotton glyoxysomal enoyl-CoA hydratase, *FEBS Lett.,* 129, 139, 1981.
35. **Frevert, J. and Kindl, H.,** Purification of glyoxysomal acetyl-CoA acyltransferase, *Hoppe-Seyler's Z. Physiol. Chem.,* 361, 537, 1980.
36. **Gerhardt, B.,** Higher plant peroxisomes and fatty acid degradation, in *Peroxisomes in Biology and Medicine,* Fahimi, D. and Sies, H., Eds., Springer-Verlag, Berlin, 1987, 141.
37. **Miernyk, J. A. and Trelease, R. N.,** Control of enzyme activities in cotton cotyledons during maturation and germination. IV. β-Oxidation, *Plant Physiol.,* 67, 341, 1981.
38. **Donaldson, R. P. and Fang, T. K.,** β-Oxidation and glyoxylate cycle coupled to NADH:cytochrome c and ferricyanide reductases in glyoxysomes, *Plant Physiol.,* 85, 792, 1987.
39. **Kleiter, A. and Gerhardt, B.,** unpublished data, 1990/1991.
40. **Hutton, D. and Stumpf, P. K.,** Fat metabolism in higher plants. LXII. The pathway of ricinoleic acid catabolism in the germinating castor bean (*Ricinus communis* L.) and pea (*Pisum sativum* L.), *Arch. Biochem. Biophys.,* 142, 48, 1971.
41. **Engeland, K. and Kindl, H.,** Purification and characterization of a plant peroxisomal Δ^2,Δ^3-enoyl-CoA isomerase acting on 3-*cis*-enoyl-CoA and 3-*trans*-enoyl-CoA, *Eur. J. Biochem.,* 196, 699, 1991.
42. **Behrends, W., Thieringer, R., Engeland, K., Kunau, W.-H., and Kindl, H.,** The glyoxysomal β-oxidation system in cucumber seedlings: identification of enzymes required for the degradation of unsaturated fatty acids, *Arch. Biochem. Biophys.,* 263, 170, 1988.
43. **Li, J., Smeland, T. E., and Schulz, H.,** D-3-Hydroxyacyl coenzyme A dehydratase from rat liver peroxisomes. Purification and characterization of a novel enzyme necessary for the epimerization of hydroxyacyl-CoA thioesters, *J. Biol. Chem.,* 265, 13629, 1990.

44. **Engeland, K. and Kindl, H.,** Evidence for a peroxisomal fatty acid β-oxidation involving D-3-hydroxyacyl-CoAs. Characterization of two forms of hydro-lyase that convert D-(–)-3-hydroxyacyl-CoA into 2-*trans*-enoyl-CoA, *Eur. J. Biochem.,* 200, 171, 1991.
45. **Kunau, W.-H. and Dommes, P.,** Degradation of unsaturated fatty acids. Identification of intermediates in the degradation of *cis*-4-decenoyl-CoA by extracts of beef liver mitochondria, *Eur. J. Biochem.,* 91, 533, 1978.
46. **Dommes, V., Baumgart, C., and Kunau, W.-H.,** Degradation of unsaturated fatty acids in peroxisomes. Existence of a 2,4-dienoyl-CoA reductase pathway, *J. Biol. Chem.,* 256, 8259, 1981.
47. **Berger, R. G., Dettweiler, G. R., Kollmannsberger, H., and Drawert, F.,** The peroxisomal dienoyl-CoA reductase pathway in pineapple fruit, *Phytochemistry,* 29, 2069, 1990.
48. **Yang, S.-Y., Cuebas, D., and Schulz, H.,** 3-Hydroxyacyl-CoA epimerases of rat liver peroxisomes and *Escherichia coli* function as auxiliary enzymes in the β-oxidation of polyunsaturated fatty acids, *J. Biol. Chem.,* 261, 12238, 1986.
49. **Kolattukudy, P. E.,** Cutin, suberin, and waxes, in *The Biochemistry of Plants,* Vol. 4, Stumpf, P. K., Ed., Academic Press, New York, 1980, 571.
50. **Giovanelli, J. and Stumpf, P. K.,** Fat metabolism in higher plants. X. Modified β-oxidation of propionate by peanut mitochondria, *J. Biol. Chem.,* 231, 411, 1958.
51. **Hatch, M. D. and Stumpf, P. K.,** Fat metabolism in higher plants. XVIII. Propionate metabolism by plant tissues, *Arch. Biochem. Biophys.,* 96, 193, 1962.
52. **Stumpf, P. K.,** Lipid metabolism, in *Plant Biochemistry,* 3rd ed., Bonner, J. and Varner, J. E., Eds., Academic Press, New York, 1976, 427.
53. **Halarnkar, P. P., Wakayama, E. J., and Blomquist, G. J.,** Metabolism of propionate to 3-hydroxypropionate and acetate in the lima bean *Phaseolus limensis, Phytochemistry,* 27, 997, 1988.
54. **Gerbling, H. and Gerhardt, B.,** Propionyl-CoA generation and catabolism in higher plant peroxisomes, in *Biological Role of Plant Lipids,* Biacs, P. A., Gruiz, K., and Kremmer, T., Eds., Plenum Press, New York, 1989, 21.
55. **Gerbling, H. and Gerhardt, B.,** Abbau von 2-Oxoisovaleriansäure und Propionyl-CoA in Peroxisomen, *Eur. J. Cell Biol.,* 54 (Suppl. 32), 74, 1991.
56. **Laties, G. G. and Hoelle, C.,** The α-oxidation of long-chain fatty acids as a possible component of the basal respiration of potato slices, *Phytochemistry,* 6, 49, 1967.
57. **Poston, J. M.,** Coenzyme B_{12}-dependent enzymes in potatoes: leucine 2,3-aminomutase and methylmalonyl-CoA mutase, *Phytochemistry,* 17, 401, 1978.
58. **Gerbling, H. and Gerhardt, B.,** Peroxisomal degradation of branched-chain 2-oxo acids, *Plant Physiol.,* 91, 1387, 1988.
59. **Gerbling, H. and Gerhardt, B.,** Degradation of 2-oxoisocaproate in higher plant peroxisomes, *Biol. Chem. Hoppe-Seyler,* 372, 530, 1991.
60. **Gerbling, H. and Gerhardt, B.,** unpublished data, 1991.
61. **Hutton, D. and Stumpf, P. K.,** Characterization of the β-oxidation system from maturing and germinating castor bean seeds, *Plant Physiol.,* 44, 508, 1969.
62. **Bortman, S. J., Trelease, R. N., and Miernyk, J. A.,** Enzyme development and glyoxysome characterization in cotyledons of cotton seeds, *Plant Physiol.,* 68, 82, 1981.
63. **Olsen, L. J. and Harada, J. J.,** Biogenesis of peroxisomes in higher plants, in *Molecular Approaches to Compartmentation and Metabolic Regulation,* Huang, A. H. C. and Taiz, L., Eds., American Society of Plant Physiologists, Rockville, MD, 1991, 129.
64. **Ni, W. and Trelease, R. N.,** Post-transcriptional regulation of catalase isozyme expression in cotton seeds, *Plant Cell,* 3, 737, 1991.
65. **Gerdes, H.-H. and Kindl, H.,** Gene response upon illumination in forming mRNA encoding peroxisomal glycollate oxidase, *Biochim. Biophys. Acta,* 949, 195, 1988.
66. **Pistelli, L., De Bellis, L., and Alpi, A.,** Peroxisomal enzyme activities in attached senescing leaves, *Planta,* 184, 151, 1991.
67. **Frevert, J., Köller, W., and Kindl, H.,** Occurrence and biosynthesis of glyoxysomal enzymes in ripening cucumber seeds, *Hoppe-Seyler's Z. Physiol. Chem.,* 361, 1557, 1980.

68. **Behrends, W., Birkhan, R., and Kindl, H.,** Transition form of microbodies. Overlapping of two sets of marker proteins during the rearrangement of glyoxysomes into leaf peroxisomes, *Biol. Chem. Hoppe-Seyler,* 371, 85, 1990.
69. **Chasan, R.,** Lipid transfer proteins: moving molecules?, *Plant Cell,* 3, 842, 1991.
70. **Bünning, J. and Gerhardt, B.,** unpublished data, 1990.
71. **Gerbling, H. and Gerhardt, B.,** Carnitine-acyltransferase activity of mitochondria from mung-bean hypocotyls, *Planta,* 174, 90, 1988.
72. **Kirsch, T., Rojahn, B., and Kindl, H.,** Diphosphatase related to lipid metabolism and gluconeogenesis in cucumber cotyledons: localization in plasma membrane and etioplasts but not in glyoxysomes, *Z. Naturforsch.,* 44c, 937, 1989.
73. **Gerhardt, B. and Fischer, K.,** unpublished data, 1991.
74. **Mettler, J. J. and Beevers, H.,** Oxidation of NADH in glyoxysomes by a malate-aspartate shuttle, *Plant Physiol.,* 66, 555, 1980.
75. **Landolt, R. and Matile, P.,** Glyoxisome-like microbodies in senescent spinach leaves, *Plant Sci.,* 72, 159, 1990.
76. **Papke, I. and Gerhardt, B.,** Metabolism of acetyl-CoA in nonglyoxysomal peroxisomes, *Biol. Chem. Hoppe-Seyler,* 372, 538, 1991.
77. **Verniquet, F., Gaillard, J., Neuburger, M., and Douce, R.,** Rapid inactivation of plant aconitase by hydrogen peroxide, *Biochem. J.,* 276, 643, 1991.
78. **Hicks, D. B. and Donaldson, R. P.,** Electron transport in glyoxysomal membranes, *Arch. Biochem. Biophys.,* 215, 280, 1982.
79. **Tolbert, N. E.,** Metabolic pathways in peroxisomes and glyoxysomes, *Annu. Rev. Biochem.,* 50, 133, 1981.
80. **Heldt, H. W. and Flügge, U. I.,** Subcellular transport of metabolites in plant cells, in *The Biochemistry of Plants,* Vol. 12, Davies, D. D., Ed., Academic Press, New York, 1987, 49.
81. **Yu, C. and Huang, A. H. C.,** Conversion of serine to glycerate in intact spinach leaf peroxisomes: role of malate dehydrogenase, *Arch. Biochem. Biophys.,* 245, 125, 1986.
82. **Schwitzguebel, J.-P. and Siegenthaler, P.-A.,** Purification of peroxisomes and mitochondria from spinach leaf by Percoll gradient centrifugation, *Plant Physiol.,* 75, 670, 1984.
83. **Luster, D. G. and Donaldson, R. P.,** Orientation of electron transport activities in the membrane of intact glyoxysomes isolated from castor bean endosperm, *Plant Physiol.,* 85, 796, 1987.
84. **Donaldson, R. P. and Fang, T. K.,** β-Oxidation and glyoxylate cycle coupled to NADH: cytochrome c and ferricyanide reductase in glyoxysomes, *Plant Physiol.,* 85, 792, 1987.
85. **Panter, R. A. and Mudd, J. B.,** Carnitine levels in some higher plants, *FEBS Lett.,* 5, 169, 1969.
86. **McNiel, P. H. and Thomas, D. R.,** Carnitine content of pea seedling cotyledons, *Phytochemistry,* 14, 2335, 1975.
87. **Wood, C., Noh Hj Jalil, M., Ariffin, A., Yong, B. C. S., and Thomas, D. R.,** Carnitine short-chain acyltransferase in pea mitochondria, *Planta,* 158, 175, 1983.
88. **Burgess, N. and Thomas, D. R.,** Carnitine acetyltransferase in pea cotyledon mitochondria, *Planta,* 167, 58, 1986.
89. **Wood, C., Noh Hj Jalil, M., McLaren, J., Yong, B. C. S., Ariffin, A., McNeil, P. H., Burgess, N., and Thomas, D. R.,** Carnitine long-chain acyltransferase and oxidation of palmitate, palmitoyl coenzyme A and palmitoylcarnitine by pea mitochondria preparations, *Planta,* 161, 255, 1984.
90. **Thomas, D. R. and Wood, C.,** The two β-oxidation sites in pea cotyledons. Carnitine palmitoyltransferase: location and function in pea mitochondria, *Planta,* 168, 261, 1986.
91. **Thomas, D. R. and Wood, C.,** Oxidation of acetate, acetyl-CoA and acetylcarnitine by pea mitochondria, *Planta,* 154, 145, 1982.
92. **Gerbling, H., Gandour, R. D., Moore, T. S., and Gerhardt, B.,** Carnitine-acyltransferase activity of plant mitochondria, in *Plant Lipid Biochemistry, Structure and Utilization,* Quinn, P. J. and Harwood, J. L., Eds., Portland Press, London, 1990, 181.

93. **McNiel, P. H. and Thomas, D. R.,** The effect of carnitine on palmitate oxidation by pea cotyledon mitochondria, *J. Exp. Bot.,* 27, 1163, 1976.

94. **Thomas, D. R. and McNiel, P. H.,** The effect of carnitine on oxidation of saturated fatty acids by pea cotyledon mitochondria, *Planta,* 132, 61, 1976.

95. **Wood, C., Burgess, N., and Thomas, D. R.,** The dual location of β-oxidation enzymes in germinating pea cotyledons, *Planta,* 167, 54, 1986.

96. **Thomas, D. R., Wood, C., and Masterson, C.,** Long-chain acyl CoA synthetase, carnitine and β-oxidation in pea-seed mitochondria, *Planta,* 173, 263, 1988.

97. **Maier, U., Fischer, K., and Gerhardt, B.,** unpublished data, 1991.

98. **Masterson, C., Wood, C., and Thomas, D. R.,** β-Oxidation enzymes in the mitochondria of *Arum* and oilseed rape, *Planta,* 182, 129, 1990.

99. **Gerhardt, B.,** unpublished data, 1986.

100. **Miernyk, J. A., Thomas, D. R., and Wood, C.,** Partial purification and characterization of the mitochondrial and peroxisomal isozymes of enoyl-coenzyme A hydratase from germinating pea seedlings, *Plant Physiol.,* 95, 564, 1991.

101. **Martin, R. O. and Stumpf, P. K.,** Fat metabolism in higher plants. XII. α-Oxidation of long chain fatty acid, *J. Biol. Chem.,* 234, 2548, 1959.

102. **Hitchcock, C. H. S. and James, A. T.,** The mechanism of α-oxidation in leaves, *Biochim. Biophys. Acta,* 116, 413, 1966.

103. **Hitchcock, C. H. S. and Morris, L. J.,** The stereochemistry of α-oxidation of fatty acids in leaves, *Eur. J. Biochem.,* 17, 39, 1970.

104. **Shine, W. E. and Stumpf, P. K.,** Fat metabolism in higher plants. Recent studies on plant α-oxidation systems, *Arch. Biochem. Biophys.,* 162, 147, 1974.

105. **Cheesbrough, T. M. and Kolattukudy, P. E.,** Alkane biosynthesis by decarbonylation of aldehydes catalyzed by a particulate preparation from *Pisum sativum, Proc. Natl. Acad. Sci. U.S.A.,* 81, 6613, 1984.

106. **Galliard, T. and Matthew, J. A.,** The enzymic formation of long-chain aldehydes and alcohols by α-oxidation of fatty acids in extracts of cucumber fruit (*Cucumis sativus*), *Biochim. Biophys. Acta,* 424, 26, 1976.

107. **Hitchcock, C. and Rose, A.,** The stereochemistry of α-oxidation of fatty acids in plants: the configuration of biosynthetic long-chain 2-hydroxy acids, *Biochem. J.,* 125, 1155, 1971.

108. **Laties, G. G., Hoelle, C., and Jacobson, B. S.,** α-Oxidation of endogenous fatty acids in fresh potato slices, *Phytochemistry,* 11, 3403, 1972.

109. **Theologis, A. and Laties, G. G.,** Membrane lipid breakdown in relation to the wound-induced and cyanide-resistant respiration in tissue slices. A comparative study, *Plant Physiol.,* 66, 890, 1980.

110. **Kolattukdy, P. E.,** Lipid-derived defensive polymers and waxes and their role in plant-microbe interaction, in *The Biochemistry of Plants,* Vol. 9, Stumpf, P. K., Ed., Academic Press, New York, 1987, 291.

SECTION V

GENERAL CONSIDERATIONS

The two chapters in this section have been written to emphasize two general areas. The emphasis in the preceding chapters has been aimed at describing the metabolism of the lipids, but much of this research has been driven by known physiological effects or responses of the lipids. Chapter 18 summarizes the role of lipids in a variety of physiological responses of plants ranging from stresses to hormones.

Chapter 17, on the other hand, is dedicated to emphasizing some aspects of lipid metabolism outside the usual areas of study. In particular, most investigations are performed on photosynthetic or lipid storage tissues, but other organs synthesize lipids too (e.g., roots). This chapter is aimed at filling gaps in these areas.

Chapter 17

METABOLISM IN NONPHOTOSYNTHETIC, NONOILSEED TISSUES

Salvatore A. Sparace and Kathryn F. Kleppinger-Sparace

TABLE OF CONTENTS

0-8493-4907-9/93/$0.00+$.50
© 1993 by CRC Press Inc.

I. INTRODUCTION

Plant lipid metabolism has been studied in a number of different plant tissues. However, as apparent in the preceding chapters, most of what is known concerning the basic metabolic pathways and regulatory mechanisms involved is derived from studies of a relatively few plant tissues. The leaves and oilseeds of higher plants and the plastids therein are most often emphasized. The reasons for this include the physiological prominence of leaves and their role in photosynthesis and the great economic importance of oilseeds. Researchers have now realized that lipid metabolism in leaves and oilseeds is, in many ways, highly specialized and often uniquely regulated in accordance with the physiological functions of these tissues. The purpose of this chapter is to direct the focus toward lipid metabolism in nonphotosynthetic plant systems other than leaves and oilseeds. This review will emphasize fatty acid and glycerolipid metabolism in nonphotosynthetic plastids.

Lipid metabolism and the pathways involved have been extensively reviewed in preceding chapters as well as earlier reviews.[1-11] They are, for the most part, similar in both photosynthetic and nonphotosynthetic tissues. However, there are some noteworthy trends and differences. Although these tissues both contain approximately the same levels of phospholipid on a per gram dry weight basis, nonphotosynthetic tissues have a higher proportion of phospholipid compared to galactolipid.[5] The increased proportion of phospholipids reflects both a decrease in the amount of galactolipids in those cell membrane fractions which contain both types of lipid,[5] as well as the decreased prominence of the plastid in the cells of some nonphotosynthetic tissues.[4] The corresponding membranes or organelles isolated from both photosynthetic and nonphotosynthetic systems contain the same characteristic glycerolipid compositions. Mitochondria uniquely contain cardiolipin,[2,5,61] plastids contain galactolipids and sulfolipid (sulfoquinovosyldiacylglycerol; SQDG), but much lower levels of phosphatidylethanolamine (PE), while the endoplasmic reticulum (ER) contains higher levels of PE than other membranes.[5] One of the more unusual aspects of the ER isolated from various nonphotosynthetic tissues is its increased capacity for triacylglycerol (TAG) biosynthesis and its plasticity for synthesizing the unusual fatty acids esterified to TAG or utilized for the formation of epicuticular waxes.[5] As will be discussed later, the needs and energy requirements of the nonphotosynthetic tissues can differ from each other and from that of the photosynthetic tissue. This in turn can ultimately affect and regulate membrane lipid compositions and metabolism.

II. METABOLISM IN ORGANELLES

A. PLASTIDS
1. Fatty Acid Biosynthesis in Plastids

Early in the study of plant lipid metabolism, following the intitial report by Smirnov[12] that spinach chloroplasts could synthesize fatty acids from acetate,

TABLE 1
Enzymes and Reactions of Fatty Acid Biosynthesis from Acetate in Plastids[7]

Acetyl-CoA synthetase	Acetate + ATP + CoA-SH → Acetyl-CoA
Acetyl-CoA carboxylase	Acetyl-CoA + HCO_3 + ATP → Malonyl-CoA
Acetyl-CoA:ACP transacetylase	Acetyl-CoA + ACP → Acetyl-ACP + CoA
Malonyl-CoA:ACP transacylase	Malonyl-CoA + ACP → Malonyl-ACP + CoA
β-Ketoacyl synthetase I	Acyl(C_2-C_{14})-ACP + Malonyl-ACP → β-Ketoacyl(C_4-C_{16})-ACP + CO_2 + ACP
β-Ketoacyl synthetase II	Palmitoyl-ACP + Malonyl-ACP → β-Ketostearoyl-ACP
β-Ketoacyl-ACP reductase	β-Ketoacyl-ACP + $NADPH_{(2)}$ → β-Hydroxyacyl-ACP + $NADP^+$
β-Hydroxyacyl-ACP dehydrase	β-Hydroxyacyl-ACP → 2-Enoyl-ACP + H_2O
Enoyl-ACP reductase	2-Enoyl-ACP + $NADH_{(2)}$ → Acyl-ACP + NAD^+

plastids emerged as an organelle of central importance. They are generally acclaimed to be the only subcellular site for *de novo* fatty acid biosynthesis in all plant tissues.[13] As such, plastids provide fatty acids for the biosynthesis of a variety of acyl lipids, particularly the membrane and storage glycerolipids, that are synthesized not only within the plastid, but also in such organelles as the endoplasmic reticulum, mitochondria, and perhaps developing oil bodies.[3,8,11,14]

The process of *de novo* fatty acid biosynthesis in all types of plastids is thought to occur by a series of virtually identical enzymatic steps. As recently summarized by Stumpf,[15] starting from acetate, the process involves the enzymes and sequence of repeating reactions shown in Table 1 until palmitoyl-acyl carrier protein (ACP) or stearoyl-ACP are synthesized. *In vitro* studies of fatty acid biosynthesis from acetate in isolated plastids normally insure that the substrate and co-factor requirements of each of the enzymic reactions listed are either directly or indirectly satisfied. As shown, *de novo* fatty acid biosynthesis is an energy-dependent process. ATP is required for the synthesis of both acetyl coenzyme A (CoA) and malonyl-CoA by acetyl-CoA synthetase and acetyl-CoA carboxylase, respectively.[15] Similarly, NADPH and NADH are required in the β-ketoacyl-ACP reductase and 2-enoyl-ACP reductase steps, respectively, in both *de novo* fatty acid biosynthesis and the desaturation of stearoyl-ACP[15] (not shown in Table 1). This means that ATP and reduced nucleotides, as well as CoA, divalent cations, and bicarbonate, must be supplied to promote maximum rates of fatty acid biosynthesis.

The degree to which these co-factors are required varies from one type of plastid to another. Table 2 illustrates the *in vitro* dependency of several different types of plastids on relatively standard exogenously supplied co-factors. As shown, the dependency on these co-factors ranges from virtually complete for ATP in pea root plastids[16] and daffodil chromoplasts[18] to slightly inhibitory for Mn^{2+} in pea root plastids[16] and for NADPH in both pea root plastids and

TABLE 2
Co-Factor Requirements for Fatty Acid Biosynthesis from Acetate in Plastids from Several Plant Tissues

Cofactor deleted	Pea root plastids[16]	Developing castor bean leukoplasts[17]	Daffodil chromoplasts[18]	Avocado mesocarp plastids[19]	Soybean suspension plastids[20]	Spinach chloroplasts[21]	Lettuce chloroplasts[22]
Complete control activity, nmol/h/mg[a]	26	54	10	110	1	150	576
				% of control activity			
ATP	1	62	<1	10	26	60	50
CoA	3	91	49	55	9	39	17
HCO_3^-	18	—[b]	<<17	33	—	83	27
Mg^{2+}	43	76	—	*	—	89	58
Mn^{2+}	120	94	*	17	—	86	99
Mg^{2+}, Mn^{2+}	15	—	—	—	—	59	—
NADH	42	36	92	36	61	*	*
NADPH	107	32	94	82	100	*	106
NADH, NADPH	19	—	66	—	—	*	—

Note: An asterisk is used to indicate that this cofactor is not part of the standard reaction mixture.

[a] Rates of acetate incorporation into fatty acids are per milligram of plastid protein for nonphotosynthetic plastids and per milligram of chlorophyll for illuminated chloroplasts.

[b] Not tested.

lettuce chloroplasts.[16,22] This variability can be explained in terms of the abilities of different plastids to retain these co-factors during isolation, as well as their capacities to synthesize additional co-factors during the isolation and incubation periods.

The formation of malonyl-CoA by acetyl-CoA carboxylase is considered to be the first committed step in the process of *de novo* fatty acid biosynthesis.[23] As such, the enzyme represents an important regulatory step that has received considerable attention. An important mode of regulation of acetyl-CoA carboxylase in all plastids appears to be the energy charge of the cell[23] and perhaps the plastid itself. In chloroplasts, however, in addition to changes in the availability of ATP, this enzyme also appears to be regulated by light-induced changes in stroma pH, Mg^{2+} concentration, and the availability of reducing equivalents[23,24] by mechanisms similar to those of several enzymes of photosynthetic carbon metabolism.[25] The existence of mechanisms similar to these latter modes of regulation of acetyl-CoA carboxylase in nonphotosynthetic plastids remains to be determined.

a. Metabolic Source of Energy for Fatty Acid Biosynthesis

One of the more important concerns regarding fatty acid biosynthesis in nonphotosynthetic plastids as compared to chloroplasts is the physiological or metabolic source of the high energy co-factors. In chloroplasts, ATP and reduced nucleotides are thought to be derived from photophosphorylation and NAD(P)H:ferredoxin reductase during photosynthetic electron transport.[15] This is supported by well-documented observations that illumination is an essential requirement for maximum rates of fatty acid synthesis in isolated chloroplasts with dark incubations generally resulting in the inhibition of fatty acid biosynthesis by 85 to virtually 100%.[12,21,22,26] In contrast, in nonphotosynthetic plastids, ATP and NADH are thought to be derived from glycolytic metabolism, while NADPH is generated in the pentose phosphate pathway. There is considerable evidence that suggests that nonphotosynthetic plastids possess essentially full complements of the enzymes of both these pathways.[27-30] The relative activities of the enzymes of the glycolytic and pentose phosphate pathways in several types of nonphotosynthetic plastids are shown in Table 3. The enzymes of these pathways that are specifically involved in providing the high energy co-factors required for fatty acid biosynthesis are phosphoglycerate kinase and pyruvate kinase which both provide ATP, glyceraldehyde-3-phosphate dehydrogenase which provides NADH, and glucose-6-phosphate dehydrogenase and 6-phosphogluconate dehydrogenase which both generate NADPH. In and of themselves, the activities of these enzymes generally appear sufficient to provide at least a portion of the ATP and reduced nucleotides required for fatty acid biosynthesis in nonphotosynthetic plastids. However, as recently pointed out by Frehner et al.[28] the activities of one or more of the enzymes of glycolytic and pentose phosphate metabolism are so low that they preclude the efficient or complete operation of the entire pathways. In plastidic glycolysis, examples of such restrictive enzymes are

TABLE 3
Activity of Enzymes of Glycolysis and the Pentose Phosphate Pathway in Several Types of Nonphotosynthetic Plastids

	Wheat endosperm amyloplasts[27]	Sycamore suspension amyloplasts[28]	Cauliflower bud plastids[29]	Developing castor bean leukoplasts[30]
		(nmol/min)[a]		
Glycolysis				
Hexokinase	—	0.76	40	—
Phosphoglucomutase	2.79	—	300	31
Hexose phosphate isomerase	38.04	—	1100	120
Phosphofructokinase	0.17	0.40	70	11
Aldolase	1.14	9.54	15	123
Triose phosphate isomerase	771.82	24.49	1600	—
Glyceraldehyde-3-phosphate dehydrogenase	6.55	0.62	190	27
Phosphoglycerate kinase	28.45	1.52	900	9
Phosphoglycerate mutase	39.38	0.03	0—30	150
Enolase	27.00	3.42	85	34
Pyruvate kinase	8.17	0.63	100	330
Pentose phosphate pathway				
Glucose-6-phosphate dehydrogenase	—	0.14	100	0
6-Phosphogluconate dehydrogenase	—	4.49	220	320
Ribose-5-phosphate isomerase	—	—	2000	—
Ribulose-5-phosphate 3-epimerase	—	—	300	—
Transketolase	—	7.18	500	70
Transaldolase	—	6.42	250	10

[a]　Rates of activity shown are per milliliter of protoplast lysate for wheat endosperm amyloplasts, per gram fresh weight of protoplasts for sycamore suspension amyloplasts, and per milligram of protein for cauliflower and castor bean plastids.

commonly glyceraldehyde-3-phosphate dehydrogenase in amyloplasts from sycamore cell suspension cultures[28] and phosphoglycerate mutase which is barely detectable in both the sycamore amyloplasts[28] and cauliflower bud plastids.[29] Similarly, in the pentose phosphate pathway, glucose-6-phosphate dehydrogenase is very low in sycamore amyloplasts[20] and absent in both developing[30] and germinating[31] castor bean endosperm leukoplasts. The restrictive activities of these enzymes would imply that these plastids must interact with the cytosolic compartment in order to maintain high rates of carbon flux through these pathways. This interaction is most likely at the level of the phosphate/triose-phosphate translocator of the plastid envelope.

The phosphate/triose-phosphate translocator has been most extensively studied in chloroplasts.[24] However, it is likely that similar translocators exist in nonphotosynthetic plastids.[32-34] During photosynthesis, the triose-phosphate translocator normally functions in the export of fixed carbon, in the form of triose phosphates, from the chloroplast to the cytosol for further metabolism.[35] However, the direction of triose transport is variable, depending of the type of plastid and its physiological status.[35] Under *in vitro* conditions, the phosphate translocator can be used as part of a shuttle mechanism which will promote ATP synthesis (via phosphoglycerate kinase) and nucleotide reduction (via glyceraldehyde-3-phosphate dehydrogenase) in the absence of either photophosphorylation in chloroplasts or complete glycolytic metabolism in nonphotosynthetic plastids. This mechanism is commonly referred to as the triose-phosphate or dihydroxyacetone phosphate (DHAP) shuttle[33] (Figure 1). As shown, this shuttle requires an external supply of a suitable triose (DHAP is shown), inorganic phosphate, and oxaloacetate and depends on portions of either Calvin cycle or glycolytic metabolism, the dicarboxylate translocator, and plastidic malate dehydrogenase. The DHAP shuttle has been used to bypass the light/ATP-dependent steps in photosynthetic CO_2 fixation[36] and fatty acid and sulfolipid biosynthesis[37,38] in isolated dark-incubated chloroplasts. Most recently, Kleppinger-Sparace et al.[34] have used the DHAP shuttle to promote fatty acid biosynthesis in pea root plastids in the absence of exogenously supplied ATP. In their study, they showed that the DHAP shuttle could promote 41% of the ATP control activity. Complete activity was restored when the shuttle was supplemented with ADP. Their observations suggest that the triose-phosphate translocator may be an effective means whereby nonphotosynthetic plastids can interact with the cytosolic compartment to support intraplastidic ATP synthesis. Similarly, this mechanism bypasses the rate-limiting step in other plastids at phosphoglycerate mutase.

Further evidence that glycolytic intermediates can be metabolized by nonphotosynthetic plastids to provide energy for fatty acid biosynthesis comes from studies with daffodil chromoplasts[39] and castor bean leukoplasts.[40] In daffodil chromoplasts, DHAP, glyceraldehyde-3-phosphate, and 3-phosphoglycerate each can almost completely substitute for ATP.[39] 2-Phosphoglycerate and phosphoenolpyruvate in daffodil chromoplasts[39] and phosphoenolpyruvate in castor bean leukoplasts[40] all give approximately 2.5 times greater activity

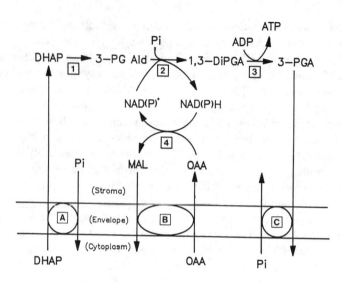

FIGURE 1. The DHAP shuttle mechanism for the generation of intraplastidic ATP. 3-PGald — 3-phosphoglyceraldehyde; 1,3-DiPGA — 1,3-phosphoglyceric acid; 3-PGA — 3-phosphoglyceric acid; MAL — malic acid; OAA — oxaloacetic acid; 1 — dihydroxyacetone phosphate reductase; 2 — glyceraldehyde-3-phosphate dehydrogenase; 3 — phosphoglycerate kinase; 4 — malate dehydrogenase; A and C — triose/phosphate translocator; B — dicarboxylate translocator. (Adapted from Werden, K., et al., *Biochim. Biophys. Acta*, 396, 276, 1975).

than ATP. Similar observations have been made with pea root plastids. However, in the latter case, all of these glycolytic intermediates, except for phosphoenolpyruvate, promote only 5 to 25% of the ATP control activity, while phosphoenolpyruvate generally results in 50 to 60% of the ATP control activity.[69] In any case, these observations all suggest that both phosphoglycerate kinase and pyruvate kinase may indeed function to provide the ATP required for fatty acid biosynthesis, although the extent to which they serve this purpose varies from one type of plastid to another.

The role of glycolytic and pentose phosphate metabolism in providing reduced nucleotides for fatty acid biosynthesis in nonphotosynthetic plastids is less well-documented. Although the key enzymes most likely involved (glyceraldehyde-3-phosphate, glucose-6-phosphate, and 6-phosphogluconate dehydrogenases; Table 2) are commonly found in these plastids, little effort has been made to determine if their activities can promote fatty acid biosynthesis in the absence of exogenously supplied nucleotides. Such studies would be difficult because plastid envelopes are not highly permeable to the substrates for these enzymic reactions and the plastids must first be lysed to maximize their activities.[26] Unfortunately, the latter treatment is not conducive to high rates of fatty acid biosynthesis, since intact plastids are essential for achieving such rates.[41] However, shuttle-stimulated fatty acid biosynthesis in pea root plastids in the absence of ATP is still highly dependent on exogenously supplied reduced nucleotide, particularly NADH.[69]

b. Metabolic Source of Carbon

The metabolic source of carbon for *de novo* fatty acid biosynthesis in both photosynthetic and nonphotosynthetic plastids has received considerable attention. Although it is generally accepted that acetyl-CoA is the direct precursor for fatty acid biosynthesis,[3,13] there appears to be a multiplicity of mechanisms that are involved in the formation of acetyl-CoA. Traditionally, radiolabeled acetate has been and continues to be used as a tracer for both *in vivo* and *in vitro* studies of fatty acid biosynthesis in plants. Acetate is readily accessible to plastids where it is activated to acetyl-CoA by acetyl-CoA synthetase, which is restricted to the plastid.[42] In earlier studies, free acetate was thought to originate in the mitochondrion of photosynthetic plant cells through the action of two mitochondrial enzymes. Pyruvate dehydrogenase formed acetyl-CoA, which was hydrolyzed to free acetate by acetyl-CoA hydrolase, and the resulting acetate freely diffused to the chloroplast.[43] A more direct source of acetyl-CoA is via plastidic pyruvate dehydrogenase. The enzyme has been found not only in chloroplasts,[44-46] but also such nonphotosynthetic plastids as developing castor bean leukoplasts,[47] daffodil flower petal chromoplasts,[48] and cauliflower bud plastids.[29] This enzyme, coupled to the capacity for glycolytic metabolism in nonphotosynthetic plastids as discussed earlier, represents an efficient and direct source of the acetyl-CoA required for fatty acid biosynthesis in these plastids. Alternatively, for plastids that have restricted glycolytic carbon flow, pyruvate derived from cytoplasmic glycolysis could also serve as a source of carbon for fatty acid biosynthesis. A number of studies have shown that exogenously supplied radiolabeled pyruvate is readily utilized for fatty acid biosynthesis in isolated plastids.[49-52] Most recently, Smith et al.[50] showed that malate was superior to either pyruvate or acetate as a substrate for fatty acid biosynthesis in leukoplasts from developing castor bean endosperm. Their results indicate that NAD(P)-linked malic enzyme may be an important source of pyruvate and eventually acetyl-CoA for fatty acid biosynthesis in these plastids.

2. Glycerolipid Composition and Biosynthetic Capacities

In general, nonphotosynthetic plastids are comparable to chloroplasts in terms of their glycerolipid compositions and capacities for glycerolipid biosynthesis. Galactolipids (mono- and digalactosyldiacylglycerol; MGDG and DGDG) and SQDG are glycolipids characteristic of all plastids and represent 60 to 80% and 5 to 6% of the total lipid, respectively (Table 4). Phosphatidylglycerol (PG) and phosphatidylcholine (PC) are the principal phospholipids, each comprising approximately 10 to 20% of the total glycerolipid. Phosphatidylinositol (PI) and PE generally represent less than 5% of the total lipid. The lipid composition of nonphotosynthetic plastids is thought to be most related to chloroplast envelopes. The reason for this is that the former plastids are usually lacking well-developed thylakoid membranes and, consequently, are essentially composed only of envelope lipids. Pea root plastids are comprised of virtually the same lipids as those plastids shown in Table 4. However,

TABLE 4
Glycerolipid Composition of Plastids from Several Plant Tissues

Lipid	Potato tuber amyloplasts[57]	Daffodil chromoplasts[18]	Cauliflower bud plastids[29]	Beet chloroplasts[58]	Spinach chloroplasts[59]	
					Envelopes	Thylakoids
			Mole %			
Triacylglycerol	1	0[a]	0	0	0	0
Monogalactosyldiacylglycerol	14	63	31	44	22	51
Digalactosyldiacylglycerol	45	18	29	25	32	26
Sulfoquinovosyldiacylglycerol	5	5	6	6	5	7
Phosphatidylglycerol	2	10	8	10	8	9
Phosphatidylcholine	16	2	19	10	27	3
Phosphatidylinositol	5	1	5	2	1	1
Phosphatidylethanolamine	5	<1	2	4	<1	0

Note: Above corrections for digalactosyl diacylglycerol in beet chloroplasts vs. spinach chloroplast envelopes.

[a] Values of 0 were either as reported or assumed from data provided in each citation.

they do have some significant differences. Total galactolipid is greatly reduced, while phospholipid is increased. In addition, the root plastids contain as much as 15% TAG.[70] These differences may be related to the physiological origins of the plastids. All of the plastids in Table 4 are derived from tissues with leaf or stem developmental origins and may thus have physiological characteristics more closely associated with photosynthetic plastids as compared to root plastids.

Nonphotosynthetic plastids can synthesize essentially all of their membrane glycerolipids. With acetate as a precursor, the products of fatty acid biosynthesis (acyl-ACPs) are used by two consecutive acyltransferases to acylate glycerol-3-phosphate to form lysophosphatidic acid (LPA) and phosphatidic acid (PA), respectively. Subsequently, PA is metabolized in several directions to eventually form those glycerolipids comprising each plastid. The details of the pathways and mechanisms involved in this metabolism are the subjects of earlier chapters and other reviews.[3,8,11,13] Table 5 summarizes the *in vitro* capacities of several types of plastids to synthesize various lipids from radiolabeled acetate. As might be expected with this precursor, free fatty acids and thioesters (acyl-ACPs and acyl-CoAs) are frequently major products. Significant (but somewhat variable) amounts of radiolabeled glycerolipid intermediates and products also accumulate. These are typically PA, diacylglycerol (DAG), PC, and PG. As before, pea root plastids are somewhat unique in that they are also capable of synthesizing the TAG that they contain. It should be emphasized, however, that the proportions of each labeled lipid product accumulated are greatly dependent on the availablility of various co-factors of fatty acid and glycerolipid biosynthesis in the *in vitro* incubation mixtures as discussed earlier and that the variability seen from one plastid to another is largely due to the different incubation conditions used in each system. As a result, key intermediates or products of lipid metabolism may or may not accumulate. For example, the accumulation of free fatty acids is greatly favored when CoA is not provided, as shown for developing safflower cotyledon plastids. Similarly, when CoA is omitted from reaction mixtures for daffodil chromoplasts and spinach chloroplasts, the accumulation of free fatty acids increases from the values shown to 66 and 71%, respectively.[18,54] The addition of glycerol-3-phosphate promotes glycerolipid biosynthesis in both pea root plastids and spinach chloroplasts, while galactolipids are not synthesized unless UDP-galactose is provided.[54,70] Similar patterns of glycerolipid biosynthesis and co-factor dependency are obtained when glycerol-3-phosphate is the radioactive precursor.[29,55,56,70] In marked contrast, there are no apparent reports of SQDG biosynthesis in nonphotosynthetic plastids, although the occurrence and synthesis of sulfolipids have been reported in intact nonphotosynthetic systems.[4] Benson[60] stresses the importance of the location of SQDG in photosynthetic tissue in implicating the role of carbon fixation in the biosynthesis of the head groups of the sulfolipid and galactolipids. Thus, with the exception of SQDG and perhaps PE, the results summarized here indicate that nonphotosynthetic plastids are essentially fully competent in glycerolipid biosynthesis and, fur-

TABLE 5
Glycerolipid Biosynthesis from Acetate in Plastids from Several Plant Tissues

	Pea root plastids[34]	Daffodil chromoplasts[18,39]	Developing safflower cotyledon plastids[53]	Spinach chloroplasts[54]
			Distribution of radioactivity(%)[a]	
Triacylglycerol	8	0	*	0
Diacylglycerol	15	14	2	17[b]
Monoacylglycerol	2	0	0	0
Phosphatidic acid	29	14	8	1
Phosphatidylglycerol	8	*	*	2
Phosphatidylcholine	17	18	6	1
Phosphatidylinositol	2	*	*	*
Free fatty acids	4	29	81	49
Acyl-CoAs/ACPs	15	25	3	27

[a] Data shown are recalculated from the indicated references and do not include lipids representing less than 1% of the total radioactivity (indicated by "*").

[b] Data for spinach chloroplasts do not include 3% SQDG.

ther, that plastidic lipid metabolism is regulated, at least in part, by co-factor availability.

B. OTHER ORGANELLES

In the 1960s and early 1970s, work with purified organelles isolated from a variety of tissues indicated that glycerolipid synthesis occurred not only in the plastid as just described, but also in the ER or microsomal fraction and in the mitochondria.[8] By 1973, the ER was credited as the site of *de novo* glycerolipid synthesis and capable of supplying all of its own glycerolipids including PA, cytidine diphospho-diacylglycerol (CDP-DAG), DAG, TAG, PC, PE, PG, PI, and phosphatidylserine (PS).[6,8] The ER is now known to also synthesize the family of polyphosphoinositol lipids. Likewise, the mitochondrion is also known to synthesize its own glycerolipids. Details of these reactions are discussed in earlier chapters and in previous reviews.[3,8,14,41] Data is notably sparse for the peroxisomal, Golgi, nuclear envelope, and plasma membrane fractions concerning both their lipid composition and ability or inability to synthesize lipids. Evidence to date suggests that these cellular fractions are not involved in *de novo* glycerolipid biosynthesis in plants,[4,8,9] although acylation of glycerol-3-phosphate by the Golgi membrane fraction can occur to a similar extent as the microsomal fraction.[2] A major problem in such studies aimed at the localization of enzymes of lipid metabolism continues to be obtaining sufficiently purified organelle fractions free of contaminates.[4,8] At present, the plastid, mitochondria, and ER-enriched microsomal fractions are relatively easy to purify to greater than 98% purity.[2,8] However, the peroxisomal, Golgi, and plasmalemma fractions still remain difficult to purify, especially from the same starting plant material[2,4] (see Chapters 47, 51, and 54 in Reference 61). Despite the many advances with molecular genetic techniques, few studies as yet address gene expression of lipid biosynthesis during differentiation or development or that specifically of nuclear, mitochondrial, or plastid origin. Whether or not all of the enzymes of glycerolipid metabolism are present in the mitochondrion, ER, and plastids isolated from the different nonphotosynthetic tissues is thus speculative.

As indicated earlier for plastids, the lipid composition of other membranes and organelles isolated from photosynthetic vs. nonphotosynthetic tissues are quite similar (Table 6). Each membrane fraction contains a characteristic glycerolipid composition, supporting the concept that the membrane composition is genetically determined.[2] The mitochondrion and the plastid (see Section A.2) isolated from all tissues thus far show the most consistency in the composition of their membranes, especially the composition of the inner membranes. The membranes in contact with the cytosolic compartment, i.e., the plasmalemma, ER, and peroxisomes, are all very similar to one another and appear the most diverse in composition when comparing between different tissues. This is due, in part, to the difficulty in their purification as mentioned earlier. Within this concept, even the outer membranes of both the plastid and the mitochondrion

TABLE 6
Percent Composition of Membrane Phospholipids of Selected Nonphotosynthetic Subcellular Fractions

Tissue	Cytoplasmic/plasmalemma (± tonoplast)		ER (microsomal)					Mitochondria															
								Intact						Outer membrane			Inner membrane			Peroxisome/glyoxysome			
	(1)	(2)	(1)	(2)	(3)	(4)	(5)	(1)	(2)	(7)	(8)	(3)	(9)	(1)	(2)	(8)	(1)	(2)	(3)	(1)	(2)	(10)	(3)
PA	0	0	3	0	—	6	2	5	2	0	0	—	0	0	0	0	0	0	0	0	0	—	—
PC	38	66	45	50	47	35	42	43	44	37	43	35	37	53	42	54	27	41	41	61	48	52	55
PE	41	22	33	36	30	28	27	30	32	38	35	38	42	25	24	30	29	37	37	20	46	48	29
PG	0	0	1	8	4	10	14	3	3	2	3	2	2	10*	10	5	25*	3	3	15	0	0	3
BPG	0	0	1	t	3	7	5	8	7	13	13	19	13	12	3	0	19	14	14	0	0	0	2
PI	15	13	16	6	14	14	10	7	11	8	6	4	6	*	21	11	*	5	5	4	6	0	9
PS	—	—	1	0	2	3	1	—	—	2	—	*	0	*	*	0	0	0	0	0	2	—	—

Note: Data were recalculated from the indicated references without glycolipid and neutral lipids to normalize for phospholipid in the ER, peroxisome, and plasmalemma. * = PG + PI + PS cochromatographed and measured together; BPG = bisphosphatidylglycerol; t = trace.

[a] Numbers in parentheses are the following:
1. Potato tubers[5]
2. Cauliflower buds[5]
3. Castor bean endosperm[68]
4. Lupin roots[5]
5. *Vicia* roots[5]
6. Cauliflower buds[5]
7. Cauliflower buds[2]
8. Sycamore suspension cells[2]
9. Pea leaves[2]
10. Potato tubers[2]

show more similarity to the membranes of the cytosol and show the most variability, albeit small, between tissues for these organelles. To date, the data support the concept in nonphotosynthetic tissues that the ER is the source of *de novo* glycerolipids for cell membranes such as the plasmalemma, tonoplast, and peroxisome. More analyses of all cell membranes and organelles, especially those isolated from the same tissue, are necessary before definitive conclusions can be drawn.

From the compositional analysis and the knowledge of the location of the lipid biosynthetic pathways, one can conclude there are three biochemically and genetically distinct membrane glycerolipid synthesizing mechanisms (see Reference 2). The relative contribution of each in nonphotosynthetic tissues remains to be determined. However, within this context, of these three, the most plastic and easily modulated is that of the cytosolic compartment. Since the plastid and the mitochondrion are the most protected, containing a double membrane, the demands on these systems and the need for modification to the membranes because of tissue and environmental differences may not be so extreme. Thus, the diversity of glycerolipid products from the ER, including TAG, may result from a greater demand on the system with fewer constraints and a need for greater modification of membrane structure. Even so, the galactolipid content of microsomal, plasmalemma, and nuclear membranes as reported by others[4,61] is quite high in view of Douce et al.'s work which indicates that galactolipid synthesis is confined exclusively to the plastid.[2] For this reason, the data in Table 6 do not include galactolipids and neutral lipids in order to normalize the amounts of phospholipids. It is still debatable whether or not the high levels of galactolipid in these membranes can be attributed entirely to plastidic contamination,[2,61] since electron micrograph and marker enzyme studies suggest lower levels of contamination than can be expected from the proportion of galactolipid present.[4,62] Modification of the lipid composition of membranes in the cytosolic compartment through cytosolic lipases, lipid transfer proteins, and other mechanisms whose biochemical regulatory properties vary with the metabolism of the tissue may partially explain why these lipid compositions differ.[2,4,8]

Much of the work with microsomes is derived from the studies of oil-rich tissues and is covered in more detail in the preceding chapters. Of interest here are the similarities and differences in the capability of the microsomal fraction isolated from the various tissues for both the synthesis of the various lipids and the regulation of biosynthetic activities. Research to date is still lacking on how glycerolipid synthesis is regulated both at the biochemical and genetic level.[1,8] Future research will hopefully address the genetic aspects even more and whether the reactions of lipid metabolism are tightly regulated or coordinately turned on and off as necessary to satisfy the needs of the different tissues.

The concept of three lipid-synthesizing mechanisms would fit well within the context of the theory of endosymbiotic origin of both the plastid and the mitochondrion (see References 4, 5, and 10). The capability of the mitochondrion for synthesis of its own glycerolipids appears universal,[2,4,8,9] and its autonomy for the synthesis of some glycerolipids such as PG and cardiolipin is evident.[8] The similarity of the mitochondrial membrane lipid composition and its biosynthetic capabilities among all tissues may reflect a uniformity in its function, metabolism, and genome in various tissues.[2,4] Although this hypothesis has been addressed for plastids from the lipid perspective,[10] it has not been adequately addressed for the mitochondria from the plant lipid perspective. In nonphotosynthetic tissues, especially meristematic where mitochondria can be equally or more prevalent than plastids, the importance of mitochondrial lipid metabolism is becoming apparent.[2,4] The similarities and differences between the plastidic and mitochondrial lipid metabolism make this question of origin pertinent[8,63] (see Chapter 6 on acyltransferases), as well as whether the lipid metabolic pathways are under the control of the nuclear, mitochondrial, or plastid genomes. Studies to date assume that plant mitochondria must import the fatty acid moieties and glycerol backbone.[1,8,9] If these were once autonomous organisms, it would be surprising that they have lost both the ability to supply their own glycerol backbone and fatty acid moiety, while plastids have retained both capabilities. Worth mentioning is a paper describing ACP-dependent *de novo* fatty acid synthesis in mitochondria isolated from the fungus *Neurospora crassa*.[55] Future work may elucidate a similar pathway in higher plants.

While *de novo* synthesis of fatty acids occurs exclusively in the plastids in plants tested to date,[15] the mitochondria and ER are credited with freely modifying preexisting fatty acids, esterified primarily to phosphatidylcholine and perhaps other phospholipids, to suit their needs.[3,8,11,13-15] Reactions covering the modifications of these fatty acids are discussed in detail in the preceding chapters. In nonphotosynthetic tissues, the emphasis to date has been on describing the enzymatic steps of fatty acid elongation, desaturation, or oxidation. More recently, this emphasis is changing to regulation of lipid biosynthesis. One aspect of regulation is in substrate availability, biochemical modulation of enzyme activity, as covered under plastid lipid metabolism. Another aspect is genetic regulation. Attention turned to genetic analysis of the regulatory control of these reactions with the discovery of the fatty acid desaturase (fad) mutants in *Arabidopsis* by 1987.[63] This topic is comprehensively reviewed by others.[1,65] Analysis of the fatty acids of the major phospholipids in both leaf and root tissues indicated the 18:1 and 18:2 desaturases on the ER were similarly affected in both photosynthetic and nonphotosynthetic tissues by the *fad2* and *fad3* mutations, respectively.

III. INTERACTION BETWEEN ORGANELLES

Very little work has been done with nonphotosynthetic tissues concerning the interactions between organelles for glycerolipid biosynthesis. However, in photosynthetic tissues, such interactions are well-known. Acyl-CoAs that originate in the plastid as a result of fatty biosynthesis are utilized by both the ER and mitochondria to acylate glycerol-3-phoshate to form PA and initiate the process of glycerolipid biosynthesis.[8,10,15,63] Glycerolipids synthesized in the plastid are generally referred to as "prokaryotic" lipids, while those synthesized outside of the plastid (primarily in the ER) are referred to as "eukaryotic" lipids.[66] This distinction is made on the basis of the fatty acids esterified to position *sn*-2 of the glycerol moiety and the different specificities of the 1-acyl-glycerol-3-phosphate acyltransfersases of different cellular compartments. In the plastid, this acyltransferase is specific for C_{16} fatty acids, while the corresponding enzyme of the ER is specific for C_{18} fatty acids. These topics are covered in more detail in Chapter 6.

It is thought that the DAG moiety of PC synthesized by the ER can be used for glycerolipid biosynthesis in plastids through a reversal of an envelope-bound CDP-choline:glyceride transferase. Such a source of DAG is extremely important for galactolipid biosynthesis in the so called "18:3 plants". These plants have very low PA phosphatase activity and thus cannot generate sufficient DAG to support galactolipid biosynthesis. The galactolipid synthesized by this interaction would thus be considered eukaryotic. In contrast, "16:3 plants" have relatively higher PA phosphatase and can thus generate much of the DAG required for the synthesis of a prokaryotic species of galactolipid. However, these plants do rely on the extraplastidic compartment for the synthesis of a portion of their lipids with the eukaryotic configuration as described for 18:3-plants.[10,66]

Some evidence from the analysis of root, stem, seed, and leaf tissue from a variety of plants and from the fad mutants of *Arabidopsis* suggests that the concepts of eukaryotic vs. prokaryotic and 18:3- vs. 16:3-plants may also apply to nonphotosynthetic tissues.[1,66] However, recent work by Frentzen et al.[63] indicates that these issues will be more complicated in nonphotosynthetic tissues. This is largely due to the more significant contribution of mitochondrial lipid relative to the total cellular glycerolipid pool, since the plastid is not necessarily the dominant organelle in these tissues.[4] Also, the specificities of the 1-acyl-glycerol-3-phosphate acyltransferase from potato tuber and pea leaf mitochondria both are such that they give rise to eukaryotic-type lipids, although they occur in a "prokaryotic" compartment. The results of Frentzen et al.[63] further suggest that the mitochondrial enzymes from photosynthetic tissue such as pea leaves display different specificity from those isolated from nonphotosynthetic tissues such as potato tuber. Given the lack of appropriate studies, it is premature to generalize the extent to which these hypotheses apply to nonphotosynthetic tissues. This is not surprising, since biogenesis of mitochondria is under the control of genes from both the plant nucleus and the mitochondria and since chloroplast DNA sequences are known to be present in the mitochondrial genomes of various plant species.[67] One may speculate that in green tissue the chloroplast may exert an overriding control of the lipid composition in part by DNA transfer between chloroplast, mitochondria, and nuclear genomes. In nonphotosynthetic tissue, this plastidal control would be less evident. More work is clearly needed in these areas.

IV. SUMMARY

Lipid composition and metabolism in nonphotosynthetic tissues and organelles is reminiscent of those in photosynthetic tissues. Differences in the lipids isolated from these tissues reflect the different requirements of each tissue as achieved by differences in the regulation of lipid metabolism. While the lipids isolated from leaves reflect primarily the composition of the chloroplast, the same cannot be said for other tissues. In nonphotosynthetic tissues, the predominant lipids are PC and PE, not the galactolipids as is the case in leaves. As in photosynthetic tissues, the plastid of nonphotosynthetic tissues plays a central role in providing fatty acids for glycerolipid biosynthesis. These plastids rely on their own glycolytic metabolism to provide much of the energy and carbon required for fatty acid biosynthesis. Although the ER, mitochondria, and plastids of essentially all tissues are all variously known to synthesize many of their component glycerolipids, insufficient work has been done with these organelles in various tissues at various developmental stages to determine whether the full complement of reactions is always present or whether certain reactions are lost during tissue differentiation. Few new comprehensive analyses of plant tissues and organelles have been made since 1980. More accurate data for the actual *in vivo* membrane compositions for all cell membranes would help clarify this issue and provide an indication of the needs of different tissues and organelles for the various lipid-synthesizing reactions. Surprisingly little information is available describing possible interactions between organelles of nonphotosynthetic tissues for glycerolipid assembly. Similarly, the extent to which the concepts of prokaryotic vs. eukaryotic metabolism and 16:3- vs. 18:3-plants may be applicable to nonphotosynthetic systems is not presently known. It is likely that future studies with these tissues will reveal novel patterns in the regulation of plant lipid metabolism.

REFERENCES

1. **Browse, J. and Somerville, C.,** Glycerolipid synthesis: biochemistry and regulation, *Annu. Rev. Plant Physiol. Plant Mol. Biol.,* 42, 467, 1991.
2. **Douce, R., Alban, C., Bligny, R., Block, M. A., Coves, J., Dorne, A.-J, Journet, E.-P., Joyard, J., Neuburger, M., and Rebeille, F.,** Lipid distribution and synthesis within the plant cell, in *The Metabolism, Structure and Function of Plant Lipids,* Stumpf, P. K., Mudd, J. B., and Nes, W. D., Eds., Plenum, New York, 1987, 505.
3. **Harwood, J.,** Lipid metabolism in plants, *Crit. Rev. Plant Sci.,* 8, 1, 1989.
4. **Kates, M.,** Glycolipids, phosphoglycolipids and sulfoglycolipids, in *Handbook of Lipid Research,* Hanahan, D. J., Ed., Plenum Press, New York, 1992, 235.

5. **Mazliak, P.,** Glyco- and phospholipids of biomembranes in higher plants, in *Lipids and Lipid Polymers in Higher Plants,* Tevini, E. and Lichtenthaler, H. K., Eds., Springer-Verlag, New York, 1977, 48.

6. **Mazliak, P.,** Lipid metabolism in plants, *Annu. Rev. Plant Physiol.,* 24, 287, 1973.

7. **Mikolajczyk, S. and Brody, S.,** De novo fatty acid synthesis mediated by acyl-carrier protein in Neurospora crassa mitochondria, *Eur. J. Biochem.,* 187, 431, 1990.

8. **Moore, T. S., Jr.,** Phospholipid biosynthesis, *Annu. Rev. Plant Physiol.,* 33, 235, 1982.

9. **Mudd, J. B.,** Phospholipid biosynthesis, in *The Biochemistry of Plants, Lipids: Structure and Function,* Vol. 4, Stumpf, P. K., Ed., Academic Press, New York, 1980, 249.

10. **Mudd, J. B., Andrews, J. E., Sanchez, J., Sparace, S. A., and Kleppinger-Sparace, K. F.,** Biosynthesis of chloroplast lipids, in *Frontiers of Membrane Research in Agriculture,* St. John, J. B., Berlin, E., and Jackson, P. C., Eds., Rowman and Allanheld, Ottowa, 1985, 35.

11. **Roughan, P. G. and Slack, C. R.,** Cellular organization of glycerolipid metabolism, *Annu. Rev. Plant Physiol.,* 33, 97, 1982.

12. **Smirnov, B. P.,** The biosynthesis of higher acids from acetate in isolated chloroplasts of *Spinacia oleracea* leaves, *Biokhimiya,* 25, 545, 1960.

13. **Stumpf, P. K.,** Fatty acid biosynthesis in higher plants, in *Fatty Acid Metabolism and Its Regulation,* Numa, S., Ed., Elsevier, Amsterdam, 1984, 155.

14. **Stymne, S. and Stobart, A. K.,** Triacylglycerol biosynthesis, in *The Biochemistry of Plants,* Vol. 9, Stumpf, P. K. and Conn, E. E., Eds., Academic Press, New York, 1987, 175.

15. **Stumpf, P. K.,** The biosynthesis of saturated fatty acids, in *The Biochemistry of Plants,* Vol. 9, Stumpf, P. K. and Conn, E. E., Eds., Academic Press, New York, 1987, 121.

16. **Stahl, R. J. and Sparace, S. A.,** Characterization of fatty acid biosynthesis in isolated pea root plastids, *Plant Physiol.,* 96, 602, 1991.

17. **Miernyk, J. A. and Dennis, D. T.,** The incorporation of glycolytic intermediates into lipids by plastids isolated from the developing endosperm of castor oil seeds (*Ricinus communis* L.), *J. Exp. Bot.,* 34, 712, 1983.

18. **Kleinig, H. and Liedvogel, B.,** Fatty acid synthesis by isolated chromoplasts from the daffodil: [^{14}C]acetate incorporation and distribution of labeled acids, *Eur. J. Biochem.,* 83, 499, 1978.

19. **Weaire, P. J. and Kekwick, R. G. O.,** The synthesis of fatty acids in avocado mesocarp and cauliflower bud tissue, *Biochem. J.,* 146, 425, 1975.

20. **Nothelfer, H. G., Barckhaus, R. H., and Spener, F.,** Localisation and characterization of the fatty acid synthesizing system in cells of *Glycine max* (soybean) suspension cultures, *Biochim. Biophys. Acta,* 489, 370, 1977.

21. **Nakamura, Y. and Yamada, M.,** Fatty acid synthesis by spinach chloroplasts. I. Property of fatty acid synthesis from acetate, *Plant Cell Physiol.,* 16, 139, 1975.

22. **Stumpf, P. K. and James, A. T.,** The biosynthesis of long-chain fatty acids by lettuce chloroplast preparations, *Biochim. Biophys. Acta,* 70, 20, 1963.

23. **Heller, A., Bambridge, H. E., and Slabas, A. R.,** Plant acetyl-CoA carboxylase, *Biochem. Soc. Trans.,* 14, 565, 1986.

24. **Sauer, A. and Heise, K.-P.,** Regulation of acetyl-coenzyme A carboxylase and acetyl-coenzyme A synthetase in spinach chloroplasts, *Z. Naturforsch.,* 39, 268, 1984.

25. **Buchanan, B. B.,** Role of light in the regulation of chloroplast enzymes, *Annu. Rev. Plant Physiol.,* 31, 341, 1980.

26. **Browse, J., Roughan, P. G., and Slack, C. R.,** Light control of fatty acid synthesis and diurnal fluctuations of fatty acid compositions of leaves, *Biochem. J.,* 196, 347, 1981.

27. **Entwistle, G. and ap Rees, T.,** Enzymic capacities of amyloplasts from wheat (*Triticum aestivum*) endosperm, *Biochem. J.,* 255, 391, 1988.

28. **Frehner, M., Pozueta-Romero, J., and Akazawa, T.,** Enzyme sets of glycolysis, gluconeogenesis, and oxidative pentose phosphate pathway are not complete in nongreen highly purified amyloplasts of sycamore (*Acer pseudoplatanus* L.) cell suspension cultures, *Plant Physiol.,* 94, 538, 1990.

29. **Journet, E.-P. and Douce, R.,** Enzymic capacities of purified cauliflower bud plastids for lipid synthesis and carbohydrate metabolism, *Plant Physiol.,* 79, 458, 1985.

30. **Simcox, P. D., Reid, E. E., Canvin, D. T., and Dennis, D. T.,** Enzymes of the glycolytic and pentose phosphate pathways in proplastids from the developing endosperm of *Ricinus communis* L., *Plant Physiol.,* 59, 1128, 1977.

31. **Nishimura, M. and Beevers, H.,** Subcellular distribution of gluconeogenetic enzymes in germinating castor bean endosperm, *Plant Physiol.,* 64, 21, 1979.

32. **Borchert, S., Große, H., and Heldt, H. W.,** Specific transport of inorganic phosphate, glucose-6-phosphate, dihydroxyacetone phosphate and 3-phosphoglycerate into amyloplasts from pea roots, *FEBS Lett.,* 253, 183, 1989.

33. **Emes, M. J. and Traska, A.,** Uptake of inorganic phosphate by plastids purified from the roots of *Pisum sativum* L., *J. Exp. Bot.,* 38, 196, 1987.

34. **Kleppinger-Sparace, K. F., Stahl, R. J., and Sparace, S. A.,** Energy requirements for fatty acid and glycerolipid biosynthesis from acetate by isolated pea root plastids, *Plant Physiol.,* 98, 723, 1992.

35. **Flügge, U.-I. and Heldt, H. W.,** Metabolite translocators of the chloroplast envelope, *Annu. Rev. Plant Physiol. Plant Mol. Biol.,* 42, 129, 1991.

36. **Werden, K., Heldt, H. W., and Mlovancev, M.,** The role of pH in the regulation of carbon fixation in the chloroplast stroma. Studies on CO_2 fixation in the light and dark, *Biochim. Biophys. Acta,* 396, 276, 1975.

37. **Sauer, A. and Heis, K.-P.,** On the light dependence of fatty acid synthesis in spinach chloroplasts, *Plant Physiol.,* 73, 11, 1983.

38. **Kleppinger-Sparace, K. F. and Mudd, J. B.,** Biosynthesis of sulfoquinovosyldiacylglycerol in higher plants: the incorporation of $^{35}SO_4$ by intact chloroplasts in darkness, *Plant Physiol.,* 84, 682, 1987.

39. **Kleinig, H. and Liedvogel, B.,** Fatty acid synthesis by isolated chromoplasts from the daffodil. Energy sources and distribution patterns of the acids, *Planta,* 150, 166, 1979.

40. **Boyle, S. A., Hemmingsen, S. M., and Dennis, D. T.,** Energy requirements for the import of protein into plastids from developing endosperm of *Ricinus communis* L., *Plant Physiol.,* 92, 151, 1990.

41. **Harwood, J. L.,** The synthesis of acyl lipids in plant tissues, *Prog. Lipid Res.,* 18, 55, 1979.

42. **Kuhn, D. N., Knauf, M., and Stumpf, P. K.,** Subcellular localization of acetyl-CoA synthetase in leaf protoplasts of *Spinacia oleracea* leaf cells, *Arch. Biochem. Biophys.,* 209, 441, 1981.

43. **Stumpf, P. K.,** Biosynthesis of saturated and unsaturated fatty acids, in *The Biochemistry of Plants,* Vol. 4, Stumpf, P. K. and Conn, E. E., Eds., Academic Press, New York, 1980, 177.

44. **Camp, P. J. and Randall, D. D.,** Purification and characterization of the pea chloroplast pyruvate dehydrogenase complex, *Plant Physiol.,* 77, 571, 1985.

45. **Heise, H.-P. and Treed, H.-J.,** Regulation of acetyl coenzyme A synthesis in chloroplasts, in *The Metabolism, Structure and Function of Plant Lipids,* Stumpf, P. K., Mudd, J. B., and Nes, W. D., Eds., Plenum Press, New York, 1987, 505.

46. **Liedvogel, B. and Bauerle, R.,** Fatty acid synthesis in chloroplasts from mustard (*Sinapsis alba* L.) cotyledons: formation of acetyl-CoA by intraplastid glycolytic enzymes and a pyruvate dehydrogenase complex, *Planta,* 169, 481, 1986.

47. **Reid, E. E., Thompson, P., Lyttle, R. C., and Dennis, D. T.,** Pyruvate dehydrogenase complex from higher plant mitochondria and proplastids, *Plant Physiol.,* 59, 842, 1977.

48. **Liedvogel, B. and Kleinig, H.,** Fatty acid synthesis in isolated chromoplasts and chromoplast homogenates. ACP stimulation, substrate utilisation and cerulenin inhibition, in *Biochemistry and Function of Plant Lipids,* Mazliak, P., Benveniste, P., Costes, C., and Douce, R., Eds., Elsevier, Amerstam, 1980, 107.

49. **Murphy, D. J. and Leech, R. M.,** The pathway of [^{14}C]bicarbonate incorporation into lipids in isolated photosynthesizing spinach chloroplasts, *FEBS Lett.,* 88, 192, 1978.

50. **Smith, R. G., Gauthier, D. A., Dennis, D. T., and Turpin, D. H.,** Malate- and pyruvate-dependent fatty acid synthesis in leukoplasts from developing castor endosperm, *Plant Physiol.,* 98, 1233, 1992.

51. **Yamada, M. and Nakamura, Y.,** Fatty acid synthesis by spinach chloroplasts. II. The path from PGA to fatty acids, *Plant Cell Physiol.,* 16, 151, 1975.

52. **Yamada, M. and Usami, Q.,** Long chain fatty acid synthesis in developing castor bean seeds. IV. The synthetic system in proplastids, *Plant Cell Physiol.,* 16, 879, 1975.

53. **Browse, J. and Slack, C. R.,** Fatty-acid synthesis in plastids from maturing safflower and linseed cotyledons, *Planta,* 166, 74, 1985.

54. **Roughan, P. G., Holland, R., and Slack, C. R.,** The role of chloroplasts and microsomal fractions in polar-lipid synthesis from [1-^{14}C]acetate by cell-free preparations from spinach (*Spinacia oleracea*) leaves, *Biochem. J.,* 188, 17, 1980.

55. **Alban, C., Joyard, J., and Douce, R.,** Comparison of glycerolipid biosynthesis in non-green plastids from sycamore (*Acep pseudoplantanus*) cells and cauliflower (*Brassica oleracea*) buds, *Biochem. J.,* 259, 775, 1989.

56. **Mudd, J. B. and Dezacks, R.,** Synthesis of phosphatidylglycerol by chloroplasts from leaves of *Spinacia oleracea* L. (spinach), *Arch. Biochem. Biophys.,* 209, 584, 1981.

57. **Fishwick, M. J. and Wright, A. J.,** Isolation and characterization of amyloplast envelope membranes from *Solanum tuberosum, Phytochemistry,* 19, 55, 1980.

58. **Wintermans, J. F. G. M.,** Concentrations of phosphatides and glycolipids in leaves and chloroplasts, *Biochim. Biophys. Acta,* 44, 49, 1960.

59. **Douce, R., Holtz, R. B., and Benson, A. A.,** Isolation and properties of the envelope of spinach chloroplasts, *J. Biol. Chem.,* 248, 7215, 1973.

60. **Benson, A. A.,** The plant sulfolipid, *Adv. Lipid Res.,* 1, 387, 1963.

61. **Packer, L. and Douce, R.,** Eds., *Methods in Enzymology,* Vol. 148, Plant Cell Membranes, Academic Press, New York, 1987, 1.

62. **Briskin, D. P., Leonard, R. T., and Hodges, T. K.,** Isolation of the plasma membrane: membrane markers and general principles, in *Methods in Enzymology, Vol. 148, Plant Cell Membranes,* Packer, L. and Douce, R., Eds., Academic Press, New York, 1987, 542.

63. **Frentzen, M., Neuburger, M., Joyard, J., and Douce, R.,** Intraorganelle localization of the mitochondrial acyl-CoA:sn-glycerol-3-phosphate O-acyltransferase and acyl-CoA:1-acyl-sn-glycerol-3-phosphate O-acyltransferase from potato tubers and pea leaves, *Eur. J. Biochem.,* 187, 395, 1990.

64. **Somerville, C. R., McCourt, P., Kunst, L., and Browse, J.,** Mutants of Arabodopsis deficient in fatty acid desaturation, in *The Metabolism, Structure and Function of Plant Lipids,* Stumpf, P. K., Mudd, J. B., and Nes, W. D., Eds., Plenum Press, New York, 1987, 683.

65. **Ohlrogge, J. B., Browse, J., and Somerville, C. R.,** The genetics of plant lipids, *Biochim. Biophys. Acta,* 1082, 1, 1991.

66. **Roughan, P. G. and Slack, C. R.,** Glycerolipid synthesis in leaves, *Trends Biochem. Sci.,* 9, 383, 1984.

67. **Newton, K. J.,** Plant mitochondrial genomes: organization, expression and variation, *Annu. Rev. Plant Physiol. Plant Mol. Biol.,* 39, 503, 1988.

68. **Donaldson, R. P. and Beevers, H.,** Lipid composition of organelles from germinating castor bean endosperm, *Plant Physiol.,* 59, 259, 1977.

69. **Qi, Q. and Sparace, S. A.,** unpublished observations, 1992.

70. **Xue, L. and Sparace, S. A.,** unpublished observations, 1992.

Chapter 18

RESPONSES OF LIPID METABOLISM TO DEVELOPMENTAL CHANGE AND ENVIRONMENTAL PERTURBATION

Guy A. Thompson, Jr.

TABLE OF CONTENTS

0-8493-4907-9/93/$0.00+$.50
© 1993 by CRC Press Inc.

I. INTRODUCTION

The preceding chapters of this volume have discussed metabolic pathways for the formation and the degradation of all major plant lipids. These pathways function cooperatively in plant tissues to provide structural, energy storage, signaling, and other lipid types as needed by the cells.

The primary emphasis in the earlier chapters has generally been on the biochemistry, especially the enzymology, of the pertinent pathways. In the following discussion, I have chosen to consider some of the same pathways from a more physiological perspective. I have selected for consideration a number of situations, such as seed development, tissue senescence, and low temperature acclimation, in which major changes in lipid metabolism occur within a relatively short time period, and I have highlighted the lipid pathways most active under these circumstances. The responses which have been observed provide some appreciation for the capabilities and the limitations under which plant lipid metabolism operates.

Opportunities to study the regulation of the lipid pathways abound in these settings. In many cases, significant progress has already been made in identifying the rate-limiting factors and in understanding the physiological consequences of these control mechanisms.

Tracing metabolic interrelationships of lipids within the plant cell poses a unique challenge. Unlike animal cells, which fabricate most of their lipids at a single location, namely, the endoplasmic reticulum, plant cells produce very significant quantities of lipids both in the endoplasmic reticulum and within plastids. The proportion of total cell lipid made at each location can vary markedly as the cell experiences developmental changes or environmental stress. The fluctuating interplay between these two centers of lipid metabolism is becoming recognized as an important factor in determining the cellular response and will be a recurring theme in the examples given below. In the face of natural perturbations which affect a plant, often without advance warning, lipid metabolism is clearly capable of remarkably versatile responses to modify the organism in a beneficial way.

II. DEVELOPMENT OF PHOTOSYNTHETIC CAPABILITY

One of the unique developmental processes in plants is the assembly of photosynthetically capable chloroplasts. In most species, light is required for the assembly of functional thylakoid membranes. Seedlings form chloroplasts from preexisting proplastids as the young leaves are exposed to light. The major factors triggering rapid thylakoid membrane expansion are the light-induced (through phytochrome) synthesis of key thylakoid proteins and light-driven conversion of protochlorophyllide to the membrane stabilizing product, chlorophyll.[1] Experimental studies of chloroplast development often utilize etiolated seedlings because of the extensive thylakoid formation that takes place within preexisting etioplasts following illumination. In one such study, etiolated leaves of 7-d dark-grown seedlings formed nearly half of their full complement of photosynthetic units together with the complete machinery for carrying out photosynthesis during the first 8 h of illumination.[2] Greening is generally complete within 24 to 28 h.

This sudden diversion of the plant's biosynthetic capacity has a profound effect on lipid metabolism. The synthesis of fatty acids, which takes place exclusively in the plastids of higher plants, can function in the preilluminated proplastid or etioplast, but in leaves or isolated chloroplasts is stimulated as much as fivefold by light.[3] The possibility that this increased activity is due to enhanced gene expression has been tested by analyzing the cellular content of proteins involved in fatty acid synthesis.[4] It has been shown, for example, that the levels of acyl carrier protein and its mRNA experience little change as etiolated seedlings undergo greening. The light stimulation of fatty acid synthesis in *Chlamydomonas* whole cells was apparently not associated with changes in substrate or ATP levels and was not prevented by inhibitors of protein synthesis.[5] Available evidence supports the concept that increased fatty acid synthesis in illuminated chloroplasts is effected principally by some relatively direct interaction of the biosynthetic pathway with the photosynthetic apparatus. The action spectrum of fatty acid synthesis stimulation in *Chlamydomonas* was very similar to the photosystem I absorption spectrum, and the stimulatory effect of light was not observed in photosystem I mutants.

The exact nature of this interaction is not clear at present. A possible activation of acetyl coenzyme A (CoA) carboxylase by photosystem I-produced ATP has not been ruled out because of the difficulty of measuring chloroplast levels of a short half-life molecule like ATP. A more direct energizing effect via the ferredoxin-thioredoxin system[6] is another possibility. Whatever the mechanism, it is clear that the greatly increased fatty acid synthesis that accompanies the early stages of chloroplast development is essential for producing the extra complement of thylakoid lipids needed at this time.

Another factor which may expedite thylakoid membrane assembly is the more direct access that lipid-forming pathways in developing chloroplasts have to the early products of photosynthesis. Developing chloroplasts from cells near the leaf bases of barley seedlings showed a much greater capacity for synthesizing fatty acids and isoprenoids from acetyl-CoA made *in situ* within the chloroplast from photosynthetically derived phosphoenolpyruvate than did mature chloroplasts from older regions of the leaves.[7] It appears that upon maturation the chloroplast becomes unable to supply adequate quantities of the substrates needed for carotenoid and fatty acid synthesis, probably due to a decrease in activity at the pyruvate decarboxylation-dehydrogenation step. As a result, importation of acetate and isopentenyl diphosphate from the cytosol is necessary to sustain optimal production of fatty acids and carotenoids, respectively (Figure 1).

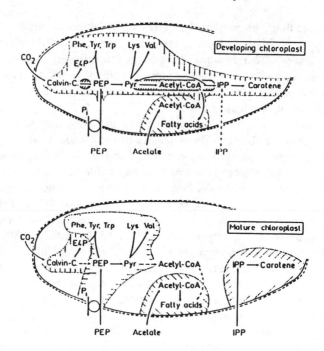

FIGURE 1. Model to illustrate the change from the autonomous $C_3 \rightarrow C_2$ metabolism of developing chloroplasts to the specialized metabolism of mature chloroplasts undergoing a division of labor with extraplastidic systems. Pathways with reduced flux are indicated by dashed arrows. (From Heintze, A., et al., *Plant Physiol.*, 93, 1121–1127, 1990. With permission.)

The bulk of available evidence indicates that isoprenoid biosynthesis in plant cells is highly compartmentalized. It appears that the formation of geranylgeranyl diphosphate ($C_{20}PP$), the key precursor of diterpenes, carotenoids, and the phytol moiety of chlorophyll, is limited to plastids, while farnesyl diphosphate ($C_{15}PP$), precursor of sesquiterpenes and sterols, is restricted to the cytoplasm/endoplasmic reticulum (ER) compartment (see Figure 2 in Chapter 13).[8] The regulation of these pathways in developing cells is difficult to decipher because many of the reactions do not persist *in vitro*. During the etioplast to chloroplast transition in mustard seedlings, the activity of a soluble phytoene synthase unaccountably declines until, in chloroplasts from primary leaves, it is no longer detectable. In contrast, chlorophyll synthase from the same tissues remains active. As it is known that carotenoid synthesis continues throughout chloroplast development *in vivo*, the changing properties of phytoene synthase are postulated to reflect a transformation of the enzyme into a form more tightly bound to the chloroplast membrane.[8]

The flux of geranylgeranyl diphosphate into competing pathways is also influenced by the availability of other substrates. In daffodil or pepper chromoplast preparations, addition of chlorophyllide diverted geranylgeranyl diphosphate away from carotenoid synthesis and into chlorophyll synthesis (reviewed by Gray[9]).

III. SEED DEVELOPMENT

The seeds of many plant species store energy in the form of oil, mainly triglycerides. Rapeseed (*Brassica napus*) and safflower (*Carthamus tinctorius*) are typical of agriculturally important crops in having approximately 40% of their seed weight as oil. The oils are synthesized during a relatively short period of seed development. For example, developing cotyledons within a safflower seed lay down about 50% of their storage triglycerides during a brief span of 16 to 20 d. The triglycerides are synthesized by enzymes of the ER and accumulate in discreet oil bodies of 0.5 to 1 μm diameter, which are surrounded by a half-unit membrane composed of proteins and phospholipids.[10]

It has been assumed that the great enhancement of triglyceride synthesis in developing seeds is fueled by increased fatty acid synthesis. This assumption is borne out by findings that the amount of active acyl carrier protein[11] and the activities of acetyl-CoA carboxylase[12] and enoyl-ACP reductase[13] all increase during this period. It is difficult to evaluate the full significance of the expanding capacity of seeds for lipid synthesis in the presence of so many other developmental changes, such as the dramatic rise in storage protein levels (see discussion by Ohlrogge et al.[4]), but persistently amplified fatty acid synthesis appears to be responsible for the steady accumulation of triglycerides in the period immediately following the completion of cell division.

Placement of the fatty acids on a triglyceride is nonrandom. In rapeseed oil, erucic acid (22:1) is concentrated in the *sn*-1 and *sn*-3 position, while the *sn*-2 position is occupied by oleate and linoleate. In many different types of developing seeds, the *sn*-2 position is the major site of linoleate formation, but in phosphatidylcholine rather than triglycerides per se. Oleate cycles rapidly into the *sn*-2 positions of phosphatidylcholine, where it is desaturated and usually recycled back, as linoleate, to the fatty acyl-CoA pool by the reverse reaction of acyl-CoA:lysophosphatidylcholine acyltransferase.[14] Thus, fatty acids for triglyceride synthesis are drawn primarily from the phosphatidylcholine pool itself and from the Co-A-bound pool of linoleate spun off from its phosphatidylcholine site of origin (Figure 2).

The prevalence of this trafficking of fatty acids through the phosphatidylcholine *sn*-2 position varies greatly, being more pronounced in safflower than in rapeseed.[15] In the latter species, the reduced conversion of oleate to linoleate by this pathway may favor the observed accumulation of 18:1 and its elongation to erucic acid (22:1).

The distribution of fatty acids among different lipid classes of the developing seed tissues is under tight control. Triglycerides may contain uncommon fatty acids, such as the erucate of rapeseed, but these fatty acids are seldom incorporated into membrane lipids, even in the same tissue (Table 1). Likewise, triglycerides of *Cuphea lutea* developing embryos contain 75% short-chain (C_{10}–C_{12}) fatty acids, while phospholipids of the embryos have only 7% of the short-chain acids.[18] The basis for this strong specificity is not certain, although in some seeds lysophosphatidic acid acyltransferase has been implicated.[17]

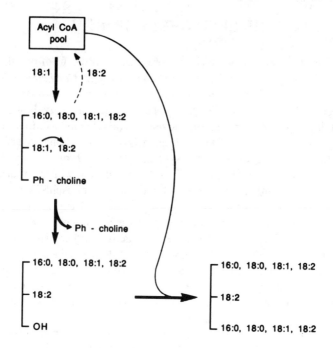

FIGURE 2. Scheme depicting the probable formation of linoleate (18:2) at the *sn*-2 position of triglycerides.

This enzyme seems to prefer placing a medium-chain length fatty acid on the *sn*-2 position of a lysophosphatidic acid already containing a fatty acid in the same size range on the *sn*-1 position. Long-chain fatty acids are paired in the same manner by the enzyme selecting substrates of matching chain length.

Alternatively, there may be as yet unrecognized slight differences in the intracellular location of phospholipid and triglyceride synthesis, with the two processes utilizing distinct acyltransferase isoforms, each with its own substrate specificity.

IV. SEED GERMINATION AND SEEDLING DEVELOPMENT

Those dicot seeds which have significant triglyceride reserves, sometimes exceeding 50% of their dry weight,[18] typically store them in the cotyledons and/or the hypocotyl, while the metabolically equivalent storage sites in monocot seeds are the endosperm and the scutellum (Figure 3). The metabolic pathways by which triglycerides are degraded and utilized for seedling growth are described in Chapter 14. I shall restrict my discussion to the dynamics by which the seed is transformed from a metabolically inert body into a vigorous processor of lipids into substrates for growth.

In the dormant seed, triglycerides are present as cytoplasmic droplets enveloped by a single monolayer of phospholipids and associated proteins

TABLE 1
Utilization of [¹⁴C]Oleate in the Developing Cotyledons of *Brassica campestris* (var. Bele)

Lipid and incubation time	Radioactivity (dpm × 10⁻³ per 30 cotyledon pairs)[a]			
	18:1	18:2	20:1	22:1
Triglyceride				
40 min	205(68)	tr[b]	54(18)	42(14)
200 min	363(63)	tr	92(16)	121(21)
Phosphatidylcholine				
40 min	247(95)	13(5)	0	0
200 min	229(82)	42(5)	8(3)	0

Note: Developing cotyledons were incubated with [¹⁴C]oleate, and after the desired time interval the radioactivity in the component fatty acids of the phosphatidylcholine and triglyceride were determined. Data from Reference 15.

[a] Figures in parenthesis are the relative distribution of radioactivity in the complex lipid.
[b] tr — trace.

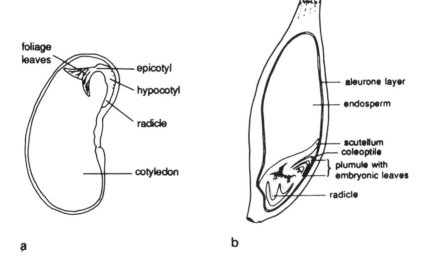

foliage leaves

epicotyl

hypocotyl

radicle

cotyledon

aleurone layer

endosperm

scutellum
coleoptile
plumule with embryonic leaves

radicle

a   b

FIGURE 3. Cross-sectional diagrams of typical dicot (a) and monocot (b) seeds.

termed oleosins.[19] Apart from stabilizing the oil bodies, oleosins may assist in targeting lipases to the triglycerides when hydrolysis of the lipids commences. Being constituted of predominantly unsaturated fatty acids, the oils are quite fluid and readily accessible to the lipases acting at the oil body surface.

The active phase of triglyceride degradation typically begins after germination proper has taken place and continues to fuel postgerminative growth. Fatty acids released from oil bodies by lipase action are transferred to nearby

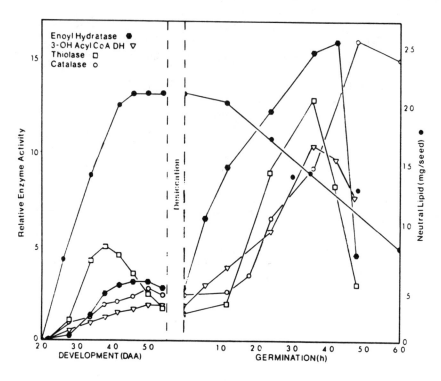

FIGURE 4. Changes in β-oxidation enzyme activities and neutral lipid content in cotyledons of maturing and germinated cotton seeds. (From Trelease, R. N. and Doman, D. C., in *Seed Physiology*, Vol. 2, Murray, D. R., Ed., Academic Press, Sydney, 1984, 201–245. With permission.)

glyoxysomes which are capable of degrading them by β-oxidation and converting the resulting acetyl-CoA to succinate.[20] Ultimately, the succinate gives rise to glucose through the reversal of glycolysis.

The rising phase of storage lipid metabolism may be charted by following the decline of lipid mass, but this assay is relatively insensitive when the quantity of lipid stores is large. Alternatively, the rise of lipase activity and the activities of enzymes further utilizing the fatty acids can be assayed to estimate the degradative potential. The enzymes of β-oxidation are relatively active in seeds even prior to germination, presumably to assure that free fatty acids do not accumulate. Their activity rises sharply during the postgerminative period of lipolysis, as illustrated in Figure 4 for cotton seeds. It is not clear what factors stimulate the activities of these enzymes, but constant removal of the soluble degradation products through the embryonic axis appears to be necessary to prevent feedback inhibition and decreased mobilization of the lipid reserves (reviewed in Reference 18).

In some instances, lipid stores can be drawn upon prior to seed germination. Seeds of certain plants require a period of physiological maturation which occurs at warm temperatures. In the freshly shed ash seed, this change is accompanied by a proliferation of ER and Golgi apparatus and a depletion of lipid reserves, shown in one assay to fall from 20 to 4% of the embryo's dry weight.[21,22] Utilization of the lipid allows starch to be synthesized and deposited in the tissue's developing plastids. Following this developmental process in ash, a period of cold stratification is necessary before dormancy of the embryo is broken.

In a study of hazel seed, total lipase activity and isocitrate lyase activity in the embryonic axis and cotyledons increased significantly during stratification at 5°C, as compared with seeds maintained at 20°C, and the activity increases there were correlated with a loss of storage lipids.[23] However, the majority of hazel seed lipid mobilization took place in the postgerminative period.

Relatively little is known regarding molecular mechanisms of lipase regulation in seeds. Effects of light and hormones such as gibberellin on whole seeds and excised seed organs are discussed in Chapters 14 and 17.

The degradation of free fatty acids by lipoxygenases may have physiological importance during seed germination. The role of lipoxygenases is complicated by the fact that several isozymes of the enzyme are present in quantities that vary independently during the germination process. For instance, the lipoxygenase activity in soybean cotyledons, which was high in ungerminated seeds, fell sharply during the first 24 h after germination, then increased until 3 d postgermination and again declined until by day 9 only 13% of the original activity remained.[24] The changes in total activity could be accounted for by a steady decline in three lipoxygenase isozymes and a rise after 24 h in the expression of different isozymes.

There is no unequivocal evidence for a direct role of lipoxygenase in the germination process. It has been proposed that the increased polarity of the lipoxygenase products may speed disruption of membranes of the lipid storage organs, thereby helping to mobilize the reserves needed for seedling growth.[25] Products of lipoxygenase action, such as jasmonic acid, can also have a profound effect on growth (see below).

V. SENESCENCE AND FRUIT RIPENING

Plant senescence as a phenomenon is much more complex than a simple aging of tissues. It is an active, energy-requiring process in which gene expression is induced and protein synthesis is necessary.[26] It may properly be considered as the final stage of programmed plant development.

In some situations, senescence takes place only in very specific tissues, such as flower petals, while at other times an entire annual plant may senesce.[27] The senescence of annual plants is extremely complex, often being coordinated with a diversion of assimilated nutrients from vegetative growth to the development of reproductive parts. However, in order to minimize complications, experimental studies on plant senescence generally involve specific tissues in which endogenous factors induce degenerative processes leading to death.

Despite the wealth of descriptive information available on the subject, the precise molecular factors triggering the onset of senescence are not understood. At the cellular level, one of the earliest manifestations of senescence is the progressive loss of membrane integrity, leading to an increasing leakiness of the cell and its organelles. The plasma membrane and ER are among the structures first affected, as are the thylakoid membranes of the chloroplast. In contrast, mitochondrial structure and function remain relatively intact until the terminal phase of senescence.

Changes in lipid metabolism that accompany senescence have recently been reviewed by Thompson.[28] Many of the reported studies have employed cell types such as the petals of cut flowers or the cotyledons of developing seedlings, which progress rapidly through a sequence of predictable physiological and metabolic events. Among the first detectable signs of approaching senescence is a rapid decrease in the bulk lipid fluidity of a tissue's isolated microsomes, as inferred by X-ray diffraction, fluorescence depolarization, or ESR.[28] Such changes are apparent during or just prior to the climacteric-like rise in ethylene production and, in at least some instances, involve membrane lipid phase separations. The factors which induce these changes are unknown, but packing defects in the membrane lipid bilayer that attend them may be responsible for the marked decrease in phospholipid content that follows, since lipids exposed at the phase boundaries are known to be highly susceptible to phospholipase attack.[29]

As an illustration of the lipid changes accompanying senescence, the phospholipid content of microsomal membranes from senescing carnations dropped from a value of 247.7 nmol of phospholipid per milligram of protein in relatively young (stage II) petals to 178.5 nmol of phospholipid per milligram of protein in stage IV petals.[30] The concurrent increase (by as much as threefold) in the sterol to phospholipid ratio as senescence proceeded was to some extent responsible for an increased rigidity of the membranes.[28]

All classes of carnation microsomal phospholipids decreased to an approximately equal extent, with degradation occurring, at least in part, through the action of phospholipase D.[31] However, within each lipid class, there was a selective degradation of individual molecular species (Figure 5). This is probably only one of several mechanisms employed to reshape the phospholipid composition in the intact senescing tissue.

In the microsomal membranes of senescing carnation flowers and in some other systems as well, observers have attributed a loss of unsaturated fatty acids to selective action by lipoxygenase (Fobel, M., M.S. thesis, University of Waterloo, Ontario, 1986, quoted by Thompson[28]). Fluorescent lipid peroxidation products were also detectable in increasing amounts in microsomal membranes as bean cotyledon senescence proceeded through a 9-d period[32] (Figure 6). While examples of lipoxygenase acting on phospholipid-bound fatty acids are known, e.g., in reticulocytes,[33] the enzyme ordinarily acts upon free fatty acids, and in senescing plant tissues its principal substrates are probably free fatty acids released from complex lipids by a combination of hydrolytic enzymes.

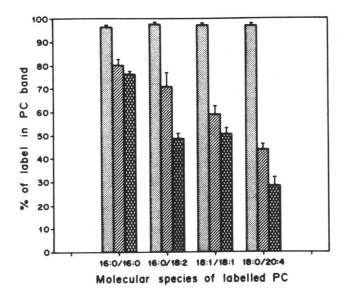

FIGURE 5. Degradation of radiolabeled molecular species of phosphatidylcholine (16:0*/16:0*, 16:0/18:2*, 18:1*/18:1*, and 18:0/20:4*) by microsomal membranes isolated from the petals of stage IV carnation flowers. See Reference 31 for details. Means ± SE are shown; n = 4 separate experiments. Stippled bars, 0 h control; hatched bars, 1.5 h of reaction; cross-hatched bars, 3 h of reaction. (From Brown, J.H., Paliyath, G., and Thompson, J.E., *J. Exp. Bot.*, 41, 979–986, 1990. With permission.)

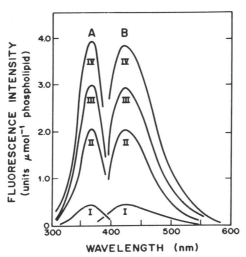

FIGURE 6. Excitation (A) and emission (B) spectra for fluorescent peroxidation pigments in microsomal membranes from senescing bean (*Phaseolus vulgaris*) cotyledons. I, II, III, and IV identify spectra for membranes from 2-d-old, 4-d-old, 7-d-old, and 9-d-old cotyledon tissue, respectively. (From Thompson, J. E., in *Senescence and Aging in Plants*, Noodén, L. D. and Leopold, A. C., Eds., Academic Press, San Diego, CA, 1988, 51–83. With permission.)

Plant lipoxygenase acts primarily on either the 9- or 13-position of linoleate and linolenate.[25] The hydroperoxides that are formed under aerobic conditions (Figure 7, step 3) may be further metabolized to fatty aldehydes and alcohols, which, incidentally, are largely responsible for the odor associated with freshly cut grass, and to the hormone-like substances 12-oxo-*trans*-10-dodecenoic acid (traumatin), a compound involved in wound healing,[34] or jasmonic acid, a promoter of senescence.[25] Under anaerobic conditions, Reaction 3 cannot occur and the fatty acid radicals are often released from the lipoxygenase to form dimers and oxodienoic acids (step 3a).

In comparison to the relative wealth of data on microsomal responses during senescence, little is known regarding the rather different process of thylakoid senescence. The bulk lipid fluidity of thylakoids from *Phaseolus vulgaris* primary leaves did not decrease as senescence progressed.[35] Whereas the formation of free radicals would be predicted as the integrity of the electron transport system is deranged, no clear role for lipid hydroperoxides or other lipid metabolites has been implicated in the senescence-induced breakdown of thylakoid chlorophyll and protein.

Fruit ripening has many features in common with senescence in vegetative tissues, but special metabolic changes, such as the hydrolysis of cell wall polysaccharides and the accumulation of pigments, are often most characteristic of the ripening process. Until the onset of ripening, a tomato fruit contains no lycopene, but during the subsequent period of color development it can form more than 1 mg of carotenoids per day. The later stages of carotenoid pigment formation in fruit appear to be catalyzed by enzymes distinct from those operating in the photosynthetic tissues, where β-carotene is a major pigment.[9] This conclusion is based partly on the observation that fruit pigmentation mutants have a normal carotenoid distribution in their leaves. In addition, the synthesis of carotenoid-forming enzymes in fruit occurs only during the ripening process. For example, the rise in phytoene synthetase in pepper fruits ripened by treatment with chemical agents was shown to stem from *de novo* synthesis of the enzyme.[36]

VI. EFFECTS OF LOW TEMPERATURE STRESS

Changes in lipid metabolism are thought to be largely responsible for the ability of plants to acclimate to low environmental temperatures. A rise in the level of fatty acid unsaturation and other more subtle structural lipid alterations increase the fluidity of membranes in "cold hardened" plants to a value approaching that found in plants grown at much higher temperatures. In the past, we could only surmise that this temperature-induced fluidization had survival value. However, current findings (see below) strongly confirm the essential role of lipid change. Most higher plants, algae, and cyanobacteria seem capable of temperature-induced lipid modification. The cyanobacterium *Anabaena variabilis* responds to chilling stress by first desaturating palmitate esterified to the *sn*-2 position of membrane glycolipids.[37] This is followed by

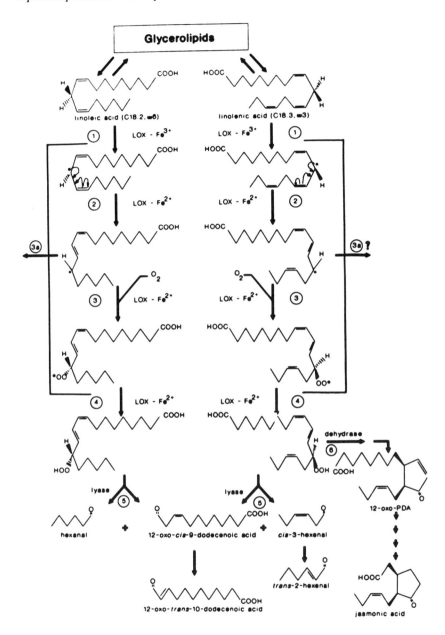

FIGURE 7. The lipoxygenase (LOX) pathway. Linoleic and linolenic acids released by lipases are the principal substrates. Intermediates that are bound to the LOX enzyme are enclosed in brackets. After formation and release of the final hydroperoxide products and oxidation of LOX back to the Fe^{+3} form (step 4), the fatty acid hydroperoxides are metabolized by hydroperoxide lyase to C_6 and C_{12} products (step 5), or by hydroperoxide dehydrase to a product that can be ultimately converted into jasmonic acid. See Reference 25 for details. (From Hildebrand, D.F., *Physiol. Plant.*, 76, 249–253, 1989. With permission.)

TABLE 2
Composition of Major Fatty Acids of Total Membrane Lipids in Various Strains of *Anacystis nidulans* R2-SPc

Strain	Fatty acid (mole %)					
	16:0	16:1 (9)	16:2 (9,12)	18:0	18:1 (9)	18:2 (9,12)
Wild type	51	36	0	3	6	0
Transformant with pUC303[a]	51	37	0	3	5	0
Transformant with pUC303/*des*A[a]	47	29	5	5	2	6

[a] The wild type was compared with transformants made using the shuttle vector alone (pUC303) or using the shuttle vector containing the *Synechocystis* Δ^{12} desaturase (pUC303/*des*A). See Reference 38 for details.

increased desaturation of the glycolipid fatty acids bound at the *sn*-1 position. Both reactions are suppressed by inhibitors of protein synthesis, indicating that an induced synthesis of fatty acid desaturase underlies the compositional change.

The resistance of certain other cyanobacteria to chilling is achieved by the same basic mechanism with slight modifications. In low temperature stressed *Synechocistis* PCC 6803, glycerolipid-bound oleate is converted to linoleate and linolenate by newly induced desaturases. In contrast, *Anacystis nidulans* forms only monounsaturated fatty acids and predictably is very sensitive to low temperature. The mitigating role of lipid desaturation was clearly demonstrated when the gene for the Δ^{12} desaturase of *Synechocystis* was cloned and introduced into *A. nidulans*.[38] The transformant not only contained lipids with polyunsaturated fatty acids (Table 2), but was also more resistant than the wild type to low temperature injury (Figure 8).

Low temperature acclimation in green algae also involves increasing desaturation of membrane glycerolipid fatty acids. *Dunaliella salina* cells shifted from 30 to 12°C exhibited measurable increases in microsomal phospholipid fatty acid unsaturation within 12 h, whereas the principal rise in chloroplast unsaturation was initiated only after 36 h.[39,40] The earliest chilling-induced lipid modifications in microsomes involved the retailoring of phosphatidylethanolamine and phosphatidylglycerol molecular species so as to redistribute preexisting fatty acids. This yielded an increase in the number of species having two unsaturated acyl chains. As time progressed, retailoring of phospholipids, especially phosphatidylglycerol, was also prominent in chloroplasts, with the most pronounced changes occurring between 36 and 60 h. During this same time period, a dramatic alteration in the threshold temperature causing thermal denaturation, as measured by chlorophyll fluorescence, also took place. The data suggested a programmed response of *D. salina*

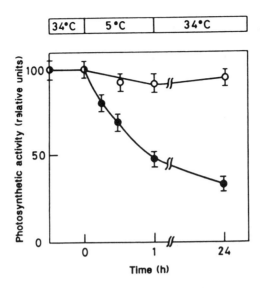

FIGURE 8.   Effect of low temperature on the photosynthetic activity of *Anacystis nidulans* R2-SPc transformed using the unmodified shuttle vector pUC303 (●) or using the shuttle vector containing the *Synechocystis* Δ^{12} desaturase (pUC303/*des*A) (○). The activity before exposure to 5°C corresponded to about 190 μmol O_2 per milligram of chlorophyll per hour for both transformants. Each point represents the mean ± SD obtained in three independent experiments. (From Wada, H., Gombos, Z., and Murata, N., *Nature*, 347, 200–203, 1990. With permission.)

lipid metabolism to chilling, with membrane fluidity being increased first through the redistribution of existing phospholipid acyl chains via molecular species retailoring. This short-term membrane fluidization was followed by a net increase in the level of certain unsaturated fatty acids, especially linolenate, which was effected before cell growth resumed after 100 h at 12°C.

The factors responsible for initiating these chilling responses are not known. Increased phospholipid molecular species retailoring may be controlled in *D. salina* by microsomal acyl hydrolases found to increase more than tenfold in activity during 60 h of chilling.[41] Unlike most plant acyl hydrolases, this one was relatively specific for phosphatidylethanolamine and phosphatidylglycerol, the lipid classes most actively modified at low temperature.

Recent research on low temperature responses in higher plants might be appropriately divided into two categories, namely, that utilizing low but above-freezing temperatures and that dealing with below-freezing conditions. The lipid metabolic responses are fairly similar in both cases, but the stress imposed by subfreezing temperatures is often compounded by dehydration of the cells. Potentially damaging cell shrinkage results when water is drawn out of the cell to dilute the hyperosmotic fluid phase remaining there, as many of the extracellular water molecules are utilized for ice crystal formation.

The lipid composition of its plasma membrane is thought to have an important bearing on a plant's ability to withstand subfreezing temperatures.[42] The plasma membrane contains a surprisingly high quantity of nonphospholipids. In rye

TABLE 3
Lipid Composition of Plasma Membrane Isolated from Leaves of Nonacclimated and Acclimated Rye Seedlings

Values are Expressed as Mole Percent of
Total Lipid ± Standard Error of 4 to 7 Determinations

Lipid fraction	Nonacclimated	Acclimated
	(mol% + SE)	
Free sterol	32.7 ± 1.7	44.4 ± 2.6
Steryl glucoside	15.1 ± 1.1	5.7 ± 0.6
Acylated steryl glucoside	4.3 ± 0.3	1.1 ± 0.1
Glucocerebroside	16.2 ± 1.2	6.8 ± 0.8
Phospholipid	31.7 ± 2.1	41.9 ± 1.6
	(μmol/mg)	
Lipid/protein	3.38 ± 0.4	3.49 ± 0.5

Note: Data from Reference 43.

plasma membranes free sterols, sterol glucosides, and acylated sterol glu-cosides account for more than 50 mol% of the structural lipids[43] (Table 3). Glucocerebrosides, particularly those containing C_{22}–C_{24} hydroxy fatty acids linked to trihydroxy long-chain bases, make up another 16 mol% of the lipids.[44] Cold hardening of rye leads to a number of plasma membrane compositional changes, of which a decline in glucocerebroside content from 16 to 7 mol% is especially worthy of note.[43] This decrease, along with an increase in phospholipids, with which glucocerebrosides are poorly miscible, may be, in part, responsible for the increased tolerance to osmotic damage observed to develop in the freezing-stressed plasma membrane following cold hardening.

The cells of many plant species are damaged when exposed to above-freezing temperatures in the 0 to 15°C range.[45] Certain species can acclimate to chilling stress, while others cannot. Acclimation is usually associated with increasing fatty acid unsaturation, but progress in understanding the molecular factors regulating these changes has been slow. Recent work with lipid mutants of *Arabidopsis thaliana* clarified certain aspects of the process. For example, Kunst et al.[46] analyzed an *Arabidopsis* mutant deficient in the chloroplast fatty acid desaturase which normally introduces a double bond at the ω-9 position of palmitate linked to monogalactosyldiacylglycerol (MGDG). Although the mutation generated a large increase in MGDG palmitate and a reduction in unsaturated C_{16} fatty acids, possible deleterious consequences of this defect on thylakoid physical properties were reduced by the enhanced accumulation of MGDG assembled using unsaturated fatty acids made in the ER by a pathway utilizing C_{18} polyunsaturates. This capacity for enhanced lipid importation into chloroplasts whose endogenous supply of lipids has been interrupted is an enlightening example of the logistical flexibility that plants can muster to achieve physiological balance. More recent data from *Arabidopsis*[47] and *Dunaliella*[48] suggest that chloroplast-derived polyunsaturated fatty acids, perhaps in the form of intact diacylglycerols, may also flow in the opposite direction, i.e., from chloroplast to ER, under certain conditions.

This same *Arabidopsis* mutant that exhibited an overall reduction in the unsaturation of chloroplast lipids and grew normally at low temperature grew more rapidly than wild-type *Arabidopsis* at temperatures above 28°C. Furthermore, it contained chloroplasts which were unusually resistant to high temperature inactivation of photosynthetic electron transport.[49]

The concept that nonoptimal membrane fluidity, and particularly lipid phase separation, causes damage by rendering membranes leaky to ions and other solutes has inspired extensive study of the most prevalent chloroplast phospholipid, phosphatidylglycerol. This is because two major molecular species, *sn*-1,2-dipalmitoyl phosphatidylglycerol and *sn*-1-palmitoyl,2-(*trans*Δ^3)hexadecenoyl phosphatidylglycerol, undergo thermotropic phase transitions at 41° and 32°C, respectively — much higher than the phase transition temperatures of other common lipids. Interestingly, these two molecular species together accounted for 26 to 65% of chloroplast phosphatidylglycerol in 9 species of chilling-sensitive plants, but only 14% or less in 12 chilling-resistant species.[50] The likelihood of a direct cause and effect relationship between chilling tolerance, lipid physical changes, and phosphatidylglycerol composition was diminished when several exceptions to the general rule were found.[51] Furthermore, calorimetric studies[52] failed to show a correlation between chilling sensitivity of photosynthesis and the detection of a phase separation of bulk membrane lipids.

It may be asking too much to expect a simple relationship between lipid physical properties and physiological responses. High melting molecular species of phosphatidylglycerol might influence chilling sensitivity more subtly (and more effectively) by interaction with membrane proteins than by perturbation of membrane fluidity per se.[53] Such an association with proteins may explain the apparent stabilization of the chloroplast photosystem II light harvesting complex by high levels of phosphatidylglycerol (PG) molecular species containing *trans*-Δ^3-hexadecenodic acid. In wild-type and mutant *Chlamydomonas*[54] and in winter rye (*Secale cereale*) grown at different temperatures,[55] the proportion of the photosystem II light harvesting complex of proteins that is found in thylakoid extracts as the oligomeric (functional) form is positively correlated with the content of *trans*-Δ^3-hexadecenoate. Replenishment of *trans*-Δ^3-hexadecenoate-containing phosphatidylglycerol in a *Chlamydomonas* mutant lacking this phospholipid could be achieved by exposure of the cells to liposomes containing the deficient PG. The restoration led not only to a recovery of the oligomeric form of the photosystem II light harvesting complex, but also a reappression of thylakoid membranes, which had been largely destacked in the mutant.[56] When PG containing only palmitate was incubated with the mutant cells, little tendency to recover wild-type thylakoid organization was observed.

The presence of this unusual fatty acid, *trans*-Δ^3-hexadecenoate, which occurs in no other phospholipid besides PG, is not essential for photosynthesis, as evidenced by the normal growth of a *trans*-Δ^3-hexadecenoate-deficient *Arabidopsis* mutant under standard cultural conditions. But the stabilizing effect it provides, even in *Arabidopsis*, appears to be of considerable value under some conditions, such as high temperature.[49] An intriguing possibility for which there is as yet no direct evidence is that the functioning of photosystem II and the biosynthesis of *trans*-Δ^3-hexadecenoate are closely linked. Recent experimentation has shown that the *trans*-Δ^3 double bond is formed not in the chloroplast envelope where all *cis* desaturases are located,[57,58] but rather in the thylakoid membranes.[58] The *trans*-Δ^3 desaturase resembles, in some respects, another thylakoid-associated desaturase, namely, the enzyme responsible for placing *trans* double bonds in carotenoids.[59]

VII. EFFECTS OF SALINITY STRESS

Plant species vary sharply in their capacity for adaptation to saline environments. In plants which can adapt, lipid compositions have been shown to change after growth at different levels of salinity, and there is a conviction among workers in the field that these changes enhance the plant's ability to grow optimally under those particular conditions. Although the fluidity of artificial lipid bilayers can be strongly affected by differing concentrations of cations,[60] much less is known about a natural membrane's perturbation by salts than by temperature. This is due, in part, to technical difficulties in making appropriate measurements. For instance, only the cell's plasma membrane can be considered to be exposed directly to the known salt concentration in the external medium, and therefore this is the pertinent membrane for study. Furthermore, in dealing with multicellular organisms, the true concentration of a particular ion is difficult to measure even in the extracellular medium.

An extreme case of adaptation to salt stress is found in the halotolerant green alga *Dunaliella salina*. This organism can withstand levels of salinity ranging from 0.86 to 4.3 *M* NaCl.[61] Because it offsets the osmotic effects of high external NaCl by maintaining an appropriately high cytoplasmic glycerol concentration, only its plasma membrane is exposed to high salt. Analyses of plasma membranes isolated from *D. salina* grown in 0.85, 1.7, and 3.4 *M* NaCl medium disclosed only modest distributional changes among the polar lipid classes.[62] The largest difference was a 10% increase in the proportion of diacylglycerol trimethylhomoserine in the plasma membranes of 3.4 *M* NaCl-grown cells. This unusual lipid is not widespread in nature, but it is a major constituent of certain nonhalotolerant algae, such as *Chlamydomonas reinhardtii*,[63] and therefore is not likely to play a special role in acclimation to saline conditions. In this case, even the dedicated lipid chemist must grudgingly assume the primacy of glycerol metabolism in overcoming the stress associated with excessive salinity.

However, in its response to sudden osmotic stress, *D. salina* does employ lipid changes of another sort. Within seconds following a dilution from 1.7 to 0.85 *M* NaCl, 30% of the cells' plasma membrane-localized phosphatidylinositol 4,5-bisphosphate (PIP$_2$) was hydrolyzed to yield inositol trisphosphate.[64] A biphasic rise in plasma membrane diacylglycerol accompanied the PIP$_2$ hydrolysis, with the short-lived initial burst yielding characteristic molecular species confirming that PIP$_2$ cleavage was its source.[48] A PIP$_2$-specific phospholipase C was also found to be highly enriched in the *D. salina* plasma membrane.[65] Stimulation of the phospholipase C activity by GTP analogs provided another indication that this pathway transduces signals across the *D. salina* plasma membrane by a GTP-binding protein-mediated pathway closely resembling that known to activate protein kinase C and mobilize Ca^{2+} in animal cells[66] (Figure 9). At the present time, the membrane receptor triggering the signaling event has not been identified, and the metabolic consequences of signaling are still obscure, despite the discovery of a specific protein that is phosphorylated following hypoosmotic shock[48] and a Ca^{2+}-stimulated protein kinase in *D. salina*.[67] The principal point of confusion is that this and most other plant protein kinases do not require lipid for activity and are therefore unlike the protein kinase C that characteristically terminates the PIP$_2$-mediated signaling pathway in animals.[66] Of interest is the recent finding that oat (*Avena sativa*) root plasma membrane has an associated protein kinase which differs in its amino acid sequence from protein kinase C, but resembles it in having a lipid requirement.[68] The data suggested that this protein kinase is very sensitive to partial proteolysis and loss of lipid dependence during isolation.

The signaling mechanism described above does not function in *D. salina* exposed to the opposite type of stress, namely, hyperosmotic conditions.[67] However, evidence for the widespread presence (often for undetermined function) of PIP$_2$-mediated signaling in higher and lower plants continues to accumulate.[70,71] An example not too far out of the present context involves ion-mediated volume control of the motor organ (pulvinus) responsible for leaf movement in the legume *Samanea saman*. The swelling and shrinking of the pulvinar cells that control folding of the leaflets require light and are preceded by an active turnover of PIP$_2$.[72]

In view of the rapid worldwide salinization of agricultural lands, understanding the molecular factors underlying salt tolerance in higher plants is of extreme importance. Some data have been gathered, but most findings have been made using whole plant tissues with their inevitable mixture of functionally different cell types and organelles. A recurring observation is that the roots of salt-tolerant plants contain more sterols and sterol derivatives than do salt-sensitive species. This tendency was found in a comparison of the salt-sensitive bean (*Phaseolus vulgaris* L. cv Saxa), the less salt-sensitive barley (*Hordeum vulgaris* L. cv. Wisa), and the salt-tolerant sugar beet (*Beta vulgaris*

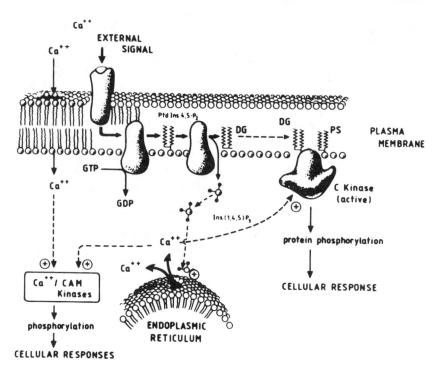

FIGURE 9. Basic pathway for phosphatidylinositol 4,5-bisphosphate-mediated transmembrane signaling. DG — diacylglycerol; PS — phosphatidylserine; C Kinase — protein kinase C; Ins (1,4,5)P$_3$ — inositol 1,4,5-trisphosphate; CAM Kinases — calmodulin-activated protein kinases. (From Einspahr, K. J. and Thompson, G. A., Jr., *Plant Physiol.*, 93, 361–366, 1990. With permission.)

L. cv. Kawemono[73]). Other lipid constituents, such as fatty acids, varied in amount also, but the significance of these changes is difficult to assess when comparing tissues from entirely different species.

A comparison that is perhaps more meaningful was made of root lipids from five grape varieties differing in the extent to which their leaves accumulated chloride.[74] The most salt-sensitive variety, which also accumulated the most chloride, had a surprisingly low content of sterols (9% of total lipids vs. approximately 20% in the other varieties), in keeping with the results described above.[73] Phosphatidylcholine was present in reduced amounts in the more salt-sensitive varieties, and MGDG displayed the opposite trend. A further noteworthy difference was the high level of very long-chain fatty acids, especially lignoceric acid ($C_{24:0}$), in phospholipids of the salt-tolerant plants.

A root's tolerance to salinity may depend, in part, upon the kind of sterol in its membranes. The sitosterol to cholesterol ratio decreased with increasing salinity in roots of *Plantago maritima*, a halophytic species, mainly due to a decrease in the sitosterol level.[75] The authors suggested that cholesterol, which is thought to be more effective than sitosterol in limiting permeability, may take a leading role in regulating ion fluxes. High sitosterol to stigmasterol ratios were correlated with efficient Cl⁻ exclusion by three citrus rootstock varieties.[76]

Many of the studies described here are a decade or more old. Although other similar reports have appeared, major advances in understanding the mechanisms of lipid influence on salt tolerance seem to await the development of experimental systems more amenable to analysis at the cellular and molecular levels.

VIII. RESPONSES TO WOUNDING AND PATHOGEN ATTACK

Many if not all plants can generate responses to counteract pathogen invasion and physical damage to their tissues. Recent studies have made it clear that both local responses at the immediate site of injury and systemic responses are possible. The formation of physical barriers to obstruct pathogen invasion at a wound site itself may involve suberization, lignification, callose formation, etc.[77] Protection of a plant tissue against attack at a specific point can also be achieved by the induced synthesis in adjacent cells of low molecular weight antimicrobials termed "phytoalexins".[78] A more systemic protection can be achieved in some plant species through the induced formation of proteinase inhibitor proteins in tissues distant from the wound as well as in its vicinity.[79] The proteinase inhibitors are thought to reduce the activity of pathogen-secreted proteolytic enzymes.

Lipid metabolism is involved in these defense responses in various ways. Lipids may participate in the initial signal transduction process in which the presence of an external threat is communicated within the cell. Efforts have been made to determine whether fungal elicitors can trigger the PIP_2 transmembrane signaling system recently found[70,71] to be present in plants. Phosphatidylinositol was observed to be degraded during the elicitation of phytoalexin production in cultured carrot cells.[80] However, there was no change in the content of PIP_2, the immediate precursor of the signaling second messengers, following an elicitor challenge to soybean or parsley cells.[81] Further study is required to clarify the physiological significance of inositol phospholipids in elicitor-induced signaling. A possibly valuable clue was discovered recently when compounds known to interact with GTP-binding proteins were shown to modulate the elicitation response in French bean cells.[82] GTP-binding proteins take part in the inositol phospholipid-mediated pathway, but also in pathways generating other second messengers, such as cAMP.

The fatty acid-derived compound jasmonic acid plays a fascinating signaling role in the defense response of some plants. The action of 13-lipoxygenase on linolenic acid initiates the series of reactions leading to the formation of jasmonic acid as well as other products[83] (Figure 7). After it is formed, jasmonic acid is converted to its methyl ester, which is volatile. Exposure of Solanaceae and Fabaceae species to an atmosphere containing methyl jasmonate led to the accumulation of proteinase inhibitor proteins within as few as 5 h.[84]

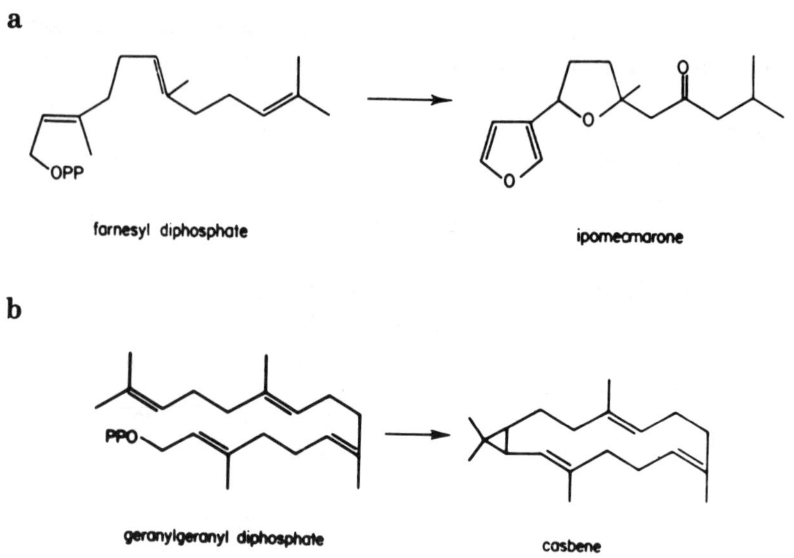

a

farnesyl diphosphate

ipomeamarone

b

geranylgeranyl diphosphate

casbene

FIGURE 10. The formation of ipomeamarone from farnesyl diphosphate (a) and casbene from geranylgeranyl diphosphate (b).

It was shown that proteinase inhibitor protein synthesis could be induced in tomato seedlings simply by placing them in a chamber with sagebrush, *Artemisia tridentata*, a plant whose leaves produce methyl jasmonate.[84]

Conventional polyunsaturated long-chain fatty acids may have a direct signaling role. Fungal invasion by *Phytophthora infestans* appears to be accompanied by a release of arachidonic and eicosapentaenoic acids from the mycelium. These acids elicit the accumulation of several sesquiterpenoid phytoalexins in solanaceous plants.[85]

It has been difficult to quantify the production of phytoalexins in elicited plant tissues because only cells near the point of damage or in the immediate vicinity of fungal mycelial growth are stimulated.[86] However, in those cells which are so stimulated, radioimmunoassay for certain types of phytoalexins showed that they are formed in dramatically higher levels near an infection site.[87]

Many representatives of the chemically heterogeneous phytoalexin family are themselves lipids. For example, the sesquiterpene ipomeamarone (Figure 10a) is produced from farnesyl diphosphate by sweet potato (*Ipomoea batatas*) responding to infection by the fungal pathogen *Ceratocystis fimbriata*,[88] and synthesis of the bicyclic diterpene casbene from geranylgeranyl diphosphate (Figure 10b) by castor bean (*Ricinus communis*) seedlings is greatly stimulated following infection with *Rhizopus stolonifer*[88] or exposure of seedlings to partially degraded cell wall products of the type thought to be formed during fungal attack.[89] Certain fatty acid derivatives have also been found to possess phytoalexin activity (reviewed by Ebel[78]).

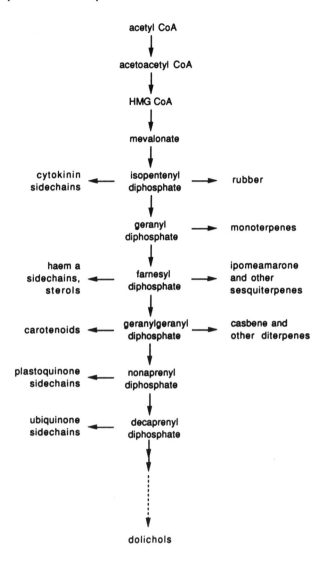

FIGURE 11. Overall scheme for the formation of various isoprenoids from acetyl-CoA.

Very little is known regarding the factors regulating the induced synthesis of lipid phytoalexins. Two enzymatic steps have been proposed as possible regulatory sites. One involves 3-hydroxy-3-methylglutaryl-CoA reductase, long known as the key regulatory enzyme controlling the biosynthesis of all isoprenoid compounds. It might seem unlikely that phytoalexin synthesis, which can be markedly induced under conditions where the formation of sterols and other isoprenoids continues at a steady rate, could be governed by an enzyme as early in the pathway as hydroxymethylglutaryl-CoA reductase. Yet considering that isoprenoid biosynthesis takes place in at least two compartments of the plant cell, the chloroplast and the endoplasmic reticulum[90] (also see Section II), it is possible that distinct regulatory mechanisms may operate in different parts of the same cell. This interpretation would be supported by the findings of Yang et al.[91] that two different isogenes for 3-hydroxy-3-methylglutaryl-CoA reductase could be induced in potato slices. Interestingly, mRNA from one of the genes increased following exposure to the pathogenic bacterium *Erwinia carotovora*, while mRNA representing the other gene increased after wounding.

Secondary sites of metabolic regulation may exist at later branch points leading from the principal isoprenoid elongation pathway. Treatment of tobacco cells with elicitor stimulated the expression of several isoforms of a sesquiterpene cyclase which converts farnesyl diphosphate to the phytoalexin 5-epi-aristolochene.[92] Activation of this alternative pathway for farnesyl diphosphate might have little effect on the metabolism of other isoprenoids (see Figure 11), particularly if it were coupled with an increased hydroxymethylglutaryl-CoA reductase activity.

In one study of fungal elicitor action on tobacco cells,[93] the stimulation of sesquiterpene formation was paralleled by a large decrease in the incorporation of isoprenoid precursors into sterols. Under these conditions, sesquiterpene cyclase activity rose from an undetectable level to a maximum within 10 h, while squalane synthetase declined to less than 15% of control values by 7 h after elicitor addition. In contrast, hydroxymethylglutaryl-CoA reductase became elevated only briefly and had returned to control levels by 10 h. Presumably, the reduced squalene synthetase activity effectively released enough farnesyl diphosphate to fuel the increased production of sesquiterpenes.

REFERENCES

1. **Rüdiger, W. and Schock, S.,** Chlorophylls, in *Plant Pigments*, Goodwin, T. W., Ed., Academic Press, New York, 1988, 1–59.
2. **Bradbeer, J. W.,** Development of photosynthetic function during chloroplast biosynthesis, in *The Biochemistry of Plants*, Vol. 8, Hatch, M. D. and Boardman, N. K., Eds., Academic Press, New York, 1980, 423–467.
3. **Browse, J., Roughan, P. G., and Slack, C. R.,** Light control of fatty acid synthesis and diurnal fluctuations of fatty acid composition in leaves, *Biochem J.,* 196, 347–354, 1981.
4. **Ohlrogge, J. B., Browse, J., and Somerville, C. R.,** The genetics of plant lipids, *Biochim. Biophys. Acta,* 1082, 1–26, 1991.
5. **Picaud, A., Creach, A., and Trémolières, A.,** Studies on the stimulation by light of fatty acid synthesis in *Chlamydomonas reinhardtii* whole cells, *Plant Physiol. Biochem.,* 29, 441–448, 1991.
6. **Huppe, H. C., de Lamotte-Guéry, F., Jacquot, J.-P., and Buchanan, B. B.,** The ferredoxin-thioredoxin system of a green alga, *Chlamydomonas reinhardtii, Planta,* 180, 341–351, 1990.

7. **Heintze, A., Gorlach, J., Leuschner, C., Hoppe, P., Hagelstein, P., Schulze-Siebert, D., and Schultz, G.,** Plastidic isoprenoid synthesis during chloroplast development, *Plant Physiol.,* 93, 1121–1127, 1990.

8. **Kleinig, H.,** The role of plastids in isoprenoid biosynthesis, *Annu. Rev. Plant Physiol. Plant Mol. Biol.,* 40, 39–59, 1989.

9. **Gray, J. C.,** Control of isoprenoid biosynthesis in higher plants, *Adv. Bot. Res.,* 14, 25–91, 1987.

10. **Wanner, G., Formanek, H., and Theimer, R. R.,** The ontogeny of lipid bodies (spherosomes) in plant cells, *Planta,* 151, 109–123, 1981.

11. **Ohlrogge, J. B. and Kuo, T.-M.,** Control of lipid synthesis during soybean seed development: enzymatic and immunochemical assay of acyl carrier protein, *Plant Physiol.,* 74, 622–625, 1984.

12. **Turnham, E. and Northcote, D. H.,** Changes in the activity of acetyl-CoA carboxylase during rape-seed formation, *Biochem. J.,* 212, 223–229, 1983.

13. **Slabas, A. R., Sidebottom, C. M., Hellyer, A., Kessell, R. M. J., and Tombs, M. P.,** Induction, purification and characterization of NADH-specific enoyl acyl carrier protein reductase from developing seeds of oil seed rape (*Brassica napus*), *Biochim. Biophys. Acta,* 877, 271–280, 1986.

14. **Stymne, S. and Stobart, A. K.,** Evidence for the reversibility of the acyl-CoA: lysophosphatidylcholine acyltransferase in microsomal preparations from developing safflower (*Carthamus tinctorius*) cotyledons and rat livers, *Biochem. J.,* 223, 305–314, 1984.

15. **Griffiths, G., Stymne, S., and Stobart, K.,** Biosynthesis of triacylglycerol in plant storage tissue, in *Proceedings World Conference on Biotechnology for the Fats and Oil Industry,* Applewhite, T. H., Ed., American Oil Chemists Society, Champaign, IL, 1988, 23–29.

16. **Singh, S. S., Nee, T. Y., and Pollard, M. R.,** Acetate and mevalonate labeling studies with developing *Cuphea lutea* seeds, *Lipids,* 21, 143–149, 1986.

17. **Somerville, C. R. and Browse, J.,** Plant lipids: metabolism, mutants, and membranes, *Science,* 252, 80–87, 1991.

18. **Trelease, R. N. and Doman, D. C.,** Mobilization of oil and wax reserves, in *Seed Physiology,* Vol. 2, Murray, D. R., Ed., Academic Press, Sydney, 1984, 201–245.

19. **Huang, A. H. C., Qu, R., Lai, Y. K., Ratnayake, C., Chan, K. L., Kuroki, G. W., Oo, K. C., and Cao, Y. Z.,** Structure, synthesis and degradation of oil bodies in maize, in *Compartmentation of Plant Metabolism in Non-Photosynthetic Tissues,* Emes, M. J., Ed., Cambridge University Press, Cambridge, 1991, 43–58.

20. **Huang, A. H. C., Trelease, R. N., and Moore, T. S., Jr.,** *Plant Peroxisomes,* Academic Press, New York, 1983.

21. **Villiers, T. A.,** Cytological studies in dormancy. I. Embryo maturation during dormancy in *Fraxinus excelsior*, *New Phytol.,* 70, 751–760, 1971.

22. **Villiers, T. A.,** Cytological studies in dormancy. III. Changes during low-temperature dormancy release, *New Phytol.,* 71, 153–160, 1972.

23. **Li, L. and Ross, J. D.,** Lipid mobilization during dormancy breakage in oilseed of *Corylus avellana*, *Ann. Bot.,* 66, 501–505, 1990.

24. **Kato, T., Ohta, H., Tanaka, K., and Shibata, D.,** Appearance of new lipoxygenases in soybean cotyledons after germination and evidence for expression of a major new lipoxygenase gene, *Plant Physiol.,* 98, 324–330, 1992.

25. **Hildebrand, D. F.,** Lipoxygenases, *Physiol. Plant.,* 76, 249–253, 1989.

26. **Noodén, L. D.,** Postlude and prospects, in *Senescence and Aging in Plants,* Noodén, L. D. and Leopold, A. C., Eds., Academic Press, San Diego, CA, 1988, 499–517.

27. **Noodén, L. D.,** The phenomena of senescence and aging, in *Senescence and Aging in Plants,* Noodén, L. D. and Leopold, A. C., Eds., Academic Press, San Diego, CA, 1988, 1–50.

28. **Thompson, J. E.,** The molecular basis for membrane deterioration during senescence, in *Senescence and Aging in Plants,* Noodén, L. D. and Leopold, A. C., Eds., Academic Press, San Diego, CA, 1988, 51–83.

29. **Wilschut, J. C., Regts, J., Westenberg, H., and Scherphof, G.,** Action of phospholipases A$_2$ on phosphatidylcholine bilayers. Effects of the phase transition, bilayer curvature, and structural defects, *Biochim. Biophys. Acta,* 508, 185–196, 1978.

30. **Brown, J. H., Chambers, J. A., and Thompson, J. E.,** Acyl chain and head group regulation of phospholipid catabolism in senescing carnation flowers, *Plant Physiol.,* 95, 909–916, 1991.

31. **Brown, J. H., Paliyath, G., and Thompson, J. E.,** Influence of acyl chain composition on the degradation of phosphatidylcholine by phospholipase D in carnation microsomal membranes, *J. Exp. Bot.,* 41, 979–986, 1990.

32. **Pauls, K. P. and Thompson, J. E.,** Evidence for the accumulation of peroxidized lipids in membranes of senescing cotyledons, *Plant Physiol.,* 75, 1152–1157, 1984.

33. **Schewe, T., Rapoport, S. M., and Kuhn, H.,** Enzymology and physiology of reticulocyte lipoxygenase: comparison with other lipoxygenases, *Adv. Enzymol. Mol. Biol.,* 58: 191–311, 1986.

34. **Zimmerman, D. C. and Coudron, C. A.,** Identification of traumatin, a wound hormone, as 12-oxo-*trans*-10-dodecenoic acid, *Plant Physiol.,* 63, 536–541, 1979.

35. **McRae, D. G., Baker, J. E., and Thompson, J. E.,** Evidence for the involvement of the superoxide radical in the conversion of 1-aminocyclopropane-1-carboxylic acid to ethylene by pea microsomal membranes, *Plant Cell Physiol.,* 23, 375–383, 1982.

36. **Camara, B.,** Terpenoid metabolism in plastids. Site of phytoene synthetase activity and synthesis in plant cells, *Plant Physiol.,* 74, 112–116, 1984.

37. **Murata, N.,** Low-temperature effects on cyanobacterial membranes, *J. Bioenerg. Biomembr.,* 21, 61–75, 1989.

38. **Wada, H., Gombos, Z., and Murata, N.,** Enhancement of chilling tolerance of a cyanobacterium by genetic manipulation of fatty acid desaturation, *Nature,* 347, 200–203, 1990.

39. **Lynch, D. V. and Thompson, G. A., Jr.,** Microsomal phospholipid molecular species alterations during low temperature acclimation in *Dunaliella, Plant Physiol.,* 74, 193–197, 1984.

40. **Lynch, D. V. and Thompson, G. A., Jr.,** Chloroplast phospholipid molecular species alterations during low temperature accclimation in *Dunaliella, Plant Physiol.,* 74, 198–203, 1984.

41. **Norman, H. A. and Thompson, G. A., Jr.,** Activation of a specific phospholipid fatty acid hydrolase in *Dunaliella salina* microsomes during acclimation to low growth temperature, *Biochim. Biophys. Acta,* 875, 262–269, 1986.

42. **Steponkus, P. L. and Lynch, D. V.,** Freeze/thaw-induced destabilization of the plasma membrane and the effects of cold acclimation, *J. Bioenerg. Biomembr.,* 21, 21–41, 1989.

43. **Lynch, D. V. and Steponkus, P. L.,** Plasma membrane lipid alterations associated with cold acclimation of winter rye seedlings (*Secale cereale* L. cv Puma), *Plant Physiol.,* 83, 761–767, 1987.

44. **Cahoon, E. B. and Lynch, D. V.,** Analysis of glucocerebrosides of rye (*Secale cereale* L. cv Puma) leaf and plasma membrane, *Plant Physiol.,* 95, 58–68, 1991.

45. **Lynch, D. V.,** Chilling injury in plants: the relevance of membrane lipids, in *Environmental Injury to Plants,* Katterman, F., Ed., Academic Press, New York, 1990, 17–34.

46. **Kunst, L., Browse, J., and Somerville, C.,** A mutant of *Arabidopsis* deficient in desaturation of palmitic acid in leaf lipids, *Plant Physiol.,* 90, 943–947, 1989.

47. **Miquel, M. and Browse, J.,** *Arabidopsis* mutants deficient in polyunsaturated fatty acid synthesis, *J. Biol. Chem.,* 267, 1502–1509, 1992.

48. **Ha, K.-S. and Thompson, G. A., Jr.,** Biphasic changes in the level and composition of *Dunaliella salina* plasma membrane diacylglycerols following hypoosmotic shock, *Biochemistry,* 31, 596–603, 1992.

49. **Kunst, L., Browse, J., and Somerville, C.,** Enhanced thermal tolerance in a mutant of *Arabidopsis* deficient in palmitic acid unsaturation, *Plant Physiol.,* 91, 401–408, 1989.

50. **Murata, N.,** Molecular species composition of phosphatidylglycerols from chilling-sensitive and chilling-resistant plants, *Plant Cell Physiol.,* 24, 81–86, 1983.

51. **Roughan, P. G.,** Phosphatidylglycerol and chilling sensitivity in plants, *Plant Physiol.,* 77, 740–746, 1985.

52. **Low, P. S., Ort, D. R., Cramer, W. A., Whitmarsh, J., and Martin, B.,** Search for an endotherm in chloroplast lamellar membranes associated with chilling-inhibition of photosynthesis, *Arch. Biochem. Biophys.,* 231, 336–344, 1984.

53. **Li, G., Knowles, P. F., Murphey, D. J., and Marsh, D.,** Lipid-protein interactions in thylakoid membranes of chilling-resistant and -sensitive plants studied by spin label electron spin resonance spectroscopy, *J. Biol. Chem.,* 265, 16867–16872, 1990.

54. **Maroc, J., Trémolières, A., Garnier, J., and Guyon, D.,** Oligomeric form of the light-harvesting chlorophyll a+b protein CPII, phosphatidyldiacylglycerol, Δ3-*trans*- hexadecenoic acid, and energy transfer in *Chlamydomonas reinhardtii* wild type and mutants, *Biochim. Biophys. Acta,* 893, 91–99, 1987.

55. **Huner, N. P. A., Krol, M., Williams, J. P., Maissan, E., Low, P. S., Roberts, D., and Thompson, J. E.,** Low temperature development induces a specific decrease in *trans*-Δ³-hexadecenoic acid content which influences LHC II organization, *Plant Physiol.,* 84, 12–18, 1987.

56. **Trémolières, A., Roche, O., Dubertret, G., Guyon, D., and Garnier, J.,** Restoration of thylakoid appression by Δ³-*trans*-hexadecenoic acid-containing phosphatidylglycerol in a mutant of *Chlamydomonas reinhardtii*. Relationships with the regulation of excitation energy distribution, *Biochim. Biophys. Acta,* 1059, 286–292, 1991.

57. **Schmidt, H. and Heinz, E.,** Desaturation of oleoyl groups in envelope membranes from spinach chloroplasts, *Proc. Natl. Acad. Sci. U.S.A.,* 87, 9477–9480, 1990.

58. **Ohnishi, M. and Thompson, G. A., Jr.,** Biosynthesis of the unique *trans*-Δ³- hexadecenoic acid component of chloroplast phosphatidylglycerol: evidence concerning its site and mechanism of formation, *Arch. Biochem. Biophys.,* 288, 591–599, 1991.

59. **Sandmann, G., Clarke, I. E., Bramley, P. M., and Böger, P.,** Inhibition of phytoene desaturase — the mode of action of certain bleaching herbicides, *Z. Naturforsch.,* 39c, 443–449, 1984.

60. **Traüble, H. and Eibl, H.,** Molecular interactions in lipid bilayers, in *Functional Linkage in Biomolecular Systems,* Schmitt, F. O., Schneider, D. M., and Crothers, D. M., Eds., Raven Press, New York, 1975, 59–90.

61. **Ginsburg, M.,** *Dunaliella:* a green alga adapted to salt, *Adv. Bot. Res.,* 14, 93–183, 1987.

62. **Peeler, T. C., Stephenson, M. B., Einspahr, K. J., and Thompson, G. A., Jr.,** Lipid characterization of an enriched plasma membrane fraction of *Dunaliella salina* grown in media of varying salinity, *Plant Physiol.,* 89, 970–976, 1989.

63. **Eichenberger, W. and Boschetti, A.,** Occurrence of 1(3),2-diacylglyceryl-(3)-0-4'-(N,N,N-trimethyl)-homoserine in *Chlamydomonas reinhardi, FEBS Lett.,* 88, 201– 204, 1978.

64. **Einspahr, K. J., Peeler, T. C., and Thompson, G. A., Jr.,** Rapid changes in polyphosphoinositide metabolism associated with the response of *Dunaliella salina* to hypoosmotic shock, *J. Biol. Chem.,* 263, 5775–5779, 1988.

65. **Einspahr, K. J., Peeler, T. C., and Thompson, G. A., Jr.,** Phosphatidylinositol 4,5-bisphosphate phospholipase C and phosphomonoesterase in *Dunaliella salina* membranes, *Plant Physiol.,* 90, 1115–1120, 1989.

66. **Michell, R. H., Drummond, A. H., and Downes, C. P.,** Eds., *Inositol Lipids in Cell Signaling,* Academic Press, New York, 1989, 534.

67. **Guo, Y. and Roux, S. J.,** Partial purification and characterization of a Ca^{2+} dependent protein kinase from the green alga, *Dunaliella salina, Plant Physiol.,* 94, 143–150, 1990.

68. **Schaller, G. E., Harmon, A. C., and Sussman, M. R.,** Characterization of a Ca^{2+} and lipid-dependent protein kinase associated with the plasma membrane of oat, *Biochemistry,* 31, 1721–1727, 1992.

69. **Einspahr, K. J., Maeda, M., and Thompson, G. A., Jr.,** Concurrent changes in *Dunaliella salina* ultrastructure and membrane phospholipid metabolism after hyperosmotic shock, *J. Cell Biol.,* 107, 529–538, 1988.

70. **Einspahr, K. J. and Thompson, G. A., Jr.,** Transmembrane signaling via phosphatidylinositol 4,5-bisphosphate hydrolysis in plants, *Plant Physiol.,* 93, 361–366, 1990.

71. **Drøbak, B. K.,** Plant signal perception and transduction: the role of the phosphoinositide system, *Essays Biochem.,* 26, 27–37, 1992.

72. **Morse, M. J., Crain, R. C., and Satter, R. L.,** Light-stimulated inositolphospholipid turnover in *Samanea saman* leaf pulvini, *Proc. Natl. Acad. Sci. U.S.A.,* 84, 7075– 7078, 1987.

73. **Stuiver, C. E. E., Kuiper, P. J. C., and Marschner, H.,** Lipids from bean, barley and sugar beet in relation to salt resistance, *Physiol. Plant.,* 42, 124–128, 1978.

74. **Kuiper, P. J. C.,** Lipids in grape roots in relation to chloride transport, *Plant Physiol.,* 43, 1367–1371, 1968.

75. **Erdei, L., Stuiver, C. E. E., and Kuiper, P. J. C.,** The effect of salinity on lipid composition and activity of Ca^{2+} and Mg^{2+} stimulated ATPases in salt-sensitive and salt-tolerant Plantago species, *Physiol. Plant.,* 49, 315–319, 1980.

76. **Douglas, T. J. and Walker, R. R.,** 4-Desmethylsterol composition of citrus rootstocks of different salt exclusion capacity, *Physiol. Plant.,* 58, 69–74, 1983.

77. **Mauseth, J. D.,** *Plant Anatomy,* Benjamin Cummings, Menlo Park, CA, 1988, 367.

78. **Ebel, J.,** Phytoalexin synthesis: the biochemical analysis of the induction process, *Annu. Rev. Phytopathol.,* 24, 235–264, 1986.

79. **Graham, J. S., Hall, G., Pearce, G., and Ryan, C. A.,** Regulation of synthesis of proteinase inhibitors I and II mRNAs in leaves of wounded tomato plants, *Planta,* 169, 399–405, 1986.

80. **Kurosaki, F., Tsurusawa, Y., and Nishi, A.,** Breakdown of phosphatidylinositol during the elicitation of phytoalexin production in cultured carrot cells, *Plant Physiol.,* 85, 601–604, 1987.

81. **Strasser, H., Hoffmann, C., Grisebach, H., and Matern, U.,** Are polyphosphoinositides involved in signal transduction of elicitor-induced phytoalexin synthesis in cultured plant cells?, *Z. Naturforsch.,* 41C, 717–724, 1986.

82. **Bolwell, G. P., Coulson, V., Rodgers, M. W., Murphy, D. L., and Jones, D.,** Modulation of the elicitation response in cultured French bean cells and its implication for the mechanism of signal transduction, *Phytochemistry,* 30, 397–405, 1991.

83. **Anderson, J. M.,** Membrane-derived fatty acids as precursors to second messengers, in *Second Messengers in Plant Growth and Development,* Boss, W. F. and Morré, D. J., Eds., Alan R. Liss, New York, 1989, 181–212.

84. **Farmer, E. E. and Ryan, C. A.,** Interplant communication: airborne methyl jasmonate induces synthesis of proteinase inhibitors in plant leaves, *Proc. Natl. Acad. Sci. U.S.A.,* 87, 7713–7716, 1990.

85. **Bloch, C. B., De Wit, P. J. G. M., and Kuć, J.,** Elicitation of phytoalexins by arachidonic and eicosapentaenoic acids: a host survey, *Physiol. Plant Pathol.,* 25, 199–208, 1984.

86. **Stanford, A. C., Northcote, D. H., and Bevan, M. W.,** Spatial and temporal patterns of transcription of a wound-induced gene in potato, *EMBO J.,* 9, 593–603, 1990.

87. **Muesta, P., Hahn, M. G., and Griesbach, H.,** Development of a radioimmunoassay for the soybean phytoalexin glyceollin I, *Plant Physiol.,* 73, 233–237, 1983.

88. **Oba, K., Takematsu, H., Yamashita, K., and Uritani, I.,** Induction of furano-terpene production and formation of the enzyme system from mevalonate to isopentenyl pyrophosphate in sweet potato root tissue injured by *Ceratocystis fimbriata* and by toxic chemicals, *Plant Physiol.,* 58, 51–56, 1976.

89. **Jin, D. F. and West, C. A.,** Characteristics of galacturonic acid oligomers as elicitors of casbene synthetase activity in castor bean seedlings, *Plant Physiol.,* 74, 989–992, 1984.

90. **Thompson, G. A., Jr.,** *The Regulation of Membrane Lipid Metabolism,* 2d ed., CRC Press, Boca Raton, FL, 1992, 144.

91. **Yang, Z., Park, H., Lacy, G. H., and Cramer, C. L.,** Differential activation of potato 3-hydroxy-3-methylglutaryl coenzyme A reductase genes by wounding and pathogen challenge, *Plant Cell*, 3, 397–405, 1991.
92. **Vögeli, U., Freeman, J. W., and Chappell, J.,** Purification and characterization of an inducible sesquiterpene cyclase from elicitor-treated tobacco cell suspension cultures, *Plant Physiol.*, 93, 182–187, 1990.
93. **Vögeli, U. and Chappell, J.,** Induction of sesquiterpene cyclase and suppression of squalene synthetase activities in plant cell cultures treated with fungal elicitor, *Plant Physiol.*, 88, 1291–1296, 1988.

SECTION VI

INDEX

INDEX